D0549106

Light Building Construction

Popular Science Books offers a wood identification kit that includes 30 samples of cabinet woods. For details on ordering, please write: Popular Science Books, P.O. Box 2033, Latham, N.Y. 12111.

Light Building Construction

Gordon A. Sanders

Reston Publishing Company, Inc.
A Prentice-Hall Company
Reston, Virginia 22090

Library of Congress Cataloging in Publication Data

Sanders, Gordon A.
Light building construction.

1. Building. I. Title.
TH145.S29 1985 690 84-2045
ISBN 0-8359-4032-2

Cover photo courtesy of Georgia Pacific.

Copyright 1985 by
Reston Publishing Company, Inc.
A Prentice-Hall Company
Reston, Virginia 22090

All rights reserved.
No part of this book may be reproduced
in any way, or by any means,
without permission in writing
from the publisher.

10 9 8 7 6 5 4 3 2

Printed in the United States of America.

Editor's Note: When possible and practical, British Imperial measurements are also given in Metric throughout this text.

TABLE OF CONTENTS

PREFACE

This book is intended to provide a broad-based construction education for students considering careers in the light construction industry. In addition to the practical, proven framing and construction techniques outlined in this book, chapters have also been presented dealing with relevant topics such as project planning methods, use of tools and fasteners, safety procedures, dealing with utilities, and building for energy efficiency.

No single book can be truly comprehensive in its treatment of the construction trade. Personal experience plays too large a role in the development of good builders. For this reason, lists of research assignments and reference materials have been included at the end of each chapter for students who wish to become more knowledgeable in areas of personal interest.

It is sincerely hoped that the material and references contained in this text stimulate interest and fresh thinking among those pursuing careers in any aspect of the light building trade.

1

PLANNING

1.1 GENERAL OBJECTIVES

This chapter introduces a complete building project planning method. This method is compatible with Performance Evaluation and Review Technique (PERT) and Critical Path Method (CPM), two standard planning methods, and does not require computer programming. Special emphasis is placed on the planning of projects using established principles of construction and easily understood building terms.

1.2 SPECIFIC OBJECTIVES

After studying this chapter and conducting the suggested readings and research, the reader should be able to

1. produce a work breakdown structure for a small building project,
2. define all tasks or activities for a small building project,
3. list the activities of a project and identify workers or worker groups responsible for them, and
4. develop a simple network program which graphically describes the construction operation.

The reader should also be able to conduct further studies with respect to

1. scheduling (estimating the duration of jobs, reviewing and checking the critical path—a systematic diagram ordering essential construction activities within specific time frames—and revising the plan), and
2. applying the planning process method to larger and more complex projects.

1.3 CHAPTER FORMAT

To illustrate a building planning process, a number of assumptions will be made about a small building project. With the assumptions as background, three different project plans will be presented, not only to suggest how the building may be constructed but also to show how different approaches to building in general may be developed. With three different plans (there could be more than three), it is possible to observe the need for defining project objectives and for producing a work breakdown structure. The builder is encouraged to think in terms of defining tasks or activities associated with the building project and to list and to order the activities. The summary of this chapter will be in the form of a network diagram.

The chapter will then close with suggested further research. This research should not be viewed as being merely supplementary to the course of study in this book, but as an integral part of the entire process of learning about building construction.

1.4 PLANNING A BUILDING PROJECT

To build, one must plan. A simple building requires a simple plan. A more complex building, with many elements brought together on a system basis, requires a more complex plan. To bring order to the planning process, one must first know what is to be built and generally how it is to be constructed. Knowledge and experience are needed to help sort identifiable building activities into a sequence of events understood by those involved with the job.

With a general idea or concept of what is to be done and in what order, as well as in what time period, and at what cost, a builder or contractor can make sure that everyone involved can effectively do his or her part. Everyone benefits: the owner, the contractor, and the builders.

1.5 PLANNING A GARAGE BUILDING PROJECT

Garages are built in many sizes and shapes and at various cost levels. For the purpose of this chapter, the planning is for a simple structure which will have the construction elements of a larger light frame building. A set of assumptions is stated for the project and then three plans are developed for project completion.

Assumptions:

1. The building is to be a single detached garage, designed to match the existing house nearby.
2. Before the concrete is placed and is sufficiently cured, there is limited on-site storage for materials.
3. The foundation is to be a concrete slab, as is the driveway.
4. The construction is to be wood framing to accommodate an overhead door, a side entry door, and side windows.
5. The gable roof is to be finished with asphalt shingles and fitted with a rainwater drainage system.

6. The building is to be serviced with electrical power, natural gas, and a cold water tap.
7. The exterior finish is to be stucco and cedar siding to match the house. Painted trim will also match the house.
8. Although time is usually a factor, a limit is not set.
9. Costing (reducing of costs) is also a factor, but will be considered later, in Chapter 6.

A projected likeness of the garage just described is shown in Figure 1-1.

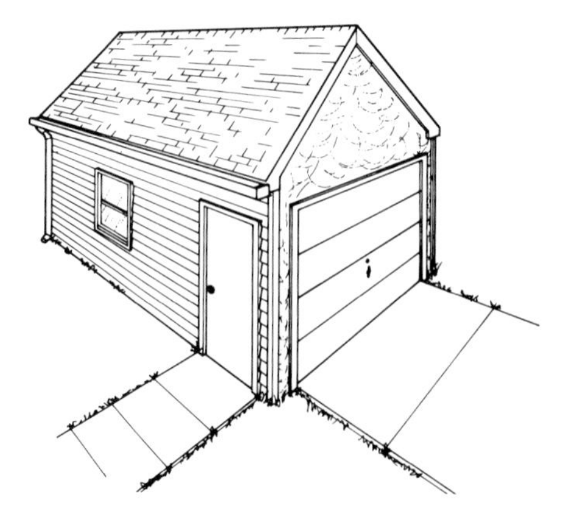

FIGURE 1-1. A hypothetical drawing of a garage as described in the Chapter 1 garage project.

Plan 1

The owner will do the planning, all the construction activities except installing and hooking up the utility services, all the scheduling and controlling, and make all the payments for materials and utility services.

The owner will

1. design the garage, the driveway, and the landscaping;
2. obtain the permits to build;
3. prepare the site for pouring concrete, dig trenches for services, do the concrete formwork, set the reinforcing steel, and mix and place the concrete;
4. frame the building: the walls, the roof, the cornice, and the rough openings for doors and windows;
5. finish the building: the roof and cornice with sheathing and trim, the walls and roof with building paper, the walls with stucco wire and cedar siding, the roof with shingles;
6. provide for the utility services to make installations for electricity, gas, and water;
7. place vapor barrier and insulation (optional);
8. install the doors, windows, and roof drainage system;
9. paint exposed wood;
10. stucco the building using a two-coat system to match the house;
11. clean up the site;
12. arrange for building inspections;
13. landscape; and
14. make all payments.

The owner must be physically able and also willing to prepare the site, haul materials, mix concrete, make arrangements for utilities, frame and finish the building, install doors and windows, clean up, and landscape. This is basically the do-it-yourself option.

Plan 2

The owner will act as contractor, subcontracting certain project activities.

The owner/contractor will

1. plan the project with respect to design, cost estimates, and building permits;

2. schedule events and manage the project so that it is completed within a reasonable amount of time and in terms of estimated costs;
3. engage subcontractors;
4. clean up and landscape; and
5. pay the subcontractors and suppliers of materials.

Plan 2.1 (an expansion of Plan 2)

1. plan the project with respect to design, cost estimates, and building permits;
2. engage a subcontractor to prepare the site and place the concrete complete with provisions for services;
3. make arrangements for and schedule the rough-servicing of the project for water, gas, and electricity;
4. frame and sheath the building—supplying doors and windows, wall framing materials, hardware, prebuilt roof trusses, and sheathing;
5. install doors and windows and trim materials;
6. apply building papers, shingles, and stucco wire;
7. arrange a subcontract for electrical work (wiring), gas-fitting, and cold water tap installation;
8. arrange a subcontract for stuccowork base and finish coats to match existing house;
9. paint, clean up, and landscape;
10. inspect the work done by subcontractors;
11. arrange for building inspectors according to permit regulations; and
12. pay the subcontractors and the suppliers of materials.

This plan allows the owner to act as contractor by subcontracting for site work, concrete work, utility services, and stuccowork, and as his or her own carpenter. In a sense, the owner in this case is subcontracting for himself as a carpenter and landscaper, while hiring other parties for the rest of the work to be contracted out. This is a compromise between a complete do-it-yourself job and having subcontractors do all of the work.

Plan 2.2 (An alternate expansion of Plan 2)

The owner/contractor will

1. plan the project with respect to general design and building permits;
2. subcontract site preparation and concrete work;
3. subcontract with respect to a garage package (have the subcontractor do the rest of the construction work in one job: framing, sheathing, stuccowork, installing hardware, and finishing);
4. subcontract with respect to each of the utility services (water, gas, electrical);
5. clean up and landscape; and
6. arrange for building inspections, money hold-backs to subcontractors, restitution, and final payments to subcontractors.

With this plan the owner is mainly a contractor, although he or she still plans the project, schedules it to some degree, and does the site cleanup and landscaping.

Plan 3

The owner will engage a contractor to provide a single garage which in general matches the construction and finish of the existing house.

With this plan the owner gives a contractor authority to build as that contractor sees fit. The burden is on the contractor to provide satisfactory service to the owner, at an agreed-upon price. The contractor is responsible for planning, scheduling, controlling, and making payments to subcontractors, if they are engaged. The owner's main decision in this case is to determine which contractor to select to do the job.

1.6 SAMPLE CONSTRUCTION ACTIVITY—FRAMING

With a general plan in mind about building a project, such as a single garage, it is possible to isolate and examine any one construction activity. Just as the overall construction process of building a garage has been broken down into its component activities, so too can each individual activity be further broken into its own basic requirements. This is one way of taking a good cross-sectional view of a construction job and seeing how that job is done from beginning to end.

Consider framing as one activity. Wall framing and roof framing are then subactivities done by the same or different persons.

To frame the walls someone must make decisions about

1. the plate layouts which determine the size of the building, stud size, and stud spacings which should allow sheathing to fit without much cutting, the widths of door and window openings, and corner details; and
2. the studding lengths which determine the height of the walls, the vertical placements of doors and windows, and cornice details.

A shop drawing may be used to provide details of the building for exact measurement, especially in the case of the cornice details which must usually be known before the studding length is finally determined. The cornice details can be drawn full-scale if the slope of the roof is known and when fixed measurements for selected doors and windows are considered. In the case of the single garage, the measurements for the rough opening of the car-entry door are crucial. They must be at hand when determining stud length.

Wall framing vertical measurements may be summarized (collected) on a shop drawing and a story rod. The story rod is simply a long 1″ × 1″ (19 mm × 19 mm) pole used to mark off vertical distances in construction.

To frame the roof, the top plate dimensions and the slope and shape of the roof must be known. If the roof is to match the roof of an existing house, one must know how to transfer measurements, cut rafters, assemble trusses, and construct cornices. A gable roof may be less complicated than a hip roof, if that is a consideration (see Chapter 12).

The shop drawing of the cornice will provide details with respect to overhang, fascias (single or double), windblocks or lookouts, and soffit (underside of overhangs). It can also be used to indicate the relationship of the soffit to the trim for windows and doors.

Design features which must be known by the framer are wall thickness, soffit width, gable-end details, insulation type and width, truss span, and services. Consideration is given to strengths of materials, wind and snow loads, temperature range, and other local environmental aspects.

The framing activity, from a contractor's point of view, includes time and cost factors. The contractor must answer the following questions. What is the least costly method of framing in terms of what is designed? What is the best way to communicate design features to the framer? How is the framing activity scheduled to fit in with other activities such as concrete placing and electrical wiring?

1.7 SAMPLE CONSTRUCTION ACTIVITY—CONCRETE WORK

A second individual construction activity needed to complete our hypothetical garage building project is concrete work. Like framing, concrete work can be broken down into several subactivities. If the owner/contractor decides to place the concrete for the driveway and garage subfloor foundation, a list of subactivities, subject to local conditions might consist of the following:

1. preparing the site (removal of topsoil, construction of forms, placement of sand/gravel base and reinforcing steel),
2. depositing concrete (premix or on-the-job mix),
3. consolidating the concrete,
4. finishing the concrete,
5. removing forms, and
6. curing and protecting the newly laid concrete.

The technology of portland cement concrete is complex. Knowledge and experience with regard to mix design and placement are essential if an activity such as concrete work is to be successful (see Chapter 8).

1.8 COMPARISON OF FRAMING AND CONCRETE WORK

Now, instead of vertically breaking down the multilayered aspects of construction activity, it can be useful for determining work time and cost factors to horizontally compare different individual activities. The two sample activities—framing and concrete work—are quite different; yet, both are essential for the building project. The materials, wood and concrete, are not alike and represent end products of two separate technologies which are brought together, along with the products of other technologies, in one job.

The two construction activities of framing and concrete work are different with respect to

1. material production,
2. material receiving and storing,
3. material handling and fastening,
4. tools and equipment for working with them,
5. fire risk,
6. environmental relationships and durability,
7. cost,
8. worker hazards, and
9. worker skills and knowledge.

Wood and concrete are both relatively common materials which are generally accepted as useful for construction. Think of wood as being a complex chemical substance and the setting of concrete as a complex chemical substance as well as a physical event. Observe how the two materials, along with others, can be put together, following the rules of a specific design, to make a functional and aesthetically pleasing structure.

1.9 SPECIALIZATION

Some persons focus on one or more technologies in the construction field as career choices. An architect, a carpenter, or a mason are all specialists. The specialist may or may not focus on a field as narrow as the framing and concrete work we spoke of as individual activities and technical aspects of building a garage. The specialist may also *not* necessarily work with just one material; however, specializing usually means having in-depth knowledge and skills related to only part of the whole spectrum of construction activities.

Other persons decide to become the generalists of the construction field, taking a broader all-encompassing approach toward building design and research. In a sense, the general contractor is or should be a skilled generalist because he or she must have enough knowledge about all aspects of construction to ensure that the entire job is done properly and cost-effectively.

1.10 BUILDING AS A MULTI-ACTIVITY PROGRAM

More than the two activities of framing and concrete work are needed in the construction of a building, even a small one such as a single garage. The usual or standard activities may be listed in order, keeping in mind that some subactivities often take place concurrently, especially if more than one person is working.

An initial trial list of activities for the garage project might consist of the following:

1. site preparation,
2. concrete work,
3. services,

4. framing,
5. finish work,
6. cleanup and landscaping.

When subactivities are added, the list becomes larger and more complex because of overlapping time frames and possible time constraints. A trial list of activities with subactivities might be expanded as follows:

1. site preparation (activity),
   - clearing/storing topsoil (subactivity),
   - removing trees and roots,
   - removing rock outcropping,
   - trenching and/or excavating, and
   - compacting soil.
2. concrete work,
   - completing form work,
   - hauling/storing sand, cement, gravel,
   - supplying water,
   - placing sand/gravel base,
   - roughing-in services,
   - placing reinforcing steel,
   - depositing concrete,
   - finishing concrete, and
   - curing and protecting finished concrete.
3. provision of services,
   - roughing-in for cold water, gas, and electricity (overhead or underground),
   - completing service lines for plumbing, gas, and electrical (after framing), and
   - finishing service work such as installing cold water tap, gas appliances, and electrical fixtures.
4. framing/sheathing,
   - walls,
   - roof, and
   - cornice.
5. closing-in and finishing (cladding), and
   - sheathing (interior),
   - building papers,
   - shingles,
   - trim, siding, stucco wire, stuccowork,
   - insulation/vapor barrier (optional),
   - installing doors and windows, and
   - painting.
6. cleanup and landscaping,
   - removing forms,
   - cleaning and piling forms,
   - backfilling,
   - grass seeding or turfing,
   - hauling refuse, and
   - planting shrubbery, flower beds, etc. (optional).

This interim list could be expanded further with subsubactivities, such as floating and troweling concrete under the finishing concrete subactivity. The list could also be expanded with nonproduction type work which is, nevertheless, absolutely necessary, such as garage design, securing permits, itemizing subcontracted activities as well as those that are not, inspections, payments, supervision, and management.

1.11 PLANNING A PROJECT

Because there are so many activities and subactivities involved in a building project, it is important for the builder to perceive the job in a methodical way. Haphazard planning and scheduling will often prove to be dysfunctional. However, even before exact scheduling and time frames for completing activities are determined, it is feasible to develop a systematic building plan by

1. defining the project objectives,
2. producing a work breakdown structure,
3. listing and defining all activities,
4. delegating those persons responsible for these activities, and
5. constructing a network diagram (flow chart).

1.11.1 Development of a Network Program for a Single Garage Project

The project objective in this case is to build a single garage adjacent to and similar in design to the owner's house. What is called a project breakdown structure is often a useful preliminary step to take before drawing a full network program. A breakdown structure (Figure 1-2) organizes activities and some subactivities to show how they contribute to the project objectives. They are not ordered, as in a full-scale network program, to show relationships between subactivities. For example, a breakdown structure will not show the relationship between sheathing and the roughing-in of services. Neither will a breakdown structure necessarily have to show all of the minor tasks associated with the construction of a single garage. By drawing a breakdown structure, a builder is able to order the priorities of the job and quite simply, see on paper what was previously only a mental picture. Most importantly, a breakdown structure lays out activities in a way that allows them to be worked into the order of a network program.

A network program is much more complex than a breakdown structure. It is also a valuable tool for a builder to have, which is why it is so important that the builder is capable of putting one together. One possible building program for the garage building project as it appears in a network diagram is shown in Figure 1-3. This diagram looks complicated, but once a builder learns how to read one, working out new ones will seem easy.

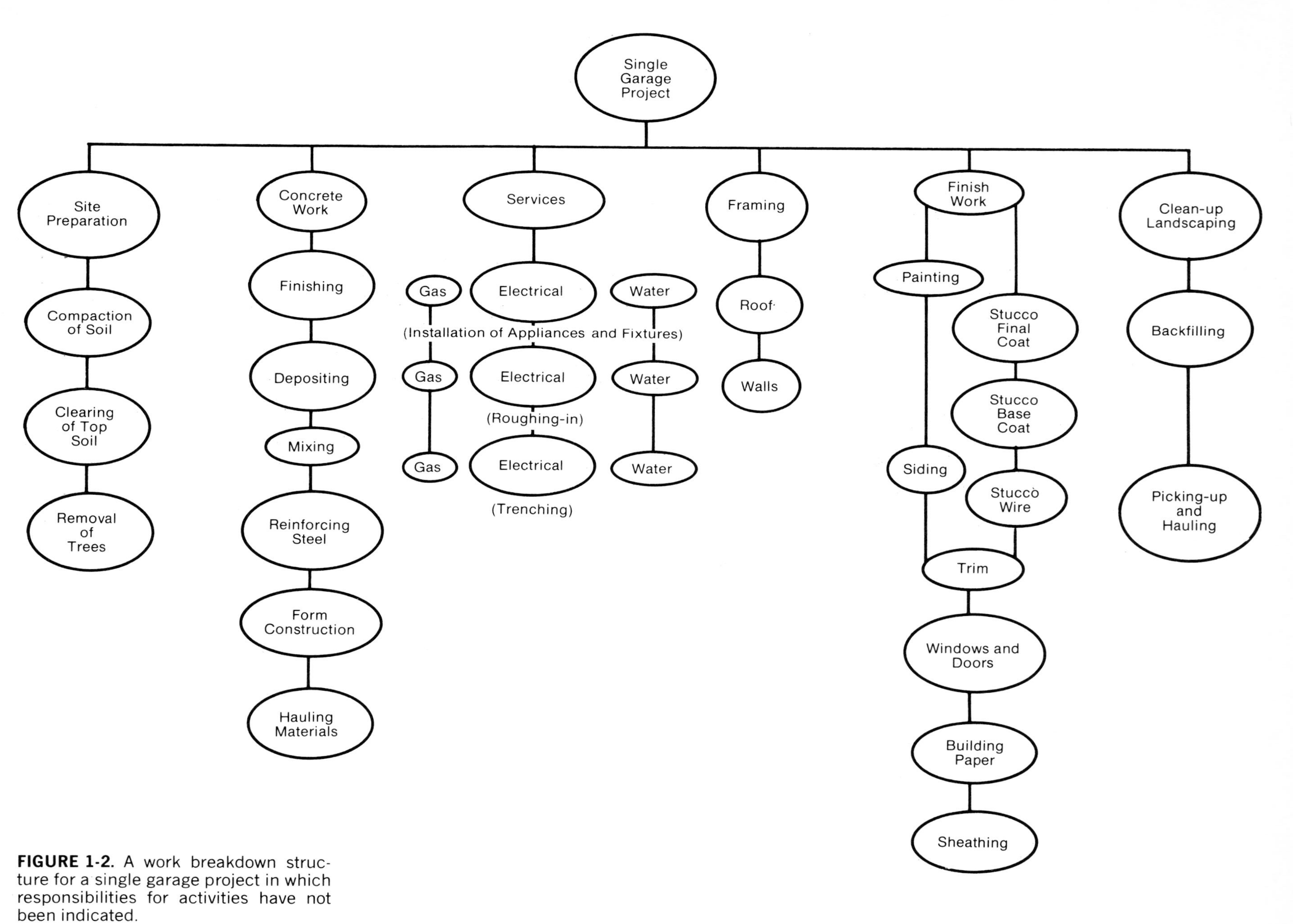

FIGURE 1-2. A work breakdown structure for a single garage project in which responsibilities for activities have not been indicated.

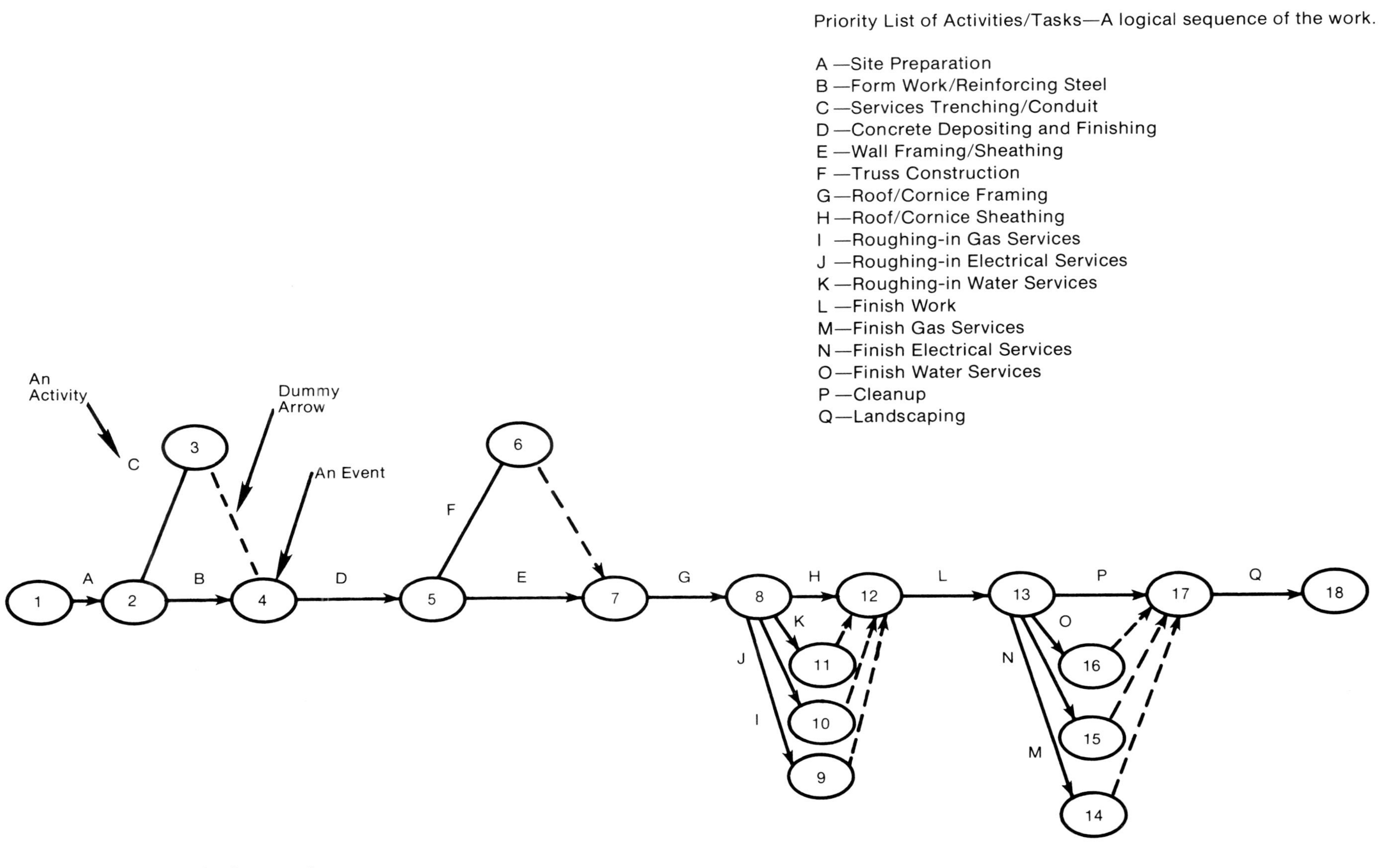

FIGURE 1-3. A network diagram for a garage building project.

In Figure 1-3, the numbered circles represent events. For example, Circle 1 denotes the beginning of the project before any work has been done. Circle 2 represents the point at which site preparation is completed and formwork/placement of reinforcement steel and services trenching/laying of conduit is ready to begin. In other words, the numbered circles represent defined progress points during the course of the job. The lettered arrows, represent the actual tasks or activities being done. For example, Arrow A refers to the act of preparing the site; work actually done during this process. The last activity done is represented by Arrow Q—landscaping—and the last circle, in this case Circle 18, represents a completed garage building project. A dummy arrow, as shown in Figure 1-3, simply means that at completion of an activity and the arrival at an established event, the flow of activities reverts to the main line of work. If time were considered, the dummy arrow represents no time allotted and no activity completed.

The entire network diagram can be explained as follows:

1 2
Site preparation must be done before formwork commences.

2 3, 2 4
Formwork/reinforcing steel and trenching/conduit work must be done before concrete is deposited and finished.

4 5
Concrete is finished before wall framing/sheathing and truss construction commences. (Concrete might set for three days or more before framing activities start on the slab. See Chapter 8.)

5 6, 5 7
Wall framing and truss construction are completed before roof/cornice framing.

7 8
Roof cornice framing is finished before roof/cornice sheathing and roughing-in of services.

8 9, 8 10, 8 11, 8 12
Roof/cornice sheathing and roughing-in for gas, electrical, and water services may all go on at once and before finish work commences.

12 13
Finish work is completed before cleanup and installation of service appliances/fixtures.

13 14, 13 15, 13 16, 13 17
Cleanup and installation of all appliances/fixtures may go on at once and must be finished before landscaping is done.

17 18
When landscaping is done, the project is completed. The objective of building a single garage has been achieved.

1.11.2 Further Uses of the Network Diagram

The network diagram as shown in Figure 1-3 is a display of the activities/tasks to be accomplished, in a correct sequence, to meet the building program objectives. It is not yet a program of work, as there are no time estimates or resource allocations of labor and materials.

If time estimates were added for each activity, the project life could then also be estimated. As suggested, it would make a difference if concrete depositing and curing took three days or seven days or if the contractor decided to start framing in 24 hours on green concrete.

When a time estimate is given, then a *critical path* (CP) becomes evident. The critical path compresses many of the activities already mentioned in inclusive time periods. Activities that overlap in reference to a time frame still require multiple sets of actual work goals, but these goals will be achieved at the same time. This will be reflected in a critical path diagram. In Figure 1-3 the critical path would probably appear as follows:

1 2 4 5 7 8 12 13 17 18

The path might change if the contractor decided to procure a garage package or decided to make doors and windows instead of buying manufactured units. By the same token, it might change if roof trusses are obtained as prebuilt units rather than waiting for a concrete slab or driveway to become a working area on which to build them.

It is often in the interest of both the contractor and the owner to have the project completed in the shortest time period possible, provided that building standards are high. Developing and using a network program is an excellent way of planning ahead to determine how long a project and each individual phase of that project will take. Having a clear network diagram is also a good way for a contractor to show the building owner how and in what time frame a job will be done. Finally, estimating activity duration implies knowledge and experience with respect to building. Such knowledge and experience comes to those persons who have a desire to become fully involved, and who become both active in the field and intellectually involved with the intricacies of the building process.

1.12 RESEARCH

1. Visit a library where there are manuals and handbooks regarding PERFORMANCE EVALUATION AND REVIEW TECHNIQUES (PERT) and CRITICAL PATH METHOD (CPM) and review them.
2. Read about the reasons for the development of PERT and CPM and how they relate.
3. Read about computations associated with network data.
4. Inquire about computer project analysis and control systems. Request permission to see a typical report.
5. Inquire about a program of studies and experiences which will enable one to become a project manager.

1.13 REFERENCES

1.13.1 *IBM System/370 Project Analysis and Control System General Information Manual.* New York: IBM Corporation, Technical Publications, Department 824, (GH19 1055-3), 1979.

1.13.2 Spencer, M. *Elements of Project Management: Plan, Schedule, and Control.* New Jersey: Prentice-Hall, Inc.

2

TOOLS AND MACHINERY

It is essential that builders and all those closely associated with the building trade know the uses and operation of tools and equipment utilized in construction work. They must also understand the cost effectiveness of investing in one tool as opposed to another and be fully aware of the maintenance and safety procedures for each tool.

2.1 GENERAL OBJECTIVE

The prime objective of this chapter is to describe the safe operation, care, and maintenance of the numerous machines and tools used in light construction. The operation of tools will be discussed in regard to both efficiency and safety, factors that are synonymous in the trade. A more complete presentation of light building construction safety principles is presented in Chapter 4.

2.2 SPECIFIC OBJECTIVES

The material in this chapter should give the reader

1. general background information on the various tools and machines commonly used in light building construction, including workshop tools;
2. the ability to select tools according to the needs of a particular job, while considering cost factors;
3. a sense of the importance safety factors play in the use of tools;
4. an introduction to tool habits and maintenance procedures; and
5. a feeling for ways to fully integrate tool and machine operation into a construction job that is both efficient and safe.

2.3 USING MACHINES AND TOOLS SAFELY

How machinery and tools are used will greatly determine both the productivity and safety at the jobsite. Workers, as well as the builder or contractor, should be aware of the need for safe machine operation.

For example, one of the most frequent jobsite hazards is created when machine guards and safety devices are removed from equipment for the sake of "easier" operation. Both builders and workers face serious consequences when they remove safety devices from equipment. Workers jeopardize their own safety and future. Employers who remove safety equipment subject their workers to increased danger and risk facing the high cost of job related accidents.

2.3.1 Tool Habits

"A place for everything and everything in its place" is just common sense. It is not possible to do a fast, efficient job if half the time is spent looking around for each tool needed. The following rules will make work easier on the jobsite or in a workshop:

1. Keep each tool in its proper storage place; a tool is useless if it cannot be found.
2. Keep all the tools in good condition. Protect them from rust, nicks, burrs, and breakage.
3. Keep the tool allowance complete. Maintain an inventory of all tools. Replace or repair worn-out or damaged tools.
4. Use each tool only on the job for which it was designed. If the wrong tool is used to make an adjustment, the results will probably be unsatisfactory. For example, if a socket wrench that is slightly too big is used, the corners of the wrench or nut can be rounded off.
5. Keep tools within easy reach when working and where they cannot fall on the floor or machinery. Avoid placing tools anywhere above machinery, equipment, or other workers. Serious damage will result if the tool falls into another piece of machinery after the equipment is energized.
6. Never use damaged tools. A battered screwdriver may slip and spoil the screw slot, damage other parts, or cause a painful injury. A gauge or square strained out of shape will result in inaccurate measurements.
7. Heavy equipment—backhoes, bulldozers, front-end loaders, trucks, etc.—should be properly maintained. All mechanicals, electrical, and hydraulic systems should be regularly checked and properly repaired if necessary. Safety procedures involving the use of any heavy equipment used in light building construction should always be followed.

Remember, the efficiency of a construction worker and the tools that worker uses are determined to a great extent by the condition in which those tools are kept. Always keep tools clean and free from dirt, grease, and foreign matter. After use, return tools promptly to their proper place in the toolbox, tool truck, or work trailer.

Maintenance and use of heavy equipment require equal care and organization. Safe machine operating conditions depend on the detection of potential hazards and on taking immediate action to remedy them. Management personnel should develop checklists regarding equipment maintenance. These checklists will then serve as standard guidelines for foremen, operators, mechanics, and so on. In fact, all construction personnel, as well as the student of construction, should approach equipment maintenance from the point of view of an inspector. Using the appropriate inspection specifications when checking machinery is a good way to make sure it is operated safely and for the student to learn exactly how a piece of equipment should and does work.

Ensuring adequate and uninterrupted work space is another important consideration when operating heavy equipment. The builder must coordinate planning to allow heavy equipment to operate with adequate room to maneuver and without the operators having to worry about other workers on the jobsite getting in their way. For the construction student to be able to plan for work space requirements necessitates knowledge about heavy equipment, their capabilities, and their limitations.

2.4 TYPES OF TOOLS AND MACHINERY

The construction process requires a wide variety of tools and machinery, and construction tools are constantly being improved with newer models that are lighter, stronger, and made from better materials. New specialty tools devised to do very specific jobs have been developed over the years. The builder must keep up-to-date with what is available in the marketplace to make construction work easier and more profitable. Builders should always think in terms of their job requirements and resulting tool needs. A requirement list can then be prepared for each particular jobsite.

The tools used in light construction can be divided into four broad groups based on their design principles.

1. **Hand tools.** These are numerous and include the following subcategories:

   - striking tools,
   - cutting tools,
   - turning tools,
   - holding tools,
   - measuring tools, and
   - miscellaneous tools and safety equipment.

2. **Power-operated hand tools.** Included in this group are electric hand tools (circular saws, portable drills, sanders, etc.); compressed-air equipment (jackhammers, tampers); powder-actuated tools (fastening tools); and small self-powered pieces of equipment (chain saws).
3. **Stationary power tools and equipment.** This group includes standard shop equipment such as table saws, radial saws, jointers, planers, band saws, stationary sanders, drill presses, and bench grinders. Also included in this general group is the gasoline-powered generator or power plant for powering electric tools on an isolated jobsite and small, relatively portable jobsite heaters.
4. **Mobile power equipment.** Included in this category are loaders, bulldozers, backhoes, graders, rollers, dump trucks, utility trucks, and trucks with mounted welding outfits and/or winches.

Careful selection of equipment and tools for the jobsite is mandatory if a building project is to be economically viable. This selection of tools is done on the basis of the activities to be performed, by or for the persons responsible for each segment of a job. Contractors who do every phase of the building progress in-house—that is, no work is done by subcontractors—will usually either have all the equipment needed to complete a job or specialty equipment will sometimes be rented or leased. Individual workers, especially skilled craftsmen, will often have their own tools even when they are working for someone else. Subcontractors are generally equipped with more specialized tools and machinery. For example, an excavator needs at least a front-end loader, a dump truck, and probably a bulldozer and, depending on the job requirements, may also need a clearing crew equipped with chain saws and/or a backhoe for trenching and foundation digging. At outside job-related shops (for millwork and cabinet making, for example), equipment is usually self-contained.

2.4.1 Hand Tools

Although this section is devoted to hand tools, remember that the worker's most valuable tools are his or her own hands. The worker must protect them as well as the rest or his or her body. The most important rule is to stay alert and concentrate on the job. In addition, hands should be protected from injury by following applicable safety rules for each and every tool. Most importantly, only use hand tools for the job they were designed to accomplish.

Striking Tools

In this group are such common tools as hammers, mallets, sledges, axes, and hatchets.

Hammers, Mallets, and Sledges

A toolbox would not be complete without at least one hammer. Several different styles are commonly used (Figure 2-1). The carpenter's or nail hammer is designed for one purpose—to drive or pull nails. Note the names of the various parts of the hammer shown in Figure 2-2. The carpenter's hammer has either a curved or straight claw. The face may be either bell-faced or plain-faced, and the handle may be made of wood or steel.

The mallet is a short-handled tool used to drive wooden handled chisels, gouges, and wooden pins. Mallet heads are made from a soft material, usually wood, rawhide, or rubber.

The sledge is a steel headed, heavy-duty tool that can be used for a number of purposes. Short-handled sledges are used to drive bolts, driftpins, and large nails and to strike cold chisels and small hand rock drills. Long-handled sledges are used to break rock and concrete, to drive spikes, bolts, or stakes and to strike rock drills and chisels.

Safety Precautions. Hammers are dangerous tools when used carelessly and without consideration. Some important things to remember when using a hammer or mallet are as follows:

- Do not use a hammer handle for bumping parts in assembly, and never use it as a pry bar. Such abuses will cause the handle to split, and a split handle can produce bad cuts or pinches. When a handle splits or cracks, do not try to repair it by binding with string or wire. Replace it!
- Make sure the handle fits tightly on the head.
- Do not strike a hardened steel surface with a steel hammer. Small pieces of steel may break off and injure someone in the eye or

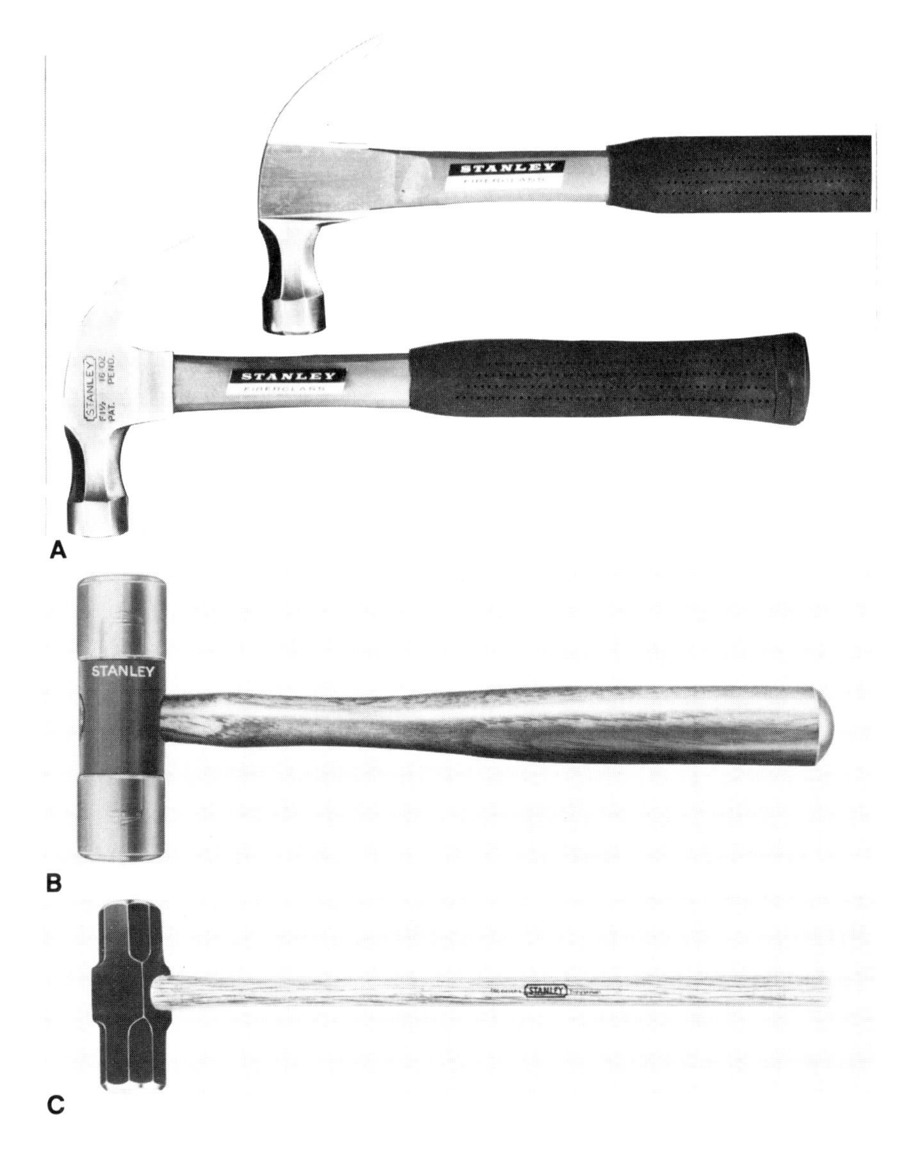

FIGURE 2-1. (A) Carpenter's hammers, (B) soft-faced mallet, and (C) sledge hammer.

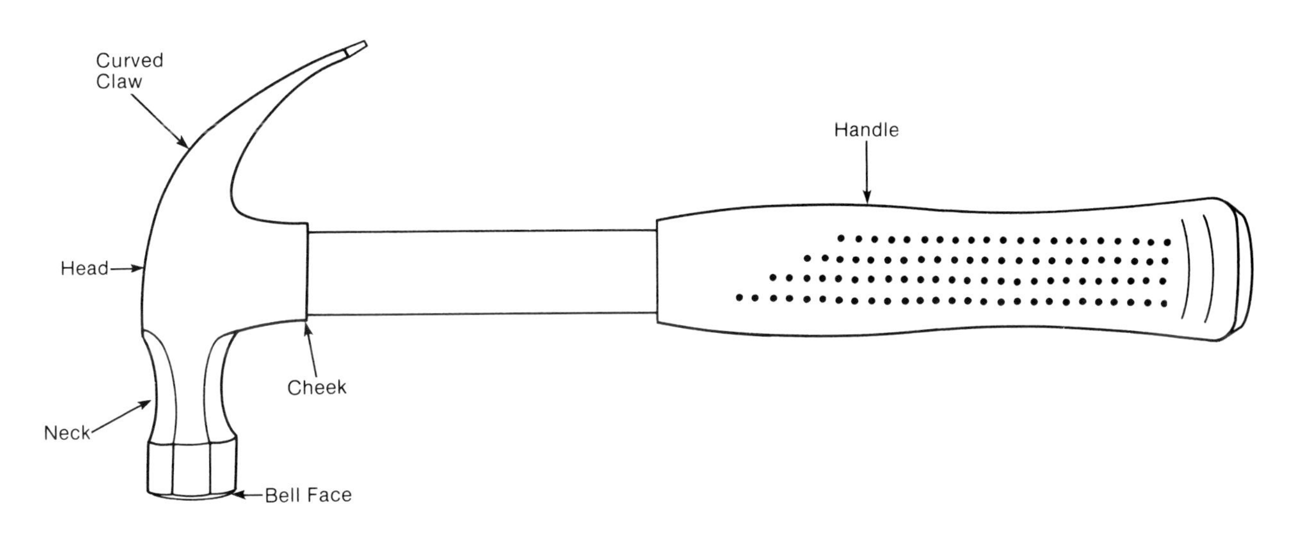

FIGURE 2-2. Parts of a carpenter's hammer.

damage the workpiece. It is, however, permissible to strike a punch or chisel directly with the hammer because the steel in the heads of punches and chisels is slightly softer than that of the hammerhead.

Axes and Hatchets

Axes are made in various patterns and head configurations (Figure 2-3). Their heads are usually forged from carbon tool steel, and the blades or bits are heat treated.

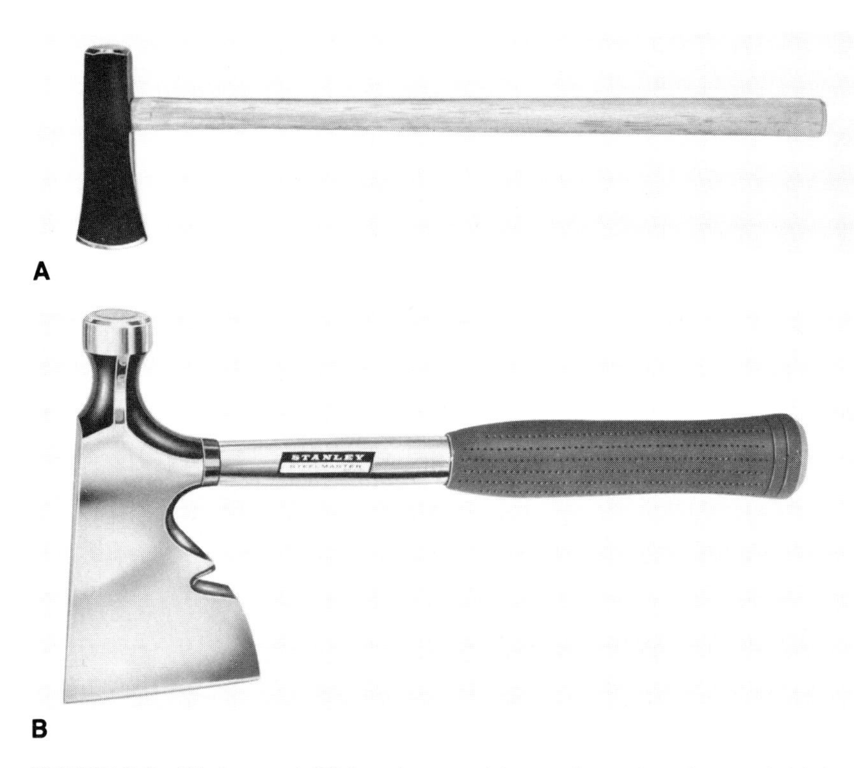

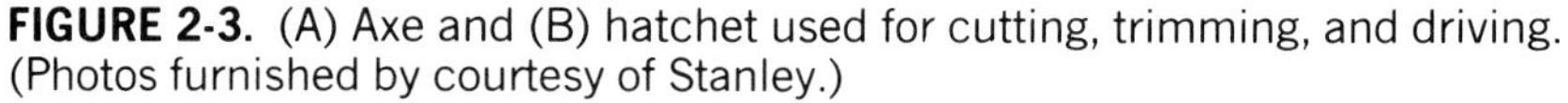

FIGURE 2-3. (A) Axe and (B) hatchet used for cutting, trimming, and driving. (Photos furnished by courtesy of Stanley.)

Hatchets are made in an even greater variety of patterns, since specific types are intended for use in various trades: carpenters, roofers, dry wall installers, etc. Their heads are usually forged from carbon tool steel and heat treated.

A single-bit axe may be used to cut, trim, and drive wooden stakes. Double-bit axes are used for felling, trimming, cutting stakes, etc. Hatchets are used for cutting, splitting, trimming, and hewing, and nails and stakes may be driven with the striking face.

The cutting edges of axes and hatchets are designed to cut wood and equally soft materials. They should never be struck against metal, stone, or concrete.

Cutting Tools

This group includes both woodcutting and metal cutting tools.

Handsaws

Although gradually being replaced by portable power-operated saws, such as circular saws and jigsaws, handsaws (Figure 2-4) are still extremely important for a wide variety of carpentry work. The main

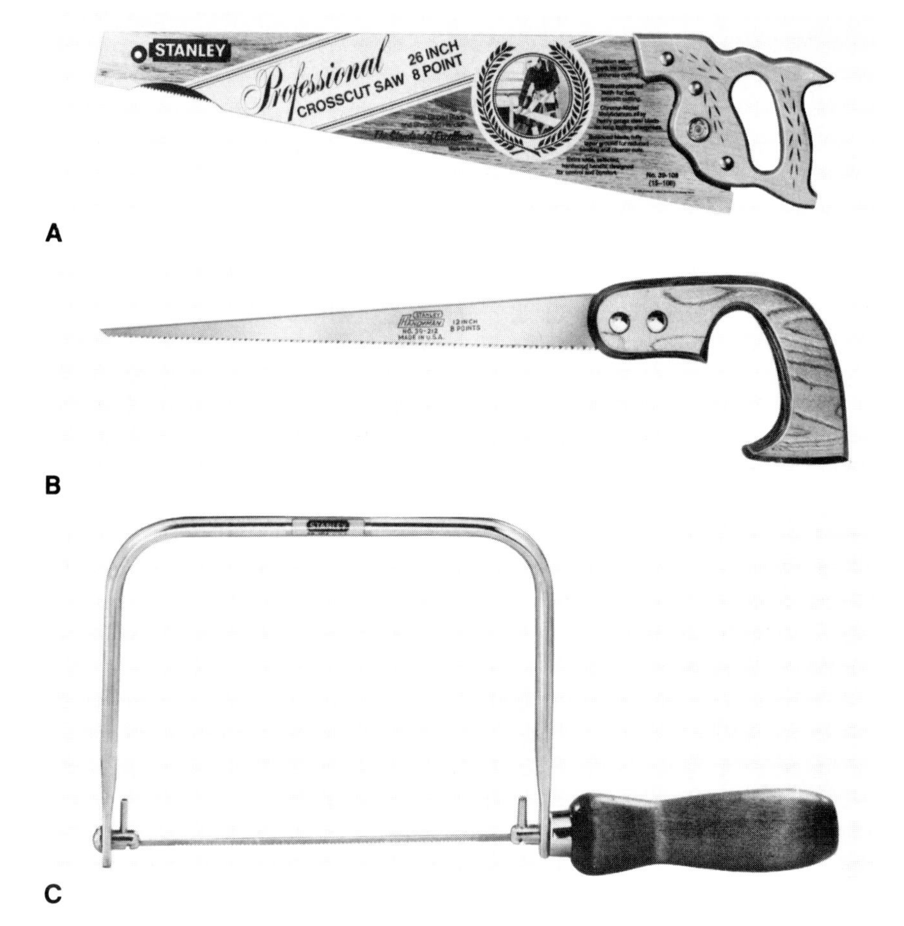

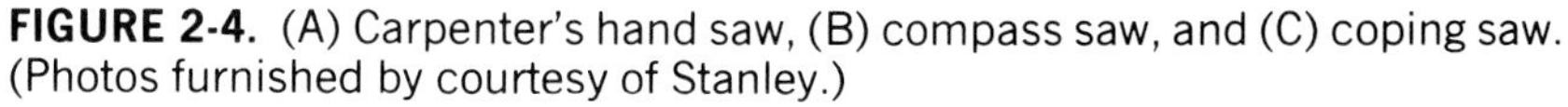

FIGURE 2-4. (A) Carpenter's hand saw, (B) compass saw, and (C) coping saw. (Photos furnished by courtesy of Stanley.)

woodworking handsaws designed for general cutting are ripsaws and crosscut saws. Ripsaws are used for cutting with the grain and crosscut saws are for cutting across the grain.

There are also a number of saws used for special purposes. The backsaw is a crosscut saw designed for sawing a perfectly straight line across the face of a piece of stock. A heavy steel backing along the top of the blade keeps the blade straight. It is used for cutting miters and other joints.

The compass saw is a long, narrow, tapering ripsaw designed for cutting out circular or other nonrectangular sections from within the margins of a board or panel. A hole is bored near the cutting line to start the saw. A keyhole saw is simply a finer, narrower compass saw.

The coping saw is used to cut along curved lines and to make circles and interior cutouts. It can also be used to shape the ends of moldings and to cut exact patterns of ornamental scroll designs. It is usually used to cut thin pieces of wood, including plywood.

Hacksaws

Hacksaws (Figure 2-5) are used to cut metal that is too heavy for snips or boltcutters. Metal bar stock can be cut readily with hacksaws. Hacksaws with tungsten-carbide blades or a metalcutting handsaw should be used for very hard materials, such as stainless steel or titanium. Newer, semiportable power hacksaws are becoming increasingly popular for heavy, production work.

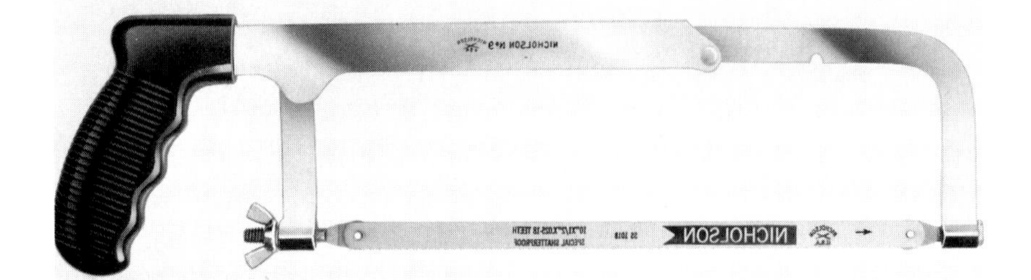

FIGURE 2-5. Hacksaw used for cutting light metals.

Hacksaw blades are made of high-grade tool steel, hardened and tempered. There are two types, the all-hard and the flexible. All-hard blades are hardened throughout, whereas only the teeth of the flexible blades are hardened.

In cutting with a hacksaw, the rigidity of the work is as important as that of the blade. The workpiece should be held steady in some way and positioned to engage the maximum number of sawteeth during the cut. A cut should be started off the edge (on the flat, for example) never against it. In other words, do not start a cut on a sharp corner.

Hand hacksaws are designed to cut on the forward or push stroke (away from the handle). Only rarely do conditions necessitate a pull cutting stroke. But in either case, make certain that the blade is inserted with the teeth pointing in the direction of the cutting stroke (the teeth point away from the handle of the hacksaw). The one possible exception, for which the blade can be dragged on the backward stroke, is when nonferrous metals of the softer varieties are being cut. A light backward drag, with no more pressure applied than the weight of the tool itself, is extremely helpful to alleviate clogging.

Handsaw and Hacksaw Safety. A saw that is not being used should be hung up or stowed in a toolbox. Before using a saw, be sure there are no nails or other edge-destroying objects in the line of the cut. When sawing out a strip of waste, do not break out the strip by twisting the saw blade. This dulls the saw and may spring or break the blade.

Be sure that the saw will go through the full stroke without striking the floor or some other object. Also, be sure to support the waste side of the board. This will help to prevent splitting and an uneven cut.

Planes

The plane (Figure 2-6) is the most extensively used wood shaving tool. While most of the lumber used in light construction is dressed on all four sides, planes must be used when performing jobs such as fitting doors and sash and interior trim work.

Bench and block planes are the most commonly used and are designed for general surface smoothing and squaring. There are three types of bench planes: the smooth plane, the jack plane, and the jointer plane. All are used primarily for shaving and smoothing with the grain.

The smooth plane is, in general, a smoother only; it will plane a smooth, but not an especially true, surface in a short time.

The jack plane is the general jack-of-all-work of the bench plane group. It can take a deeper cut and plane a truer surface than the smooth plane. The jointer plane is used when the planed surface must meet the highest requirements with regard to trueness.

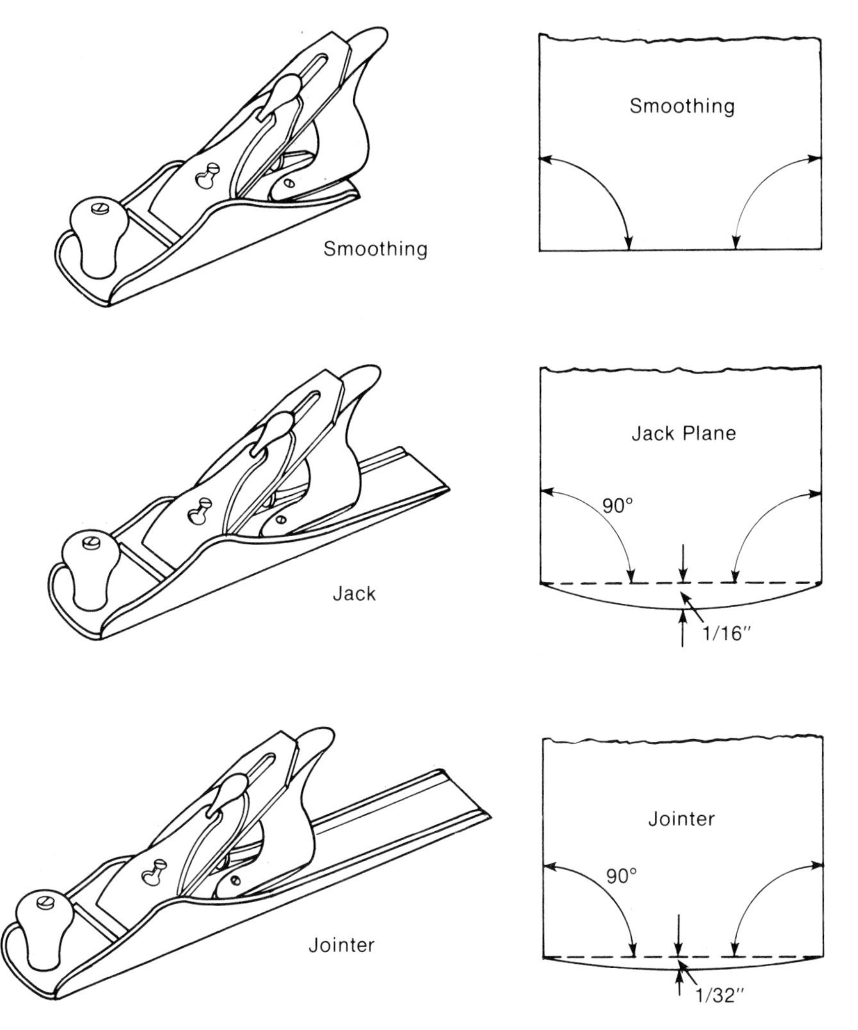

FIGURE 2-6. Planes used for wood shaving.

A block plane is a less heavy-duty tool. It is usually held at an angle to work and is used chiefly for the cross-grain squaring of endstock. It is also useful, however, for smoothing all place surfaces on very small work.

Wood Files and Rasps

For smoothing edges and small shaping problems, wood files and rasps (Figure 2-7) are essential tools. The cut of a file—that is, the angle and spacing of its teeth—determines its best use. Single-cut files have teeth running diagonally in parallel lines; double-cut files have two sets of crossing parallel lines, making a grid. Files are classified as coarse, bastard, second, and smooth, each (in order) representing a smaller distance between the lines and therefore creating a smoother finish on the work.

Rasps that have deeper, coarser cuts than files have triangular teeth rather than parallel lines of teeth like files. The rasp cut also differs from both the single and double cuts of files in that the teeth are individually formed and disconnected from each other. A rasp takes away wood particles at a rapid rate. Yet its shape—a square edge, a flat back, and a curved face—makes it useful on almost any surface. Many shaping and smoothing operations are also performed with another

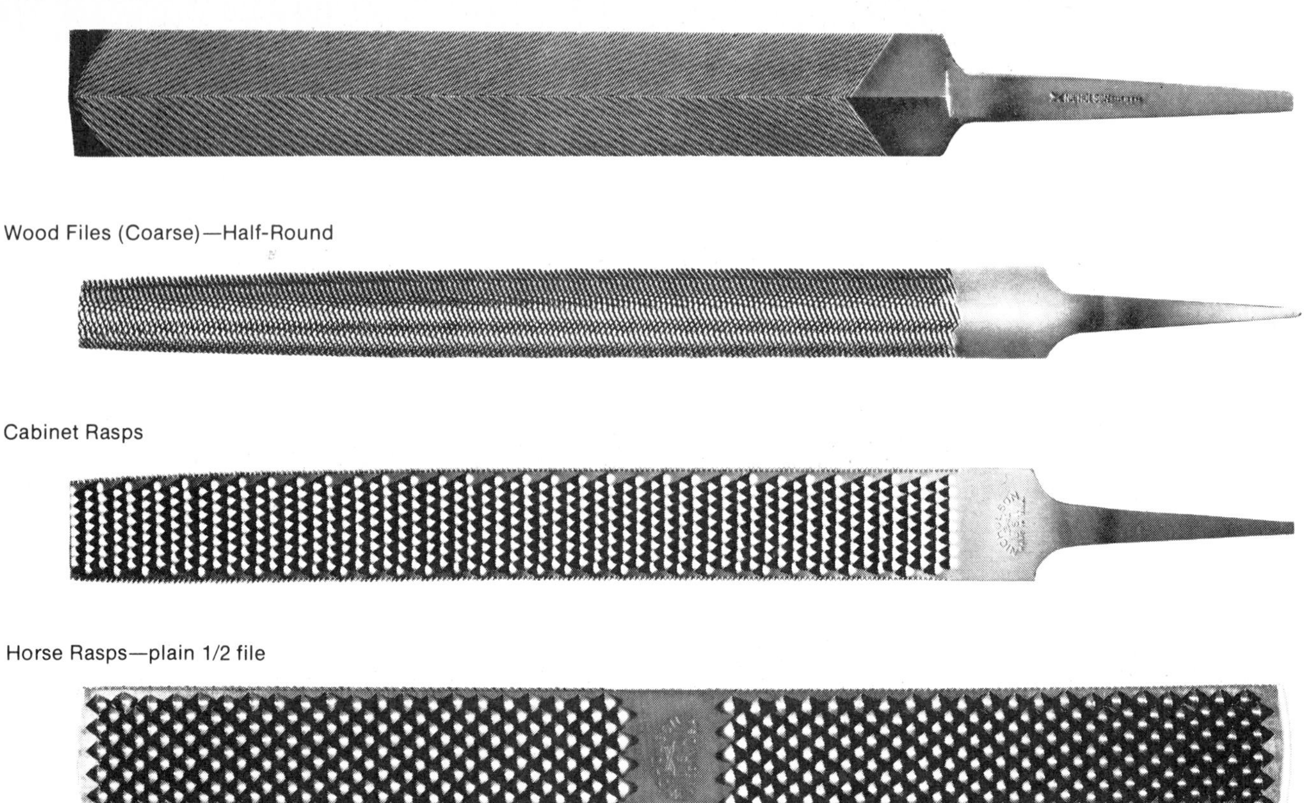

FIGURE 2-7. Wood files and rasps.

tool manufactured in a variety of shapes. Known as *Surform tools*, they are used for shaping, trimming, and forming wood, plastics, and soft metals. With a surface that looks a little like a roughened colander and a little like a file, Surform tools have many tiny cutting teeth that work like individual planes. All Surform tools have replaceable blades, both flat and curved. Special blades are made for cutting metals in the same fashion. An appropriate Surform tool can be used for almost any shaping problem; they are especially valuable for finishing any intricate shape.

Wood Boring Bits

When working with wood, it will frequently be necessary to bore holes. Auger bits and a variety of braces and drills are used for boring holes for screws, dowels, and hardware; as an aid in mortising (cutting a cavity in wood for joining members); in shaping curves; and for many other purposes (Figure 2-8).

Auger bits are made in either single or double twist. The most commonly used is the single-twist bit, which is best for fast, easy boring.

Expansive auger bits have adjustable cutters for boring holes of different diameters. They also have adjustable cutting blades which can be set to bore holes of any diameter within their range. Another wood-boring bit, the Forstner bit, does not have a feed screw and must be fed by pushing while turning. Because of the absence of a feed screw,

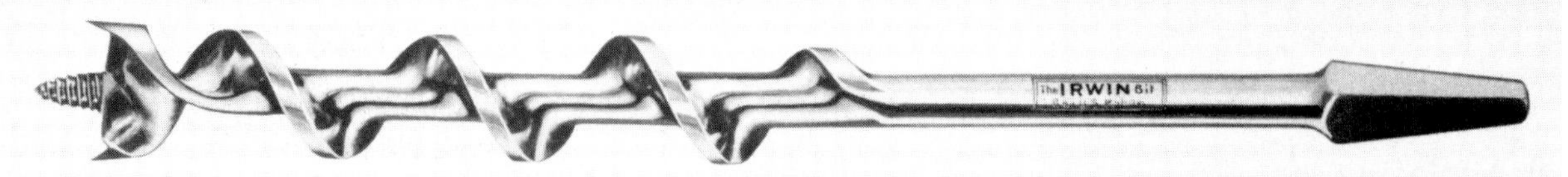

FIGURE 2-8. Auger bit used in wood boring. (Photo furnished by courtesy of Irwin Auger Bit Company.)

Forstner bits will bore holes almost through a board without defacing the opposite side of the board. To center or start a Forstner bit, scribe a circle the size of the hole with dividers, and press the rim of the Forstner bit into it.

Another bit, the gimlet bit, is used to bore holes through very thick timbers and planks sometimes used in construction work (especially for post and beam construction).

Braces are used to hold the auger bit and give enough leverage to turn the bit. Several braces are still used: the plain, the ratchet, the corner, and the short brace (Figure 2-9). All braces have an adjustable chuck that receives bits from 1/4″ to 1-1/8″ (6 mm to 29 mm) in diameter and will also receive most special bits, such as the expansive auger bit, Forstner bit, wood twist drill, and gimlet bit.

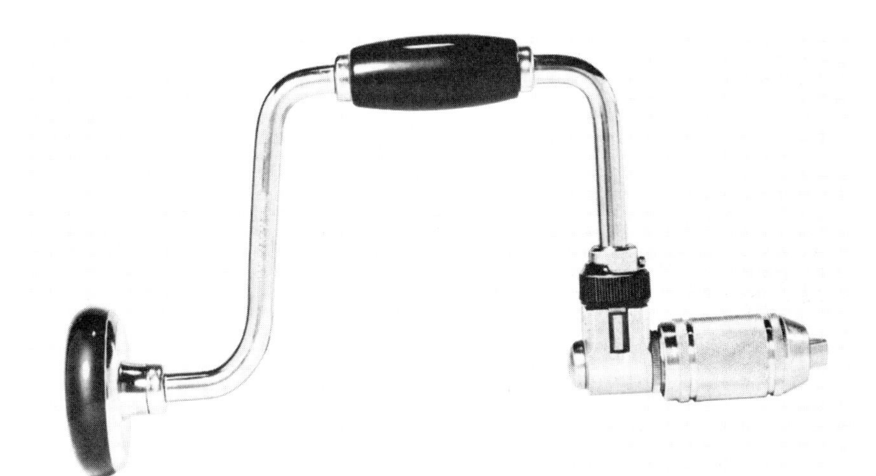

FIGURE 2-9. Braces used to secure auger bits.

Chisels

There are two main groups of chisels: woodcutting (Figure 2-10) and metal cutting (Figure 2-11). Depending upon whether they are intended for rough cutting or finishing work, woodcutting chisels are classified as either framing or finishing chisels. Light-duty wood chisels are fitted with wood or plastic handles. These chisels can be worked by hand or struck with a soft-faced mallet. Never use a steel hammer or sledge.

Heavy-duty wood chisels are forged from a single piece of tool steel. Designed for rougher work, these all-metal tools can be struck with a hammer or small sledge. The two most popular all-steel wood chisels are the flooring and the electrician's varieties. As their name implies, flooring chisels are used to cut through flooring and flooring nails. They can also be used to cut through wood shingles and clapboard sidings. Electrician's chisels are simply larger versions of the flooring

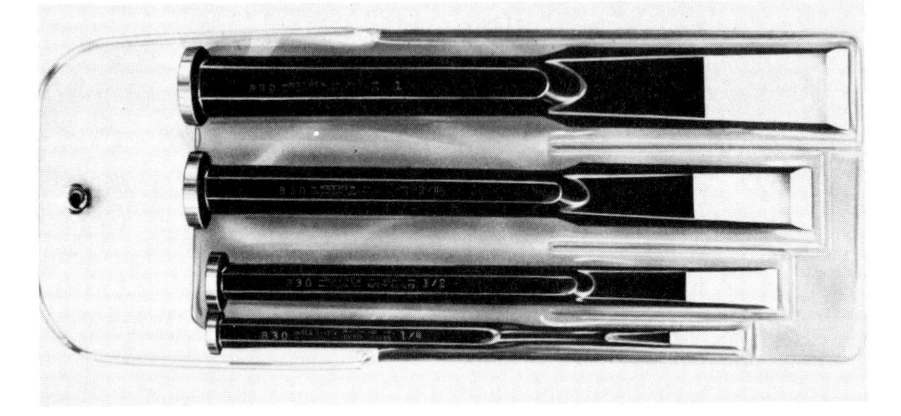

FIGURE 2-10. Wood cutting chisels. Never strike a wooden-handled tool with a steel striking tool. (Photo furnished by courtesy of Mayhew Steel Products, Inc.)

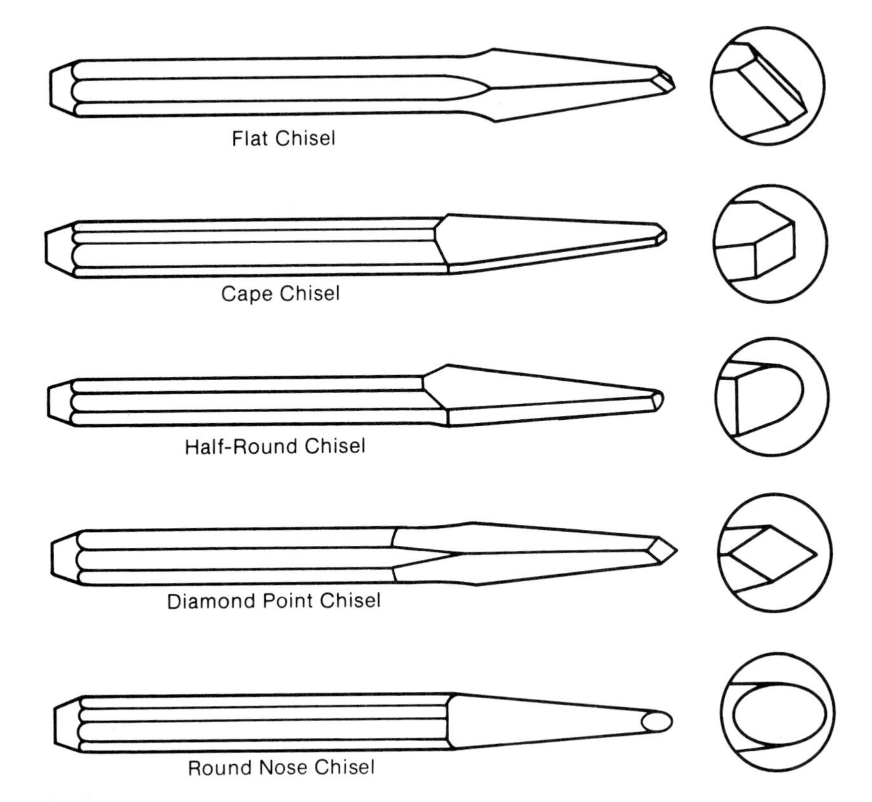

FIGURE 2-11. Metal cutting or cold chisels.

chisel. They are used to cut channels in beams or joists for Romex or armored electrical cable.

Metalcutting chisels are designed for cutting and chipping cold metals such as steel, cast and wrought iron, aluminum, brass, and copper. Also called cold chisels, these tools cut any metal softer than the material from which the chisel is made.

Snips and Shears

Snips (Figure 2-12) and shears are used for cutting sheet metal and steel of various thicknesses and shapes. Normally, the heavier or thicker materials are cut by shears.

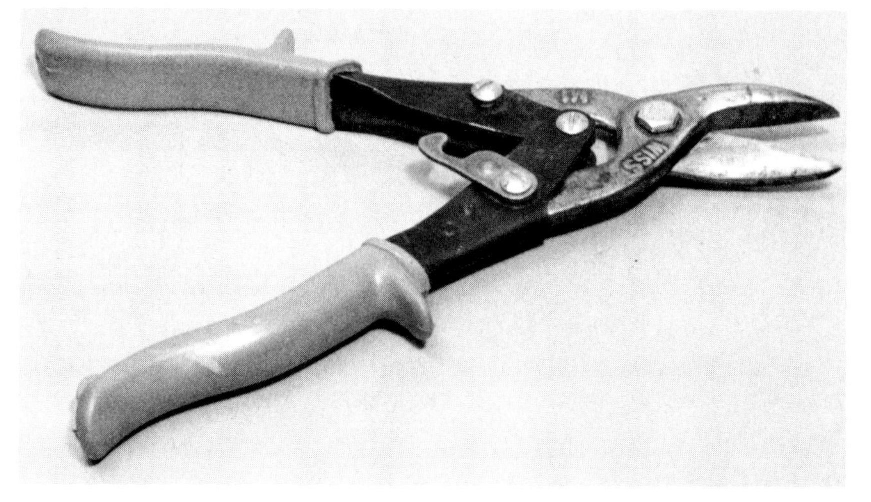

FIGURE 2-12. Hand or tin snips.

One of the handiest tools for cutting light (up to 1/16″ [2 mm] thick) sheet metal is the hand or tin snip. Straight hand snips have blades that are straight and cutting edges that are sharpened to an 85° angle. Tin snips will also work on slightly heavier gauges of soft materials such as aluminum alloys.

Snips will not remove any metal when a cut is made. There is danger, though, of causing minute metal fractures along the edges of the metal during the shearing process. For this reason, it is better to cut just outside the layout line.

Cutting extremely heavy-gauge metal always presents the possibility of springing the blades. Once the blades are sprung, hand snips are useless. Therefore, when cutting heavy material use the rear portion of the blades. This procedure not only avoids the possibility of springing the blades, but also creates greater cutting leverage.

Turning Tools

This group consists of screwdrivers and wrenches.

Screwdrivers

Not only are screwdrivers (Figure 2-13) one of the most basic of tools, they are also one of the most abused. Screwdrivers are designed for one task only—to drive and remove screws. A screwdriver should not be used as a pry bar, a scraper, a chisel, or a punch.

When using a screwdriver, it is important to select the proper size and slot design so that the blade fits the screw slot properly. This prevents burring the slot and reduces the force required to hold the driver in the slot. Always keep the shank perpendicular to the screwhead.

Besides the standard slot design, recessed screws are available in various other shapes including the Phillips, Frearson (Reed and Prince), and newer Torq-Set types. Other special screwdrivers for recessed-type screws are the clutch-head and Pozidriv. An offset screwdriver may be used where there is not sufficient vertical space for a standard or recessed screwdriver.

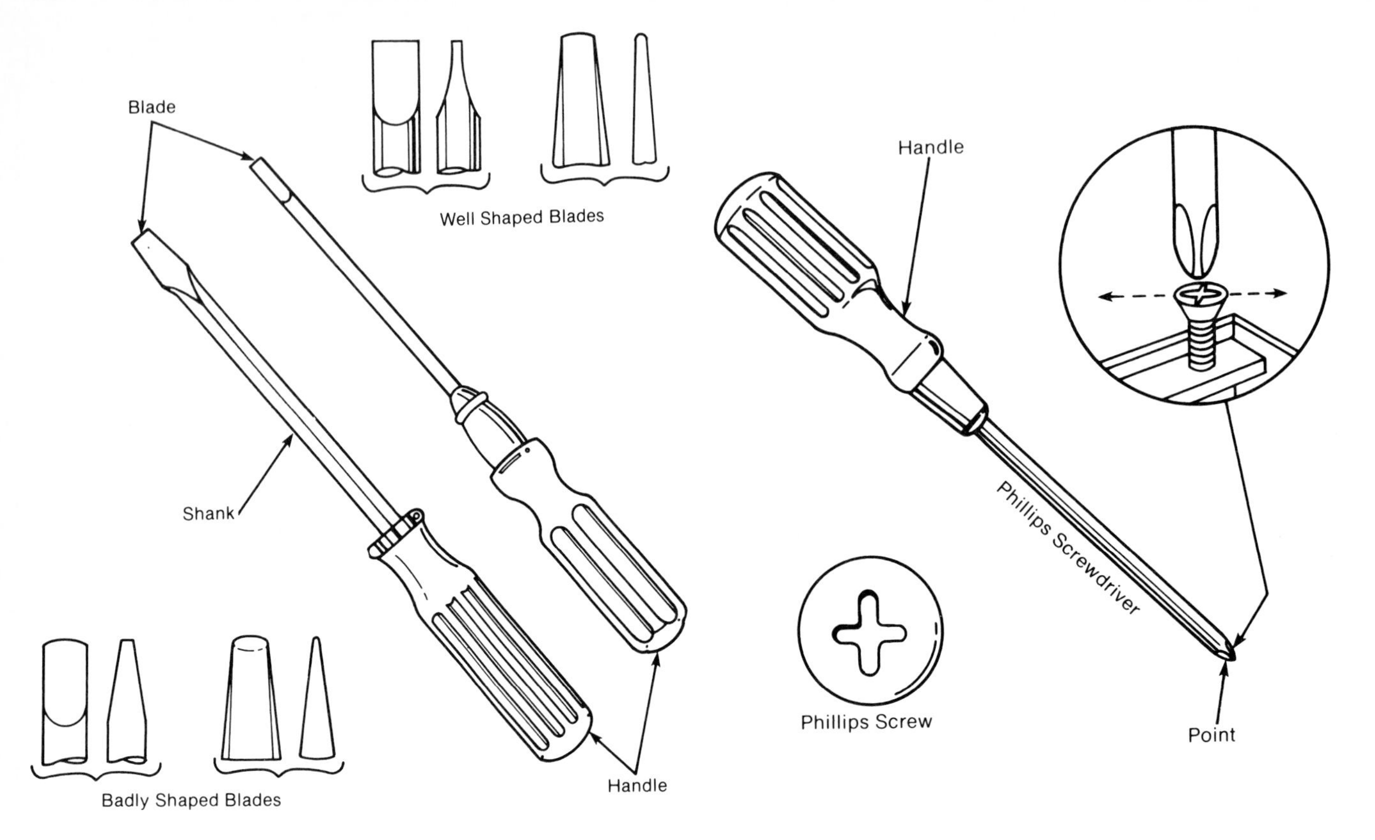

FIGURE 2-13. Screwdrivers used in light construction work.

Wrenches

Wrenches are used to exert a twisting force for loosening or tightening bolt heads, nuts, studs, pipes, and other threaded materials. Special wrenches designed for specialty jobs are usually nothing more than variations of the basic wrench design (Figure 2-14).

Solid, nonadjustable wrenches with openings in one or both ends are called open-end wrenches. Box wrenches completely surround or box a nut or bolt head and are less likely to slip off the work than open-end

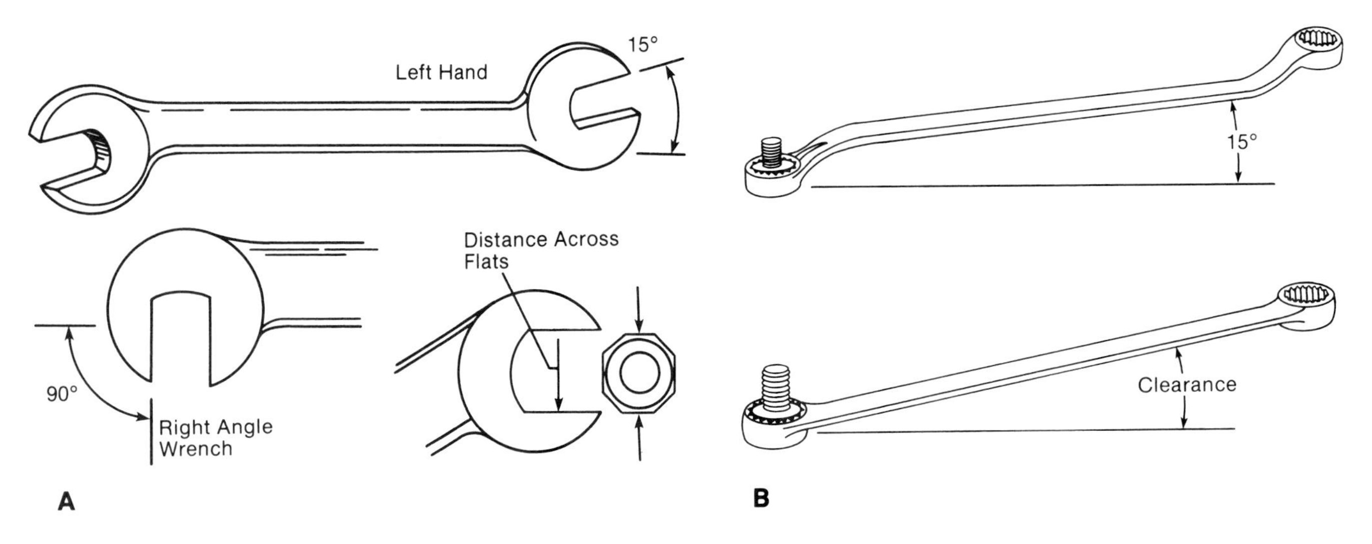

FIGURE 2-14. (A) Open-end and (B) box design wrenches.

wrenches. Socket wrenches consist of a special wrench handle with any number of interchangeable wrench heads, while adjustable wrenches have a single open-end wrench that can be adjusted to various bolt sizes.

Pipe wrenches, chain pipe wrenches, and strap wrenches are all designed to rotate or hold pipes and other round or cylindrical work (Figure 2-15). Internal pipe wrenches that fit on the inside of pipes rather than the outside are also available.

FIGURE 2-15. Pipe wrenches used to rotate or hold pipes or other round objects.

Holding Tools

Holding tools can be placed in either of two basic categories: hand-held holding tools (usually some type of pliers) and vises and clamps.

Pliers

Pliers are made in many styles and sizes and are used to perform many different operations (Figure 2-16). They are used for cutting purposes as well as holding and gripping small articles in situations where it may be difficult or impossible to use one's hands.

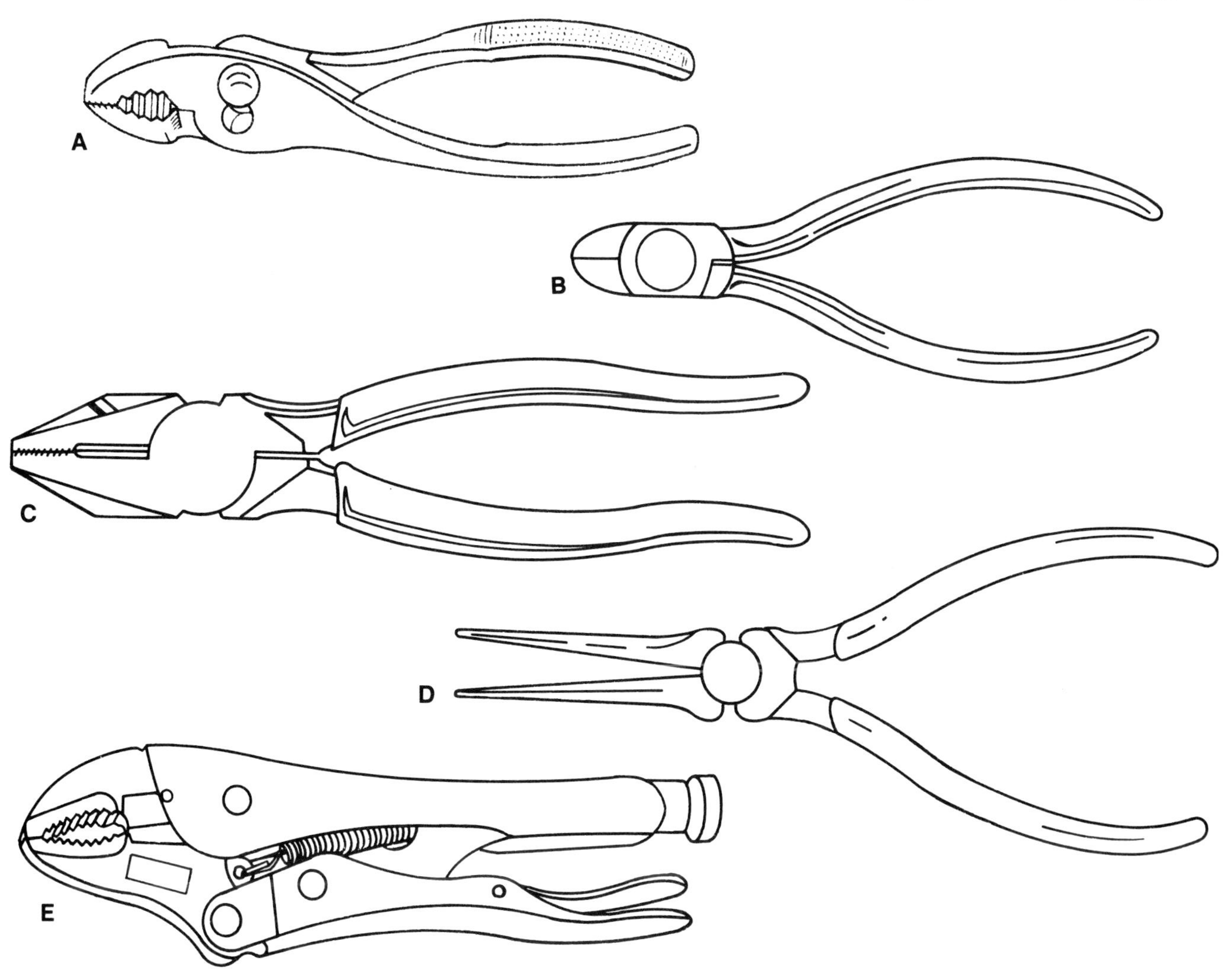

FIGURE 2-16. Various types of pliers: (A) slip-joint, (B) diagonal cutting, (C) side cutting, (D) long-nose, and (E) locking wrench.

Vise grip pliers are especially useful. They can be used for holding objects regardless of their shape. A screw adjustment in one of the handles makes them suitable for several different sizes. The jaws of vise grips may have standard serrations or a clamp-type jaw. Vise grip pliers have an advantage over other types of pliers because they can be clamped onto an object and they will stay. This will leave a worker's hands free to do other tasks. Vise grip pliers should be used with care since the teeth in the jaws tend to damage the object on which they are clamped. They should not be used on nuts, bolts, tube fittings, or other objects which must be reused.

Vises and Clamps

Vises (Figure 2-17) are used for holding work when it is being planed, sawed, drilled, shaped, or riveted. The machinist's bench vise, bench and pipe vise, and pipe vise are common designs.

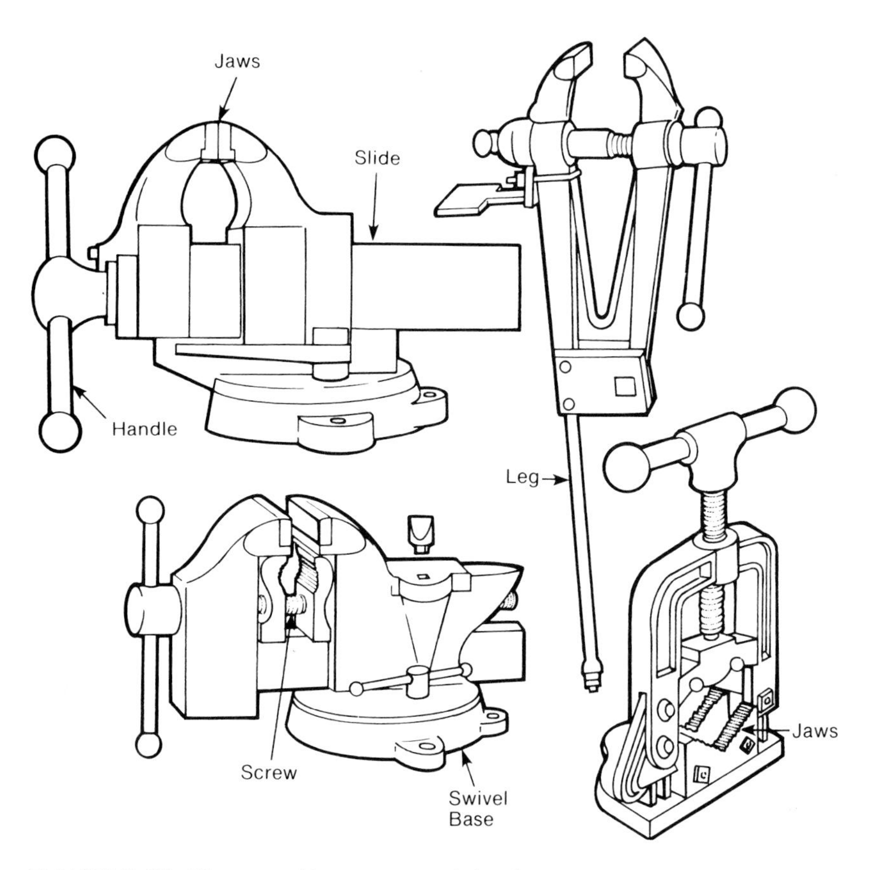

FIGURE 2-17. Vises used to secure work in place.

Clamps are used for holding work that cannot be satisfactorily held in a vise due to its size and shape. They are also used in locations where a vise would be impractical. The most common types of clamps used in light building construction are C-clamps, hand screw clamps, bar clamps, and pipe clamps (Figure 2-18).

Measuring Tools

Every step of the construction process relies on tolerances of one degree or another. Needless to say, this requires a great deal of measuring and a variety of measuring tools. These layout and marking tools

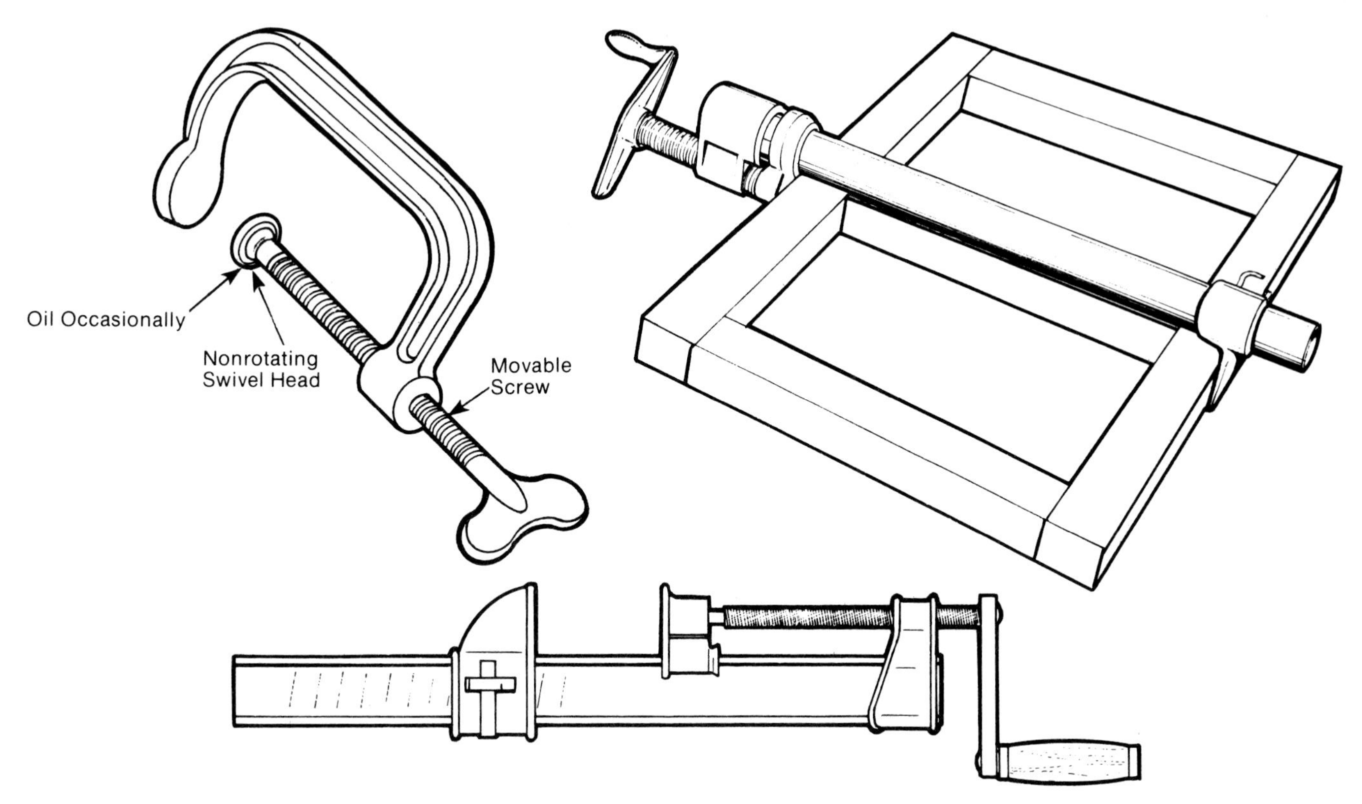

FIGURE 2-18. Clamps.

include such items, as the framing square, combination square, try square, sliding T-level, dividers, scriber, trammel points, marking gauge, mortise gauge, butt gauge, chalk line, and straightedge. Another group of measuring tools are the leveling and plumbing type, which include the carpenter's level, line level, level sights, plumb bob, and builder's levels. Finally, light construction work requires such common measuring devices as the folding rule, spring-loaded pocket tapes, and rolled steel tape, as well as the more specialized story rod.

Framing Square

The standard framing square, also called a carpenter's square or rafter square, is one of the most basic layout tools (Figure 2-19). It consists of a long, wide member called the blade and a shorter member called the tongue, which forms a right angle with the blade. The outer corner where the blade and tongue meet is called the heel. The face of the square is the side one sees when the square is held with the blade in the left hand and the tongue in the right hand, with the heel pointed away from the body. The blade is usually 24″ (610 mm) long and 2″ (51 mm) wide, and the tongue varies from 14″ to 18″ (356 mm to 457 mm) long and is 1-1/2″ (38 mm) wide, as measured from the heel. The outer and inner edges of the tongue and the blade, on both face and back, are graduated in inches and millimeters.

The framing square is used most frequently to find the length of the hypotenuse (longest side) of a right triangle when the lengths of the other two sides are known. This is the basic problem involved in, for example, determining the length of a rood rafter, a brace, or any other member which forms the hypotenuse of an actual or imaginary right triangle.

FIGURE 2-19. The standard framing square or carpenter's square.

Figure 2-20 shows how the framing square is used to determine the length of the hypotenuse of a right triangle when the other sides are each 12″ (305 mm) long. Plane a true, straight edge on a board and set the square on the board so as to bring the 12″ (305 mm) mark on the tongue and the 12″ (305 mm) mark on the blade even with the edge of the board. Draw the pencil marks shown in the second view. The distance between these marks, as measured along the edge of the board, is the length of the hypotenuse of a right triangle with the other sides each 12″ (305 mm) long. This distance, which is called the bridge measure, will be just a shade under 17″ (432 mm).

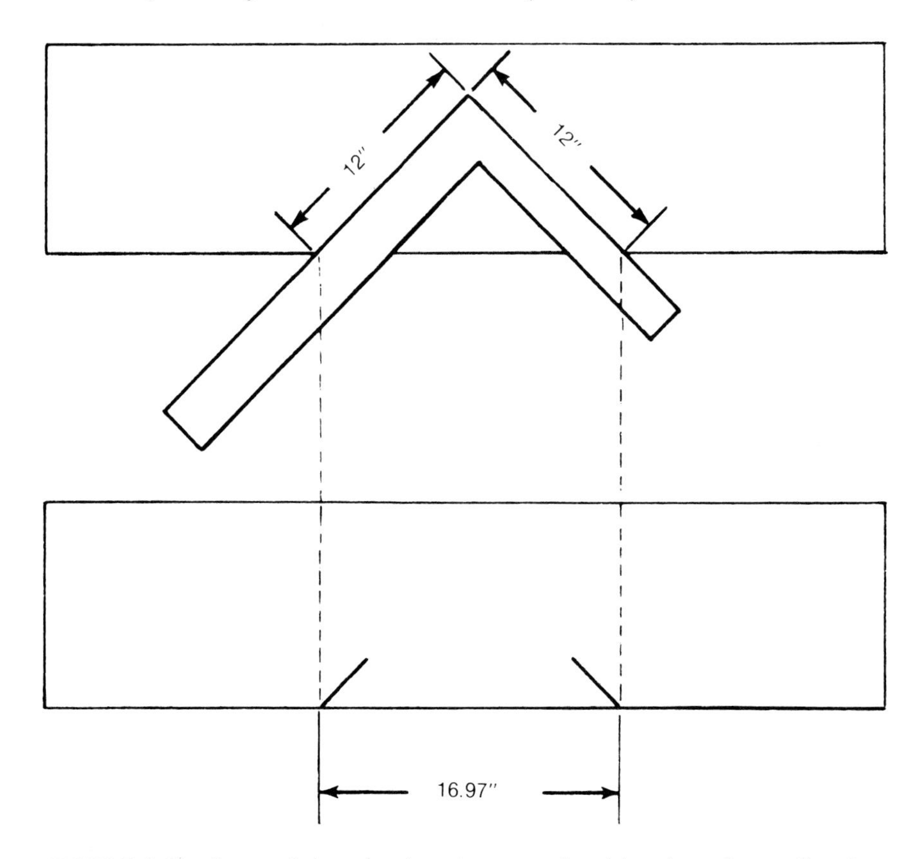

FIGURE 2-20. Determining the hypotenuse of a triangle using a framing square.

Actually, the problems that can be solved with a good framing square are so varied that books have been written on the square alone. Thus, the carpenter or builder who desires to take full advantage of the square's capacities for solving a whole host of construction problems should obtain and study one of the books on the square. However, one of the most common uses of the framing square—laying out risers and stringers—is illustrated in Figure 2-21.

Try Square

The try square (Figure 2-22) consists of two parts, a thick wood or metal handle and a thin steel blade, at right angles to each other. Most try squares are made with the blades graduated in inches and fractions of an inch. In metric measure they are graduated in millimeters. The blade length varies from 3″ to 12″ (76 mm to 305 mm). This square is used for testing the squareness of ends and edges of stock. Held upside down on a flat surface, it may be used to check for warp; held against

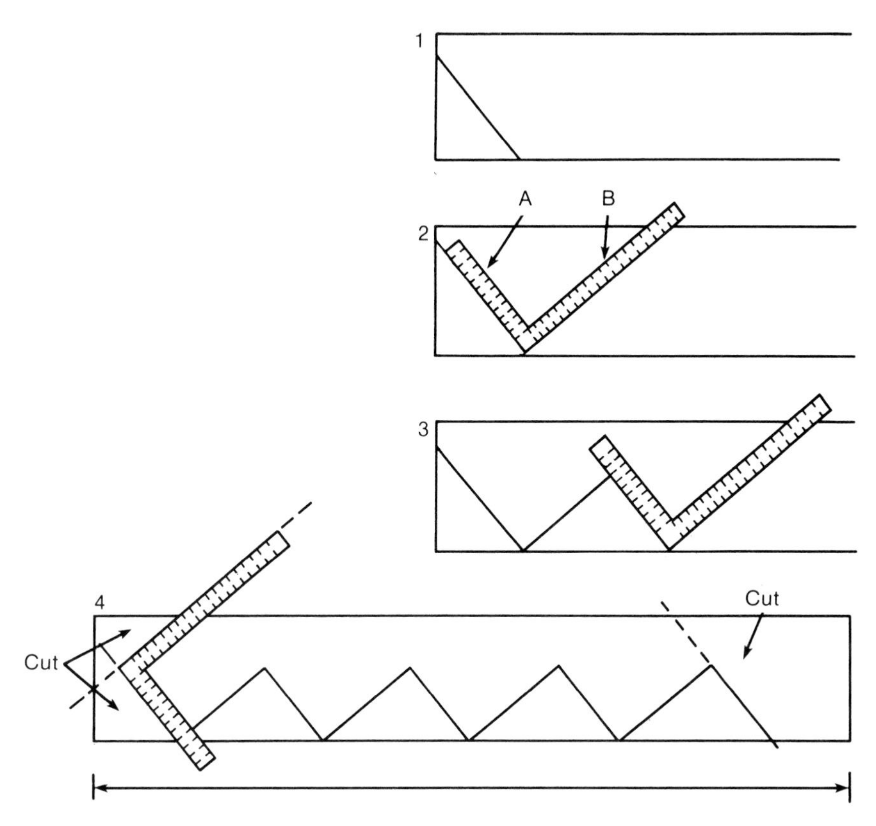

FIGURE 2-21. Laying out risers and stringers using a framing square. (1)Mark the tread size on the width of the board, the riser on one side. Connect the points to form a triangle. (2) Lay the framing square on the line; mark the riser height at A, the tread width at B. (3) Place the square with the point at the edge of the stringer, and the riser width meeting B. Draw the lines, and repeat to the end of the board. (4) The dotted lines indicate the cutoff points for the top and bottom of the stringer.

FIGURE 2-22. The try square used in testing for squareness (Photo furnished by courtesy of Stanley.)

an inside corner, it may be used to check a right angle. It can also be used to measure sections not exceeding the length of the blade and to scribe across the face of stock.

A try-miter square is similar to the conventional try square except that it has a cross beam that permits its use as a quick mitering guide, in either the right- or left-hand position, in addition to its squaring function.

Combination Square

The combination square (Figure 2-23) looks a lot like a try square, but it can perform more tasks. The square head may be adjusted to any position along the 12″ (305 mm) blade and clamped securely in place.

FIGURE 2-23. The combination square is a multi-functional measuring tool.

The combination square can serve as a depth gauge, height gauge, or marking gauge. Two of the faces of the head are ground at right angles to each other and a third face at 45°. This means that the combination square can be used as an inside and outside try square, as well as a miter square. A small spirit level is built into the head for checking whether surfaces are plumb, and a small scriber is usually housed in a hole in the end of the head for marking layout lines. Of course, without the square head, the square can be used as a straightedge.

Some combination squares have a center head that can slide onto the blade in place of the square head. This is a V-shaped member designed so that the center of the 90° V lies exactly along one edge of the blade. This attachment is useful when locating the exact center of round stock.

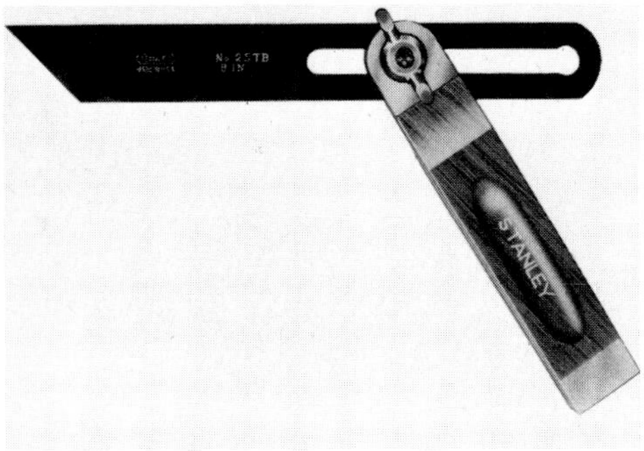

Sliding T-Bevel

FIGURE 2-24. The sliding T-bevel. (Photo furnished by courtesy of Stanley.)

The sliding T-bevel (Figure 2-24) is an adjustable try square with a slotted, beveled blade. The blade is normally 6″ or 8″ (152 mm or 203 mm) long. The sliding T-bevel is used for laying out angles other than right angles and for testing constructed angles such as bevels.

To adjust a sliding T-bevel to a desired setting, loosen the blade screw at the round end of the handle just enough to permit the blade to slide along its slot and to rotate with little friction.

To set the blade at a 45° angle, hold the handle against a framing square with the blade intersecting equal gradations on the tongue and blade of the square (Figure 2-25A), or hold the bevel against the edges of a 45° drafting triangle (Figure 2-25B). When using a drafting triangle for setting a sliding T-bevel, a different size of triangle must be used for each different setting. A 45° angle can also be set by using the squaring head of a combination set (Figure 2-25C).

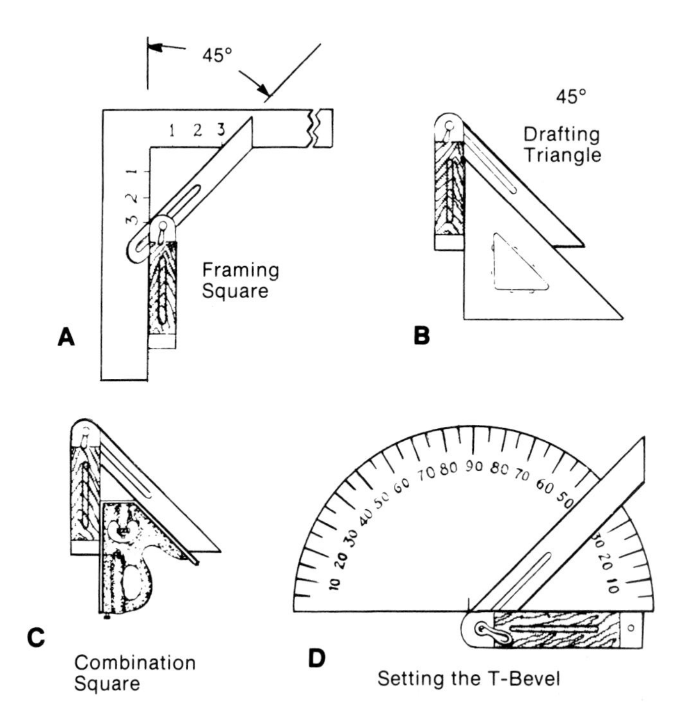

FIGURE 2-25. Methods of setting the T-bevel at 45°: (A) framing square, (B) drafting triangle, (C) combination square, and (D) protractor.

A sliding T-bevel can be set to any desired angle with a protractor. Loosen the blade screw as before. Hold the bevel with its blade passing through the gradation selected and the center of the protractor (Figure 2-25D).

To test a chamfer or bevel for trueness, set the T-bevel to the required angle, and hold the handle to the working face of the stock being tested. Face a source of light, and with the blade brought into contact with the surface to be tested, pass the blade along the length of the surface. The appearance of light between the blade and the surface of the stock indicates where the angle is not correct.

A related device, the angle divider, is a double-bevel tool that functions in somewhat the same manner as a T-bevel in marking off identical angles on several pieces of work. It is also used to take off and divide angles for a miter cut in one operation or to find the center of two pieces of wood joined at an angle. The handle is graduated on the back for laying off 4-, 5-, 6-, 8-, and 10-sided work. The square blade may also be used as a try square.

Dividers

Dividers have many uses. They are absolutely necessary for scribing a line when matching a workpiece (perhaps a wall panel) to an irregular edge such as brick or stone masonry (Figure 2-26). Dividers can also be used to accurately step off a measurement several times, as well as to scribe circles and arcs (Figure 2-27).

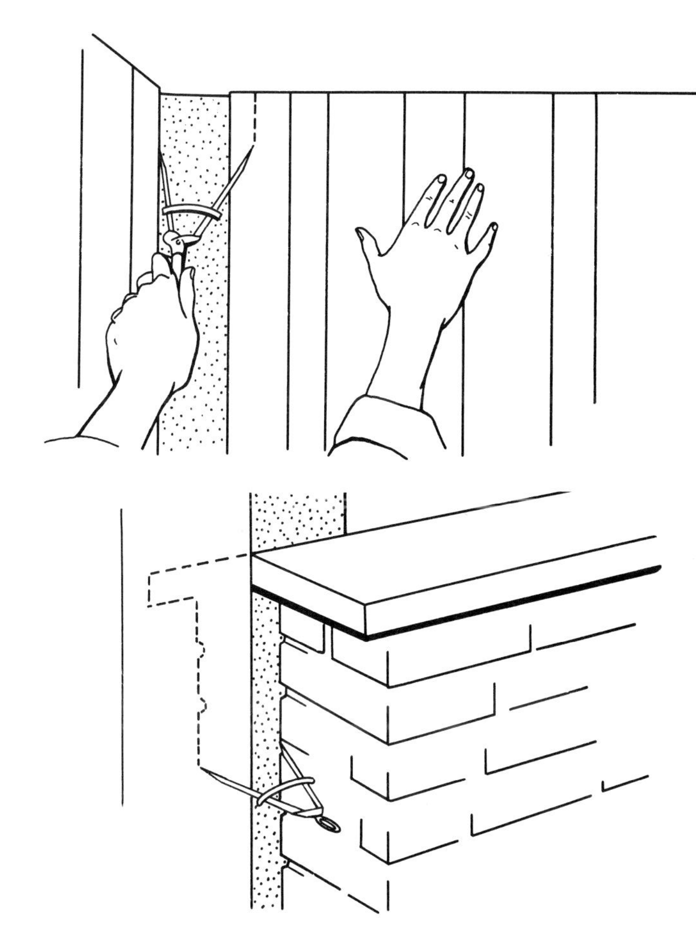

FIGURE 2-26. Using dividers to transcribe a cutting line.

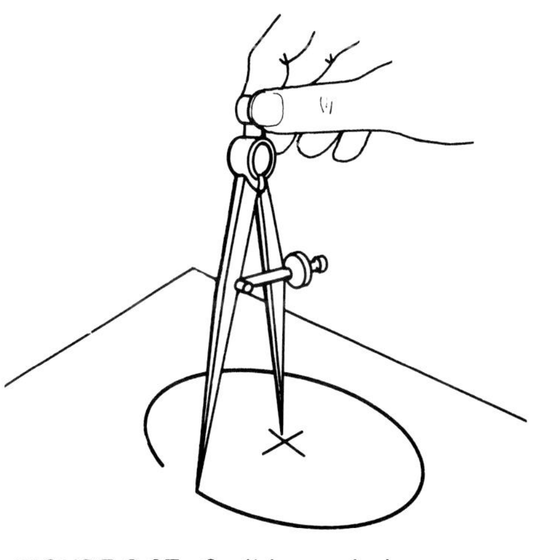

FIGURE 2-27. Scribing a circle.

Trammel Points

Trammel points (Figure 2-28) are used to lay out the distance between two points and to scribe circles beyond the capacity of dividers. Some trammel points are made to slide over a metal or wooden beam, while others are in the form of C-clamps and fasten to one side of a straight board. Some trammel points are also designed with a special socket for a carpenter's pencil for marking.

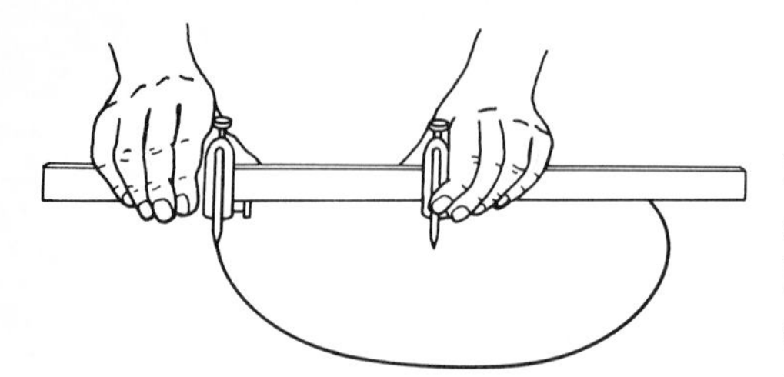

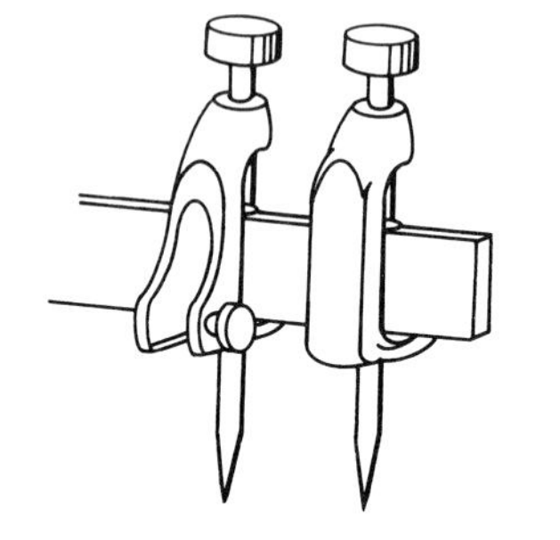

FIGURE 2-28. Trammel points.

Marking Tools

There are several specialized marking tools used in construction besides the more common pencils, chalk, pens, and crayon.

The scriber (Figure 2-29) is used for marking lines on wooden stock and can also be used to mark metal. It is especially useful when measuring with a rule. Centers can be located by using it to draw two intersecting lines. A bent end is convenient for marking the insides of cylindrical objects or partially closed recesses. Keep the scriber sharp, and use it like a pencil, with only enough pressure to make a clear mark.

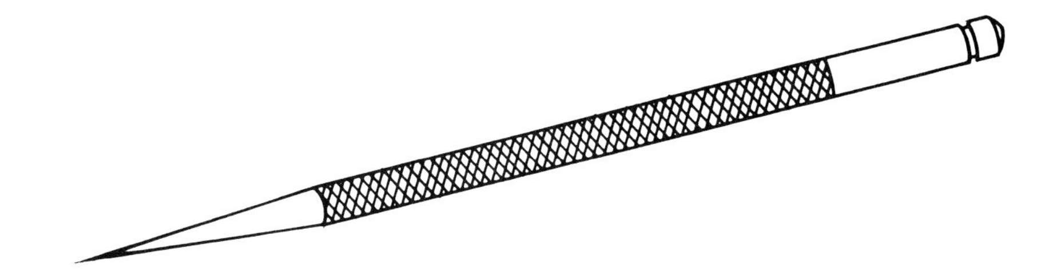

FIGURE 2-29. Scriber.

The awl (Figure 2-30), or scratch awl as it is often called, is a sharp-pointed piece of steel, like an ice pick, which is used to score cutting lines on metal or wood. The awl is also used for making pilot holes for screws and small nails.

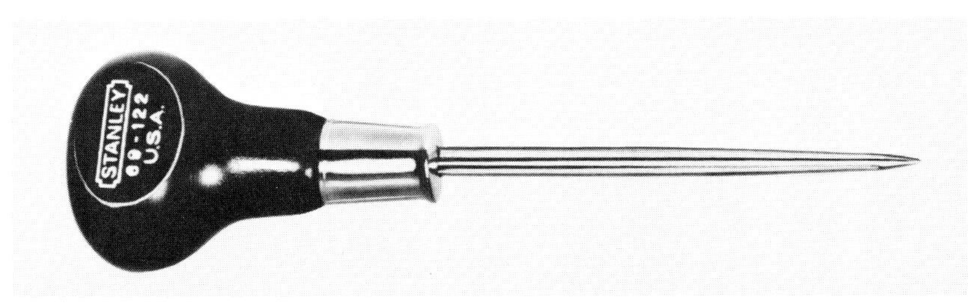

FIGURE 2-30. Awl.

Marking Gauges

Marking gauges (Figures 2-31A and 2-31B) are used to mark off guidelines parallel to an edge, end, or surface of a piece of wood or metal. They are made of metal or wood and consist of a graduated beam about 8″ (203 mm) long on which a head slides. The head can be fastened at any point on the beam by means of a thumbscrew. The thumbscrew presses a brass shoe tightly against the beam and locks it firmly in position. The steel pin or spur that does the marking projects from the beam about 1/16″ (2 mm).

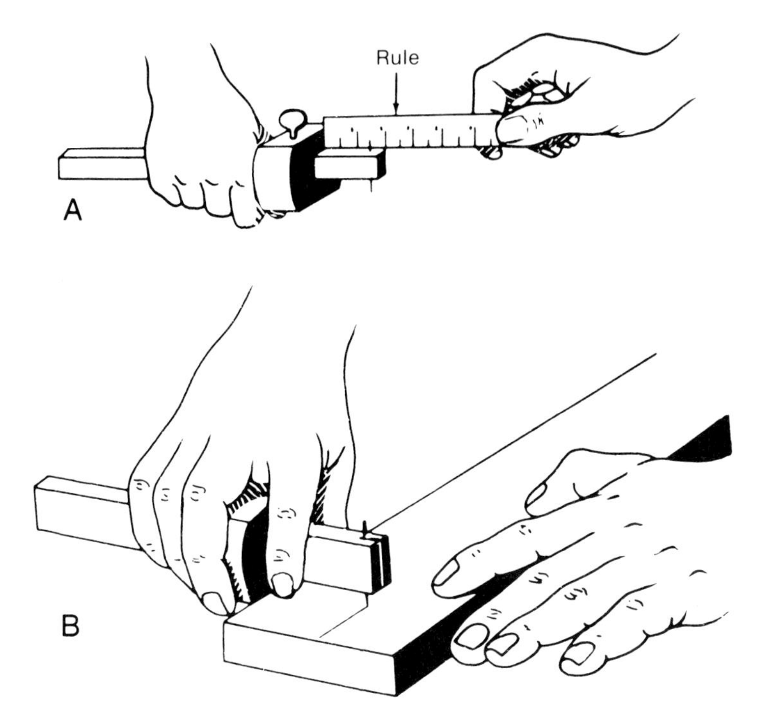

FIGURE 2-31. (A) Setting the marking gauge, and (B) using the gauge.

Mortise Gauge.

The mortise gauge is simply a variation of the marking gauge, but it has two beams instead of one and in place of pins it has two tiny sharp-edged wheels to make the appropriate marks. One of the beams or arms is set to mark the distance from the edge of the workpiece to the edge of the mortise. The other beam is then set to mark the width of the mortise. This gauge is useful for marking off both mortises or the corresponding tenons this way.

Butt Gauges

Butt gauges are really designed for only one job—to lay out the gains for butt hinges. The gauge is simply a metal block with two sliding arms. One arm has one marker attached; the other arm has two. The arm with one marker is used to set the depth of the hinge gain. The other markers set the distance from the back edge of the door that the hinge will reach and the distance from the inside face of the door rabbet that the jamb leaf of the hinge will reach.

Chalk Line

Long, straight lines between distant points are marked by snapping a chalk line (Figure 2-32). The line is first chalked by holding the chalk in the hand and drawing the line across it several times. The line is then stretched between the points, pulled taut, and snapped. For an accurate line, never snap the chalk line for a distance greater than 20′ (6.1 m). The chalk line is one of many essentially simple tools which are so necessary for construction work.

FIGURE 2-32. Chalk line.

Straightedge

Some form of solid straightedge is a standard, basic construction tool. Usually a straightedge can be made on the worksite from job to job to meet the specific requirements of a particular situation. The job at hand will determine the straightedge dimensions; however, they are usually from 2′ to 4′ (.6 to 1.2 m). When making any straightedge, preferably out of hardwood, the edges must be dressed perfectly straight and should be parallel to each other.

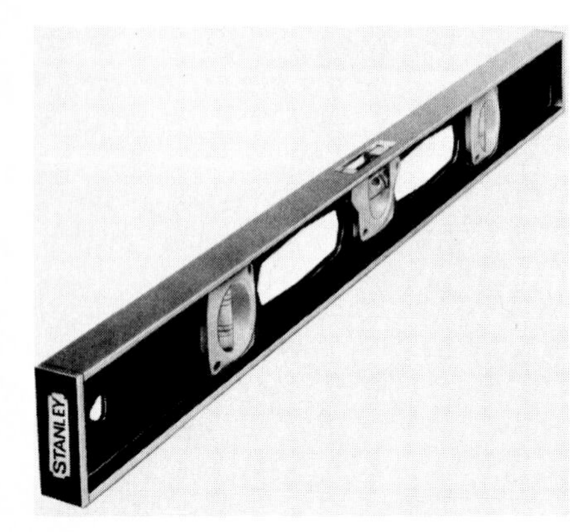

FIGURE 2-33. Carpenter's level.

Carpenter's Level

The carpenter's or spirit level (Figure 2-33) is usually of either a 24″ (610 mm) wood block or metal I-beam construction with true surface edges. While this type of level may have two or more vials, there are two important bubble tubes in it. One is in the middle of one of the long edges. The other is at right angles to this and parallel to the end of the level. The bubble tubes are slightly curved glass vials nearly filled with alcohol. As a bubble of air in such a tube will rise to the highest point, the bubble will be in the middle of the tube only when the tube is in a horizontal position. Scratch marks at equal distances from the middle of the tube mark the proper position of the bubble when the surface on which the tube rests is level.

The carpenter's level is used to determine whether a surface is level or plumb. Level usually describes a horizontal surface which, throughout its extent, lies on a line corresponding to that of the horizon. Plumb means vertical, or at right angles to level.

To test for the level of a surface, lay the carpenter's level on the surface and see where the bubble comes to rest. If the surfaces are level and the level is in adjustment, the bubble will come to rest exactly between the two scratch marks. Turn the level end-for-end and re-check. The bubble should come to rest in the same place. If it does not, raise the end of the surface being tested, which is toward the low end of the bubble, until it checks level.

To check for plumb, set the long side of the carpenter's level against the upright to be tested, and use the bubble which is set in the end in the same way as just described. Again, turn the level end-for-end to ensure accuracy.

Mason's levels are very similar in design to carpenter's levels, except that they are longer—usually 48″ to 80″ (1,219 mm to 2,032 mm) long—and usually have more bubble vials.

Level Sights. Level sights (Figure 2-34) fit on a carpenter's level and are used in conjunction with it for sighting and aligning grades, walls, fences, and other surfaces which must be level.

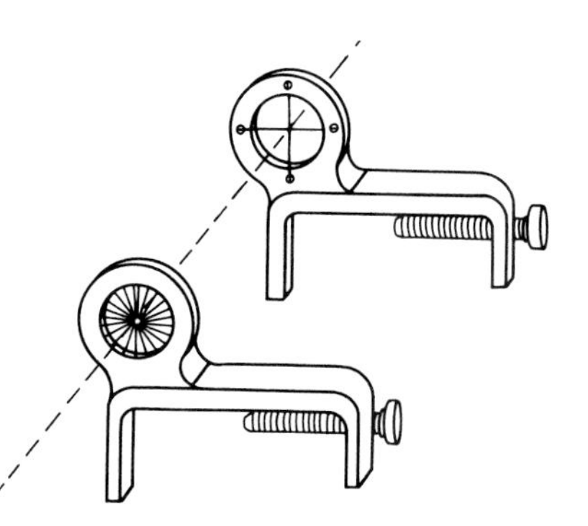

FIGURE 2-34. Level sights.

Line Level

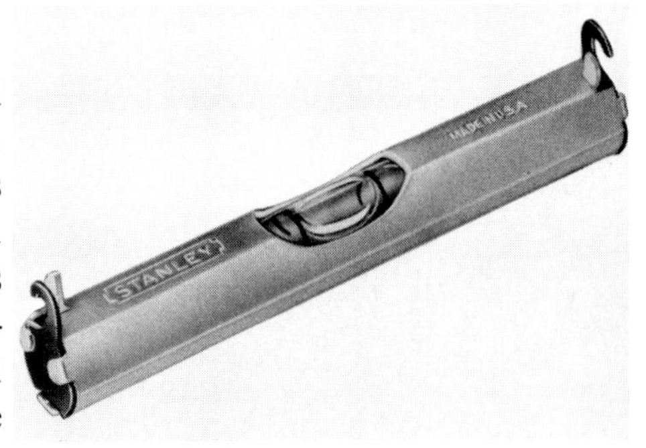

FIGURE 2-35. Line level.

The line level (Figure 2-35) consists of a bubble tube set into a metal case with a hook at each end, permitting it to be hung on a line or cord. The line level is used to test whether or not a line or cord is level. It is particularly useful when the distance between two points to be checked for level is too long to permit the use of a mason's level or carpenter's level. However, the line level has a disadvantage in that at long distances the line has a tendency to sway. To use the level, stretch a cord between the two points that are to be checked for level. Hang the line level on the cord, and adjust the line height until the bubble is in the middle of the tube.

Torpedo Level

Smaller than the carpenter's level, the torpedo level is nevertheless a useful addition to a tool inventory. It has two advantages over the larger models: first it has a bevel bubble for checking 45° angles, and second, it has a grooved base for unassisted positioning on shafts, pipes, and similar rounds.

Level Care. Levels must be checked for accuracy. This is readily accomplished by placing the level on a true horizontal surface and noting the vial indication. Reverse the level end-for-end. If the bubble appears on one side of the gradations on the first reading and on the other side for the second reading, the level is out of true and must be adjusted.

Do not drop a level or handle it roughly. To prevent damage, store it in a rack or other suitable place when it is not in use.

Plumb Bob

A plumb bob (Figure 2-36) is a pointed, tapered, brass or bronze weight that is suspended from a cord to determine the true vertical or plumb line to or from a point on the ground. The plumb bob has many uses, but it is most frequently used in building to determine true verticality for erecting the uprights and corner posts of framework. It is also used for transferring and lining up points when surveying a lot.

A plumb bob is a precision instrument and must be cared for as such. If the tip becomes bent, the cord from which the bob is suspended will not show the true plumb line over the point indicated by the tip. For this reason many plumb bobs have a detachable tip, so that if the tip should become damaged it can be replaced without replacing the entire instrument.

FIGURE 2-36. Plumb bob.

Leveling Instruments

In light construction, either the builder's level or the transit level is commonly used to lay out foundations (Figure 2-37). The builder's level is not as sophisticated an instrument as the transit level and neither is as precise as an engineer's level which can be used in place of either, although it is rarely required for general work.

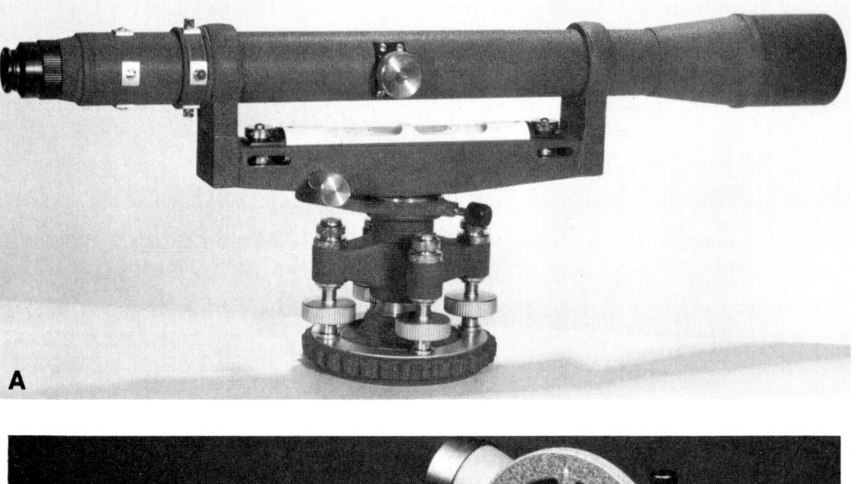

B

FIGURE 2-37. (A) Builder's or dumpy level and (B) transit levels. (Photos furnished by courtesy of Teledyne Gurley.)

The builder's level, also called a dumpy level, is a precision instrument that is used for locating points in a horizontal plane. Most levels consist of a telescope with a focusing knob resting on a leveling base plate. The level is equipped with a spirit level and four leveling screws so that it can be properly adjusted. Some levels are equipped with a horizontal circle graduated from 0° to 360°. The level is most often used to determine elevations and horizontal angles, but it can be used for grading, leveling, and aligning fences and driveways.

The transit level is very similar to the builder's level, except that it can pivot in a vertical plane. This added feature allows the transit level to measure vertical as well as horizontal angles. The transit level is also very useful in plumbing a vertical wall.

Both the builder's level and the transit level are mounted on a tripod, which is usually constructed of specially treated hardwoods that will not bend or warp even under adverse field conditions. The tripod is available with either fixed or extension legs. Extension legs allow lengthening or retracking for easier handling.

If either instrument is placed over a hub or if a plumb bob is attached to the bottom of the instrument, the plumb bob should be properly located before the instrument is leveled.

To level the instrument, the horizontal clamp screw must be loosened; then the telescope is positioned directly over two leveling screws. To level the telescope, one leveling screw should be turned in a clockwise direction while the other leveling screw is being turned counterclockwise. These leveling screws should be turned uniformly to center the bubble of the spirit level.

After the bubble is centered, the telescope must be rotated 90° and positioned directly over the other two leveling screws. These two leveling screws are then adjusted until the telescope is level.

After the second leveling operation, the telescope can be rotated to any desired angle and still be level. But, it is good practice to check the spirit level before each reading.

Care of Leveling Instruments. Since the builder's level and the transit level are precision pieces of equipment, there are certain rules that should be followed:

- The instrument should be kept clean and dry and placed in a carrying case when not in use.
- To keep the instrument from getting wet when in use, a plastic cover should be carried and used as needed.
- The leveling screws should be snug, but not overtight. Overtightening could warp the leveling base plate.
- The instrument should always be carefully carried in both hands and not slung over the shoulder.

Taking Level Sightings. Once the instrument is leveled, it is possible to take level sightings with the use of a leveling rod. A leveling rod is a thin wooden measuring rod calibrated to take precision vertical measurements to 0.01′ or to 0.005 (Figure 2-38). Some rods are equipped with a sliding target which is adjusted until the horizontal line on the rod aligns with the horizontal cross-hair of the level. The target is then locked into position and a reading is taken. To take a vertical measurement with a builder's or transit level

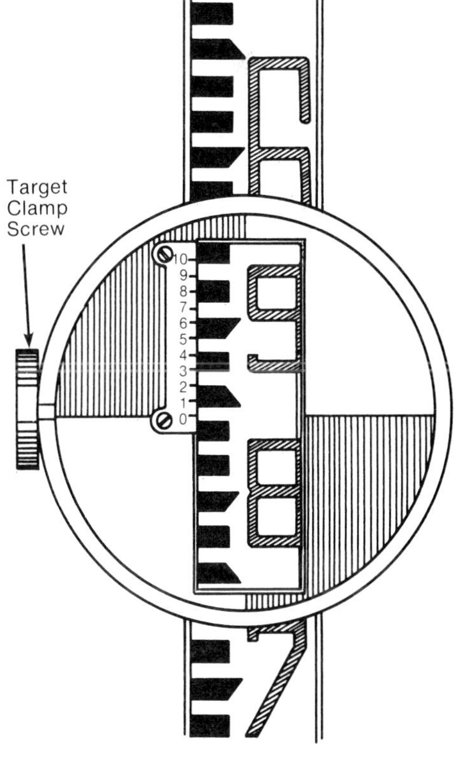

FIGURE 2-38. Leveling rod set for target reading of less than 7′ (2.1 m).

1. Have an assistant hold the leveling rod at some convenient distance away from the level (approximately 50′ [15.24 m]).
2. Aim the level at the rod by sighting along the top.
3. Sight through the eyepiece and adjust the focus by turning the focusing knob. Adjust the instrument until the object is centered. Tighten the horizontal motion clamp screw and adjust with the horizontal motion tangent screw until the vertical cross-hair is centered on the rod.
4. Have an assistant adjust the rod target up or down the rod until the horizontal cross-hair coincides with the horizontal line on the rod target. The assistant can now take the reading indicated on the rod.

This reading is known as the height of instrument (h.i.) for that position. For example if the reading is 3.12′ (1.0 m), it would mean that the horizontal cross-hair of the level is that distance above the ground.

If the leveling rod is moved to a second position and a second reading is obtained, the difference in elevation between the first and second positions can be determined. For example, if the second reading is 6.78′ (2.1 m), the difference in elevation would be 6.78′ − 3.12′ = 3.66′ (1.1 m).

Story Rods

Story rods are a final measuring device commonly used in light construction. Unlike many of the other instruments used, story rods are for vertical measurements only. They are derived from the idea of laying off on a thin wooden rod or piece of lath that heights of critical horizontal members of a wall between successive floors. They are used by carpenters to lay off the heights for concrete bucks for basement windows in the cribbing forms. They are also used for establishing the levels for mud sills and header joists. When constructing the cribbing walls, the carpenter stands the story rod on top of the concrete footings and transfers to the cribbing walls the heights at which different members will be placed. In this manner, all the windows may be set at the same height. It is a handy and accurate method, and it saves time in measuring for each window separately.

A story rod is also used by carpenters when hanging a number of similar type doors (for the same side of opening). A thin, straight, dry lath called a door strap is used. It is usually about 3/8″ (10 mm) wide and is the exact length of the door. On the rod the positions for the hinges and the door lock are accurately marked off. The rod is then used to transcribe the markings onto the doorframes and onto the edges of the doors. Any doors may then be hung in any frame and the doors with the best grain may be placed where they will be most seen. Another advantage is that the carpenter may use one room as a temporary work place and confine wood chippings to one area.

Kitchen cabinet makers use a story rod on which they mark all heights of the cabinet. A similar measuring device is used for the horizontal measurements of cabinets. A rod is also used to set the height of all valances for drapes. This is important because manufacturers market drapes in standard lengths.

Another area where story rods make a builder's life a lot easier is masonry work. Masons lay off on a story rod the height of all courses of concrete block or bricks and their mortar joints. In this way, two masons working on the same wall may lay up the corners and both check for uniformity of height by checking with the same story rod. Thus, they may keep all courses parallel to each other and to the top of the concrete footings.

A straight story rod of square or rectangular section can be used to record four sets of data by marking one set on each face of the rod.

Other Measuring Tools

Besides the relatively more specialized measuring tools already mentioned, common rules and tapes are popular workhorse tools in the construction business, especially useful when extremely precise measurements are not critical.

Of all measuring tools, the simplest and most common is the steel rule. This rule is usually a yard or meter in length, although other lengths are available.

A metal or wooden folding rule is often used in construction work for general measuring purposes. These folding rules are usually 2′ to 6′ (.6 m to 1.8 m) long.

Steel or fiberglass tapes are made from 6′ to about 300′ (1.8 m to 91.4 m) in length. The shorter lengths are frequently made with a curved cross section so that they are flexible enough to roll up, but remain rigid when extended. Longer, flat, rolled steel or fiberglass tapes require support over their full length when measuring or the natural sag will cause an error in reading. These tapes are usually wound up by hand using a small insert knob.

Miscellaneous Tools and Safety Equipment

This broad heading covers an assortment of essential or commonly used tools that do not fall into the other categories. These include ripping and pry bars; mechanical staplers, tackers, and nailers; trowels and other masonry tools; paint brushes and paint rollers; sharpening stones; and safety or protective equipment.

Ripping Bars

Ripping, wrecking, pry, and nail bars (Figure 2-39) are used to dismantle structures that have already served their purpose, to pull fasteners, to dig, to break construction materials, and to open heavy-duty packaging.

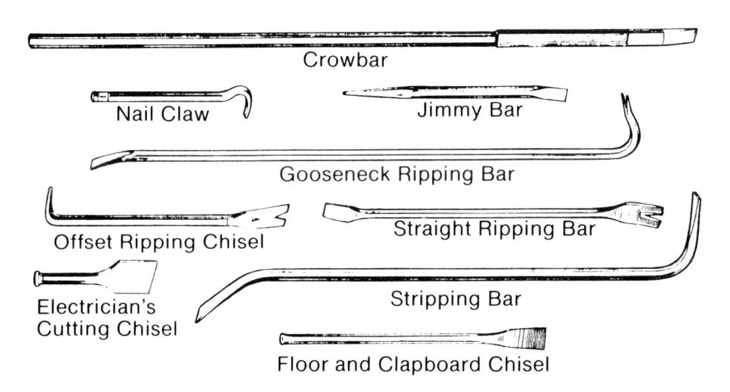

FIGURE 2-39. Ripping bars.

In using any bar, the principle hazard is encountered when the tool slips, causing the user to fall or the bar to fall on someone's foot. This is generally due to improper placing of the bar under the object to be moved or the bar not readily gripping the object.

Staplers, Tackers, and Nailers

Today, mechanical staplers, tackers, and nailers (Figure 2-40) perform a variety of operations formerly done by hand. They provide an efficient method of attaching insulation, roof material, building paper, screening, underlayment, weather stripping, upholstery material, ceiling tile, and many other products.

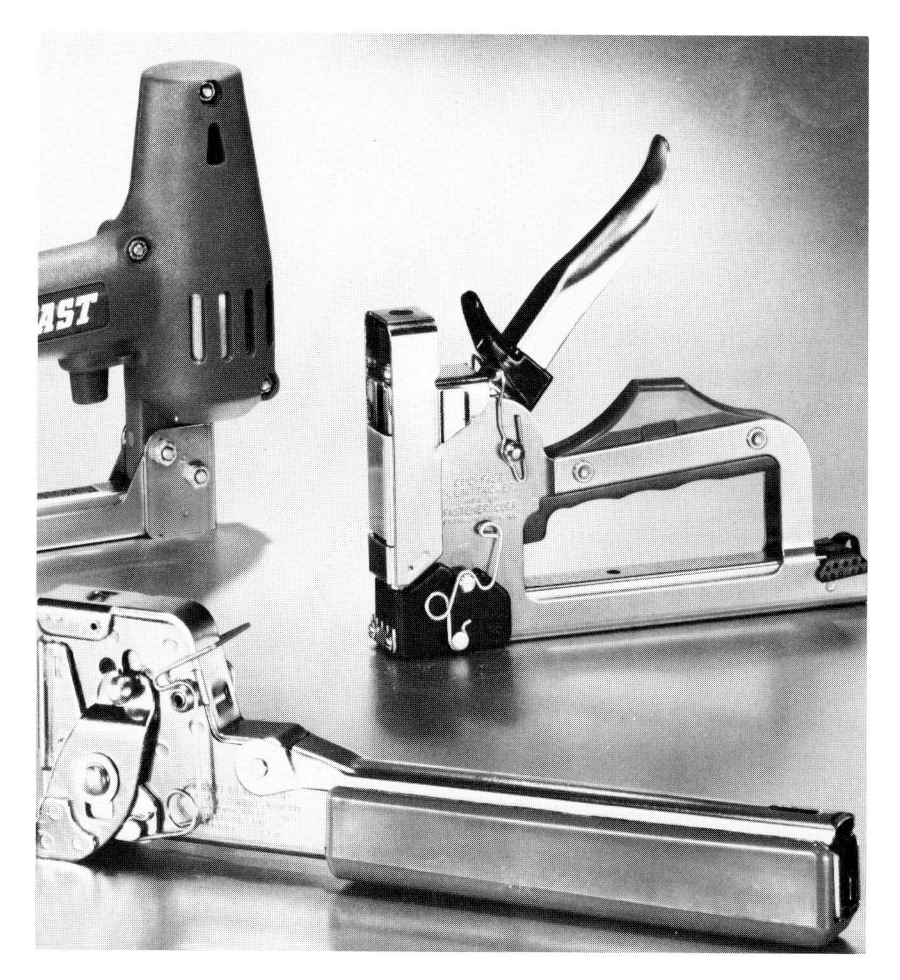

FIGURE 2-40. Staplers, tackers, and nailers.

Tackers and staplers are both designed to drive staples. In most cases, the maximum length of tacker staples will be approximately 1-1/6″ (30 mm). They are used in upholstering and in fastening building paper, asphalt shingles, and some types of paneling. The fasteners used in staple guns can be up to 1-7/12″ (40 mm) in length, suitable for securing paneling, subflooring, and sheathing. Nailers are designed to drive nails up to about 3-11/12″ (100 mm) in length and 1/6″ (4 mm) in diameter. They can be used almost anywhere nails of that size are required.

Safety Equipment

Although personal protection equipment is not precisely a category of tools, it is just as valuable to a construction team. Tools enable construction workers to do their various types of work, while protective equipment helps to ensure that the work is done safely. On many construction sites all or some of the items listed here will be mandatory. In such a case, their use is a must with no questions asked and no exceptions allowed.

FIGURE 2-41. Hard hat.

Hard Hats. Hard hats or protective helmets (Figure 2-41) come in a variety of shapes. They must be made of tough polyethylene or polycarbonate, one of the toughest hat materials yet developed. Many construction workers have had their lives saved because they wore a protective hat. When a falling object strikes the hat, the shock-absorbing suspension capabilities minimize damage to the wearer's head.

Regular hard hats are required to have a degree of insulation resistance so that general personnel are protected from accidental head contacts with electrical circuits and equipment at comparatively low voltages (less than 2,200 volts). Electrical workers require special insulating safety helmets with a much greater capacity to protect against high voltages.

Safety Goggles. Proper eye protection is essential for all construction site personnel (Figure 2-42). Many construction operations such as sawing, drilling, and grinding produce flying chips of wood or metal. Sparks are often a problem during grinding operations. If welding is done on the site, precautions must be taken to protect workers' eyes from the ultraviolet radiation produced during arc welding. Use proper eye protection during all eye-hazard operations.

Remember, eye damage can be excruciatingly painful and costly. No one on a construction site can afford a serious eye injury.

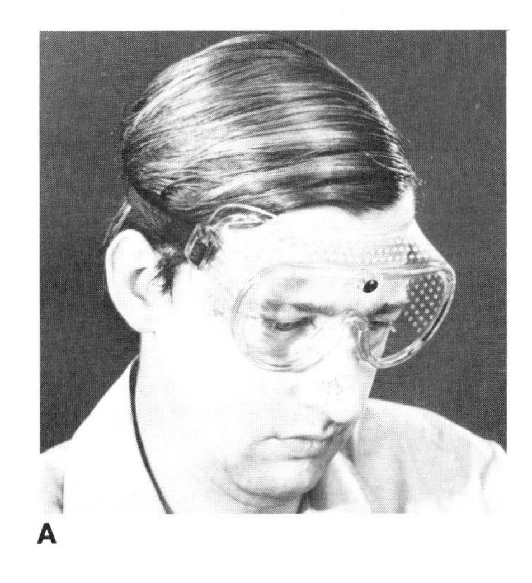

A

B

FIGURE 2-42. (A) Protective goggles and (B) protective glasses.

Gloves. Gloves should be used whenever a worker is required to handle rough, scaly, or splintery objects. Special flameproof gloves are designed for gas and electric welding to limit danger from sparks and other hot flying objects. Personnel in the electrical fields are usually required to wear insulating rubber gloves. Extreme care must be taken when gloves must be worn.

Safety Shoes. Safety shoes are designed to limit damage to a worker's toes from falling objects. A steel plate is placed in the toe area of such shoes so that the toes are not crushed if an object impacts there.

2.4.2 Power-Operated Hand Tools

Power-operated hand tools, also commonly known as portable power tools, have gradually been replacing manually powered hand tools as first-line equipment in the construction business. Portable power tools have been designed in most cases to perform a task previously done by hand. In a sense, then, these tools can be thought of as replacements for hand tools. They work faster and at much higher speeds than is possible with hand work, and they do finer work, making sharper, smoother, and more accurate cuts.

Portable Power Saws

The three main types of electric portable power saws are circular saws, saber saws, and reciprocating saws.

Circular Saws

The workhorse of the portable power saw group in the light construction business is the portable circular saw (Figure 2-43), which performs the functions of crosscut and rip handsaws. In general, it saves 50% to 90% in terms of time for most operations, as well as a proportionate amount of energy. The circular saw can be used for crosscutting operations (against the grain), ripsawing (with the grain), miters, bevel sawing, dadoes, grooves, rabbets, and even plunge or pocket cuts in plywood. In addition, a circular saw equipped with the appropriate blade or cutoff wheel can be used to cut nonferrous metals, iron, steel, and masonry materials.

FIGURE 2-43. Portable circular saw.

The portable electric circular saw is a right-handed tool and is used very much like a handsaw. The difference is that the cut is made in the opposite direction: the blade cuts from the bottom to the top of the work, and the speed of the machine does the cutting. Therefore, the workpiece is always placed with the good side down. Hold the saw with the right hand and the work with the left hand. Make sure that the work is safely supported and in a position that allows comfortable sawing. Never allow a worker to stand directly behind the saw when operating it. The operator should stand off to one side, and keep the other (left) hand well out of the line of cut. The workpiece should never be held with a finger underneath it. The circular saw will always have a lower blade guard that will not return until the blade is clear of the work material.

Circular Saws Safety. The operator should always observe the following safety rules.

1. Always wear goggles or a face shield while using the saw.
2. Hold the saw firmly with one hand. Care should be taken that the saw does not break away and thereby cause injury. If guiding the saw with two hands, keep the second hand on the motor housing, not near the blade.
3. Unplug the saw and inspect the blade at frequent intervals and always after it has locked, pinched, or burned. Remember to never use a dull saw blade. Pinching or binding indicates a dull blade.
4. Do not overload the saw motor by pushing too hard or cutting stock that is too heavy. Let the blade do its own cutting.
5. Before using the saw, carefully examine the material to be cut, and free it of nails and other metal substances. Avoid cutting into or through knots as much as possible.
6. All circular saws should be equipped with guards that will automatically adjust themselves to the workpiece when in use, so that none of the teeth protrude above the workpiece.
7. Check the shoe (flat bottom of saw) and bevel-angle adjustments to be certain that they are tight. Plug the cord into a grounded outlet (unless the saw is double insulated), and be sure that the cord will not become tangled in the workpiece or the blade.

Saber Saw

The saber saw (Figure 2-44), sometimes called a portable jigsaw or bayonet saw, does not have the cutting speed or power of the circular saw. However, it more than makes up for this by its versatility. It can handle almost any operation of which the circular saw is capable, except dadoing and rabbeting. A saber saw can be used for crosscutting, ripping, mitering, beveling, and for plunge cutting. In addition it is extremely useful for making large or small arcs, circles, scrolls, and any other curved, intricate, or irregular cuts, including zigzags and pattern cuts.

FIGURE 2-44. Portable saber saw.

Blades for saber saws are of four general types: woodcutting blades, metal cutting blades, toothless tungsten-carbide blades, and knife blades. The last type slices through a variety of soft materials, such as rubber, leather, cork, and insulation board.

Reciprocating Saws

Often referred to as a plunge-cutting or all-purpose saw, the reciprocating saw (Figure 2-45) resembles an overgrown saber saw with the blade projecting from the front of the tool instead of from the bottom. It has a long blade that may be up to 12″ (305 mm) in length. Equipped with the proper type of blade, it will cut through wood, metal, and plaster. The blade can be mounted in four to six different positions to permit cutting in almost any direction, and it is ideally suited for plunge cutting in walls, floors, and ceilings. Although larger and heavier than a saber saw, the reciprocating saw can be used for scroll cutting as well as straight-line cutting, and most models will cut flush up against walls or other obstructions. They can even be used for pruning large shrubs and cutting off tree limbs, although contractors might prefer a light chain saw for this type of work.

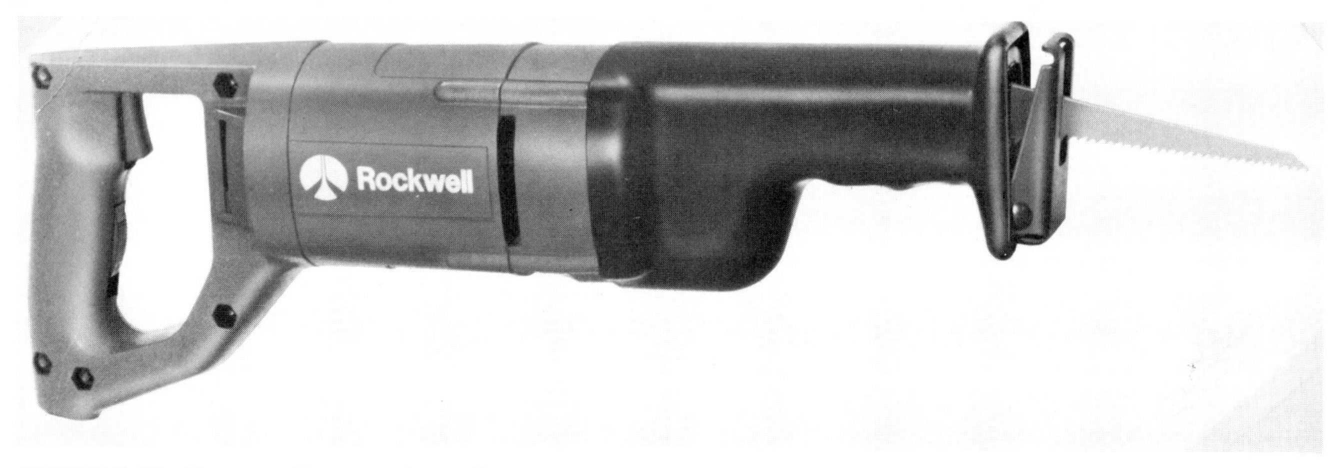

FIGURE 2-45. Plunge cutting or reciprocating saw.

Routers

The portable electric router (Figure 2-46) is another very versatile power tool. It is especially designed for advanced carpentry and cabinetmaking. In its modern efficiency, it not only makes decorative edges on counters, trims plastic laminates, and cuts outlines for inlay jobs, but it also does many typical jobs occurring in light construction. For instance, it is a tool that can cut grooves and rabbets for cabinet-work, mortises for door hinges, fancy edges for bookcases and shelving, grooves for storm windows and weather stripping, and sink openings in countertops. It can also duplicate shapes of intricate design, cut circles and ovals with perfectly smooth edges, and round corners in all sorts of work.

FIGURE 2-46. Portable router.

Portable Power Drills

Electric drills (Figure 2-47) with bits for boring wood or drilling metal are a huge improvement over the old hand drills and braces. Portable drills come in more varieties and are made by more manufacturers than any other power tool. Essentially a drill consists of a motor, a trigger switch to control the motor, and a chuck into which the shafts of twist drills, bits, and other attachments are inserted. With a variety of accessories, a portable electric drill can also grind, sand, drive or remove screws, chisel wood, mix paint, trim hedges, and perform many other tasks. Not every drill, depending on its speed and power, is capable of doing all of these jobs; however, the portable power drill is generally an extremely versatile tool.

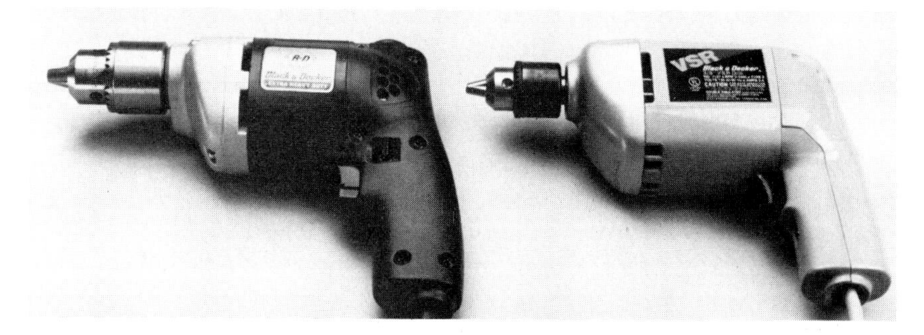

FIGURE 2-47. Portable power drills.

Portable Power Sanders

Portable power sanders are fast working tools used for smoothing rough wooden surfaces. Belt sanders (Figure 2-48A) have an abrasive belt that travels over two drums, one driven by an electric motor. These sanders are not true finishing sanders. They work best with coarse abrasive belts for removing old paint, reducing stock thickness, and final sizing for doors and windows.

Disk sanders (Figure 2-48B) spin on a motorized shank. All disk sanders have a circular rubber pad mounted at right angles to the drive spindle of the motor. The abrasive disks are attached by a flange and screw that thread into the spindle. Disk sanders remove stock rapidly and roughly, and they are excellent for removing paint or shaping wooden pieces.

Pad or orbital sanders (Figure 2-48C) are light-duty sanders ideal for extremely smooth finishing jobs. They move the abrasive paper with a very light action to produce a smooth finish. Always sand with the grain to avoid scratching the wood surface.

A

B

Power Screwdrivers

There are several types of power screwdrivers or fastener drivers on the market that drive screws quickly. Industrial-type screwdrivers or fastener drivers usually have an easy reversing action and have 0 to 1,000 rpm no-load ratings. A few fastener drivers operate at 2,500 rpm, but these are used for drill-point fasteners. They have easy-change bit holders to save time on the job and manual or automatic torque-control clutches to ensure proper seating of all types of fasteners—slotted screws, Phillips screws, hex-head screws, cap screws, nuts, etc.

C

FIGURE 2-48. Portable power sanders: (A) belt, (B) disk, and (C) pad designs. (Photo B furnished by courtesy of 3M.)

Electric Hammer-Drills

The electric hammer-drill is really a multipurpose electric drill (Figure 2-49). That is, by shifting its collar, the tool can easily be converted from a rotary drill for drilling holes in metal and wood to a percussion drill for making holes in concrete, brick, and tile. In the hammering mode, a typical hammer-drill can deliver about 34,000 blows per

FIGURE 2-49. Electric hammer-drill.

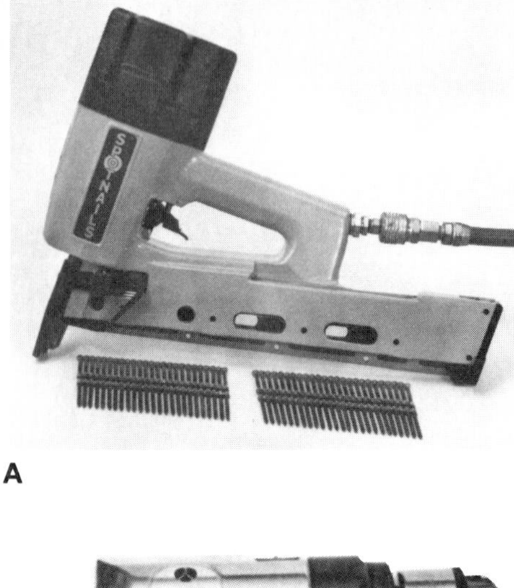

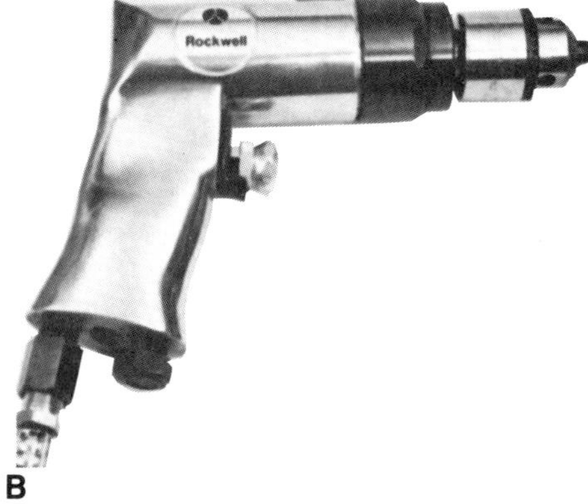

FIGURE 2-50. Pneumatic tools using compressed air as a power source: (A) nailer and (B) air drill.

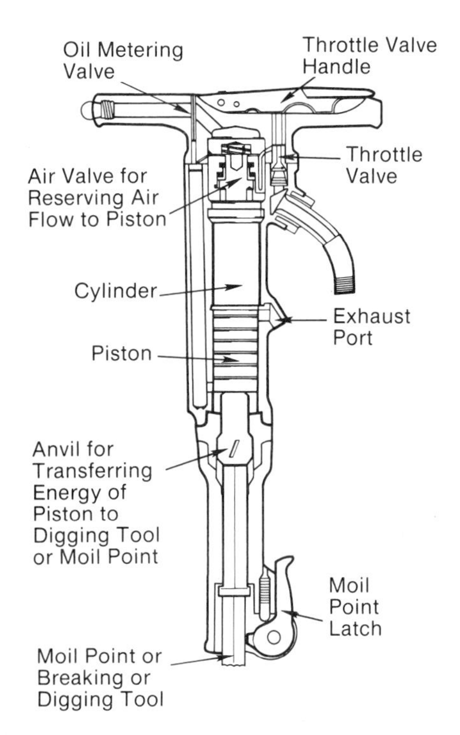

FIGURE 2-51. Jackhammer.

minute. With the proper attachments this tool can drive and remove screws, nuts, and bolts. It can also mortise, chisel, gouge, shape, and remove scale, rust, plaster, tile, and putty. The operation of this tool is basically the same as that of the electric drill.

Compressed-Air Equipment

The most common tools powered by compressed air, also called pneumatic tools, are staplers, tackers, nailers, and impact wrenches, plus jackhammers and tampers. In addition, even circular saws and wood drills are available as air-powered tools. Air-powered staplers, tackers, nailers, and impact wrenches perform the same functions and are similar in design to their electrically-powered and mechanical counterparts. Pneumatic tools are being used more often in light building construction because, where applicable, they have some advantages over the electrically-powered ones. They allow more flexibility, run cooler, have the advantage of variable speed and torque, are generally lighter in weight, require fewer repairs and less maintenance, and are safer because they eliminate potential fire hazards caused by the sparking of electric tools in close, dusty areas. The staplers, tackers, and nailers (Figure 2-50) are particularly useful for securing shingles, shakes, and siding.

Jackhammers (Figure 2-51), which have no electrical equivalent, are used when it is necessary to break up the surface for excavating at the building site or to smash old foundations, basements, and driveways that are to be replaced. Pneumatic tampers, rammers, and vibratory compactors are all used for compacting sand, backfill, and subgrades. A pneumatic tamper makes it possible to do this type of work quickly and easily, as opposed to laborious hand tamping.

Powder-Actuated Tools

Many times in light construction work it is necessary to attach wood or some other material to masonry or steel. Sometimes this is accomplished by means of brackets, adhesives, or anchor bolts. However, often what will be needed will be a special fastener capable of penetrating such a hard surface and a tool with enough power to force the fastener into position.

This task is often accomplished by powder-actuated tools (Figure 2-52), which are pistol-like in appearance and carry small cartridges similar to a .22 caliber rifle cartridge. The tool is placed against the material to be fastened, a trigger is pulled which fires an explosive charge, driving one of several types of special pins and fasteners—headed, eyed, threaded, or knurled shank—through the material and into the steel, concrete, or other masonry material. Cartridges of different explosive force are available for use with fasteners of different length and materials of varying density. The explosive power of a cartridge is denoted by a color code. Green cartridges are most powerful, followed in decreasing order by yellow, red, purple, gray, and brown.

Powder-actuated tools can be very useful for some applications, but extreme care should be taken when they are used on a job. The pins and fasteners are driven at a bullet-like velocity; these tools are potentially

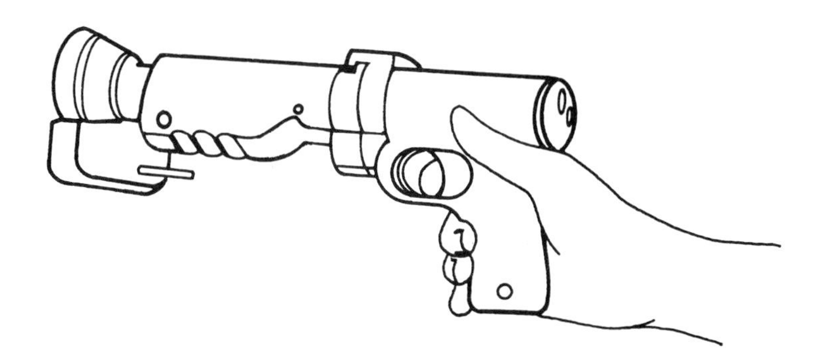

FIGURE 2-52. Powder-actuated tools.

lethal if they should accidentally misfire or are used improperly. The operator of a powder-actuated tool should keep their free hand clear and should always wear safety goggles or a mask.

Chain Saws: Self-Powered Equipment

Chain saws (Figure 2-53) are extremely useful during surveying operations and for clearing when laying out and staking a building site lot. They are also used for such obvious tasks as tree cutting in the initial stages of site clearing and for cutting trees and brush down to size so they can be hauled away. Chain saws are also good to have around a jobsite for any number of rough cutting jobs, although they are obviously not used for actual carpentry.

FIGURE 2-53. Chain saw.

Chain Saw Safety

A chain saw can be dangerous if not handled properly. Much of the man-hour loss incurred during clearing and cleanup operations on a jobsite is caused by chain saw accidents. Unlike the blade on a circular saw, the chain on a chain saw is exposed. Leg cuts and gashes caused by a chain saw are fairly common. Builders should make sure that workers follow all safety rules when using a chain saw.

1. Always hold the saw with both hands, making sure that the thumbs are hooked around the handles and that the grip is tight.
2. Be careful not to let the end of the blade bump into any objects not being cut.
3. Be sure that other workers are a safe distance away from the operator and the saw, and make sure everyone is out of the way when trees are being felled.
4. Turn the saw off when carrying it from one area to another, even if more cutting needs to be done.
5. Never use metal wedges or an ax to hold cuts open.
6. Never set a chain saw on the ground with the engine running.
7. Do not start a chain saw in the same spot in which it has been refueled.
8. Wear short sleeves or otherwise close-fitting, snug clothing when working a chain saw.
9. Only make cuts within the capability of the saw.

FIGURE 2-54. Table saw.

FIGURE 2-55. The radial arm saw.

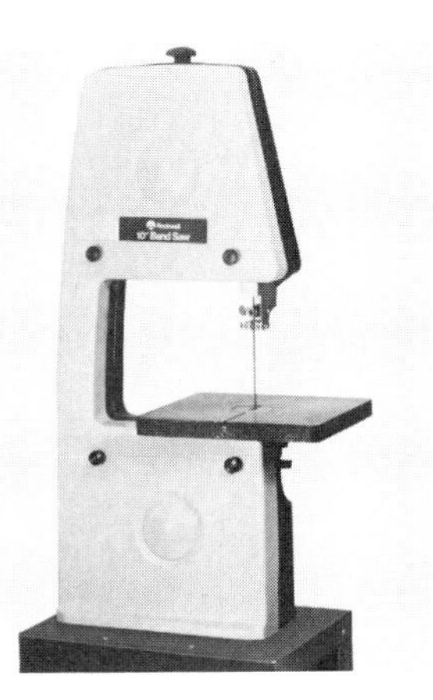

FIGURE 2-56. Stationary band saw.

2.4.3 Stationary Power Tools and Equipment

Many of the tools in this group will be permanently installed in a workshop rather than transported and used on the jobsite. The standard stationary shop tools such as table and radial saws, bandsaws, jointers, planers, stationary sanders, drill presses, and bench grinders are all in this category.

Table Saw

The table saw (Figure 2-54) can make crosscut, miter, bevel, compound angle, and rip cuts. With accessories, it can also dado, shape, and sand workpieces. With the correct abrasive blade, it cuts particleboard, composition board, metals, plastics, and stone.

The table saw is simply a circular blade driven by a motor. The blade projects through a slot in the top of the table. A fence or miter gauge is used to guide workpieces through the blade. Elevating and blade tilting mechanisms position the combined movable motor-arbor-cutter assembly for crosscut, rip, miter, bevel, and compound angle cutting.

The table saw is superior for cutting straight lines and is especially useful for production rip sawing. It is slightly less expensive than a radial arm saw. Its major disadvantage compared to the radial saw is that the cutter is beneath the pattern line where the operator cannot see the cut being made.

Radial Saw

The other shop production saw is the radial saw, also sometimes called a radial arm saw or overhead saw (Figure 2-55). The radial saw can perform all the functions of the table saw. With other accessories, however, the radial saw can perform additional functions that the table saw cannot. These include routing, surface planing, drilling, and drum sanding.

The blade of the radial saw is above the workpiece; whereas, the blade of the table saw is below the workpiece. This is its greatest advantage. With the cutting done from the top of the workpiece, the operator can see what is being cut; therefore, it is easier to follow the pattern line. This results in more accurately cut workpieces with less scrap from inaccurately cut lumber.

Because of its versatility, the radial saw is often recommended as the first bench power tool for a workshop. The basic saw can cut quickly, accurately, and conveniently. Because workpieces are placed lengthwise on the worktable for both the crosscut and ripping operations, the radial saw can be located against the wall in the workshop.

Band Saw

The band saw (Figure 2-56) is used in the workshop to cut long curved lines, intricate shapes, scrolls, curlicues, inside contours, and straight lines in thick wood workpieces or thin metals. It is also used to resaw thick boards into thinner boards. The band saw cuts much heavier stock than a scroll saw and does it faster and smoother.

The band saw is useful for rough cutting wood that is to be carved, for cutting tapered and cured workpieces, and for precutting wood to be turned on a lathe. It makes intricate cuts like a scroll saw, but does not have the capacity (as the scroll saw has) for making internal cuts in a workpiece.

FIGURE 2-57. Scroll saw or stationary jigsaw. (Photo furnished by courtesy of Rockwell International: Power Tool Division.)

Scroll Saw

The scroll saw (Figure 2-57), also known as a stationary jigsaw, has the ability to cut the curves and sharp corners of designs and patterns and to rough saw larger pieces by means of a fine reciprocating blade. Used correctly, it is a precision tool for intricate work including fine frets, inlays, puzzles, marquetry work, and other ornamental objects of great variety. Although primarily a freehand saw, the jigsaw can be adapted with simple shop-built jigs for making straight line and circular cuts. The blade can be rotated 90° left or right on some models for ripping.

The jigsaw is also the only bench power tool that can make cuts inside a workpiece without leaving a saw kerf. It cuts patterns in wood, plastic, thin and soft metals, and in paper, leather, and cloth when special methods are used. Although simple and safe to use, the scroll saw is not intended for heavy production cutting, a job better suited for a table, radial, or band saw.

FIGURE 2-58. Power jointer.

Jointer

The jointer (Figure 2-58) does the work of a hand plane; however, the jointer does it faster and better. The jointer gets its name from its primary function—planing board edges smooth for joining together. The jointer can also plane board surfaces. It is capable of planing boards smooth up to widths that are almost two times as wide as the knife length. The jointer is used to make rabbet cuts for doors, window frames, and drawers. It can also bevel edges, cut chamfers and tenons, make a variety of taper cuts, and take the warp out of a board.

FIGURE 2-59. Planer. (Photo furnished by courtesy of Rockwell International: Power Tool Division.)

Planer

The planer (Figure 2-59) is a specialty power tool designed for heavy-duty production work. It is used to clean rough lumber and to smoothly plane wood to the desired thickness. Other tools similar to the thickness planer, but more elaborate and expensive, are the planer-molder and the planer-molder-saw.

The planer is self-feeding. Its maximum cut is 1/8″ (3 mm) per pass; it accepts workpieces 12″ (305 mm) wide by 4″ (102 mm) thick and turns out smooth planed wood. Workpieces less than 1/16″ (2 mm) thick or 6″ (152 mm) in length cannot be planed.

The planer-molder performs both planing and molding operations in a single pass of the workpiece through the machine. A planer-molder-saw combines a sawing operation with the other two operations. These power fed tools trim rough lumber into moldings, trim, frames, and other finer work.

FIGURE 2-60. Shaper used for decorative work.

Shaper

One of the primary functions of the shaper (Figure 2-60) is to decoratively shape wood for decorative molding; frames; straight, curved, and irregular shaping; fluting; beading and scalloping; and the cutting of sash and door moldings. Another primary function of the shaper is to cut wood in preparation for joining two workpieces together; it cuts tongue-and-groove joints, drop-leaf joints, matched shaping, glue joints, and rabbets. The shaper also cuts cabinet door lips and can be used like a jointer to plane narrow workpiece edges. Basically, the shaper does the same work as the portable router, but the shaper head cannot be moved around freehand for carving as is sometimes done with a router.

Drill Press

The main function of the drill press (Figure 2-61) is to turn drills and boring bits for cutting holes in workpieces of many types of material. The drill press is sturdily built to provide accurate drilling at the proper speed to suit the cutting tool and the material. While the portable power hand drills used at a jobsite are handy, the drill press allows work to be done with much finer tolerances.

Secondary functions of the drill press are sanding, routing, shaping, mortising, and making dovetails. Drill presses are not meant to take the place of routers, shapers, and sanders. Instead, they are used with the necessary attachments if these other tools are not available. Drill presses do not run at as high speeds as the router or shaper, but the speeds are adequate for light-duty routing and shaping.

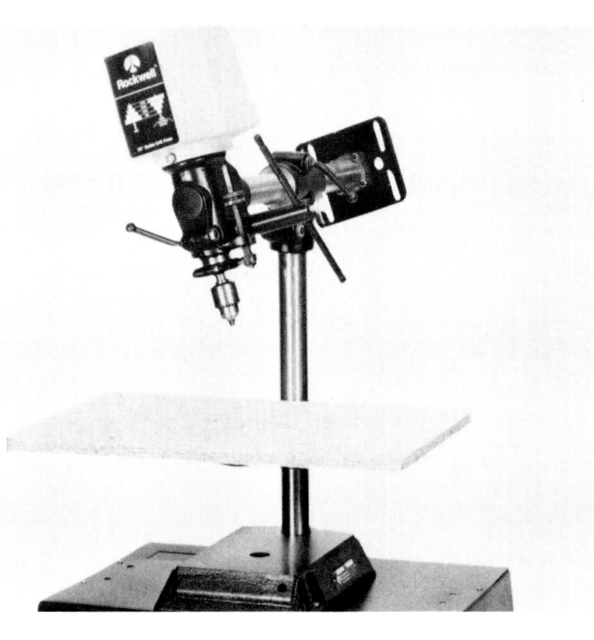

FIGURE 2-61. Stationary drill press.

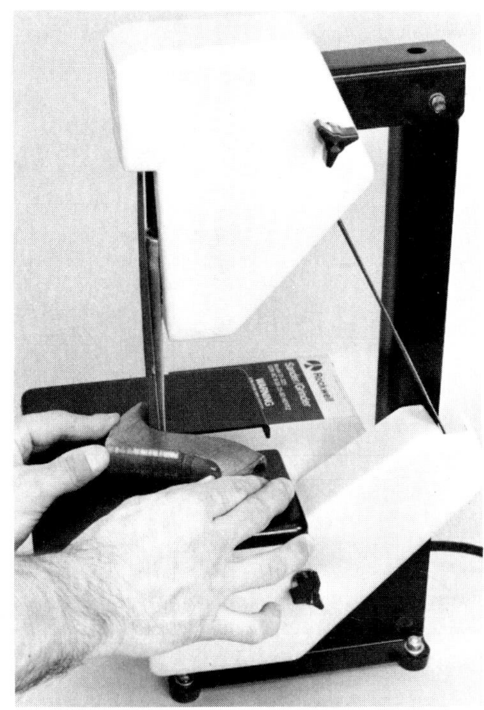

FIGURE 2-62. Sander-grinder.

Sander-Grinder

The sander-grinder (Figure 2-62) is a relatively new tool which utilizes a continuously turning belt to perform a variety of operations that include sanding, grinding, polishing, sharpening, and deburring. It is a valuable tool for finishing very small or contoured workpieces and for sharpening many of the tools used in light building construction. The sander-grinder works well on nearly all materials including wood, metal, plastic, composition materials, leather, rubber, and ceramics.

Air Compressors

Portable air compressors (Figure 2-63) have many uses in light building construction. The compressor motor can be either electric or gasoline powered; for constant pressure, a pressure regulator must be included in the rig. Air compressors are used perhaps most widely for spraying paint. Paint sprayers are particularly useful for large areas and are ideally suited for the heavy production work common to the construction business. Spraying is much faster than brushing or rolling, and although some paint will likely be wasted through overspraying, the savings in time and effort will more than compensate.

In addition to paint sprayers, portable air compressors are used to power a variety of portable air tools (pneumatic tools), caulking guns, and sandblasters. The sandblaster accessory for a compressor provides a method of removing paint, dirt, rust, and scale from castings, stone, brick, ceramic, concrete, metal, plastic, and wood. In addition to heavy-duty cleaning, it can be used to etch plastic, glass, metal, or stone—all with ordinary sand.

FIGURE 2-63. Portable air compressor.

Electric Generators

One of the more valuable pieces of building equipment is the portable electric generator, also called an electric power plant (Figure 2-64). These gasoline-driven generators are used to supply electricity at building sites where there is no other available supply of power. They generally furnish 120 or 240 volts or both and are available with capacities of between 2,000 and 7,500 watts. A wide range of portable power tools can be run off of one of these electric generators. Without them, a builder has little option but to strictly use hand tools at any building site without commercial power available. Take note, however, that the generator must be of sufficient capacity to handle the individual starting load of any motor it is to power.

FIGURE 2-64. Portable electric generator.

2.4.4 Mobile Power Equipment

Mobile power equipment includes all large pieces of equipment normally associated with the construction business. Dump trucks, flatbed trucks, or smaller trucks with winches or heavy-duty welding outfits can be used for a variety of jobs during almost every phase of a construction job. Other pieces of self-powered equipment such as graders and rollers are usually used for more specific types of work.

However, the most common types of heavy machinery (and most typical of light building construction) are the machines used to clear the land and to make the proper cuts and fills. Although the type of machinery use on a job depends on the site, location, and the size of the job, the most common are backhoes, wheel loaders, track-type loaders, and bulldozers. These machines are used to prepare the jobsite, excavate for the basement, dig the footings, backfill around the foundation, and level the jobsite when the construction has been completed. Not every builder has all of these expensive types of machinery. Sometimes, one piece of machinery (a loader, for example) can sufficiently do some of the work best done by another.

FIGURE 2-65. Backhoe. (Photo furnished by courtesy of Caterpillar Tractor Co.)

Backhoes

A backhoe (Figure 2-65) is used to excavate below the surface of the ground. This machine is sometimes also called a hoe, back shovel, or pull shovel. It operates by first digging and then pulling the load toward the power unit. It can be used to dig trenches and excavate for basements and footings. The buckets are designed to load easily and dump cleanly. If the backhoe is operated in cohesive soils, the bucket can be equipped with a cleaner bar. Fitted to the inside of the bucket, the bar automatically operates when the bucket is opened for dumping, forcing all soil out of the bucket.

The backhoe can also be equipped with ripper teeth for ripping frozen ground, asphalt, or other hard materials. The teeth are fitted with reusable drive pins that can easily be removed and replaced.

FIGURE 2-66. Loader. (Photo furnished by courtesy of Caterpillar Tractor Co.)

Loaders

The two basic types of loaders, also known as front-end loaders, are the wheel loader and the track-type loader (Figure 2-66). The machine's primary purpose is to load excavated soil into trucks; however, it can be used to excavate loose soils, and if a multipurpose bucket is used, a loader can strip topsoil, bulldoze, clean up debris, move lumber, and carry pipe.

Some loaders are equipped with automatic bucket controls, which allow the operator to set the bucket at both a preset height and a digging angle. The bucket then stops automatically at the desired height, empties the load, and returns the bucket to the present digging angle.

The capacity and reach of a loader vary with the manufacture and type of loader.

Bulldozers

A bulldozer (Figure 2-67) is a tractor-driven machine that has a broad, blunt blade mounted in front of it. The conventional bulldozer has the blade mounted perpendicular to the line of travel so it can push the dirt forward. An angledozer has its blade mounted at an angle so that it can push the dirt forward and to one side. Some blades can be adjusted to perform as either bulldozers or angledozers.

Bulldozers can be classified as either crawler tractors or wheel tractors. In most cases a crawler tractor is more efficient in soft and muddy soils. The undercarriage of the crawler tractor is designed with wide-track shoes and a long track form. The wheel tractor, designed for articulated steering and fast maneuvering, operates more efficiently on firm ground.

FIGURE 2-67. Bulldozer. (Photo furnished by courtesy of Caterpillar Tractor Co.)

2.5 RESEARCH

1. Establish a hypothetical light building project. For example, use the one-car garage discussed in Chapter 1. For a construction crew of a given number of workers (four would be appropriate), determine the minimum number of tools and machinery that would allow the project to be completed correctly. Determine what special skills or training the workers would need to operate these tools.
2. Talk to various craftsmen and other working professionals in the construction business (carpenters, masons, plumbers, electricians, cabinetmakers, etc.). Ask them what their most important tools are. Get a feel for the tools they favor. Do most craftsmen prefer portable electric tools over hand tools? In what situations is this true? How do they rate the efficiency of various hand and power tools? Do they believe certain hand tools are becoming absolute?

What shop tools do they find most important? Sort this information; then do a cost analysis, comparing prices for comparable tools. Create a hypothetical basic inventory of tools and machinery for a light building construction company with a crew of ten workers. Create separate inventories for jobsite tools and equipment and workshop tools. Do a cost analysis of these inventories. Remember, enough tools must be available so that each worker can accomplish his or her specific tasks. After listing the inventories, check with construction people to see if you have forgotten anything.

3. Find out what special training and/or licensing is required in your state and locality for operating heavy mobile construction equipment, such as bulldozers, front-end loaders, and backhoes.
4. Do some reading on tool maintenance. Learn standard procedures for keeping construction tools and equipment in safe running order. Visit a tool sharpening shop and observe how saw blades and other cutting tools are kept sharp.
5. Make a brief outline determining procedures for the maintenance and storage of tools and machinery at a jobsite. Consider the problems of vandalism and theft.
6. Find out more about air-powered equipment. What are the special problems and advantages associated with its use? Locate and read national, state, and local safety regulations for using air-powered tools. Write your own set of safety rules and give reasons for each rule.
7. Learn more about power tools. Study the yellow pages of the telephone book to find retailers and wholesalers. Visit these outlets to see the tools and to collect specification sheets. Note prices, warranties, and repair services.
8. Compare the pros and cons of buying or leasing heavy construction equipment. Consider such factors as work loads and amount of time the machinery will be used, optimum use of working capital, maintenance, depreciation, and financing.

2.6 REFERENCES

2.6.1 Drake, George R. *Everyone's Book of Hand and Small Power Tools.* Virginia: Reston Publishing Company, Inc., 1977.

2.6.2 Kratfel, Edward R. and George R. Drake. *Modern Shop Procedures.* Virginia: Reston Publishing Company, Inc., 1974.

2.6.3 Scharff, Robert. *The Complete Book of Home Workshop Tools.* New York: McGraw-Hill, Inc., 1979.

2.6.4 Scharff, Robert. *Your Chain Saw.* California: McCulloch Corporation, 1980.

3

FASTENERS AND FASTENING METHODS USED IN LIGHT CONSTRUCTION

By its very nature, construction involves the joining together of numerous separate components to form a finished structure. The fasteners and fastening methods used in light building construction must be capable of securely joining a wide variety of construction materials, such as wood, man-made wood products, concrete and masonry, metals, and chemical based materials such as insulation board.

Nails, screws, and bolts of numerous sizes and designs have long been the mainstay fasteners in the light building industry. While these fasteners remain extremely important, newer fastening methods are becoming more common. Structural wood connectors and clips are now being used in many framing operations, and special adhesives and glues are rapidly becoming one of the most popular and easy methods of securing floor, wall, and ceiling coverings.

3.1 SPECIFIC OBJECTIVES

Before a building project can begin, the various fastening needs of the job must be identified and subsequent decisions made concerning the types of fasteners and fastening methods to be used. After reading this chapter, the student should be familiar with the most common fasteners and fastening methods used in light construction and be capable of making sound decisions concerning the uses, advantages, and disadvantages of each type. In addition, the material in this chapter should enable the reader to

1. understand the relative strengths of various fastening methods,
2. select the proper type of fastener or fastening method for any job associated with light building construction,
3. be capable of safely applying any of these fasteners or fastening methods, and
4. research the cost factors associated with the use of various fasteners.

3.2 NAILS AND NAILING METHODS

Despite the influx of new fastening materials, nails remain, by far, the most frequently used fasteners in construction work. Nails achieve their holding power when they displace material (usually wood fibers) from their original position. The pressure exerted by this material as it attempts to return to its original position creates the holding force.

Nails are most commonly used in wood and wood based materials. Specialty nails can be used in masonry and light metals, such as aluminum siding. Nails are usually made of steel, but rust resistant nails made of aluminum, galvanized steel, and copper are available for outside construction such as siding, exterior trim, and shingle work. Unpainted, non-corrosive nails are also used as finishing nails when working with redwood and other woods commonly left exposed to the elements. Other nail designs are especially suited for fastening roofing, drywall, insulation board, and wood paneling. Smaller nails and brads are designed for finish and trim work.

3.2.1 Nail Sizes

The lengths of the most commonly used nails are designated by the "penny" system. The abbreviation for the word *penny* is the letter "d." So the expression "a 2d nail" means a two-penny nail. The penny sizes and corresponding length and thicknesses (in wire gauge sizes) are listed in Table 3-1. As the penny size increases, the thickness of a nail increases and the number of nails per pound (kg) decreases.

Nails larger than 20d are called spikes and are generally designated by their length in inches or millimeters. Nails smaller than 2d are designated in fractions of an inch or millimeters instead of the penny system.

In ascending order according to size, nails are referred to as brads, finishing nails, casing nails, box nails, common nails, and spikes (Figure 3-1). Common nails are used most often for the widest variety of fastening work in the light construction industry. Common nails

TABLE 3-1 COMMON NAILS AND FINISHING NAILS

Common Nails				
Penny Size	**Length in inches (mm)**	**Diameter Gauge Number**	**Diameter of Head in inches (mm)**	**Approx. Number per Pound (kg)**
2d	1 (25)	15	11/64 (4)	850 (385.6)
3d	1-1/4 (32)	14	13/64 (5)	560 (254)
4d	1-1/2 (38)	12-1/2	1/4 (6)	320 (145.1)
5d	1-3/4 (44)	12-1/2	1/4 (6)	260 (117.9)
6d	2 (51)	11-1/2	17/64 (7)	185 (83.9)
7d	2-1/4 (57)	11-1/2	17/64 (7)	150 (68)
8d	2-1/2 (64)	10-1/4	9/32 (7)	106 (48)
9d	2-3/4 (70)	10-1/4	9/32 (7)	90 (40.8)
10d	3 (76)	9	5/16 (8)	70 (31.8)
12d	3-1/4 (83)	9	5/16 (8)	60 (27.2)
16d	3-1/2 (89)	8	11/32 (9)	48 (21.8)
20d	4 (102)	6	13/32 (10)	30 (13.6)
30d	4-1/2 (114)	5	7/16 (11)	24 (10.9)
40d	5 (127)	4	15/32 (12)	18 (8.2)
50d	5-1/2 (140)	3	1/2 (13)	14 (6.4)
60d	6 (152)	2	17/32 (13)	11 (5)

TABLE 3-1 COMMON NAILS AND FINISHING NAILS (Continued)

Finishing Nails				
Penny Size	Length in Inches (mm)	Diameter Gauge Number	Diameter of Head Gauge Number	Approx. Number per Pound (kg)
2d	1 (25)	16-1/2	13-1/2	1,350 (612.3)
3d	1-1/4 (32)	15-1/2	12-1/2	850 (385.6)
4d	1-1/2 (38)	15	12	600 (272.2)
5d	1-3/4 (44)	15	12	500 (226.8)
6d	2 (51)	13	10	300 (136)
8d	2-1/2 (57)	12-1/2	9-1/2	195 (88.5)
10d	3 (64)	11-1/2	8-1/2	120 (54.4)
16d	3-1/2 (70)	11	8	90 (40.8)
20d	4 (76)	10	7	62 (28.1)

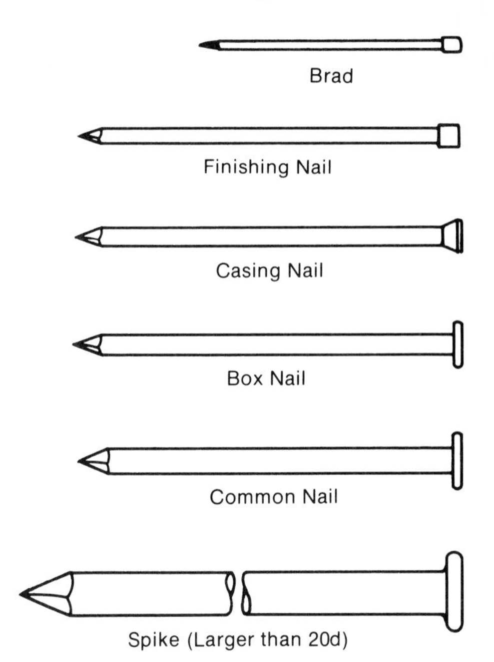

FIGURE 3-1. Nail Varieties.

have relatively large, flat heads and fairly heavy shanks. They are intended for any type of rough work where the heads are driven flush with the surface. Almost all wood framing work is done with common nails, although the newer structural connectors discussed later in this chapter are also used to some extent.

Spikes are larger versions of common nails and are used to hold large wooden members where great fastening strength or load bearing strength is required. Box nails are slim versions of common nails and are used in assembly of lighter stock. Finishing nails are slightly slimmer in the shank than common nails, but they have a small, ball-shaped head that can easily be sunk below the surface with a nail set and concealed with putty.

Casing nails are similar to finishing nails, but casing nails have a tapered head and a somewhat heavier body. They may be either countersunk below the surface or driven flush with the material. They are used mainly in trimwork. Small finishing nails less than 1″ (25 mm) are called brads.

3.2.2 Other Nail Types

Duplex or double-headed nails, (Figure 3-2A) are used for nailing temporary structures such as scaffolds and concrete forms, which will eventually be disassembled. They are also frequently used for attaching string lines. Double-headed nails are simply driven in up to their lower head; the second or top head can be used to easily withdraw the nail when disassembly is required.

Barbed nails and deformed shank nails of both annular-grooved and spiral-grooved designs offer superior withdrawal resistance (Figure 3-2B). Annular-grooved nails offer the finest resistance to withdrawal loads and are commonly used in the construction of pole-type framed buildings or with special building clips and fasteners. Very small, short annular-grooved nails are often used for fastening interior paneling, tongue-and-grove plywood flooring, some types of hardboard, and thin sheets of plywood. Their extra withdrawal resistance helps prevent nail popping, a common problem when fastening large sheets of heavy, pliable material.

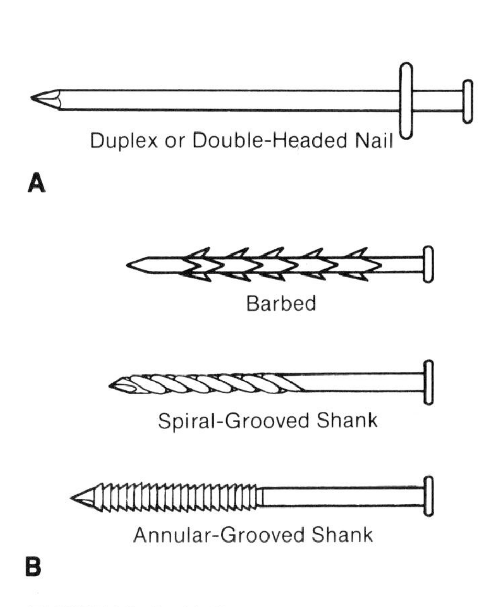

FIGURE 3-2. Nails with superior withdrawal resistance.

3.2.3 Nailing Practices

The type of wood being used in a construction affects the strength of any joint because the resistance offered by the wood to the withdrawal of nails varies. Usually, the denser and harder the wood, the greater is the inherent nail-holding ability, assuming the wood does not split.

The size, quantity, and placement of nails all have a marked effect on the strength of a joint. Thus, more nails are required in woods of medium holding power than in woods of high holding powers. The resistance of nails to withdrawal increases almost directly with their diameter; if the diameter of the nail is doubled, the holding strength is doubled, providing the nail does not split the wood when it is driven. The lateral resistance of nails also increases as the diameter increases.

The moisture content of the wood at the time of nailing is extremely important for good nail holding. If plain-shank nails are driven into wet wood, they will lose about three-fourths of their full holding ability when the wood becomes dry. Thus, the most important rule in obtaining good nail-holding ability and strong joints is to use well-seasoned wood.

The splitting of wood by nails greatly reduces their holding ability. Even if the wood is split only slightly around the nail, considerable holding strength is lost. Because of hardness and texture characteristics, some woods split more in nailing than others do. The heavy, dense woods, such as maple, oak, and hickory, split more in nailing than do the lightweight woods, such as basswood, spruce, balsam, and white fir. In general, splitting will be more of a problem in finishing work than in general carpentry and framing work.

Predrilling is a good practice when working with dense woods, especially when large diameter nails are used. The drilled hole should be about 75% of the nail diameter. Woods without a uniform texture, like Southern yellow pine and Douglas fir, split more than the uniform-textured woods, such as Northern and Idaho white pine, sugar pine, or ponderosa pine. In addition to predrilling, the most common method of reducing splitting is to use small diameter nails. The number of small nails must be increased to maintain the same gross holding strength as with larger nails. Slightly blunt-pointed nails have less tendency to split wood than do sharp-pointed nails. Sharp nails may be blunted with a hammer before driving; too much blunting, however, results in a loss of holding ability.

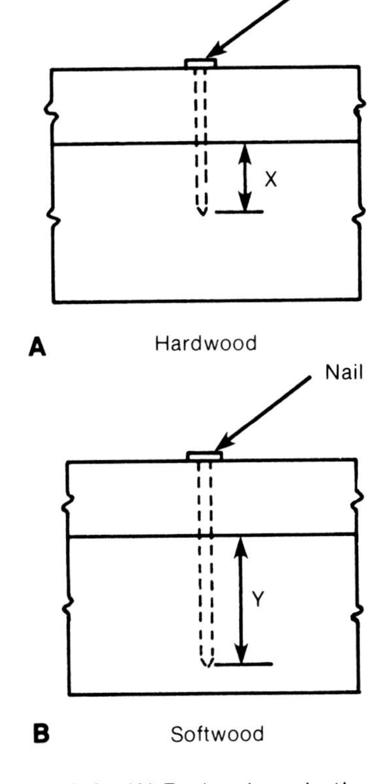

FIGURE 3-3. (A) For hardwoods, the nail penetration (X) in the bottom piece should be one-half the length of the nail; (B) for softwoods, the penetration (Y) into the bottom piece should be two-thirds the length of the nail.

Selecting Nail Length

There is a simple rule to follow when selecting the nail lengths for both rough (framing) and finish carpentry. The rule varies for hardwoods and softwoods. Suppose that pieces A and B are to be nailed together. For hardwoods, the nail penetration (X) into the bottom piece should be one-half the length of the nail. For softwoods, the penetration (Y) into the bottom piece should be two-thirds the length of the nail. Thus, the thickness of the top piece determines the required nail length (Figure 3-3).

When plywood is being fastened, the choice of nail depends on the thickness. For 5/8″ (16 mm), use 8d. For 1/2″ (13mm) and 3/8″

(10mm), 6d is best. These are both finishing nails. If heavier nails are required, use a casing nail. A 6″ (152mm) spacing is about right for most plywood work.

Increasing Holding Power

There are several ways to increase the holding power of nails. Common nails used in rough carpentry work are often clinched (Figure 3-4A). In other words, if a portion of a nail extends beyond the surface of the wood, that portion may be bent over. If common nails are clinched, they will have about 45% greater holding power than corresponding nonclinched nails. The clinched portion of a nail should be at least 1/4″ (6mm) long; longer clinched portions do not provide stronger joints. Clinching should be done perpendicular to the grain of the wood.

Other methods of nailing joints can increase their strength. A number of nails at a joint strengthens it, but they should be staggered, making it less likely for the wood to split down the grain (Figure 3-4B). Where possible, driving the nails in at oblique angles, called skewing, creates a sounder connection and reduces the possibility of splitting. These angles, however, should be slight, and alternate nails should be angled in opposite directions (Figure 3-4C).

Toenailing (Figure 3-4D) is a standard technique used for the nailing of butt joints in situations where it is impossible or impractical to nail from the cross member into the end of the butting member. The nail goes at an angle through the butting member into the cross member. At least two nails are required—frequently more—to make any kind of a solid joint. It is often difficult to get nails started in toenailing. For best results place the point in position and give the nail a gentle tap as if to drive it straight into the member. Then, angle it properly and continue nailing. The small start made by that first tap will keep the nail in the proper position. Because the angle uses up more of the nail's length than straight nailing, it is best to use the longest nail that will not break through or split the wood.

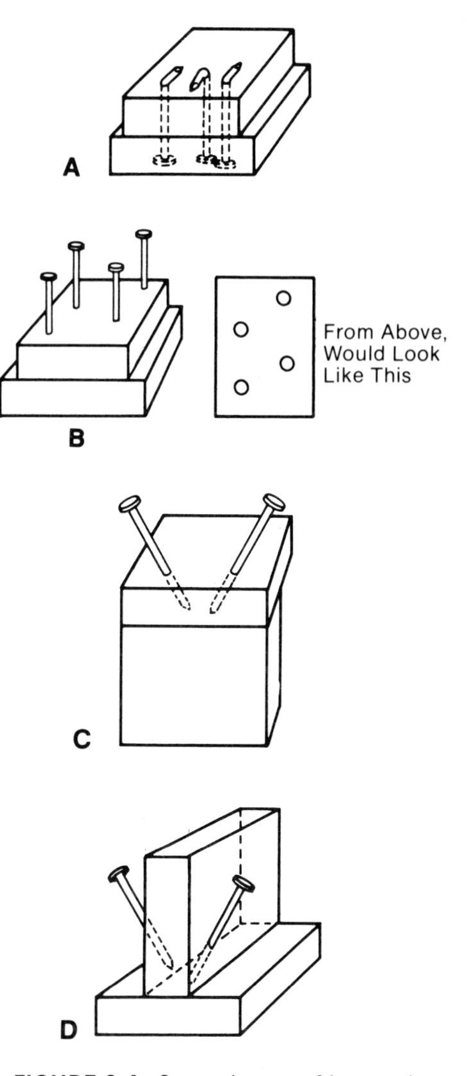

FIGURE 3-4. Several ways of increasing the strength of nail joints: (A) clinching, (B) staggering nails, (C) skewing or driving in nails at a slight angle, (D) toenailing at butt joints.

3.3 SCREWS

Screws offer several advantages over the use of nails in some types of fastening jobs. For their size they have much greater holding power, are neater in appearance, and can be tightened to a required holding power easier than adding nails. They offer the additional advantage of being easily disassembled and reassembled without damage to the work, but take longer to install than nails and are generally more trouble with which to work. For these reasons, screws are primarily used for finish carpentry and cabinetwork.

3.3.1 Screw Types and Sizes

Wood screws are the primary screws used in light construction work. Most wood screws are made from steel or brass and are available in flathead, roundhead, and ovalhead designs (Figure 3-5). Flatheads are the most popular. They can be countersunk flush with the surface in routine assembly or sunk below the surface and filled for finer ap-

FIGURE 3-5. Types of heads on wood screws.

pearance. The roundhead can also be driven in a counterbored hole for concealment, but it is most often left exposed, especially in exterior work. The ovalhead requires a partial contersink and is ordinarily used with a special washer in joints that are commonly disassembled. As mentioned in Chapter 2, screws can also be grouped according to the design of the screwdriver grip in the screw head. The standard single slot head is still the most widely used, but Phillips head, Forge-Set, and Frearson head screws are frequently used when greater driving power is required to install the screw.

In addition to the wide variety of wood screws, machine screws, cap screws, setscrews, and self-tapping sheet metal screws are used in light building construction for fastening hardware, heating and plumbing fixtures, electrical outlets, etc. Special masonry screws are also available for fastening materials to concrete and masonry.

The size of an ordinary wood screw is indicated by the length of the screw and its diameter (gauge) (Figure 3-6). Screw length is always measured from the screw point to the point of greatest head diameter. Body diameters are designated by gauge numbers. Gauge sizes and lengths for standard steel and brass screws are shown in Table 3-2.

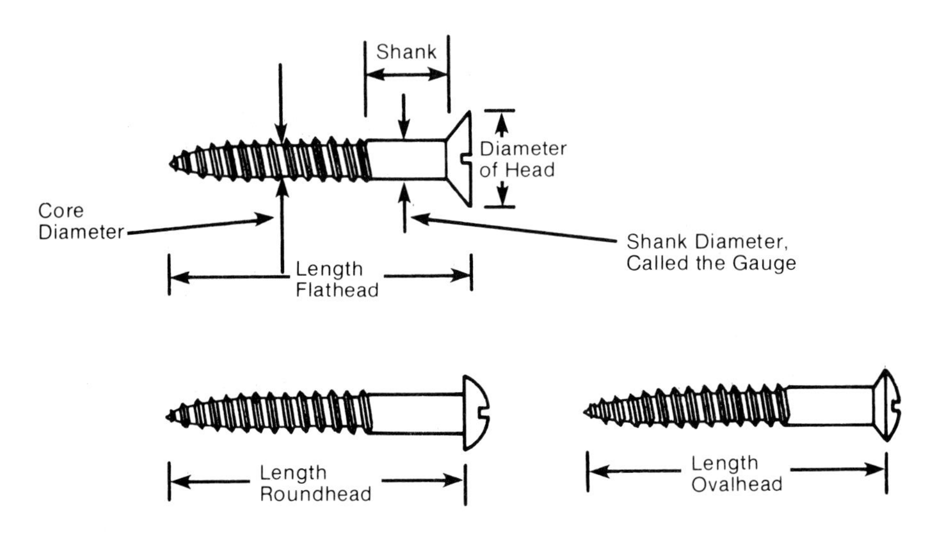

FIGURE 3-6. Nomenclature of wood screws.

3.3.2 Working With Screws

Screws always require a starting hole or pilot hole. Failure to use a starting hole will make the screw difficult to drive and may cause splitting. In softwoods, starting holes can be made with an awl when the screw being driven is small. In most cases starting holes are drilled.

Because the shank of a screw is wider in diameter than its threaded portion, a starting hole requires two distinct diameters, one wide enough to accommodate the shank (the shank hole) and one narrow enough to exert force on the threaded portion of the screw (the pilot or lead hole) (Figure 3-7). The proper size drill bits used to create the correct shank and pilot holes for various screws are listed in Table 3-2.

The depth of the pilot hole should be approximately one-half the length of the threaded portion of the screw; the depth of the shank hole usually matches the length of the shank.

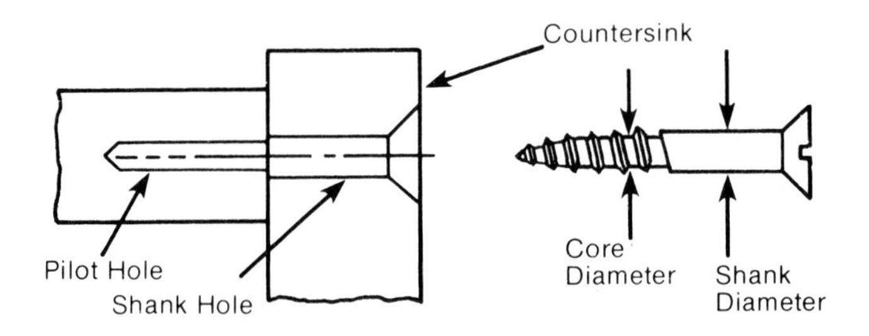

FIGURE 3-7. Starting holes and countersink for a flathead wood screw.

TABLE 3-2 STANDARD STEEL AND BRASS SCREWS

Length Inches (mm)	Gauge	
	Steel Screws	Brass Screws
1/4 (6)	0 to 4	0 to 4
3/8 (10)	0 to 8	0 to 6
1/2 (13)	1 to 10	1 to 8
5/8 (16)	2 to 12	2 to 10
3/4 (19)	2 to 14	2 to 12
7/8 (22)	3 to 14	4 to 12
1 (25)	3 to 16	4 to 14
1-1/4 (31)	4 to 18	6 to 14
1-1/2 (38)	4 to 20	6 to 14
1-3/4 (44)	6 to 20	8 to 14
2 (51)	6 to 20	8 to 18
2-1/4 (57)	6 to 20	10 to 18
2-1/2 (64)	6 to 20	10 to 18
2-3/4 (70)	8 to 20	8 to 20
3 (76)	8 to 24	12 to 18
3-1/2 (89)	10 to 24	12 to 18
4 (102)	12 to 24	12 to 24
4-1/2 (114)	14 to 24	14 to 24
5 (127)	14 to 24	14 to 24

Gauge Number	Actual Size	Decimal	Approx. Fraction	Drill or Auger Bit Size in Inches (mm)		
				Shank Hole	Pilot Hole	Countersink Hole
0	○	0.060	1/16	1/16 (2)		
1	○	0.073	5/64–	3/32 (2)		
2	○	0.086	5/64+	3/32 (2)	1/16 (2)	3/16 (5)
3	○	0.099	3/32	1/8 (3)	1/16 (2)	4/16 (6)
4	○	0.112	7/64	1/8 (3)	1/16 (2)	4/16 (6)
5	○	0.125	1/8	1/8 (3)	3/32 (2)	4/16 (6)
6	○	0.138	9/64	5/32 (4)	3/32 (2)	5/16 (8)
7	○	0.151	5/32–	5/32 (4)	1/8 (3)	5/16 (8)
8	○	0.164	5/32+	3/16 (5)	1/8 (3)	6/16 (10)
9	○	0.177	11/64	3/16 (5)	1/8 (3)	6/16 (10)
10	○	0.190	3/16	3/16 (5)	1/8 (3)	6/16 (10)
11	○	0.203	13/64	7/32 (6)	5/32 (4)	7/16 (11)
12	○	0.216	7/32	7/32 (6)	5/32 (4)	7/16 (11)
14	○	0.242	15/64	1/4 (6)	3/16 (5)	8/16 (13)
16	○	0.268	17/64	9/32 (7)	7/32 (6)	9/16 (14)
18	○	0.294	19/64	5/16 (8)	1/4 (6)	10/16 (16)
20	○	0.320	21/64	11/32 (9)	9/32 (7)	11/16 (17)
24	○	0.372	3/8	3/8 (10)	5/16 (8)	12/16 (19)

Countersinking is drilling a special hole to contain the head of the screw. In this way the screw can be driven down flush to the work surface (Figure 3-8A). The countersunk hole should be fully formed in hardwoods. In softwoods, it is better to stop short of the full hole depth so that the force of the screw being driven can complete the hole.

Counterboring is drilling a hole which will allow the screw to set completely beneath the finished surface (Figure 3-8B). Counterbored holes are usually filled with wood putty or wood plugs. Counterboring

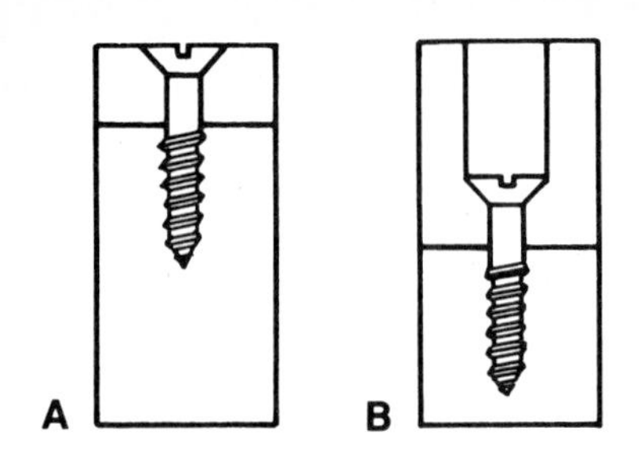

FIGURE 3-8. Counterbored screw with a plug filler.

makes it possible to attach thick stock using shorter screws. Always be sure to use a screwdriver with a blade narrower than the diameter of the counterbored hole. Screws driven at an angle will seat properly if the screwhole is counterbored to provide a flat surface for the screwhead. Countersinking and counterboring operations are done with specially designed drill bits or by using several standard drill bit sizes.

3.4 BOLTS

Where they are appropriate, bolts are probably the strongest wood fasteners used in light building construction. They are generally heavy-duty connectors used in places in which nails and screws cannot handle the demanded load. For example, wooden decks should always be bolted, not nailed, to the side of a house. Backheaders which support overhanging roofs should also be bolted to the building.

Bolts are often used to connect wood members to other building materials. A fine example is the use of bolts to secure the sill plate to the foundation (Figure 3-9). These bolts, usually 1/2″ (13mm) in diameter and 12″ to 18″ (305 mm to 457 mm) in length are set head-down into the wet concrete or are suspended by straps from the foundation forms before the concrete is poured. After the concrete dries and sets, the sill plates are dropped over the shaft of the bolts protruding from the concrete and are bolted fast. (More will be said concerning this procedure in Chapter 15.)

Unlike a wood screw, a bolt does not thread into the wood, but it is passed through a prebored hole and is held in place by a bolt tightened by means of a wrench to create a much greater torque or holding power.

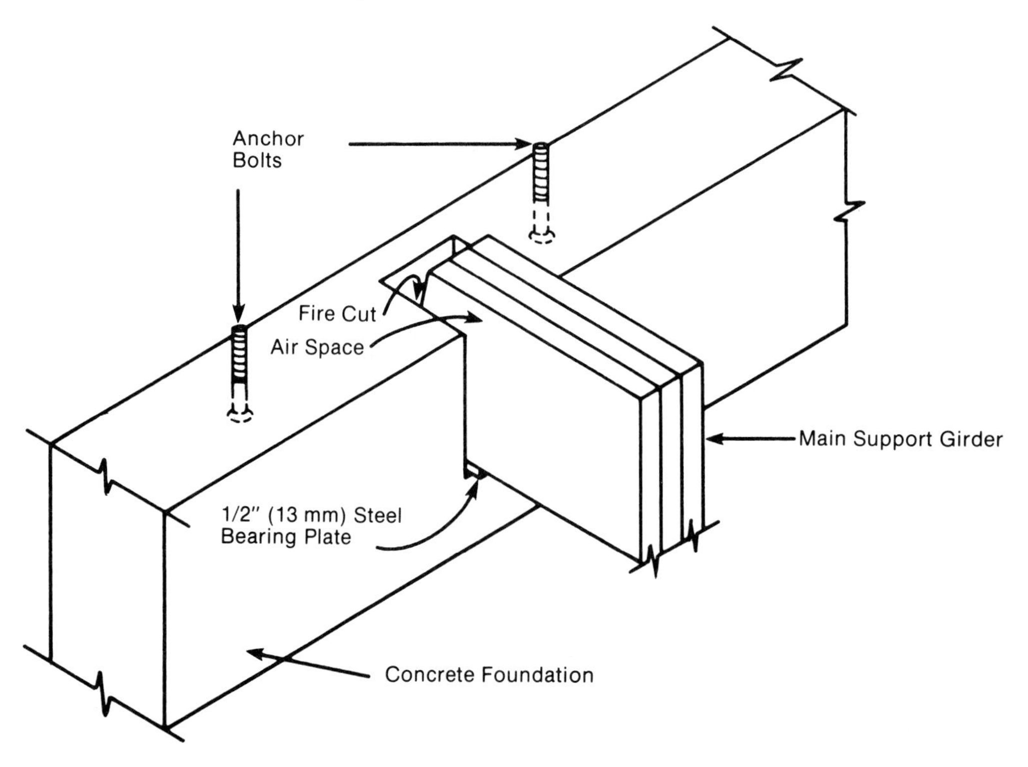

FIGURE 3-9. Protruding anchor bolts for tying the sill plate to the foundation.

3.4.1 Bolt Types

The most common bolts used in light construction (Figure 3-10) are carriage and machine bolts ranging from 3/4″ to 20″ (19 mm to 508 mm) in length and 3/16″ to 3/4″ (5 mm to 19 mm) in diameter. Smaller stove bolts range from 3/8″ to 4″ (10 mm to 102 mm) in length and 1/8″ to 3/8″ (3 mm to 10 mm) in diameter. Lag bolts of various sizes are also used. These are sometimes referred to as lag screws.

The heavy-duty carriage bolt has a square shoulder below a rounded head; the square section is imbedded in the wood as the bolt is tightened. This in turn keeps the bolt from turning as the nut is tightened on the other side of the joint. The machine bolt has a hexagon head or a square head which is held with a wrench to prevent it from turning when the nut is tightened. Stove bolts have a slotted head and are tightened with a screwdriver, not a wrench. They are available in a flathead style and can be countersunk and filled much like a screw. Lag bolts are driven with a wrench, but are for use where only the head will be accessible. In other words, they are driven into wood like a screw, but with the force of a wrench. They are uniquely suited for joints where neither carriage bolts nor machine bolts can be used. Carriage, machine, and stove bolts should be used with a washer at the nut end; for machine bolts and lag bolts a washer under the head will give better bearing strength against the wood. There are other types of bolts as well: masonry bolts with expansion anchors, eyebolts, U-bolts, turnbuckles, and so on.

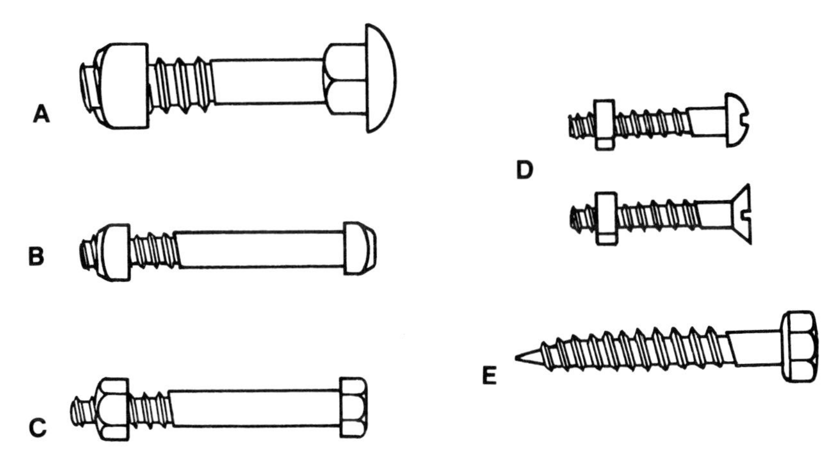

FIGURE 3-10. Bolts used in light building construction: (A) carriage bolt, (B) machine bolt with square head and (C) hexagon head, (D) stove bolts, and (E) lag bolt.

3.4.2 Working With Bolts

The use of bolts is simple. Position the mating material pieces properly and bore or drill through them with a bit that matches the size of the bolt. Counterbore if it is desirable to conceal the head. However, when installing bolts, be certain that the grip length of the bolt is correct. The grip length is the length of the unthreaded portion of the bolt shank. Generally, the grip length should be equal to the thickness of the material that is being bolted together. Under most conditions, no more than one thread should bear on the material, and the threaded portion of the shank should be showing through the nut. If demountability is a factor, wing nuts can be used instead of regular hexagonal or square nuts. Wing nuts are even easier to remove and can be tightened with pliers.

When fastening into masonry—brick, stone, concrete—it is best to use masonry bolts with metal expansion shields. Although there are several types of bolt arrangements available, all work on the same principle that the shield expands to grip the sides of a predrilled hole as the bolt is driven home (Fig. 3-11). It is important to drill the hole to fit the size of the metal insert before expansion.

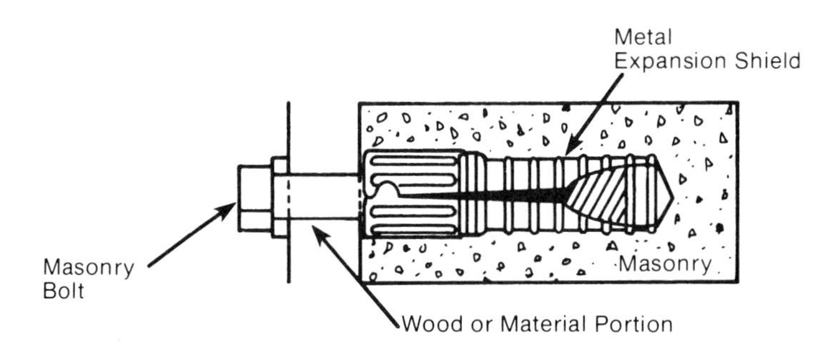

FIGURE 3-11. Masonry bolt with metal expansion shield.

3.5 STRUCTURAL WOOD CONNECTORS AND CLIPS

In the past several years, a wide variety of structural wood connectors and clips have been designed to help eliminate certain structural flaws inherent to more traditional fastening methods. These special fasteners either completely replace nails, screws, and bolts or are used in combination with these fasteners to form a stronger joint. Such fasteners are becoming more and more popular in all types of framing and production work involving fastening wood to wood, concrete to wood, drywall to wood, and other material combinations. Brief summaries of the most useful connectors and clips are given in the following paragraphs.

Anchor Clips. Although anchor clips have a variety of uses, they are primarily used to anchor wooden footing plates or sill plates to concrete foundations (Figure 3-12). Made of 16 gauge steel, anchor clips are available in two sizes—a smaller size for embedding in concrete or around a single masonry block and a larger size for fitting around two masonry blocks. These clips can also be used to secure sheathed wall or finished walls or as a rafter tie.

To install, the lower arms of the clip are bent around the blocks and/or embedded into the concrete. The upper arms are opened, wrapped around the plate, and secured to it with 8d common nails.

Post Base Clips. Post base clips (Figure 3-13) are used to anchor and support 4″ x 4″ (89 mm x 89 mm) post on concrete. The bottom of the clip is tied to the concrete foundation with an anchor bolt or masonry nails. The post rests on a spacer plate that protects the wood against surface moisture, and it is secured to the clip with barbed nails driven through the slots in the clip base.

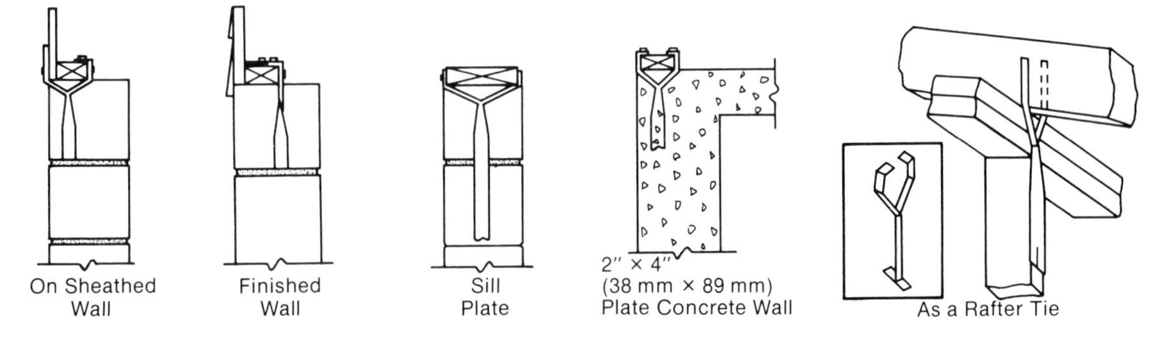

FIGURE 3-12. Various uses of anchor type structural connectors.

Beam Clips. These conectors (Figure 3-14) provide superior structural strength for anchoring the top of a post or column to a beam. Two clips per post are normally used. Beam clips eliminate the need for awkward toenailing. To install, simply nail the clip onto the post or column, position the beam into the clip, and nail it fast with barbed nails for superior withdrawal resistance.

Bridging Clips. These easily installed clips serve as a substitute for traditional wood bridging (Figure 3-15). Although they lack the rigid strength of solid bridging, they eliminate the usual cutting, mitering, and nailing, and they are strong enough to prevent squeaky floor problems. No nails are required. The straight end of the clip is hammered into the floor joist; the pre-bent end is pushed up to a 45° angle and then hammered into place.

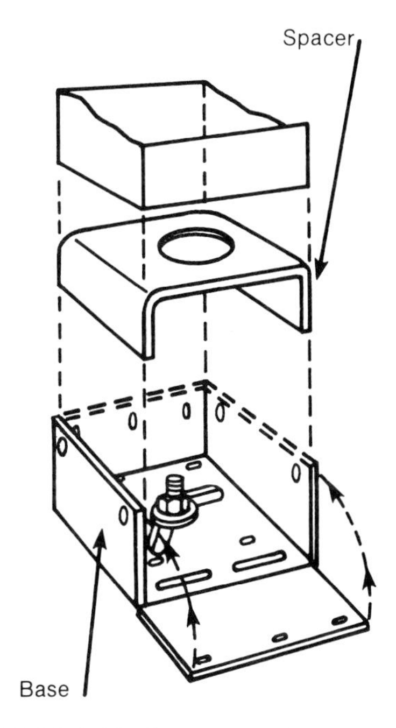

FIGURE 3-13. Post base clips used for tying 4" x 4" (89 mm x 89 mm) posts to a concrete foundation.

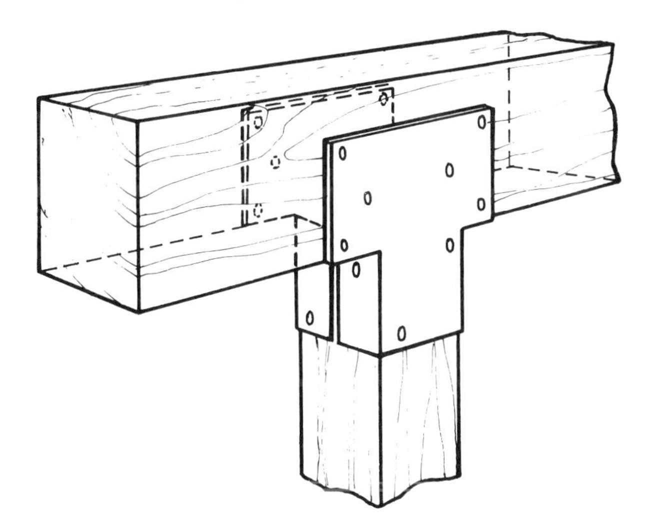

FIGURE 3-14. Beam-type clip for anchoring the top of a post or column to a beam.

Framing Clips. Framing clips are used for securing a variety of two-and three-way ties (Figure 3-16) in floor, roof, and wall framing. They provide excellent structural strength and eliminate the need for toenailing. They can be adjusted for both left- and right-handed joints and can accommodate a variety of nailing patterns and the use of lag bolts. To install a framing clip, the slotted angle is bent to fit the joint and the clip is nailed into place. Again, barbed nails work best with this clip.

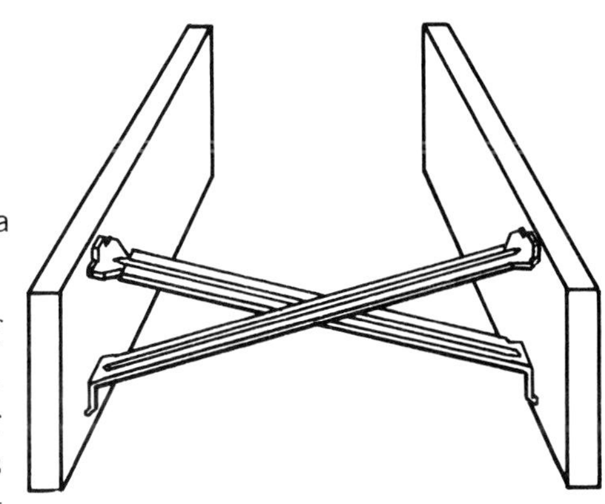

FIGURE 3-15. Bridging type connectors can be used as a substitute for traditional wood bridging in some cases.

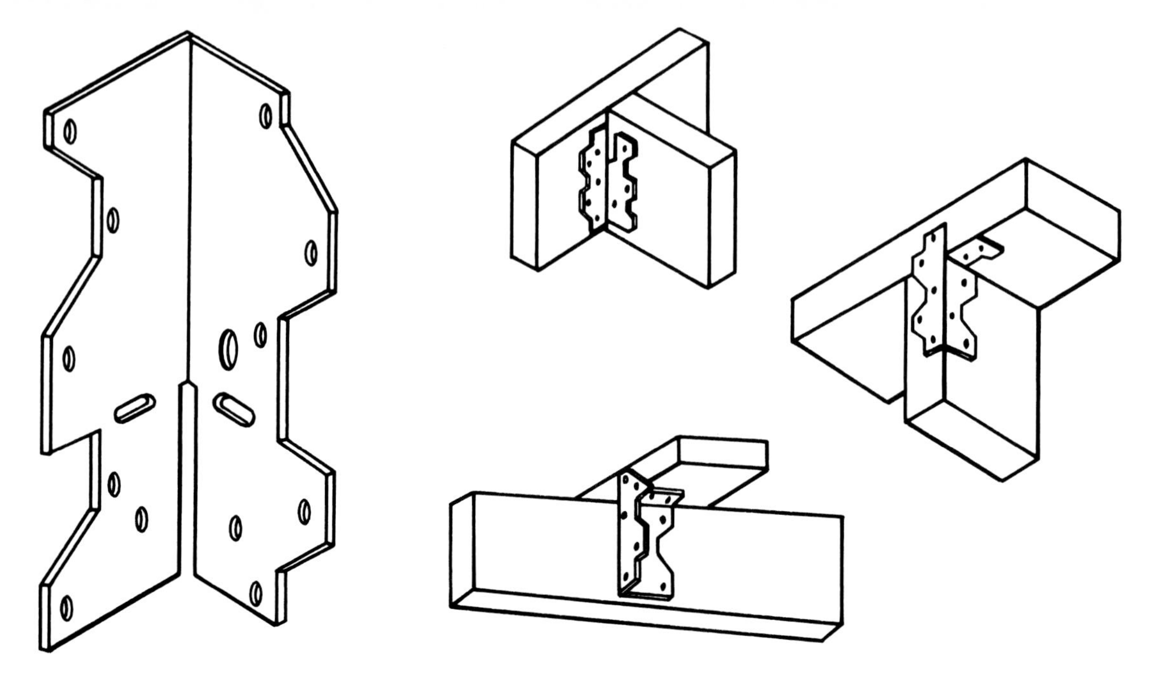

FIGURE 3-16. Framing type clips are used for a variety of two- and three-way joints. Note how they eliminate the need for awkward toenailing.

Storm Clips. These clips provide a simple, secure method of anchoring trusses or rafters to plates and studs to eliminate wind uplift problems. Two clips should be used per truss or rafter, one on each end (Figure 3-17).

Joist Clips. Joist clips or hangers (Figure 3-18) ensure that floor and ceiling joists are properly attached to headers and beams. This widely used clip eases alignment and nailing problems and increases the strength of the joint when compared to toenailing, end nailing, and the use of ledger strips. Both single and double joist clips are available.

To install this connector, measure and mark for location on the header, nail the clip to the joist, line the joist up on the header, and drive in the self contained speed nails to temporarily keep it in place. Drive in additional 10 gauge barbed nails to permanently hold the joist in place. For double joists, use heavier nails, such as 16d common nails.

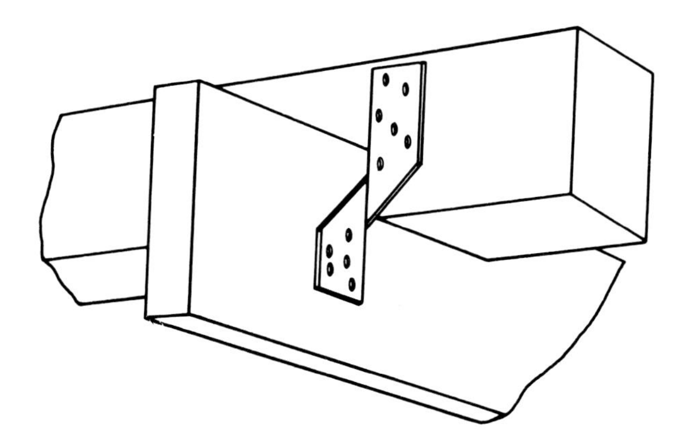

FIGURE 3-17. Storm clips for anchoring trusses or rafters to plates or studs.

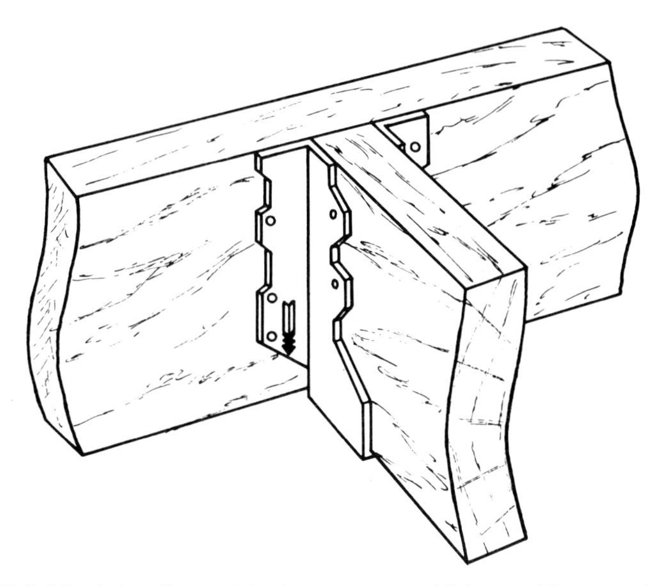

FIGURE 3-18. Joist clips or joist hangers are widely used for securing joists to headers and beams.

Purlin Clips. These structural wood connectors speed up framing, layout, and installation. By using purlin clips to tie together roof trusses, purlin members can be placed between trusses flush with the top of each truss (Figure 3-19). This eliminates the need for notching and provides solid bracing for the trusses. Both single and double purlin clips are available.

When installed, the purlin clip is nailed to the top of the rafter at the spacing required. A sectional purlin or blocking member is set into the side slots of the clip and nailed fast with 8d common nails.

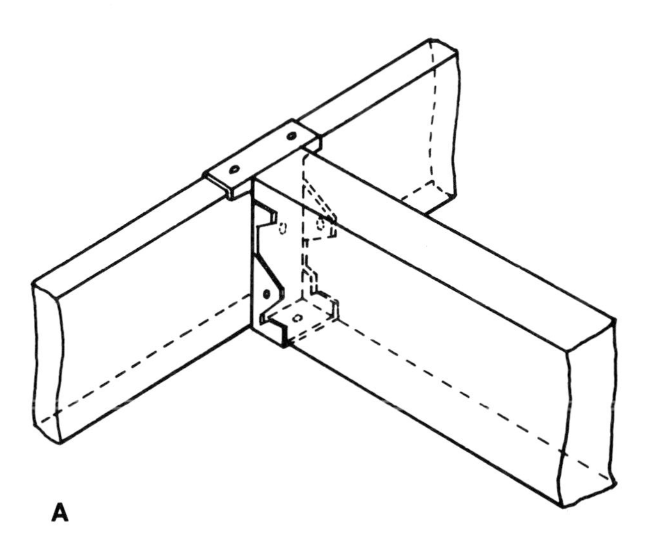

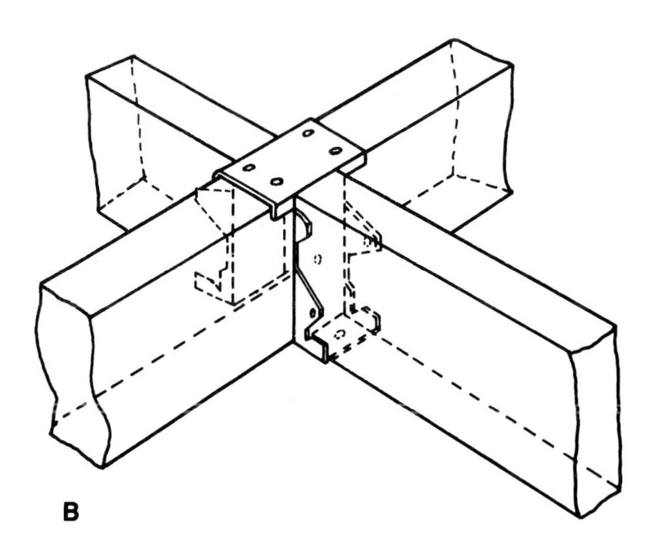

FIGURE 3-19. (A) Single purlin clip; (B) saddle purlin clip.

Panel Clips. Panel clips (Figure 3-20A) are designed to provide vertical strength and a snug, tight fit between panels of plywood roof sheathing. Their chief advantage is that they eliminate the need for edge blocking to support the plywood panels whose edges fall between rafters in a roof layout.

Dry Wall Clips. Dry wall connectors (Figure 3-20B) can be used as a structural backup for drywall and similar materials. They eliminate the need for extra studs and ceiling backup materials. They are easily attached to existing studs with self-contained speed nails and are usually located no less that 16″ (406 mm) on center.

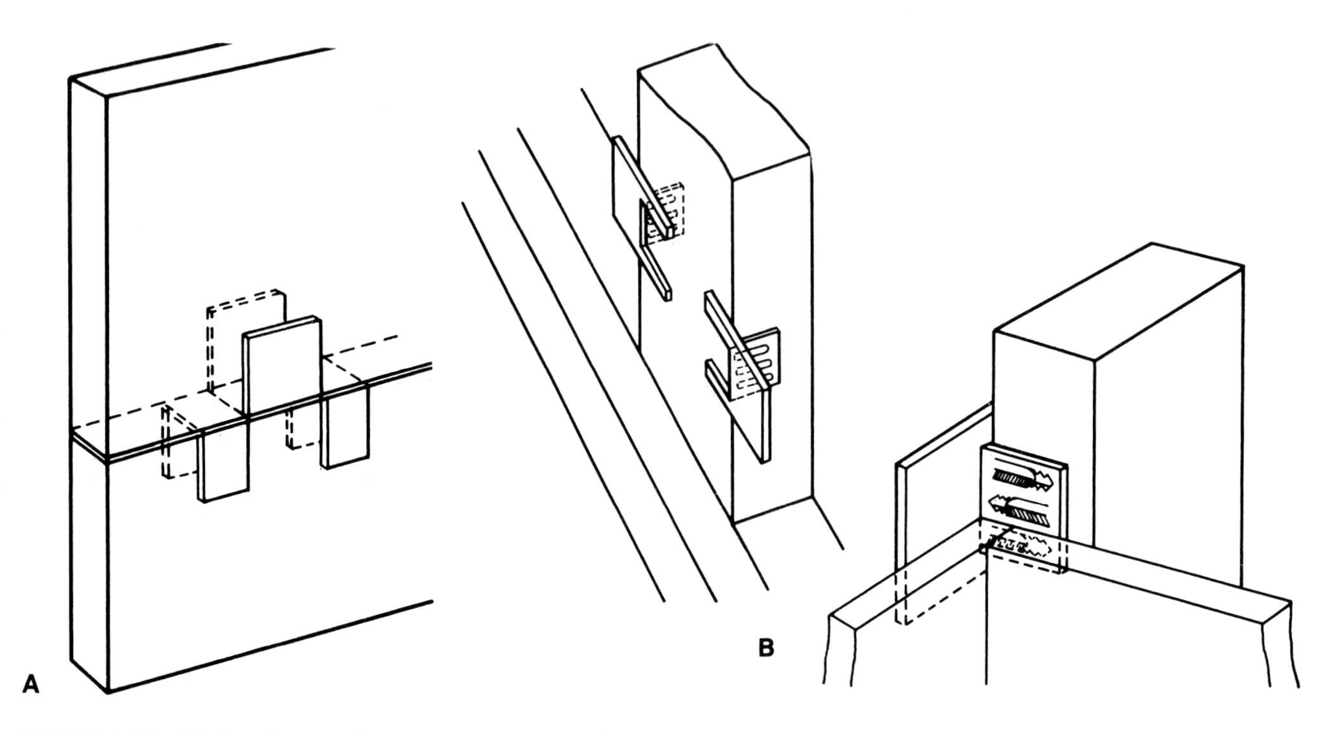

FIGURE 3-20. (A) Panel type clips to support plywood roof sheathing. (B) Drywall clips provide a structural backup for drywall and similar materials.

Other Specialty Clips. *Top plate ties* (Figure 3-21A) can be used to secure joints on the top plate or cap plate of walls and partitions. These plates hold rows of self-contained claw-like barbs that have sufficiently more holding power than 16d nails. The plate is simply placed over the joint and hammered in.

Using *truss clips* or plates (Figure 3-21B) is an increasingly popular method of fastening truss assemblies. They have great holding power and eliminate the need for lapped joints or gussets.

3.6 ADHESIVES AND GLUES FOR LIGHT CONSTRUCTION

Traditionally, adhesives and glues played a minor role in the light construction industry. They were commonly used to install ceramic tile and for a number of light holding jobs such as cabinet and furniture making. Advances in the adhesive industry have greatly enlarged the use of adhesives and glues in all phases of modern construction. In many cases, they have become the ideal fastening method for securing floor, wall, and ceiling materials to the building frame. Specially formulated adhesives and glues have been developed to fasten nearly all of the materials used in light construction projects: wood and wood based materials, drywall, metals, foam insulation board, tile, and carpeting.

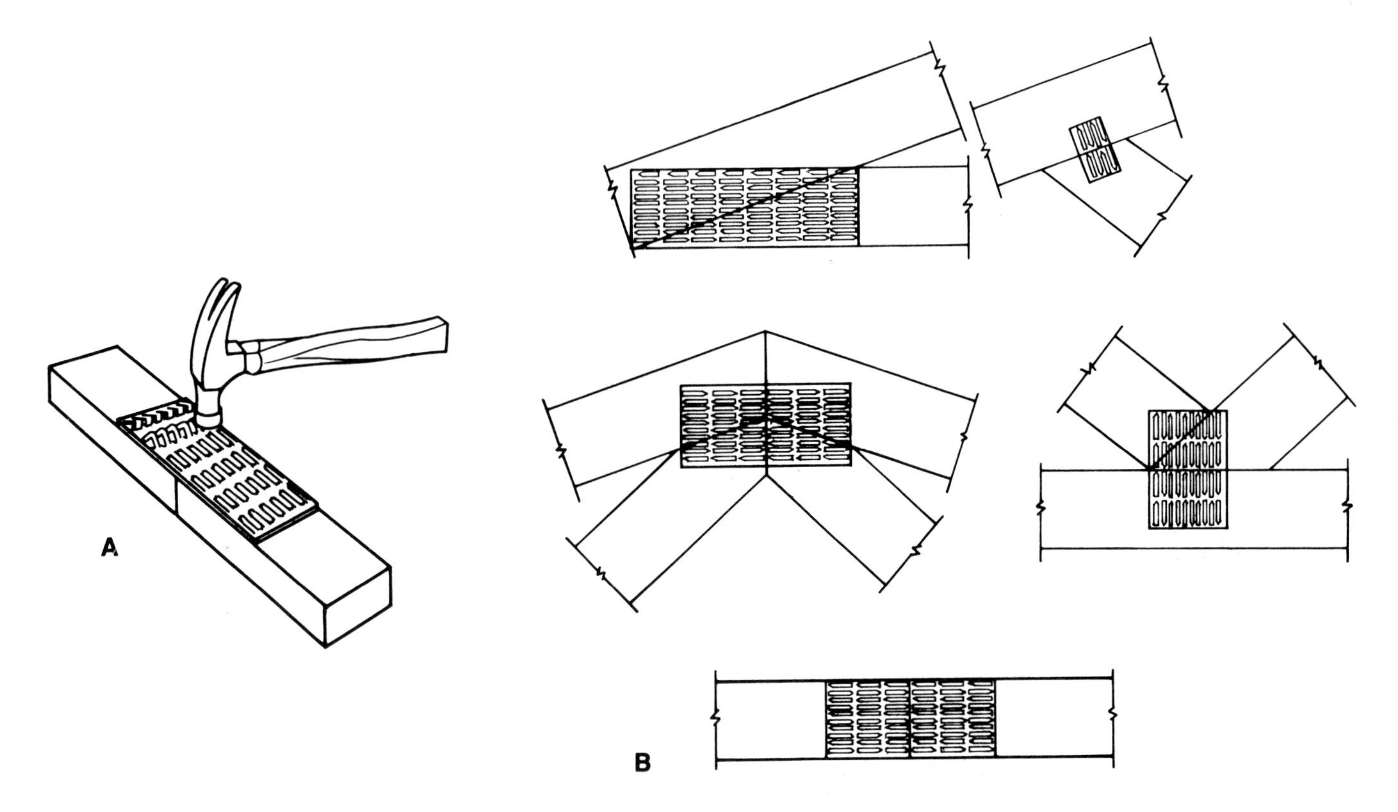

FIGURE 3-21. (A) Truss clips are hammered into place using self-contained barbs; (B) various truss assembly joints secured with truss clips, also known as truss plates or truss connectors.

Modern general construction adhesives (Figure 3-22) dry exceptionally fast (10 to 30 minutes) at normal temperatures (above 60° F). In cold temperatures, drying will be somewhat retarded. Do not apply them at temperatures below 10° F. Many formulations can be used for both interior and exterior applications. The adhesive can be gun extruded (using a caulking gun) or applied with a spatula or putty knife; some are designed for trowel applications. Their high degree of

FIGURE 3-22. Some of the more popular "construction" types of adhesives available to both the construction industry and the do-it-yourselfer.

initial tack allows light pieces to be bonded with only momentary pressure. Exposure of the adhesive to air, before assembly, will increase this initial grab. The allowable open time is regulated by the temperature, humidity, and air circulation and the basic characteristics of the formulation. Many of the modern construction adhesives are nonflammable and nontoxic, while others contain volatile hexane, acetone, and toluene. Always follow all manufacturer's safety and installation instructions. Since some construction adhesives attack polystyrene foams, always check the manufacturer's recommendations on the container before using it on this material.

The following are general instructions for the application of most construction adhesives.

1. Remove dust and other foreign matter from surfaces to be bonded.
2. There are two types of installations depending upon the type of wall surface with which you are working.
   - Studs, joists, or furring strips—Apply the proper bead (1/8″ to 3/8″ [3 mm to 10 mm]) as determined by surfaces to be bonded (1/8″ [3 mm]) where there is contact, 1/4″ (6 mm) for most applications, and 3/8″ (10 mm) for bridging gaps and where the surface is irregular) (Figure 3-23). A spaghetti or serpentine pattern is recommended for large contact areas, generally on 12′ to 16′ (3.7 m to 4.9 m) centers.
   - Existing walls of plywood, plaster, plasterboard, etc.—Solid backing walls of this type generally present the best surfaces for wall covering materials. In this case, trowelable grade construction or tileboard adhesives should be used and the wall covering should be troweled over the entire surface (wall covering surface) with the recommended notched spreader.
3. Apply to one surface only.
4. Press substrates firmly in order to wet the total area to be bonded. Adhesive should spread to the edges of the desired contact area.
5. Use supplemental fasteners or weight to hold substrates in position until adhesive sets.

FIGURE 3-23. Applying panel adhesive to studs.

Metal Framing and Structural Adhesives

These products frequently use a neoprene base. They are designed to meet the growing demands of galvanized steel and aluminum framing members both in floor and wall systems. Many develop exceptional strength when bonding plywood, gypsum wallboard, hardboard, and paneling to these metal surfaces. They remain permanently resilient through wide temperature variations from -40° F to 200° F (-40° C to 93.3° C) Initial tack is quite high. The creep resistance is also unusually high on some formulations.

Panel Adhesives

These adhesives, frequently called panel and plywood adhesives or mastics, can be used to bond panels, drywall, hardboard, corkboard, bulletin boards, or chalkboards to masonry, studs, drywall, or concrete. That is, paneling of all types can be bonded tightly to studs, drywall, masonry, or furring strips with these adhesives and requires few or no nails for a smooth, one-step installation. This eliminates nail pops and patching, plus removes the danger of marring prefinished panels with hammer marks. They also overcome structural deficiencies, fill gaps, and bridge minor framing defects. Usually dispensed from cartridges, most panel adhesives will not drip from beams or sag from vertical surfaces.

Foamboard Adhesives

Plastic foams, both polystyrene and polyurethane, are being used more frequently today in all types of construction; however, they can present some problems if the incorrect adhesive is applied. This is particularly true of the common insulation varieties of polystyrene foams, which are quite susceptible to attack from certain solvents found in many panel and construction types of adhesives. Therefore, be absolutely certain that the adhesive used is designated by the manufacturer as being satisfactory for use on polystyrene foams. Certain solvents, however, can be used and there are some solvent based adhesive formulations on the market that are perfectly satisfactory. Those with an emulsion (water) base are generally the best choice, since water will not bother the foam.

Polyurethane foams provide as much insulation value as a greater thickness of polystyrene would. Currently, the largest use for polyurethanes in the home construction area is in simulated wood beams for ceilings. Polyurethane foams have a great deal of resistance to most of the solvents used in adhesives. Therefore, a good construction or panel adhesive will work with the polyurethanes.

To distinguish between polystyrene and polyurethane consider that urethanes are generally of a yellowish color and have a smaller or finer pore structure. Polystyrenes, on the other hand, are generally white or blue. They may or may not have a small or fine pore structure, as the "bead" boards are merely polystyrene beads fused together.

Gypsum Drywall Adhesives

These adhesives are specially formulated for bonding gypsum drywall to wood or metal studs and for laminating drywall panels. They greatly increase wall strength, reduce sound transmission, and improve wall appearance overall. Those made with an elastomeric base contain no asphalt or tar and can be painted over without fear of stain-through. A high solids content makes them an excellent gap filler that bridges irregularities. Properly applied, they stay in place without sagging and allow ample time for alignment of each drywall panel. Only minimum supplemental nailing is required on walls or ceiling, thereby reducing nail pops.

Tileboard and Wallboard Adhesives

These adhesives are specifically designed for use on large sheets of prefinished hardboard, plywood, and wallboard. Smooth spreading and easy to use, tileboard adhesives form a high-strength bond to wood, masonry, poured concrete, plaster, and gypsum wallboard, and they are highly water-resistant. (Some are even waterproof.) They usually have a working or open time of approximately 30 minutes. The open time, however, may also depend upon other variables. High ambient temperatures, low humidity conditions or dry air, and high wind conditions will all shorten the working time. On the other hand, lower temperatures and high humidity conditions or moist air will tend to lengthen the working time.

Ceiling Tile Adhesives

These adhesives are specially formulated for the installation of acoustical ceiling tile. They have ultra-high cohesive strength. Cohesive strength is the ability of an adhesive to adhere within itself. This allows the adhesive to hold the ceiling tile in place without any sagging. Quick green grab is another desirable trait of a ceiling tile adhesive; it must keep the tile in place without a lot of unnecessary mechanical fasteners. Sometimes the characteristics of high cohesive strength and quick grab are desired in certain wallboard installations. Because of this, some manufacturers market these adhesives as ceiling tile and wallboard adhesives.

Subfloor Adhesives

These adhesives were specifically developed for bonding plywood directly to floor joists, but have now developed into a general purpose use category in many cases (Figure 3-24). Most of these multipurpose formulations achieve exceptional bonds to woods, concrete, gypsum wallboard, insulation board, and similar construction surfaces. Their high degree of initial tack allows light pieces to be bonded with only momentary pressure. The allowable open time is regulated by the

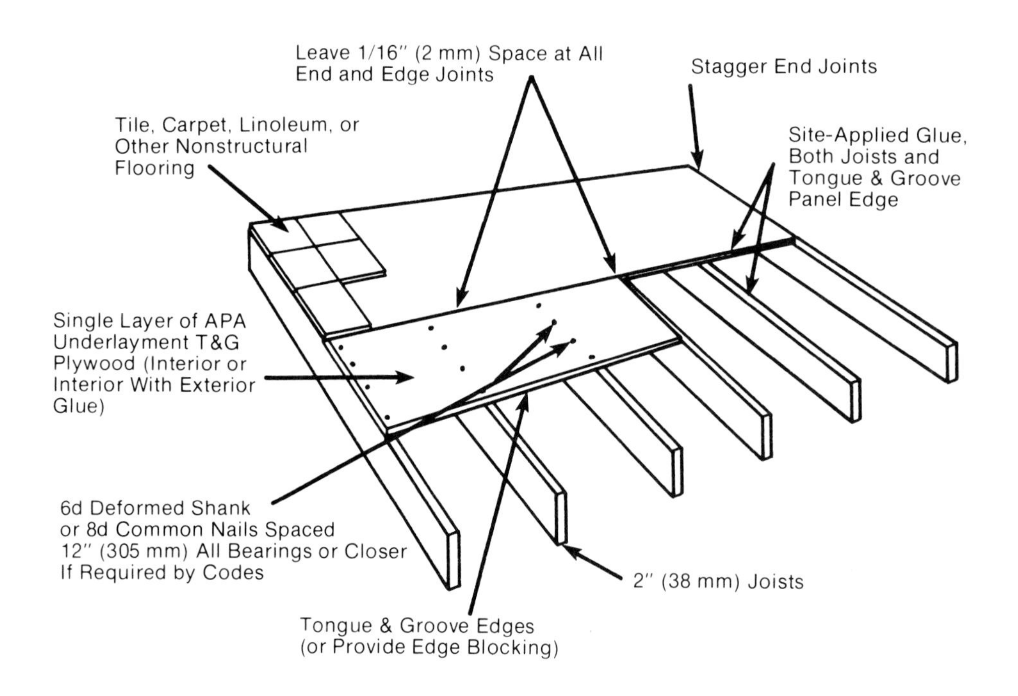

FIGURE 3-24. APA glued floor system.

temperature, humidity, and air circulation. The porosity of the substrate or adherend will also have an effect on the drying rate, depending upon the ability of these surfaces to absorb solvent from the system. Irregularities and voids in substrates, such as framing, are readily bridged with subfloor adhesive, providing a more solid backup surface.

The most common methods for applying subfloor mastics is by extrusion, either in cartridge form or specialized extrusion equipment.

Floor Tile Adhesives

Most multipurpose floor tile adhesives are water resistant both indoors and out after completely drying. They are suitable for vinyl tile, vinyl asbestos tile, vinyl roll goods, rubber tile, asphalt tile, and carpeting. In addition, they may be used on floor surfaces such as plywood, particleboard, and smooth, dry concrete either below, on, or above grade, providing that there is no moisture or hydrostatic pressure in the floor. Most of these multipurpose adhesives permit adjustment of the flooring material after they are applied.

Cove Base Adhesives

These resin base formulations are designed specifically for the easy installation of rubber and vinyl cove base materials. They develop excellent adhesion on wall surfaces such as plaster, gypsum wallboard, wall papers, plywood and hardboard paneling, ceramic tile, tileboard, etc.

Carpet Adhesives

These adhesives were developed specifically for installation of residential and commercial carpeting with impregnated, laminated foam rubber backing, or plain jute backing. They provide quick bond between carpet and floor surfaces of wood, plywood, particleboard, and concrete; and it can be used above, on, or below grade level.

Wood Floor Mastics

These products are specially designed for the installation of wood parquet, strips, or planks. They may be applied over smooth, dry concrete, plywood, particleboard, or hardboard surfaces. The use of a latex emulsion system for wood flooring is not recommended.

Ceramic Floor Tile Adhesives

These solvent and latex adhesives are similar to the ceramic adhesives used for walls, except that the floor adhesives generally dry more rapidly than wall types.

3.7 RESEARCH

1. Outline the various stages of a relatively simple light construction project, such as a large garage or storage building, and state the types of fasteners and fastening methods that can be used at each stage of the project.

2. Visit a major building supply dealer and review the most modern fasteners and adhesives available on the market. Note the advantages and disadvantages of each, and draw up general cost comparisons.
3. Work up a cost sheet for the fasteners used in research question 1.
4. Do further research into the area of adhesives and glues used in construction work. Fully outline their properties, applications, and safety points.

3.8 REFERENCES

3.8.1 Miller, Robert S., *Adhesives and Glues, How to Choose and Use Them.* Ohio: Franklin Chemical Industries, 1980.

3.8.2 The Panel Clip Company, *Build-it With Clips.* Michigan, 1979.

4

CONSTRUCTION SAFETY PRINCIPLES AND PRACTICES

A good place to work is also a safe place to work. To make a jobsite safe requires the conscious cooperation of everyone on the job. Safety consciousness implies a positive attitude toward the building tasks and toward other workers. It also requires a considerable effort to learn and to be aware of safe limits with respect to one's health, to equipment, and to materials.

4.1 GENERAL OBJECTIVES

The primary intent of this chapter is for the reader to be able to recognize a safe work environment and be able to create the work conditions that will optimize both productivity and safety.

To fully benefit from the material in this chapter, it will be helpful for the reader to experience certain practical construction activities. Although not strict prerequisites, the reader who has worked on or has taken field trips to light construction jobsites and who has had some building experience using handtools and some machine tools will better appreciate the material presented here. It is also important for the reader to be interested in learning how people work together safely during the construction process.

4.2 SPECIFIC OBJECTIVES

The material in this chapter should enable the reader to

1. help create the conditions for a relatively safe building site or workshop;
2. write a set of guidelines for a personal on-the-job safety program;
3. be able to respond to a jobsite or workshop accident or emergency;
4. be aware of the various access structures and hoisting techniques common to light building construction;
5. conduct research and write specifications for a light or heavy-duty wood scaffold which meets local requirements;
6. research and write a set of safety rules for the safe operation of any common jobsite tool or piece of machinery, including heavy equipment, such as front-end loaders or backhoes;

7. research and write a set of shop rules for the safe operation of a radial saw or table saw, jointer, and band saw, or any other standard shop tool; and
8. plan a typical woodworking shop and organize it in an efficient and safe manner, determining which tools will be necessary and making a list of power tool safety device requirements.

4.3 A SAFETY PROGRAM

If a program is an organized sequence of events, then a safety program is an intentionally organized program which encourages safety for an individual, safety for a person working as a member of a group, and safety for a group generally.

Organizing means planning for positive outcomes; however, if something useful is expected then something worthwhile must go into the making of the plan. The input must produce benefits to all concerned: the employees, the builder, and the owner. The employees should expect gainful employment, relatively safe working conditions, and satisfaction from work well done and from creative building experiences. The builder should expect to get the job done with minimal risks and without lawsuits and budget overruns. The owner expects good value for the money spent and general satisfaction with respect to the craftsmanship displayed. All should benefit from a positively planned safety program.

For the initial stages of a construction job, some elements of a safety program would involve the site preparation, the working climate, and first aid. The layout of the site is important to avoid on-site confusion. Provision must be made for working space around the actual project space to allow and to control movements of persons, equipment, and materials. Water, power, and fuel services must be safely provided where they are needed. Security with respect to materials and equipment as the project develops will also be required. The builder must create a setting in which work can regularly be done systematically and safely.

4.3.1 Relevant Literature

Although much of what is called jobsite safety involves first and foremost common sense, by builder and worker alike, many exact rules and regulations have been enacted in recent years. Legislation concerning on-the-job safety has become an ongoing exercise; therefore, it is important for builders to be aware of new safety legislation as it becomes effective.

A good way to review and update information concerning jobsite safety rules is to read the relevant government publications. Because occupational, health, and safety acts and regulations are always changing, keeping in touch with this type of material will enable the builder to know precisely what safety rules are in force. However, checking the relevant government publications will also give the builder a broad background in the entire area of jobsite safety.

In addition, the existence of the current government documents and standards helps to delineate the responsibilities of persons involved in the building process and can create a greater general feeling of security for those on the job. Probable spin-offs of a greater concern about safety by the builder are greater productivity and an increased sense of worker satisfaction.

The relevant information is found in the publication of legislative acts, including current updates, and in codes, manuals, reports, and specifications. They are in print and to some degree available at appropriate computer terminals. This information is produced by the government and by private standard setting organizations.

Another way for the builder to become informed is by visiting libraries to find and read materials. Decide which materials to obtain for a personal or company reference library from regional, state, or national sources. From such publications one may obtain further information from the references given. One must be selective in terms of personal interests and applicability of materials for a given locality.

The builder should read relevant materials to search not only for general information, but for limits as to what may be safely constructed on the job without professional approvals. When reading data sheets, say for metal scaffolding or for form ties, determine the qualifications of those persons who have approved the designs. Make note of which standards setting organization has given its stamp of approval.

4.3.2 Safety Considerations and the Worker

Workers go to the jobsite with certain expectations and some questions.

1. To what extent does the builder know about and comply with local, regional, and national standards of occupational health and safety?
2. How does the builder plan to make the jobsite safe against local weather patterns, vandals, and roaming children who might get hurt?
3. Does the builder know enough about safe loads, scaffolds, forms, strengths of structural members, and green concrete?
4. Will the builder encourage good housekeeping on the site? What are the standards for guardrails, handrails, illumination, ventilation, shoring, and storage of hazardous materials?
5. Under what conditions should one be willing to work? Under what conditions should one not be willing to stay on the job?
6. If one were injured or became ill on the job, what are the first aid provisions?
7. How can one positively influence the creation of a safe building climate? What are everyone's responsibilities?

Such questions may be answered at jobsite safety meetings. However, to obtain meaningful answers good questions need to be formulated and asked. This requires regular dialogue about safety matters between all members of a construction team on an ongoing and informal basis.

4.3.3 First Aid

Adequate first aid capabilities are a vital aspect of jobsite safety programs. A builder who is prepared for job related accidents benefits not only workers who are injured and need immediate treatment, but also the construction team as well. Better on-the-job treatment can lessen the severity of injuries, decrease worker recovery time, and help to keep worker downtime to a minimum.

There are standard regulations governing jobsite requirements for first aid and medical services in both the United States and Canada. For example, a U.S. Department of Labor, Occupational Safety and Health Administration (OSHA) regulation, revised in February, 1983, requires all construction employers to meet the following standards:

1. The employer shall ensure the availability of medical personnel for advice and consultation on matters of occupational health.
2. Provisions shall be made prior to commencement of the project for prompt medical attention in case of serious injury.
3. In the absence of an infirmary, clinic, hospital, or physician's office (which is reasonably accessible in terms of time and distance to the jobsite) that is available for the treatment of injured employees, a person who has a valid certificate in first aid training from the U.S. Bureau of Mines, the American Red Cross, or equivalent training that can be verified by documentary evidence, shall be available at the worksite to render first aid.
4. First aid supplies approved by the consulting physician shall be easily accessible when required. The first aid kit shall consist of materials approved by the consulting physician in a weatherproof container with individual sealed packages for each type of item. The contents of the first aid kit shall be checked by the employer before being sent out on each job and at least weekly on each job to ensure that the expended items are replaced.
5. Proper equipment for prompt transportation of the injured person to a physician or hospital, or a communication system for contacting necessary ambulance service, shall be provided.
6. The telephone numbers of the physicians, hospitals, or ambulances shall be conspicuously posted.

Regulations governing first aid in Canada are similar. Such regulations regarding medical services and first aid are statements designating responsibilities of principal contractors, employers, and employees. For a particular jobsite all must know the regulations that apply to local codes, as well as those that are current in the region or state.

4.3.4 Personal Safety Program

Of course, it is important for the contractor, builder, supervisor, or job foreman to have a wide-ranging and positive conception of occupational health in general and jobsite safety in particular. It is equally important, however, for every worker on a job to realize the importance of working productively and safely. The following list is not meant to be a point by point account of how builders or workers must conduct themselves when on a job. Needless to say, conditions change from job to job. Rather, it suggests a general safety-conscious state of mind, desirable in anyone involved in building construction. Not every item listed here will necessarily apply directly to every member of a construction crew; however, it is the responsibility of the builder to make everyone aware of their safety responsibilities just as the builder must give everyone their work responsibilities.

1. Develop a positive attitude toward the job and others on the job. Care for oneself and for others. Be alert. Be well mentally and physically, and be the person able to resolve conflicts.
2. Show by your attire that you are ready to look after yourself, and expect others to do the same for themselves and you. Clothes should fit and be weather related. Use appropriate protective gear: headwear, footwear, eye protectors, ear-drum protectors, and face-shields, and know how and why they are used. Remove watches and jewelry so they will not be caught in a machine. Discard or repair torn garments that might wind into a machine or cause tripping or falling.
3. Control long hair that otherwise might cause one to get caught in a machine.
4. Keep tools sharp and in good repair. Tools in good condition are easier to use and tend to cause fewer accidents.
5. Understand the design characteristics of tools and equipment. Be a specialist in a field of work by being knowledgeable with respect to cutting and fastening actions, to high pressure/tension lines (air, water, natural gas, electricity), to materials (strengths, spalling, shattering, breaking) and to hazardous materials.
6. Study safe movements of persons, mobile machines, and hoisting. Be space conscious in terms of what others are doing when you move about (Figure 4-1). Anticipate what might happen. Insist by personal behavior that a high standard of safety prevails.
7. Be medically fit. Know your strengths and weaknesses, especially your physical limitations in terms of sight (including color blindness, hearing, agility, and allergies). Know about lifting limits and how to lift (Figure 4-2).

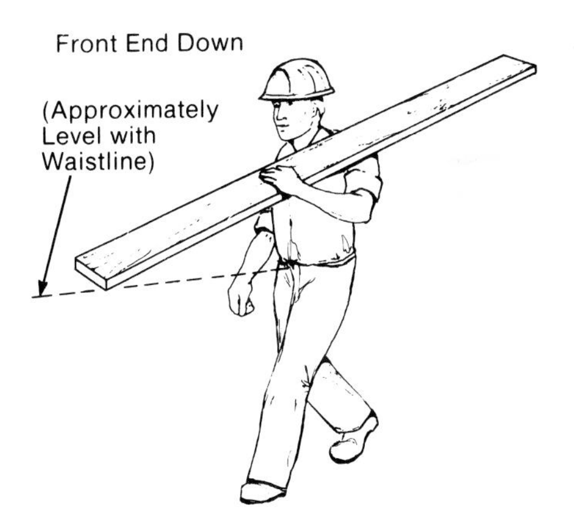

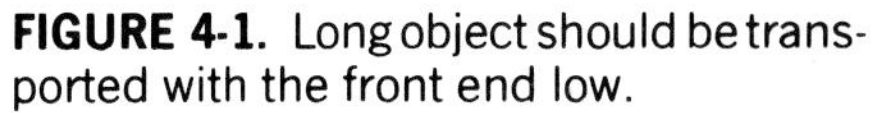
FIGURE 4-1. Long object should be transported with the front end low.

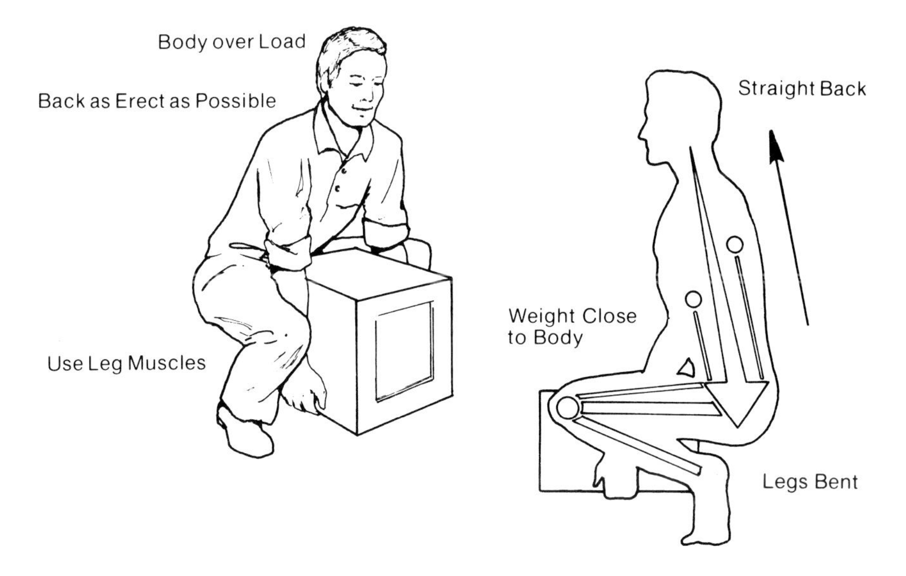

FIGURE 4-2. Make sure you use the correct movements when lifting.

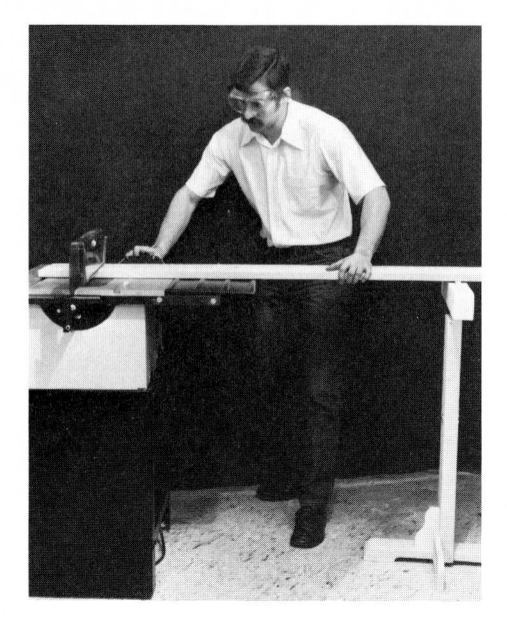

FIGURE 4-3. Be safety conscious when performing light building construction work.

8. Assess your work talents and develop them while on the job.
9. Show a responsible and constructive attitude toward the job and others on the job. Observe how negative feelings may permeate a group and engender carelessness. Know how one's feelings affect others.
10. Recognize general hazardous conditions and know how to cope by using technical knowledge (Figure 4-3). Know about safe noise levels, explosion potentials, toxic fumes, machine characteristics (speeds, cutting actions, feeds, guards, kickbacks, shutdowns), construction faults (forming, framing, pouring), green concrete, high pressure lines, and welding flash. Know local regulations about making and using pits, trenches, hoists, scaffolds, and ladders at a construction site. Insist on adequate access and exit provisions.
11. Develop good housekeeping habits to prevent slipping, tripping, and falling, or being hit by falling materials (Figure 4-3). Know about toeboards, guardrails, and shields.
12. Read national occupational safety and health regulations, and keep in tune with changes.

4.4 LADDERS AND SCAFFOLDING

The construction and use of ladders and scaffolding require special care because the potential for creating a hazardous work situation can be high. Both of these are usually characterized as access structures; they give workers access to the building they are constructing. From a standpoint of safety and work efficiency, builders should always strive to build and use access structures as safe and well-constructed as the building on which they are working.

4.4.1 Ladders

Ladders have a multiplicity of uses in the construction business. Primarily, they are used in the early stages of a light construction project, including form construction. OSHA stipulates that "Except where either permanent or temporary stairways or suitable ramps or runways are provided, ladders...shall be used to give safe access to all elevations." In any case, ladders must be carefully constructed, used, and maintained. Materials for side rails (vertical members), rungs (horizontal members), and cleats (supporting pieces) must not be taken for granted, but carefully checked for defects before use.

Various types of ladders, both wooden and aluminum, are used in light construction, including stepladders, single or straight ladders, extension and pushup ladders, and safety rolling ladders (Figure 4-4). A stepladder is a portable ladder which opens out, sawhorse fashion, for self support. The maximum permissible height for a stepladder is 16′ (4.9 m). Stepladders must always be used fully opened and they should not be used as regular working platforms. Safety rolling ladders usually consist of three or four steps mounted on rollers.

A fixed ladder is one which is fastened to a structure in a more or less permanent manner. Top, bottom, and intermediate fastenings must be used as required. The rails of a fixed ladder must extend at least 3′ (.9 m) above the top landing. If the landing at the top requires passing

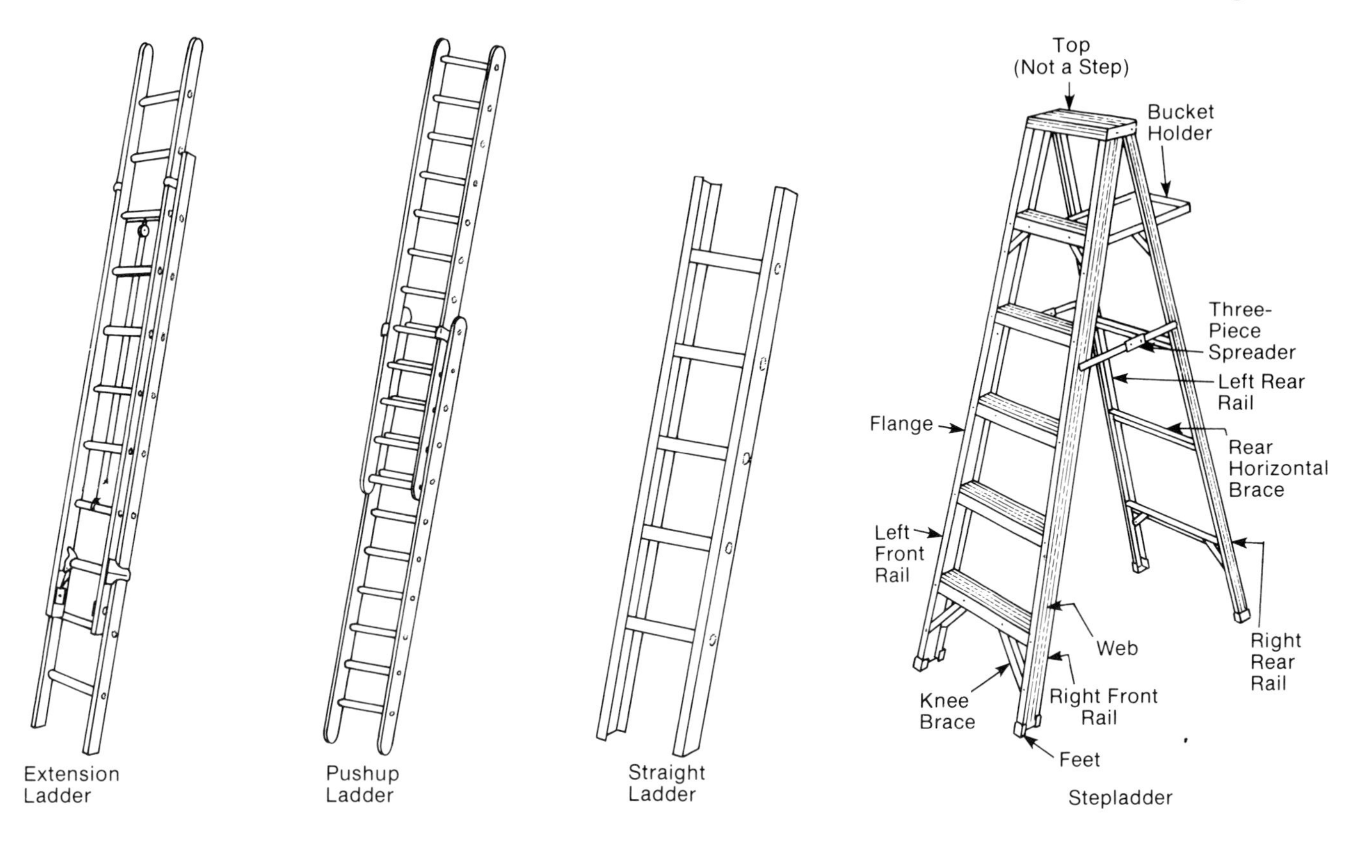

FIGURE 4-4. Typical ladder types.

between the rails, rungs above the landing must be removed (Figure 4-5). Landing platforms should be provided where a person must step a greater distance than 14″ (356 mm) from the ladder to a floor, roof, or landing.

Most requirements for a ladder are satisfied with the portable single and extension types. A single ladder has, as its name implies, a single section which should not exceed 30′ (9.1 m) in length. Careful consideration must be given to the placement of a ladder at a safe angle against the wall or other fixed object to be scaled.

The clearance space in back of a single ladder should always be sufficient to obtain a secure foothold on the rungs. A back clearance of at least 6″ (152 mm) is recommended. The clearance space in front of the ladder should be such that it will not be necessary to assume a cramped or unnatural position when climbing. A front clearance of at least 30″ (762 mm) is recommended.

If a single ladder is to be used on smooth floors, concrete walks, or sloping surfaces, make sure it is equipped with a nonslip base or that other suitable means are provided to prevent displacement while in use. Single portable ladders not constructed for use as sectional ladders must not be spliced together to form a longer ladder.

An extension ladder is one consisting of two sliding sections that can be adjusted to different heights. No extension ladder may contain more than two sections, nor may it be extended to more than 60′ (18.3 m). An extension ladder must be constructed to bring the rungs of overlapping parts of sections opposite each other when the ladder is locked in an extended position. For extension lengths of up to 38′ (11.6 m), the

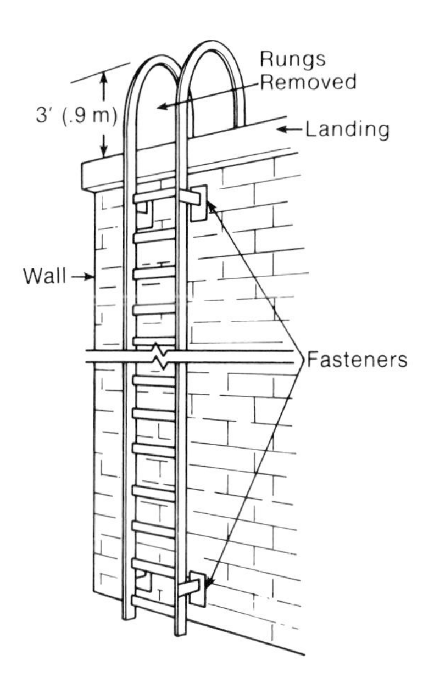

FIGURE 4-5. A fixed ladder must extend at least 3′ (.9m) above the landing.

minimum section overlap should be 3′ (.9 m); for lengths from 38′ to 44′ (11.6 m to 13.4 m) it should be 4′ (1.2 m); and for lengths from 44′ to 60′ (13.4 m to 18.3 m) it should be 5′ (1.5 m).

Sturdy single ladders are often made on a jobsite. The wood should be seasoned, straight-grained stock. It must be free from knots that would weaken the ladder. For ladders of up to a length of 16′ (4.9 m), use 2″ × 4″ (38 mm × 89 mm) stock; ladders of greater length should be built from 2″ × 6″ (38 mm × 140 mm) stock.

Ladders can be reinforced in several ways. One way is to run a heavy, galvanized wire in a precut groove along the underside of each side rail. The ends of the wire must be wrapped around the ends of the rails and anchored on nails (Figure 4-6A). The rungs can be reinforced by running a piece of rod with threaded ends along the back of each rung and through holes bored in the side rails. A washer and nut are tightened on the ends of the reinforced rod against the outside of the side rails (Figure 4-6B).

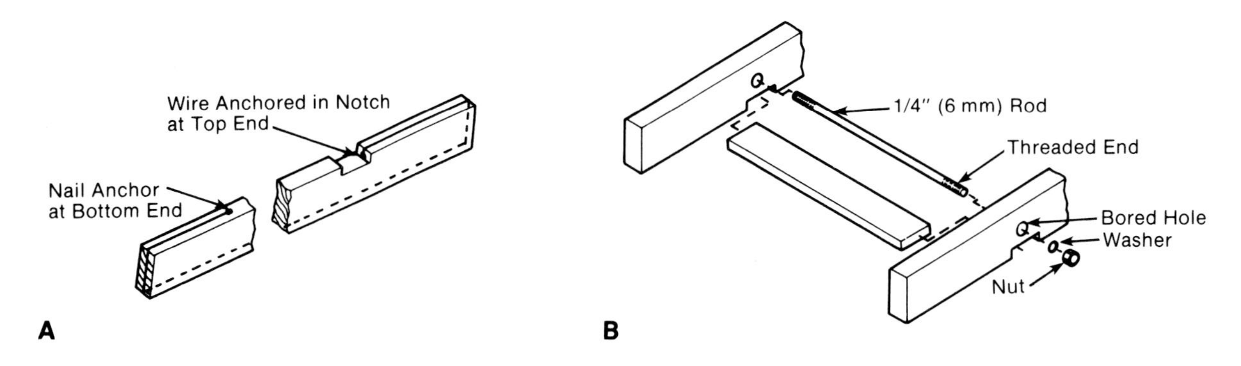

FIGURE 4-6. Ladder reinforcement: (A) wire reinforcement on the rails, (B) rod or bar reinforcement behind rungs.

Ladder Safety Rules

Ladders should always be inspected before they are used. A ladder with parts missing, with bent or cracked sides or rungs, and those made with faulty material should be condemned. Badly worn and weathered aluminum ladders and wooden ladders with rotten spots should not be used because they are subject to breaking and can cause a serious accident. Ladders with rough spots, such as protruding metal fastenings, screws, and nails, should be repaired or reconstructed to prevent injuries.

When putting a ladder into its final position, make sure the bottom is firmly footed. If the ground is soft or if the ladder does not rest squarely on both bottom legs, a board may be placed under the feet. If the bottom of the ladder has some type of nonslip grips or surfaces, they should be cleaned of mud or debris before putting up the ladder. A ladder should always be placed at a safe angle against the wall. A good rule is to place the base of the ladder about one-fourth as far out from the upper support as the length of the ladder (Figure 4-7). The upper end of a nonfixed ladder should not extend more than 2′ (.6 m) above the upper support.

The following are a few additional safety precautions and tips that apply to ladders in general:

- Ladders should be inspected at regular and frequent intervals.

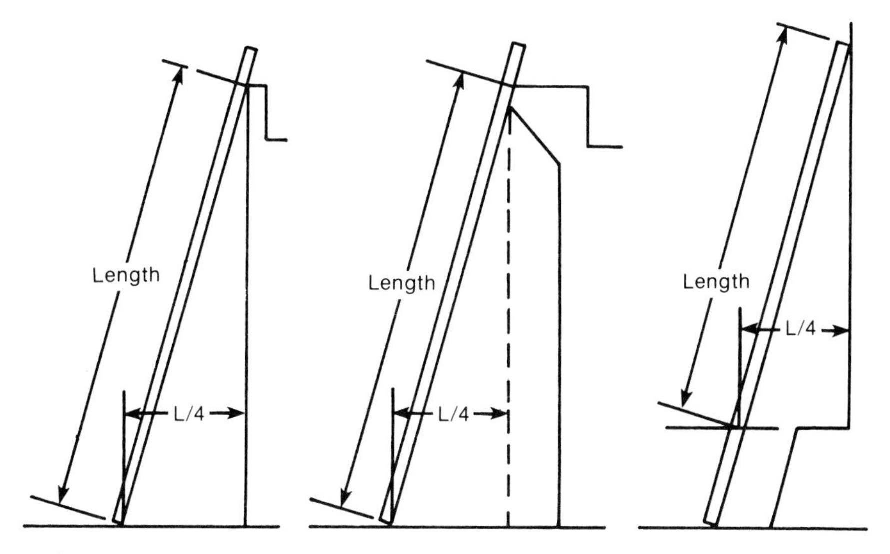

FIGURE 4-7. Correct angles for ladders.

- Ladders with weakened, broken, or missing treads, rungs, or cleats, or broken or splintered side rails should not be used.
- Ladders should be kept coated with a clear shellac or other transparent material or treated with linseed oil. Painting with opaque or pigmented paint should be avoided because it can cover defects.
- Separate ladders for ascending and descending should be provided in a building construction of more than two stories in height or where traffic is heavy.
- Where a ladder is installed wide enough to permit traffic in both directions at the same time, a center rail should be provided. One side of the ladder should always be for moving up, the other side for down.
- Ladders used in passageways, driveways, or thoroughfares should be guarded by barricades or guardrails. Doors which open adjacent to portable ladders should be locked or otherwise blocked or guarded while the ladder remains in use.
- Ladders should be placed so that the rails have a secure footing and a substantial support at or near the top.
- Ladders should not be placed against sash, window panes, or unstable supports such as loose boxes or barrels. The use of ladders during a storm or in a high wind should be avoided unless absolutely necessary, in which case the ladder should be securely lashed into position.
- If a ladder is to be placed against a window frame, a board should first be spiked across the side rails at the top.
- Ladders should not be placed or used in elevator shafts or hoistways. Should such a procedure be necessary, the ladders should be protected from objects at higher elevations.
- Ladders should not be left standing, especially on the outside, for long periods of time unless securely anchored at both top and bottom.
- Ladders constructed of metal should not be used near electricity. Metal ladders should not be used within 4′ (1.2 m) of any electrical wiring or equipment.
- When ascending or descending a ladder, always face the ladder.

- No one should go up or down a ladder without the free use of both hands. If handling material, a rope should be used.
- Before attempting to climb a ladder, remove oil or grease from the soles of shoes and do not allow paint, oil, or grease to accumulate on ladder rungs or rails.
- Lubricate locks and pulleys on extension ladders, keep fittings tight, and replace worn rope.
- Single portable ladders over 30′ (9.1 m) in length should not be used.
- Fixed ladders should be securely held in place by top, bottom, and intermediate fastenings as required.
- Sloping ladders which require climbing on the underside of the ladder should not be used.

4.4.2 Scaffolding

Many light construction type jobs require the use of scaffolding to work in areas beyond the normal reach of a worker on the ground. Scaffolding is usually comprised of rigid, sturdy supports holding an elevated platform. The platform is used by workers with their tools and building materials as an elevated work surface. Scaffolding can be made of wood or metal. Wooden scaffolds are usually made to order on a jobsite, while metal scaffolds are manufactured items, usually formed into one of a variety of patented systems or styles.

Types of Scaffolding

There are actually a great number of scaffolding types, but the differences between many are relatively minor, related mostly to a specialized adaptation for a particular type of work (Table 4-1). The most common wooden scaffold types used in light building construction are single pole scaffolding and double pole or independent pole scaffolding. Swinging scaffolds are also sometimes used for relatively light work. Several types of prefabricated patent scaffolding are available. The types most commonly used for light construction are usually categorized as sectional scaffolding. Also appropriate for many jobs is tubular clamp-type scaffolding.

TABLE 4-1 TYPES OF SCAFFOLDS

Type	Selected Descriptive Statement
1. Single Pole	Every wooden putlog in single pole scaffold shall be reinforced.
2. Double or Independent Pole	Independent pole scaffolds shall be set as near to the wall of the building as practicable.
3. Sectional	Drawings and specifications for all frame scaffolds over 125′ (38.1 m) in height above the base plates shall be designed by a registered professional engineer.
4. Tubular Steel Clamp Type	Bearers shall be at least 4″ (102 mm) but not more than 12″ (305 mm) longer than the post spacing or runner spacing.

TABLE 4-1 TYPES OF SCAFFOLDS (Continued)

Type	Selected Descriptive Statement
5. Lean-to	The use of shore or lean-to scaffolds is prohibited.
6. Manually Propelled Mobile	The height shall not exceed four times the minimum base dimension.
7. Mason's Adjustable Multiple-Point Suspension	The scaffold shall be provided with hoisting machines that meet the requirements of Underwriters' Laboratories or Factory Mutual Engineering Corporation.
8. Outrigger	Outrigger beams shall extend not more than 6′ (1.8 m) beyond the face of the building.
9. Swinging	Two-point suspension scaffolds shall be suspended by wire, synthetic, or fiber ropes capable of supporting at least six times the rated load.
10. Stone Setter's Adjustable Multiple-Point Suspension	The scaffold shall be capable of sustaining a working load of 25 lb (11.3 kg) per 1 ft^2 (.093 m^2) and shall not be overloaded.
11. Single-Point Adjustable Suspension	All power-operated gears and brakes shall be enclosed.
12. Carpenters' Bracket	The bracket shall consist of a triangular wood frame.
13. Bricklayers' Square	The squares shall not exceed 5′ (1.5 m) in width and 5′ (1.5 m) in height.
14. Horse	Horse scaffolds shall not be constructed or arranged more than two tiers or 10′ (3 m) in height.
15. Needle-Beam	Each employee working on a needle-beam scaffold shall be protected by a safety belt and lifeline.
16. Plasterers', Decorators', and Large Area	These shall be constructed in accordance with the general requirements set forth for independent wood pole scaffolds.
17. Interior Hung	This shall be hung or suspended from the roof structure or ceiling beams.
18. Ladder Jack	This shall be limited to light duty.
19. Window Jack	This shall be used only for the purpose of working at the window opening through which the jack is placed.
20. Roofing Brackets	These shall be constructed to fit the pitch (slope) of the roof.

TABLE 4-1 TYPES OF SCAFFOLDS (Continued)

Type	Selected Descriptive Statement
21. Crawling Boards or Chicken Ladders	A firmly fastened lifeline at least 3/4″ (19 mm) diameter rope, or equivalent, shall be strung beside each crawling board for a handhold.
22. Float or Ship	Each employee shall be protected by an approved safety lifebelt and lifeline.
23. Form	All scaffolds shall be designed and erected with a minimum safety factor of 4 computed on the basis of the maximum rated load.
24. Pump Jack	Pump Jack brackets, braces, and accessories shall be fabricated from metal plates and angles.

Pole Scaffolds

Pole scaffolds are probably the most common type of scaffolds constructed on the jobsite. Single pole scaffolds have one rear leg per section and are tied to the building on the facing side (Figure 4-8). Double pole scaffolds are also called independent pole scaffolds because they are not braced against the building but are completely

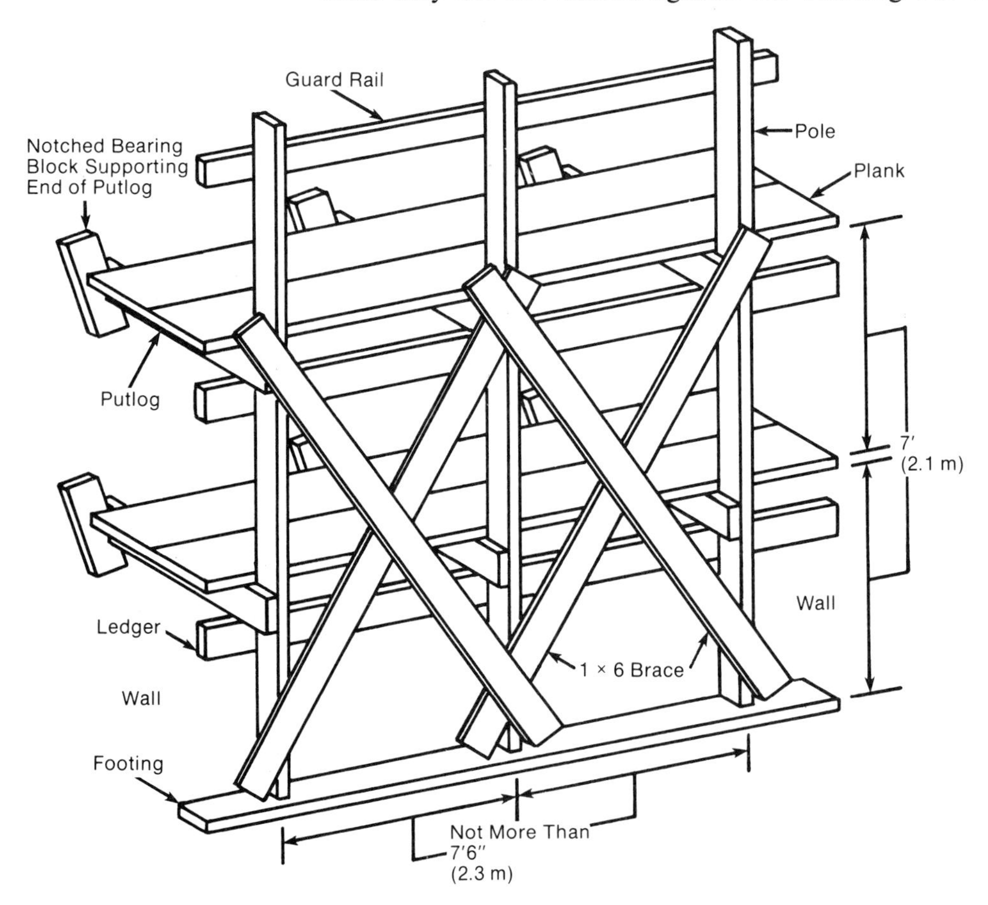

FIGURE 4-8. Single pole light-duty scaffolding.

self-standing. They have two legs per section, front and back (Figure 4-9). Single pole scaffolds fastened to a wall are less expensive to build but prevent free movement of both the scaffolding and the siding between the scaffold and the wall. Independent scaffolds can be made stronger and constructed to carry heavy loads of materials and workers.

Aside from their style, pole scaffolds are also divided according to their proposed use. Heavy-duty or heavy-trade scaffolds are used by bricklayers, stone masons, concrete block workers, etc. Light-duty or light-trade scaffolds are for carpenters, painters, plasterers, etc. The primary difference between them is the stock used in their construction.

The poles on a job-built pole scaffold should not exceed 40′ (12.2 m) in height. If higher poles than this are required, the scaffolding must be designed by an engineer.

For a heavy-duty (25 to 75 lb per 1 ft^2 [11.3 to 34 kg per 0.09 m^2]) single pole scaffold, the minimum dimensions are as follows:

Poles: 24′ (7.3 m) or less, 2 × 6 (38 mm × 140 mm); 24′ to 40′ (7.3 m to 12. 2 m), doubled 2 × 4 (doubled 38 mm × 89 mm)

Putlogs: doubled 2 × 4 (doubled 38 mm × 89 mm) or 2 × 8 (38 mm × 184 mm) on edge

Ledgers: 2 × 8 (38 mm × 184 mm)

Braces: 1 × 6 (19 mm × 140 mm)

Planking: 2 × 10 (38 mm × 235 mm)

Guardrails: 2 × 4 (38 mm × 89 mm)

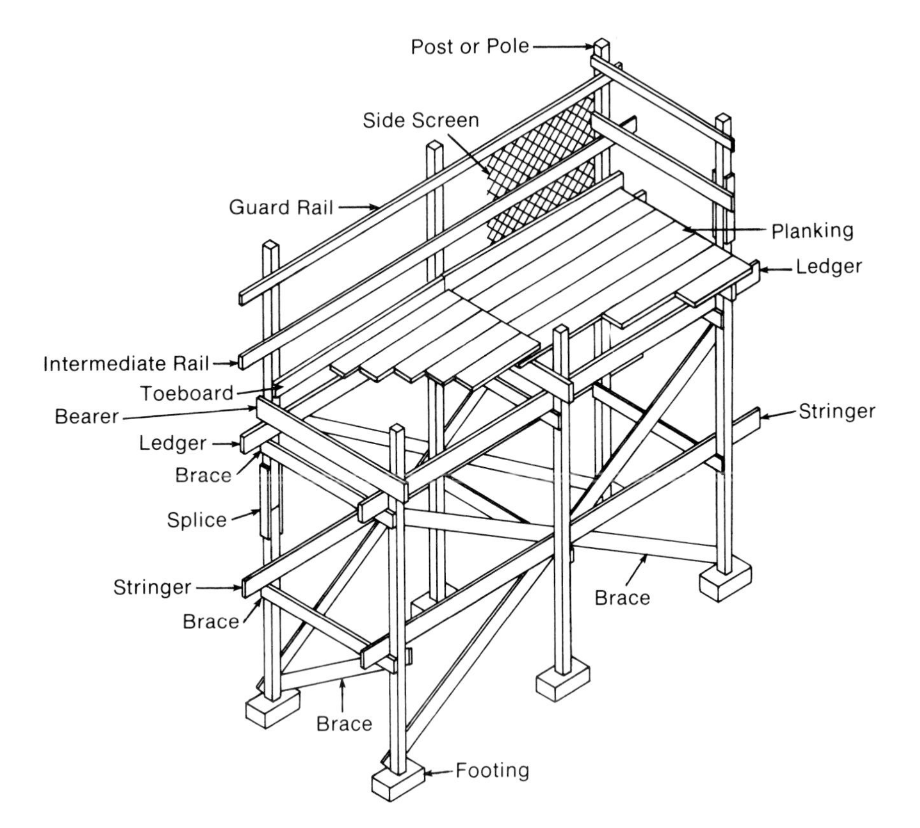

FIGURE 4-9. Double pole or independent pole scaffolding.

For a light-duty (not over 25 lb per 1 ft^2 [11.3 kg per 0.09 m^2]) scaffold, either single pole or double pole, the minimum dimensions are as follows:

Poles: 24′ (7.3 m) or less, 2 × 4 (38 mm × 89 mm); 24′ to 40′ (7.3 m to 12.2 m), 2 × 6 (38 mm × 140 mm)

Putlogs: 2 × 6 (38 mm × 140 mm) on edge

Ledgers: 2 × 6 (38 mm × 140 mm)

Braces: 1 × 4 (19 mm × 89 mm)

Planking: 2 × 10 (38 mm × 235 mm)

Guardrails: 2 × 4 (38 mm × 89 mm)

For a heavy-duty, double pole scaffold, the minimum lumber dimensions are as follows:

Poles: 24′ (7.3 m) or less, 2 × 6 (38 mm × 140 mm); 24′ to 40′ (7.3 m to 12.2 m); for a load from 25 to 50 lb per 1 ft^2 (11.3 kg to 22.7 kg per 0.09 m^2), double 2 × 4 (double 38 mm × 89 mm); for load from 50 to 75 lb per 1 ft^2 (22.7 to 34 kg per 0.09 m^2), double 2 × 6 (double 38 mm × 140 mm)

Putlogs: 2 × 8 (38 mm × 184 mm) on edge

Ledgers: 2 × 8 (38 mm × 184 mm)

Braces: 1 × 6 (19 mm × 140 mm)

Planking: 2 × 10 (38 mm × 235 mm)

Guardrails: 2 × 6 (38 mm × 140 mm)

The longitudinal maximum pole spacing for a light-duty scaffold is 7′6″ (2.3 m). For a heavy-duty scaffold it is 7′ (2.1 m).

The transverse maximum pole spacing for light or heavy-duty double pole scaffolds with poles up to 24′ (7.3 m) is 6′6″ (1.9 m). For a light-duty double pole scaffold with poles 24′ to 40′ (7.3 m to 12.2 m), the transverse maximum pole spacing is 7′ (2.1 m). For a heavy-duty scaffold with poles 24′ to 40′ (7.3 m to 12.2 m), the transverse maximum spacing is 10′ (3 m).

For a single pole light or heavy-duty scaffold, the pole spacing from the wall should be from 3′ to 5′ (.9 m to 1.5 m).

For a light-duty scaffold the maximum ledger vertical spacing is 7′ (2.1 m). For a heavy-duty scaffold the maximum ledger vertical spacing is 4′6″ (1.4 m).

To build all types of pole scaffolds, poles must be set up perfectly plumb. The lower ends of poles must not bear directly on a natural earth surface. If the surface is earth, a board footing 2″ (51 mm) thick and from 6″ to 12″ (152 mm to 305 mm) wide (depending on the softness of the earth) must be placed under the poles.

If poles must be spliced, splice plates must not be less than 4′ (1.2 m) long, not less than the width of the pole wide, and each pair of plates must have a combined thickness not less than the thickness of the pole. Adjacent poles must not be spliced at the same level.

A ledger must be long enough to extend over two pole spaces, and it must overlap the poles at the ends by at least 4″ (102 mm). Ledgers must be spliced by overlapping and nailing at poles—never between poles. If platform planks are raised as work progresses upward, the ledgers and putlogs on which the planks previously rested must be left in place to brace and stiffen the poles. For a heavy-duty scaffold, ledgers must be supported by cleats, nailed or bolted to the poles, as well as by being nailed themselves to the poles. Also, be sure the poles are well braced on the ends and the back, as shown in Figure 4-10.

A single putlog must be set with the longer section dimension vertical, and putlogs must be long enough to overlap the poles by at least 3″ (76 mm). They should be both face-nailed to the poles and toenailed to the ledgers. When the inner end of the putlog butts against the wall (as it does in a single pole scaffold), it must be supported by a 2 × 6 (38 mm× 140 mm) bearing block not less than 12″ (305 mm) long, notched out the width of the putlog and securely nailed to the wall. The inner end of the putlog should be nailed to both the bearing block and the wall. If the inner end of the putlog is located in a window opening, it must be supported on a stout plank nailed across the opening. If the inner end of the putlog is nailed to a building stud, it must be supported on a cleat of the same thickness as the putlog, nailed to the stud.

A platform plank must never be less than 2″ (51 mm) thick. Edges of planks should be close enough together to prevent tools or materials from falling through the opening. A plank must be long enough to extend over three putlogs, with an overlap of at least 6″ (152 mm), but not more than 12″ (305 mm). In addition, a heavy-duty type scaffold must have a 1″ × 8″ (19 mm × 184 mm) toeboard nailed onto the poles at the outer edges of the planking to prevent bricks, stones, or other material from being inadvertently kicked over the edge, creating a hazard for the workers below.

The stock specifications and dimensions given here are relatively standard; however, it would be a good idea to check local building codes or safety regulations to make sure.

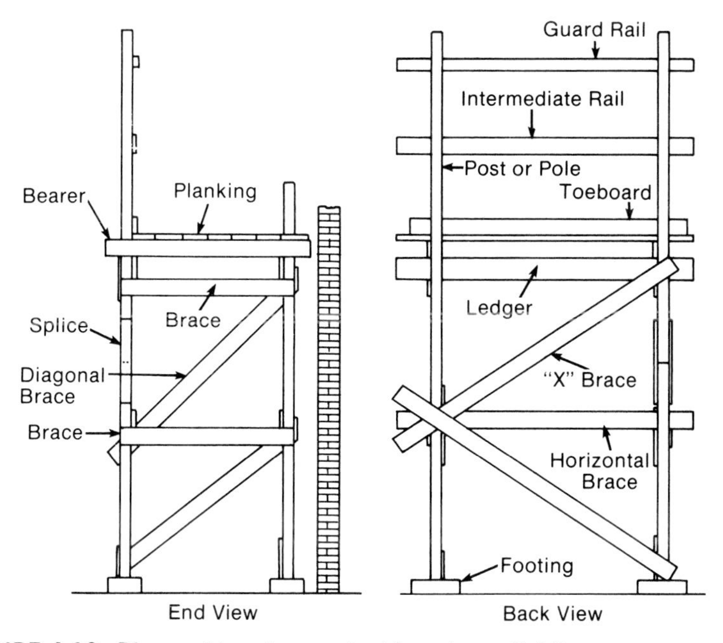

FIGURE 4-10. Diagonal bracing on double pole scaffolding.

Swinging Scaffold

While pole scaffolding is supported on the ground, swinging scaffolding is suspended from above. The simplest type of swinging scaffold is one which consists simply of a stout plank (minimum thickness 2″ [51 mm]) with a couple of transverse horns nailed or bolted to the underside, near the ends. The stage hangs from a couple of lines, (minimum size 2″ [51 mm]) which lead up and over or through some supporting device (such as a pair of shackles secured to outriggers at the roof line) and back to the stage.

The method of bending a bowline to a stage by means of a scaffold hitch is shown in Figure 4-11A. A stage provides a convenient means for working downward (painting down a wall from roof line to ground level, for instance), but since workers cannot hoist themselves aloft on a stage, it cannot be used for working upward.

When the rig shown in Figure 4-11B is hooked to the tackles, it can be moved up or down at will simply by heaving in or slacking out on the tackles. The two projecting timbers to which the tackles will be attached are called outriggers.

The swinging type of scaffold must always be secured to the building or structure to prevent it from moving away and causing workers to fall. Where swinging scaffolds are suspended adjacent to each other, planks should never be placed so as to form a bridge between them.

This type of scaffold, while versatile for work in which an entire completed wall or story must be painted or finished, should not be used for any heavy-duty work and should always be equipped with the proper guardrails and toeboards.

Sectional and Clamp-Type Scaffolding

Several types of patent independent sectional scaffolding are available for simple and rapid erection (Figure 4-12). A common model is

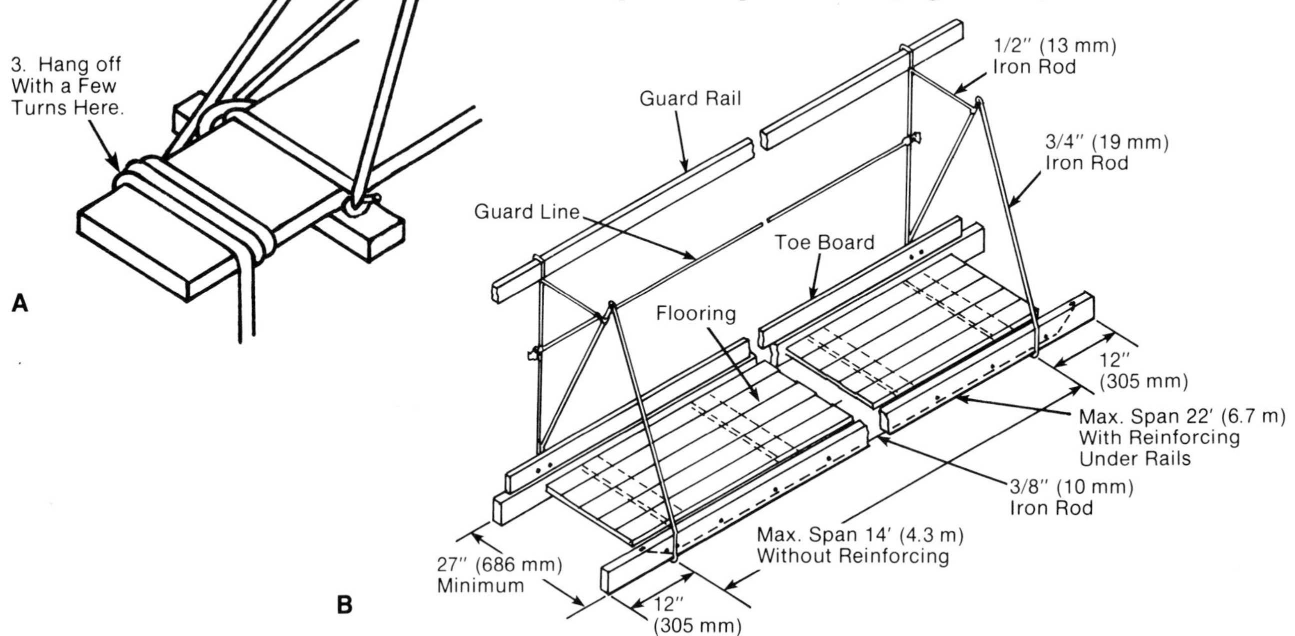

FIGURE 4-11. (A) Making a scaffold hitch; (B) a typical swinging scaffold.

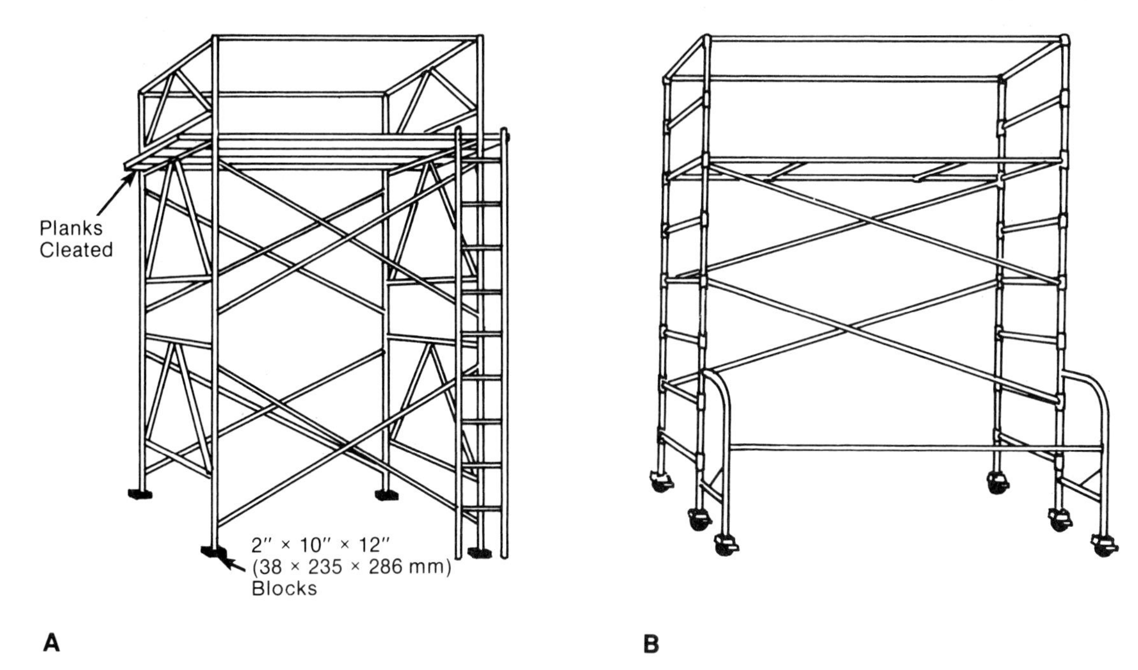

FIGURE 4-12. Two types of sectional scaffolding: (A) set (can be secured to building with #12 double-wrapped wire); (B) rolling (minimum dimensions of base should not be less than one-third the height of the scaffold).

sometimes called double pole built-up scaffolding, which like wooden double pole scaffolding is completely independent of the building. The scaffold uprights are braced with diagonal members (Figure 4-13) and the working level is covered with a platform of planks. All bracing must form triangles and the base of each column requires adequate footing plates for the bearing area on the ground. This patented steel scaffolding is usually erected by placing the two uprights on the ground and inserting the diagonal members. The diagonal members have end fittings which permit rapid locking in position. The first tier is set on steel bases on the ground. A second tier is placed in the same manner on the first tier, with the bottom of each upright locked to the top of the lower tier. A third and fourth upright can be placed on the ground level and locked to the first set with diagonal bracing. The scaffolding can be built as high as desired, but high scaffolding should be tied into the main structure.

Tubular steel clamp-type scaffolding is shown in Figure 4-14. Clamp tubular scaffolding is found on many construction jobs because its design permits it to be shaped to any contour and it goes up rapidly. The safety rules given in Figure 4-15 are recommended by the Scaffolding, Shoring, and Forming Institute when installing and using steel clamp-type scaffolding.

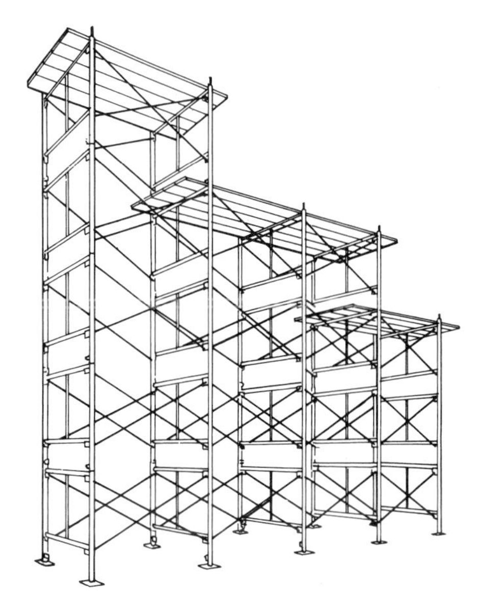

FIGURE 4-13. Note diagonal bracing in double pole built-up scaffolding.

4.4.3 General Scaffolding Safety

The following scaffolding safety precautions must be observed by all persons working on scaffolds or tending other persons who are working on scaffolds.

- Standard scaffolds suitable to the work at hand must be provided and used. The use of weak makeshift substitutes or ladders is prohibited.

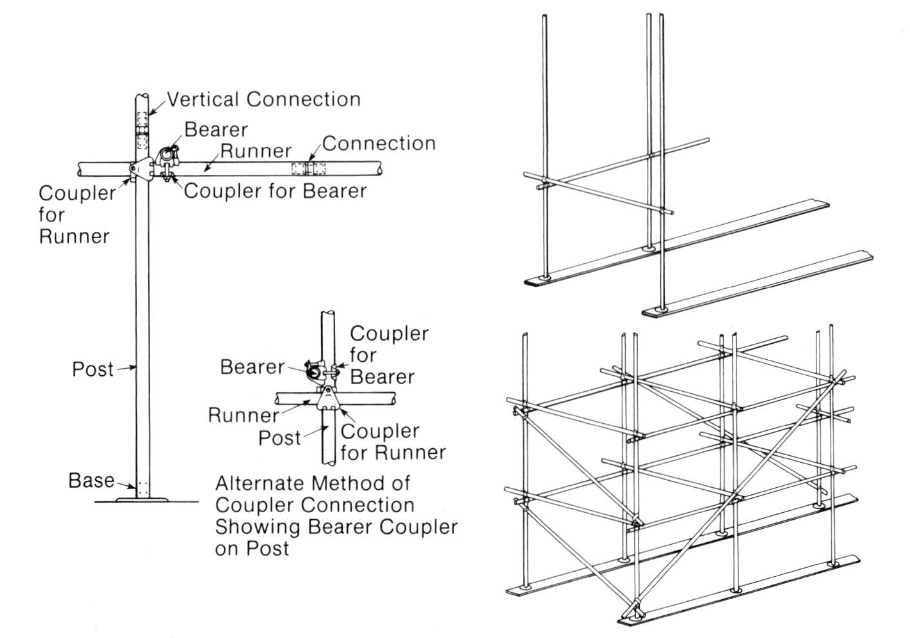

FIGURE 4-14. Tubular steel clamp-type scaffolding.

- All scaffolds must be maintained in a safe condition, and a scaffold must not be altered or disturbed while in use. Personnel must not be allowed to use damaged or weakened scaffolds.
- Structural members, supporting lines and tackles, and other scaffold equipment must be inspected daily before work on scaffolding is started.
- When personnel working on a scaffold are directly below others working above, the ones below must be sheltered against possible falling objects by a protective covering. Those below must also wear protective headgear.
- If the frequent presence of personnel directly under a scaffold is unavoidable, a protective covering must be set up under the scaffold. A passageway or throughfare under a scaffold must have both overhead and side protection.
- Access to scaffolds must be by standard stairs or fixed ladders only.
- The erection, alteration, and dismantling of scaffolds must be done under the supervision of workers who are experienced in scaffold work. When scaffolding is being dismantled, it should be cleaned and made ready for storage or use. Scaffolding that is not ready for use should never be stored.
- Work on scaffolds should be stopped during storms or high winds or when scaffolds are covered with ice or snow. Materials should not be left on scaffolding during high winds.
- Unstable objects such as barrels, boxes, loose brick, or building blocks must not be used to support scaffold planking.
- No scaffold may be used for the storage of materials in excess of those currently required for the job.
- Tools not in immediate use on scaffolds must be stowed in containers to prevent tools lying around from being knocked off. Tool containers must be lashed or otherwise secured to the scaffolds.
- Scaffolds must be kept clear of accumulations of tools, equipment, materials, and rubbish. If part of a scaffold must be used as a loading

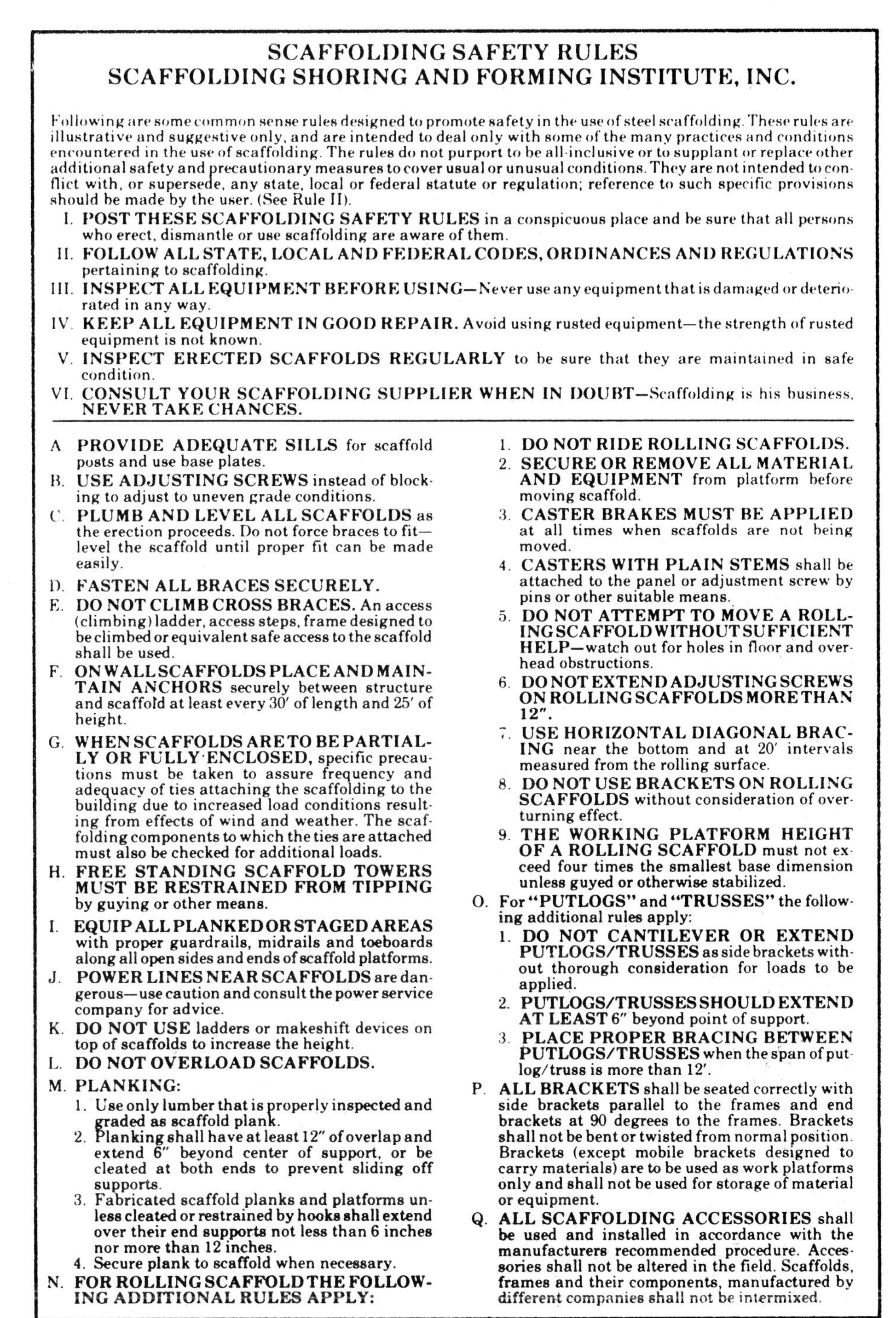

SCAFFOLDING SAFETY RULES
SCAFFOLDING SHORING AND FORMING INSTITUTE, INC.

Following are some common sense rules designed to promote safety in the use of steel scaffolding. These rules are illustrative and suggestive only, and are intended to deal only with some of the many practices and conditions encountered in the use of scaffolding. The rules do not purport to be all-inclusive or to supplant or replace other additional safety and precautionary measures to cover usual or unusual conditions. They are not intended to conflict with, or supersede, any state, local or federal statute or regulation; reference to such specific provisions should be made by the user. (See Rule II).

I. **POST THESE SCAFFOLDING SAFETY RULES** in a conspicuous place and be sure that all persons who erect, dismantle or use scaffolding are aware of them.
II. **FOLLOW ALL STATE, LOCAL AND FEDERAL CODES, ORDINANCES AND REGULATIONS** pertaining to scaffolding.
III. **INSPECT ALL EQUIPMENT BEFORE USING**—Never use any equipment that is damaged or deteriorated in any way.
IV. **KEEP ALL EQUIPMENT IN GOOD REPAIR.** Avoid using rusted equipment—the strength of rusted equipment is not known.
V. **INSPECT ERECTED SCAFFOLDS REGULARLY** to be sure that they are maintained in safe condition.
VI. **CONSULT YOUR SCAFFOLDING SUPPLIER WHEN IN DOUBT**—Scaffolding is his business. **NEVER TAKE CHANCES.**

A. **PROVIDE ADEQUATE SILLS** for scaffold posts and use base plates.
B. **USE ADJUSTING SCREWS** instead of blocking to adjust to uneven grade conditions.
C. **PLUMB AND LEVEL ALL SCAFFOLDS** as the erection proceeds. Do not force braces to fit—level the scaffold until proper fit can be made easily.
D. **FASTEN ALL BRACES SECURELY.**
E. **DO NOT CLIMB CROSS BRACES.** An access (climbing) ladder, access steps, frame designed to be climbed or equivalent safe access to the scaffold shall be used.
F. **ON WALL SCAFFOLDS PLACE AND MAINTAIN ANCHORS** securely between structure and scaffold at least every 30′ of length and 25′ of height.
G. **WHEN SCAFFOLDS ARE TO BE PARTIALLY OR FULLY ENCLOSED,** specific precautions must be taken to assure frequency and adequacy of ties attaching the scaffolding to the building due to increased load conditions resulting from effects of wind and weather. The scaffolding components to which the ties are attached must also be checked for additional loads.
H. **FREE STANDING SCAFFOLD TOWERS MUST BE RESTRAINED FROM TIPPING** by guying or other means.
I. **EQUIP ALL PLANKED OR STAGED AREAS** with proper guardrails, midrails and toeboards along all open sides and ends of scaffold platforms.
J. **POWER LINES NEAR SCAFFOLDS** are dangerous—use caution and consult the power service company for advice.
K. **DO NOT USE** ladders or makeshift devices on top of scaffolds to increase the height.
L. **DO NOT OVERLOAD SCAFFOLDS.**
M. **PLANKING:**
  1. Use only lumber that is properly inspected and graded as scaffold plank.
  2. Planking shall have at least 12″ of overlap and extend 6″ beyond center of support, or be cleated at both ends to prevent sliding off supports.
  3. Fabricated scaffold planks and platforms unless cleated or restrained by hooks shall extend over their end supports not less than 6 inches nor more than 12 inches.
  4. Secure plank to scaffold when necessary.
N. **FOR ROLLING SCAFFOLD THE FOLLOWING ADDITIONAL RULES APPLY:**
  1. **DO NOT RIDE ROLLING SCAFFOLDS.**
  2. **SECURE OR REMOVE ALL MATERIAL AND EQUIPMENT** from platform before moving scaffold.
  3. **CASTER BRAKES MUST BE APPLIED** at all times when scaffolds are not being moved.
  4. **CASTERS WITH PLAIN STEMS** shall be attached to the panel or adjustment screw by pins or other suitable means.
  5. **DO NOT ATTEMPT TO MOVE A ROLLING SCAFFOLD WITHOUT SUFFICIENT HELP**—watch out for holes in floor and overhead obstructions.
  6. **DO NOT EXTEND ADJUSTING SCREWS ON ROLLING SCAFFOLDS MORE THAN 12″.**
  7. **USE HORIZONTAL DIAGONAL BRACING** near the bottom and at 20′ intervals measured from the rolling surface.
  8. **DO NOT USE BRACKETS ON ROLLING SCAFFOLDS** without consideration of overturning effect.
  9. **THE WORKING PLATFORM HEIGHT OF A ROLLING SCAFFOLD** must not exceed four times the smallest base dimension unless guyed or otherwise stabilized.
O. For **"PUTLOGS"** and **"TRUSSES"** the following additional rules apply:
  1. **DO NOT CANTILEVER OR EXTEND PUTLOGS/TRUSSES** as side brackets without thorough consideration for loads to be applied.
  2. **PUTLOGS/TRUSSES SHOULD EXTEND AT LEAST** 6″ beyond point of support.
  3. **PLACE PROPER BRACING BETWEEN PUTLOGS/TRUSSES** when the span of putlog/truss is more than 12′.
P. **ALL BRACKETS** shall be seated correctly with side brackets parallel to the frames and end brackets at 90 degrees to the frames. Brackets shall not be bent or twisted from normal position. Brackets (except mobile brackets designed to carry materials) are to be used as work platforms only and shall not be used for storage of material or equipment.
Q. **ALL SCAFFOLDING ACCESSORIES** shall be used and installed in accordance with the manufacturers recommended procedure. Accessories shall not be altered in the field. Scaffolds, frames and their components, manufactured by different companies shall not be intermixed.

FIGURE 4-15. Typical scaffold safety rules for use of tubular steel built-up scaffolding.

or landing stage for materials, the scaffold must be additionally braced and reinforced at and around the landing stage area.

- Throwing objects to or dropping them from scaffolds is absolutely prohibited. Handlines must be used for raising or lowering objects which cannot be passed hand-to-hand. Also, materials being hoisted onto a scaffold should have a legline.
- A standard guardrail and toeboard should be provided on the open side of all platforms 5′ (1.5 m) or more above ground; otherwise, safety belts tied off to safety lines must be used.

- If the space between the scaffold and building is more than 18″ (457 mm), a standard guardrail should be erected on the building side.
- No person should remain on a rolling-type scaffold while it is being moved.
- No person should lean on or against or stand or sit on any guardrail or guardline.
- A scaffold must *never* be overloaded. Scaffolds are often categorized according to the type of work and load they can accommodate. These categories are: (1) extra heavy-duty, (2) heavy-duty, (3) intermediate category between light and heavy, and (4) light-duty. The maximum uniform safe working load per 1 ft^2 (.09 m^2) of platform for each of these categories is as follows:
  Extra heavy-duty (stone masons)—75 lb (34 kg)
  Heavy-duty (stone setters and bricklayers)—50 lb (22.7 kg)
  Intermediate (stucco workers and plasterers)—30 lb (13.6 kg)
  Light-duty (carpenters and miscellaneous)—25 lb (11.3 kg)

To get the load per 1 ft^2 (.09 m^2) of platform of a pile of materials on a platform, divide the total weight of the pile by the number of square feet (square meters) of platform it covers. For strength of plank see Table 4-2.

4.5 HOISTING

Some type of hoist will be needed to lift work materials which are applied to a building—siding, bricks, plaster, etc.—from a scaffold or

TABLE 4-2 STRENGTH OF PLANK FOR SCAFFOLDING

	Span in Feet (M)							
	4 (1.2)	6 (1.8)	8 (2.4)	10 (3.0)	12 (3.7)	14 (4.3)	16 (4.9)	18 (5.5)
Size in Inches (mm)	**Safe Load in Pounds (kg)**							
2 × 8 (38 × 184)	400 (181)	260 (118)	190 (86)					
2 × 10 (38 × 235)	500 (227)	325 (147)	235 (107)	180 (82)				
2 × 12 (38 × 286)	600 (272)	390 (177)	285 (129)	220 (100)	165 (75)			
3 × 10 (57 × 235)	1245 (565)	820 (372)	600 (272)	465 (211)	370 (168)	305 (138)		
3 × 12 (57 × 286)	1495 (678)	980 (445)	720 (327)	555 (252)	445 (202)	370 (168)	300 (136)	
3 × 14 (57 × 337)	1740 (789)	1145 (519)	840 (381)	650 (295)	520 (236)	430 (195)	350 (159)	290 (131)

The loads given are in pounds (kg) concentrated at the center of the span.

The above loads are for fir or spruce planks in first class condition. For yellow pine planks in first class condition, add 10% to the above allowable loads.

Safe loads given are based on finished planks (as is for metric).

used on a roof. Many times materials can be moved by hand, but heavier items require a mechanical hoist (Figure 4-16). There are several kinds of hoisting equipment available. A forklift or the front-end loader on a tractor or dozer are useful for relatively low lifts. Small mechanical or powered lifts are also useful if the necessary height and weight are not too great. For reaching greater heights and lifting a heavier load, especially on bigger construction jobs, small elevator-type lifts are used. Although not usually necessary in the field of light building construction, a crane may also be occasionally necessary.

Another important aspect of hoisting, whether by hand or mechanical, is the method of hitching the work material to be lifted. The most common way of attaching a load to a lifting hook is to put a sling around the load and hang the sling on the hook (Figure 4-17). A sling may be made of rope, line, wire, or wire rope with an eye in each end. Wire rope or rope with an eye in each end is usually called a strap or endless sling (Figure 4-18). When a sling is passed through its own bight or eye or shackled or hooked to its own standing part so that it tightens around the load like a lasso when the load is lifted, the sling is said to be choked. The hitch may be called a choker as shown in Figures 4-17 and 4-18. A two-legged sling which supports the load at two points is called a bridle (Figure 4-19).

FIGURE 4-16. A typical mechanical hoist.

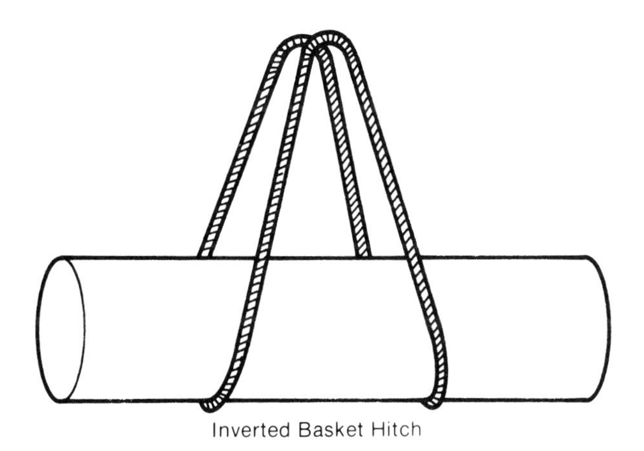

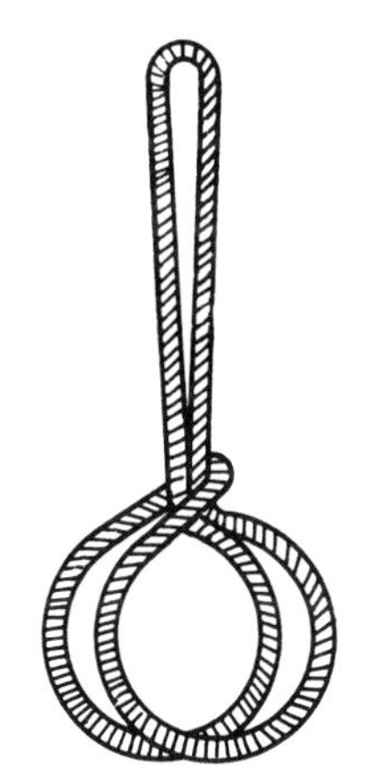

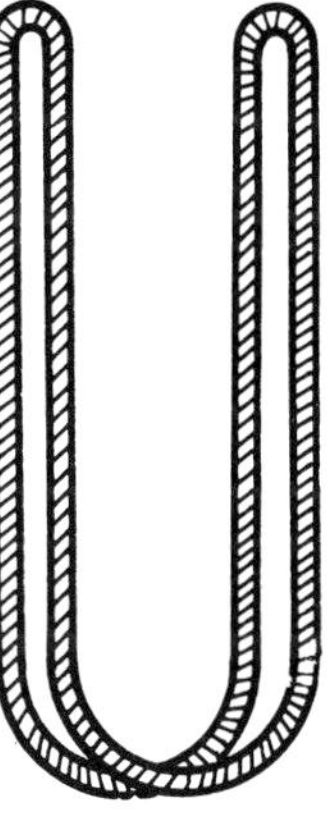

FIGURE 4-17. Ways of hitching on a sling.

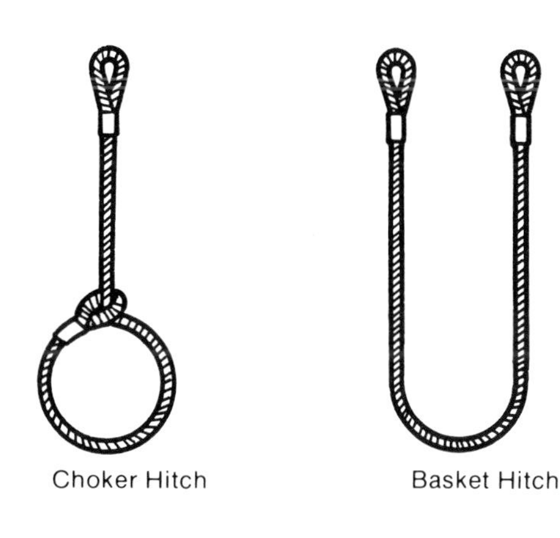

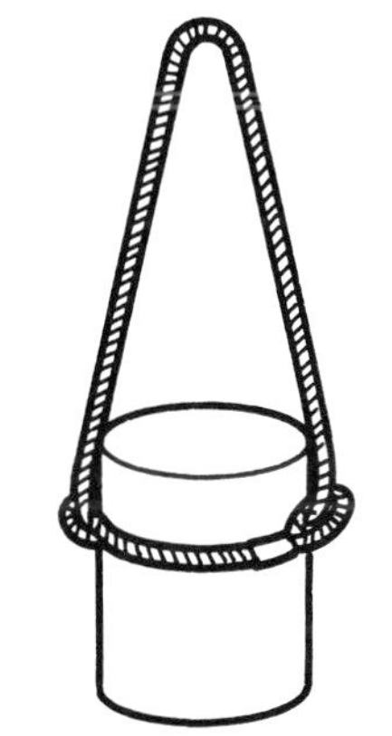

FIGURE 4-18. Ways of hitching on a strap.

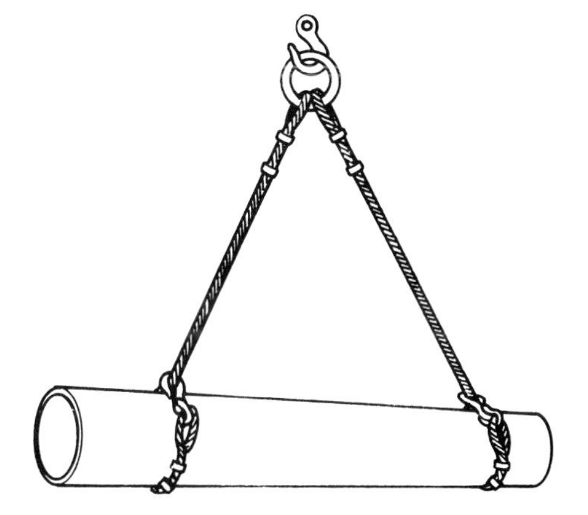

FIGURE 4-19. A typical bridle hitch.

4.5.1 Hoisting Safety

Hoisting always requires a great deal of care to avoid both injury and damage to work materials. The following safety rules must be observed by everyone engaged in a hoisting operation:

- The person in charge of hitching must know the safe working load of the rig and the weight of every load to be hoisted. The hoisting of any load heavier than the safe working load of the rig is absolutely prohibited.
- When a cylindrical metal object such as a length of pipe, a gas cylinder, or the like is hoisted in a choker bridle, each leg of the bridle should be given a turn around the load before it is hooked or shackled to its own part. The purpose of this is to ensure that the legs of the bridle will not slide together along the load, thereby upsetting the balance and possibly dumping the load.
- The point of strain on a hook must never be at or near the point of the hook.
- Before hoisting, the person in charge must be sure that the load will balance evenly in the sling.
- As the load leaves the ground, the person in charge must watch carefully for kinked or fouled lines or slings. If any are observed, the load must be lowered at once for clearing.
- Tag lines must be used to guide and steady a load whenever there is a possibility that the load might get out of control.
- Before any load is hoisted, it must be inspected carefully for loose parts or objects which might drop as the load goes up.
- All workers must be cleared from and kept out of any area that is either under a suspended load or over which a suspended load may pass.
- Whenever materials are placed on scaffolding or other work areas, shoring must be provided to prevent tipping of the load or shifting of the materials.
- When slings are being heaved out from under a load, all workers must stand clear to avoid a backlash and also to avoid a toppling or a tip of the load which might be caused by fouling of a sling.
- *Never* walk or run under a suspended load.

4.6 TOOL AND MACHINERY OPERATION SAFETY

The tools and machinery commonly used in light building construction have been described in Chapter 2. Knowing how and why the tools used in light building construction work the way they do is an excellent way to help ensure that they are used efficiently and safely.

Knowing how to use and adapt tools and machinery in the safest and most productive way is a learned skill that should be implanted within the entire construction team. For example, a builder should schedule and record inspections of all tools and machinery at frequent and regular intervals. This is particularly true of the protection guards and protective devices, which have been so unfortunately neglected in the past by many builders for one reason or another. In addition, however, a machine or tool operator should be able to recognize a damaged, worn-out, or otherwise flawed tool and should bring the matter to the foreman's or builder's attention. Almost nothing has more potential danger on a jobsite than using a power saw with very dull blades, a misfiring powder-actuated tool, or some other tool which is not working quite the way it should be.

Too often workers continue to use damaged or flawed tools and the foreman or builder allows them to do so for reasons ranging from saving time to lack of concern. Both the workers' safety and possibly their future are at stake. Perceived shortcuts in tool use and disregard for malfunctioning machinery often prove to be extremely costly, both for the worker and the builder.

Safe machine operating conditions depend on the detection of potential hazards and in taking immediate action to remedy them. Machine and tool checklists can be used to formulate a general set of rules for their use. Many times a tool operator may, by experience, learn of hazards associated with a certain tool which is not included in an operator's manual. The builder should include this type of information in a company tool dossier. In addition, an operator may also, after years of experience, learn how to perform a special job or how to make a special jig, which increases the value of a tool. Again, the builder should record this learned technique so that it does not leave his or her company with one innovative operator.

A safe work environment also includes the proper dispensation of working space. This is particularly true in regard to tool use, and even more so with power tools. The builder must learn to allocate sufficient and undisturbed working space for tool operation to the extent that the operation can be done most efficiently and with complete safety.

Care and maintenance of machines is equally important not only for increased production, but also for good workmanship and safe operations. Badly maintained machines are a hazard. Care and maintenance means keeping machines in design condition or as close to it as is possible.

Ensuring safe working conditions on a jobsite requires of a builder a concerned attitude; the willingness to spread this attitude throughout the company; knowledge of machinery and tools and how they work; a regular program of tool care; maintenance, and repair; and at least informal guidelines for operating the variety of tools and machinery that are so essential to construction work. More precise or formal guidelines are usually even more helpful.

4.6.1 Safety Guidelines for Tool and Machine Operation

A good place to start is to use and collect manufacturer's instruction manuals for operators and safety tip literature. Seek other published material concerning effective and safe ways to use tools. The books and magazine articles on this subject are numerous. In addition, personal and collective company experience can add a lot of information to the published material. There are numerous ways to put this safety information in some type of systematic guidelines.

One possibility is to create a series of logical equations about the nature and operation of individual tools or machines. For example, in Tables 4-3A and 4-3B a series of general statements are made about power saws. These statements are neither all-inclusive, nor do they differentiate between the various types of power saws. Together, these tables represent a formulation for safe power saw operations. The first part of the formulation, as shown in Table 4-3A, sets forth in summary form a set of concepts or ideas about the operation of a tool or piece of equipment, in this case power saws as a group. These concepts can also be clarified by descriptive statements. In this example, the explanatory statements are relatively short; however, they may be as long as is found necessary to make the meaning of the concept clear.

The second part of the formulation is logical based on the statements listed in Table 4-3A. Table 4-3B consolidates these ideas into "if-then" statements regarding the principles of operation. Study the two tables. Table 4-3B is entitled Principles of Power Sawing, but this is merely a tentative statement of principle which can be refined and expanded. The experiences of each and every construction company, the tools *they* use, and innovations in the field of tools generally will affect safety guidelines.

Attempting to create a statement of principle will provoke numerous immediate questions. In this case, how does one know if a power saw is well designed and built? Where are the performance test results? What are the criteria for properly positioning and installing a stationary power saw? How is a power saw properly maintained? The answers to the first questions are assumed to be embodied in the stamp of approval attached by a recognized underwriters laboratory and in statements of guarantee or warranty by the manufacturer. In addition, purchasing

TABLE 4-3A POWER SAWING

Concept	Descriptive Statement
1. Cutting action	Knife or chisel or combination.
2. Crosscut/ripping	Across/with the wood grain.
3. Tooth shape	Angled, beveled, triangular (with combinations, particularly for reciprocating saws; cutting actions.
4. Tooth set	Swage, flat-ground, hollow-ground, thin-rim, carbide-tipped.
5. Feeding of stock for ripping	Pushing stock into the saw. Consider, without trying the operation, the effect of feeding with the saw as to how the saw might catch the wood stock and throw it forward. In case of a radial saw, the throwing action would probably cause the arbor to bend and the stock to be thrown at very high speed, at great danger to any other person nearby.
6. Guarding	Safe covering, over and under the blade, exposing the cutting edges only to the stock pushed into it.
7. Splitting	Keeping the kerf open by means of a fixed knife to prevent pinching or stalling.
8. Holding down/back guarding	To avoid freehand sawing.
9. Pushing	Use of specially shaped pushsticks to keep hands clear.
10. Grounding	Avoiding electrical shocks.
11. Safe switching	Placement of switches for safe starting, stopping, and shutdown.
12. Peripheral speed of saw	Design limits, measured in feet or meters per minute.
13. Anti-kickback fingers	Allows only outfeed movement.
14. Sawdust exhausting	Directs sawdust/takes it away from the workspace.

TABLE 4-3B PRINCIPLES OF POWER SAWING

If
1. The power saw is well designed and built, and 2. It is properly positioned and installed, and 3. It is properly maintained...
And if the operator
1. Uses the safety devices (fence, guards, splitter, etc.), 2. Uses the appropriate blade, 3. Feeds clean stock properly, 4. Applies safe work loads to the machine (according to design specifications), 5. Selects materials properly, as to kind, thickness, and length, 6. Wears proper protective clothing and gear, 7. Pays strict attention to each operation being performed...
Then
1. That operator may safely use the power saw.

agents should know about such statements and what they mean or imply for a purchaser.

However, in a truly safe construction environment, all members of the construction team should have enough knowledge about each type of equipment used to have developed a sound idea as to its operating capacities, need for maintenance, etc. This knowledge will be based on experience and the formal safety guidelines laid down by a construction company.

For machines that produce shavings rather than sawdust, another set of concepts can be defined and a statement of principle formulated. Tables 4-4A and 4-4B list operational concepts for jointing and planing and a principle of power jointing and planing. These lists could be

TABLE 4-4A HAND-FED JOINTING AND THICKNESS PLANING

Concept	Descriptive Statement
1. Cutting action	Chisel-type blades set in revolving heads/cylinders.
2. Infeed	Infeed tables set for depth of cut.
3. Outfeed	Outfeed tables usually set to precisely accommodate the depth of cut, therefore, set higher than the infeed table.
4. Guarding	Cutting head covered and table opening as small as possible.
5. Stock size	Short pieces being fed in may be pressed into the jointer table opening then kicked back, out of the operator's hands. Results: flying stock and hands going into the cutters!
	Thin pieces and loose knots may break up and be thrown back at the operator.
6. Peripheral speed	Design limits, measured in feet or meters per second.
7. Blade sharpening/setting	Hollow-ground blades set parallel with tables.
8. Shavings and exhaust	Shavings collector to keep workspace and air clean.

TABLE 4-4B PRINCIPLES OF POWER JOINTING AND PLANING

If
1. The machine is well designed and built, positioned, installed, and maintained...
And if the operator
1. Uses the safety devices (guard, fence, exhaust duct),
2. Feeds clean stock properly,
3. Applies safe work loads to the machine according to design specifications,
4. Selects materials properly (as to kind, thickness, and length),
5. Wears proper protective clothing and gear (including noise mufflers),
6. Pays attention to each operation being performed...
Then
1. That operator may safely use the planer or jointer.

further subdivided and planers and jointers considered separately, just as the power saw category could be subdivided into lists for table saws, radial saws, band saws, scroll saws, circular saws, saber saws, and so on.

Writing sets of concepts for other woodworking power tools allows a builder or the student of construction to formulate additional statements of principles. Each machine will have its own design features, working characteristics, and safety devices to consider. Table 4-5 gives some ideas for preparing the "if-then" type of formulation for a few other woodworking tools.

TABLE 4-5 PARTICULARITIES WITH RESPECT TO SOME WOODWORKING MACHINES

Bandsaw
1. Blade straightness.
2. Blade tension.
3. Wheel alignment.
4. Blade breakage and safe position of operator.
Lathe
1. Guards between centers over long pieces of stock.
2. Safe positions of operator for in-board and out-board work.
3. Use of cutting chisels.
4. High speed bearings.
5. Cutting speeds.
Router
1. Cutting actions, feed.
2. Guards, fence, guides.
3. Noise levels.
4. Matching cutters to materials.
Sander or Sanding Attachment (Drum, Disk, Belt)
1. Guards (against sanding fingers).
2. Exhausting (against allergy problems from dust).

The safe and efficient operation of other stationary and portable power tools and even some hand tools could be aided by a formulation of this type of safety guideline. In addition to providing a basis for accepted work rules, such a formal guideline could easily serve as part of the training given to new workers or foremen. For example, Tables 4-6A and 4-6B lay out safety guidelines for pneumatic tools—a category of tools a new worker may be unfamiliar with. The more unusual, specialized, or dangerous a tool or group of tools is, the more reason for a construction company to draw formal guidelines for its use. A good example is powder-actuated tools; a sample formulation is given in Tables 4-7A and 4-7B.

Tools and equipment can also be cross-formulated. For example, besides grouping tools in regard to their design or use, they might be grouped together in regard to some other common denominator—amount of training a worker needs before operating a type of tool, the energy source used, and so on. In this vein, jobsite electrical equipment—all tools and machinery powered by electricity—are formulated in Tables 4-8A and 4-8B.

TABLE 4-6A JOBSITE PNEUMATIC POWER TOOL OPERATION

Concept	Descriptive Statement
1. Safety Standards	National, regional, local.
2. Compressor	Operation, safe limits, hoses, safety valves.
3. Tool action	Nailing, stapling, sealing, cleaning, painting, concrete ramming.
4. Fasteners	Nails, staples (automatic fastener feed).
5. Hoses, pipes	Connectors, sizes, bleeding.
6. Safety devices	Whip restraining, couplings, clips, retainers, pressure reducers, diffusers.
7. Training	Industrial safety practices.
8. Hazards	Hose whipping, air blast on skin, high velocity particles, tool failure under high pressures.
9. Inspection	Maintenance, records.

TABLE 4-6B PRINCIPLES OF USING JOBSITE PNEUMATIC POWER TOOLS

If
1. The tool is well designed, well built, inspected, and maintained,
And if
1. The manufacturer's safe operating pressure instructions are heeded, 2. Safety clips or retainers are securely installed (as required) and maintained, 3. The tool has a muzzle safety device when required, 4. The tool is secured to the hose, 5. The tool is properly hoisted and lowered, 6. An air blast is not directed at the skin, 7. Protective clothing and gear are worn,
Then
1. The operator may safely use the tool.

TABLE 4-7A POWDER-ACTUATED FASTENING TOOLS

Concept	Descriptive Statement
1. Action of tool	Direct or indirect: powder to fastener.
2. Velocity of fastener	Measured in feet or meters per second, color coding of charges.
3. Fastening	Applications/specifications.
4. Hazards	Particles (spalling), ricochets, noise, perforations, explosive atmospheres, and failure to fasten.
5. Testing/operation	Loading, positioning, firing, materials.
6. Guards/shields	Ricochets, spalling.
7. Maintenance	Cleaning, storing.
8. Protective clothing	Hardhats, safety goggles, face shields.
9. Job safety	Procedures, precautions/certification of operators.

TABLE 4-7B PRINCIPLES OF POWDER-ACTUATED TOOL OPERATION

If
1. The tool, charges, and fasteners meet all the requirements of the respective national standards,
And if the operator
1. Has been trained in the operation of the tool, 2. Selects the charge and fastener appropriate for the materials to be fastened, 3. Wears proper protective clothing and gear, 4. Uses the correct shield or guard, 5. Avoids firing into a spalled area or into a flammable atmosphere...
Then
1. That operator may safely use the powder-actuated tool.

TABLE 4-8A JOBSITE ELECTRICAL EQUIPMENT OPERATION

Concept	Descriptive Statement
1. Safety standards	National Electrical Code, Regional Acts/Regulations.
2. Electrical power circuits	Exposed, concealed.
3. Load ratings	Specifications, voltage, amperage, power.
4. Lockout of circuit	De-energized, inoperative.
5. Ground-fault protection	Interruptors, grounding program.
6. Phase	Single, Three.
7. Grounding	Cord sets, plug/receptacle, continuity, resistance, bonding, double insulation, testing, records.
8. Overcurrent protection	Fuses, circuit breakers.
9. Transformer	Access, voltages, hazards.
10. Hazardous locations	Classes.
11. First aid	Shock, electrocution, fibrillation.

TABLE 4-8B PRINCIPLES OF USING JOBSITE ELECTRICAL EQUIPMENT

If
1. The piece of equipment meets specifications of the National Electrical Code and Regional Electrical Protection Acts/Regulations,
2. The employer has provided an assured grounding conductor program,
3. The employer is satisfied that the employee is knowledgeable with respect to the operation of the equipment,
4. Exposed noncurrent-carrying metal parts of fixed electrical equipment are grounded,
Then
1. The operator may safely operate that equipment.

Finally, these tables might require special subdivisions. The formulation for mobile power equipment (Tables 4-9A and 4-9B) could be further broken into tables for different types of equipment. The skills and safety factors involved in digging a deep trench around buried pipes, cables, and wires with a backhoe are far different from those required to smooth crushed stone with a roller.

The tables shown here are in many respects limited with regard to both concepts listed and principles derived therefrom. However, they should serve as a model for the builder determined to create a safe workplace. In any case, this particular type of formulation is by no means the only way to set formal safety guidelines. Every builder may have his or her own way of setting up guidelines. What is important is for the builder to understand how construction tools and machinery operate and are properly used and how to implant both safe operation principles and a "safety" attitude in the whole construction team.

TABLE 4-9A JOBSITE MOBILE POWER EQUIPMENT OPERATION

Concept	Descriptive Statement
1. National rules	Acts, Regulations.
2. Safety standards	Government, Society, Institute, Manufacturer.
3. Safety systems	Braking, lighting, signaling, alarm, coupling, starting, controls.
4. Safety devices	Seatbelts, fenders, rollover protective structures, guards, canopies.
5. Ratings	Ratios, capacities, altitudes, limitations.
6. Site clearing	Access, hazards, clearing of trees, rock and earth.
7. Excavating	Buried services, opening the excavation, pilings, shoring, pumping, angle of repose, bracing, underpinning, draining, ramps, walkways, ladders.
8. Trenching	Cave-ins, shoring, sloping, sheeting, stress, angle or repose, jacks, unstable soil.
9. Lifting, hoisting	Specifications, limitations.
10. Augers	Hazards, infeed, outfeed, dust.
11. Fuels/lubricants	Flash points, storage, handling.
12. Training, education	Care and maintenance, mechanisms, operating, protection, inspection.

TABLE 4-9B PRINCIPLES OF JOBSITE MOBILE POWER EQUIPMENT OPERATION

If
1. The machine is well designed, well built, inspected, and maintained,
And if the operator
1. Uses the machine for the purpose it was designed, 2. Complies with the manufacturer's specifications and limitations applicable to the operation, 3. Obtains information from a professional engineer competent in the field, when manufacturer's specifications are not available, 4. Is trained and authorized to use the machine, 5. Is protected from environmental hazards, 6. Has access to first aid and life-saving equipment, 7. Uses good judgment with respect to building site conditions,
Then
1. That operator may safely use the piece of equipment.

4.7 RESEARCH

1. For ladder placement, what does a 1 to 4 ratio mean? If the placement limits are horizontal and vertical, why is 1 to 4 best?
2. Why are lean-to scaffolds prohibited?
3. For one- and two-story light construction projects in local areas, what are the per unit foot (meter) costs for wooden single pole and independent pole scaffolds and for sectional and clamp-type scaffolds? Make comparisons for buying, building, or renting.
4. Make a sketch of each type of scaffold discussed in this chapter.
5. For a given light construction worksite job (carpenter, electrician, plumber, or other), list a complete set of personal gear and itemize a cost statement.
6. Make a carefully scaled sketch of a woodworking shop with various woodworking machines in the plan. Indicate safe working spaces, direction of feed, direction of possible kickbacks, and size of stock to be accommodated. Also, given the equipment located in this shop, list the operations that could or could not be done there.
7. Write a personal set of operating rules for each of the machines shown in the drawing for Research Question 6.
8. Obtain a power saw operator's manual (table saw, circular saw, radial saw, etc.) complete with detailed specifications and operating instructions. Using the information, update or revise Table 4-4B.
9. By means of a sketch, describe the probable safe standing position for a band saw operator.
10. By means of a sketch for a particular model machine, describe how straight jointer blades are hollow-ground and set parallel with the tables.

11. Also in regard to a jointer, use a sketch to explain what happens to the edge of a piece of wood stock when the outfeed table is set slightly higher than the blades and then what happens when it is slightly lower than the blades.
12. Visit a regional occupational health and safety library or some other location where safety information can be obtained. Acquire a copy of a safety report on jobsite operations of mobile power equipment. From it make a comprehensive list of reasons for accidents. Look for recommendations by the report writer for improving the accident rate. Indicate how you think this accident rate might be improved. Delineate responsibilities for achieving improvement.
13. Other areas that require attention in the field of occupational health and safety are electrical/electronic hazards, chemical hazards, and design of safety equipment. Investigate progress and problems in these areas.
14. Pretend you are at home getting ready for a typical day at the construction site. How should you prepare yourself mentally? What about your appearance? What should it reflect? Now allow your imagination to transport you to the site. How should you behave there? What are your responsibilities? What types of things should you constantly be aware of? When answering these questions, consider how your actions will affect others, not only yourself.
15. What safety precautions should always be taken before and while climbing a ladder? Take into account the condition and position of the ladder as well as climbing style.
16. How much weight per square foot can the platforms of each of the following categories of scaffolds safely support: extra heavy-duty, heavy-duty, intermediate, and light-duty?
17. To become better acquainted with the different ways of hitching work that must be hoisted, get a piece of large diameter, lightweight plastic pipe that you can easily handle and a piece of rope or nonplastic clothesline in which a loop, or "eye," has been tied in each end. Practice making the various hitches around the pipe, then lifting the pipe with the rope to see the effect. Note where the hitches must be positioned to prevent the load from tipping or slipping.
18. Radial arm saws are frequently used as primary tools at house construction sites because they simplify the trimming and shaping of hard-to-handle components such as studs and rafters. On a sketch of a particular model, show how the blade should be positioned for the following cutting actions: neutral crosscut, in-rip, out-rip, and simple miter cut. Review the various and diverse operations that can be performed by a radial arm saw along with what type of blade should be used for each particular job.
19. Go to a local store and ask the salesperson to explain the various types of air compressors available. What is the difference between

the diaphragm type and piston type models? Which type and size model would be most practical for use at a construction site? Next look at the selection of pneumatic tools and accessories. Which would be most useful to you in the present? How about the future? Compare the pressure ranges required by the tools with the maximum pressures that can be produced by the different compressors.

20. What types of operations can be performed by a router? How can the depth of cut be altered? Review the types of bits available for use with a router. How could each one be of use at the construction site?

4.8 REFERENCES

4.8.1 Drake, George R. *The Complete Handbook of Power Tools.* Virginia: Reston Publishing Company, Inc., 1977.

4.8.2 *Getting the Most Out of Your Band Saw.* Pennsylvania: Rockwell International, 1978.

4.8.3 Hand, Jackson. *Modern Woodworking.* Virginia: Reston Publishing Company, Inc., 1975.

4.8.4 *How to Choose and Use an Air Compressor.* Ohio: Campbell-Hausfeld, 1979.

4.8.5 OSHA Safety and Health Standards. *Construction Industry.* Washington, D.C.: U.S. Department of Labor Occupational Safety and Health Administration, 1983.

4.8.6 *Rigging for Commercial Construction.* Virginia: Reston Publishing Company, Inc., 1983.

4.8.7 *Using a Router and Router Bits.* North Carolina: Vermont American, 1981.

5

MEASUREMENT

Proper measurement is as important as any single aspect of building construction. The contractor or builder must be fully able to make jobsite or shop measurements and to translate these into shop drawings. The builder must also be able to make constant on-the-job measurements as part of the building process and to be able to translate an architect's plans into a physical structure, another task requiring the ability to measure and to anticipate the changes that take place during the construction process.

5.1 GENERAL OBJECTIVES

This chapter is primarily intended to give the builder the basic framework of construction measurement. How to make measurements and drawings, recognize tolerances, use measuring devices, and deal with construction materials in terms of their standard sizes are all aspects of construction measurement which must be mastered by the builder.

A complication involving construction measurement is the current official attempt in North America to change from the British Imperial System to the Metric System. This process of change is hardly completed, and in the United States it has barely begun. The builder in the 1980s is caught between these systems of measurement. For this reason, most measurements in this book are shown in both the British Imperial System and the Metric System.

5.2 SPECIFIC OBJECTIVES

On reading this chapter and conducting the suggested research, one should be able to

1. convert linear measurements from or to Metric readings and change British Imperial common fractions to Metric (decimal) fractions;
2. apply recognized tolerances with respect to those linear measurements;
3. read and use British Imperial and Metric measurements of area and volume;
4. take jobsite and in-shop measurements;

5. relate the concepts of contour lines, bench marks, and elevations as they pertain to a typical jobsite;
6. make shop drawings;
7. make and read an elementary story rod;
8. make and read a layout board for a small cabinet;
9. write a table of measurement tolerances for jobsite work;
10. relate jobsite angular measurement terms;
11. state sizes of typical jobsite materials; and
12. read and use measurement conversion tables for various construction materials.

5.3 MEASUREMENT SYSTEMS

The two predominant systems of measurement in the world have been the British Imperial and the Metric. Currently, a major shift seems to be taking place in America from British Imperial measure to Metric, mainly because the Metric System is a decimal system and its units of measurement are more consistent and easier with which to work.

The modern Metric System is the International System of Units (Le Systeme International d'Unites) or SI. Each country using this system has created its own style guide.

Industries still using the British Imperial System have often shifted from using common fractions to decimal fractions to assure accurate measurement statements, particularly for computer applications. In building construction both common fractions and decimal fractions are used. A major problem has been how to interrelate the different units of British Imperial measure such as miles, yards, feet, and inches.

The units of the Metric System, however, are well related internally. To illustrate the differences in the relationship of linear units of measurement within the two systems, the following two examples are taken from Table 5.1.

Example 5.1: (British Imperial) 1 mi = 320 rods = 1,760 yd = 5,280′ = 63,360″.

Example 5.2: (Metric) 1 km = 1,000 m (10^3 m) = 100,000 cm (10^5 m) = 1,000,000 mm (10^6 mm).

In example 5.1 the relating units are many: 320, 5.5, 3, and 12. In Example 5.2 the focus is on the number 10. Needless to say, this is generally the great difference between both systems. British Imperial measure lacks common or even related dimensions of measure, while all Metric measure is based on standard divisions always being multiples of ten. Table 5-1 shows the measurement units common to light building construction work and their Metric counterparts.

Over the years in most countries around the world, the scientific communities, the chemical industries, and the military have each adapted the Metric System. In North America this is also partially the case. But many other industries are still using the British Imperial system only, although some have changed linear measurements in cumbersome common fractions to decimal fractions. However, even

TABLE 5-1 COMMON UNITS FOR LIGHT CONSTRUCTION MEASUREMENTS

British Imperial			Metric		
Relationship	Abbreviation	Unit	Unit	Abbreviation	Relationship
		Length			
1 mi = 1,760 yd	mi	mile	kilometer	km	1 km = 1,000 m
1 yd = 3′	yd	yard	meter	m	1 m = 10 dm = 100 cm
1′ = 12″	ft	foot	decimeter	dm	1 dm = 10 cm
	in.	inch	centimeter	cm	1 cm = 10 mm
			millimeter	mm	
		Area			
1 sq mi = 640 a	sq mi	square mile	square kilometer	km^2	1 km^2 = 1 ha
1 a = 43,560 sq ft	a	acre	hectare	ha	1 ha = 10,000 m^2
1 sq yd = 9 sq ft	sq yd	square yard	square meter	m^2	1 m^2 = 100 dm^2 = 10,000 cm^2
1 sq ft = 144 sq in.	sq ft	square foot	square decimeter	dm^2	1 dm^2 = 100 cm^2
	sq in.	square inch	square centimeter	cm^2	
		Volume			
1 cu yd = 27 cu ft	cu yd	cubic yard	cubic meter	m^3	1 m^3 = 1,000 dm^3
1 cu ft = 1,728 cu in.	cu ft	cubic foot	cubic decimeter	dm^3	1 dm^3 = 1,000 cm^3
	cu in.	cubic inch	cubic centimeter	cm^3	
			kiloliter	kl	1 kl = 1,000 l
			liter	l	1 l = 1,000 ml
			milliliter	ml	
		Mass			
1 short ton = 2,000 pounds	—	ton	tonne	t	1 t = 1,000 kg
1 pound = 16 oz = 7,000 grains	lb	pound	kilogram	kg	1 kg = 1,000 g
	oz	ounce	gram	g	1 g = 1,000 mg
			milligram	mg	
		Time			
1 day = 24 hr	—	day	day	d	1 d = 24 h
1 hr = 60 min	hr	hour	hour	h	1 hr = 60 min
1 min = 60 sec	min	minute	minute	min	1 min = 60 s
	sec	second	second	s	
		Temperature (for water)			
freezing point is 32°F	°F	degree Fahrenheit	degree Celsius	C	freezing point is 0°C
boiling point is 212°F					boiling point is 100°C

here problems exist. Examples 5.3 and 5.4 which follow indicate the lack of consistency in the application of British Imperial measure.

Example 5.3: 5-3/8″ + 1/16″

This measurement can be taken with a standard tape measure used by construction workers.

Example 5.4: 5.375″ + 0.0625″

This measurement cannot be taken with a common tape measure. It can, however, be obtained with a micrometer or vernier caliper for automotive or other engineering applications. In other words, in British Imperial measure the measuring tools often seem to prescribe the measuring technique.

The construction worker is sometimes obliged to use decimal fractions, especially in roof framing calculations because framing squares have tables with measurements expressed in decimal fractions. However, the tape measure used for applying the calculated measurements is usually graduated in common fractions (1/16 or 1/8, etc.).

The following set of calculations will illustrate the incongruities of construction measurement and means to deal with this problem. The set will also emphasize the importance of the concept of "nearest" (the nearest 1/16″ or ± 1/16″ as a level of tolerance, for example).

5.3.1 Calculations in British Imperial Measure

If, on the framing square tables, the unit line length for a rafter is 18.76″ and seven steps are required for the rafter layout, the calculated line length of the rafter is 7 × 18.76″ or 131.32″. As it is good practice to check the stepped-off length with a taped measurement, say with a tape graduated in 1/16″, another calculation is required. The measurement of 131.32″ is equivalent to 10.943′, but this is not readily measured with a standard tape measure. The measurement 131.32″ is also equivalent to 10′ + 11″ + 0.32″, with the first two parts of the equivalency being measurable. The problem is to convert the 0.32″ to the nearest 1/16″. The rule is to multiply by the desired denominator: 0.32″ × 16 = 5.12″. This figure signifies the nearest 1/16″, in this case, 5/16. In other words, 131.32″ = 10′ + 11-5/16″ to the nearest 1/16″ (that is, 5.12 is nearer to 5 than to 6).

It is possible to generalize from this set of calculations about decimal to common fraction conversions as shown in the following examples.

Example 5.5: The nearest 1/8″ for 0.5″ would be 0.5 × 8 = 4; 0.5″ = 4/8″. The nearest 1/16″ would be 0.5 × 16 = 8; 0.5″ = 8/16″.

Example 5.6: The nearest 1/8″ for 0.52″ would be 0.52 × 8 = 4.16; 0.5″ = 4/8″. The nearest 1/16″ for 0.52 would be 0.52 × 16 = 8.32; 0.52 = 8/16″. The nearest 1/64″ for 0.52″ would be 0.52 × 64 = 33.28; 0.52″ = 33/64″.

Then the general rule is to set the measurement tolerance level by selecting the multiplier (8, 16, 64, etc.). That multiplier should, in turn, be determined by the required tolerance for a specific job.

Example 5.7: The nearest 1/8″ for 0.57″ would be 0.57 × 8 = 4.56; 0.57″ = 5/8″.

Example 5.8: The nearest 1/16″ for 0.57″ would be 0.57 × 16 = 9.12; 0.57″ = 9/16″.

Take note the degree of error in reading 5/8″ rather than 9/16″ on a tape graduated in 1/16″. This degree of error may be significant depending on the specific tolerance required.

A construction worker is obligated to measure according to information given on drawings. A draftsperson, who may provide the information, is responsible for loss of time and for measurement errors if drawings are not readily interpreted in the field. Therefore, for interests of efficiency and accuracy, the on-the-job worker needs a completely integrated measurement system.

5.3.2 Metric Measurement

Where the Metric System is being used in the light building construction industry, the millimeter is rapidly evolving as the standard unit of linear measurement and of tolerances. Longer distances, especially for site and foundation layout, are often measured in meters and fractions thereof, but even there the millimeter is being used more as the basic measuring unit. Because the millimeter seems like such a small unit, measuring a building in millimeters may seem strange at first, especially for someone used to the British Imperial System.

There are, however, definite advantages in using the millimeter as a basic unit. It eliminates the use of a combination standard, such as is usual in the British Imperial System where many construction measurements are in feet, inches, and fractions of inches. Measurements in millimeters require no conversions within the system and fractions are usually not required because millimeters are such a small unit to begin with. Even where computations leave decimal fractions of millimeters, rounding off to the nearest millimeter is rarely a problem within the tolerances usually required in light building construction. In addition, millimeters can be used as a basic unit for all types of construction measurement from site layout to cabinetry. Again, this eliminates cross-conversions and assures fine tolerances on even the greatest measured distances. Furthermore, the millimeter is an ideal unit for computer calculations. For example, when linear measurements for a building are given in millimeters, measurements of area, volume, and load can be obtained from standard computer programs.

In general, the system of measurement most readily adaptable to programming in the computer age seems to be the Metric. This becomes increasingly important as computers are utilized more frequently by light construction people ranging from architects, engineers, estimators, and draftspersons to materials suppliers and inventory controllers. Computer calculations are commonly being made with respect to cost estimates, end price, strengths of materials, strengths of built-up units (such as trusses and laminated arches), and inventories, as well as many other applications. The Metric System will grow in importance hand-in-hand with computer applications for a wide variety of traditional jobs associated with construction work.

When Metric measure is being used, the tape, rule, or other measuring device should obviously be graduated in millimeters, both for actual measuring purposes and to facilitate easier calculating. For example, in Figure 5-1 the unit line length of a particular rafter is

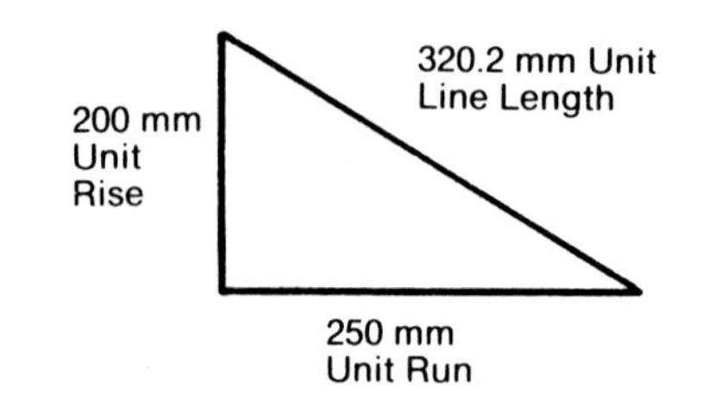

FIGURE 5-1. Unit line length measurement for a typical rafter application.

indicated as 320.2 mm. If eight steps are required to determine the entire length, then

$$8 \text{ steps} \times 320.2 \text{ mm} = 2{,}561.6 \text{ mm}$$

The measurement 2,561.6 mm can be rounded off to 2,562 mm as standard procedure and, in fact, would be read directly from the tape as such.

For extremely long distances and especially large areas, using millimeters or square millimeters may be undesirable because the numbers involved become so large as to be unmanageable. Therefore, the meter (and fractions thereof, depending on the tolerance necessary) is frequently used as the unit to express linear measurements of greater length, and the square meter is usually the standard metric unit for measuring area: floor square, foundation area, and so on. A building's area will usually be expressed in square meters (m^2) as shown in the following example.

Example 5.9: A building measuring 5,500 mm by 7,900 mm has an area of 43.45 m^2. Even though the millimeter is not used as the base unit here, the metric system, by its very nature, makes internal conversions simply a matter of moving decimal places.

Much of the resistance to the Metric System in North America is the result of tradition and misconception. Although Americans are used to the British Imperial System of measurement, it has some distinct disadvantages in relation to the Metric System. Interrelationships between units give the Metric System coherency (meaning each part is related logically to other parts) and a rationality (an orderly progression), both of which are lacking in the British Imperial System. The British Imperial System and all other previous systems of measurement were largely composed of arbitrary units with an assortment of unrelated names and irregular progression from one size to the next. Of course, in the Metric System every unit can be multiplied or divided by a factor of 10 to get larger units (multiples) or smaller units (submultiples).

What is most important about Metric measurement in the construction industry today is how it is affecting the way in which various materials are estimated for size and volume, ordered and purchased, and used on the jobsite. In addition, it is necessary to view the use of the British Imperial and Metric systems within the context of a traditional period. A reasonable assumption is that the British Imperial System will in the future no longer be used.

Take note, however, that several customary units of measure in the United States will remain the same, although none of these are specifically related to construction work. Electricity will still be measured in amperes and watts, and time will be measured in seconds, minutes, and hours, rather than in Metric standardized seconds. Contractors and construction workers will continue to be paid in dollars and cents, which is already a Metric measure.

In the following sections, the dual system of British Imperial and Metric measure will be described for each of the most common construction materials, including: wood, metals, plumbing materials, electrical materials, concrete, glass, and insulation.

5.4 WOOD

Wood products can be classified as softwood lumber, hardwood lumber and flooring, panel products, doors and windows, laminated sheets, pallets, containers, poles and piling, or shingles and shakers. Traditional methods of sizing construction wood products have always been somewhat irrational. Now with the introduction of the Metric System, it can be said that lumber sizing is in a period of flux. Traditionally, lumber in the United States and Canada is identified in British Imperial measure by nominal sizes. However, the nominal size of a piece of lumber is larger than the actual dressed finished dimensions. The common widths and thicknesses of lumber in their nominal and dressed dimensions with the Metric conversions are shown in Table 5-2.

TABLE 5-2 NOMINAL AND DRESSED LUMBER DIMENSIONS WITH METRIC EQUIVALENTS

			Minimum Dressed					
			Dry			Green		
Nominal Thickness (in.)	Exact (mm)	Rounded (mm)	(in.)	Exact (mm)	Rounded (mm)	(in.)	Exact (mm)	Rounded (mm)
1	25.4	25	3/4	19.1	19	25/32	19.8	20
1-1/4	31.8	32	1	25.4	25	1-1/32	26.2	26
1-1/2	38.1	38	1-1/4	31.8	32	1-9/32	32.51	33
2	50.8	51	1-1/2	38.1	38	1-9/16	39.7	40
2-1/2	63.5	64	2	50.8	51	2-1/16	52.4	52
3	76.2	76	2-1/2	63.5	64	2-9/16	65.1	65
3-1/2	88.9	89	3	76.2	76	3-1/16	77.8	78
4	101.6	102	3-1/2	88.9	89	3-9/16	90.5	91
5	127.0	127	4-1/2	114.3	114	4-5/8	117.5	118
6	152.4	152	5-1/2	139.7	140	5-5/8	142.9	143
7	177.8	178	6-1/2	165.1	165	6-5/8	168.3	168
8	203.2	203	7-1/4	184.1	184	7-1/2	190.5	191
10	254.0	254	9-1/4	235.1	235	9-1/2	241.3	242
12	304.8	305	11-1/4	286.2	286	11-1/2	292.1	292
14	355.6	356	13-1/4	337.1	337	13-1/2	342.9	343
16	406.4	406	15-1/4	386.6	387	15-1/2	393.7	394

The fact that lumber does not even equal its nominal British Imperial measurement by the time it is used by a builder further complicates the eventual change to metric measurement. The situation in North America represents several steps in the evolution from British Imperial measure to Metric and from the use of nominal size identification to actual size. Canada is further along this evolutionary path than is the United States; however, even in Canada the Metric System is only partially in use. Canada does now use metric terminology, but the conversions are considered "soft," meaning that the lumber is cut the same as it used to be, but that the British Imperial measurements are simply converted to the metric and the lumber is labeled as such. However, rather than still maintaining both nominal and actual sizes, as was previously done, only actual Metric sizes are used. This alone can be considered a giant step forward in the drive to rationalize construction measurement.

Nevertheless, the Canadian softwood lumber industry will continue to produce existing sizes and lengths of softwood lumber until a conversion to Metric sizes and lengths can be coordinated with the U.S. softwood lumber industry. During this interim period, the Canadian softwood lumber industry is prepared to market the existing sizes and lengths in Metric terms in Canada, while continuing to market in the British Imperial terminology in the United States.

As Metric measurement is gradually introduced into the U.S. construction industry, the use of nominal sizes for lumber will disappear there as well. This means that the surfaced dimension lumber now called a 2″ × 4″, which actually measures 1-1/2″ × 3-1/2″, will in future only be referred to as a 38 mm × 89 mm piece of lumber. In other words, Metric nomenclature for width and thickness is being developed, based on the actual sizes of lumber, rounded to the nearest millimeter.

The Metric nomenclature for surfaced dimension lumber, boards, and worked or patterned lumber is based on the seasoned S4S (surfaced four sides) size whether shipped seasoned or unseasoned. Metric nomenclature for rough lumber depicts the actual size at the time of manufacture and applies whether shipped seasoned or unseasoned. Timbers which are 4-1/2″ (114 mm) or more in least dimension are surfaced unseasoned; therefore, the Metric nomenclature is based on the unseasoned surfaced size.

After a consolidation of the Metric System in the United States and Canada, lumber sizes may eventually be changed again to conform with more regular and rational Metric readings.

Lengths of lumber will be measured in meters (m), expressed to the second decimal place. Generally, the lengths will also be soft conversions of existing British Imperial lengths. However, four new precision and trimming (PET) stud lengths may be provided to suit new wall heights, which will change because of metric panel sizes. This will be more efficient than the previous practice of producing 11 to 15 different British Imperial PET stud lengths.

Volume will be measured by the cubic meter (m^3), expressed to the third decimal place. It is based on the Metric nomenclature. Small quantities of lumber, however, may be sold by the piece or by the lineal meter. A few items such as bevel siding will be sold on the basis of coverage in square meters, and mouldings and trim will be sold by the lineal meter.

During the period when the Canadian softwood lumber industry is operating in Metric terminology and the U.S. remains on the British Imperial System, there will be a need by some to convert softwood lumber volumes and prices from foot board measure (FBM) to cubic meters and vice versa. British Imperial volumes of lumber are based on nominal sizes while Metric volumes are based on actual size; therefore, conversion factors are needed for each item and size of softwood lumber.

For additional information about softwood lumber sizes, obtain current data from American and/or Canadian lumber standards associations for

1. surfaced boards, shiplay and center match;
2. surfaced finished;
3. surfaced dry industrial clears;
4. surfaced stepping;
5. surfaced flooring, ceiling and partition;
6. surfaced siding;
7. rough lumber; and
8. surfaced dimension lumber and timbers.

Construction Modules for Framing. Actually, in construction, soft conversion from traditional to metric sizes is of relatively little value in itself and could be a waste of time unless there is a significant reason for the conversion. However, in the building industry there will be a major change in basic modular standards for framing, incorporating the advantages that are inherent in the Metric System.

The traditional module of construction in the British Imperial System is 4″ (102 mm). This module provides for 16″ (406 mm) and 24″ (610 mm) framing layouts for studding, joists, and rafters/trusses and 48″ (1,219 mm) layouts for softwood and hardwood panels. Panels measuring 4′ × 8′ (1.2 m × 2.4 m) are fastened to framing without cutting, if and when it is possible to do so. The module is thus applicable to framing foundation forms, subfloors, floors, walls, ceilings, and roofs. These spacings also accommodate insulation rolls and batts.

Under the Metric System, the construction industry is adopting, or has adopted, the 100 mm module. Stud, joist, and truss spacings are 300 mm, 400 mm, and 600 mm, and Metric framing square dimensions reflect the adoption. Panel sizes have been changed in width and length from 1,220 mm × 2,440 mm to 1,200 mm. Precut stud lengths account for a change to generally lower ceiling heights.

Traditional paneling dimensions and their Metric counterparts are shown in Table 5-3.

TABLE 5-3 TRADITIONAL PANELING DIMENSIONS AND THEIR METRIC COUNTERPARTS

Paneling Widths/Lengths*	Thicknesses*
4′ = 1,200 mm	1/8″ = 3 mm
8′ = 2,400 mm	1/4″ = 6 mm
9′ = 2,700 mm	3/8″ = 9 mm
10′ = 3,000 mm	1/2″ = 12 mm
12′ = 3,600 mm	5/8″ = 15 mm
14′ = 4,200 mm	3/4″ = 18 mm
	1″ = 25 mm

*Metric measurements are rounded off.

5.5 METAL

Metals commonly used in the field of light building construction are aluminum, copper, iron, and steel. Measurement of these materials has traditionally been, if anything, more confusing than measurements for wood, simply because there has never been consistent standards throughout the industry. Metal materials have in the past frequently been quoted in feet, inches, common fractions, or in pounds. In addition, metal materials also have had a gauge designation, a term describing common metal thicknesses. As a gauge number increases, the metal thickness decreases. The greatest problem of using these gauge designations is that each segment of the metal industry has their own gauge system.

Even before the introduction of the Metric System, certain trends seemed to be apparent. Common fractions had already largely been replaced by decimal fractions in the British Imperial System and attempts were made to standardize gauge sizes. In fact, this standardization will probably eventually be achieved under the auspices of the Metric System. Specifically, the more than 20-odd gauge systems now being used will be replaced by one Metric gauge size. The old profusion of gauge systems has always caused a great deal of confusion. With the metric standards, all gauge designations are eliminated and metal thicknesses or diameters—wire, for example—are given in millimeters. Table 5-4 gives the more common gauge sizes of nonferrous and ferrous metals and the preferred millimeter thicknesses.

The general evolution of metal's measurement involves a switch away from the use of common fractions and away from the use of gauge sizes altogether other than actual Metric gauge measurement. The obvious improvement is that it will be easier to work with actual measurements in millimeters rather than to think in terms of a gauge

TABLE 5-4 COMPARISON OF COMMON GAUGE SIZES

Gauge Number	Manufacturer's Standard Gauge for Sheet Metal (in.)	U.S. Standard Gauge for Iron and Sheet Steel (in.)	American Standard Wire Gauge or Brown & Sharpe Gauge (in.)	ISO/R388 (International System of Equivalents) (in.)	ISO/R388 (International System of Equivalents) (mm)	Suggested Rounded Modules
10	.1345	.1350	.1019	.0984	2.500	2.4
11	.1196	.1205	.0907	.0882	2.240	2.2
12	.1046	.1055	.0808	.0787	2.000	2.0
13	.0897	.0915	.0720	.0709	1.800	1.8
14	.0747	.0800	.0641	.0630	1.600	1.6
15	.0673	.0720	.0571	.0552	1.400	1.4
16	.0598	.0625	.0508	.0492	1.250	1.2
17	.0538	.0540	.0453	.0441	1.120	1.1
18	.0478	.0475	.0403	.0394	1.000	1.0
19	.0418	.0410	.0359	.0354	0.900	0.90
20	.0359	.0348	.0320	.0315	0.800	0.80
21	.0329	.0317	.0285	.0280	0.710	0.70
22	.0299	.0286	.0253	.0248	0.630	0.60
23	.0269	.0258	.0226	.0221	0.560	0.55
24	.0239	.0230	.0201	.0197	0.500	0.50
25	.0209	.0204	.0179	.0177	0.450	0.45
26	.0179	.0181	.0159	.0158	0.400	0.40

TABLE 5-5 FRACTIONAL INCHES INTO DECIMALS AND MILLIMETERS

Inch	Decimal Inch	Millimeter	Inch	Decimal Inch	Millimeter
1/64	0.015625	0.396875	33/64	0.515625	13.096875
1/32	0.03125	0.79375	17/32	0.53125	13.49375
3/64	0.046875	1.190625	35/64	0.546875	13.890625
1/16	0.0625	1.5875	9/16	0.5625	14.2875
5/64	0.078125	1.984375	37/64	0.578125	14.684375
3/32	0.09375	2.38125	19/32	0.59375	15.08125
7/64	0.109375	2.778125	39/64	0.609375	15.478125
1/8	0.125	3.175	5/8	0.625	15.875
9/64	0.140625	3.571875	41/64	0.640625	16.271875
5/32	0.15625	3.96875	21/32	0.65625	16.66875
11/64	0.171875	4.365625	43/64	0.671875	17.065625
3/16	0.1875	4.7625	11/16	0.6875	17.4625
13/64	0.203125	5.159375	45/64	0.703125	17.859375
7/32	0.21875	5.55625	23/32	0.71875	18.25625
15/64	0.234375	5.953125	47/64	0.734375	18.653125
1/4	0.25	6.35	3/4	0.75	19.05
17/64	0.265625	6.746875	49/64	0.765625	19.446875
9/32	0.28125	7.14375	25/32	0.78125	19.84375
19/64	0.296875	7.540625	51/64	0.796875	20.240625
5/16	0.3125	7.9375	13/16	0.8125	20.6375
21/64	0.328125	8.334375	53/64	0.828125	21.034375
11/32	0.34375	8.73125	27/32	0.84375	21.43125
23/64	0.359375	9.128125	55/64	0.859375	21.828125
3/8	0.375	9.525	7/8	0.875	22.225
25/64	0.390625	9.921875	57/64	0.890625	22.621875
13/32	0.40625	10.31875	29/32	0.90625	23.01875
27/64	0.421875	10.715625	59/64	0.921875	23.415625
7/16	0.4375	11.1125	15/16	0.9375	23.8125
29/64	0.453125	11.509375	61/64	0.953125	24.209375
15/32	0.46875	11.90625	31/32	0.96875	24.60625
31/64	0.484375	12.303125	63/64	0.984375	25.003125
1/2	0.50	12.7	1	1.00000	25.4

number which must be translated. Before the Metric System is fully established, it will be necessary to be able to make the conversion from specific gauge designation to common and decimal fractions in the British Imperial System to Metric fractions which are, by their nature, decimal fractions. Table 5-5 gives the conversions of common fractional inches into British Imperial decimals, fractions, and millimeters.

Lengths of threaded fasteners will be the same as the actual screws, or other fastener, except that they will be soft-converted to millimeter designations, but the diameter-pitch sizes will be hard-converted. Twenty-five new Metric diameters replace 59 different sizes and thread combinations utilized under the current inch system. The coarse and fine thread distinction will be eliminated and a new single thread will fall somewhere in between the two.

Many other metal building items will undergo a soft conversion into Metric units. For example, the common ten-penny (10 d) nail which is about 3″ long will still be made that same length, but will then be 76.2 mm long in Metric. Rounded off, the nail could be 75 mm and may eventually be referred to as a "75 mm" nail (see Chapter 3 for nail conversions).

5.6 PLUMBING MATERIALS

When planning a light construction building, one must first know which plumbing fixtures will be installed, according to the expected occupancy of the building. In other words, the plumbing subsystem must be developed step by step, much as the rest of the building. After those decisions are made, roughing-in measurements are determined with the aid of information from the fixture manufacturer. When construction begins, the initial plumbing work is the roughing-in, with a general plan in mind for piping, water supply, and disposal. Of course, the piping installers and the fixture installers must agree with respect to the roughing-in measurements and to the total number of fixture units. A typical set of roughing-in measurements is shown in Figure 5-2.

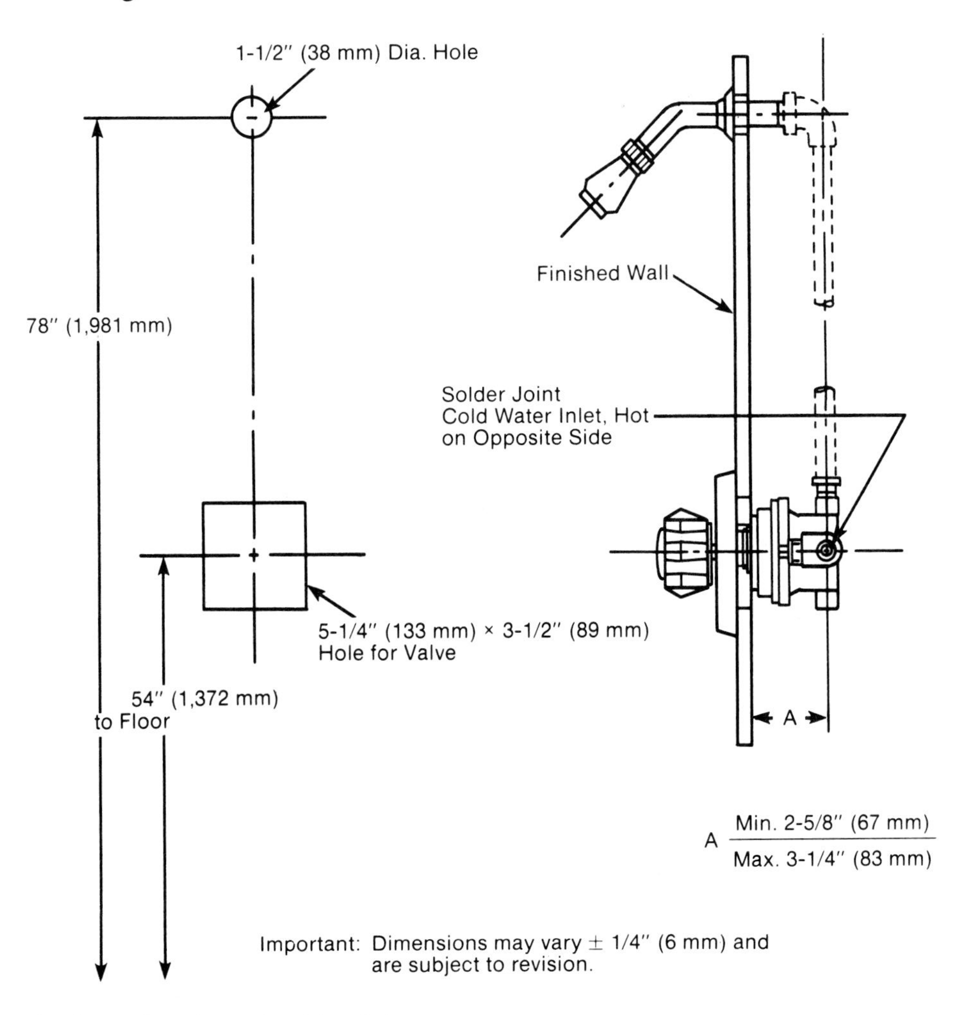

FIGURE 5-2. A typical set of roughing-in dimensions for a plumbing fixture.

When reading roughing-in measurements, particular attention must be given to references to the floor, finished or rough, or to some other point such as the bath rim. Story rod methods of sorting measurements are useful for this type of roughing-in work. If all vertical measurements can be marked on a rod, then most of the ultimate plumbing measurement decisions are under control.

Metric conversions involving materials associated with roughing-in and plumbing work pose no specific problems for the builder. Pipe and tubing sizes will remain at their present size, but will be soft-converted into approximate Metric equivalents. For instance, a 1/2" pipe will be labeled roughly as 12 mm, but will still be the traditional 1/2" pipe. Other Metric approximations of the actual British Imperial measurements for pipes and fittings are shown in Table 5-6.

TABLE 5-6 ACTUAL BRITISH IMPERIAL MEASUREMENTS OF PLUMBING PIPES AND FITTINGS AND THEIR METRIC APPROXIMATIONS

inch	mm	inch	mm	inch	mm	inch	mm
1/4	6	3-1/2	90	18	450	54	1,350
3/8	9	4	100	20	500	60	1,500
1/2	12	5	125	21	525	66	1,650
5/8	15	6	150	24	600	72	1,800
3/4	18	7	175	27	675	78	1,950
1	25	8	200	30	750	84	2,100
1-1/4	30	10	250	33	825	90	2,250
1-1/2	40	12	300	36	900	96	2,400
2	50	14	350	39	975	108	2,700
2-1/2	65	15	375	42	1,050	120	3,000
3	75	16	400	48	1,200		

5.7 ELECTRICAL SUPPLIES

Similar to the planning for a plumbing subsystem, the planning for an electrical subsystem begins with a firm list of hardware and appliance specifications. This list should again reflect occupancy needs balanced by capital and operating costs. The list can be in the form of common wiring charts or tables developed from carefully reading and marking a set of building plans. These charts are frequently organized into sections by circuits. As a practical feature; for example, a bedroom or living room may be divided into two electrical circuits or be on two different circuits, to avoid problems should one fail. This sort of decision is made up-front and is reflected in the electrical list or chart.

A logical extension of a wiring chart is to list the individual circuit power requirements. If the set of circuits is already known and sized, the size of the service panel or panels can also be determined. The planning sequence for the electrical subsystem moves from (1) hardware and appliance specifications to (2) electrical power requirements to (3) panel(s) size.

A typical service panel specification for a light building construction project might be as follows:

1. Sizing—100 amperes.
2. Circuit capacity—24 or 32.
3. Voltages—110/220.

In all cases, such measurements must comply with local and national codes.

As with plumbing, the roughing-in measurements are critical for ensuring an electrical subsystem that functions properly and for creating an aesthetically pleasing appearance. These roughing-in measurements include the following:

1. service panel for placement;
2. meter placement;
3. placement of boxes for lighting, switching, and receptacles;
4. determining relationships between electrical circuitry hardware and finished walls and ceilings;
5. determining arrangements for electrical circuitry hardware, switches, and receptacles in reference to safety and aesthetic considerations.

Local and national codes help to ensure that wiring and electrical appliance hookups are adequate and safe, but they have much less effect on the convenience and aesthetic value inherent in an electrical subsystem. The architect and the builder are responsible for creating a subsystem that ensures maximum efficiency, plus agreeable appearance within the bounds of safety codes.

The introduction of the Metric System should not cause a drastic change in construction electrical work. First of all, the current units used for measuring electricity and for determining the electrical use of appliances and so on, will be retained. The builder should know what these units are and how they apply to construction planning and application. For example,

1. One newton is the force which when applied to a mass of 1 kg gives it an acceleration of 1 m per second squared ($1\ N = 1\ kg\ m/s^2$).
2. One joule is the work done when the point of application of a force of 1 N is displaced through a distance of 1 m in the direction of the force (1 J = 1 Nm).
3. One watt is the power available when the energy of 1 J is expended in 1 second (1 w = 1 J/s). The watt is the electrical power unit and is exactly the same in both British Imperial and Metric systems of measurement. A kilowatt, the unit often utilized to express electrical use, is 1,000 w.

Other measuring units are amperes (A) for measuring voltage, ohms for measuring electrical resistance, and coulombs (c) for measuring quantities of electricity.

The builder should understand the common units of electrical measure so that he or she is fully aware of how much electricity a completed building will use during a stated period of time. Electricity is measured and charged on a kilowatt per hour basis by electric companies. The builder should be aware of the potential for electric use in a house he or she is building. For example, if a house is equipped with an electrical heating system or electric hot water heater, the builder should be able to tell the building owner the approximate cost of this facility at current rates. He or she can rely on the hourly electrical use of some typical appliances to make such determinations. By using such data, a builder should be able to develop an electric usage table, on a monthly basis, for a particular building project.

A different aspect of electrical installation which will be affected by Metric conversion is wire sizes. Wires are presently measured in a variety of ways. Diameters of electric wires are usually measured in mils (a mil is 1/1,000″) and given a gauge number to identify it. Wires are also measured according to the area they cover in circular mils (CM). A circular mil is obtained by squaring the diameter of the wire in mils. In addition, large quantities of wire are bought and sold by weight (pounds per 1,000′) and length (feet per pounds). Eventually, gauge numbers will be replaced by a straight Metric decimal measurement, such as is the case with other types of metal. Weight gauge number will be replaced by a straight Metric decimal measurement, like other types of metal. Weight and length will be measured in kilograms and meters instead of pounds and feet.

5.8 CONCRETE

Measurement pertaining to the design and control of concrete mixtures seems to be shifting rapidly from the British Imperial to the Metric System. Data presentations in the industry are now in either British Imperial/Metric or Metric only. Some of the changes in measurement from British Imperial to Metric which affect work with concrete mixtures are described in Table 5-7.

TABLE 5-7 TYPICAL MEASUREMENT CHANGES FROM BRITISH IMPERIAL TO METRIC MEASURE FOR MIXING CONCRETE

Item		Relationship
	Volumes	
Aggregate (sand/gravel)		
British Imperial:	cubic yard	1 cu yd = 0.76455 m³
Metric:	cubic meter	1 m³ = 1.30795 cu yd
Cement		
British Imperial:	1 bag (1 cu ft)	
Metric:	(see Mass)	
Water		
British Imperial:	1 gal	1 U.S. gal = 0.8327 Imperial gal
Metric:	1 l	1 cu ft = 7.4805 U.S. gal
		1 cu ft = 6.25 Imperial gal
		1,000 l = 1 kl

TABLE 5-7 TYPICAL MEASUREMENT CHANGES FROM BRITISH IMPERIAL TO METRIC MEASURE FOR MIXING CONCRETE (Continued)

Item			Relationship
	Mass		
Aggregate			
British Imperial:	1 ton		1 tonne = 1.1 tons
Metric:	1 tonne (t)		= 1,000 kg
Cement:			
British Imperial:	(see Volumes)		
Metric:	40 kg (bag)		25 bags (40 kg = 1,000 kg = 1 tonne = 1.1 tons)
	30 kg (bag of masonry cement)		33.333 bags (30 kg = 1,000 kg = 1 tonne = 1 ton)
Water			
Metric:	1 kg		1 kg of water has a volume of 1 l
	Common Tests		
Slump			
British Imperial:	inches		
Metric:	mm (increments of 10 mm)		
Test Cylinder			
British Imperial:	6″ × 12″		
Metric:	150 mm × 300 mm		
	Water/Cement Ratio (by weight)		
Set 1			
W/C Ratio	Time	**Psi* (Pounds per square inch)**	**MPa* (Megapascals)**
0.4	3 days	580	4.0
0.5	3 days	363	2.5
0.6	3 days	145	1.0
Set 2			
W/C Ratio	Time	Psi*	MPa**
0.4	28 days	4,786	33.0
0.5	28 days	3,625	25.0
0.6	28 days	2,756	19.0

*Approximate values.
**1 MPa = 145.04 psi.

For design and large volume mixing, the trend seems to be from measurements by volume to measurements by mass. In other words, ingredient materials are weighed in kilograms and metric tonnes or pounds and tons (the Metric is much more useful for this purpose) before mixing rather than being measured by volume afterwards, in terms of so many cubic feet, cubic yards, or gallons.

The data in Set 1 and Set 2 of Table 5-7 shows how the water/cement ratio affects the ultimate strength of concrete and how this will be expressed in metric measure. As will be described later in Chapter 11, the less the amount of water, the better (within limits).

Yield of concrete is another subject related to the effects of mixing water, cement, and the other ingredients of concrete. It is expressed as the volume of concrete produced per 40 kg of cement. Note that kilogram is a unit of mass, not volume.

A convincing argument for proportioning by mass rather than by volume is shown in Figure 5-3. The surface moisture in the fine aggregate (sand) used in concrete and mortar causes considerable bulking, the amount of bulking varying with the amount of moisture and the light aggregate grading. However, Figure 5-3 shows that light surface moisture will increase the mass of dry, fine sands by about 5%, but at the same time will increase the volume of the sand by as much as 38%. Consider the implications of that information for the buyers and sellers of sand and mixers of concrete, and it will not be difficult to see why measurement by mass is becoming the standard.

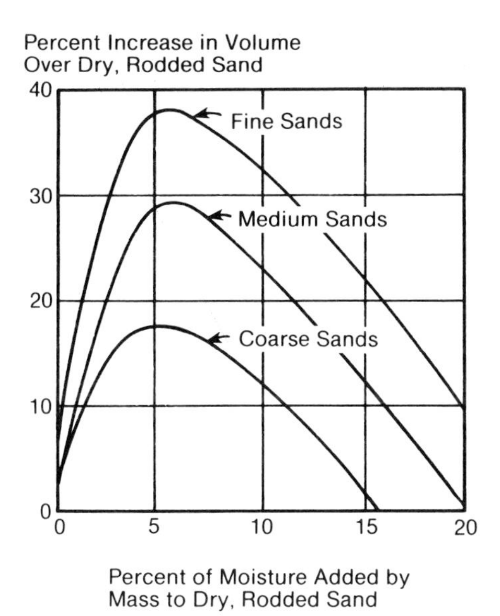

FIGURE 5-3. Graph indicating how the addition of moisture can increase the volume of dry, rodded sand.

5.9 GLASS

In light building construction work, glass is installed as fixed glazing; in a wood or metal framing, such as in a window sash or in sliding doors; or for decorative purposes such as surface-mounted mirrored feature walls.

Much like the rest of the construction industry, glass measurement is slowly being changed from British Imperial to Metric measure. Measurements for glass are thickness, width, and length with tolerances ranging from ± 1/120″ (± .008″) (± 0.2 mm) to ± 1/25″ (± .04″) (± 1.0 mm) for thickness and ± 1/25″ (± .04″) (± 1.0 mm) to ± 1/8″ (± .12″) (± 3.0 mm) for the other measurements. Glass thickness is also usually specified according to perimeter measurements as indicated by the general thicknesses shown in Table 5-8 and specifications for glass thicknesses in doors, as shown in Table 5-9. Quite simply, the greater the perimeter of a piece of glass, the greater its thickness must be.

TABLE 5-8 SPECIFICATIONS FOR GLASS THICKNESS PER PERIMETER SIZE

Minimum Glass Thickness of Inner and Outer Panes	Limiting Glass Size		
		Factory-Sealed Double Glazing	
	Sash Type or Fixed Glazing	Fused Edges	Other Than Fused Edges
.08″ (2 mm)	120.1″ (3,051 mm) perimeter	180.1″ (4,575 mm) perimeter	150.0″ (3,810 mm) perimeter
.12″ (3 mm)	168.1″ (4,270 mm) perimeter	252.2″ (6,406 mm) perimeter	210.2″ (5,339 mm) perimeter
.16″ (4 mm)	240.2″ (6,101 mm) perimeter	360.2″ (9,149 mm) perimeter	300.2″ (7,625 mm) perimeter
.20″ (5 mm)	280.3″ (7,120 mm) perimeter	420.5″ (10,681 mm) perimeter	350.4″ (8,900 mm) perimeter
.24″ (6 mm)	50 ft^2 (4.65 m^2)	112.9 ft^2 (10.5 m^2)	78 ft^2 (7.25 m^2)
.32″ (8 mm)	No limit	No limit	No limit

TABLE 5-9 SPECIFICATIONS FOR GLASS THICKNESS IN DOORS PER PERIMETER SIZE OF GLASS

Minimum Glass Thickness	Maximum Perimeter
.08″ (2 mm)	78.7″ (2,000 mm)
.12″ (3 mm)	118.1″ (3,000 mm)
.16″ (4 mm)	157.5″ (4,000 mm)
.20″ (5 mm)	177.2″ (4,501 mm)
.24″ (6 mm)	No limit

Width and length measurements are important design features. For example, when glass is installed in a sash, the assembly measurements of width and length are especially vital. This is particularly true because of the different ways in which a sash is constructed. The lower rail of a sash is often of a size different from the top rail and the two side pieces. The size of the glazing must determine the size of the sash or confusion will result. Also, some decorative glass panels have patterns which suggest or demand either vertical or horizontal placement. When specifying glass size, write the horizontal measurement first and the vertical measurement second.

If the windows and doors used in a building are factory or shop assembled, then the outside lengths and widths must be known before wall framing can begin. With those measurements determined, specific roughing-in allowances—usually about 1/4″ (6 mm) each way—may be made when doing the plate layouts and making vertical markings with the story rod for cripple studs. Roughing-in allowances provide space within the stud framing for leveling and plumbing and for weatherproofing the assembly.

5.10 INSULATION

Heat is transferred by means of conduction, convection, and radiation. In the United States the insulating values of component materials are sometimes measured by making reference to their U-values. A U-value is simply the specific rate at which a material conducts heat. However, to be in accordance with Metric procedure, the measurements of heat transfer, in terms of insulation values for light construction, are made here with reference to units of thermal resistance (the reciprocal of a material's U-value). These units are called the R-value of a material in the British Imperial System and the RSI value in the Metric System. Using such a unit is also logical because it gives the precise information needed to determine the insulating qualities of a material. A higher R-or RSI-value means that a material does not readily allow heat to pass through it and is, in fact, a good insulator.

Thermal resistance then is the reciprocal of thermal conductance. By measuring heat flow through a material, one can calculate thermal resistance. In the Metric System, thermal resistance, sometimes also called thermal insulance, is expressed in square meters times degree Celsius per watt. Traditionally, the procedure for estimating the heat flow through a material has been to assume that this heat flow per unit surface area can be expressed as $Q = U(T_i) - T_o)$, where U is the heat loss coefficient and T_i and T_o are interior and outside temperatures. The heat flow is in Btu/hr/ft² of surface. The heat transfer coefficient represents conductance, so its reciprocal is determined ($R = 1/U$) to arrive at the thermal resistance of a material per unit surface area. A constant ratio exists between the R system and the RSI system of thermal resistance: RSI = 0.176 R.

For the British Imperial System, a useful range of R-values is from 0.1 to 30.0; for the Metric System, the equivalent RSI range is 0.018 to 5.283. In other words, an R-value of 30.0 is equivalent to an RSI-value of 5.283; whereas, an R-value of 10.0 is equivalent to an RSI-value of 1.761. R-value to RSI-value equivalents are given in Table 5-10, and RSI-value to R-value equivalents are given in Table 5-11.

TABLE 5-10 R-VALUES TO RSI-VALUES

R-Value (ft² × h × °F)/Btu	0.0	0.1	0.2	0.3	0.4	0.5	0.6	0.7	0.8	0.9
	RSI-Value (m² × °C)/W									
0.0	0.000	0.018	0.035	0.053	0.070	0.088	0.106	0.123	0.141	0.158
1.0	0.176	0.194	0.211	0.229	0.247	0.264	0.282	0.299	0.317	0.335
2.0	0.352	0.370	0.387	0.405	0.423	0.440	0.458	0.475	0.493	0.511
3.0	0.528	0.546	0.564	0.581	0.599	0.616	0.634	0.652	0.669	0.687
4.0	0.704	0.722	0.740	0.757	0.775	0.792	0.810	0.828	0.845	0.863
5.0	0.881	0.898	0.916	0.933	0.951	0.969	0.986	1.004	1.021	1.039
6.0	1.057	1.074	1.092	1.109	1.127	1.145	1.162	1.180	1.198	1.215
7.0	1.233	1.250	1.268	1.286	1.303	1.321	1.338	1.356	1.374	1.391
8.0	1.409	1.426	1.444	1.462	1.479	1.497	1.515	1.532	1.550	1.567
9.0	1.585	1.603	1.620	1.638	1.655	1.673	1.691	1.708	1.726	1.743
10.0	1.761	1.779	1.796	1.814	1.832	1.849	1.867	1.884	1.902	1.920
11.0	1.937	1.955	1.972	1.990	2.008	2.025	2.043	2.060	2.078	2.096
12.0	2.113	2.131	2.149	2.166	2.184	2.201	2.219	2.237	2.254	2.272
13.0	2.289	2.307	2.325	2.342	2.360	2.377	2.395	2.413	2.430	2.448
14.0	2.466	2.483	2.501	2.518	2.536	2.554	2.571	2.589	2.606	2.624
15.0	2.642	2.659	2.677	2.694	2.712	2.730	2.747	2.765	2.783	2.800
16.0	2.818	2.835	2.853	2.871	2.888	2.906	2.923	2.941	2.959	2.976

TABLE 5-10 R-VALUES TO RSI-VALUES (Continued)

R-Value (ft² × h × °F)/Btu	0.0	0.1	0.2	0.3	0.4	0.5	0.6	0.7	0.8	0.9
	RSI-Value (m² × °C)/W									
17.0	2.994	3.011	3.029	3.047	3.064	3.082	3.100	3.117	3.135	3.152
18.0	3.170	3.188	3.205	3.223	3.240	3.258	3.276	3.293	3.311	3.328
19.0	3.346	3.364	3.381	3.399	3.417	3.434	3.452	3.469	3.487	3.505
20.0	3.522	3.540	3.557	3.575	3.593	3.610	3.628	3.645	3.663	3.681
21.0	3.698	3.716	3.734	3.751	3.769	3.786	3.804	3.822	3.839	3.857
22.0	3.874	3.892	3.910	3.927	3.945	3.962	3.980	3.998	4.015	4.033
23.0	4.051	4.068	4.086	4.103	4.121	4.139	4.156	4.174	4.191	4.209
24.0	4.227	4.244	4.262	4.279	4.297	4.315	4.332	4.350	4.368	4.385
25.0	4.403	4.420	4.438	4.456	4.473	4.491	4.508	4.526	4.544	4.561
26.0	4.579	4.596	4.614	4.632	4.649	4.667	4.685	4.702	4.720	4.737
27.0	4.755	4.773	4.790	4.808	4.825	4.843	4.861	4.878	4.896	4.913
28.0	4.931	4.949	4.966	4.984	5.002	5.019	5.037	5.054	5.072	5.090
29.0	5.107	5.125	5.142	5.160	5.178	5.195	5.213	5.230	5.248	5.266
30.0	5.283	—	—	—	—	—	—	—	—	—

TABLE 5-11 RSI-VALUES TO R-VALUES

RSI-Value (m² × °C)/W	.00	.05	.10	.15	.20	.25	.30	.35	.40	.45
	R-value (ft² × h × °F)/Btu									
0.00	0.000	0.284	0.568	0.852	1.136	1.420	1.703	1.987	2.271	2.555
1.00	5.678	5.962	6.246	6.530	6.814	7.098	7.382	7.666	7.950	8.234
2.00	11.357	11.640	11.924	12.208	12.492	12.776	13.060	13.344	13.628	13.912
3.00	17.035	17.319	17.603	17.887	18.171	18.454	18.738	19.022	19.306	19.590
4.00	22.713	22.997	23.281	23.565	23.849	24.133	24.417	24.701	24.984	25.268
5.00	28.391	—	—	—	—	—	—	—	—	—

RSI-Value (m² × °C)/W	.50	.55	.60	.65	.70	.75	.80	.85	.90	.95
	R-value (ft² × h × °F)/Btu									
0.00	2.839	3.123	3.407	3.691	3.975	4.259	4.543	4.827	5.110	5.394
1.00	8.517	8.801	9.085	9.369	9.653	9.937	10.221	10.505	10.789	11.073
2.00	14.196	14.480	14.764	15.047	15.331	15.615	15.899	16.183	16.467	16.751
3.00	19.874	20.158	20.442	20.726	21.010	21.294	21.578	21.861	22.145	22.429
4.00	25.552	25.836	26.120	26.404	26.688	26.972	27.256	27.540	27.824	28.108
5.00	31.231	—	—	—	—	—	—	—	—	—

The working ranges of values given in Tables 5-10 and 5-11 are adequate for discussing specific building insulation materials and for total building insulation values for some, particularly warmer, regions of the continent. For some inland and more northerly regions, however, the ranges should, to meet improving building standards, be increased to at least R-50 and possibly as high as R-80 in the British Imperial System and from 9 to 14 in the RSI. Working with such high

R-value or RSI-value figures is especially the case for attic or ceiling insulation applications.

5.11 BUILDING LEVELS

Although the concept of building elevations is further developed in Chapter 9, an introduction here in terms of differences in relative elevations rather than actual elevations will be helpful. When buying or constructing a building, there are many decisions to make, including those regarding appearance, location, and cost. Location is important. If the location turns out to be a bad one (resulting in chronic and severe drainage problems, for example), the best built house in the world may be a bad value.

To make decisions about a building location, one must be acquainted with the topography of the vicinity. Visual checks are necessary, but they should always be supplementary with readings of maps prepared by surveyors. These maps will carry contour lines showing like-elevations with the intervals between lines as close or as far apart as meets the needs of the builder. In most cases (especially where a site has a steep grade), intervals representing a smaller distance, 5′ (1.5 m), rather than a greater one, 25′ (7.7 m), are much more helpful and allow the builder to more carefully plot the site. These surveyor's maps should also indicate the direction and depth of sewers.

Consider the sites S_1, S_2, and S_3 in Figure 5-4. Here the intervals represent a difference of 10 units (which will differ depending on the

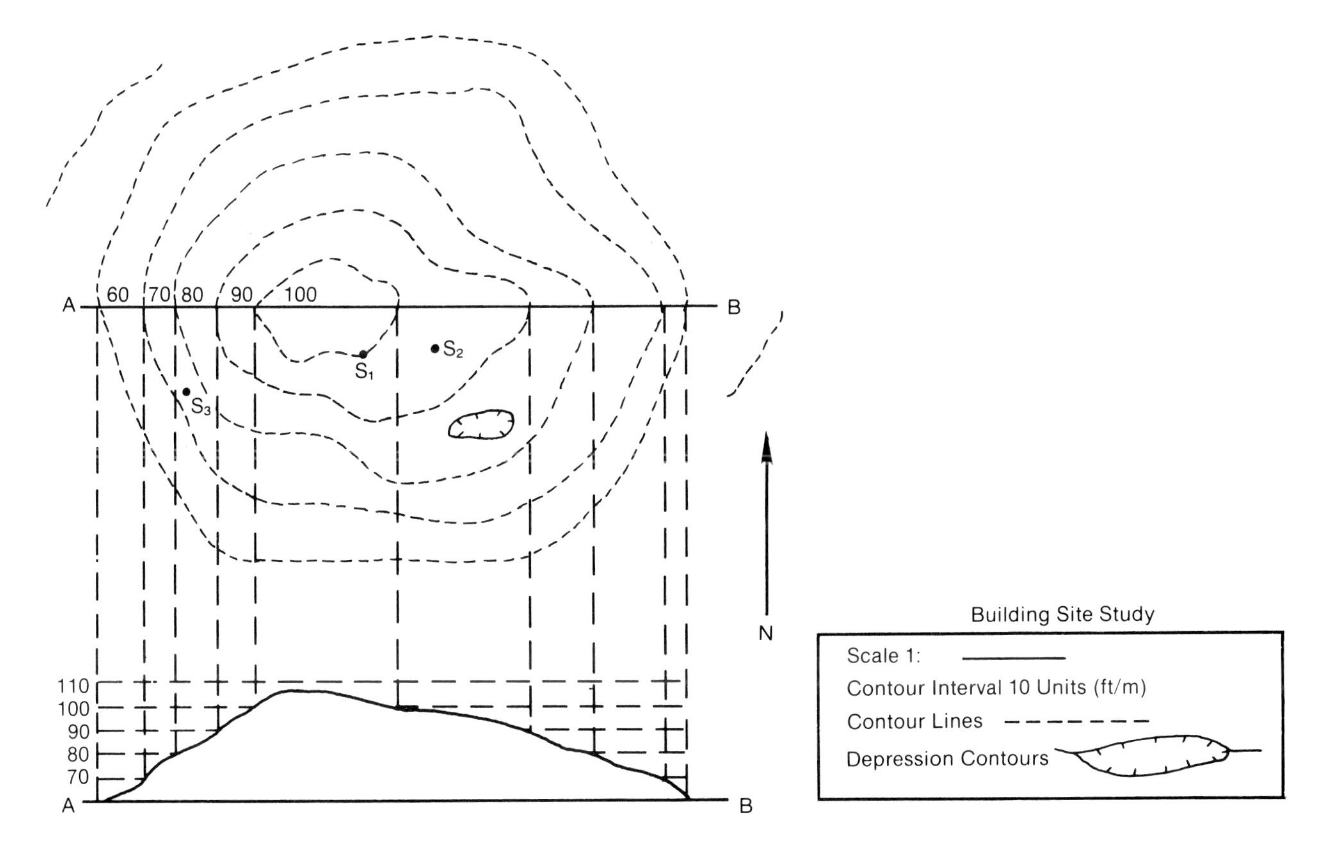

FIGURE 5-4. A building site contour map.

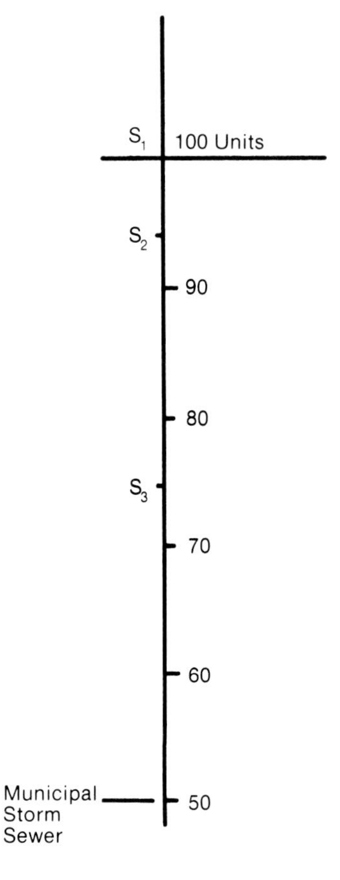

FIGURE 5-5. Surveyor's map showing various site measurement heights in relation to the municipal storm sewer level.

measurement system used). Site S_1 is higher than S_2 by less than 10 units and higher than S_3 by more than 20 units. The slopes at S_1 and S_2 seem southeast and at S_3 mainly southwest. If the contour lines are close together, the slope is steep. The rate of slope in degrees is called the grade or gradient. If the lines are far apart, the slope is relatively gentle. By looking at Figure 5-4, it becomes immediately clear that landscaping problems at S_3, where there is a steeper slope, will be quite different from those at S_2 which is on a much flatter piece of ground.

If this map were a municipal map also showing the depth and direction of the local sewer lines, the further information about the three potential building sites could be gathered and useful decisions rendered. The respective site measurements above the municipal storm sewer, as shown in Figure 5-5 would be 50 units for S_1, +44 units for S_2, and +23 units for S_3. If the measuring units used were inches, all of the sites would be at risk; they would probably be too low for hookups and proper drainage. However, if the units used were feet or meters, all of the sites would probably be safe with or without basements, especially S_1 and S_2. Also, if a two-story building with a basement were built at S_1, the ridge of the building would be higher than the hilltop. A building located at S_3, however, would be lower and somewhat protected from winds from the north.

If auxiliary contours were shown on the map in Figure 5-4 at 5 units rather than 10, then the information necessary for making a site selection would be further enhanced. When the decision has been made to proceed at a certain site, say at S_1, based on all available information, another map, such as shown in Figure 5-6, can be drawn to a larger scale with additional elevations marked.

One way of setting these building levels at a site is to use a story rod as illustrated in Figure 5-6. The story rod can be stood on a completed concrete footing and used to carry the critical measurements from above the footing. These measurements include contour marks, bench marks, first and second finished floor levels, landing levels, concrete forms heights, and window and door levels. As many detail vertical measurements as desired can be marked on the rod, which can then becomes a control device for the elevation measurements of that particular building (in this case on site S_1). The markings for site S_1 would, of course, be of little use on sites S_2 and S_3.

Another way of setting building levels, in lieu of a story rod, is to take a set of recorded measurements from a particular site reference point—a benchmark—by means of a builder's level and surveying instruments (see Chapter 9).

5.12 IN-SHOP MEASUREMENTS

Shop and site drawings are the working tools of the construction business. They are the intermediate guidelines between the draftsman's drawings and the actual construction. Shop and site drawings are also frequently used to make adaptations to the original plans when actual conditions on a site do not conform precisely to those plans.

5.12.1 Shop Drawings

Shop drawings are done with shop tools on plywood surfaces or building paper, rather than with drafting instruments. With sharp, hard-leaded pencils, awls, scribers, bevels, and straightedges (see Chapter 2), one can make full-sized detail drawings or scaled drawings. Scaling is usually controlled by the given graduations on a framing square or roofing square.

Shop or site drawings must be carefully done to exactly, not approximately, solve construction problems. They help to determine the following:

1. cutting measurements, often in the form of cutting lists, to given tolerances, especially for cabinetwork and stairway construction;
2. angular measurements for compound cuts, hoppers, purlins, windblocks, rafters, and so on;
3. form sizes, to produce a given sized product; and
4. construction details, such as for effective jointing, moulding, and so on.

A shop drawing should be able to tell a worker exactly how certain work will be done; the draftsman's drawings, on the other hand, are often more general in describing what is to be accomplished in terms of design.

A traditional British Imperial measure square is useful for 1/12 full-scale drawings where 1″ = 1′ and 1/12″ represents 1″. Some squares have 1-1/2″ equalling 1′, in which case a 1/8 full-scale drawing is produced. Full-scale drawings of wood joint details, for example, may be measured to a tolerance of ±1/64″. A shop drawing of a 24′ × 36′ building drawn 1/12 full-scale would measure 24″ × 36″. A common rafter of line length 19′7″ would measure 19-7/12″. The accuracy of this measurement would relate to the 1/12″ graduations, not to the 1/64″ tolerance already mentioned for full-scale drawings.

Of course, a Metric square lends itself to making 1/10 full-scale drawings. For example, 100 mm on a drawing might represent 1,000 mm of a building measurement. A building 8,540 mm × 10,980 mm could be drawn 854 mm × 1,098 mm. Full-scale drawings of construction details may be accurate within a tolerance of ±1 mm or ±2 mm, depending on the graduations of the square used.

Common examples of shop drawings are small-scale drawings of building sites, buildings, and roof framing members and full-scale drawings of intricate construction details on stairs, the cornice, and particularly cabinet joints, which describe how the construction will be done. In other words, if a person makes a scaled drawing of a project, that person should know precisely how to actually construct it. A list of preferred scales for building drawings is shown in Table 5-12.

When taking a measurement with a framing square, be careful of parallax error, caused by the thickness of the square. Always read and mark measurements "straight down." Use a sharp pencil and a clean, bright surface, when possible, on which to mark. By being careful one may stay within a tolerance of ±1/16″ or ±1 mm.

Careful measurements with dividers or calipers are important when stepping off a story rod for stairs (Figure 5-7). If there is an error in the

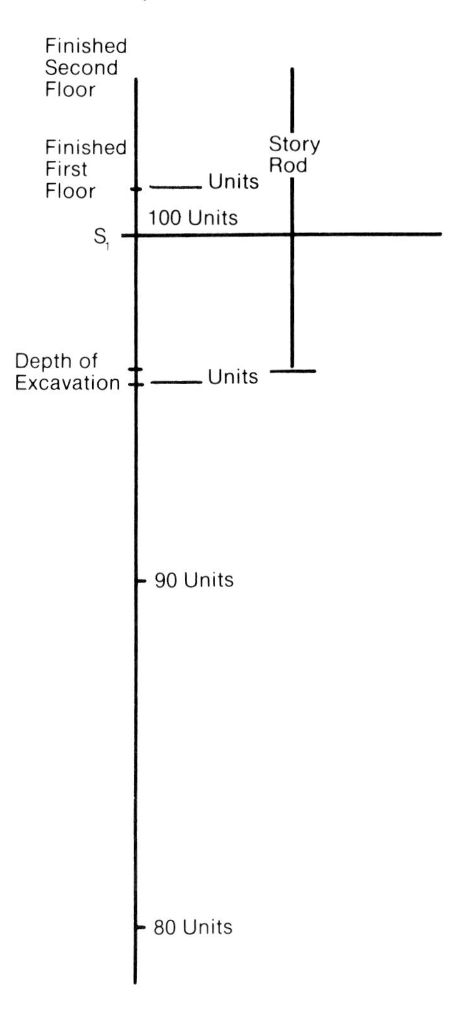

FIGURE 5-6. Using the site drawing and the story rod technique to set various building levels.

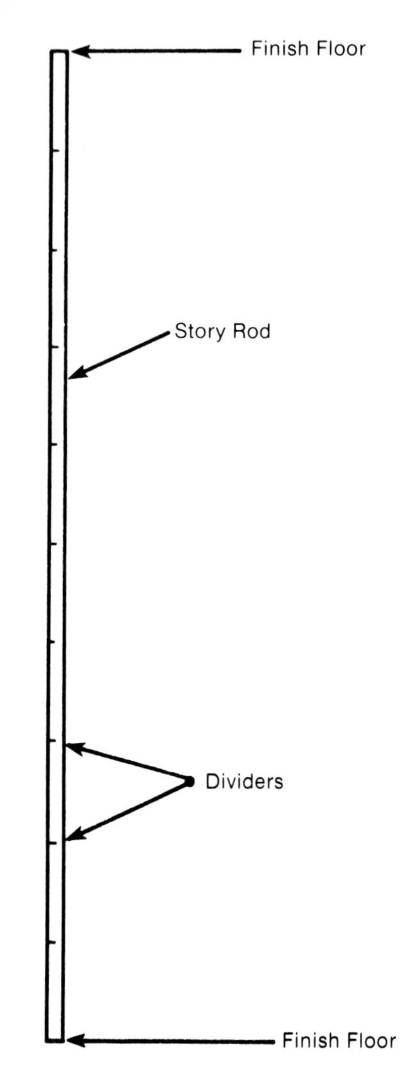

FIGURE 5-7. Stepping off a story rod for stairs using dividers.

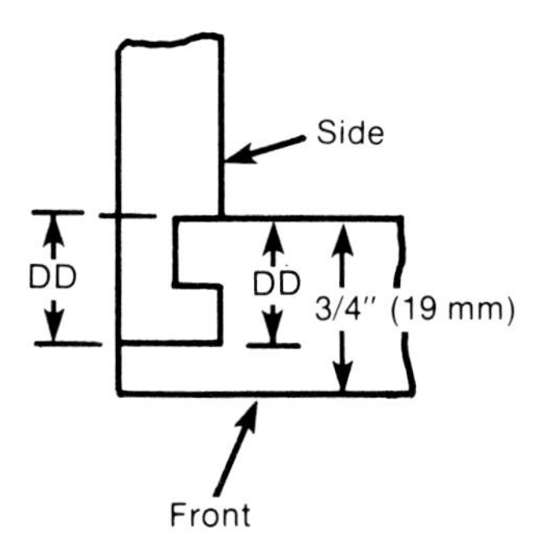

FIGURE 5-8. Joint details from a layout board can be drawn to a larger scale to ensure proper visual inspection and accuracy.

setting, it will be multiplied by the number of risers. If the error is 1/16″ or 1 mm, then for 11 risers the aggregate error will respectively be 11/16″ or 11 mm.

When preparing a shop layout board for cabinet construction, equal care should be given to marking measurements from a steel tape. All measurements are marked on the board as actual cutting measurements. Lengths, widths, and depths are all given. For a really accurate layout board, joint details should be included, although they may also be detailed to a different scale (Figure 5-8). From such a set of measurements, a cutting list can be written in preparation for "breaking out" materials.

Of course, not all measurement tolerances need to be the same. The margin of error for rougher cuts can be greater than for cabinetry, stair construction, and joinery. For example, when precutting wood framing stock (rafters, trusses, studs, etc.), tolerances may be slightly less exact, say ±1/8″ or ±2 mm. Nevertheless, the tolerance level should always be established on each individual site for all types of work before any cutting begins.

5.12.2 Moisture Level Readiness

Generally, when working with unfinished wood, one should expect changing moisture content to cause variations in stock measurements. Usually thickness and width measurements change proportionately more than lengths. Moisture levels and the wood measurements they affect are, for the most part, of more concern to cabinetmakers, millworkers, painters, and wood finishers than to framers. However, testing instruments are available for measuring moisture content whenever excessive moisture could throw off the measurement tolerance level. Such testing removes some of the guesswork with respect to when materials should be worked. It is important that moisture level standards should be recognized both by persons writing material specifications and by those who work with wood.

5.12.3 Drawing and Reading a Diagonal Scale

A British Imperial measure or Metric diagonal scale, carefully drawn, may be used to measure to hundredths of a unit and to obtain a reading with dividers to two decimal places. A diagonal scale can be helpful for a wide variety of fine tolerance construction measurement. This type of shop scale should be drawn to a convenient number of units in length, with one additional unit divided into 10 equal spaces (Figure 5-9).

For an easy reading of the scale, number the spacings. There should be 10 of each. Draw the diagonal lines 0-1, 1-2, and so on. Observe the set of data added to line "8-9" in Figure 5-9. When reading across, one dot is one-tenth of one-tenth of a unit "short" of the next one down. On the fourth line down, the reading for the example measurement is 2.34 units. On the sixth line down, the reading is 2.56 units.

TABLE 5-12 PREFERRED SCALES FOR BUILDING DRAWINGS

Drawing	Recommended Scales	Use	Former Scales (Ratios)
Block plan	1:2,000 1:1,000 1:500	To locate the site within the general district.	1″ = 200′ (1:2,400) 1″ = 100′ (1:1,200) 1″ = 40′ (1:480)
Site plan	1:500 1:200	To locate building work, including services and site works, on the site.	1″ = 40′ (1:480) 1/16″ = 1′ (1:192)
Sketch plans General Location drawings	1:200 1:100 1:50	To show the overall design of the building. To indicate the juxtaposition of rooms and spaces and to locate the position of components and assemblies.	1/16″ = 1′ (1:192) 1/8″ = 1′ (1:96) 1/4″ = 1′ (1:48)
Special area Location drawings	1:50 1:20	To show the detailed location of components and assemblies in complex areas.	1/4″ = 1′ (1:48) 1/2″ = 1′ (1:24)
Construction details	1:20 1:10 1:5 1:1	To show the interface of two or more components or assemblies for construction purposes.	1/2″ = 1′ (1:24) 1″ = 1′ (1:12) 3″ = 1′ (1:4) Full size (1:1)
Range drawings	1:100 1:50 1:20	To show, in schedule form, the range of specific components and assemblies to be used in the project.	1/8″ = 1′ (1:96) 1/4″ = 1′ (1:48) 1/2″ = 1′ (1:24)
Component and assembly details	1:10 1:5 1:1	To show precise information of components and assemblies for workshop manufacture.	1″ = 1′ (1:12) 3″ = 1′ (1:4) Full size (1:1)

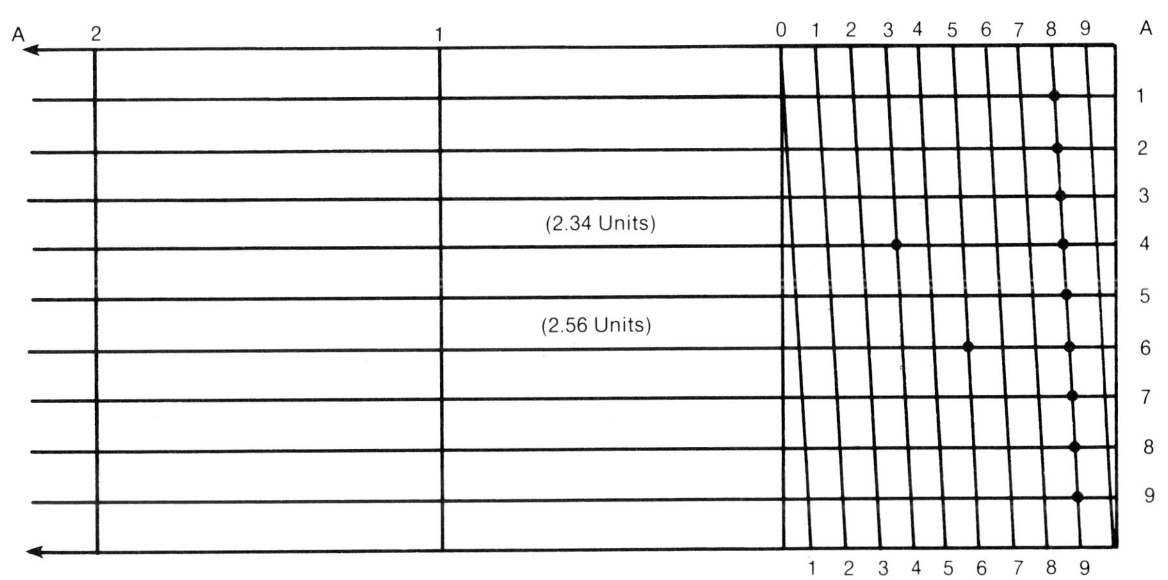

FIGURE 5-9. A typical diagonal scale.

5.13 MEASUREMENT RELATIONSHIPS

A common understanding of definitions and terms is extremely important for construction work. When all specific definitions can be agreed upon in advance, then far better communications will take place between architects, draftsmen, and in-shop and jobsite workers. Some measurement terms that are very common to light construction work are pitch, slope, slope triangle, equilateral triangle, equiangular triangle, isosceles triangle, hexagon, square, grade, elevation, bearing, and ratio. Most of these are common math or measurement terms and need not be elaborated on here. Some of the terms more specifically related to construction will, however, be further explained.

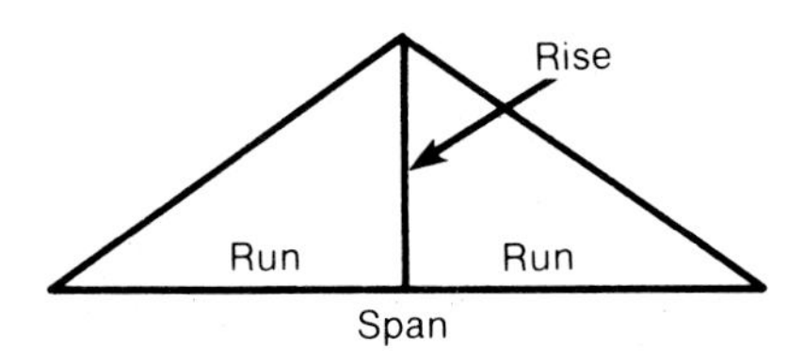

FIGURE 5-10. Pitch = rise/span.

Pitch is usually defined as the relationship of vertical rise to horizontal span (Figure 5-10). For example:

$$\text{Pitch} = \frac{\text{Rise}}{\text{Span}}$$

$$\frac{1}{3}\text{P} = \frac{8'\ (2.4\ \text{m})}{24'\ (7.3\ \text{m})}$$

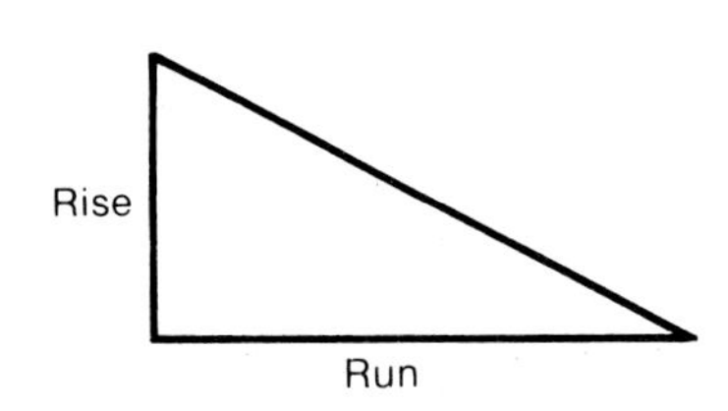

FIGURE 5-11. Slope = rise/run.

Slope, on the other hand, is the relationship between rise and run (Figure 5-11). In this case,

$$\text{Slope} = \frac{\text{Rise}}{\text{Run}}$$

$$\frac{2}{3}\text{S} = \frac{8'\ (2.4\ \text{m})}{12'\ (3.7\ \text{m})}$$

A slope triangle simply indicates where the slope formula is applicable. Figure 5-12 shows two slope triangles. If the rise of slope triangle A is 8 units and the run 16 units and if the run of slope triangle B is 8 units, then

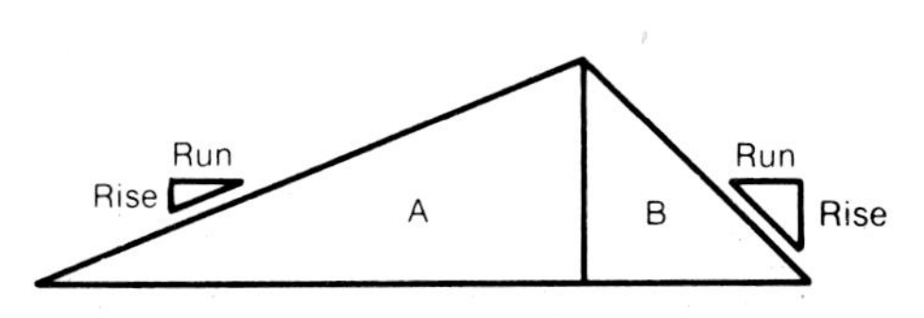

FIGURE 5-12. Typical slope triangles.

$$\text{Slope A} = \frac{\text{Rise}}{\text{Run}} = \frac{8}{16} = \frac{1}{2} = 0.5$$

$$\text{Slope B} = \frac{\text{Rise}}{\text{Run}} = \frac{8}{8} = 1$$

Because the term *pitch* is difficult to apply when a roof is not regular, as for example in Figure 5-12, the term is falling into disuse. Obviously, the term *slope* can be applied to regular or irregular roofs because it is a relationship involving the angles of only one section of the roof at a time.

The term *grade* pertains to the rate of ascent or descent of a road or the trench for a sewer line. It also has a more general use with landscapers who must assume surface drainage by grading or in a sense sloping the ground from a given point on a building. Grades are measured as a percentage and in degrees, in the latter case the same as an angle in a triangle (Figure 5-13).

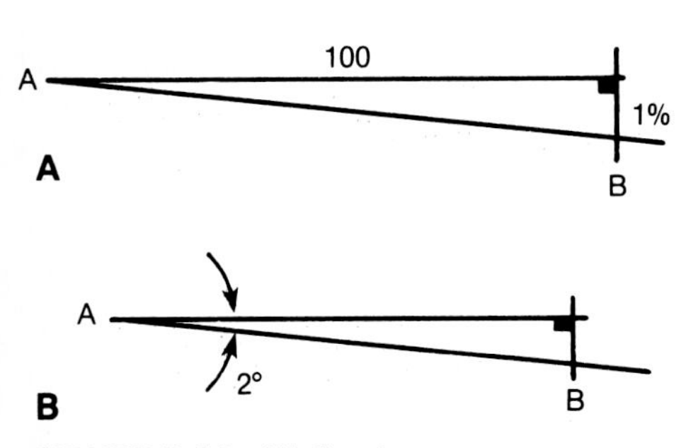

FIGURE 5-13. (A) Grade as a percentage and (B) grade measured as the degrees of an angle.

The expression of slope generally encourages the use of nondimensional ratios, as shown in Table 5-13. Such expressions can be applied not only to roads and sewer line trenches, but also to roofs, ramps, and stairs.

TABLE 5-13 EXPRESSION OF SLOPE

Ratio ($\frac{y}{x}$)	Angle	Percentage (%)
Shallow Slopes		
1:100	0° 34′	1
1:67	0° 52′	1.5
1:57	1°	1.75
1:50	1° 09′	2
1:40	1° 26′	2.5
1:33	1° 43′	3
1:29	2°	3.5
1:25	2° 17′	4
1:20	2° 52′	5
1:19	3°	5.25
Slight Slopes		
1:17	3° 26′	6
1:15	3° 48′	6.7
1:14.3	4°	7
1:12	4° 46′	8.3
1:11.4	5°	8.75
1:10	5° 43′	10
1:9.5	6°	10.5
1:8	7° 07′	12.5
1:7.1	8°	14
1:6.7	8° 32′	15
1:6	9° 28′	16.7
1:5.7	10°	17.6
1:5	11° 19′	20
1:4.5	12° 30′	22.2
1:4	14° 02′	25
Medium Slopes		
1:3.7	15°	26.8
1:3.3	16° 42′	30
1:3	18° 26′	33.3
1:2.75	20°	36.4
1:2.5	21° 48′	40
1:2.4	22° 30′	41.4
1:2.15	25°	46.6
1:2	26° 34′	50
1:1.73	30°	57.5
1:1.67	30° 58′	60
1:1.5	33° 42′	67
1:1.33	36° 52′	75
1:1.2	40°	84
1:1	45°	100
Steep Slopes		
1.2:1	50°	119

TABLE 5-13 EXPRESSION OF SLOPE (Continued)

Ratio ($\frac{y}{x}$)	Angle	Percentage (%)
Steep Slopes		
1.43:1	55°	143
1.5:1	56° 19′	150
1.73:1	60°	173
2:1	63° 26′	200
2.15:1	65°	215
2.5:1	68° 12′	250
2.75:1	70°	275
3:1	71° 34′	300
3.73:1	75°	373
4:1	75° 58′	400
5:1	78° 42′	500
5.67:1	80°	567
6:1	80° 32′	600
11.43:1	85°	1143
∞	90°	∞

The elevation of an object is usually measured vertically from a local reference point known as a bench mark. Surveying instruments, levels, and story rods are used to determine as many measurements as needed from a given bench mark. Bench marks are simply common reference points for making vertical measurements.

Builders must frequently locate building sites from a given map. In some instances, bearing of a site (essentially its points on a compass) may be needed to find the correct spot. A bearing reading first lists whether the site lies to the north or south and then how many degrees it is east or west of the north-south axis (Figure 5-14).

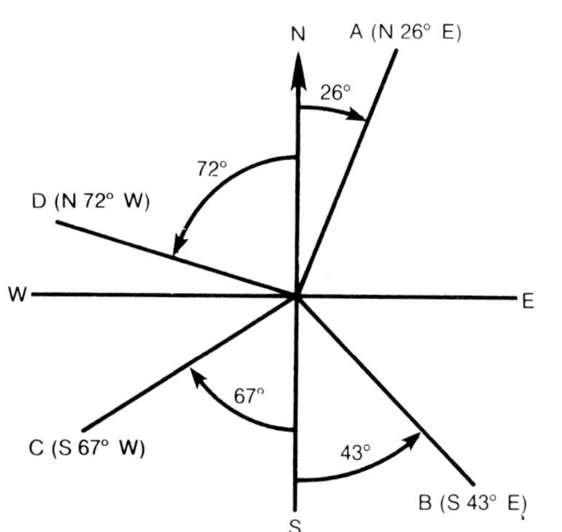

FIGURE 5-14. Example of bearing used to locate a building site on a given map.

5.14 RESEARCH

1. Obtain the building plans and building supply list for a small light construction job (for example, a project similar to the one-car garage discussed in Chapter 1). Determine how a breakdown of materials was decided upon. Consider the implications of planning the project in a Metric frame of reference rather than British Imperial measure. How will this affect the building supply list and the way materials are ordered and used?
2. Consider possible eventual hard-conversion standards in Metric for lumber sizes. What might be good common lumber sizes for strictly the Metric System? Make reference to the Metric modules for building, future insulation requirements, possible future framing and building styles (with increased truss assembly use, for example), passive solar energy possibilities, as well as other engineering, cost, and aesthetic factors which could determine the Metric sizes. What are current standard Metric lumber sizes elsewhere in the world?
3. Study several building plans. Convert insulation U-values or R-values for each building—floors, walls, roofs, and so on—into

RSI-values. Consider possible RSI measurement standards for building insulation, taking into consideration current and potential future regulations governing insulation.

4. Obtain building level data for a particular site. Incorporate the following steps:
   - From the appropriate municipal office obtain a map which identifies the site that you wish to study. Obtain the legal description of the property.
   - From the engineer's office, request detailed information for that property with respect to levels of sewer lines, roads, curbs, and service lines such as water, gas, and power.
   - Using an appropriate scale, show on a drawing how the different elevations relate for that site.
   - Using the same scale, show how those elevations relate to the proposed (or existing) building including excavation, footings, outside door sill, and finished floor.
   - Identify a bench mark on the property. Show its position on the scaled drawing.
   - Obtain a long story rod. Assuming that it would stand on top of the building footing, mark on it the bench mark, the outside door sill, and the finished first floor.
   - Roll the stick and mark on the adjacent face the bottom and top of the concrete form. Add other construction details if you wish. Be sure that the marks relate positively to one another.
   - As you conduct the study, make notes with respect to how measurements are taken in the field. Determine whether there is a universal, coordinated system, and consider improvements in present measurement practices.
5. Using building terminology, label shop drawings that describe developments regarding cuts for purlins, windblocks, and roof sheathings. Be specific about types of joints and names of angles. Incorporate the following steps:
   - Define terms.
   - See typical on-the-job assemblies.
   - Conduct research with respect to drawing practices.
   - Make the shop drawing complete with sets of specifications, including slope(s).
   - Pick up the developed angular measurements and transfer them to standard stock.
   - Consider how compound cuts would be made.
   - Consider how a table of compound cuts might be developed.
6. Prepare a layout board for an existing small cabinet from which a cutting list of materials might be created. Incorporate the following steps:
   - Refer to Figure 5-8 and to Chapter 22.
   - Select a small cabinet to measure. Make a sketch of it.
   - Prepare a layout board.
   - Mark the necessary measurements on the board. Do not dimension.
   - Organize and take-off a cutting list.
   - On a scale drawing, show how the materials would be broken out if plywoods were to be used or rough stock machined.

- Keep the layout board for reference purposes. Come back to it in a week or so to determine whether you can read it without seeing the actual cabinet.

7. Discuss with someone who is knowledgeable in the field of mathematics the subfields that pertain to the construction industry. Be sure to inquire about descriptive geometry and its many applications. Find out how mathematics can expedite better, more efficient building practices.

5.15 REFERENCES

5.15.1 Granet, Irving. *Strength of Materials for Engineering Technology.* Virginia: Reston Publishing Company, Inc., 1980.

5.15.2 Hardie, Glenn M. *Construction Contracts and Specifications.* Virginia: Reston Publishing Company, Inc., 1981.

5.15.3 Herubin, Charles A. and Theodore W. Marotta. *Basic Construction Materials.* Virginia: Reston Publishing Company, Inc., 1981.

5.15.4 Kratfel, Edward R. and George R. Drake. *Modern Shop Procedures.* Virginia: Reston Publishing Company, Inc., 1974.

5.15.5 Maguire, Byron W. *Construction Materials.* Virginia: Reston Publishing Company, Inc.

5.15.6 Scharff, Robert. *Math for Construction, Workshop, and the Home.* New York: Harper & Row.

5.15.7 Sullivan, James A. *Plumbing Installation and Design.* Virginia: Reston Publishing Company, Inc., 1980.

6

FORMS OF ENERGY

Different forms of energy are available for use in light building construction. In view of the options, the builder must decide early in the planning stage which energy type to use since the building design can be greatly affected by this decision.

The builder must be knowledgeable with respect to what forms of energy are now available, how they are delivered for use, and how each may safely be used both during construction and in the completed building. He or she must be fully aware of trends toward change in the types of energy used.

As energy costs rise, these changes take place quickly. Shelved but tested systems may become economically viable or building structures may be modified to reduce need-levels for energy. Technologies with respect to building subsystems become sophisticated and interrelated. Electrical energy, for example, is used directly for lighting and for heating or in low-voltage forms to activate natural gas furnaces. But natural gas furnaces may also operate without external electrical controls.

Another example is the development of technologies to use solar energy in passive or active subsystems.

Energy is the potential force which gives one the capability to do mechanical work or to produce a change in temperature (that is, to heat or to cool).

6.1 SPECIFIC OBJECTIVES

On reading the chapter and completing the suggested research, the reader should be able to

1. identify some of the forms of energy available for use in buildings and/or building sites,
2. say generally how those forms of energy are delivered or made available on site,
3. encourage open minds and creative thinking with respect to use of energy forms,
4. indicate how energy is measured and ways of making cost comparisons of the different forms,

5. suggest some of the environmental considerations when deciding which energy forms to use in a building,
6. point out some of the practical considerations when deciding which energy forms to use,
7. encourage research and further reading with respect to vocational choices and to eventually give leadership in the field.

6.2 FORMS OF ENERGY USED IN LIGHT CONSTRUCTION BUILDING

Energy is available at building sites as solar, wind, water, wood, and possibly coal. It is brought to a site as electricity, gas, oil, and coal or wood. The following energy use points apply to the light building construction industry:

1. Energy directly from the sun is used in passive or active systems. (Nearly all energy available on earth ultimately derives from solar radiation.)
2. Wind power is sometimes used to operate pumps and generators.
3. A local water supply is the motive power for machines, including generators.
4. Wood and coal are control-burned to heat buildings.
5. Public and private power companies supply electrical power to sites. Usually it is in the form of alternating current at about 110 volts and about 220 volts. It is locally transformed from high-voltage transmission lines.
6. Portable generators, often diesel driven, provide electrical energy in the form of alternating or direct current. Generators are available in many sizes and may be the primary or the alternate (back-up) supply.
7. A fossil fuel known as propane is brought to a site and stored in tanks. The fuel is used in expanded gaseous form to heat buildings and water and to cook and to cool.
8. Fossil fuels in natural gaseous form are piped to building sites by utility authorities and used to heat buildings and water, to cook, and to cool. The mechanics of handling and using natural gas are somewhat different from propane. Care must be taken not to confuse procedures.
9. Producer gas from coal is distributed through a piping system and used for heating, cooking, and cooling.
10. Oil, a liquid fossil fuel, is delivered and stored in tanks on site. It is piped to specially designed furnaces to heat buildings and water (Figure 6-1).

Builders in the light construction industry must study the source of the energy supply, the delivery methods, and comparative costs. Consideration should also be given to environmental effects and to the technologies associated with each. Always remember, user safety is of first importance.

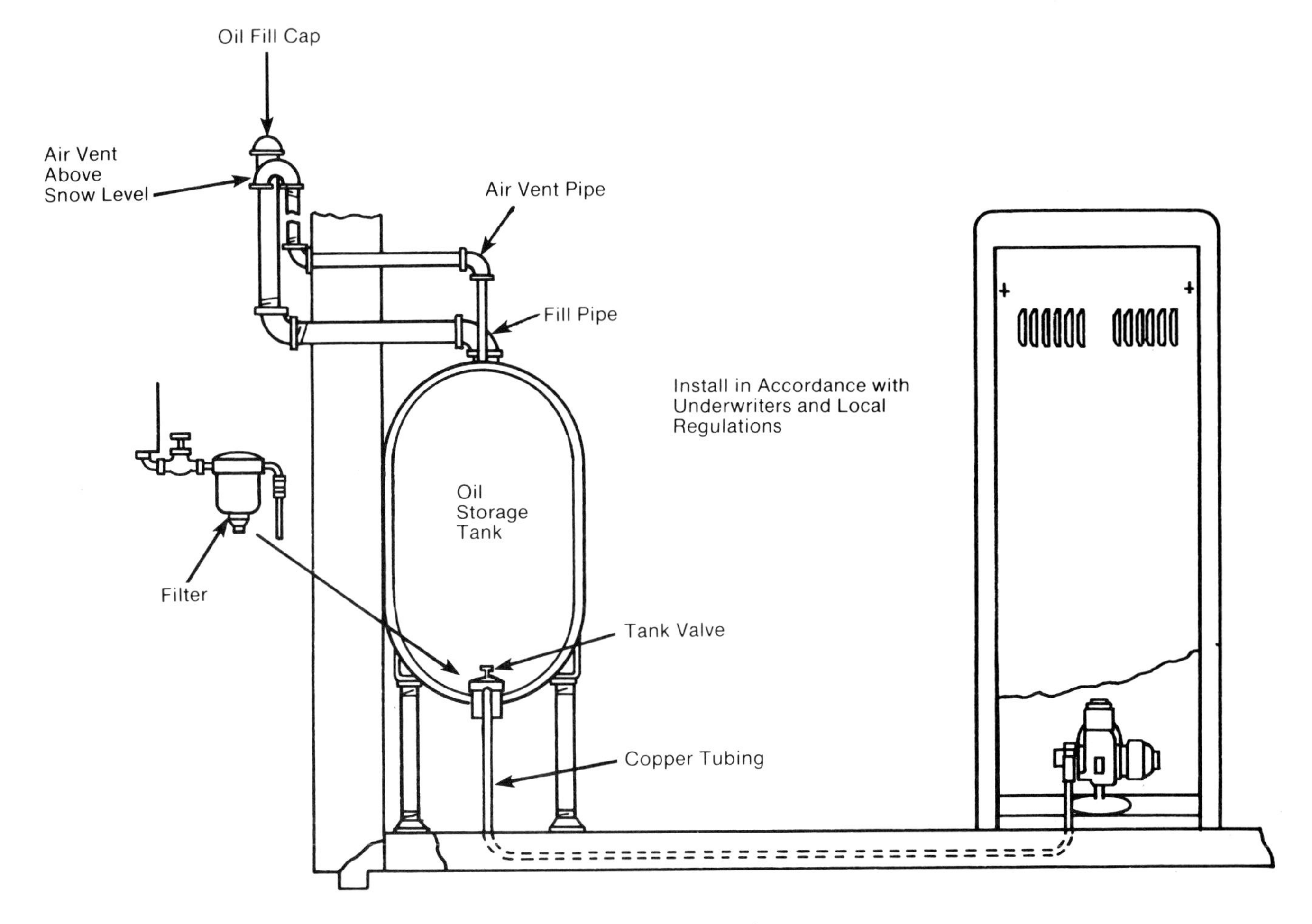

FIGURE 6-1. Typical oil storage and piping system for an oil burning furnace.

6.3 ENERGY MEASUREMENT

The quantity of energy can be expressed by a variety of equivalent units which apply to the mechanical, heat, or electrical forms of energy. Imperial units are the British thermal unit (Btu), applicable to heat energy; the kilowatt-hour (kwh), applicable to electrical energy; and the foot-pound-force (ft-lb-f), applicable to mechanical energy. In the Metric System, all forms of energy or work are expressed with one comprehensive unit, the joule (J), and its multiples and submultiples.

Table 6-1 is an energy conversion table in British Imperial units. When reading it and wishing to compare British Imperial and Metric energy units, keep in mind that 1 Btu is nearly the equivalent of 1 kJ:

1 Btu = 1.055 kJ (Kilojoule)
1 kJ = 0.948 Btu

Heat energy is the major energy form utilized in building technology. Basically, it applies to heating, cooling, and cold storage systems and insulated assemblies.

TABLE 6-1 ENERGY CONVERSION TABLE

To convert from	to	Multiply by
kilowatt hour	Btu	3,412.8
1 ton bituminous coal	Btu	25,200,000
1 bbl (35 Imperial gallons crude oil)	Btu	5,803,000
1 gallon gasoline	Btu	150,000
1 gallon #2 fuel oil	Btu	166,600
1 cubic foot natural gas	Btu	1,030
1 Mcf natural gas	Btu	1,030,000
1 therm natural gas	Btu	100,000
1 Btu natural gas	kwh	0.000293

6.4 ENVIRONMENTAL CONSIDERATIONS

Decisions with respect to which form of energy to use at a site may have environmental implications. Many builders may have feelings about the use of certain forms, but are unable to change until different technologies are developed.

6.4.1 Electricity

Electricity is generated in hydroelectric plants, fossil fuel-fired central stations, and nuclear power plants. It is carried by high-tension power lines which cross the countryside and in so doing may change the nearby environment for plants, animals, and humans. Electrical energy is delivered locally at transformed low voltages for safer usage by humans. It is clean to use in that the pollutants are produced at the generating plants and along the transmission lines. For portable generators, the pollutants are usually emissions from diesel engines.

6.4.2 Fossil Fuels

Fuel oil, natural gas, propane, and producer gas are some of the fossil fuels used at building sites. They are obtained from oil and gas wells, and, in the case of producer gas, from coal. Oil and natural gas may be moved to the general area by pipeline, rail, or truck and to the site by truck (for oil) or smaller pipelines (for natural gas). Propane is carried by rail or truck and to the site by truck.

Natural gas is usually metered as it is piped directly to a furnace or stove. Propane is stored in large tanks outside a building and then piped to heating or cooling units. Oil is also stored in tanks outside a building and piped to appliances. Producer gas is handled much the same as natural gas.

Each of the fuels is handled with great care. Carefully stated regulations are enforced in each community. The products of combustion determine the designs of vents and flues. Flash points of each determine sets of safety rules.

6.4.3 Wood

Softwoods and hardwoods are burned in stoves, furnaces, and fireplaces. Firewood is locally cut, stored, and split and is used more as prices of fossil fuels rise. Stove design is changing rapidly to more efficient units which need recharging fuel less often.

Wood fuels vary in respect to energy potential and to residues. Lined chimneys are needed to withstand accumulations of hazardous tars which may ignite to cause chimney fires. Emissions tend to be pollutants which need to be carried away by chimneys higher than the buildings.

Care must be taken to supply fresh air to the fire. In this regard, attention needs to be given to fireplace design to prevent unpleasant cold drafts in rooms when such air is being supplied. Preheating of outdoor air rather than using room air seems to be one solution to the problem. Electric fans are another means of moving warm air to the room rather than having most lost through the chimney.

6.4.4 Wind

Windmills have long been used to pump water, grind grain, and generate electricity. Some attempts have recently been made to increase the usefulness and efficiency of wind-driven generators.

6.4.5 Solar

The sun's energy reaches the earth in the form of radiation, including ultraviolet rays, infrared radiation, and visible light. Infrared radiation and visible light are the components of solar radiation which provide most of the thermal energy we receive from the sun. This is commonly referred to as solar energy.

The earth gains about 2,000 times as much energy from the sun in one day as mankind will obtain from fossil fuels in the next 20 years. Solar energy is there to be used and does not create pollutants.

6.5 PRINCIPLES AND APPLICATIONS

The following sections further outline the principles and applications of the major energy systems used in light building construction.

6.5.1 Electrical Installations

Qualified persons are designated to perform specific installation duties. On behalf of an electrical utility authority, power electricians connect high-voltage energized lines to a local transformer. Such persons may also bring lines from the transformer to the service entrance of a building. Construction electricians install the roughing-in circuitry and finish work. National and regional electrical codes and regulations are specific as to which duties may be performed by the construction and power electricians and as to standards of installation.

For reasons of public safety, electrical materials and equipment are approved by national standards associations or laboratories. Each piece is stamped to indicate approval.

The size of a service entrance panel is determined from the building plans. Power requirements are known when the purposes of the building are known and specific needs assessed. Provisions are made for lighting circuits, power circuits (for machines), and low-voltage communications circuitry. Future electrical needs must be considered when designing the original system.

Inbuilding voltages range from 24 for communications and heating controls to 110 for lighting and appliances to 220 for heating and large motors. The size (amperage) of an electrical service panel may be 100 for a building the size of a modest house. Circuits are continuity tested by qualified persons before being energized.

6.5.2 Fossil Fuels

Coal, oil, and natural gas production is monitored with great care. Strip (surface) mines and underground mines yield coals which differ with respect to hardness and energy potential. Lignite, bituminous, and anthracite are major coal types. The use of coal as a small building fuel has dramatically declined. When coal is delivered to a site for trench line thawing or for building heating, it needs protection from weather and storage where coal dust will not be a problem.

Combustion gases produced by burning coal need proper venting. Adequate oxygen supply is required for complete combustion of coal and prevention of unwanted gas buildup and minor explosions. Care is taken when adding coal as fuel so as not to smother the existing flame and to keep the fire burning.

The typical fuel oil furnace functions with a burner and an electric motor. Components are power supply, electric motor (with transformer and control box), air control, fuel oil pump, and an air blast tube fitted with a nozzle of specific size. A chimney damper allows products of combustion to escape while keeping most of the warm air in the furnace system.

A qualified person checks a furnace twice a year: for soot deposits on the firepot, heat exchanger and pipes; for stack temperatures in relation to speed of circulating fan and oil nozzle size; for stack soot (an efficient furnace will produce little or no stack soot); for carbon dioxide (CO_2) level to measure furnace thermal efficiency; and for adjustments of check draft and barometric damper. The following periodic inspections are made with respect to possible hazardous conditions: flue pipe connections, character of the flame, oil valve at the tank, oil filter and gasket, stack controller, chimney and base, burner, electrodes, oil nozzle, oil line, fan assembly, and air filter. Consult local authorities with respect to proper stack temperatures and CO_2 levels.

Gas furnace types will be liquid petroleum (probably propane) and natural gas. The two function differently but have in common automatic shut-down devices in case of power or fuel shortages. Other components are gas supply pipe, main shut-off valve, main gas line to burners, pilot gas line and air intake ports. The pilot light is usually on continuously and ignites the flame when the burner switches on. A thermostat calls for heat and for switching off.

Service persons may be qualified to handle liquid petroleum or natural gas or both types. Although building owners may be competent to start up and shut down a furnace and to change air filters and lubricate fan motors, service persons clean the main burners and adjust the flame and clean and adjust the pilot safety cutoff. They also check or replace a thermostat and measure stack temperatures. The thermal efficiency of a well-functioning gas furnace should be about 78%, so stack temperatures should not be too high.

6.5.3 Heat Pumps

The use of heat pumps in residential and light building applications is increasing. Heat pumps function as refrigerator machines which absorb heat from a low temperature source (usually the outside air) and deliver it to the building interior where it is released. A liquid refrigerant serves to absorb heat from the desired area, and an electric compressor is used to raise the pressure of the refrigerant so it will more easily release this heat at the proper time (Figure 6-2A).

Because heat pumps are equipped with valves that can reverse the flow of refrigerant, the heat pump can also act as an air-conditioner in warmer weather. When operated in this mode, the heat pump absorbs heat from the building interior and releases it to the outside (Figure 6-2B).

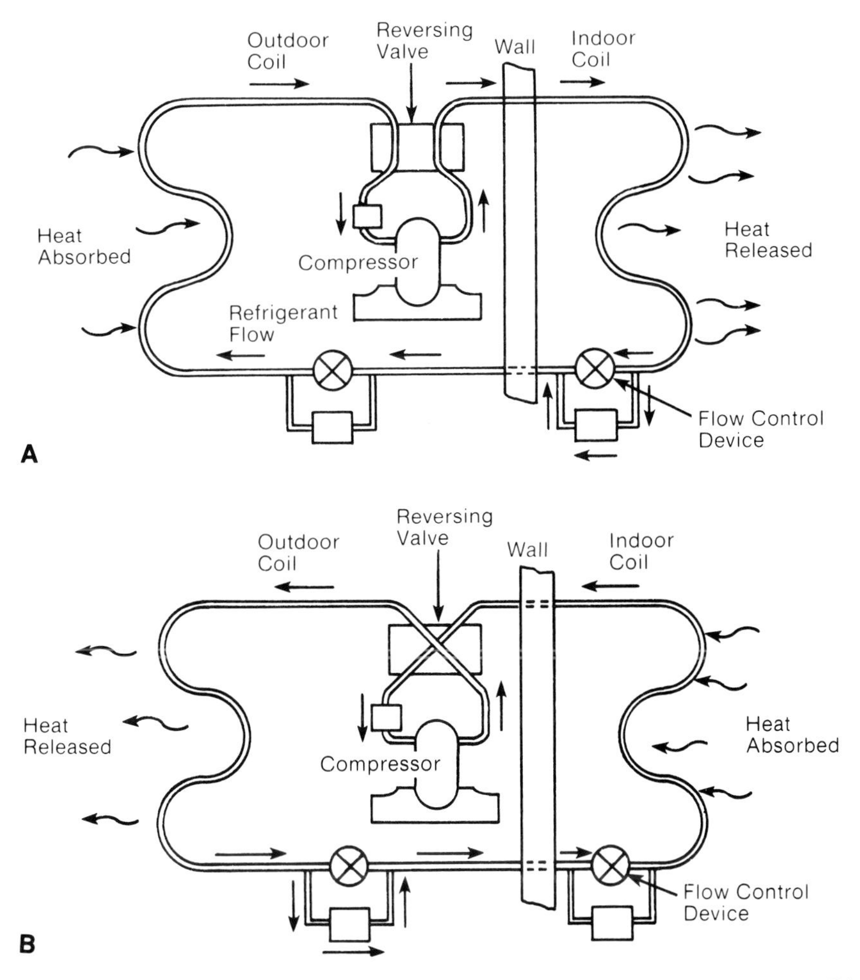

FIGURE 6-2. (A) Schematic view of a heat pump in the heating cycle and (B) view of the heat pump in the cooling cycle.

6.5.4 Solar Heating Systems

Solar heating systems applicable to the light building trade are constantly being developed, but as this occurs, actual architectural designs must be altered to accommodate this solar technology. The change to solar will be slow as long as conventional fuels remain relatively inexpensive. Economic collection of solar energy is the key to its effective use as a heating system.

Solar energy is available in three forms: direct radiation, diffuse radiation (being scattered by the atmosphere), and *albedo* (radiation reflected from the surface of the earth). On overcast days when the sun cannot be seen, solar energy is still available in the form of diffuse radiation and albedo. As the seasons change, the angle of the sun's rays to Earth also change. The seasonal imbalance that gives us winter also makes solar energy scarce when it is most in demand.

Since solar radiation is not a concentrated form of energy, the first step in designing solar architecture is to reduce energy demands by reducing heat losses. Air leaks need to be sealed and insulation levels need to be increased while providing sufficient preheated fresh air for ventilation and safe operation of appliances. Windows need to be at least double-glazed and limited to walls which allow solar radiation to warm adjoining rooms. Shielding the building from cold winter winds will also reduce heat losses.

Systems for collecting solar energy are divided into two categories: passive and active.

Passive solar energy systems collect and store solar energy primarily by using natural heat flow processes. Because of their simplicity, passive systems tend to be very reliable. They can also be inexpensive. In fact, some passive features can be incorporated into a building design without any increase in construction cost. For example, south-facing glass windows can serve two functions. They allow shortwave radiation to enter the building (Figure 6-3A) and, when temperatures drop (after dark) prevent the escape of longer wave reradiated energy (Figure 6-3B).

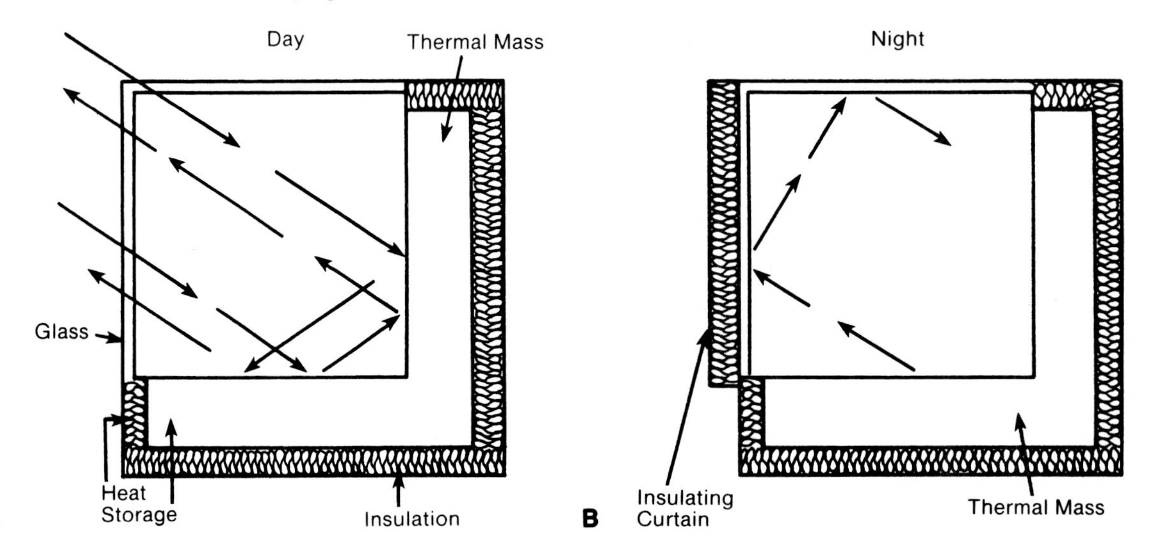

FIGURE 6-3. (A) In a passive system, sunlight enters the building through south-facing windows and is absorbed by the thermal mass of walls, floors, etc. (B) This heat is released at night. An insulating curtain eliminates loss through the windows.

If the shortwave radiation strikes rough surfaces in the building, more absorption takes place. To prevent overheating, some means of storing heat is required. Tanks of water or massive materials such as brick, stone, or concrete have very large heat capacities and are used to increase the amount of energy a building can store (Figure 6-4). When the temperature of the surrounding air drops below that of the storing materials, they begin to reradiate their heat energy.

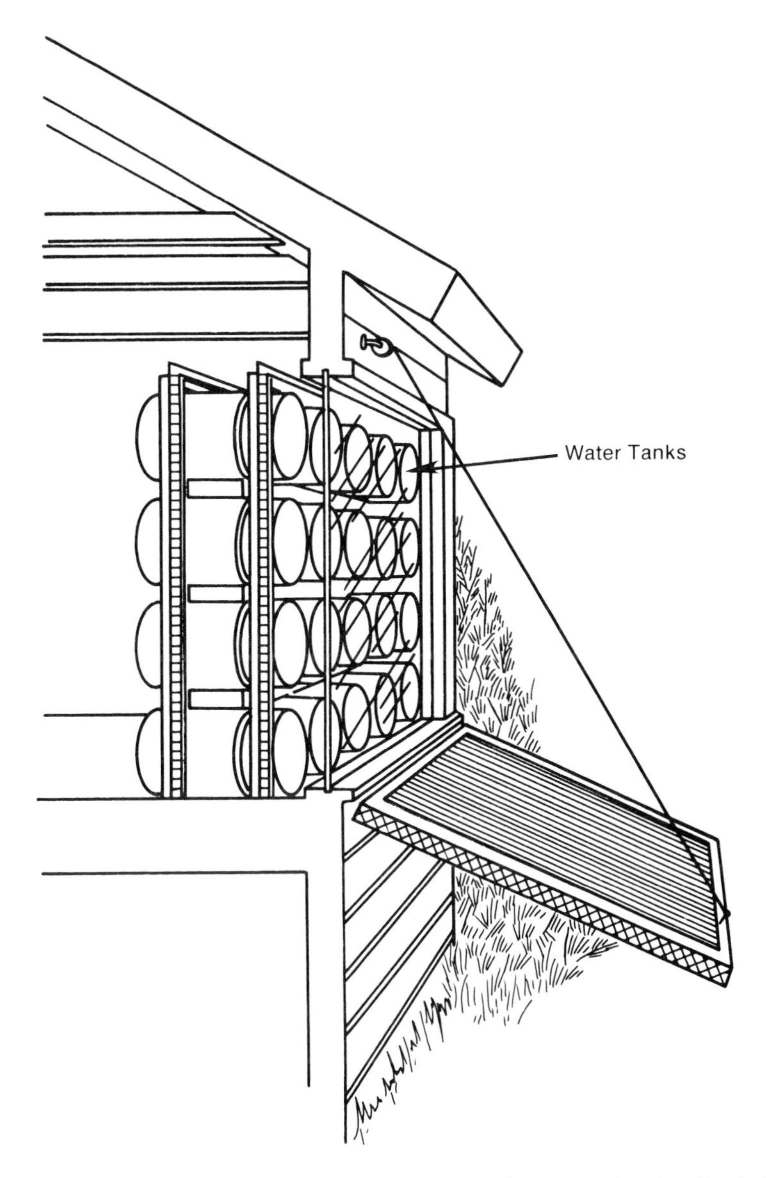

FIGURE 6-4. Tanks of water are sometimes used to store heat collected in passive systems.

Passive systems can store enough energy to be of significant benefit for up to two consecutive overcast days. A conventional furnace provides back-up heat when required. The heat storage capacity and the area of the south-facing windows must be matched to prevent overheating. Passive heating systems can be cost-effective in many regions, and by adding passive solar heating features such as windows, masonry walls, and greenhouses, they can also be visually pleasing.

Active solar energy systems collect and store solar energy primarily by using mechanical equipment such as pumps, heat exchangers, and flat plate collectors (Figure 6-5). Because they employ a number of high-performance components, active systems tend to be less reliable and much more expensive than passive systems.

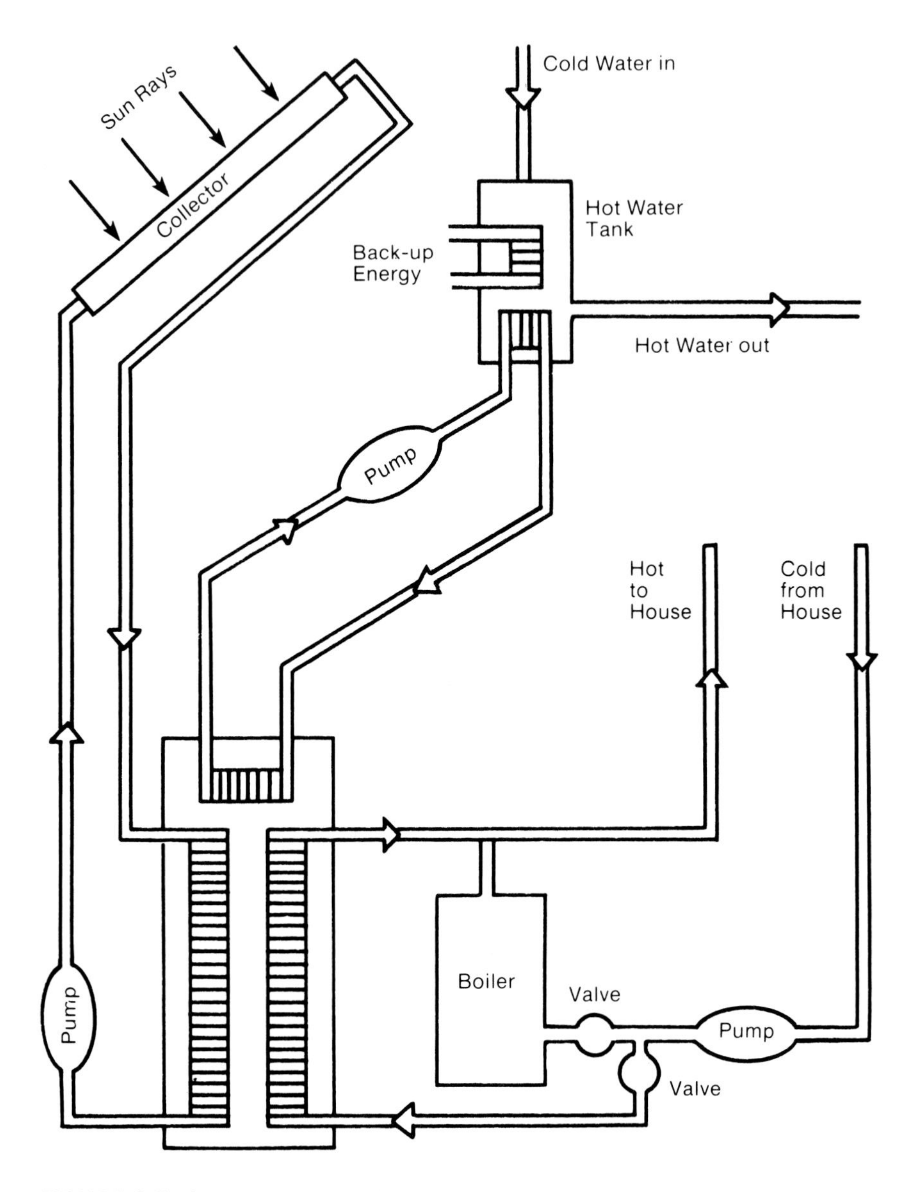

FIGURE 6-5. Schematic view of an active solar heating system that is also used to supply hot water.

In an active system, the collector admits and absorbs solar radiation. The heat energy is then transferred from the collector to a water tank or rock bed for storage. A suitable distribution system moves heat from storage into the house when activated by a thermostat.

Either air or a liquid (usually water and antifreeze) is used to carry heat energy from the collector to storage. Air systems do not cause any damage if they leak, but they do require bulky ductwork and a large storage area. Liquid systems, although more compact, can experience problems with leaking and freezing under some circumstances.

Most active systems will have a conventional heating system as a backup.

6.6 RESEARCH

1. Create a *Forms of Energy* file. As it develops, subdivide it to store topical information.
2. Collect or identify national and regional materials pertaining to energy forms used during construction and in buildings. Codes and regulations for each form of energy should be noted.
3. Identify specific vocational designations in the energy field. Determine how one prepares for, makes entry, and qualifies to work in each. Read computer software data.
4. Collect engineering/architectural data with respect to light construction applications of energy forms.
5. For your region, identify the forms of energy which are now available and which heating/power systems are most cost-effective.
6. Identify publications which state local safety regulations for installations and operations of light construction building heating/cooling/lighting systems.
7. Compare and contrast the environmental and safety pros and cons of the major forms of energy. Familiarize yourself with local and state codes that deal with permissible levels of emissions.
8. Working from a simple sketch of a building, determine what, if any, design modifications the structure would require to use each of the following heat sources: electricity, fossil fuels, heat pumps, and solar heating.
9. Review the advantages and shortcomings of both passive and active solar systems. Figure out how much each would increase the construction costs of the same building design.
10. Pick up the latest government pamphlets on ways of reducing heat losses. Identify the problems of overinsulating as well as underinsulating a building.

6.7 REFERENCES

6.7.1 Alberta Energy and Natural Resources. *Solar Energy Matters.* Alberta: Energy Conservation Branch, 9915-108 Street, Edmonton, 1980.

6.7.2 Dagostino, Frank R. *Estimating in Building Construction.* Virginia: Reston Publishing Company, Inc., 1978.

6.7.3 Energy, Mines and Resources Canada. *100 Ways to Save Energy and Money in the Home.* Ottawa: Information Canada, 1975.

6.7.4 *Fundamentals of Solar Heating.* Washington, D.C.: U.S. Department of Energy, 1978.

6.7.5 Langley, Billy C. *Comfort Heating.* Virginia: Reston Publishing Company, Inc., 1978.

6.7.6 OSHA Safety and Health Standards. *Construction Industry*. Washington, D.C.: U.S. Department of Labor Occupational Safety and Health Administration, 1979.

6.7.7 Powell, Evan and Ernest V. Heyn. *Popular Science Book of Home Heating (and Cooling)*. Virginia: Reston Publishing Company, Inc., 1983.

6.7.8 Trechsel, Heinz R. and Sheila J. Launey. *Criteria for the Installation of Energy Conservation Measures*. Washington, D.C.: U.S. Department of Commerce/National Bureau of Standards, 1981.

7

PERMITS AND UTILITIES

Builders and contractors involved in the light construction industry must be familiar with the diverse permits required before various stages of construction work can begin. Building permits and special utility permits give the builder the "right" to commence operations. Permits are issued by municipal or regional authorities that regulate the building industry and set building standards. Builders must also be familiar with the utility services offered in their area, how to integrate them into the building plan and schedule, and the advantages each utility offers to the finished project.

7.1 GENERAL OBJECTIVES

The general objective of this chapter is to provide the light building contractor with the information needed to secure building permits and utility services for a light building project. Utilities included are water, sewage disposal, electricity, and fuel.

7.2 SPECIFIC OBJECTIVES

The specific objectives of this chapter are to

1. encourage the use of network programs when planning a light construction project,
2. impress upon the student the importance of carefully weighing the advantages and disadvantages of each utility service before selecting specific services for a project,
3. encourage the study of utilities as a vocational field in itself,
4. present an entire overview of a light construction project as opposed to an in-depth look at one or two aspects of the job, and
5. help students learn about national building codes and regional regulations.

7.3 THE BUILDING PERMIT

Before any actual construction begins, a proposal for a new building permit must first be submitted to a municipal or regional planning authority. The planning authority examines the plans and specifications to determine whether the proposed building will conform to the planned pattern of the area in which it is to be built. If approved by the planning authority, the application is then passed on to a board of engineers for closer structural examination. When the plans are given the engineer's approval, a building permit is then issued upon payment of a scaled fee based on the type of building proposed.

All permits must be displayed on the jobsite. In the course of a light building project, the builder will probably have to secure several different types of building permits. Permits are required for tapping into the municipal water system or sewage lines, using water from public fire hydrants during construction, and closing roads or diverting traffic. Because the exact nature and details of such permits vary from region to region and community to community, the builder must work closely with local authorities to ensure that all work conforms to local regulations.

Each permit declares the basis on which the builder may proceed. The planning authority knows what to expect from the builder, and the builder knows his or her freedoms and restrictions. The planning authority is taking general, long-term responsibility for a building project and its surroundings. The builder is ensuring the community of his or her competency to build.

In this regard, the builder must be fully knowledgeable with respect to the water supply system, sewer or disposal field, electrical power system, fuel supplies, and all other utilities before beginning work in these areas. The builder must also follow all safety regulations and be aware of special emergency procedures that may be necessary in the event of an accident or natural disaster, such as a flash flood or severe storm.

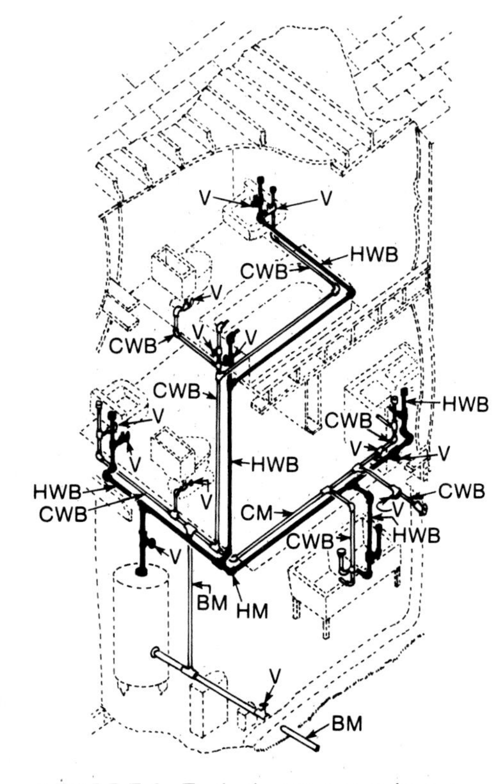

FIGURE 7-1. Typical water supply system of a residential home: BM = building main, CM = cold water main, HM = hot water main, CWB = cold water branch, HWB = hot water branch, V = valve.

7.4 WATER SUPPLY

The water supplied to buildings for human consumption and use is generally referred to as potable water. The term states that the water meets the standards set for impurities and certain permissible substances. These health standards are set by public health authorities and analyses are done by designated laboratories. Limits are set on substances such as free chlorine, copper sulphate, solids, and color. Local authorities set the number of tests to be done to ensure the public that the water is satisfactory for drinking, culinary, and domestic purposes. The term "clean water" does not make this assurance, so it is best to always use the term potable when dealing with knowledgeable persons in this field.

Water supplied to a building site may be from a municipal system, wells, or a known fresh water supply (such as springs or artesian well) (Figure 7-1).

7.4.1 Municipal Water Systems

If the water serving the building is to come from a municipal water system run by a municipal water authority, the builder needs to know specific information with respect to

1. the location, depth, and size of the water main and when a connection may be made;
2. the digging of a trench to the main from the building location and the laying of pipe of proper size (with a goose neck to allow for ground movement, to prevent pipe breakage);
3. the installations of valves at the main and at the curb;
4. the creation of a service entrance and installation of a meter;
5. the location of the sewer or sewers (Municipalities may require the water wastes from inside the building to run to the public sewer and the runoff of rain/snow outside the building to the storm sewer.);
6. the installations of rough cold water and hot water supply piping, finish plumbing, hot water heating unit, and possibly a hot water heating system (The system must withstand pressures greater than atmospheric.); and
7. the elapsed time and cost of each activity.

With this specific information, the builder can now coordinate all activities by means of a planning model, and a critical path of action begins to take shape.

7.4.2 Wells as Water Supplies

If the water source is a well, the builder assumes different responsibilities. A major one is the contracting with a driller (who obtains a permit to drill) to supply an adequate amount of water on a long-term basis. Whether the water is potable is formally determined through required analysis and reporting. Decisions as to the water being "soft" enough or too "hard" would be made before installing a pump and building a pump house.

Water tests will indicate the need to drill deeper, to drill at a different location, or to pipe water from an adjacent property or community well.

When the water supply is confirmed, installations of the following items are made: a pump (with electric motor and controls); a below frost-line trench and piping to the building(s); and piping connections to roughed-in plumbing.

The time line and cost of the activity are important to the builder. As there are usually more variables with which to contend (not being assured of a water supply before drilling), it is more difficult to draw the critical path. Estimates may be based on other area experiences.

7.5 BUILDING DRAINS AND SEWAGE SYSTEMS

In any type of building, the flow of waste water in the drainage pipes of its plumbing system is from points of higher elevation to points of lower elevation. Because gravity is the force that propels the flow of wastes, the flow can occur at atmospheric pressure. The principle parts of a waste water system that allow this flow to take place in a safe and controlled manner are traps, vents, drainage pipes, the building drain,

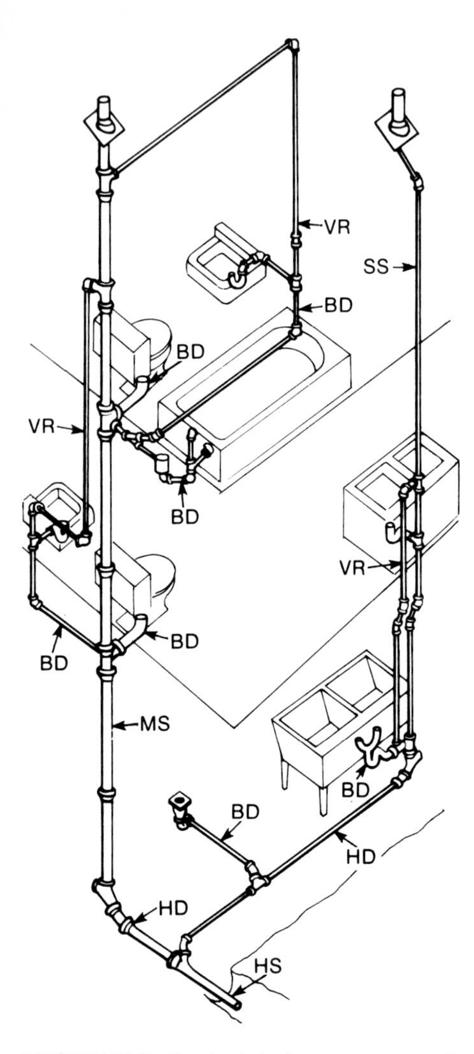

FIGURE 7-2. Typical drainage system of a residential home: MS = main (soil) stack, SS = secondary (soil) stack, BD = branch drain, HD = house drain, HS = house sewer, VR = vent run.

and the sanitary sewer (Figure 7-2). The "horizontal" pipes of a waste system are placed with sufficient slope just steep enough to ensure self-cleaning.

The drainage system must carry waste water rapidly away from the fixtures and prevent passage of air, odors, or vermin into the building. Durable drainage pipes must be airtight, watertight, and gastight; they must also be installed to withstand building vibrations. Cleanouts are installed at all logical, accessible places in the system. There must be no way for blocked sewage in the system of one building to back up into the system of another building.

Sewer trenches are as narrow as possible to minimize the amount of backfill and machine (compacting) loads on the pipes. Before backfilling, pipes are carefully bedded on the undisturbed trench bottom and joints are made secure. After inspection the backfilling is done promptly to minimize danger of trench cave-in. To ensure proper drainage of the basement floor and fixtures, the building drain is brought well under the concrete basement footing and floor. Elevations and sewer slope measurements are crucial. Remember, the building sewer must be connected to the correct sewer line—*the sanitary sewer.*

Rain and snow water from the roof is usually drained through a series of gutters and leaders into a separate sewage system, the *storm sewer.* Surface water from rain and snow also drains into the storm sewer. The storm sewer is kept apart from the sanitary system. In both cases, municipal regulations must be followed concerning hookup and materials.

7.5.1 Septic Tank Disposal Systems

Where a municipal sewage system is not available, other means of safe disposal are usually required by the local health authority. A septic tank with underground disposal of the effluent is one means of treatment (Figure 7-3).

The septic tank, possibly made of concrete or fiberglass, is a watertight container of sewage. It is sized so that the period of sewage retention is sufficient to allow septic action—a bacterial action that changes the chemistry of the sewage—and to allow outflow of effluent at about the same rate it enters the tank. The tank is about 10′ (3 m) (see local regulations) from the building and needs periodical servicing. It also needs to be accessible for cleaning and to be securely covered.

The National Building Code, as interpreted locally, establishes sizes of septic tanks connected to a building. Data is based on the number of persons to be served and is stated in terms of septic tank capacity, giving inside dimensions, liquid depth, and total depth. The code describes the construction of a tank and how it is designed to discharge.

Normally the tank discharges into a specially designed field. The effluent moves into a porous tile or seepage pit, or both so that the liquid will seep into the upper layer of surface soil to be lost to evaporation or to growing plants. The porous tile field and seepage pit are designed to suit the soil conditions. Tests may be run for standard soil percolation time to help determine the effective size of an absorption area.

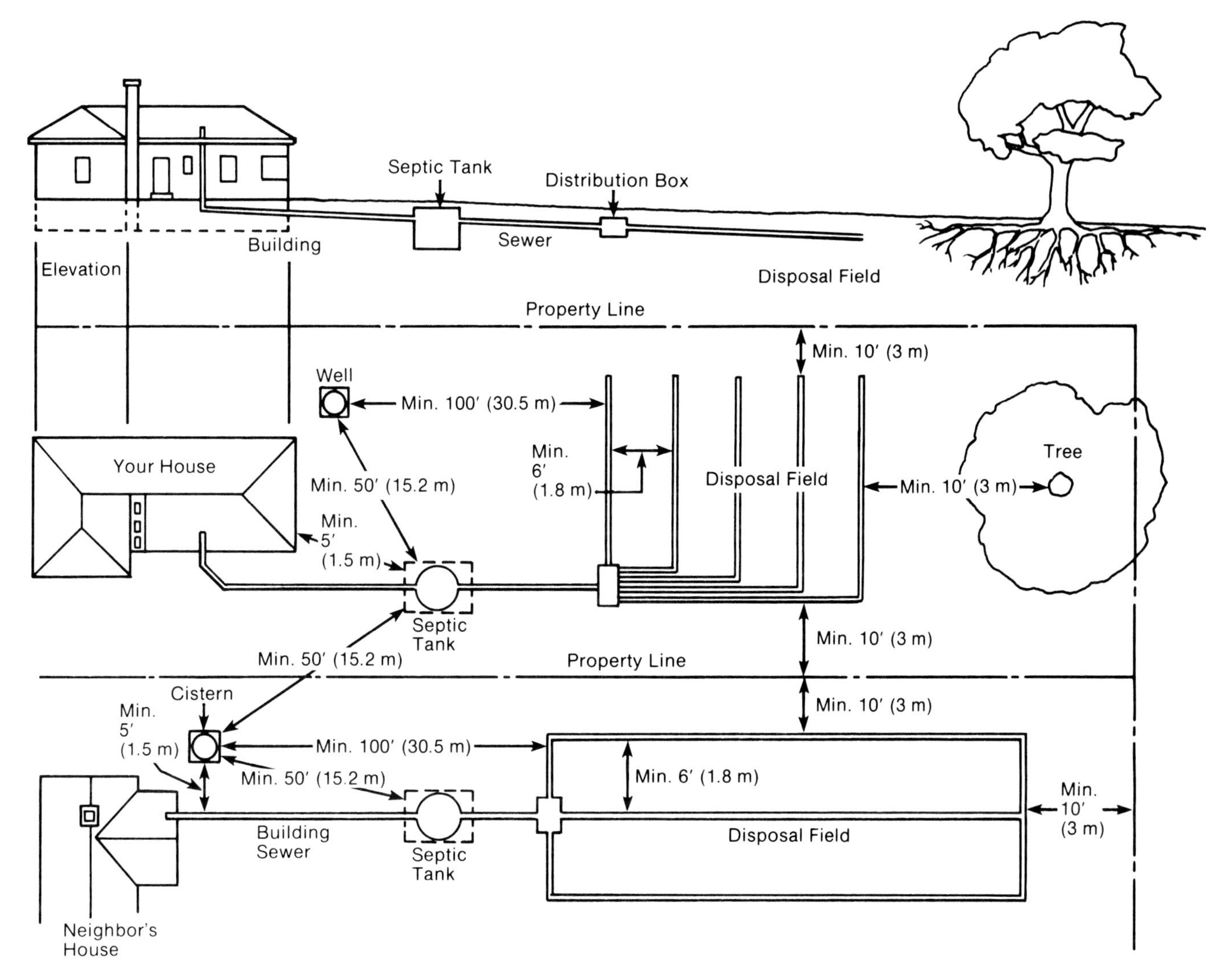

FIGURE 7-3. Septic tank disposal system.

National plumbing codes are explicit and helpful in the design of a septic tank and its field. When installing one, "it is better to be safe than sorry." Digging up a new field to enlarge it is costly, and if the problems arise in wintertime, it can be an immense problem. It is a sizeable task to place a septic tank (and possibly an automatic syphon), seepage pit, and porous tile and then to properly backfill with filter material and cover with landscaped topsoil.

Cleaning a septic tank to remove sludge is done with caution. Risks are associated with poisonous, asphyxiating, or explosive gases—sewer gases.

7.6 ELECTRICAL POWER

Electricity is one utility service that is used in virtually all light building projects. It is used as a temporary installation during construction to help power machinery and tools. The permanent electrical system is installed and connected during various stages of the entire construction project. It serves to power the lighting, small motor, and appliance requirements of the building.

In most cases, the temporary electrical service will be supplied through a hookup to an overhead distribution pole. Most permanent installations are made by running underground service lines from the building to the utility line. Permits must be secured for electrical work and hookup. Proper metering of temporary and permanent installations is required and all work must pass inspection by the proper authorities.

7.6.1 Units of Electrical Measurement

Power is the rate at which energy is used or the rate at which work is done. The units of electrical power in the British Imperial and Metric or SI Systems are the watt, the kilowatt, and the megawatt. Horsepower ratings (Imperial) are replaced with watts in SI. One electric horsepower equals 746 watts, the same for practical purposes as the mechanical power equivalents.

A watt (w) is energy per unit of time measured in joules per second (Js). Therefore:

1 Js = 1 w, and
1,000 w = 1 kw, and
1,000 kw = 1 Mw.

One horsepower (Imperial) is approximately 0.75 kw (SI) and 746 w. One Kilowatt is approximately 1-1/3 horsepower (Imperial).

7.6.2 Estimating Expected Electrical Demand

Before a builder or contractor can accurately determine the size of the electrical service panel required to fulfill the electrical demands of a particular building, he or she must gather extensive data concerning the building's intended use, organize this data, and then test this data against comparable existing installations and the prevailing standards given in the National Electrical Code.

For example, once this data has been compiled and processed, the builder may determine that a 100 ampere service is sufficient for a moderately sized modern home. Two such service lines may be required for a larger home, and the requirements for a shop where electric motors are used extensively may be even greater.

The data base collected for a project could consist of a list of individual appliances stating the exact number of such appliances and their exact wattage. With this information on hand, it is a simple task to calculate the total power load of the building. By knowing that appliances will not all be on one circuit but will be distributed on a planned basis to many circuits, say 24 or 32 for a house, some estimate can be made of a peak demand, with provision for spare circuits for adding wiring in the future.

Table 7-1 contains typical electrical demand data for a small building, such as a house. Appliances are listed with corresponding wattages and kwh/h amounts. *Note:* Watts = volts × amperes.

TABLE 7-1 HOURLY USE OF APPLIANCES*

Appliance	Wattage	kwh/h
Air conditioner		
Furnace (24,000 Btu)	2,000	2
Room (12,000 Btu)	1,000	1
Blanket (double)	200	.05
Car block heater (single unit)	450	.45
Car interior warmer	750	.8
Clothes dryer	6,000	4.6
Coffee maker	600	.6
Corn popper	1,000	.5
Deep fryer/cooker	1,500	
Frying foods		1.2
Simmering foods		.3
Dishwasher	1,500	
Double wash		.8
Single wash		.6
Rinse'n hold		.1
Electronic air cleaner	40	.04
Frypan	1,250	
Frying		.6
Simmering and baking		.45
Hair dryer		
Bonnet	325	.36
Hard hat	1,000	1
Heat lamp	250	.25
Heating tape	7 w/ft	.007/ft
Iron	1,100	.45
Kettle	1,500	1.4
Lawn mower	1,000	1
Lighting		
100 w bulb	100	.1
40 w fluorescent tube	40	.04
Microwave oven	1,200	1.2
Space heater	1,500	1.5
Sunlamp	275	.275
Television		
Black and white	120	.1
Color	250	.2
Solid-state color	120	.1
Instant "on"	55	.055
Toaster (2 slice)	1,200	1.2
Vacuum cleaner	750	.75
Waffle baker	1,100	1
Washer		
Automatic (1/2 hp)	1/2 hp	.42
Wringer (1/3 hp)	1/3 hp	.28

*Table courtesy of Alberta Agriculture.

Most circuits in a typical house are designed to carry 12 amperes or approximately 1,500 w of electricity (12 × 120 v = 1,440 w). This means that the total wattage of all the appliances and lights being used on any circuit at one time must not exceed 1,500 w. If excessive, the circuit will be overloaded and the breaker will trip, or the fuse will blow.

Each circuit must be carefully planned, and the wiring for each room must also be planned. Usually it is best to have two circuits serving different parts of a room, in case one is dysfunctional.

Service panels are covered metal boxes containing a main switch and several circuit breakers. The main switch controls all the electricity flowing into the building, and each circuit breaker controls the electricity flowing into that circuit.

Branch circuits are not all the same. Table 7-2 describes four types.

TABLE 7-2 ELECTRICAL BRANCH CIRCUITS*

Amperes	No. of Wires	Volts	Purpose
			General Purpose
15	2	120	Supplies all lighting circuits and all convenience circuits except those listed under separate and split duplex.
			Separate
15	2	120	Supplies electricity to only one duplex receptacle or two single receptacles.
			Split Duplex
15	3	115/230	Supplies kitchen counter type receptacles on separate circuits.
			Major Appliance
40	3	120/240	Range
40	3	120/240	Built-in oven
30	3	120/240	Clothes Dryer (up to 7.2 kw)
20	3	120/240	Water Heater (up to 3 kw)

*Table courtesy of Alberta Agriculture.

7.6.3 Planning the Installation

With regard to the CPM activities associated with the installation of electrical service, some of the major subactivities consist of the

1. securing of all necessary permits;
2. temporary installation, including the metering of power (for municipal or public utility),
3. planning for wiring, and providing materials and equipment,
4. installation of service entrance,
5. roughing in and testing of circuits,
6. finish work, including installation of appliances, and
7. final inspections.

7.6.4 Portable Electric Generators

In areas remote from electrical utility lines, fossil fuel driven generators are temporarily or, in the case of larger diesel driven machines, permanently installed to provide electric power to a building. Such a machine requires care and maintenance. It is leased or purchased on a warranty basis and also requires a capital outlay to obtain it and place it on a protected site. The electrical installations in such a building must be tailored to the generator's output.

7.7 FUEL SERVICE

Building project planning includes the supply and placement of a fuel system or systems. These systems supply heat energy and in some cases fuel for cooking and certain appliances. Common fuels used in modern light construction are natural gas, liquid propane, coal, heating oil, wood, and solar energy. Once a decision has been made as to the type of fuel or fuels that will be used, more complete planning is required.

Many of the major subactivities associated with each type of fuel mentioned are outlined in the following sections. These broad outlines can be easily fleshed out with further detailed planning.

7.7.1 Natural Gas

When installing a natural gas fuel system

1. Obtain the permit.
2. Contract with qualified persons to
   - trench,
   - connect at service entrance and at the main (cold water shutdown valves),
   - build an approved type chimney(s) on existing concrete pad (footing),
   - install heating units (gravity or forced air or hot water or space) and piping,
   - connect heating units to meter and gas supply, and
   - install and test controls.
3. Arrange inspections for installation, inclusive of automatic shutdown.
4. Backfill and compact the trench.
5. Receive invoices for payments.

7.7.2 Liquid Propane

When installing a liquid propane fuel system

1. Research and plan with respect to tank size, and installation requirements.
2. Obtain the permit.
3. Contract with qualified persons to
   - place a concrete tank-pad/cradle,
   - deliver and install the tank,
   - trench from tank to service entrance,

- connect at service entrance and at the tank cold water shut-down capabilities,
- build an approved type chimney(s) on existing concrete pad,
- install heating units (gravity, forced air, hot water, space) and piping,
- connect heating units to gas supply and measuring instrument, and
- install and test controls.

4. Arrange formal inspections for installation/safety tests, inclusive of automatic shutdown.
5. Backfill and compact trench.
6. Receive invoices for payments.

7.7.3 Coal

When installing a heating system fueled by coal

1. Research with respect to supply: type of coal (soft or hard), burning characteristics, availability on long-term basis, dust control, and storage cost.
2. Research with respect to heating unit(s): furnace, space heater, controls, fuel feeding, chimney, air supply, and energy efficiency (Btu/kJ).
3. Decide on units to be used.
4. Obtain permit.
5. Build storage unit (shed, hopper).
6. Contract with qualified person(s) to
   - supply equipment, materials, and duct system;
   - build an approved-type chimney (c/w lined flue);
   - install heating unit(s);
   - connect low voltage electrical controls (if required); and
   - test the system.
7. Arrange formal inspections for installation and safety tests.
8. Buy a supply of coal and have it delivered.
9. Receive invoices for payments.

7.7.4 Heating Oil

To install a heating oil fuel system

1. Research with respect to supply of oil: Btu/kJ availability, flash point, cost, safe storage.
2. Research with respect to system development: tank, tank placement, controls, piping, connections to tank and heating units, heating units, venting, air supply, ductwork, electrical power supply, and safety (including chimney cleaning).
3. Decide on the system to be used and how it is to be installed.
4. Obtain the permit. (Firefighters need to know what kind of fuel is being used in a building.)
5. Contract with qualified persons with respect to
   - supplying equipment and materials;
   - placing a tank pad and possibly a footing for heating unit(s);
   - building an approved type chimney;
   - installing the tank and heating units, and making connections;

- connecting electrical power supply and installing controls; and
- testing the system.

6. Arrange formal inspections for installation and start-up.
7. Test the system: flame characteristics, burner soot buildup, barometric damper operation, stack temperature (for efficiency test), smoke number, carbon dioxide level, and nozzle size.
8. Have the fuel tank filled with oil.
9. Receive invoices for payments.

7.7.5 Wood

Wood-burning systems are most often used to provide supplemental heat, but they can be used as the total heating system (Figure 7-4). To install a wood-burning system

1. Research with respect to supply: kinds (softwoods and hardwoods), burning characteristics, storage requirements, and availability of poles, blocks, and splits.
2. Research with respect to burning equipment and installations: furnace, space heater, stove, fireplace, efficiencies, chimney requirements, chimney fire prevention, and costs.
3. Decide on the units to be used to develop a heating system, possibly in conjunction with use of another fuel.
4. Obtain the permit.
5. Contract with qualified persons with respect to
   - supplying equipment and materials;
   - supplying wood;
   - building an approved type chimney and/or fireplace/chimney;
   - assembling and installing the heating units cold water ducts and pipes, where required; and
   - testing the system.
6. Cut or buy the wood supply and store it.
7. Arrange formal inspection and testing.
8. Receive invoices for payments.

FIGURE 7-4. Wood burning stoves are commonly used to supply supplemental heat.

7.7.6 Solar Heating System

Because of the many factors that can affect the performance of a solar heating system, such as location, alignment, materials, and climatic conditions, extreme thoroughness in planning is essential for the system's success (Figure 7-5). As with wood systems, solar heating systems are often used as a supplement to another type of heating system. When considering a solar heating system

1. Research with respect to the state of the technology: the degree to which current systems are viable and to cost effectiveness.
2. Research with respect to which system will be used: passive or active (installation costs, reliability, availability of materials and equipment, and available expertise).
3. Decide which system to use and which other heating system to use as a backup.
4. Obtain necessary permits.
5. Contract with qualified persons with respect to
   - design of system and the building,

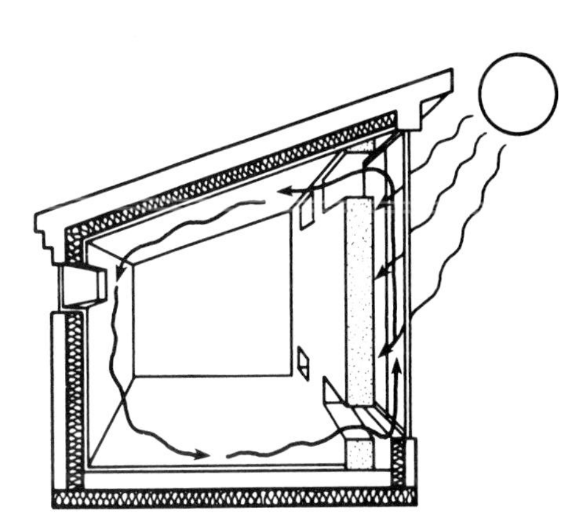

FIGURE 7-5. In a passive solar heating system, heat is collected in a thermal mass located on the south face of the structure.

- supplying equipment and materials,
- structural work to support and contain the system,
- structural work to use the system (greenhouse, swimming pool, and main building), and
- testing and regulating the system and the back-up system.

6. Arrange formal testing and inspections.
7. Receive invoices for payments.

7.8 EMERGENCY PROVISIONS FOR UTILITY SERVICES

To ensure the continued safe operation of water, sewer, electrical, and fuel utilities, certain automatic and manual controls are designed into each system. These controls offer protection against accidents and malfunctions and also give the system owners control over the utility. The major automatic and manual control devices for each utility and service are outlined below.

7.8.1 Water Supply

1. Each cold water supply requires a shut-off valve between the faucet and the in-building service pipe valve to allow repairs of the faucet without shutting down the cold water supply to the building.
2. Each hot water supply needs a shut-off valve between the faucet and hot water tank to allow repairs.
3. The water supply pipe needs an in-building main valve between the meter and the external curb valve.
4. The water supply pipe, external to the building, requires a "corporation" valve at the water main and possibly a curb valve for shutting off the water supply to a building in case of supply pipe rupture.

7.8.2 Sanitary Sewage Systems

Backflow valves are needed on drains to prevent backflow from entering public sewers and to prevent flooding of basements in case of heavy rains.

7.8.3 Electrical Systems

With respect to the electrical supply system, each circuit has a breaker or fuse, and each building has a main breaker (both controlled by a building owner) and a secondary distribution system breaker (controlled by the power authority). Internal and external controls are safety devices for part or all of a building.

7.8.4 Fuel Systems

1. A set of on-site valves for natural gas or liquid propane lines also provide protection to owners. Automatic shutdown is provided at each heating unit; in-building shutdown is provided at the meter, and the curb valve is external to the building. A comparable set of controls is provided for oil-burning systems.

2. Coal fires need proper "banking" to prevent buildup of explosive unburned gases. The amount of fuel added to a furnace or stove fire is controlled manually or by automatic feed.
3. Selection of wood is important with respect to burning characteristics and to soot/tar formation in chimneys.
4. Where coal and wood are burned, owners need to contract with persons who maintain such systems, including chimneys, or learn specifically how to fight chimney fires. Selection of fuels is important.

7.9 RESEARCH

1. Inquire locally specifically how a water supply is tested with respect to being potable. Either have a sample tested or see a typical report. Ask for an interpretation of the report.
2. Make an isometric sketch of a light construction building cold/hot water system. Identify any cross connections, and report on the effects.
3. Describe with sketches the types of wells being used in your area. Tell how the water is delivered to buildings. Inquire about making cost estimates.
4. Using sketches illustrate the principles of well water pumping for your area.
5. Report on local regulations with respect to slopes of building drain pipes, storm sewers, and sanitary sewers. What is 100% slope?
6. Using a sketch describe the functioning of a plumbing system vent.
7. Report on local regulations with respect to dry wells, cesspools, and septic tanks.
8. Report on functioning of a septic tank/seepage pit/field installation. Use quotes from one or two regulations given in the National Plumbing Code.
9. Make a labeled drawing with symbols of the electrical wiring system for a particular house. Identify the types of branch circuits.
10. Observe a natural gas installation with respect to a light construction project. Identify the subactivities. Note the sequence of events and elapsed time for each. Sketch a simplified plan for such an activity.
11. Obtain an estimate of cost of installing a liquid propane system for a small building in a rural setting: tank rental, installation, tank fill, and LP appliances.
12. Obtain information sheets/brochures for all types of wood-burning building heating units: fireplaces, stoves, furnaces, and heaters. Compare the sets of data. Write recommendations for or against usage or indicate specific limitations.

7.10 REFERENCES

7.10.1 Alberta Agriculture. *Residential Electric Wiring Guide.*

7.10.2 Domtar Construction Materials. *Construction MetriGuide.* Third edition: 1975.

7.10.3 National codes and regulations with respect to light construction:

- building;
- electricity;
- oil burning equipment;
- natural gas burning appliances and equipment;
- plumbing;
- heat, ventilation, and air-conditioning; and
- construction safety

7.10.4 Wass, Alonzo, and Gordon A. Sanders. *Materials and Procedures for Residential Construction.* Virginia: Reston Publishing Company, 1981.

8

JOB COST ACCOUNTING

While all contractors must face the problem of accurately recording the costs of their building projects, the task is particularly troublesome for the beginning builder. Without an ongoing, up-to-date system of recording financial information, there is a high risk of business failure. Complete, accurate financial records concerning a building project and its subcontracted activities form the reliable basis on which to determine profit and loss. Finally, the contractor must know the financial status of a project at all times or be subjected to foreclosure and the denial of the right to proceed.

The builder must also be able to accurately differentiate between the costs of the building project and determining the amount charged for the work done. In this chapter emphasis is placed on recording costs.

8.1 GENERAL OBJECTIVES

The main purpose of this chapter is to present a job cost accounting system designed to provide ongoing and up-to-date information for a small contracting business. The accounting principles outlined can be adapted into a PERT/CPM plan as suggested in Chapter 1.

8.2 SPECIFIC OBJECTIVES

On reading this chapter and conducting additional studies with respect to other accounting systems and cost recording forms and ledgers, the reader should

1. be ready and able to select a job cost accounting system for starting a building contracting business;
2. be able to measure the compatibility of a job cost accounting system and a formal building planning program;
3. be able to conduct further study and useful research in other accounting areas, such as payroll; and
4. be better able to analyze the accounting activity of a building program.

8.3 GUIDELINES FOR JOB COST ACCOUNTING

The accounting activity should be systematic and in harmony with the organizational structure of the job being performed. It should

1. be simple, easy to learn, and encourage, rather than discourage productivity and proficiency;
2. provide accurate and up-to-date records of costs;
3. enable the reader to immediately determine whether costs are above or below the projected cost estimate for the project or subactivity;
4. provide trial balances which can help management personnel make sound financial decisions during the course of the project;
5. be inexpensive and legible in nature;
6. be flexible to fit current and possible expansion of job requirements;
7. be designed to reduce accounting work loads; and
8. be designed to reduce accounting errors.

As mentioned previously, the cost accounting system must fit into the overall planning system outlined in Chapter 1.

8.4 COST ACCOUNTING TASKS

Any cost accounting system adopted by a builder should use generally accepted and standardized recording procedures and materials. Using a one-of-a-kind system can cause problems when dealing with other businesses, financial institutions, etc. In this regard the following procedures should be undertaken:

1. All transactions must be documented. Invoices and checks must be kept to prove costs and payments.
2. All costs and payments should be posted or recorded in journals or ledgers as they are received or as they occur.
3. All postings must be proved or checked for accuracy whenever posting for the day is completed or a journal or ledger is completed.
4. A system of accounts should appropriately be selected for debiting or crediting for each transaction. Total debits will equal total credits.
5. The system of tasks must not be inconsistent with computer data processing procedures.

8.5 A JOB COST ACCOUNTING FLOWCHART

All accounting documents such as invoices, checks, journals, and ledgers are elements of a comprehensive, standardized accounting system. This ongoing set of documents must sufficiently record all financial aspects of a job and should only be discontinued if a superior documentation system is discovered.

Because they are part of a total accounting system, these documents have formal relationships to one another and can be used to determine accounting results, such as profit margins, total credits, and total debits.

In the system described in this chapter, suppliers' invoices will be listed in three separate places: in a journal of invoices, on an accounts payable ledger, and on a job cost ledger. The invoice journal will also be used to transfer building materials from the builder's general supply inventory to a specific job. It will also be used to keep a record of unused materials that must be transferred from a specific project account back into general inventory.

Checks that are used as expenditure or disbursement documents will be recorded in two places: in a disbursement journal and on an accounts payable ledger. This accounting system and the components in it are shown graphically in Figure 8-1. This illustration establishes the relationship between suppliers' invoices and disbursement checks. It also introduces a "one-writing" system in which the recording sheets are designed so that the record is made in more than one place when it is only written out once. This "one-writing" procedure, developed by Systems Business Forms, ensures orderly posting of data without transcribing errors (see 8.14 REFERENCES).

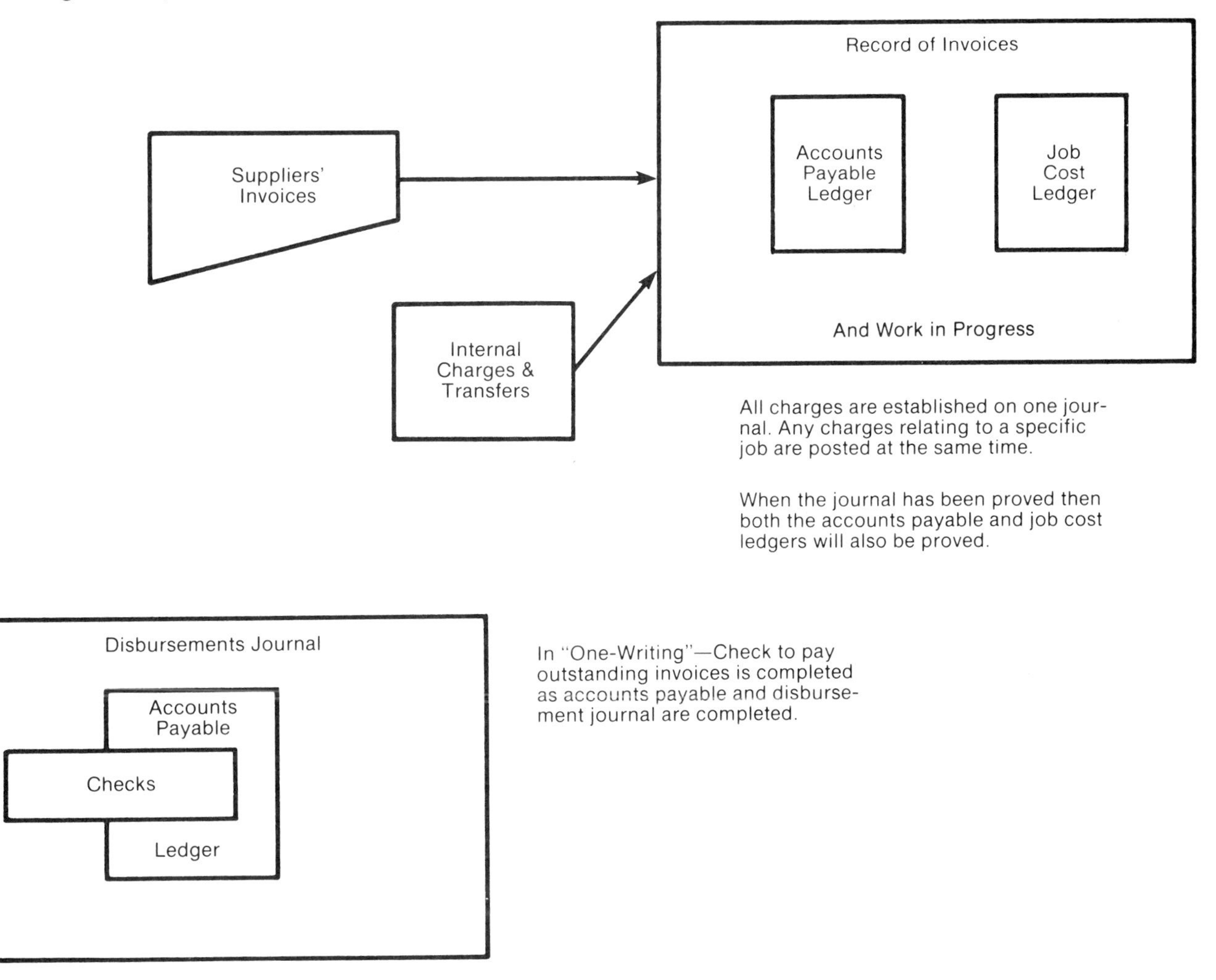

FIGURE 8-1. Flowchart of job cost accounting using the "one writing" system of accounting developed by Systems Equipment Limited. The author wishes to acknowledge Systems Equipment Limited for use of their material in this chapter.

8.6 THE ONE-WRITING JOB COST SYSTEM

Figures 8-2, 8-3, and 8-4 illustrate three accounting forms that can be used by builders to help create a job cost accounting system. Figure 8-2 illustrates a form used to record invoices and work in progress. The form is perforated to fit onto a backing board equipped with hanging pegs. Figure 8-3 illustrates an accounts payable ledger while Figure 8-4

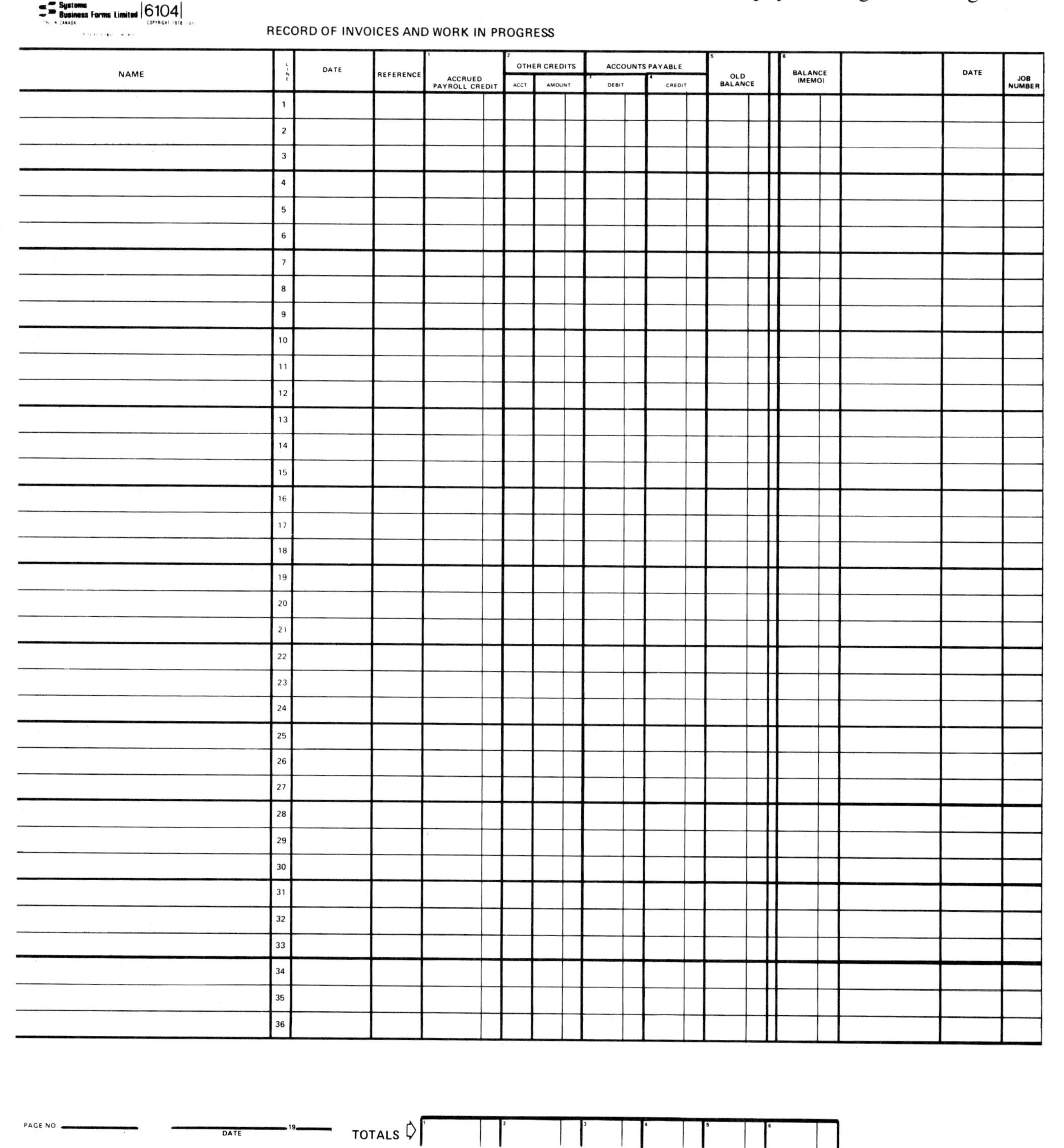

Systems Business Forms Limited 6104

RECORD OF INVOICES AND WORK IN PROGRESS

NAME	LINE	DATE	REFERENCE	1 ACCRUED PAYROLL CREDIT	2 OTHER CREDITS ACCT	OTHER CREDITS AMOUNT	ACCOUNTS PAYABLE 3 DEBIT	ACCOUNTS PAYABLE 4 CREDIT	5 OLD BALANCE	6 BALANCE (MEMO)		DATE	JOB NUMBER
	1												
	2												
	3												
	4												
	5												
	6												
	7												
	8												
	9												
	10												
	11												
	12												
	13												
	14												
	15												
	16												
	17												
	18												
	19												
	20												
	21												
	22												
	23												
	24												
	25												
	26												
	27												
	28												
	29												
	30												
	31												
	32												
	33												
	34												
	35												
	36												

PAGE NO ________ ________ DATE 19____ TOTALS ⇨ | 1 | 2 | 3 | 4 | 5 | 6 |

FIGURE 8-2. Record of invoices and work in progress form (6104). (Form furnished by courtesy of Systems Business Forms Limited.)

shows a job cost ledger. Both are also perforated to fit onto the pegged backing board. When used together as a set, the accounts payable ledger and the job cost ledger are placed over the record of invoices and work in progress form. The pegs provide for proper alignment of form lines.

Figure 8-5 illustrates how a check form is used in the posting of disbursements. As the check is written, the data is also posted on the disbursement journal and the accounts payable ledger without the risk

DESCRIPTION	WORK IN PROGRESS				11 OTHER COSTS		LINE	12	13	14	15	16	17	18	19 GEN. LEDGER	
	7 LABOUR	8 MATERIAL	9 SUB-CONTRACTS	10	ACCT	AMOUNT									ACCT.	AMOUNT
							1									
							2									
							3									
							4									
							5									
							6									
							7									
							8									
							9									
							10									
							11									
							12									
							13									
							14									
							15									
							16									
							17									
							18									
							19									
							20									
							21									
							22									
							23									
							24									
							25									
							26									
							27									
							28									
							29									
							30									
							31									
							32									
							33									
							34									
							35									
							36									

PLACE LAST JOURNAL ON BOTTOM PEG AND SHINGLE OTHER JOURNALS ON TOP TO TOTAL AMOUNTS FOR SUMMARIZING BY PERIOD.

7	8	9	10	11	12	13	14	15	16	17	18	19

FIGURE 8-2. Continued.

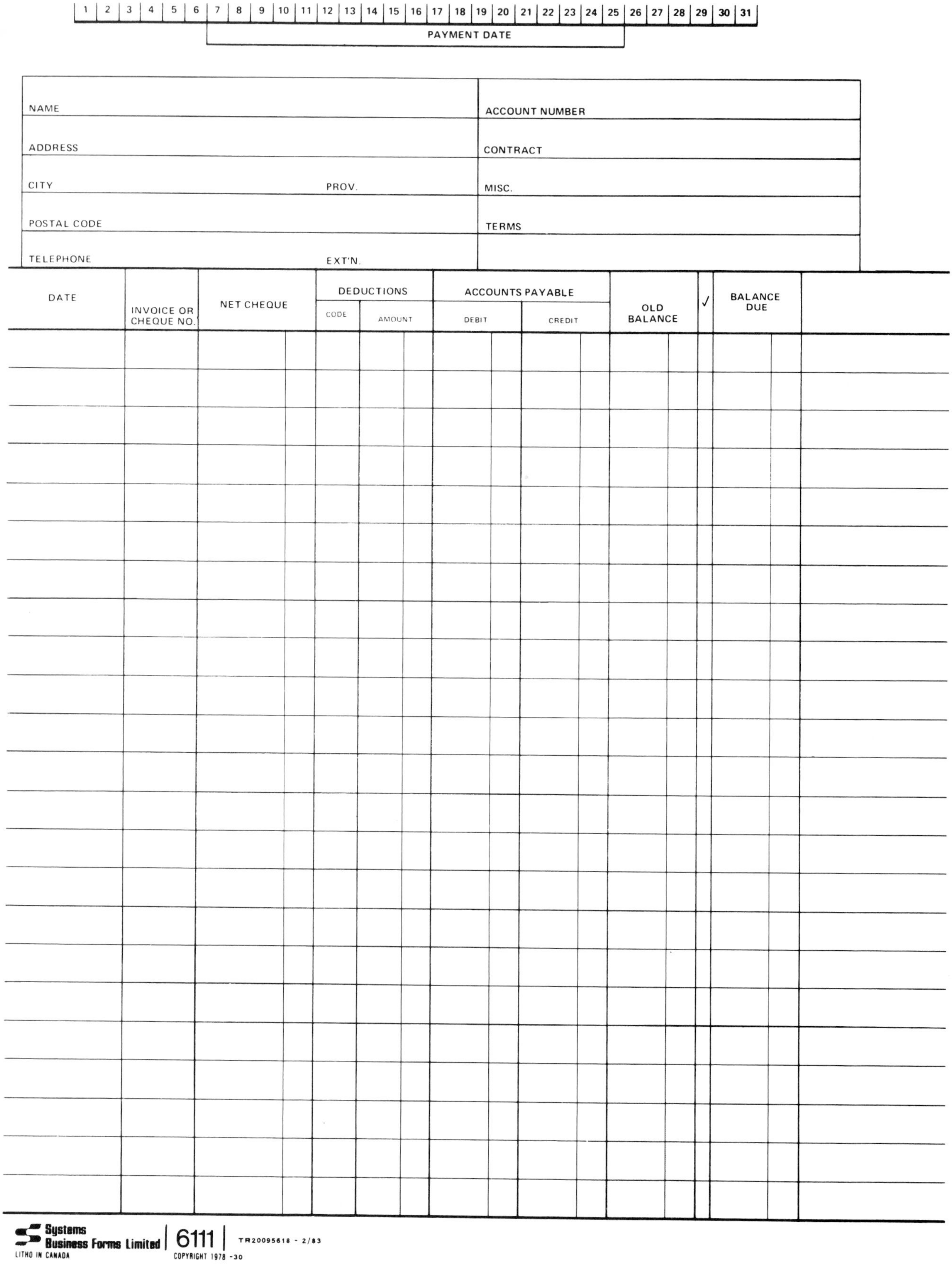

1	2	3	4	5	6	7	8	9	10	11	12	13	14	15	16	17	18	19	20	21	22	23	24	25	26	27	28	29	30	31

PAYMENT DATE

NAME		ACCOUNT NUMBER
ADDRESS		CONTRACT
CITY	PROV.	MISC.
POSTAL CODE		TERMS
TELEPHONE	EXT'N.	

DATE	INVOICE OR CHEQUE NO.	NET CHEQUE	DEDUCTIONS		ACCOUNTS PAYABLE		OLD BALANCE	✓	BALANCE DUE	
			CODE	AMOUNT	DEBIT	CREDIT				

Systems Business Forms Limited | 6111 | TR20095618 - 2/83
LITHO IN CANADA
COPYRIGHT 1978 -30

FIGURE 8-3. Accounts payable ledger (6111). (Form furnished by courtesy of Systems Business Forms Limited.)

CONTRACT PRICE CARD NO.

JOB NO.	ACCOUNT NO.
JOB NAME	
ADDRESS	
TELEPHONE	
PERSONAL CONTACT	

PROGRESS BILLINGS AND RECEIPTS					
BILLINGS			RECEIPTS		
DATE	AMOUNT	TOTAL	DATE	AMOUNT	TOTAL

DATE	JOB NUMBER	DESCRIPTION	LABOUR	MATERIAL	SUB-CONTRACTS		OTHER COSTS	
							ACCT.	AMOUNT
		ESTIMATE						

Systems Business Forms Limited | 6113 |
LITHO IN CANADA COPYRIGHT 1978 - 39
TR20099476 - 7/83

JOB COST LEDGER

FIGURE 8-4. Job cost ledger (6113). (Form furnished by courtesy of Systems Business Forms Limited.)

of a transcribing error. For a single contracting job, the set of job cost accounting forms would include the following: at least one record of invoices and work in progress form, an accounts payable ledger form for each vendor, at least one job cost ledger, at least one disbursement journal, and a number of check forms with stubs.

8.7 THE MECHANICS OF THE JOB COST SYSTEM

The objectives of the one-writing system are to keep job costs up-to-date and to maintain control over accounts payable, labor costs, mate-

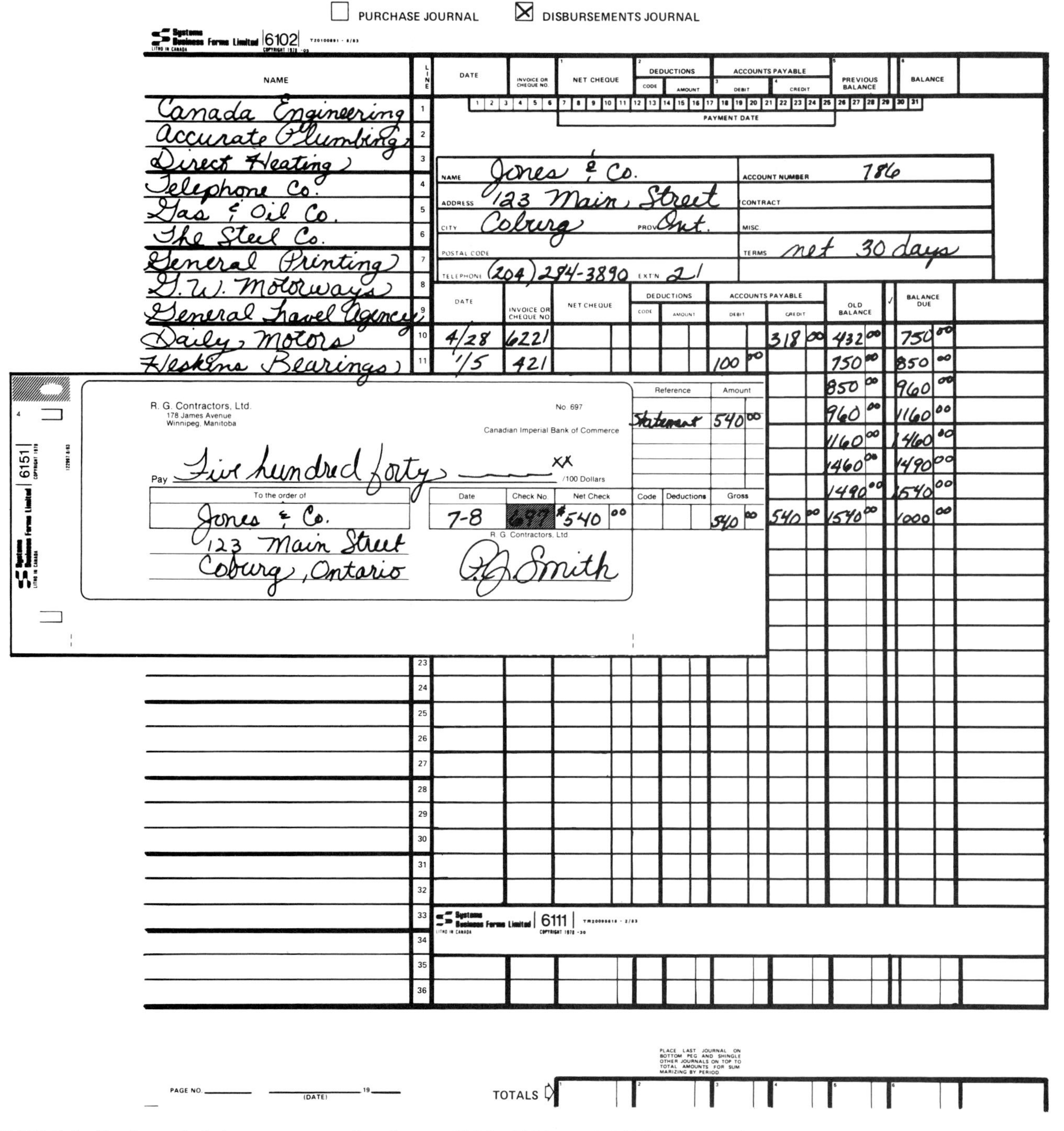
☐ PURCHASE JOURNAL ☒ DISBURSEMENTS JOURNAL

Systems Business Forms Limited 6102

NAME	LINE
Canada Engineering	1
Accurate Plumbing	2
Direct Heating	3
Telephone Co.	4
Gas & Oil Co.	5
The Steel Co.	6
General Printing	7
G.W. Motorways	8
General Travel Agency	9
Daily Motors	10
Weskins Bearings	11

NAME Jones & Co. ACCOUNT NUMBER 786
ADDRESS 123 Main Street CONTRACT
CITY Coburg PROV Ont. MISC.
POSTAL CODE TERMS net 30 days
TELEPHONE (204) 284-3890 EXTN 21

DATE	INVOICE OR CHEQUE NO	NET CHEQUE	DEDUCTIONS CODE	DEDUCTIONS AMOUNT	ACCOUNTS PAYABLE DEBIT	ACCOUNTS PAYABLE CREDIT	OLD BALANCE	BALANCE DUE
4/28	6221					318 00	432 00	750 00
7/5	421				100 00		750 00	850 00
							850 00	960 00
							960 00	1160 00
							1160 00	1460 00
							1460 00	1490 00
							1490 00	1540 00
7-8	697	540 00			540 00	540 00	1540 00	1000 00

R. G. Contractors, Ltd.
178 James Avenue
Winnipeg, Manitoba
No. 697
Canadian Imperial Bank of Commerce

Reference	Amount
Statement	540 00

Pay Five hundred forty XX/100 Dollars

To the order of
Jones & Co.
123 Main Street
Coburg, Ontario

Date	Check No	Net Check	Code	Deductions	Gross
7-8	697	$540 00			540 00

R. G. Contractors, Ltd
R. G. Smith

Systems Business Forms Limited 6151

Systems Business Forms Limited 6111

PAGE NO. (DATE) 19
TOTALS

FIGURE 8-5. Posting of disbursements using forms 6102, 6111, and 6151. (Forms furnished by courtesy of Systems Business Forms Limited.)

rial costs, subcontract costs, and all other costs which might be charged to work in progress.

An accounts payable ledger is set up for each vendor, inclusive of utility companies, and all invoices are recorded on the record of invoices. Every entry on the record of invoices will affect an accounts payable ledger and/or a job cost ledger.

It follows that no entry can be made on the record of invoices unless the left half of the journal is covered by an accounts payable ledger and/or the right half is covered by the appropriate job cost ledger. Just as no entry can be made on the journal without the presence of one or

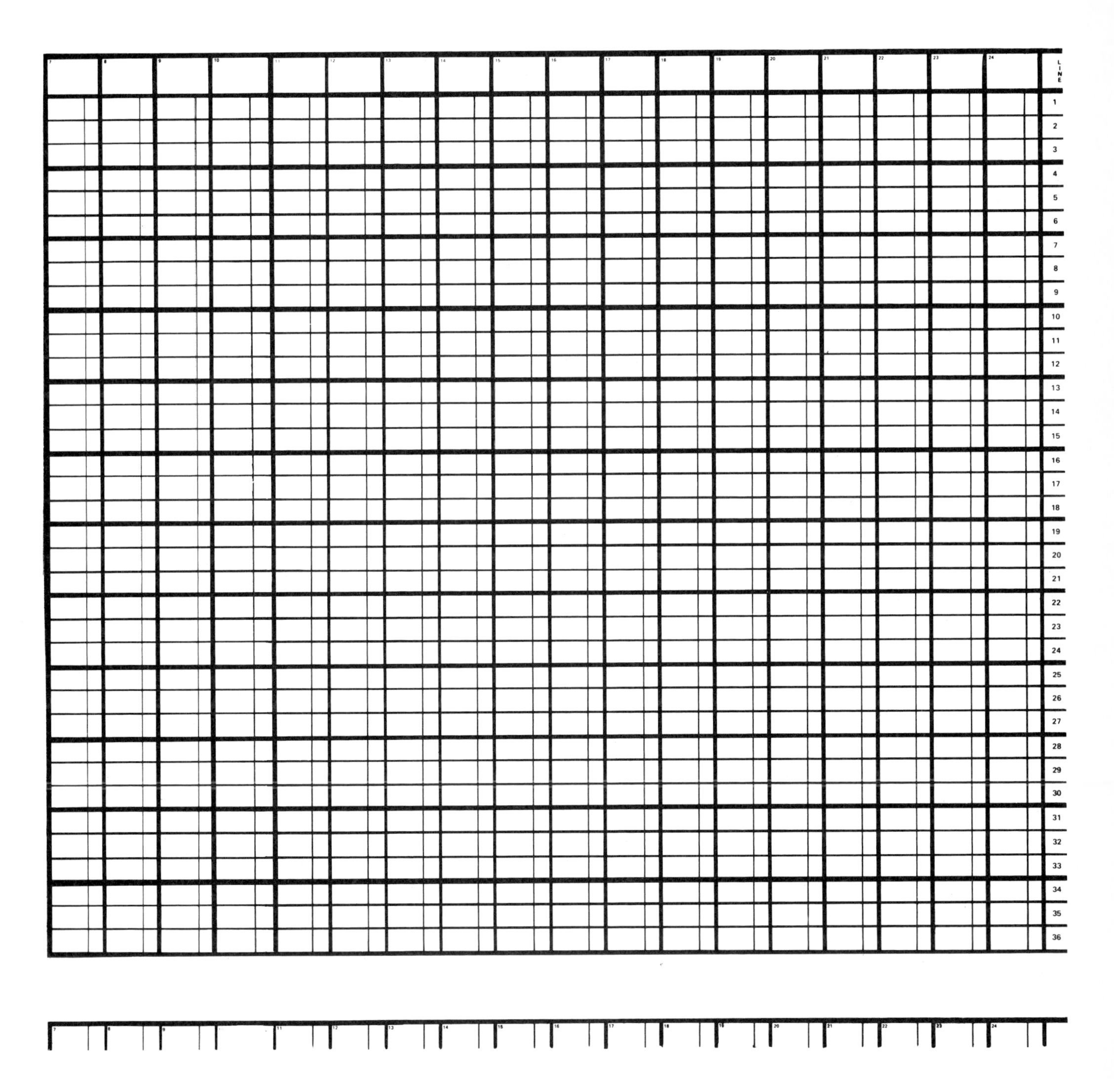

FIGURE 8-5. Continued.

both ledgers, neither can any entry be made on a ledger unless it is superimposed upon a journal.

8.8 JOURNAL CONTROL

As shown in Figure 8-2, the record of invoices and work in progress form is divided into 19 numbered columns. Some columns used to

Systems Business Forms Limited |6104|

RECORD OF INVOICES AND WORK IN PROGRESS

NAME	LINE	DATE	REFERENCE	ACCRUED PAYROLL CREDIT	OTHER CREDITS		ACCOUNTS PAYABLE		OLD BALANCE	BALANCE (MEMO)		DATE	JOB NUMBER
					ACCT	AMOUNT	DEBIT	CREDIT					
Jones & Co.	1	(1) 4-28	(2) 6221	(3)		(4)		(5) 318 50	(6) 432 00	(7) 750 00		(8)	(9)
Smith Bros.	2	4-28	6621					226 20	602 70	828 90		4/30	111
	3	4-29	2476									4/30	111
Brown Lumber Co.	4	4-28	4465					384 00	1061 50	1445 50		4/30	111
	5	4-29	L.D.S.	254 00								4/30	111
Wayne Plumbing Co.	6	4-28	A112					217 00	42 85	259 85		4/30	111
City	7	4-29						32 50	-0-	32 50		4/30	111
	8	4-29	2481									4/30	111
	9	4-29	2481									4/30	115
	10	4-30			699	129 76						4/20	111
Bell Telephone	11	4-30						93 75	0	93 75			
	12			254 00		129 76		1271 45	2139 05	3410 50			
	13												
	14												
	15												
	16												
	17												
	18												
	19												
	20												
	21												
	22												
	23												
	24												
	25												
	26												
	27												
	28												
	29												
	30												
	31												
	32												
	33												
	34												
	35												
	36												

PAGE NO. ______ ______ DATE ______19______ TOTALS

FIGURE 8-6. Typical entries on the record of invoices and work in progress form 6104. (Form furnished by courtesy of Systems Business Forms Limited.)

explain entries 1 to 19 are not numbered. Figure 8-6 shows how a contractor might renumber the columns 1 to 23 to help further break-down and describe entries made in each of the columns. A written key explaining exactly what is recorded in each of the 23 columns would be extremely helpful to those using the revised form. It must be noted, however, that these revised headings would not be used for accounting purposes; only the 19 original headings will match up with the other forms and records used in the overall accounting system. A number of

DESCRIPTION (10)	WORK IN PROGRESS 7 LABOUR (11)	8 MATERIAL (12)	9 SUB CONTRACTS (13)	10 (14)	11 OTHER COSTS ACCT	OTHER COSTS AMOUNT (15)	LINE	12 Inventory Dr. (16)	13 Inventory Cr. (17)	14 (18)	15 (19)	16 (20)	17 (21)	18 (22)	19 GEN. LEDGER ACCT	GEN. LEDGER AMOUNT (23)
							1	318 00								
Lumber		226 20					2									
#2476 from Stock		110 00					3		110 00							
Lumber		384 00					4									
Direct Labour 4/29	254 00						5									
Rough Plumbing			217 00				6									
Curb Permits					412	32 50	7									
Transf. to #115		(62 40)					8	62 40								
Transf. from 111		62 40					9		62 40							
Overhead					699	129 76	10									
							11								614	93 75
	254 00	720 20	217 00			162 26	12	380 40	172 40							93 75
							13									
							14									
							15									
							16									
							17									
							18									
							19									
							20									
							21									
							22									
							23									
							24									
							25									
							26									
							27									
							28									
							29									
							30									
							31									
							32									
							33									
							34									
							35									
							36									

PLACE LAST JOURNAL ON BOTTOM PEG AND SHINGLE OTHER JOURNALS ON TOP TO TOTAL AMOUNTS FOR SUMMARIZING BY PERIOD

7	8	9	10	11	12	13	14	15	16	17	18	19

FIGURE 8-6. Continued.

important journal control features follow (in all cases, the number reference refers to the revised 1 to 23 numbering):

1. Column 4 is used when an offsetting credit entry is required and it cannot be assigned by any other columns.
2. Column 9 contains the job number. Every entry on the journal which has been changed to a job can be traced to the particular job by its number. The repetition of the job number on the job cost ledger is visual proof that the charges have been made to the proper job.
3. Column 15 contains other costs such as fees and permits.
4. Columns 16 and 17 may be used as debit and credit columns for inventory.

8.9 TYPICAL ENTRIES ON THE JOURNAL

Ten typical journal entries are shown in Figure 8-6. Ledgers are omitted so that the entries can be seen in full. The total of all invoices is $1,271.45.

1. Invoice from Jones and Company for $318.00 chargeable to inventory. Accounts payable ledger used (but not shown in Figure 8-6).
2. Invoice from Smith Brothers for $226.20 chargeable to materials delivered to Job 111. Accounts payable ledger for Smith Brothers and job cost ledger used.
3. Materials worth $110.00 were withdrawn from inventory on requisition No. 2476 and delivered to Job 111. Job cost ledger used.
4. Brown Lumber Company invoice for $384.00 for materials delivered to Job 111. Accounts payable ledger and job cost ledger used.
5. Direct labor amounting to $254.00 charged to Job 111 from daily Labor Distribution Sheet (not shown here) for January 29. Job cost ledger used.
6. Invoice from Wayne Plumbing Company for $217.00 for completing rough plumbing on Job 111. Accounts payable ledger and job cost ledger used.
7. Invoice from the city for curb-breaking permit on Job 111. Accounts payable ledger and job cost ledger used.
8. Materials costing $62.40 were transferred from Job 111 to Job 115. This entry requires two lines to complete, on line 8 a debit is made to inventory and a credit to Job 111; on line 9, a credit is made to inventory and a debit to Job 115. Job cost ledgers for Job 111 and Job 115 are used.
9. End-of-the-month entry distributing $129.76, its share of overhead, to Job 111. Job cost ledger used.
10. Telephone Company bill for $93.75. Accounts payable ledger used.

An entry, unless it is a transfer of costs from one job to another, seldom affects more than one job. Vendors and subcontractors generally submit separate invoices for different jobs. Should an invoice be received, however, covering more than one job, one line on the left of the journal would be used to record the invoice, but as many lines as there were jobs concerned would be used on the right in the work in progress section.

8.10 USE OF THE JOURNAL AND LEDGER IN A ONE-WRITING SYSTEM

Figure 8-7 illustrates the record of invoices and work in progress form (6104) with accounts payable ledger (6111) and job cost ledger (6113) superimposed on top of it as would be the case during actual use. While the latter two forms cover and hide many of the journal entries, they illustrate how the one-writing system works. In this particular example, we are concerned with the entries of the Building Supply Co. in row 16. As you can see, columns 4, 5, and 6 of form 6104 are in alignment with the corresponding columns of form 6111. Column 8 of form 6104 (material) is also in alignment with the material column of form 6113. From Figure 8-7 you can see that entries made on forms 6111 and 6113 also appear on form 6104, the heart of the entire cost system.

8.11 CONTROL OF THE SYSTEM

Examination of the entries of Figure 8-6 indicate how costs on each job are classified. Since the journal is self-balancing (total debits equal total credits) and since the journal will be proved at least once each day of work, control over the accounts payable ledger and job cost ledger is absolute. Remember the following three points:

1. No entry can be made on a ledger unless it is on the board.
2. No entry can be made in Columns 5 and 11 through 15 except as a copy of an entry on an accounts payable ledger or job cost ledger.
3. Proving the journal proves the accounts payable ledger and job cost ledger.

Daily posting of all available costs should be insisted upon even though on some days a single posting is made. The major purpose of the system is to gain control of costs. The key to its success is that all related records be kept up-to-date and in balance at all times.

While the one-write system will not be used by all companies, it is based on sound, standardized accounting principles. A thorough understanding of this system can help the student understand other commonly used accounting systems.

8.12 POSTING INVOICES

Invoices are itemized statements furnished to a purchaser by a seller specifying the price of goods or services and the terms of sale. These important documents must be checked and approved before posting. All invoices should be checked to ensure that they are being credited to the correct job. To speed the posting of invoices, arrange the invoices in a logical order, and use the adding machine to run a tape on all invoices to be posted.

When the set of invoices is ready, post them on the job cost ledger and accounts payable ledger as they are "shingled" on the journal. As the set of invoices pertains to one job, one job cost ledger would be used while more than one accounts payable ledger would be used. There would be an accounts payable ledger for each vendor sending materials to the job.

As mentioned earlier in this chapter, all postings should be proved whenever a journal is completed or posting for the day is completed.

Systems Business Forms Limited | 6104 |

RECORD OF INVOICES AND WORK IN PROGRESS

NAME	LINE	DATE	REFERENCE	ACCRUED PAYROLL CREDIT	OTHER CREDITS ACCT	OTHER CREDITS AMOUNT	ACCOUNTS PAYABLE DEBIT	ACCOUNTS PAYABLE CREDIT	OLD BALANCE	BALANCE (MEMO)
Canada Engineering Co.	1	Dec 5/78	#626					650 00	—	650 00
Accurate Plumbing Co.	2	Dec. 8/78	#438					285 00	—	285 00
Direct Heating Co.	3	Dec 12/78	#73					120 00	300 00	420 00
Building Supply Co.	4	Dec 15/78	#4361					6000 00	—	6000 00
Telephone Co.	5	Dec 30/78	#4782					65 00	120 00	185 00
Office Supply Co.	6									
Gas & Oil Co.	7									
The Steel Co.	8									
J. R. Excavating Ltd.	9									
Direct Labour	10									
G.W. Motorways Express	11									
General Printing Co.	12									
L. & M. Painting Co.	13									
Accurate Plumbing Co.	14									
General Travel Agency	15									
Building Supply Co.	16									

PAYMENT DATE 1 2 3 4 5 6 7 8 9 10 11 12 13 14 15 16 17 18 19 20 21 22 23 24 25 26 27 28 29 30 31

NAME Building Supply Co. ACCOUNT NUMBER
ADDRESS 123 Main St. CONTRACT
CITY Cobourg, PROV ONT. MISC.
POSTAL CODE TERMS Net 30 Days
TELEPHONE (204) 284-3890 EXT'N 21

LINE	DATE	INVOICE OR CHEQUE NO	NET CHEQUE	DEDUCTIONS CODE	DEDUCTIONS AMOUNT	ACCOUNTS PAYABLE DEBIT	ACCOUNTS PAYABLE CREDIT	OLD BALANCE	✓	BALANCE DUE
15	Dec. 15/78	#3672					6000 00	—		6000 00
16	Jan. 22/78	#4161					360 00	6000 00		6360 00

Systems Business Forms Limited | 6111 |

PAGE NO. ____ DATE ____ 19__ TOTALS

FIGURE 8-7. Record of invoices and work in progress form 6104 with the accounts payable ledger 6111 and the job cost ledger superimposed on top of it as they are in use. (Forms furnished by courtesy of Systems Business Forms Limited.)

8.13 RESEARCH

1. In outline form, prepare a descriptive statement with respect to a typical system for job cost accounting.
2. Also obtain a set of one-writing forms for a job cost accounting system, and describe how it is designed to function.
3. Compare the two systems and make recommendations for use with respect to a building contractor starting a small business.

DATE	JOB NUMBER	DESCRIPTION	WORK IN PROGRESS: 7 LABOUR	8 MATERIAL	9 SUB CONTRACTS	10	11 OTHER COSTS: ACCT	AMOUNT	LINE	12 INVENTORY DR.	13 INVENTORY CR.	14 TELEPHONE	15 MTCE	16 FUEL	17 ADMIN SUPPLIES	18 OVER-HEAD	19 GEN. LEDGER: ACCT	AMOUNT
									1								234	650 00
									2									
									3									
									4	6000 00								
									5			65 00						
									6						85 00			
									7					350 00				

CONTRACT PRICE 35,000.00 CARD NO 1

JOB NO 241 ACCOUNT NO

JOB NAME SHOPPING CENTER

ADDRESS 1595 BUFFALO PLACE

TELEPHONE 297-4321

PERSONAL CONTACT J. ALLEN

PROGRESS BILLINGS AND RECEIPTS

BILLINGS: DATE	AMOUNT	TOTAL	RECEIPTS: DATE	AMOUNT	TOTAL
1-15-79	4000 00	4000 00	1-12-79	2000 00	2000 00
2-15-79	6000 00	10000 00	2-3-79	5000 00	7000 00
3-15-79	7500 00	17500 00	10-4-79	5000 00	12000 00

DATE	JOB NUMBER	DESCRIPTION	LABOUR	MATERIAL	SUB CONTRACTS		OTHER COSTS: ACCT	AMOUNT	LINE
									10
OCT 12/78	241	ESTIMATE 23500 00	750 00	8300 00	5700 00		ENG	2000 00	11
JAN 8/79	241	# 8271		1200 00					12
JAN 14/79	241	# 151			180 00				13
JAN 15/79	241	Direct Labour #	840 00						14
JAN 20/79	241	# 632					240	85 00	15
JAN 23/79	241	# 4161		360 00					16

Systems Business Forms Limited 6113 JOB COST LEDGER

FIGURE 8-7. Continued.

4. From job cost accounting, extend the research to other accounting areas such as payroll. Use a comparison flowchart such as Figure 8-7 to prepare your statement.
5. Reread Chapter 1. Indicate in writing how compatible the principles of a one-write system of job cost accounting and PERT/CPM are.
6. Visit a jobsite to observe job cost accounting firsthand. Learn of personnel requirements with respect to training and experience for such accounting.
7. Build a file of brochures and sample forms with respect to accounting systems.

8.14. REFERENCES

8.14.1 Systems Business Forms Limited. Rexdale, Ontario; Winnipeg, Manitoba; and Vancouver, British Columbia.

Job Cost Accounting—Systems Series 6300

Accounts Payable—Systems Series 6100, and

Payroll—Systems Series 2100.

9

ELEVATIONS FOR BUILDING

Any building with a basement must be set into the ground at an appropriate depth, a depth in harmony with the contours and elevations of the building site and its surrounding locality. The elevation at which the building is set must not only be aesthetically pleasing, but it also must be functional. Setting the building too low will allow water to enter windows and doorways. Setting the building at too high an elevation may result in an inordinate amount of expensive finish grading. In all cases, the building must be situated at an elevation which allows for proper drainage from the building's plumbing system to the municipal sewer or private disposal field.

In developed areas, some critical elevation measurements such as the depths of public sewers and building drains are provided by the municipal authority. Many city engineering departments employ an "official grade striker" who will come to the jobsite after all surveying work is complete. The job of the official grade striker is to set up a wooden stake peg showing the depth of the sewer line and the height of the city sidewalk curb.

In less developed areas, the builder must take more responsibility for establishing optimum levels for finished floors, finish grade, footings, building drains, building sewers, sidewalks, and roads. Building entrances and driveways are particularly important to the occupants in terms of easy and safe entrance and exit, energy conservation, and appearance.

While a builder is not normally a land surveyor, he must pay close attention to the establishment of critical levels and be insistent on maintaining them during the construction process. The excavator, for example, must leave undisturbed soil for footings, or the foundation may later move or break under pressure of the building load. Such footings must not only be on solid ground, but they must also be at the appropriate level so that normal building operations may proceed: The exit door sill is placed without conflict with the existing grade and the building drain properly meets the tile field or public sewer.

In this same regard, the builder must be aware of the exact location of all building lines. A building line is the prescribed minimum setback distance from the street or avenue fronting the lot; it is illegal to build between that line and the street. It is also the prescribed minimum distance that the structure must be kept from other buildings or boundaries of the property. A house that is built infringing on these regulations may have to be moved to its correct place on the lot. Be careful; this is very important! A builder may recover from all sorts of mistakes, but an avoidable error of this nature may force a builder out of business immediately.

It is evident that a builder requires a complete set of carefully established and accurately recorded measurements. These measurements must be easily accessible during all phases of construction, one reason a story rod is often employed by builders and contractors (see Chapter 2.) In addition, these measurements must be reliable.

To ensure accuracy and consistency in *vertical* measurement of building elevations, a bench mark is used. A bench mark is a single reference point, readily accessible during all stages of construction, from which all elevation measurements are taken.

9.1 SPECIFIC OBJECTIVES

On reading this chapter and conducting the suggested research, the reader should be able to

1. establish and use a bench mark for a light construction project,
2. select and use a set of instruments and/or devices for measuring on-site elevations,
3. record on-site elevations data for building reference purposes,
4. read and interpret a contour map for building site planning purposes, and
5. use library reference data/books/computer software in the field of surveying for vocational reasons.

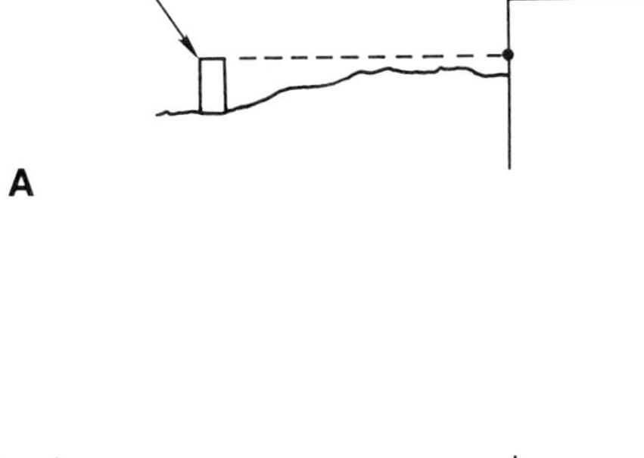

9.2 THE BENCH MARK

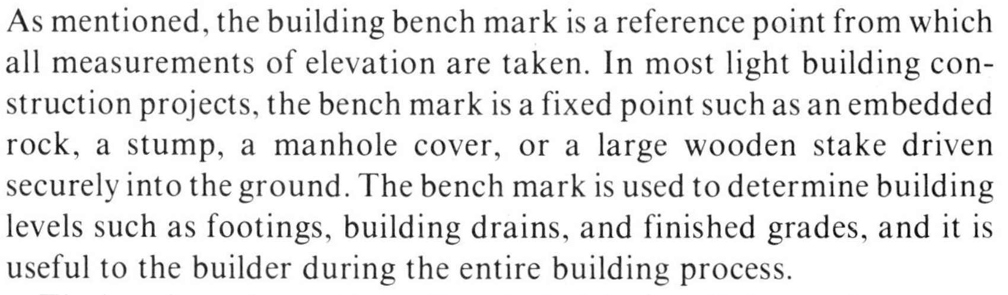

As mentioned, the building bench mark is a reference point from which all measurements of elevation are taken. In most light building construction projects, the bench mark is a fixed point such as an embedded rock, a stump, a manhole cover, or a large wooden stake driven securely into the ground. The bench mark is used to determine building levels such as footings, building drains, and finished grades, and it is useful to the builder during the entire building process.

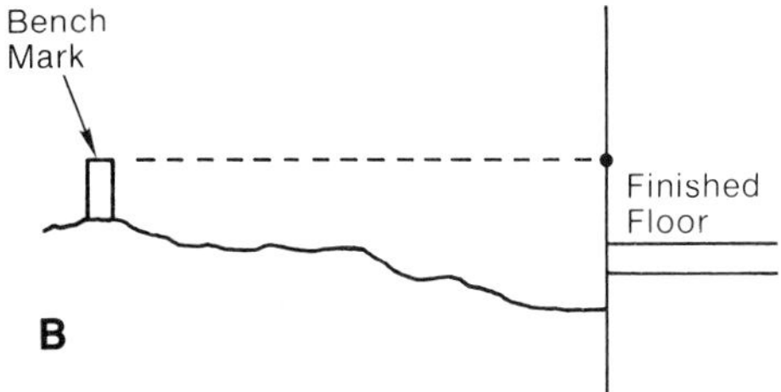

The bench mark must be well protected during all stages of construction, but need not be a permanent bench mark as required for land surveying. As shown in Figure 9-1, a bench mark can be placed in any reasonable position with respect to the actual building elevations as long as it is within easy measuring distance of the building project.

FIGURE 9-1. All vertical measurements are taken using a fixed bench mark as a reference point. As shown in examples A and B, the relative position of the bench mark to the building or building site does not matter as long as all measurements are taken and recorded accurately. Once established the bench mark should not be moved.

9.3 MEASURING EQUIPMENT

Once a bench mark has been established, elevation measurements can be made using any number of measuring and leveling tools, some more

A B

FIGURE 9-2. Precision leveling instruments: (A) transit levels and (B) builder's level.

accurate than others. The degree of accuracy required determines which tools should be used.

Precision direct leveling instruments used to measure differences in elevations are the transit level and the builder's or dumpy level (Figure 9-2). Operation of these precision leveling tools was discussed in Chapter 2. When working with precision leveling instruments, all elevation readings are taken on graduated wooden rods called leveling rods (Figure 9-3). Figure 9-4 illustrates how a builder's or dumpy level is used to determine elevations.

As indicated in Figure 9-5, a set of leveling tools, rather than instruments, can also be used for marking off and measuring less

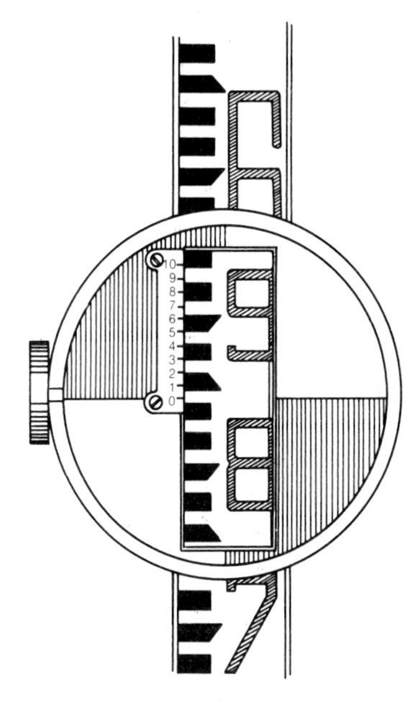

FIGURE 9-3. The leveling rod used to measure vertical elevations.

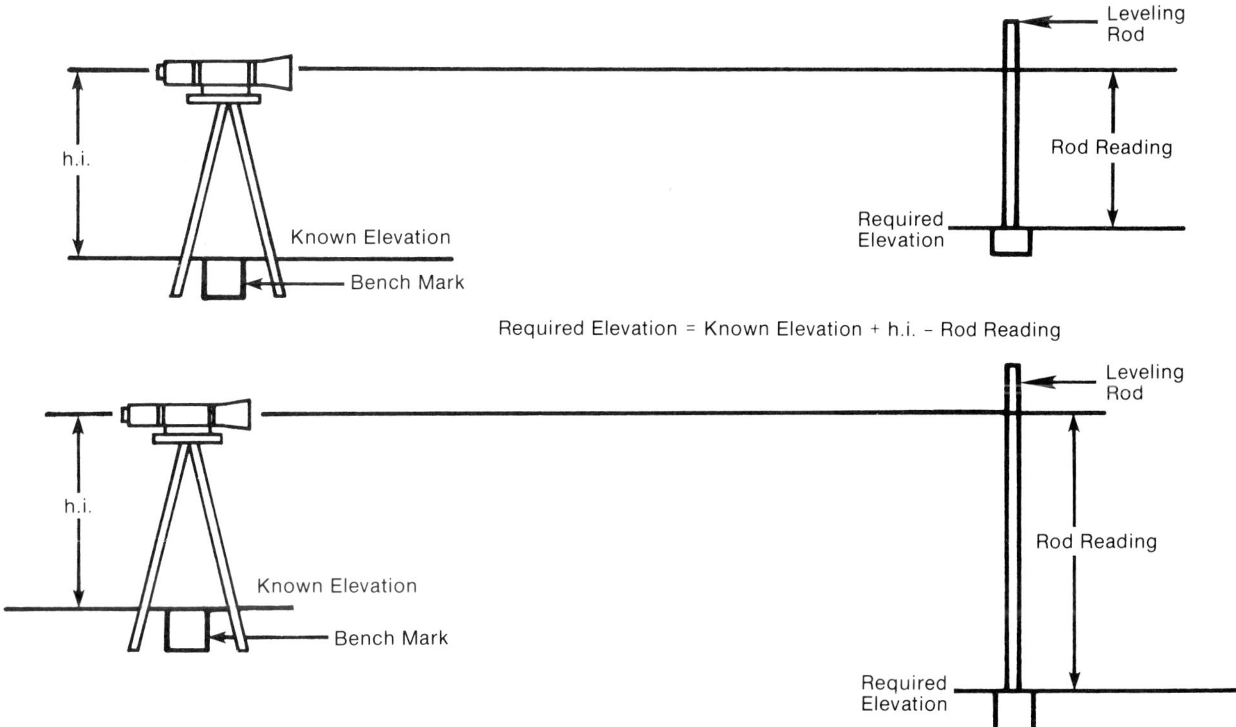

FIGURE 9-4. Using a builder's level and a leveling rod to determine elevation.

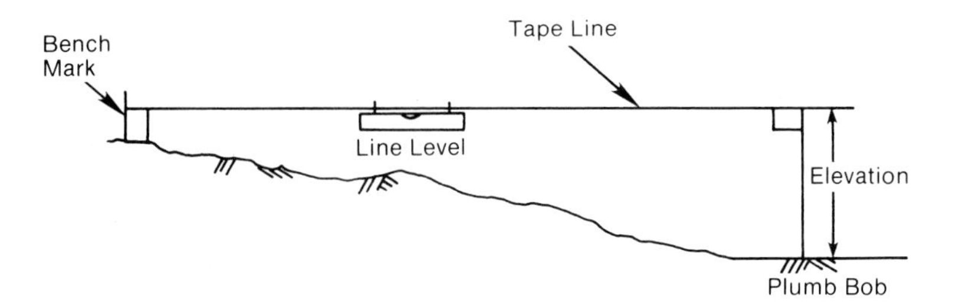

FIGURE 9-5. Using a set of measuring and leveling tools to determine an elevation.

accurate elevations. These leveling tools include the steel measuring tape, the line level, and the plumb bob. Still another measuring system involves the use of a spirit level equipped with a pair of level sights and a story rod. Figure 9-6 illustrates how these two tools are used to record elevations taken from a bench mark.

In all cases the precision levels, line level, and spirit level are used to ensure that the builder is measuring up or down from a true horizontal plane. Once a true horizontal plane is established out from the bench mark, points of equivalent elevation can be marked on batter boards, stakes, the story rod, or the building itself. Differences in elevation between two points can also be determined.

9.4 WORKING WITH CONTOUR MAPS AND LINES

Figure 9-7 illustrates a contour map of a proposed building site. Contour maps are designed to give a graphic indication of the land elevations of a given area. For this example the unit of measurement is simply stated as x units (British Imperial System or Metric System) and could be assigned any reasonable value such as 1′, 10′, 10m,

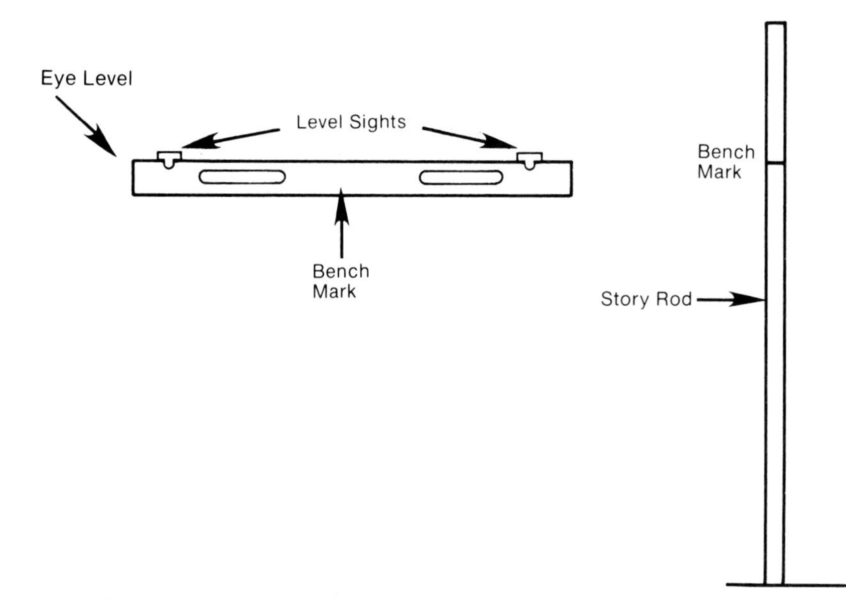

FIGURE 9-6. Using a spirit level and story rod to read an elevation.

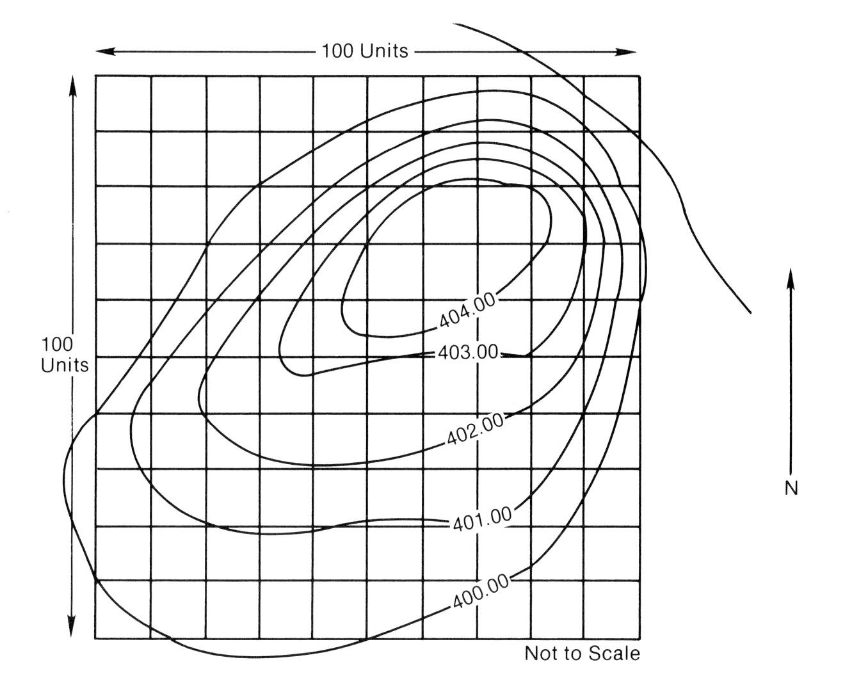

FIGURE 9-7. A contour map of a proposed building site.

etc./unit. Five contour lines are superimposed over the map. Each contour line connects points of equal elevation, and the difference in elevation between adjacent contour lines is one unit.

This hypothetical piece of land rises generally from the southwest corner to the northeast corner. It rises at least four units and then falls abruptly to the northeast corner. The more gentle slope is from the top of the hill to the south and west. Drainage of the area covered by the contour lines seems excellent.

A general overview of this site indicates no major elevation related construction problems, but before a final building decision is made a closer examination of the property is in order.

9.5 A DETAILED PROPERTY EXAMINATION

Figure 9-7 illustrates horizontal measurements and elevations taken in one unit increments. Data missing from this contour map includes information concerning adjacent pieces of land, the utility services available, vegetation, soil analysis, the direction of prevailing winds, annual rainfall, temperature range, and longitude and latitude readings. Rather than specify such detailed data for this sample problem, the reader is asked to think in terms of the data relevant to his or her own geographic area. With this in mind, the detailed examination of Figure 9-7 as a building site can continue.

One method of gaining insight as to how a proposed building would fit into the contours of the land would be to cut a piece of paper into a scale model of the building, say ten units by 20 units. This scale model can then be placed on the contour map in various locations and the relative advantages and disadvantages of each location determined.

For example, if the building is placed on the highest point on the site somewhere within the 404.00 contour line, it would be subject to maximum exposure to the elements. While this fact may not be sufficient to eliminate this position from consideration, certain home and building designs would fair better than others in terms of energy efficiency and aesthetic appearance, and it would be foolish for the builder to ignore these considerations. If the building is located at the northeast corner of the lot, the steep grade present in this area must be accommodated, probably at a higher cost per square unit of floor space. If located at the south central part of the site between contour lines 401.00 and 402.00 where the slope is more gentle, access will be easier and some shelter will be obtained from the colder northwest winds.

Drainage for the two sites is probably adequate; whereas, further detailed checks should be made for the upper area, especially for some soils that retain water.

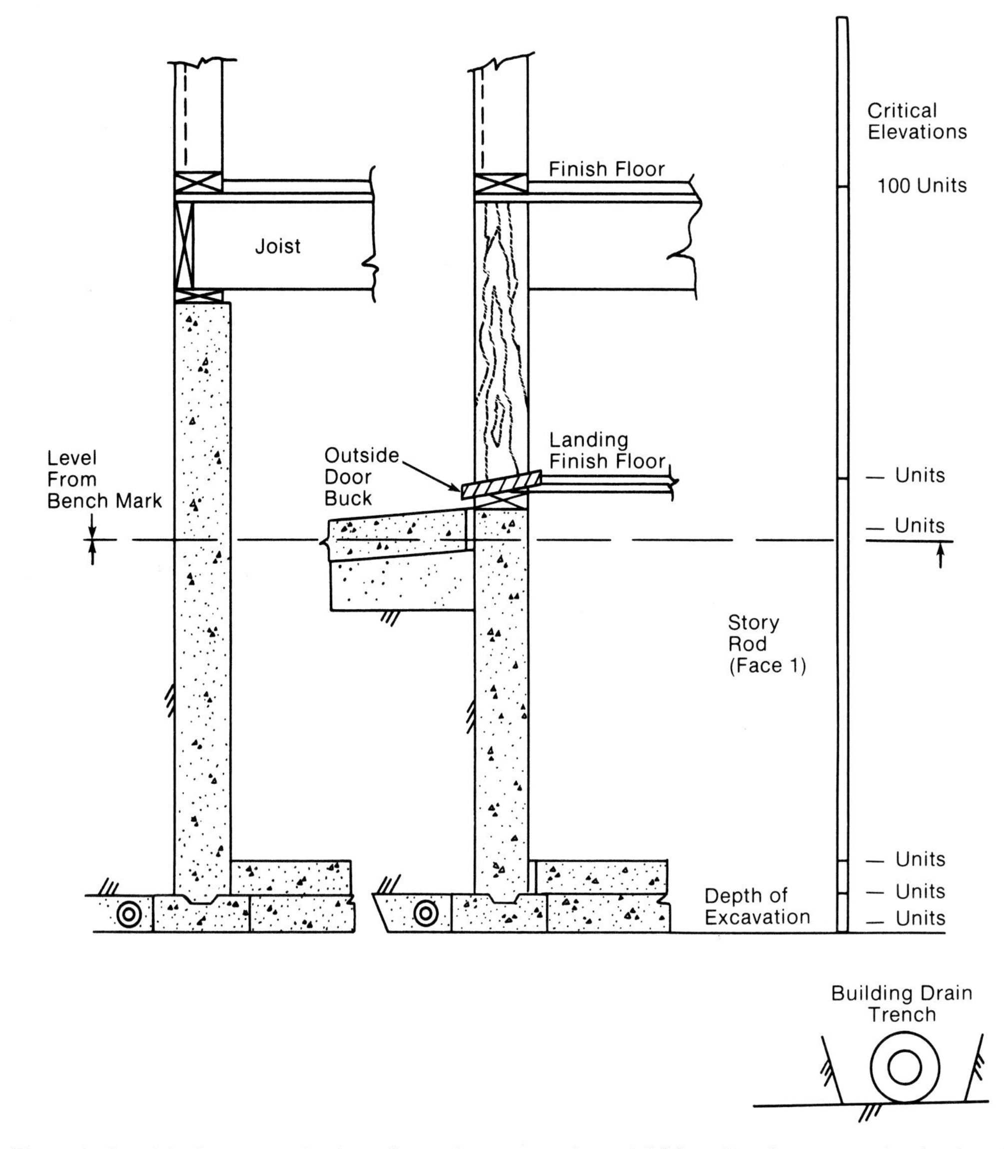

FIGURE 9-8. The relationship between the bench mark, story rod, and 100 unit reference point in the determination of building elevations.

9.6 THE BUILDING SITE

Following the selection of the general building location, the detailed elevation measurements can be taken with reference to a bench mark. Using a precision leveling instrument with a leveling rod is the preferred method of taking these measurements, but the job can also be done with a spirit level and rod.

The unit of measurement used in taking elevation measurements should be much smaller than the unit of measurement used on the contour map. A unit between one-half and one-tenth as long is usually sufficient.

It is best to use an assistant who can help set up stakes, position the leveling rod, and mark off the actual elevations on the stakes or story rod. Accurate measurement is extremely important for determining the exact location and amount of earth that must be removed so that rainwater will run around and not through the finished building. A visual inspection can help in this regard, but it is often not sufficient, especially on a side hill location or where a number of building elevations must be considered.

Key points on the jobsite are marked with wooden stakes. When extreme accuracy is required, a tack or nail is driven into the top of the stake and measurements are taken to this point. Instructions to the builder or excavator are marked on the stakes. The construction object to which the stake refers is noted with an abbreviation such as NE cor. for northeast corner or DrC for driveway curb. The distance above or below the stake is noted and preceded by the letter F for fill if the finished elevation is above the stake and C for cut if the finished elevation is below the stake. For example, C-8.5 indicates that the building object referred to, say the building drain, should be located 8.5 units below the reference stake. The excavator would now know to remove earth to this point.

9.7 CRITICAL BUILDING ELEVATIONS

The total concept of building elevations is best understood by studying a set of elements and how they interact. These elements are the bench mark, the various building sections or floors, the story rod, and the contour map. Figures 9-8 and 9-9 illustrate the relationships between these elements.

As shown in Figure 9-8, the first finished floor of the building has a critical elevation of 100.00 units as marked on the story rod. This 100.00 unit elevation reference point has been arbitrarily selected by the builder and has no relationship to the elevation units used on the contour map.

The builder can now use this 100.00 unit mark as a reference point for marking other elevations. Notice that it is very convenient to add or subtract measurements to or from this 100.00 unit base figure. So, elevations higher than the first finished floor will have unit values greater than 100.00, and those below the first finished floor level will have unit values less than 100.00. For example, the elevation of the

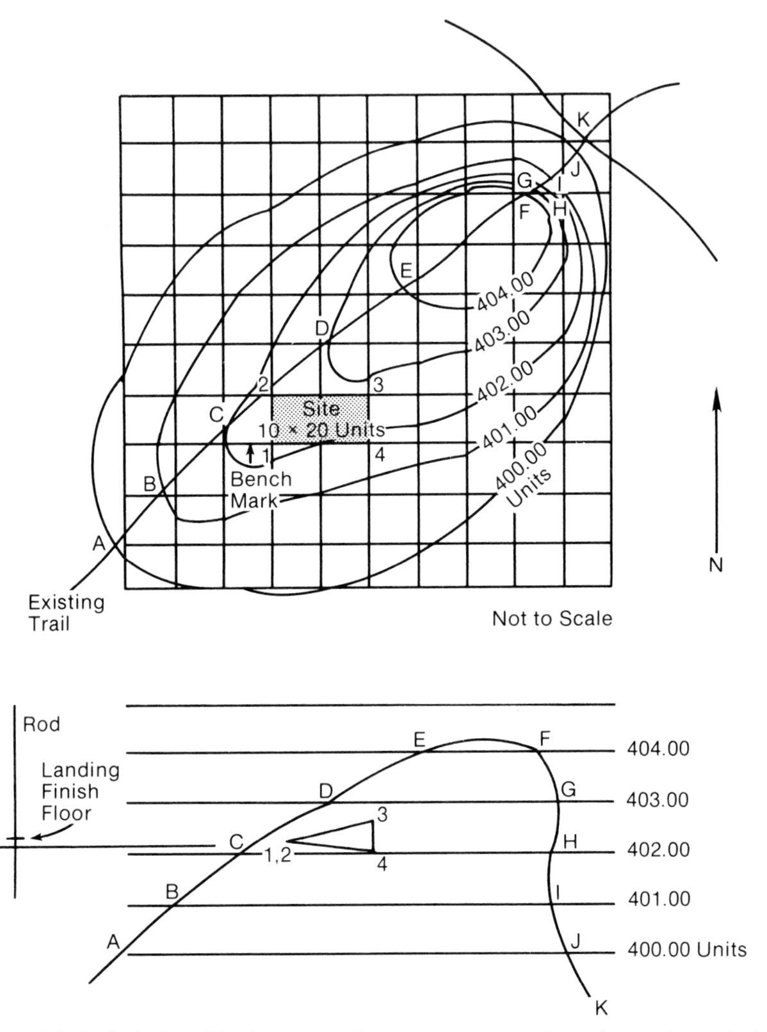

FIGURE 9-9. Relationship between the contour map, bench mark, and story rod.

second finished floor may be 108.00 units; whereas, the elevation at the base of the foundation footer may be 92.00 units.

Notice too that the bench mark level on the story rod does not correspond to any critical elevation on the building itself. But because the bench mark level is the ultimate vertical reference point, extreme care must be taken when establishing the story rod elevation relationship between the bench mark level and the first finished floor (100.00 unit) level.

The bench mark elevation can be transferred to the story rod by resting a spirit level fitted with two eye sights directly on the bench mark (Figure 9-6). If a precision leveling instrument is mounted on a tripod and then set on the bench mark, it is essential to account for the height of the tripod when transferring the bench mark measurement to the story rod (Figure 9-10).

9.8 SUMMARY

Critical elevations for a building must be determined before excavation work commences.

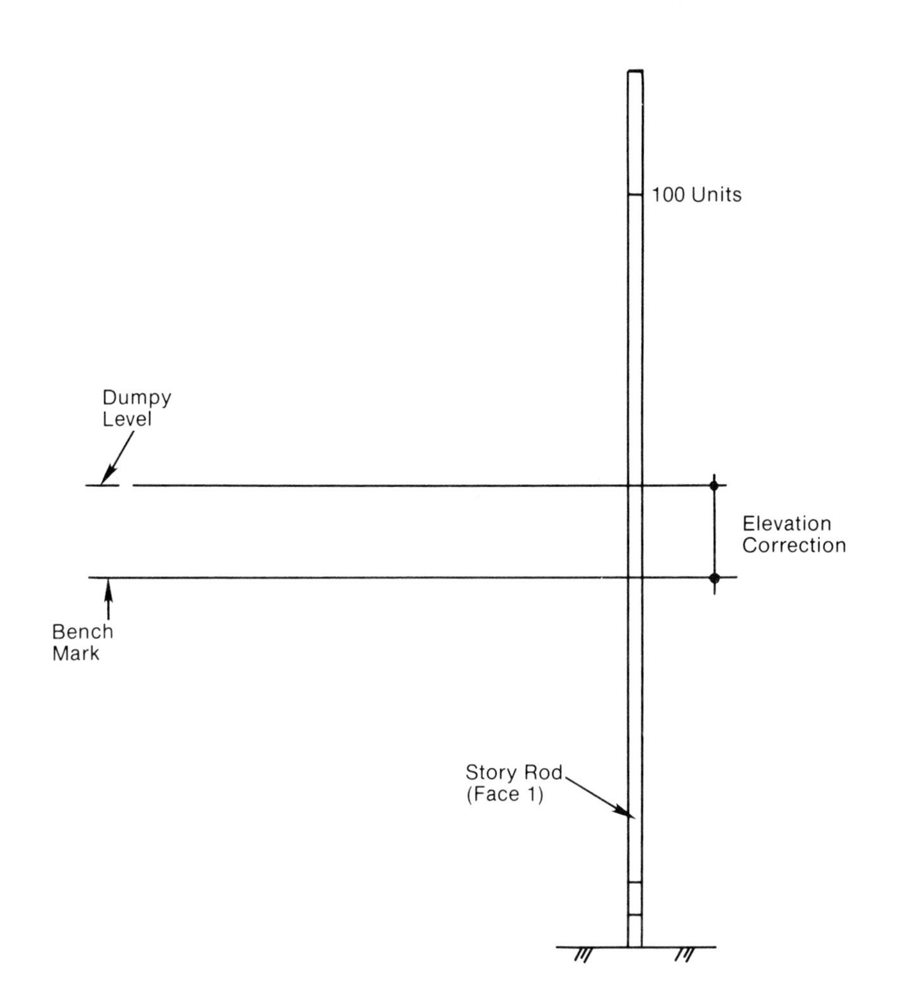

FIGURE 9-10. Transferring the bench mark level to the story rod using a builder's level. Remember to account for the height of the builder's level.

1. The site must be thoroughly studied and a bench mark established.
2. The building section must be known: the depth of footings; the height of basement framework, including construction details with respect to forms, door and window bucks, and first level floor framing.
3. The sets of critical measurements need to be stored for repeated use—stored as notes in a field book, marks on a story rod, or possibly as both.

Remember, it is practical to work with excess drainage water rather than against it.

9.9 RESEARCH

1. Obtain a contour map for your area. Identify a vacant piece of property on the map and overlay it with transparent paper with a suitable grid (British Imperial System or Metric System). Trace the contour lines. Indicate with an arrow the direction north.
2. Describe the piece of land.
3. Mark the general area where you would build a house.
4. In relation to the building site selected locate a possible bench

mark (⊤). Draw a straight line arrow through to show the general direction of rain runoff.

5. Using part of the arrow as a hypotenuse, draw a right-angled triangle to show the elevation measurement for a set of contour lines.
6. Assuming that utility services are not available at this property, mark WW for a desirable location for a water well, DF for waste drainage field, PT for propane tank or other fuel storage, and EG for electric generator. Give serious consideration to safety and to environmental factors for each location. Consult with municipal authority with respect to your plan.
7. For a given lot of land in a developed area, obtain a set of municipal elevations for house drain and public sewer.
8. Prepare a contour map (British Imperial System or Metric System) for a house lot or small fenced area. Make decisions about boundaries, the contour interval, and the bench mark. With a dumpy level (or equivalent) and a set of pegs, stake the area and then make a drawing to a given scale.
9. Comment in writing on the following statement: "For comparison purposes, all critical vertical measurements for a small building can be stored (collected) on one rod."
10. Explain how one uses surveying equipment and a field data book to collect elevations data, rather than a story rod.

9.10 REFERENCES

9.10.1 Ketchem, Milo S. *Handbook of Standard Structural Details for Buildings.* New Jersey: Prentice-Hall, Inc., 1956.

9.10.2 Stillman, W. J. *Construction Practices for Project Managers and Superintendents.* Virginia: Reston Publishing Company, Inc., 1978.

9.10.3 Untermann, Richard K. *Principles and Practices of Grading, Drainage and Road Alignment: An Ecologic Approach.* Virginia: Reston Publishing Company, Inc., 1978.

9.10.4 Wass, Alonzo. *Data Book for Residential Contractors & Estimators.* Virginia: Reston Publishing Company, Inc., 1979.

10

BUILDING AND THE ENVIRONMENT

A building project does not take place in a vacuum. A construction project will be affected by the site and surroundings in which it takes place. The local environment will in turn be altered in some way by the process and ultimate completion of the project. It is important for the builder to always keep in mind this integral relationship between the job being done and the site and surroundings in which the process is taking place.

10.1 GENERAL OBJECTIVES

This chapter describes the close association between various aspects of light construction and the environment. Foundations, footings, basements, pilings, berms, landscaping, wells, drains, and sewers usually all have an intrinsic relationship with topography and soil. In addition, completed buildings are directly influenced by earth movements, the insulation and protection provided by the earth, and by the surface and below grade drainage. The type of earth a builder will have to deal with is a primary concern, particularly as soil types differ from region to region. The different types of soil and fill encountered in construction will, therefore, be discussed here.

Another important factor to consider is the climate in which light construction will be done and how this will affect both building design and construction techniques. In addition, there are other environmental variables that must be fully discussed and understood by the prospective light construction builder. In the broadest sense, builders must always be aware of the indirect factors—soil, climate, presence of water—that will greatly affect how light building construction will be done.

10.2 SPECIFIC OBJECTIVES

This chapter should encourage the reader to:

1. appreciate the relationship between building and environment;
2. become knowledgeable about the climate, about the environment generally, and how these factors affect builder's operations;
3. respond to known soils technology when preparing a site, when backfilling, and when landscaping;

4. develop a regional soils' file; and
5. develop regional files (long-term data) with respect to water and water levels, wind, temperature, and humidity levels and earth movements.

10.3 CLIMATE AND GEOGRAPHY

Prevailing weather conditions affect changes on the earth's surface. Every region is influenced by temperature range, the amount of moisture (rain, snow, hail, vapor) it receives, and wind speeds and directions of airflow. By careful reference to the appropriate atlases, sheet maps, and geography books, it is possible to become familiar with world climate and vegetation patterns. Additional reading will be necessary for an overview of logical principles including composition of the atmosphere, solar radiation, heating of the earth's surface, heating of the air, cooling of the earth, vertical arrangements of temperature, vertical air movements, pressure, convection, deflection caused by earth's rotation, winds and pressures, humidity, condensation of moisture, evaporation, and precipitation. Also important is a basic understanding of the reasons for climatic differences on the earth including such factors as latitude, the influence of land and water, pressures and winds, cyclonic storms, ocean currents, mountain barriers, and vegetation.

The reader may question whether these areas of study are truly pertinent to the practical aspects of light construction. While not directly concerned with the actual mechanics of building, this type of background knowledge is vital for developing proper construction techniques. By knowing how to accomplish light construction projects in a variety of climates and conditions, a builder will be better able to adapt to whatever local conditions he or she encounters. In other words, a general knowledge of climates and world geography better prepares a builder for an in-depth study of any one particular region or locality.

A versatile and efficient builder should be able to do light construction anywhere and under most conditions. Hypothetically the good builder should be able to successfully accomplish a project 2,500 miles (4,023.4 km) from home by adapting general knowledge to local conditions. For that reason, a thorough understanding of background subjects—climate and geography in this case—is important in the building profession.

A builder with a strong general background can plan with a degree of confidence, knowing what to expect in terms of working conditions, project planning, selection of materials, and construction methods. Typical decisions made with respect to climate, soils, and other environmental factors are

1. temperature range and materials specifications,
2. rainfall and permeance of materials,
3. hydrostatic pressure and foundation materials,
4. water table and site materials,
5. wind and permeance of materials, and
6. earth/water ratios and compaction requirements for proper foundation work.

10.4 SOILS

To start a building, other than a portable one, a builder must give attention to the soil of the land where work will be done. Soil properties have been studied and classified in different ways by different groups of people, each for their own specific purposes. For example, objectives of classifying soils will be different for oil-field drillers, army engineers, and light construction builders. Probably the best classification system is the Unified Soil Classification System, used by the U.S. Army Corps of Engineers, Bureau of Reclamation as well as others (Table 10-1). This system places soils in 15 groups, identified by name

TABLE 10-1 THE UNIFIED SOIL CLASSIFICATION SYSTEM*

Main Group			Group Symbols	Typical Names
Coarse-Grained Soils—More than half of material is larger than No. 200 sieve size.	Gravels—More than half of coarse fraction is larger than No. 4 sieve size.	Clean Gravels—Little or no fines.	GW	Well graded gravels, gravel-sand mixtures, little or no fines.
			GP	Poorly graded gravels, gravel-sand mixtures, little or no fines.
		Gravels With Fines—Appreciable amount of fines.	GM	Silty gravels, poorly graded gravel-sand-silt mixtures.
			GC	Clayey gravels, poorly graded gravel-sand-clay mixtures.
	Sands—More than half of coarse fraction is smaller than No. 4 sieve size.	Clean Sands—Little or no fines.	SW	Well graded sands, gravelly sands, little or no fines.
			SP	Poorly graded sands, gravelly sands, little or no fines.
		Sands With Fines—Appreciable amount of fines.	SM	Silty sands, poorly graded sand-clay mixtures.
			SC	Clayey sands, poorly graded sand-clay mixtures.
Fine-Grained Soils—More than half of material is smaller than No. 200 sieve size.	Silts and Clay—Liquid limit less than 50.		ML	Inorganic silts and very fine sands, rock flour, silty or clayey fine sands with slight plasticity.
			CL	Inorganic clays of low to medium plasticity, gravelly clays, sandy clays, silty clays, lean clays.
			OL	Organic silts and organic silt-clays of low plasticity.
	Silts and Clays—Liquid limit greater than 50.		MH	Inorganic silts, micaceous or diatomaceous fine sandy or silty soils, elastic silts.
			CH	Inorganic clays of high plasticity, fat clays.
			OH	Organic clays of medium to high plasticity.
Highly Organic Soils			Pt	Peat and other highly organic soils.

*See 10.8.1.

and letter symbols. Coarse-grained soils are defined as soils with more than 50% of the material larger than the No. 200 sieve. Fine-grained soils are divided into soils with low (L) or high (H) compressibility.

B. Broms and L. Forsblad proposed another classification system (Table 10-2) to be used with reference to a mechanical compaction of soils. According to this system, the gradation of the soils in Groups 1, 2, and 3 is predetermined. For Group 4, the strength of the soil has to be measured by testing its unconfined compressive strength. The builder should have a good general working knowledge of the various soil types and how they could affect a building project.

TABLE 10-2 SOIL CLASSIFICATION SYSTEM WITH REFERENCE TO COMPACTION*

I.	Rock fill and granular soils with large stones and boulders[1]
II.	Sand and gravel[1]
	a. Well-graded
	b. Uniformly graded
III.	Silt, silty soils, etc.
	a. Silty sand, silty gravel, moraines
	b. Silt and sandy silt, clayey sand, clayey gravel
IV.	Clay
	a. Clay with low or medium strength[2]
	b. Clay with high strength[3]

[1]With less than 5% to 10% of material smaller than 0.002" (.06 mm)

[2]Unconfined compressive strength < 29 psi (.2 MPa)

[3]Unconfined compressive strength > 29 psi (.2 MPa)

*See 10.8.1.

When excavating for a foundation or when preparing a roadway or driveway, consideration should be given to the origin of the soils, their characteristics, and the type of fill materials that should be used for backfilling or for other compacting. Table 10-3 indicates the origins of soils and Table 10-4 lists special types of fill materials. Note that organic soils are normally not used as fill materials; therefore, they are usually cleared from a building site before excavation and are often stockpiled for landscaping purposes after backfilling and roadway operations are completed.

Observations in the field and compaction test results in laboratories determine the optimum water content and the corresponding maximum dry density of soils. Figure 10-1 illustrates a set of compaction curves for different types of soils. Study of such curves will indicate that through compaction the different soils may approach a condition of zero air voids—meaning no pockets of air—and that each type of soil has a different curve (shape and position on the graph).

A rough interpretation of the gravel and clay curves (without knowing what compaction method was used) is that gravel is a far better fill material because it allows water to drain off much better.

TABLE 10-3 ORIGIN OF SOILS*

Fluvial soils have properties affected by the action of flowing water.
Alluvial soils are deposited when water, transporting soil particles, flows out over a plain or when a river flows out in the sea. The coarse material is first deposited and then successively finer material.
River bed deposits, consisting of naturally washed sand and gravel, are common sources for concrete, asphalt, and road base materials.
Lake sediments vary from fine sand to clay.
Glacial deposits, moraines, and glacial tills are created by the action of the ice sheets, during glacial times covering northern and southern parts of the world, and are common, e.g., in Canada and Scandinavia.
Aeolian deposits are transported by the wind. Fine sand (dune sand) and silt (loess) are the most common.
Residual soils are the result of the weathering of the rock ground, resulting in soils varying from clayey to more or less granular types.
Organic soils consist of decomposed vegetation. They appear as peats, organic silts, and clays. Organic soils are with a few exceptions not used as fill materials.

*See 10.8.1.

TABLE 10-4 SPECIAL TYPES OF FILL MATERIALS*

Bentonite	very fine and active clay
Bank gravel	natural mixture of cobbles, gravel, sand, and fines
Basalt	a dark grey to black, dense fine-grained igneous rock
Binder	fines for filling of voids in order to increase the stability
Black cotton soil	very cohesive soil with swelling properties common in Africa, India, etc.
Caliche	soils where the grains are cemented together with lime
Chalk	white weak limestone found in, for example, England, France, and the U.S.A.
Collapsible soils	wind-transported soil with a loose structure which can break down under the influence of saturation and load (traffic or static loads)
Coral	dredged from coral banks, used as road base and fill material
Fly ash	finely divided residue obtained at combustion of coal, also called fuel ash
Granite	a very hard igneous rock of visible crystalline structure, composed essentially of quartz and feldspar

TABLE 10-4 SPECIAL TYPES OF FILL MATERIALS* (Continued)

Gumbo	sticky clay with a soapy appearance (U.S.A., Canada)
Hardpan	very hard well-graded aggregates of mineral particles
Humus	an organic material formed by the partial decomposition of vegetable matters
Laterite	a residual, more or less clayey soil originating from weathering of rock in tropical or subtropical parts of the world often with red color due to presence of iron in the rock
Limestone	rock of calcium carbonate, usually consisting of organic remains such as shells
Loam	soil containing sand, silt, and clay
Loess	wind-blown sediment, uniformly graded, grain size usually .0004″ to .002″ (.01 mm to .05 mm)
Marl	stiff marine calcareous clay
Mica	a group of mineral silicates that readily separate into thin leafs
Moraine	well-graded glacial soil containing all particle sizes from clay to blocks (boulders)
Peat	soil deposit of decayed and chemically decomposed vegetable matter
Sedimentary rock	fragments and particles of igneous rock that has been broken down, transported, and rehardened by natural forces
Shale	fissile rock with laminated structure formed by consolidation of clay
Slag (furnace slag)	stone material obtained as residue at metallurgical process
Till	morainic type of soil transported by glaciers often very heterogeneous
Tuff	fine-grained water- or wind-laid aggregate of very small mineral or rock fragments ejected from volcanoes

*See 10.8.1.

The effects of change in water content for each type of soil are clearly indicated here. Clay holds water to about 29% content; whereas, moraine and gravel do so only to an 8% to 11% content. How well a type of fill material allows water to drain away can be very important in regard to light building construction. For example, if clay is incorrectly used as backfill around a footing in an inappropriate situation, problems could result. The worst possible case would be if a heavy rainstorm occurred immediately after the clay was poured. The clay could change from a dry water content of 12% to as much as 29% if the rainfall was heavy enough. This would cause some clay to expand and move, and it might even break a newly set footing. A backfill with less

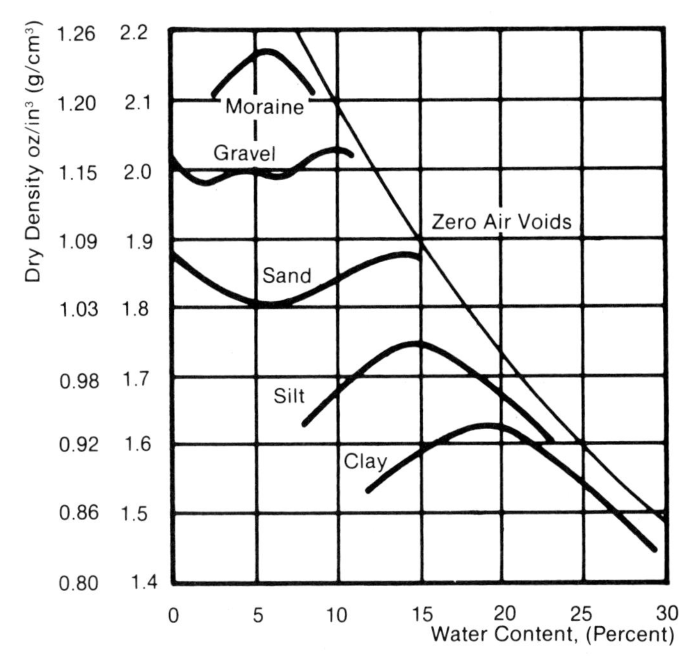

FIGURE 10-1. Laboratory compaction curves for various soil types.

capacity to become saturated with water will allow rainwater to seep away. In the case of coarse gravel, very little or no water will be absorbed and all rainwater will run off.

Great care must be taken when excavating to remove soil either to the level of the top of the footing or, in special cases, to the level of the bottom of the footing (correcting a water problem). Soil beneath should not be disturbed because it will be most difficult, if not impossible within a reasonable span of time, to compact it to its original condition.

10.4.1 Compacting Soils

When backfilling around a building, in a service trench, or in a roadway/driveway bed, compaction is desirable. Machines, such as those shown in Figure 10-2, are built for precisely this purpose. The vibratory tamper (Figure 10-2A) is a good all-around compactor and is especially useful for compacting smaller, relatively hard-to-reach areas. Tamping backfill around water or sewer hookups is a typical utilization of this machine. Vibratory plate compactors (Figure 10-2B) and the walk-behind double drum roller (Figure 10-2C) are both used for smoothing and compacting larger, flatter areas such as the gravel base of a foundation or the subbase of a driveway or parking area. The combination roller shown in Figure 10-2D is needed only when very large areas—a parking lot, for example—must be compacted. The combination roller also has the most weight and provides the greatest compaction: however, it will not usually be needed for light building construction jobs. Without compaction, backfills will need refilling after setting, and roadways/driveways may settle or settle unevenly which will cause cracking. Needless to say, not doing a good job of compacting backfill during the construction process can turn out to be cost ineffective in the long run.

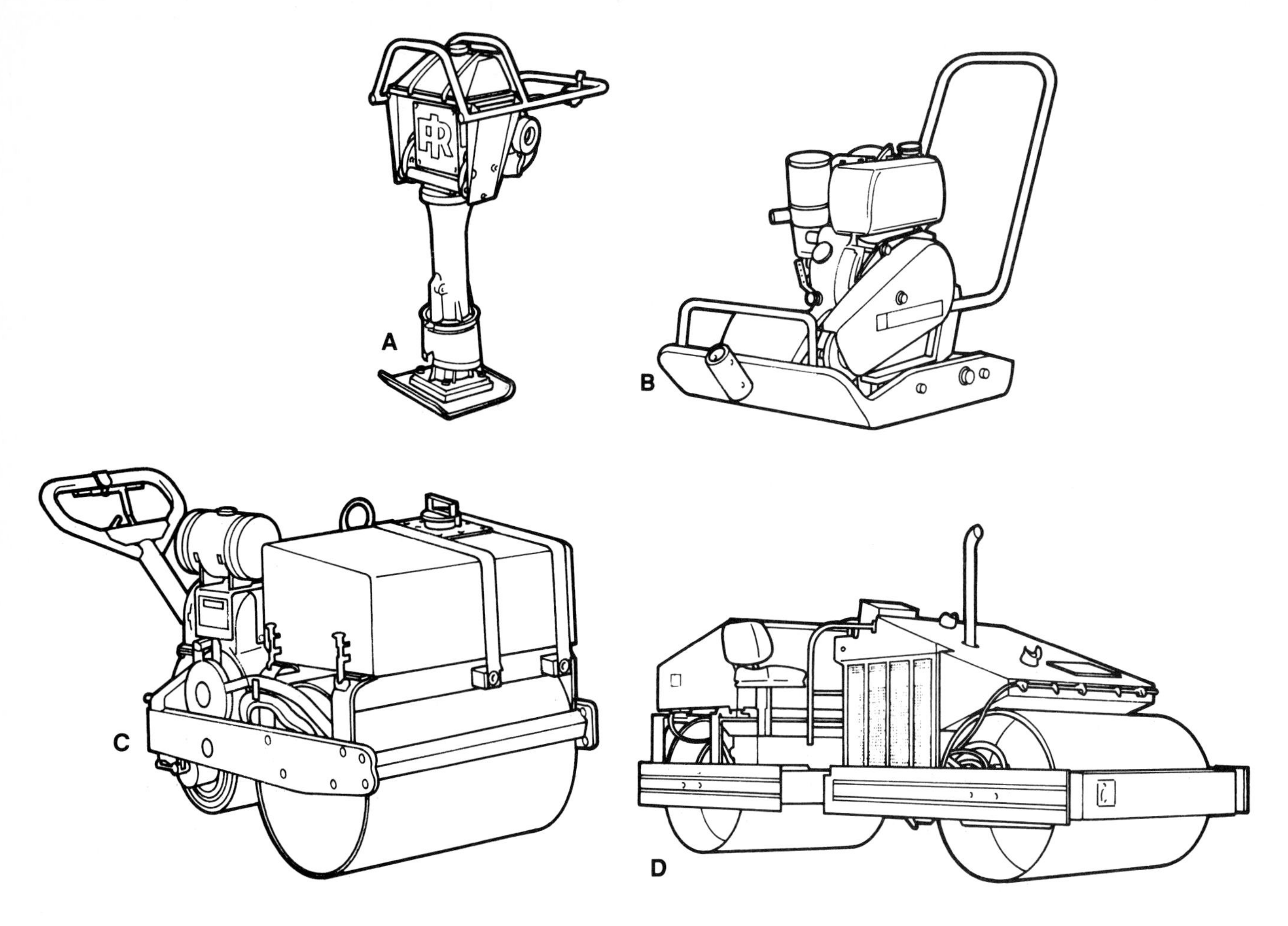

FIGURE 10-2. (A) Vibratory tamper, (B) vibratory plate compactor, (C) walk-behind double drum roller, and (D) combination roller with a vibratory drum and pneumatic tires.

10.4.2 Compacting Subsoil Conditions

An important factor affecting any compacting work is the nature of the subsoil. It is often impossible to reach a high density when compacting if the subsoil is loose or elastic as, for example, is a clay with a high water content. In such cases this layer must be consolidated by some means. Spreading lime on moist clay, a process called lime-stabilization, will draw the water from the clay, shrinking it in the process. This consolidates the subsoil base before the backfill is applied and compacted. If this is not feasible, it is necessary to build up the fill to a certain height in compacted layers before a normal degree of compaction is obtained.

The opposite effect results when soil is vibrated on a very dense base or in a confined space such as a pipe trench. The sides of the trench reflect the vibrations causing a higher degree of compaction at a greater depth than it is normally possible to obtain. The only problem this will usually cause is a need for more fill than is absolutely necessary. Remember that problems with backfill can result from using the wrong type of fill for a particular job or from failing to compact the fill the proper way.

10.4.3 Soil Volumes

For a builder, one of the practical aspects of using fill involves knowing how much fill will be needed for a job. The calculations necessary are not, however, based on a constant. Fill takes a varying amount of space, depending on the state which it is in. In other words, earth volume can be defined under the following different conditions:

1. in its natural state (in place),
2. in a loose state (on trucks, for example), and
3. after compaction.

The average values of swelling and shrinkage for different types of fill are shown in Figure 10-3. Note that individual variations caused by soil properties, water content, and so on, can be comparatively large. As shown in Figure 10-3, the compacted state in three of the four soil groups listed may have volumes smaller than in the natural state. It is also important to note that all of the soils increase in volume from the natural state to loose state.

Rock fill, stone, gravel, and sand are easier to compact than fine-grained stones. The coarse-grained soils may also be compacted in thicker layers, are better with respect to load bearing, and are not as susceptible to frost action because they allow water to more easily seep through.

The builder who desires to learn more about soils will need to understand a number of relevant engineering concepts and how they apply to the mechanics of building. Bearing capacity, plasticity, elasticity, capacity, capillary force, uniformity, pH, cohesiveness, compressibility, consolidation, graduation, gravity, stability, deformation, shear, and strength are some of the terms a builder will have to become familiar with for a more in-depth knowledge of this area.

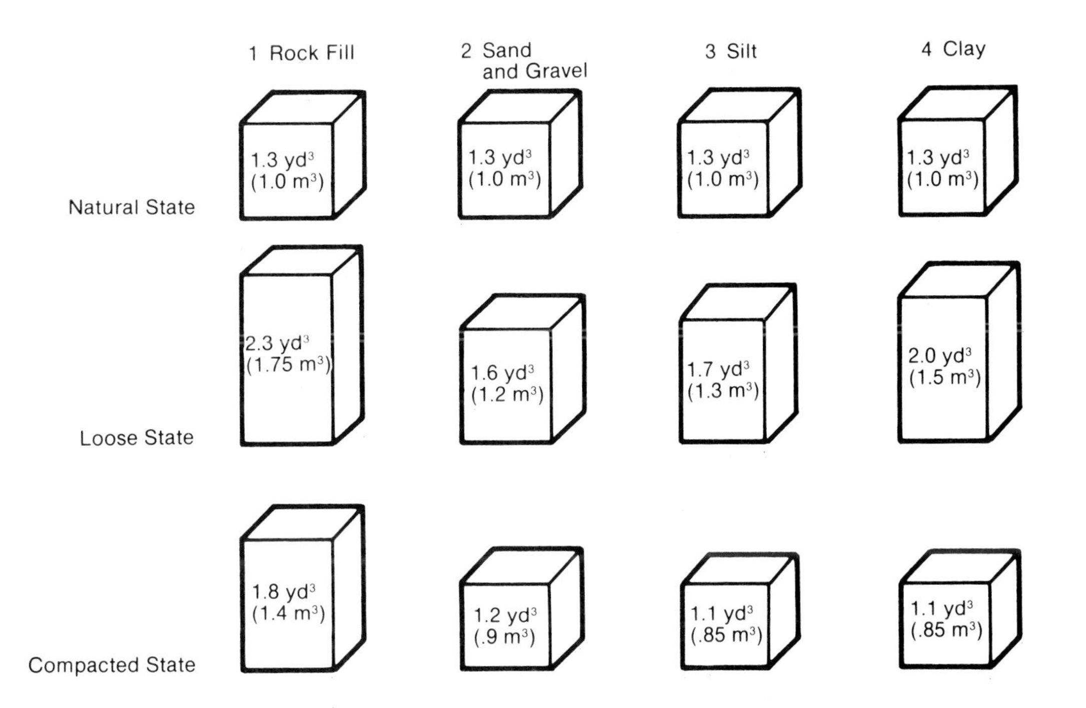

FIGURE 10-3. Volumes of different types of fill materials in natural, loose, and compacted states.

10.4.4 General Soil Investigation Guidelines

Builders who are able to read and interpret data regarding soil regions are able to deal with many building situations before they become serious problems. When possible, builders should try to avoid building on problem soils. Unfortunately, population growth and sales pressures often encourage people to buy problem lots, in which case a builder must do the best job possible under less than ideal conditions. There are, nevertheless, some general soil investigation guidelines that are always useful for sizing up the situation on any specific site.

1. Dig or bore a test hole(s) to examine site soils. A test pit can be dug by machine or by hand. An alternative is to use a hand or machine auger to bring subsoil to the surface for testing. Look for potential problems such as an excessive ground water level, organic fill, or expansive clays. A changing water level may have sufficient hydrostatic head (water pressure) to lift a building. In such cases draining of the water may be very expensive or impossible. The decision might be to avoid the site—to refuse to build. The same decision might be made where soils are very bad risks as footing soils.
2. If a water problem is a minor one, a compacted granular subbase may be provided for footings and slabs, or a capillary break (a waterproof membrane) may be installed. A common precaution, mandatory in many communities, is the installation of weeping tile connected to a drainage well or sewer line.
3. When a basement floor is included in a building, the soil beneath should either be undisturbed soil or compacted granular fill. Where a concrete slab is a floating slab such as for a garage and the excavation is incomplete, the soil beneath the slab needs to be uniform in kind and density.
4. If footings are used, they must be placed below the frost line. Freezing and thawing will heave and break a footing, and thereby shorten the effective life of a building.

10.5 OTHER ENVIRONMENTAL AND AESTHETIC CONSIDERATIONS

Besides information about soils, the builder should be in the habit of using a wide assortment of environmental data. Reasons for this are twofold. First are practical environmental considerations. For example, just as it would not be fully satisfactory to build in an area with a very poor soil base (especially without making the necessary construction adaptations), neither would it be a good idea to build in an area prone to earthquake activity without complete knowledge of the problems involved. The second consideration involves the aesthetics of a building. How well will a building fit into its environment? What kind of design will best fit the area in which the building is to be constructed? Are there historical, traditional, or scenic considerations that should or could be incorporated into the building's design? What are the aesthetic aims of the the design? Is it meant to blend into or to clash with its surroundings? How easily can the landscaping and overall setting be accomplished? These are all questions that must be asked in regard to the aesthetic value of a building.

10.5.1 Environmental Data

The National Weather Bureau, government departments of housing, forestry, and agriculture, and college and public libraries all generally have some useful long-term environmental data. Regional reports from national sources are of particular value to the builder who wants general and specific information covering long periods of time—up to 100 years. Offices of architects and engineers might provide reading lists covering local conditions.

Charts for precipitation and for wind are especially useful for building designers. In any case, care needs to be taken to obtain all possible relevant data. For example, it is not sufficient to accept data from a wind chart showing average surface wind velocities; recorded high wind velocities are also needed to provide complete wind effects.

An active environmental data file can be quite useful for a builder. The following can be used as a starter list:

1. temperature—range, averages. frost penetrations;
2. winds—velocities (highs, averages), prevailing direction (summer and winter), probability of tornadoes;
3. precipitation—rainfall/snowfall (maximums, minimums, averages), humidity levels, history of flooding; and
4. earthquake/volcano phenomena.

Having such a file allows a builder to make necessary decisions about weather resistivity levels needed in the materials and construction methods to be used for a job.

10.5.2 Design Features

With long-term weather data at hand and with local experiences from which to draw, a builder can think in terms of construction that will provide the optimum degree of comfort at the least possible cost. In addition, the builder will be prepared to address the aesthetic considerations of a job, both inside and outside a building.

By guiding himself or herself with the necessary weather data, the builder is providing vital background material for making technical decisions with respect to heating, cooling, and safety from wind and earth movements. Study of climate and other environmental factors will help to determine the proper siting of a building. The natural topography is, of course, always a prime consideration. The lay of the land on which the building will be constructed will determine the amount of clearing and excavating a builder will have to do. It may be a factor in determining the type of foundation to be built. It might also influence the size, shape, and style of a building and will usually affect the way in which landscaping is done.

Orientation of a building—the direction in which it faces—should also be influenced by a wide variety of environmental factors. For example, the effects of wind, rain, snow, sand, and dust might all influence the proper functioning of a building differently and must be considered separately in regard to orientation. Optimum exposure levels, expected moisture levels, latitude (for direction of sun's rays during the course of the year), and foreseeable noise levels are extremely important design considerations. In an era of energy shortages, minimum wind and moisture exposure and maximum exposure to the sun (south-facing exposures in the Northern Hemisphere) are vital for the creation of a cost-effective building in terms of heating and cooling. Designing a building that takes advantage of all positive aspects of a local environment and improves as much as possible the negative aspects at the most cost-efficient level is the goal of all light building construction.

To help improve on nature, attention may be given to the construction of berms to deflect noise; to screens and fences for noise reduction, security, and privacy; to retaining walls for grading/landscaping features; and to planting shrubs and trees for protection from wind and sun and for aesthetic purposes. By using the relevant precipitation data and local building codes, adequate surface drainage can be accomplished with maximum avoidance of flooding and wash-outs.

10.5.3 Landscaping

Building and landscaping need to complement one another, not be in conflict, if a functional unit is to be created. Today, it is recognized that landscaping serves a technical as well as an aesthetic function. Not only can landscaping help a building to look better, but it can also help to determine how well a building performs. A well-designed landscape, integrated into the overall design features of a building, can greatly increase the heating and cooling abilities of the building and considerably add to its comfort features. Think of landscaping as an extension of the building construction process rather than as an additional or separate operation.

As an example, planting trees is one landscaping variable that can make an environmental impact on a building in a variety of different ways. Furthermore, because trees take a considerable time to grow, planting them involves a long-term impact. Trees are frequently planted on the north-northwest side of a lot to provide a wind barrier in the winter to help lower heating costs and to prevent drafts and air infiltration. For such a purpose, softwoods and evergreens are well suited; they grow quickly and they retain their needles or leaves to provide maximum wind cover in the shortest time. Planting relatively mature evergreens as a windbreak provides the building with an immediate payback from the builder's landscaping job and gives the builder a living advertisement of practical landscaping.

The knowledgeable builder—the builder who does his or her homework—will, however, often choose deciduous trees for the south, southeast, and southwest sides of a building where these trees will provide a shading and cooling effect in the summer, but will allow the sun's rays through in the winter to maximize any passive solar effect.

10.6 SUMMARY

Materials, construction methods, and workmanship all contribute equally to the successful completion of a construction project. Appearance is affected by material shapes, horizontal and vertical lines, and such surface features as grain, smoothness/roughness, color, chamfers, flutes, and scorings. Positive or negative feelings about a building are influenced by such factors as the style of construction, bonding, pattern selection, and mortar treatment. However, materials and methods are insufficient in themselves to ensure a good construction job. Good workmanship is needed to put it all together.

Design is not usually the builder's job, even though it should be thought of as an integral part of the construction project. Builders often receive detailed specifications that set their objectives. It is then their responsibility to meet on-site problems while producing the desired building. The necessary steps must be taken for the rhythm of the project to evolve. The "place" being created—the building, the landscape, the entire entity involved in the construction project—will then be able to accommodate itself to local weather patterns, topography, and earth conditions. The footings and walls will be properly controlled for stress. The outdoor steps and driveway will look "right" and indoor stairs and other individual design features will warrant display.

10.7 RESEARCH

1. Visit the soils laboratory of a college or university to see local soil samples, methods of testing, and the resulting data sheets.
2. Visit the art department of a college or university and learn about colors and pigments, how different colors contrast and complement each other. Consider the colors of common building materials and how these colors have affected traditinal color patterns and styles. List common color patterns used by the housing industry. List potentially viable patterns which are relatively uncommon.
3. Make a list of historical and contemporary building styles and consider how each fits into a variety of potential settings: urban, rural, suburban, mountain, coast, newly-developed area, old established area with a dominant theme (Nantucket salt-box, for example), and so on.
4. Study meteorological information for several sample areas in the country. Determine how long-term weather patterns might affect the way a building project would be approached.
5. Consider various ways in which the aesthetic and utilitarian features of a building can be combined. Using south-facing glass for a beautiful view and to let in the winter sun for natural auxiliary heating is just one example of a contemporary move to maximize the cost-effectiveness of a building project. Beauty and practicality can no longer be thought of as distinct features in a building. What are other examples of this current trend? Try to think of other dual-purpose features or methods not currently being used in the building industry.

10.8 REFERENCES

10.8.1 De Chiara, Joseph, and Lee E. Koppelman. *Site Planning Standards.* New York: McGraw Hill, Inc., 1978.

10.8.2 *Earth Manual: A Water Resources Technical Publication.* United States Government Printing Office, Washington, D.C., 1980.

10.8.3 Forsblad, L. *Vibratory Soil and Rock Fill Compaction.* Solna, Sweden: Dynapak Maskin AB, P. O. Box 1103 S-17122.

10.8.4 Graf, Don. *Basic Building Data.* New York: Van Nostrand Reinhold Company, 1949.

10.8.5 Stillman, W. J. *Construction Practices for Project Managers.* Virginia: Reston Publishing Company, 1978.

10.8.6 Untermann, Richard K. *Principles and Practices of Grading, Drainage, and Road Alignment: An Ecologic Approach.* Virginia: Reston Publishing Co, Inc., 1978.

11 CONCRETE

Concrete has numerous applications in the light building construction industry. It is the principle construction material used in footings, foundations, basement floors and walls, exterior steps, walks, and driveways. Certain earth materials, water, and air are the primary ingredients of concrete. These earth materials include cement, aggregates (either fine or coarse), and other additives known as admixtures. The most common fine aggregate is sand; coarse aggregates include gravel and crushed stone of various sizes. Aggregates are referred to as inert ingredients in the concrete mixture because they serve as fillers and do not undergo a chemical change in the hardening process. Cement and water are known as active ingredients because they undergo a chemical reaction that causes the concrete to harden. Admixtures are all other materials added to the concrete to improve its quality or give it certain desirable characteristics. Mixtures of cements, water, and sand without coarse aggregates are often referred to as mortar and grout.

Working with concrete is not a simple process. If the incorrect ingredients are used or improperly mixed, severe structural difficulties can result. Quality concrete will only result if the materials used are of known quality and if the builder is well versed in their use.

11.1 SPECIFIC OBJECTIVES

This chapter is designed to give the student a basic background in concrete makeup and use. On reading the chapter the student should have

1. the ability to select the proper materials and combine them in the proper proportions to make concrete, mortar, and grout, and
2. an understanding of the methods used to place, finish, protect, and cure concrete.

11.2 CEMENT

Cement, more correctly called portland cement, is a manufactured product that forms a paste when mixed with water. Portland cement normally consists of portions of lime, silica, alumina, and iron. A relatively complex manufacturing procedure involves creating a portland cement clinker and pulverizing this clinker into a fine powder.

By definition, portland cements are hydraulic cements; this means that they set and harden by reacting chemically with water. This process, called hydration, combines cement with water to form a stonelike mass. It is important to remember that the hardening process is caused by the hydration, not by a drying out of the mix. In fact, the concrete formed by the reaction between cement and water must be kept as moist as possible during the initial hydration process. Drying out would cause a drop in water content below the amount required for the chemical reaction to work properly. Because concrete hardens and does not dry, it will harden just as well under water as it will in the air.

11.2.1 Cement Types and Uses

Different types of portland cement are manufactured to meet certain physical and chemical requirements for specific applications. The American Society of Testing and Materials (ASTM) lists five basic types of cement and three air-entrained varieties.

ASTM Type I. This type is also called normal portland cement and is a general purpose cement suitable for all uses when the special properties of the other types are not required. Type I portland cement is more generally available than are the other types of cement. It is used in pavement and sidewalk construction, reinforced concrete buildings and bridges, railways, tanks, reservoirs, sewers, culverts, water pipes, masonry units, and soil/cement mixtures. In general, it is used when concrete is not subject to special sulfate hazard or where the heat generated by the hydration of the cement will not cause an objectionable rise in temperature.

ASTM Type II. This type is a modified cement used where precaution against moderate sulfate attack is important, as in drainage structures where the sulfate concentrations in the soil or ground water are higher than normal, but not unusually severe. Type II will usually generate less heat at a slower rate than Type I. It may be used in structures of considerable size where cement of moderate heat of hydration will tend to minimize temperature rise, as in large piers, heavy abutments, and heavy retaining walls and is also used when the concrete is placed in warm weather.

ASTM Type III. Type III is a high-early-strength cement which provides high strengths at an early period, usually a week or less. Concrete made with Type III cement has a 7-day strength comparable to the 28-day strength of concrete made with Type I cement, and a 3-day strength comparable to the 7-day strength of concrete made with Type I cement. Type III cement has a higher heat of hydration and is more finely ground than Type I cement. It is used where it is desired to remove forms as soon as possible, to put the concrete in service as quickly as possible, and in cold weather construction to reduce the period of protection against low temperatures. Although richer mixtures of Type I may be used to gain high-early-strength, Type III may provide it more satisfactorily and/or more economically.

ASTM Type IV. Type IV is a low heat cement for use where the rate and amount of heat generated must be minimized. The development of strength is also at a slower rate. It is intended for use only in large masses of concrete such as large gravity dams where temperature rise resulting from the heat generated during hardening is a critical factor.

ASTM Type V. Type V is a sulfate-resistant cement used only in concrete exposed to severe sulfate action. It is used principally where soil or groundwater in contact with the concrete structure has a high sulfate content. It gains strength more slowly than Type I.

Air-Entraining Cements. Type IA, IIA, and IIIA correspond in composition to ASTM Types I, II, and III. The one difference is the addition of small quantities of air-entraining materials interground with the clinker during manufacture. These cements produce concrete with improved resistance to freeze-thaw action and to scaling caused by chemicals applied for snow and ice removal. Such concrete contains minute, well-distributed, and completely separated air bubbles.

Masonry Cements. Masonry cements are mixtures of portland cement, air-entraining additives, and supplemental materials selected for their ability to impart workability, plasticity, and water retention to masonry mortars.

Other types of portland cements include blast furnace slag cements, waterproofed cements, and plastic cements.

11.2.2 Canadian Equivalents

The Canadian Standards Association (CSA) provides in CSA Standard A5-M for five types of cement which correspond to the ASTM Types I, II, III, IV, and V. These are

- type 10 (normal),
- type 20 (moderate),
- type 30 (high early strength),
- type 40 (low heat of hydration), and
- type 50 (sulfate resisting).

Requirements for masonry cement are provided in CSA Standard A8-M.

11.2.3 Packing

Portland cement is packed in cloth or paper bags. In the United States a bag of regular cement normally weighs 94 lbs (42.6 kg) which amounts to approximately 1 ft^3 (.03 m^3). The bulk and bag weights of cement in the United States and Canada are listed in Table 11-1.

See Chapter 13 for information concerning the proper handling and storage of cement and concrete-related materials.

TABLE 11-1 WEIGHTS OF CEMENT

	United States	Canada
Bulk	1 ton (2,000 lb) (907.2 kg)	1 ton (1,000 kg)
Bag	94 lb (42.6 kg)	40 kg
Bag of masonry cement	(Weight printed on bag)	30 kg

11.3 WATER FOR MIXING

Almost any water that is drinkable and has no pronounced taste or odor can be used as mixing water for making concrete. In addition, mixing water should be free of oil, acid, heavy mineral content, high levels of sulfates, and other foreign substances. (It should be noted, however, that water suitable for mixing concrete may not be fit to drink.)

Water of questionable suitability can be used for making concrete if mortar cubes made with it have 7- and 28-day strengths equal to at least 90% of companion specimens made with suitable water. Mortar cubes should be made and tested according to the Standard Method of Test for Compressive Strength of Hydraulic Cement Mortars, American Society for Testing and Materials Designation: C109. In addition, Vicat needle tests, ASTM C191, should be made to ensure that impurities in the mixing water do not adversely shorten or extend the setting time of the cement. Excessive impurities in mixing water also may cause efflorescence, staining, or corrosion of reinforcement. Therefore, certain optional limits may be set on chlorides, sulfates, alkalies, and solids in the mixing water.

Water containing algae is unsuitable for making concrete because the algae can cause excessive reduction in strength either by influencing cement hydration or by causing a large amount of air to be entrained in the concrete. Algae may also be present on aggregates, in which case the bond between the aggregate and cement paste is reduced.

11.4 AGGREGATES

Even though aggregates are considered as inert materials acting as filler, they make up from 60% to 80% of the volume of concrete. The characteristics of the aggregates have a considerable influence on the mix proportions and on the economy of the concrete. For example, rough-textured or flat and elongated particles require more water to produce workable concrete than do rounded or cubical particles. Hence, aggregate particles that are angular require more cement to maintain the same water/cement ratio and, thus, the concrete is more expensive. For most purposes, aggregates should consist of clean, hard, strong, durable particles which are free of chemicals or coatings of clay or other fine materials that affect the bond of the cement paste.

Contaminating materials most often encountered are dirt, silt, clay, mica, salts, and humus or other organic matter that may appear as a coating or as loose, fine material. Many of them can be removed by washing, but weak, friable or laminated aggregate particles are undesirable. Sand containing organic material cannot be washed clear. Shale, stones with shale, and most cherts are especially undesirable. Visual inspection often discloses weaknesses in coarse aggregates. In doubtful cases the aggregates should be tested.

Characteristics and tests for common aggregates are outlined in Table 11-2. Table 11-3 lists materials harmful to the concrete mix, the effects of these substances, and the ASTM test that determines their presence.

TABLE 11-2 CHARACTERISTICS AND TESTS OF AGGREGATES

Characteristic	Significance	Test Designation	Requirement
Resistance to abrasion	Index of aggregate quality; wear resistance of floors, pavements	CSA A23.2-M16A CSA A23.2-M17A ASTM C131 ASTM C295 ASTM C535	Max. percentage loss of mass
Resistance to freezing and thawing	Surface scaling, roughness, loss of section, and unsightliness	CSA A23.2-M4B ASTM C295 ASTM C666 ASTM C671 ASTM C672 ASTM C682	Max. number of cycles or period of frost immunity; durability factor
Resistance to disintegration by sulfates	Soundness against weathering action	ASTM C88	Weight loss, particles exhibiting distress
Chemical stability	Strength and durability	CSA A23.2-M9A CSA A23.2-M14A CSA A23.2-M15A CSA A23.2-M, Appendix B	Max. expansion of mortar bar; aggregates must not be reactive with cement alkalies*
Particle shape and surface texture	Workability of fresh concrete	CSA A23.2-M13A ASTM C295 ASTM D339B	Max. percentage of flat and elongated pieces
Grading	Workability of fresh concrete; economy	CSA A23.2-M2A CSA A23.2-M5A ASTM C117 ASTM C136	Min. and max. percentage passing standard sieves
Bulk density and relative density	Mix design calculations; classification	CSA A23.2-M10A ASTM C29	Min. or max. relative density
Specific gravity	Mix design calculations	ASTM C127, fine aggregate ASTM C128, coarse aggregate ASTM C29, slag	—
Relative density	Mix design calculations	CSA A23.2-M6A CSA A23.2-M12A	—
Absorption and surface moisture	Control of concrete quality	CSA A23.2-M6A CSA A23.2-M11A CSA A23.2-M12A ASTM C70 ASTM C127 ASTM C128 ASTM C566	—
Compressive strength	Acceptability of fine aggregate failing other tests	CSA A23.2-M8A ASTM C39 ASTM C78 ASTM C87	Strength to exceed 95% of strength achieved with purified sand
Definitions	Clear understanding and description of aggregates	ASTM C125 ASTM C294	—

*Independent of conformance to any or all test requirements of CSA Standard A23.1-M, the Authority may accept or reject aggregate on the basis of performance of similar aggregate, from the same source, in concrete of comparable properties that has been exposed for at least 5 years to service similar to that to be encountered.

TABLE 11-3 HARMFUL MATERIALS

Substances	Effect on Concrete	Test Designation
Organic impurities	Affect setting and hardening, may cause deterioration	ASTM C40 ASTM C87
Materials finer than No. 200 (80 μm) sieve	Affect bond, increase water requirement	ASTM C117
Coal, lignite, or other lightweight materials	Affect durability, may cause stains and popouts	ASTM C123
Soft particles	Affect durability	ASTM C851
Clay lumps and friable particles	Affect workability and durability, may cause popouts	ASTM C142
Chert of less than 2.40 specific gravity	Affects durability, may cause popouts	ASTM C123 ASTM C295
Alkali-reactive aggregates	Abnormal expansion, map cracking, popouts	ASTM C227 ASTM C289 ASTM C295 ASTM C342 ASTM C586

11.4.1 Types of Aggregates

Brief descriptions of some of the natural materials of which mineral aggregates are composed are given in Descriptive Nomenclature of Constituents of Natural Mineral Aggregates (ASTM C294). Identification of the constituent materials in an aggregate may assist in recognizing the properties of the aggregate, but identification alone cannot provide a basis for predicting the behavior of the aggregate in service. All aggregates that do not have reasonably reliable service records should be tested for compliance with requirements.

Aggregates are divided into two sizes: fine and coarse. Natural sand is the most common fine aggregate. Sand used in the making of concrete should have particles ranging in size from 1/4″ (6 mm) down to dust-sized particles. Mortar sand contains only small particles and should not be used in the making of concrete.

Gravel and crushed stone are the most common coarse aggregates. Rounded or cubical pieces are far better than long sliver-like pieces (Figure 11-1). Coarse aggregates range in size from 1/4″ (6 mm) to a maximum of 1-1/2″ (38 mm). For slab applications, the largest maximum size of coarse aggregates should be equal to one-quarter of the thickness of the slab.

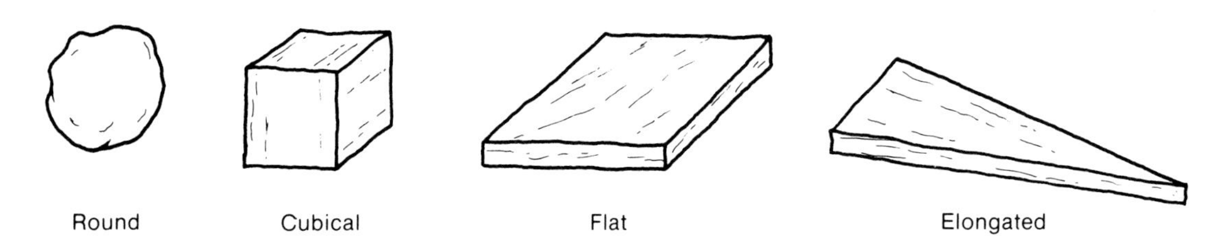
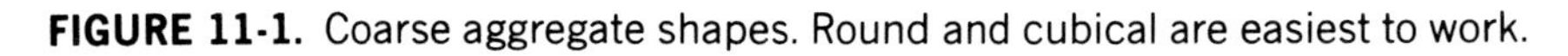

FIGURE 11-1. Coarse aggregate shapes. Round and cubical are easiest to work.

Suitable concrete aggregates have a full range of sizes but no excess amount of any one size. Large aggregates fill out the bulk of the concrete, minimizing the use of the more expensive cement. Smaller aggregates fill the spaces between the larger ones. An even distribution of sizes produces the most economical and workable concrete.

11.4.2 Water Content of Aggregates

Aggregates are also capable of containing water, the agent that combines with cement to make a paste. Figure 11-2 shows that water accounts for 14% to 18% of the bulk of a concrete mixture. These

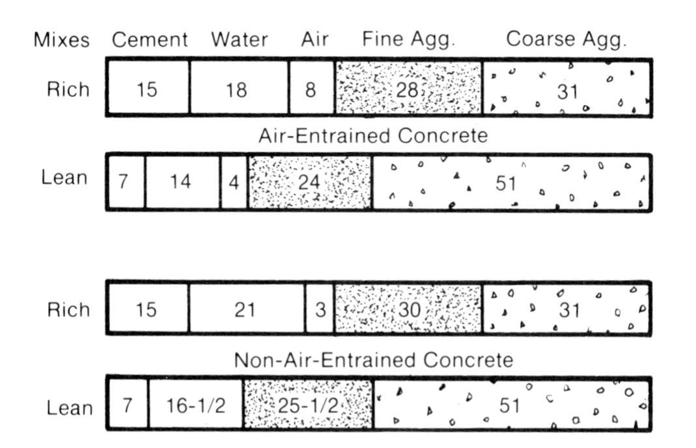

FIGURE 11-2. Proportions of concrete ingredients in percentages. Water content includes all water contained in the aggregates.

percentages of water include water added to a mixture, water content in the aggregates, and water surrounding the aggregates. Any aggregates may have a varying water content such as none (oven dry), some (because of air drying), damp, and saturated (Figure 11-3). This factor is extremely important because it must be accounted for in concrete mixing. Generally, if bone-dry aggregates are used, greater quantities of water are needed; less water is needed where damp or wet aggregates are used.

Since aggregates have the ability to retain water, they will absorb water from the paste. This, in turn, causes some separation of paste and

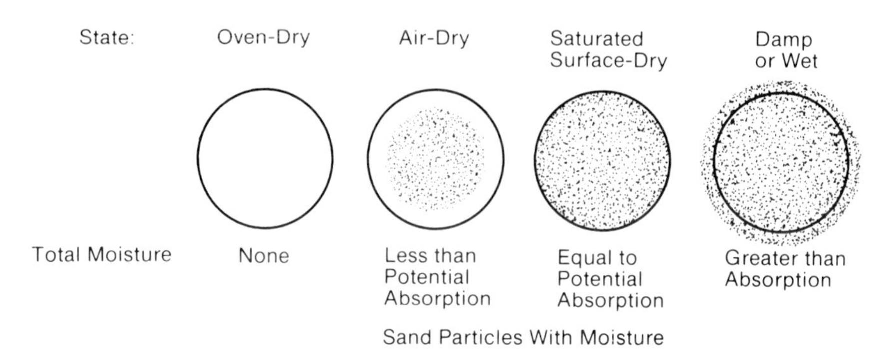

FIGURE 11-3. Range of water content in aggregates.

poor quality concrete. In addition, freezing and thawing temperatures expand and contract the water which causes expansion and contraction of the concrete and results in structural damage to the mass.

In any event, the water content of the aggregate (or the lack of it) must be accounted for in determining the mix ratio of water to cement.

11.5 ADMIXTURES

Admixtures include all materials other than portland cement, water, and aggregates that are added as ingredients to concrete, mortar, or grout immediately before or during mixing.

Admixtures can be broadly classified as follows:

1. air-entraining admixtures,
2. chemical admixtures,
3. pozzolanic-mineral admixtures, and
4. miscellaneous, such as workability agents, dampproofing and permeability-reducing agents, grouting agents, gas-forming agents, and coloring agents.

Concrete should be workable, finishable, strong, durable, watertight, and wear-resistant. These qualities can often be obtained most conveniently and economically through the proper design of the mix and the selection of suitable materials without resort to admixtures (except air-entraining admixtures when necessary).

There may be instances, however, when special properties are required such as extended time of set, acceleration of early strength, control of alkali reactivity with certain aggregates, or the reduction of heat of hydration. Although these special properties can often be obtained by selecting the appropriate type of portland cement, sometimes this is not practical, and it may be advisable to use admixtures to produce the special effects desired. In certain instances properties may be desired that can only be obtained by the use of admixtures. No admixture of any type or amount, however, should be considered as a substitute for good concreting practice.

The effectiveness of an admixture depends upon such factors as the type and amount of cement, water content, aggregate shape, gradation and proportions, mixing time, slump, and the temperatures of concrete and air.

Admixtures being considered for use in concrete should meet applicable ASTM or CSA Standards. Trial mixes should be made with the admixture and job materials at temperatures and humidities anticipated on the job. In this way the compatibility of the admixture with other admixtures and job materials as well as the effects of the admixture on the properties of the fresh and hardened concrete can be observed. The amount of admixture recommended by the manufacturer or the optimum amount determined by laboratory tests should be used.

Even though an admixture may produce concrete with the desired properties, the same results can often be obtained just as economically by changing the mix proportions or by selecting other concrete ingredients. Whenever possible, comparison should be made between the cost of changing the basic mix and the additional cost of using an admixture.

11.5.1 Air-Entraining Admixtures

One of the major advances in concrete technology in recent years has been the advent of air entrainment. The use of entrained air is recommended in concrete for most purposes. The principal reason for using intentionally entrained air is to improve concrete's resistance to freezing and thawing exposure. Air-entrained concrete is produced by using either an air-entraining cement or an air-entraining admixture during the mixing of the concrete. Unlike air entrapped in nonair-entrained concrete, which exists in the form of relatively large air voids unevenly dispersed throughout the mix, entrained air exists in the form of minute disconnected bubbles well dispersed throughout the mass.

In general, air-entraining agents are derivatives of natural wood resins, animal or vegetable fats or oils, alkali salts of sulfated or sulfonated organic compounds, and water-soluble soaps. Most air-entraining agents are in liquid form for use in the mix water. The major properties and advantages of air-entrained concrete are listed in the following paragraphs.

Workability. Entrained air improves the workability of concrete. It is particularly effective in lean mixes and in mixes with angular and poorly graded aggregates. This improved workability allows a significant reduction in water and sand content. The disconnected air voids also reduce segregation and bleeding of plastic concrete.

Freeze-Thaw Resistance. The freeze-thaw resistance of hardened concrete is significantly improved by the use of intentionally entrained air. As the water in concrete freezes, it expands, causing pressure that can rupture concrete. The entrained air voids act as reservoirs for excess water forced into them, thus relieving pressure and preventing damage to the concrete.

Resistance to Deicers. Entrained air is effective for preventing scaling caused by deicing chemicals used for snow and ice removal. The use of air-entrained concrete is recommended for all concretes that come in any contact with deicing chemicals.

Sulfate Resistance. The use of entrained air improves the sulfate resistance of concrete. Concrete made with a low water/cement ratio, entrained air, and cement having a low tricalcium aluminate content will be most resistant to attack from sulfate soil waters or seawater.

Strength. Strength of air-entrained concrete depends principally upon the voids/cement ratio. For this ratio, *voids* is defined as the total volume of water plus air (entrained and entrapped). For a constant air content, strength varies inversely with the water/cement ratio. As air content is increased, a given strength generally may be maintained by holding to a constant voids/cement ratio by reducing the amount of mixing water, increasing the amount of cement, or both. Some reduction in strength may accompany air entrainment, but this is often minimized since air-entrained concretes have lower water/cement ratios than nonair-entrained concretes having the same slump (consistency).

Watertightness. Air-entrained concrete is more watertight than non-air-entrained concrete, since entrained air inhibits the formation of interconnected capillary channels. Air-entrained concrete should be used where watertightness is desired.

11.5.2 Other Types of Admixtures

The following sections describe the properties of other commonly used admixtures.

Water-Reducing Admixtures. A water-reducing admixture is a material used for the purpose of reducing the quantity of mixing water required to produce concrete of a given consistency. These admixtures increase the slump for a given water content.

Retarding Admixtures. Retarders are sometimes used in concrete to reduce the rate of hydration in order to permit the placement and consolidation of concrete before the initial set. These admixtures are also used to offset the accelerating effect of hot weather on the setting of concrete.

Accelerating Admixtures. Accelerating admixtures accelerate the setting and the strength development of concrete.

Pozzolans. Pozzolans are siliceous or siliceous and aluminous materials which combine with calcium hydroxide to form compounds possessing cementitious properties. The properties of pozzolans and their effects on concrete vary considerably. Before one is used, it should be tested in order to determine its suitability.

Workability Agents. It is often necessary to improve the workability of fresh concrete. Workability agents frequently used include entrained air, certain organic materials, and finely divided materials. Fly ash and natural pozzalons used should conform to ASTM C618.

Dampproofing and Permeability-Reducing Agents. Dampproofing admixtures, usually water-repellant materials, are sometimes used to reduce the capillary flow of moisture through concrete that is in contact with water or damp earth. Permeability-reducing agents are usually either water-repellents or pozzolans.

Grouting Agents. The properties of portland cement grouts are altered by the use of various air-entraining admixtures, accelerators, retarders, workability agents, and so on, in order to meet the needs of a specific application.

Gas-Forming Agents. Gas-forming materials may be added to concrete or grout in very small quantities to cause a slight expansion prior to hardening in certain applications. However, while hardening, the concrete or grout made with gas-forming material has a decrease in volume equal to or greater than that for normal concrete or grout.

11.6 WATER/CEMENT RATIO AND COMPRESSION STRENGTH

Because it can be easily specified and tested, compression strength is the most often used measurement of concrete quality. For fully compacted concrete made with sound and clean aggregates, strength and other desirable characteristics are determined by the quantity of mixing water used per unit of cement. Within the normal range of strengths in concrete construction, compression strength is inversely related to the water/cement ratio (Figure 11-4).

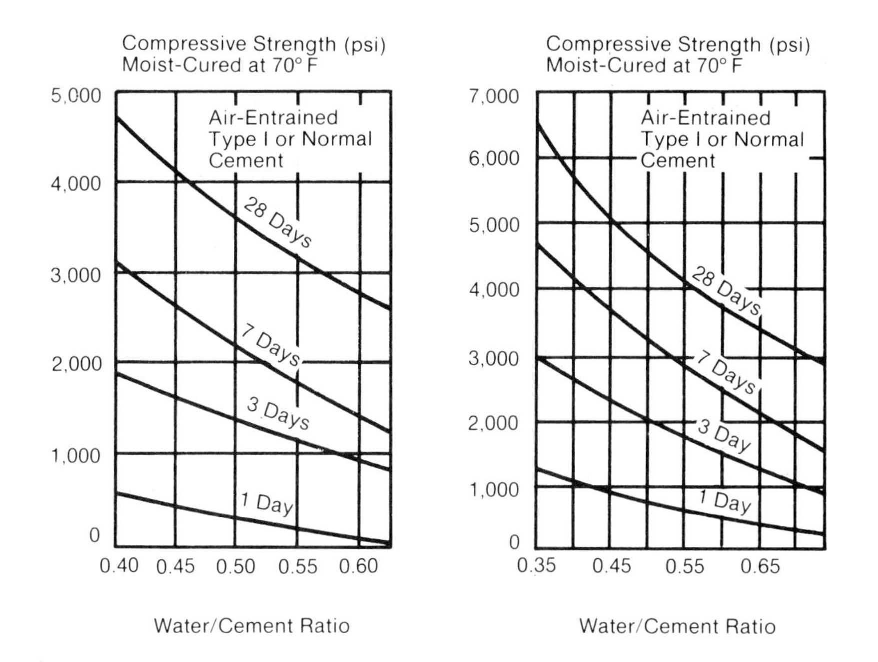

FIGURE 11-4. Typical age-strength relationships of concrete based on compression tests of concrete cylinders using Type I portland cement and moist curing at 70°F (21°C).

Approximately 2.5 gal (9.5 l) of water are required for hydration in one 94 lb (42.6 kg) sack of cement. However, such a low water/cement ratio results in a paste that does not coat all aggregates, is difficult to work with, and is low yielding in terms of concrete volume.

For these reasons, a higher water/cement ratio is used. This additional water thins the paste, allowing it to coat more aggregates. This increases the yield per sack, resulting in a more economical mix. The minimum and maximum amounts of water generally used for economical mixes range from 4 gal to 8 gal (15.1 to 30.3 l) per sack. Excessive water/cement ratios should be avoided because they weaken the paste and result in reduced compression strength.

On larger building projects where off-site mixing and on-site placing of concrete are closely supervised, the risk of mixing errors or inadvertently adding additional water to the mix is low. On smaller jobs, the risk is higher. This is especially true if workers mixing and placing concrete are tempted to add water to the mix to ease handling and placement. All workers involved in concrete work should be aware of the importance of the proper water/cement ratio.

Differences in strength for a given water/cement ratio may result from changes in the size, grading, texture, shape, strength, and stiffness of the aggregates; differences in cement types and sources; differences in air-entrained content; the presence of admixtures; and the length of curing time. In fact, the strength of the cement paste binder depends not only on the quality and quantity of the reacting components, but it is also largely controlled by the degree to which the hydration reaction is completed. Because the same concrete becomes stronger with time as long as moisture and temperature are favorable, the importance of proper curing cannot be overstressed.

11.7 PROPORTIONING CONCRETE MIXTURES

Traditionally, mix proportions for concrete were stated in numerical proportions coinciding with the relative amounts of cement, fine aggregates, and coarse aggregates used in the mix. Mix ratios of 1:2:4 and 1:2:5 are common examples.

The trial batch method involves working up several sample mixtures of different proportions. The hardened samples are then tested against the desired specifications and a practical selection is made based on these findings. Tables listing the proportions and characteristics of numerous trial batches are also available, but the proportions of mixes selected by this book method should be tested and modified as needed.

Modern proportioning methods include the weight and absolute volume methods set forth by the American Concrete Institute Committee 211, Proportioning Concrete Mixes. Weight proportioning methods estimate mix proportions by using an estimated or known weight of the concrete by unit volume. While this method is fairly simple and quick to use, a more accurate estimating procedure involves using the specific gravity values of the ingredients involved to calculate the absolute volume each will occupy in a unit volume of concrete.

Regardless of the proportioning method used, the ultimate objective in proportioning a normal concrete mixture is to determine the most economical and practical combination of readily available materials to produce a concrete that will satisfy the specification requirements under particular conditions of use. To fulfill this objective, a properly proportioned concrete mix will possess the following qualities:

1. acceptable workability of freshly mixed concrete;
2. strength, durability, and good appearance of hardened concrete; and
3. economy.

11.7.1 Properties of Fresh Concrete

Concrete must always be made with a workability, consistency, and plasticity suitable for the job conditions. Workability is the term used to describe how easy or difficult it is to place, consolidate, and finish the concrete within the forms without harmful segregation of the concrete's ingredients. Workability is difficult to measure, but can be judged by experienced batchmen, concrete finishers, and technicians.

Plasticity is the property of freshly mixed concrete that determines its ease of molding. A nonplastic or low plastic mix is difficult to mold; a medium to highly plastic mix is easily molded. The more water in a given volume of concrete, the more fluid the mix. If more aggregate is used in the same volume or if less water is added, the mix becomes more stiff. Neither very dry, crumbly mixes nor very watery, fluid mixes can be regarded as having plasticity.

Consistency is the ability of freshly mixed concrete to flow. The usual measure of consistency in concrete is its slump which is measurable in inches (millimeters). The higher the slump, the wetter the mixture. Slump is indicative of workability when assessing similar mixtures. It should not be used to compare mixes of totally different proportions. When used with different batches of the same mix, a change in slump indicates a change in consistency and in the characteristics of materials, mix proportions, or water content. Because different slumps are needed for various placements of concrete, slump is usually indicated in the concrete specifications.

When the amount of mixing water is held constant, the use of air-entrained concrete will increase slump. In lean concretes when cement content and slump are held constant, the entrainment of air means less mixing water is needed. This decrease in the water/cement ratio can actually increase concrete strength and decrease concrete permeability, a beneficial result. The strength of air-entrained concrete may equal, or nearly equal, that of nonair-entrained concrete when the cement content and slump are the same in both cases. In moderate to rich concretes, entrained air causes some reduction in strength.

Slump tests are made using a slump cone and tamping rod. The cone is filled with samples of the wet concrete which are tamped firmly down with the tamping rod and struck flush with the top of the cone. The cone is then lifted off the concrete and the amount the concrete slumps from its original height is measured (Figure 11-5). All tests should be made in accordance with ASTM C143, Slump of Portland Cement Concrete.

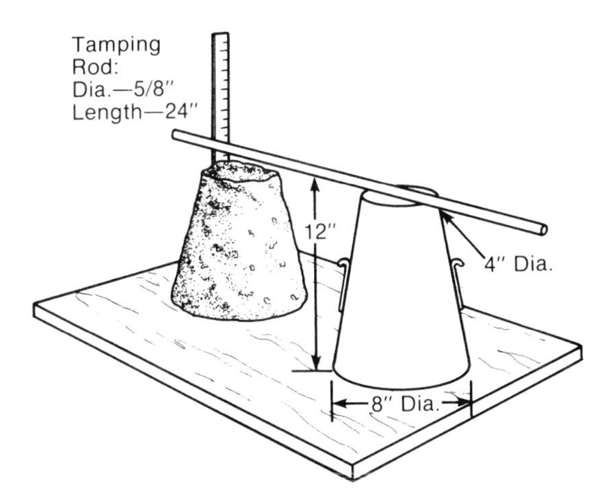

FIGURE 11-5. Measuring the slump of a concrete test sample.

Another method of measuring consistency is the ball penetration test: ASTM C360, Ball Penetration in Fresh Portland Cement Concrete. In this test, the depth to which a 30 lb (13.6 kg), 6" diameter (152 mm) hemisphere will sink into fresh concrete is measured. When calibrated for a particular set of materials, the results can be related directly to slump. The test has the advantage of being relatively simple and not requiring a molded specimen. Fresh concrete can be placed in any open container but the minimum lateral dimension of the sample must be about 18" (457 mm) and the depth at least 8" (203 mm).

11.7.2 Economy

The amount of mixing water required to produce a cubic yard (cubic meter) of concrete of a given slump is dependent on the maximum size of the aggregate. The larger the maximum size of the aggregate, the smaller the amount of water required. Because of this, it is advisable to use the largest practical maximum size of coarse aggregate. This minimizes water content and allows the cement content to also be reduced.

Using the stiffest practical concrete mix and the optimum ratio of fine to coarse aggregates also increases concrete economy. The relative costs of fine and coarse aggregates must also be considered in determining the most economical mix proportions.

11.8 TESTING CONCRETE PROPERTIES

In general, specifications for concrete and its component materials give detailed requirements for limits of acceptability. The requirements may affect (1) characteristics of the mixture such as maximum size of aggregates or minimum cement content; (2) characteristics of the cement, water, aggregates, and admixtures; and (3) characteristics of the plastic or hardened concrete such as temperature, slump, air content, or compressive strength.

Portland cements are tested for their compliance with established standards to avoid any abnormal performance such as early stiffening, delayed setting, or low strengths in the concrete.

Tests of aggregates have two major purposes: first are tests to determine the suitability of the material itself for use in concrete, including tests for abrasion, soundness, specific gravity, and petrographic and chemical analysis; second are tests to assure uniformity such as tests for moisture control and gradation of aggregates. Some tests are used for both purposes.

Tests of concrete can be similarly grouped. First are tests to evaluate the performance of available materials and to establish mix proportions. These usually are done in the laboratory and include tests for unit weight, strength, and workability. Second are tests used principally for control and include those for slump, air content, unit weight, and strength. These are also used for materials evaluation and mix proportioning.

11.9 PREPARATIONS FOR PLACING CONCRETE

Preparation prior to placing concrete includes compacting, trimming, and moistening the subgrade; erecting the forms; and setting the reinforcing steel and other embedded items securely in place. Moistening the subgrade is important, especially in hot weather to keep the dry subgrade from drawing out water from the concrete. In cold weather the subgrade must not be frozen. Snow, ice, and other debris must be removed from within the forms before concrete is placed. Where concrete is to be deposited on rock, all loose material must be removed, and cut faces should be nearly vertical or horizontal rather than sloping.

Forms should be clean, tight, accurately set, adequately braced, and constructed of or lined with materials that will impart the desired formed-surface finish to the hardened concrete (Figure 11-6). Wood forms should be moistened before placing concrete; otherwise, they will absorb water from the concrete and swell. Forms should be made for removal with minimum damage to the concrete. With wood forms, use of too large or too many nails should be avoided to facilitate removal and reduce damage. Forms should be treated with a release agent such as oil or lacquer to facilitate removal. For architectural concrete, this agent should be a lacquer or an emulsified stearate so it will be nonstaining.

FIGURE 11-6. Typical footing form used in concrete work.

Reinforcing steel should be clean and free of loose rust or mill scale when concrete is placed. Mortar coatings need not be removed from reinforcing steel and other items to be embedded, if the lift is to be completed within a few hours; loose, dried mortar, however, must be removed from items that will be encased by later lifts of concrete.

All equipment that will be used to place the concrete must be clean and in working condition. Standby equipment should be available in the event of a breakdown.

11.10 PLACING AND CONSOLIDATING CONCRETE

After the preparations are completed, concrete placement can begin. After placement is completed, consolidation can be undertaken.

11.10 Placing Concrete

Concrete will not attain its maximum possible strength, density, and uniformity unless proper methods are used to place it in the forms. Proper methods are methods which will ensure the thorough filling of all form spaces, while at the same time confining segregation to a minimum.

Concrete should be deposited in even horizontal layers and should not be puddled or vibrated into place. The layers should be from 6″ to 24″ (152 mm to 610 mm) in depth depending on the type of construction. The initial set should not take place before the next layer is added. After the concrete is in place, it should be vibrated or spaded to prevent honeycombing or air pockets in the concrete. This is particularly desirable in wall forms with considerable reinforcing. Care should be taken not to overvibrate because segregation and a weak surface may result.

There is a temptation, in the interests of time and effort, to drop the concrete from the point to which it has been transported regardless of the height of the forms, but the free fall of concrete into the forms should be reduced to a maximum of 3′ to 5′ (.9 m to 1.6 m) unless vertical pipes, suitable drop chutes, or baffles are provided.

Concrete should be placed as nearly as possible in its final position. Horizontal movement should be avoided since this results in segregation because mortar tends to flow ahead of coarser material. Concrete should be worked thoroughly around the reinforcement and bedded fixtures, into the corners, and on the sides of the forms.

On large pours so as to avoid excess pressure on forms, the rate of filling should not exceed 4′ (1.2 m) per hour measured vertically, except for columns. Placing will be coordinated so that the concrete is not deposited faster than it can be properly compacted. In order to avoid cracking during settlement, an interval of at least 4 hours, but preferably 24 hours, should elapse between completion of columns and walls and the placing of slabs, beams, or girders supported by them.

For walls, the first batches should be placed at the ends of the section. Placing should then proceed toward the center for each layer, if more than one layer is necessary, to prevent water from collecting at the ends and corners of the forms. This method should also be used in placing concrete for beams and girders. For wall construction, the inside form should be stopped off at the level of construction. Overfill the form for about 2″ (51 mm) and remove the excess just before setting occurs to ensure a rough, clean surface. Before the next lift of concrete is placed on this surface, a 1/2″ to 1″ (13 mm to 25 mm) thick layer of sand/cement mortar should be deposited on it. The mortar should have the same water content as the concrete and should have a slump of about 6″ (152 mm) to prevent stone pockets and help produce a water-tight joint.

For slabs, the concrete should be placed at the far end of the slab, each batch dumped against previously placed concrete. The concrete should not be dumped in separate piles and the piles then leveled and worked together. Nor should the concrete be deposited in big piles and then moved horizontally to its final position, since this practice results in segregation.

Always deposit the concrete at the bottom of the slope first, and proceed up the slope as each batch is dumped against the previous one. Compaction is thus increased by the weight of the newly added concrete when it is consolidated.

A

B

FIGURE 11-7. Consolidating concrete with a mechanical vibrator: (A) before vibration and (B) after vibration.

11.10.2 Consolidating Concrete

With the exception of concrete placed under water, concrete is compacted or consolidated after placing. Consolidation may be accomplished by the use of handtools such as spades, puddling sticks, and tampers, but the use of mechanical vibrators is preferred (Figure 11-7). Compacting devices must reach the bottom of the form and must be small enough to pass between reinforcing bars. Consolidation eliminates rock pockets and air bubbles and brings enough fine material to the surface and against forms to produce the desired finish.

11.11 FINISHING AND CURING CONCRETE

After the concrete has been placed and consolidated, it can be finished and cured.

11.11.1 Finishing Concrete

The concrete finishing process may be performed in many ways depending on the effect desired. Occasionally, only the correction of surface defects, such as the filling of bolt holes, is needed. Unformed surfaces may require only screeding for proper contour and elevation, or a floated, troweled, or broom finish may be required.

Screeding. The chief purpose of screeding is to level the surface of concrete slabs by striking off the excess concrete. A hand-operated strike-off board and the method of using it are shown in Figure 11-8.

FIGURE 11-8. Screeding or striking-off concrete level with the top of the form.

Floating. If a smoother surface is required than that obtained through screeding, the surface should be worked sparingly with a wood or metal float or finishing machine. This process should take place shortly after screeding when the concrete is still workable. Through floating, high spots are eliminated, low spots are filled in, and enough mortar is brought to the surface to produce the desired finish (Figure 11-9).

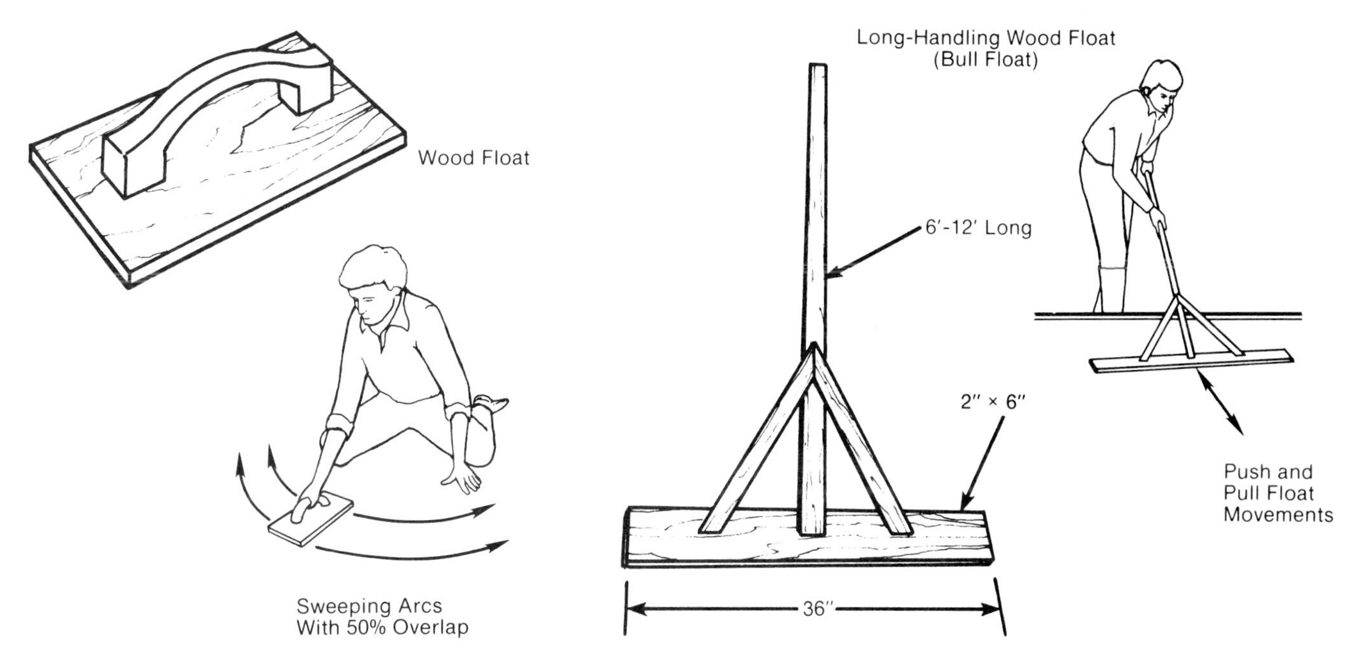

FIGURE 11-9. Floating tools and technique.

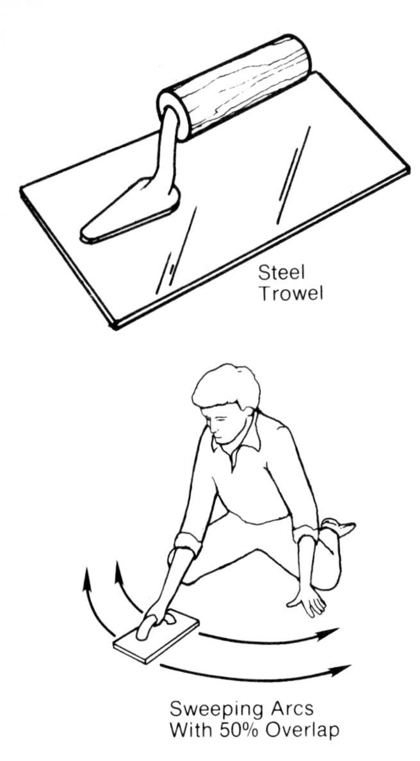

FIGURE 11-10. Steel troweling for a dense smooth finish.

Troweling. If a dense smooth finish is desired, floating must be followed by steel troweling (Figure 11-10). This occurs at some point after the moisture film or sheen disappears from the floated surface and when the concrete has hardened enough to prevent fine material and water from being worked up to the surface.

Brooming. A nonskid surface can be produced by brooming the concrete before it has thoroughly hardened. Brooming is done after the floating operation.

11.11.2 Curing Concrete

To achieve the designated age-strength relationship, concrete needs to be carefully placed, protected from the elements, and cured for the proper length of time. Protection is offered by the forms, by burlap coverings, heating units, and by the spraying on of water and protective chemicals. If placed concrete is exposed to hot sun or winds before hydration has taken place, dusting or cracks may mar the surface and weaken the concrete. Exposure to rainwater may change the water/cement ratio in the concrete mix.

Freshly placed concrete contains more than enough water to hydrate the cement completely, but if the concrete is not protected against excessive dryout, especially near the surface, the water content may drop below the level required for complete hydration.

Curing is designed to prevent surface evaporation of water during the period between beginning and final set. Concrete takes a beginning set in about 1 hour. A final set takes around 7 days. Curing is brought about by keeping the concrete surfaces continuously moist. This is usually done by spraying, ponding, or covering with wet burlap or a water retaining membrane.

Concrete made with ordinary cement should be kept moist for a minimum of 7 days and should be protected from direct sunlight for at least the first 3 days of the curing period. Forms left in place also furnish good protection against sunlight; they should, however, be loosened at the time when they might safely be removed and the space between the forms and the concrete should be flushed with water at frequent intervals to keep the concrete moist.

11.11.3 Removal of Forms

Forms should, whenever possible, be left in place for the entire curing period (about 7 days). Forms which are to be reused, however, must be stripped for reuse as soon as possible. In any event, forms must not be stripped until the concrete has hardened enough to hold its own weight and any other weight it may be carrying. The surface must be hard enough to remain uninjured and unmarked when reasonable care is used in stripping the forms.

Under ordinary circumstances, forms for various types of construction may be removed after intervals as follows

- Haunch boards (side forms) on girders beams—1 day.
- Soffits on girders and beams—7 days.
- Floor slab forms—10 days.
- Wall forms—1 day.
- Column forms—3 days.

11.12 RESEARCH

1. Inquire into the suitability of area water sources for use in the mixing of concrete. If they have not been previously tested, find out what the testing procedure is. What would happen if water was used that had organic impurities? What about clay lumps?
2. Why is it important to know the existing water content of an aggregate? What problems could arise if this factor is not taken into consideration?
3. For a hypothetical job, compare the cost of changing a concrete mix with that of using an aggregate to obtain a desired property. Would the added convenience of an aggregate outweigh any increased cost? Why?
4. Describe the advantages of the addition of entrained air to concrete. What types of jobs would particularly benefit from the use of such an admixture and why?
5. Visit a local contractor or supply company and compare the prices of the various portland cement types and admixtures.
6. What is the importance of having a proper water/cement ratio? How does it relate to the concrete's compression strength? How can a high water/cement ratio be beneficial? What happens if it becomes too high?
7. Trial mix some small batches of portland cement in 1:2:4 and 1:2:5 ratios. Is there any difference in the hardened samples?
8. Make a slump test using three batches of the same mix. Comparing the results, determine the accuracy of your proportioning.
9. What is the relationship between the amount of water needed and aggregate size?
10. What types of preparations should be taken prior to the placing of concrete, particularly concerning the treatment of forms?
11. Why shouldn't concrete be moved horizontally?
12. Visit a construction site to watch the pouring of concrete walls and slabs. What are the proper procedures? Note any variances you observe during the pouring and inquire into the reasoning behind them.
13. Take time to examine various concrete surfaces. Try to determine whether they were finished by floating or by screening.
14. Why should surface evaporation be prevented? Visit the same site for several days after the concrete has been poured to observe whether or not steps have been taken to assure a proper cure.

11.13 REFERENCES

11.13.1 *Concrete and Masonry.* Washington, D.C., Headquarters, Department of the Army, 1970.

11.13.2 DeCristoforo, R. J. *Concrete and Masonry Techniques and Design.* Virginia: Reston Publishing Company, Inc., 1975.

11.13.3 DeCristoforo, R. J. *Handyman's Guide to Concrete and Masonry.* Virginia: Reston Publishing Company, Inc., 1978.

11.13.4 Maguire, Byron W. *Masonry and Concrete.* Virginia: Reston Publishing Company, Inc., 1978.

11.13.5 Smith, R.C., T. L. Honkala, and C. K. Andres. *Masonry: Materials Design Construction.* Virginia: Reston Publishing Company, Inc., 1979.

11.13.6 Wass, Alonzo, and Gordon A. Sanders. *Materials and Procedures for Residential Construction.* Virginia: Reston Publishing Company, Inc., 1981.

12

MASONRY

Masonry is construction consisting of prefabricated or natural masonry units, such as concrete block, brick, structural clay tile, or field stone laid in various ways and patterns and joined together with mortar. The term masonry is often further defined by specifying the materials used in construction. Concrete masonry, brick masonry, and stone masonry are examples of specialized branches of the masonry trade.

A more general view of the masonry trade can be taken if careful study is given to each masonry material and the possible combinations of these materials for light construction work. A general concept of masonry should include all major branches and be extended to related activities such as parging and plastering.

Masonry units are used in the construction of bearing, partition, and screen walls; load-carrying columns, pillasters, and piers; fences and sound barriers; decorative facing work; paving; drainage systems; and specialty structures such as fireplaces, chimneys, and planters. Masonry work is also used for fire protection of structural steel building members.

Heavy masonry building segments such as walls and fences are built (carried) on carefully designed and prepared footings made of grouted bedrock or placed concrete set below the frostline to avoid heaving or breaking. Composite walls, such as brick veneer walls with concrete block backing, are usually carried on concrete walls with footings.

Columns and pillasters are heavy load-bearing members that usually require their own stronger footing, separate from wall footings. Footings for chimneys and fireplaces are also often separate from other building footings to allow for building and earth movements and to offer increased fire protection. Concrete floor slabs used in light building construction projects are poured after such footings and seldom carry heavy loads.

Masonry walls and units possess definite heat transfer and acoustical properties. They are also subject to structural stresses from materials weight, wind, earth movements, vibrations, water, snow, and freeze-thaw cycles.

12.1 SPECIFIC OBJECTIVES

After reading the chapter and conducting the research work, the student should be able to

1. define the term masonry and be able to list the major branches of the masonry trade;

2. understand the basic principles of masonry construction with regard to materials, footings, unit placement, and cold weather work;
3. define the types of stress placed on masonry walls, columns, etc., and understand how buildings are constructed to accommodate this stress;
4. understand the basic principles of heat and moisture transfer through masonry materials and the construction methods used to control this phenomenon;
5. understand the acoustical properties of masonry units and the basics of noise control in light building construction; and
6. appreciate aesthetically pleasing, technically correct masonry work.

12.2 MORTAR AND GROUT

Mortar and grout are similar products to the extent that both require a cementitious material and aggregates for their production. But mortar and grout are not the same product. They are introduced into the wall system differently, are used for different purposes, usually require a different size of aggregate, and quite often use a different type of cement.

Mortar is designed for a number of purposes, but it serves primarily to join masonry units together in an integral structure. In addition to that, it is also required to hold the units a specified distance apart; to produce tight seals between units to prevent the passage of air and moisture; to bond metal ties and anchor bolts to the steel joint reinforcement in order to integrate them into the masonry structure; to provide a bed that will help accommodate variations in the size of units; and to provide an architectural effect on exposed masonry walls through various styles of mortar joints. In order to achieve these purposes, mortar is applied to the edges of masonry units (Figure 12-1).

On the other hand, grout is designed primarily to bond masonry units and steel together in reinforced masonry walls so that they act together to resist imposed loads (Figure 12-2). It is usual to place grout

FIGURE 12-1. Mortar applied to edges of masonry units. (Photograph furnished by courtesy of Portland Cement Association.)

FIGURE 12-2. Reinforced masonry wall. (Photograph furnished by courtesy of Portland Cement Association.)

in only those cells containing steel reinforcement (Figure 12-3); in some load-bearing reinforced masonry walls, however, all cells will be filled with grout. Sometimes the cells of nonreinforced masonry walls are grouted to provide added strength (Figure 12-4), and grout may also be used to bond together the two wythes of a cavity wall (Figure 12-5). In order to achieve its purposes, grout is introduced into the cells of the units or the cavities between wythes.

FIGURE 12-3. Cells containing grouted reinforcement.

FIGURE 12-4. Cells of nonreinforced masonry wall grouted. (Photograph furnished by courtesy of Portland Cement Association.)

12.2.1 Desirable Properties of Mortar

For good workmanship and proper structural performance of concrete masonry, mortar must possess a number of important properties both while it is in its plastic stage and after it has hardened. Important properties of mortar while plastic include workability, water retentivity, and a consistent rate of hardening. Hardened mortar must have good bond, durability, good compressive strength, low volume change, and good appearance.

Workability. Workability is probably the most important quality of plastic mortar because of the effect that it has on other mortar qualities (in both the plastic and hardened stages). The following interrelated factors are considered to have an influence on workability: quality of the aggregate, amount of water used, consistency, water retentivity, settling time, mass, adhesion, and penetrability.

Mortar with good workability should spread easily on the masonry unit, cling to vertical surfaces readily, extrude from joints without dropping, and permit easy positioning of the unit without subsequent shifting due to its weight or the weight of successive courses.

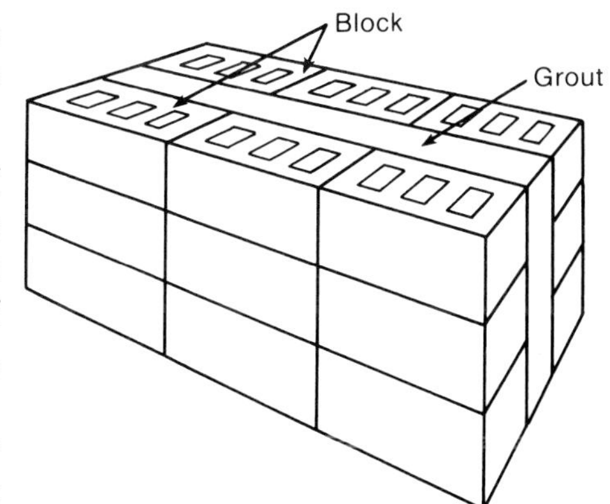

FIGURE 12-5. Grouted cavity wall.

Water Retentivity. Water retentivity is the property of mortar that resists rapid loss of mixing water to the air or to an absorptive masonry unit. Rapid water loss will cause the mortar to stiffen prematurely, making it extremely difficult to obtain good bond and weathertight joints. A mortar with good water retentivity, however, will remain soft and plastic long enough for the masonry units to be aligned, leveled, plumbed, and adjusted properly without danger of breaking the bond between the mortar and units.

Poor water retentivity could be caused by poorly graded aggregate, aggregate of too large a size, insufficient mixing, or the wrong type of cement. The level of water retentivity may be increased by an increase in mixing, the addition of an air-entraining agent or a pozzolanic material, or sometimes, the addition of more water to the mortar.

Consistent Rate of Hardening. The hardening of mortar is due to hydration—the chemical reaction between cement and water—and the rate at which this occurs determines the time required for the mortar to develop resistance to an applied load. If the mortar hardens too quickly, it interferes with the proper laying of the units and finishing of the joints. On the other hand, if hardening is too slow, work may be impeded by having to wait for the mortar to harden sufficiently so that it will not flow from completed joints.

An increase in temperature tends to speed up the hydration process. During hot weather it may be necessary to use colder water or to decrease the length of time that the mortar is exposed to the hot atmosphere. One way in which the latter may be accomplished is to lay shorter mortar beds and fewer units in advance of tooling. Conversely, in cold weather, it may be necessary to use warm mixing water or to provide heat for the surrounding atmosphere, or both.

Bond. The bond between mortar and a masonry unit is a combination of the degree of contact between the mortar and the masonry unit and the adhesion of the cement paste to the unit surface.

The degree of contact between mortar and unit is important to watertightness and tensile bond strength. Poor bond of this nature could lead to the penetration of moisture through the unbonded areas. A high degree of mortar contact with the unit is obtained with a workable and water-retentive mortar, full joints, concrete masonry units having a medium initial rate of absorption, and good workmanship.

The adhesion of the cement paste to the unit surface, or tensile bond strength, is probably the most important property of hardened mortar. Mortar must develop enough bond to withstand the tensile forces exhibited by structural, earth, and wind loads; shrinkage of concrete masonry units and mortar; and temperature changes.

The adhesion of the paste to the masonry surface depends on the following factors: mortar ingredients; workability and water retentivity of the mortar; surface texture, moisture content, and suction of the masonry units; curing conditions (such as temperature, relative humidity, and wind); and workmanship.

Durability. The durability of mortar is its ability to withstand exposure to weather and atmospheric conditions. Polluted air may cause some mortar deterioration, but the major cause is frost action on mortar saturated with water.

The deterioration of mortar in walls above grade due to frost action is not usually a major problem if the mortar has first been allowed to develop to its full potential strength. For frost damage to occur, hardened mortar must first be subjected to a high degree of water saturation. During the hardening process, the original water content of the mortar is greatly reduced. The mortar does not normally become saturated again unless the masonry is in contact with saturated soil, it is exposed to heavy rain, or the formation of horizontal ledges on the wall allows water to be held there. Under any of these conditions, the mortar may again become saturated and subsequent frost action can cause deterioration.

There are a number of steps which may be taken to help prevent such deterioration. Mortar with high compressive strength normally has good durability. Mortar containing air entrainment will withstand a great many freeze-thaw cycles. Masonry below grade should be protected by a waterproof coating on the exterior surface, and the tooling of mortar joints should not produce ledges which may collect and hold water.

Compressive Strength. The compressive strength of mortar is largely dependent on the type and quantity of cement used in the mortar mix. The compressive strength increases as the amount of cement in the mix is increased in relation to the amount of water used, provided the mix remains workable. An increase in water content, air entrainment, or lime content will reduce compressive strength.

Volume Change. The maximum shrinkage across a mortar joint is very small and, therefore, usually not troublesome. This is especially true with weaker mortars, which have greater extensibility and, so, are better able to accommodate shrinkage.

Appearance. The overall appearance of a masonry structure is affected by the appearance of the mortar joints. Factors contributing to the appearance of mortar joints include uniformity of color and shade, uniformity of texture and thickness, and the workmanship involved in tooling the joints.

The color and shade of the joints are affected by the moisture content of the masonry, admixtures in the mortar, atmospheric conditions, uniformity of proportions of the mortar mix, and the time of tooling. Good uniformity of texture depends on careful measurement of batch proportions from batch to batch and also on thorough mixing. Regular joint thickness and consistent tooling are both the results of good workmanship.

12.2.2 Types of Mortar

Mortar is composed of cement, lime, and sand in proportions varying according to the use of the mortar. The following types of mortar are proportioned on a volume basis:

Type M. This type is made with (a) 1 part portland cement, 1/4 part hydrated lime or lime putty, and 3 parts sand *or* (b) 1 part portland cement, 1 part type II masonry cement, and 6 parts sand. It is suitable for general use and is recommended specifically for masonry below grade and in contact with earth such as foundations, retaining walls, and walks.

Type S. This type is made with (a) 1 part portland cement, 1/2 part hydrated lime or lime putty, and 4-1/2 parts sand *or* (b) 1/2 part portland cement, 1 part type II masonry cement, and 4-1/2 parts sand. It is also suitable for general use and is recommended where high resistance to lateral forces is required.

Type N. This type is made with (a) 1 part portland cement, 1 part hydrated lime or lime putty, and 6 parts sand *or* (b) 1 part type II masonry cement and 3 parts sand. It is suitable for general use in exposed masonry above ground and is recommended specifically for exterior walls subjected to severe weather.

Type O. This type is made with (a) 1 part portland cement, 2 parts hydrated lime or lime putty, and 9 parts sand *or* (b) 1 part type I or type II masonry cement and 3 parts sand. It is recommended for nonload-bearing walls of solid units where the compressive stresses do not exceed 100 pounds per square inch and the masonry will not be subjected to freezing and thawing in the presence of excessive moisture.

Mortar is also classified according to its compressive strength. The compressive strength ratings required for each type of mortar are given in Table 12-1.

TABLE 12-1 COMPRESSIVE STRENGTH OF MORTAR CUBES

		Minimum Compressive Strength, psi (MPa) Average of Six 2″ (51 mm) Cubes	
Mortar Type		**Tested at 7 Days**	**Tested at 28 Days**
Laboratory	M	1,600 (11)	2,540 (17.5)
prepared	S	1,088 (7.5)	1,813 (12.5)
	N	435 (3)	725 (5)
	O	220 (1.5)	363 (2.5)
Job			
prepared	M	1,300 (9)	2,030 (14)
	S	870 (6)	1,450 (10)
	N	363 (2.5)	580 (4)
	O	145 (1)	290 (2)

Note: If the mortar does not meet the 7-day strength but meets the 28-day strength, it is acceptable. If it fails to meet the 7-day strength but reaches two-thirds of it, the contractor may continue at his own risk, pending the 28-day results.

The compressive strength, water retentivity, and workability of cement-lime mortars can be varied over wide ranges by changing the proportions of cement and lime in the mortar.

As noted, the cement is the main contributor to the strength of mortar. But, at the same time, the cement contributes to rapid setting, low water retentivity, and poor workability. On the other hand, lime in the mix contributes very little towards strength; it does, however, improve the workability and makes for better water retentivity.

The amount of water to be used on the job should be the maximum that will produce a workable consistency during construction.

12.2.3 Grout Mix Proportions

Grout is made to a consistency that will flow or pour easily into cores and cavities without segregation of the ingredients. Coarse grout, for filling larger spaces, may contain small gravel; whereas, fine grout does not. A set of ingredients (in parts by volume) for fine grout might include 1 part portland cement; 1/10 part hydrated lime; 2-1/2 parts fine, damp, and loose aggregate; and water, for desired consistency.

For coarse grout, the following set of ingredients (in parts by volume) might be used: 1 part portland cement; 1/10 part lime putty; 2-1/2 parts fine aggregate; 1 part coarse aggregate; and water, for desired consistency.

The mix proportions shown here will, in most cases, produce grout with a compressive strength of 580 psi to 2,465 psi (4 MPa to 17 MPa) in 28 days, depending on the amount of water used in the mix when tested by the standard laboratory methods. However, the in-place compressive strength of the grout will often exceed 2,465 psi (17 MPa) because normally the units will absorb some of the mixing water from the grout, and this lowering of the water/cement ratio of the grout mix will increase the compressive strength. In addition, the presence of moisture in the units surrounding the grout will ensure favorable curing conditions, which is an aid to improved strength.

12.3 MASONRY CONSTRUCTION PRINCIPLES

The usage of masonry materials is now increasing and no discussion of light building construction would be complete without a description of the fundamental methods of using them in modern buildings. The basic structure of masonry walls and their physical characteristics are discussed in this section.

12.3.1 Types of Walls

A number of methods or systems are used in the construction of masonry walls, chimneys, and pavings. The most suitable system for a project is determined by the structure's desired appearance and economic, acoustic, insulation, and strength requirements. Masonry walls may be of the following types: solid, hollow, cavity, composite, veneered, or reinforced. These classifications may overlap, but the following basic terminology (Figure 12-6) and bonding directions (Figure 12-7) remain the same.

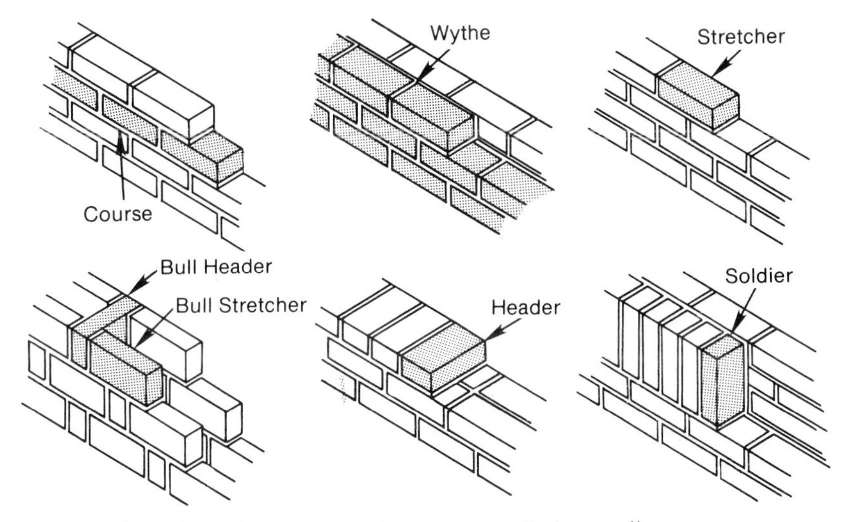

FIGURE 12-6. Basic terminology of masonry units in a wall.

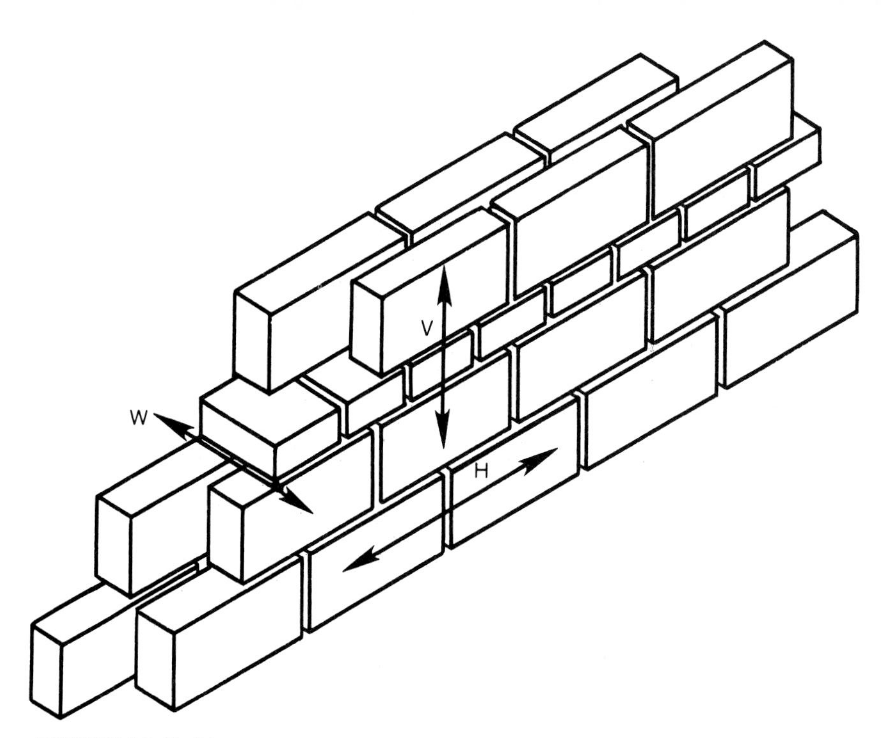

FIGURE 12-7. Masonry units bonded in three directions—horizontally, vertically, and laterally—to form a continuous mass.

Course. One of the continuous horizontal layers (or rows) of masonry which, when bonded together, form the masonry structure.

Wythe. A continuous vertical 4″ (102 mm) or greater section or thickness of masonry, such as the thickness of masonry separating flues in the chimney.

Stretcher. A masonry unit laid flat with its longest dimension parallel to the face of the wall.

Header. A masonry unit laid flat with its longest dimension perpendicular to the face of the wall. It is generally used to tie two wythes of masonry together.

Rowlock. A brick laid on its edge (face).

Bull-Stretcher. A rowlock brick laid with its longest dimension parallel to the face of the wall.

Bull-Header. A rowlock brick laid with its longest dimension perpendicular to the face of the wall.

Soldier. A brick laid on its end so that its longest dimension is parallel to the vertical axis of the face of the wall.

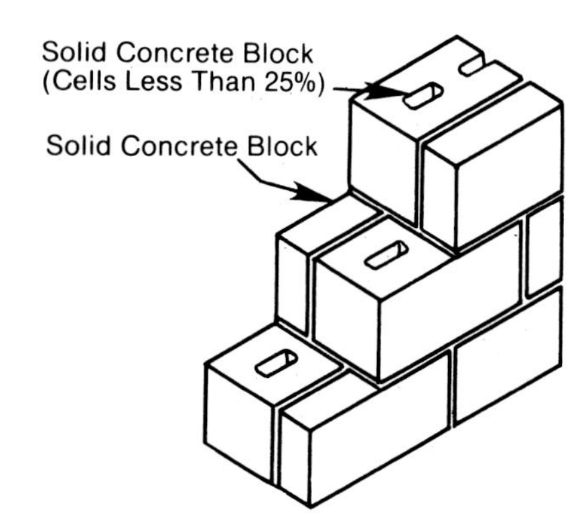

FIGURE 12-8. Solid masonry wall.

Solid Walls. Solid masonry walls are built up of solid units laid up in mortar. Since solid units are those whose net area is equal to or greater than 75% of the gross area (Figure 12-8), most solid walls will be built of brick or, in some cases, stone.

Such walls may be either single or multiple wythe. Single-wythe, solid masonry walls are generally used as partitions in a nonload-bearing capacity. Multiple-wythe, solid walls are used extensively in load-bearing construction for fire walls and in reinforced brick masonry.

Hollow Walls. Hollow masonry walls are built with hollow masonry units, which may be brick, concrete masonry, or structural tile. Hollow units are those whose net area is less than 75% of the gross area (Figure 12-9). As in the case of solid units, all horizontal and vertical edges are embedded in mortar.

Hollow walls may be used in either load-bearing or nonload-bearing capacities, depending on the type of unit and the particular circumstances.

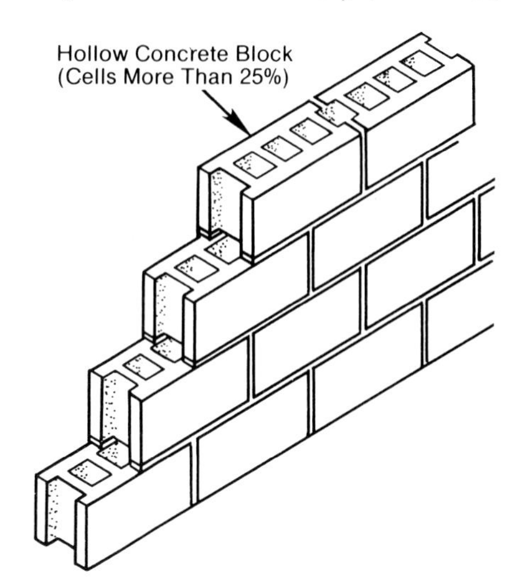

FIGURE 12-9. Hollow masonry wall.

Cavity Walls. Cavity walls consist of two wythes of masonry separated by an air space of 2″ to 3″ (51 mm to 76 mm) and joined together by horizontal metal ties for structural strength (Figure 12-10). The outside or face wythe and the backup wythe may be of similar materials, as in many cases, the face wythe will be brick or facing tile and the backup wythe will be concrete masonry or structural tile and will serve to support the vertical loads of the building.

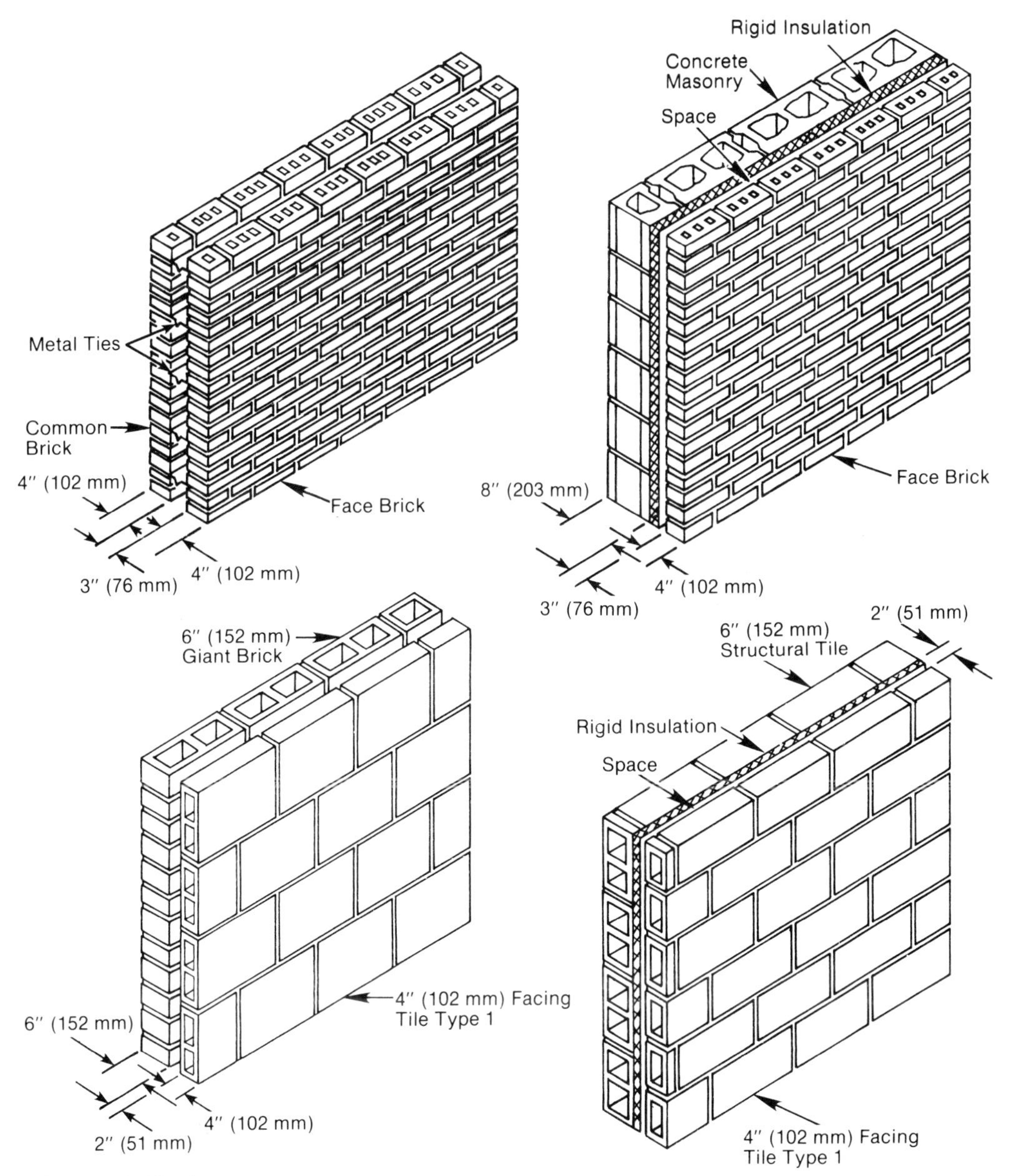

FIGURE 12-10. Cavity walls.

The important advantages of cavity walls over those of solid masonry are the positive protection against rain penetration that cavity walls can provide and the fact that insulation can be introduced into the cavity to provide thermal protection.

Composite Walls. Composite walls are multiwythe walls having at least one of the wythes dissimilar to the others in the type or grade of masonry unit or mortar used (Figure 12-11). Each wythe contributes to the strength of the wall. These walls are normally used as exterior walls

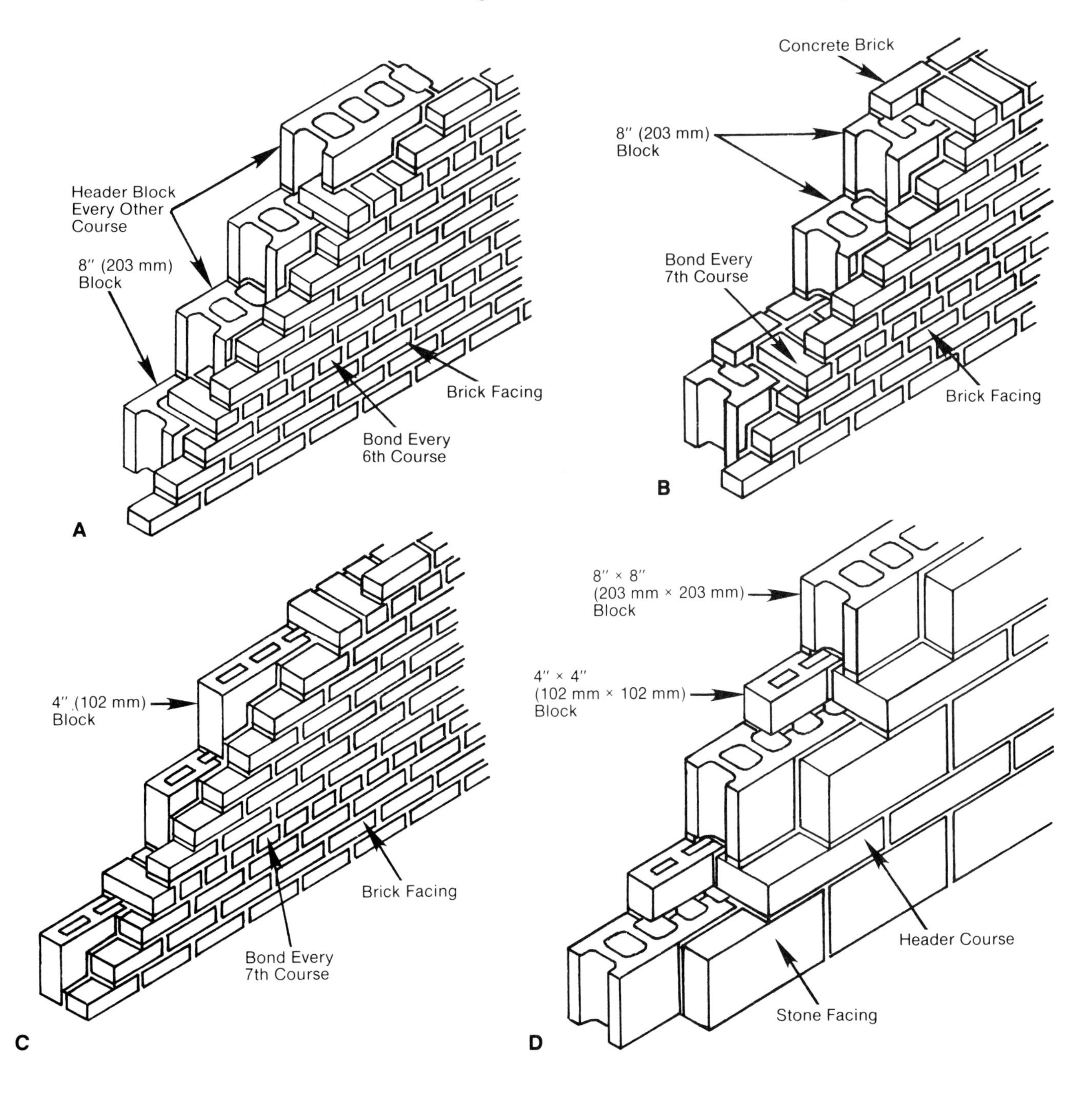

FIGURE 12-11. Composite walls: (A) 12" (305 mm) wall (block and brick, 6th course bonding); (B) 12" (305 mm) wall (block and brick, 7th course bonding); (C) 8" (200 mm) wall (block and brick); and (D) 12" (305 mm) wall (block and stone).

and will usually consist of a brick, stone, or tile face bonded to a concrete masonry or structural tile backup. Since composite walls have wythes that are bonded together, they are not considered veneered walls.

Veneered Walls. A veneered wall consists of a brick, stone, tile, or concrete masonry facing over a backup wall of wood, concrete, concrete masonry, common brick, or structural tile. The outside wythe or veneer is merely a facing and a weather protective coating; it is nonload-bearing and, in some cases, quite thin. This veneer is tied but not bonded to the backup wall (Figure 12-12). The metal ties used are flexible to allow for differential movement.

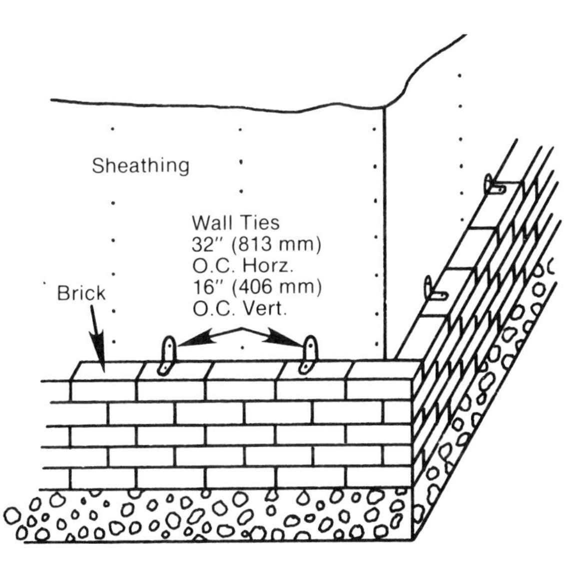

FIGURE 12-12. Veneered wall.

Reinforced Masonry Walls. When buildings are more than three stories high and are located in seismic zones or in any situation where greater resistance is required to applied loads and stresses, added strength should be provided for the walls by introducing reinforcement into them.

Reinforcing masonry walls are walls in which the interior space or spaces are filled or partially filled with grout and reinforcing steel. This may involve a single-wythe concrete masonry wall with some or all of the cells filled or a cavity wall with the space between the two wythes filled with a reinforced grout. Reinforced masonry acts very similarly to reinforced concrete, due to the flexibility added when reinforcing is introduced.

Reinforced masonry walls are generally of three types: a single-wythe wall with only some of the cells filled, a single-wythe wall fully grouted, and a cavity wall with the space between the wythes grouted (Figure 12-13).

12.3.2 Control Joints

As masonry units are placed and the mortar sets, both internal and external stresses will be exerted against the structure. Internal stresses

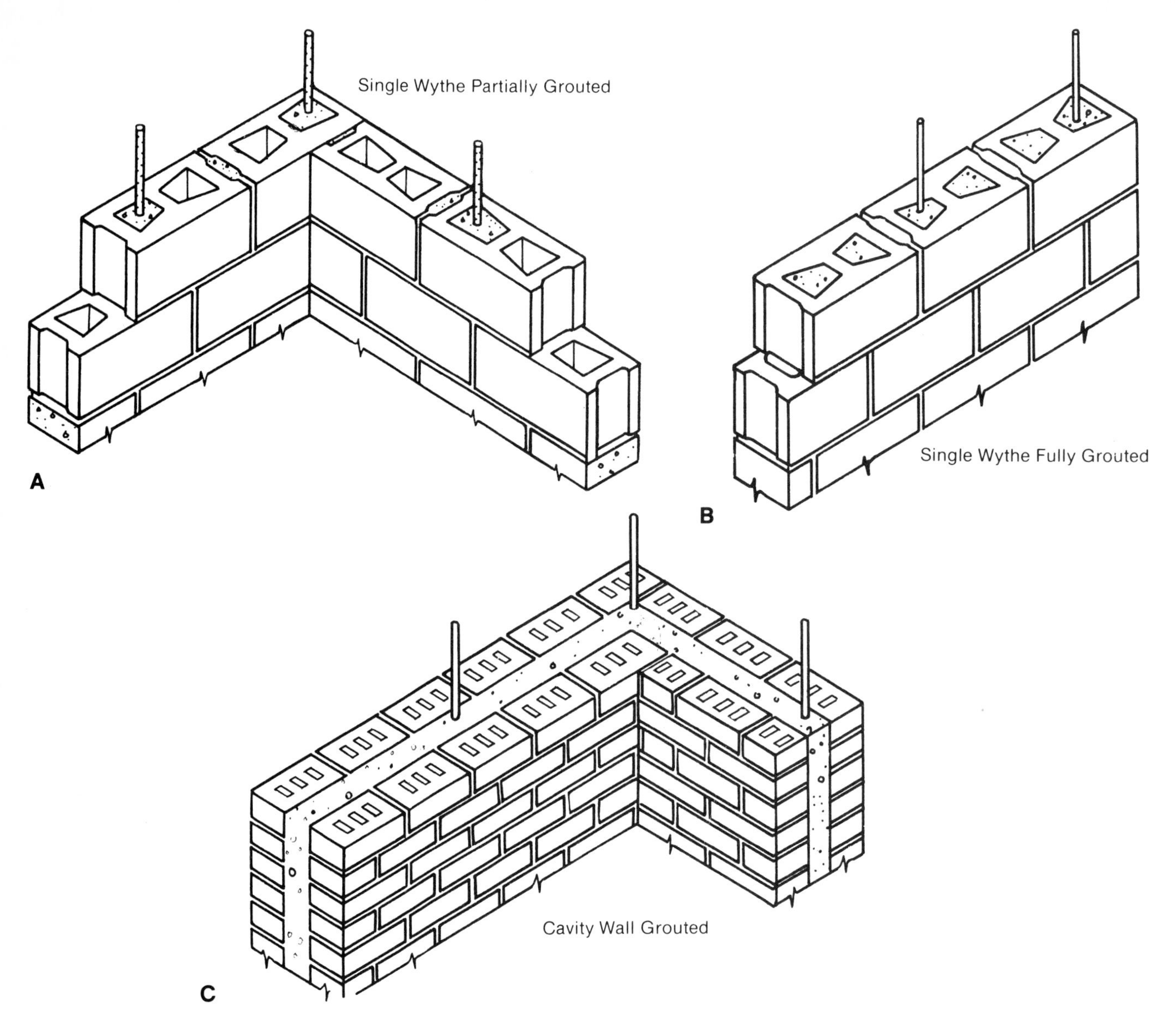

FIGURE 12-13. Reinforced masonry walls.

are created by shrinkage and by building loads. The movement of concrete masonry walls (due to temperature changes, changes in moisture content and material chemistry, and the movement of other parts of the structure) will also set up stresses within the wall. External stresses may be caused from poor footings, a water problem, an excessive number of freeze-thaw cycles, or earth movements and wind. Indications of too much stress are cracks and spallings. In extreme cases a structure will collapse.

Control joints are vertical joints built into a wall to relieve stresses by reducing restraint and permitting movement to take place. They are incorporated into a wall in locations where concentrated stress might occur, such as at locations where abrupt changes occur in wall height or in wall thickness (Figure 12-14).

12.3.3 Wall Patterns

Because of the wide choice of sizes, shapes, textures, and colors in masonry units, an endless variety of architectural effects may be obtained simply by varying the pattern in which the units are laid (Figure 12-15) and by applying different treatments to the mortar joints (Table 12-2).

12.3.4 Heat Transfer

The primary purpose of thermal insulation in exterior walls is to reduce the heat flow and to maintain desired air temperatures in winter and summer. Interior air temperature is not the only factor that affects comfort. Humidity and air movement are also important considerations. As the relative humidity is reduced, the temperature must be increased to keep the same comfort level. In winter conditions, for example, the comfort level at 68° F (20° C) and 100% relative humidity is about equivalent to 79° F (26° C) and 10% humidity.

Air movement has a significant effect on comfort. An increased airflow across the body can increase the convection losses, and when perspiration is occurring, heat is lost due to evaporation. This movement will increase the comfort in high temperatures, but it can have the reverse effect in low temperatures.

The rate of heat flow through walls, floors, ceilings, windows, and doors depends on the difference in inside and outside temperatures and the resistance to heat flow of the materials used in the element. This heat flow, from a region of high temperature to one of low temperature, occurs by conduction, convection, radiation or a combination of these. Conduction is usually given the most attention when insulating and element.

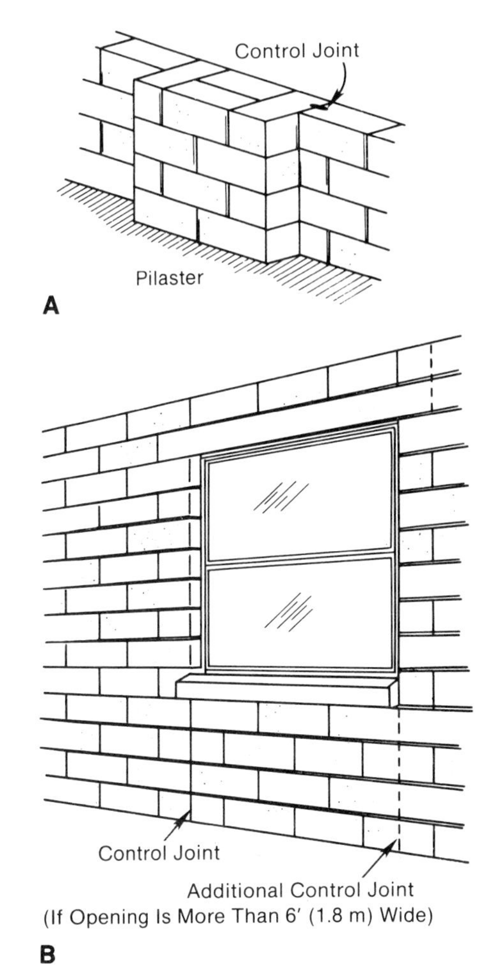

FIGURE 12-14. Control joints (A) at edges of pilaster and (B) at window opening.

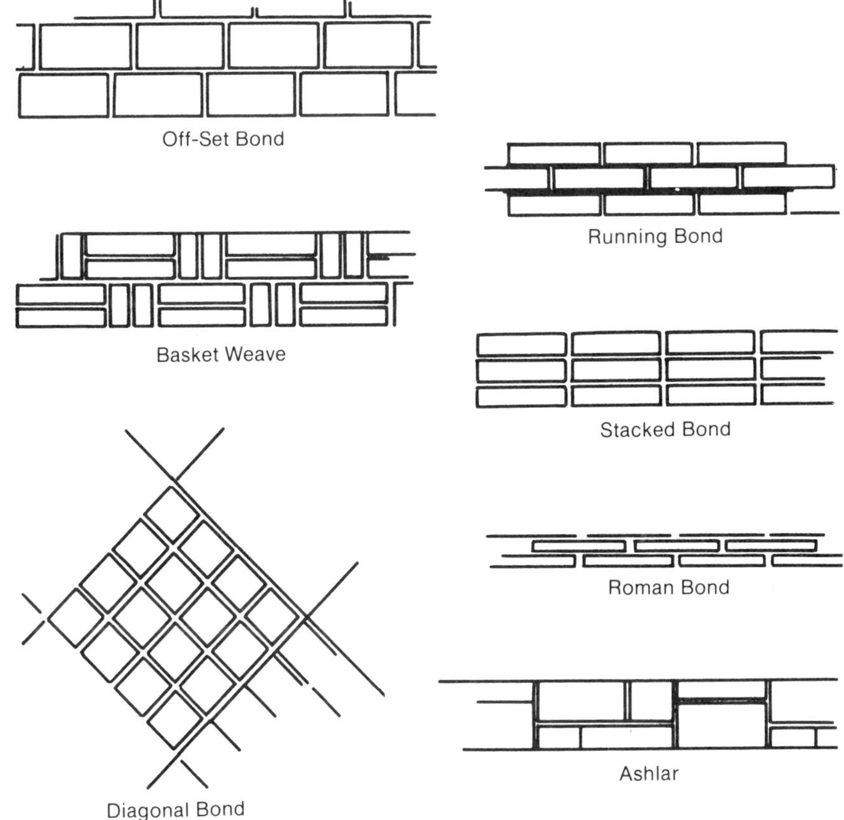

FIGURE 12-15. Wall patterns.

TABLE 12-2 TYPES OF MORTAR JOINTS

Shape	Name	Remarks
	Concave	Formed with jointing tool; resists rain penetration; very good in heavy rain and high wind areas; joints are usually kept small.
	V	
	Weathered	Regarded as a top grade troweled joint; sheds water easily.
	Flush (rough cut)	Easiest of the troweled joints to do; watertightness cannot be guaranteed.
	Struck	Not difficult to do with a trowel; a common joint but the ledge does not shed water easily.
	Raked	Formed with a tool; creates good shadow lines; is not particularly resistant to freezing, high winds or rain.
	Extruded	Mortar squeezed out is simply left as is; used for appearance only; makes for a very rough wall.
	Shaped	A raked joint done with a special tool; characteristics similar to the raked joint.
	Convex	Extruded mortar is shaped with a special tool; makes a fairly acceptable joint but is done mostly for appearance.

Other factors affecting heat flow include the following:

1. The presence of moisture in the units raises the U factors and increases heat flow.
2. Surface color of the wall determines the amount of heat absorbed.
3. Thermal bridges, such as ties, tend to increase heat flow.

The coefficient of heat flow, or U value, is essential to the designer's calculations regarding the flow of heat through walls as well as other building components. The U value expresses the total amount of heat in British thermal units (Btu) that 1 square foot of wall (or ceiling or floor) will transmit per hour for each degree Fahrenheit of temperature difference between the air on the warm and cool sides.

The R value, which is reciprocal of the U value, is a measure of the resistance that a building section, material, air space, or surface film offers to the flow of heat. The unit value for R is (° F × h × ft²) Btu. Because R values can be added directly, they are particularly useful for estimating the effect of components of a section on the total heat flow.

The U value of a wall must include the effect of air film or surface conductance for the inside of a wall (f_i) as well as its outside (f_o). Thus, a U value is calculated by taking the reciprocal of $1/f_i + 1/f_o$ + the sum of R values for each component of the wall. No meaningful results can be obtained by added or subtracted U values.

The rational method for estimating the U values by calculation was devised because of the countless types and combinations of materials used in constructing modern walls. This method is based on the heat flow resistance concept and is related to the principles governing the flow of electricity. Heat flow resistance is similar to electrical resistance. Thus, the flow of heat is inversely proportional to the heat resistance and directly proportional to the temperature difference.

By multiplying the U value of the section by the total net area in square feet (with areas of windows and doors deducted and dealt with separately) and then multiplying that answer by the design temperature difference between the inside and the outside air, the hourly heat flow through the exterior wall construction of a building can be determined. Because the surface area and temperature difference are almost always established or fixed quantities, the U value alone remains subject to design adjustments.

U values increase in walls as the moisture content increases. Figure 12-16 shows the relationships found between the conductivity, density, and moisture content of concrete, mortar, or grout. Note that conductivity in the normally dry condition is only slightly higher than in the oven-dry condition. However, in the event that concrete masonry should become saturated with water, heat transfer will increase markedly. This condition might occur in a basement wall that is not damp-proofed. It could also prevail for a short time after a heavy driving rain in walls above grade.

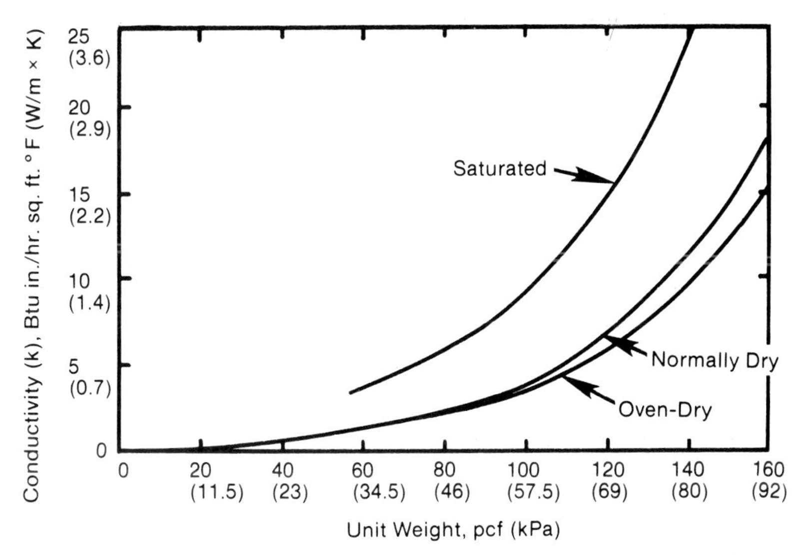

FIGURE 12-16. Conductivity of concrete, mortar, or grout as affected by density and moisture. Aggregate type influences conductivity only as it affects the resulting unit weight and moisture content.

A sealer should not be applied to the cold side of a wall since moisture travels toward that side. If this surface is sealed, the wall can approach saturation there and, thus, lose much of its insulating value. This illustrates the importance of using exterior paints that can breathe.

It should be noted that heavy construction such as concrete masonry does not respond to temperature fluctuations as rapidly as does lightweight construction, even though the U values may be identical. This is because masonry walls have a greater heat storage capacity; therefore, buildings with heavy masonry walls are sometimes designed with higher U values.

The designer must consider several factors when selecting the proper U value for a wall section. The relative importance of each of those factors usually depends upon the type of building and its purpose. For buildings such as apartments, hotels, dormitories, and houses, human comfort needs will probably govern the selection. On the other hand, process control requirements may be the primary factor in deciding the U value for commercial buildings devoted to manufacturing operations.

12.3.5 Control of Condensation

To prevent condensation or sweating on the interior surface of a wall, the surface temperature of the wall must be kept above the dew point of the room air. In Table 12-3, the dew points for various room temperatures and relative humidities are listed.

TABLE 12-3 DEW POINT TEMPERATURES

Dry Bulb or Room Temp., °F (°C)*	Dew Point, °F (°C)*, Based on Relative Humidity									
	10%	20%	30%	40%	50%	60%	70%	80%	90%	100%
40 (4)	-9 (-23)	5 (-15)	13 (-11)	19 (-7)	24 (-4)	28 (-2)	31 (1)	34 (1)	37 (3)	40 (4)
45 (7)	-5 (-21)	9 (-13)	17 (-8)	23 (-5)	28 (-2)	32 (0)	36 (2)	39 (4)	42 (6)	45 (7)
50 (10)	-1 (-18)	13 (-11)	21 (-6)	27 (-3)	32 (0)	37 (3)	41 (5)	44 (7)	47 (8)	50 (10)
55 (13)	3 (-16)	17 (-8)	25 (-4)	31 (-1)	37 (3)	41 (5)	45 (7)	49 (9)	52 (11)	55 (13)
60 (16)	6 (-14)	20 (-7)	29 (-2)	36 (2)	41 (5)	46 (8)	50 (10)	54 (12)	57 (14)	60 (16)
65 (18)	10 (-12)	24 (-4)	33 (1)	40 (4)	46 (8)	51 (11)	55 (13)	58 (14)	62 (17)	65 (18)
70 (21)	13 (-11)	28 (-2)	37 (3)	45 (7)	51 (11)	56 (13)	60 (16)	63 (17)	67 (19)	70 (21)
75 (24)	17 (-8)	31 (-1)	42 (6)	49 (9)	55 (13)	60 (16)	65 (18)	68 (20)	72 (22)	75 (24)
80 (27)	20 (-7)	36 (2)	46 (8)	54 (12)	60 (16)	65 (18)	69 (21)	73 (23)	77 (25)	80 (27)
85 (29)	23 (-5)	40 (4)	50 (10)	58 (14)	65 (18)	70 (21)	74 (23)	78 (26)	82 (28)	85 (29)
90 (32)	27 (-3)	44 (7)	55 (13)	62 (17)	69 (21)	74 (23)	79 (26)	82 (28)	86 (30)	90 (32)

*Values are rounded off.

When proper attention is given to the insulating quality of the walls and the relative humidity is controlled within reasonable limits, sweating of the interior wall surfaces is usually not a problem. Sometimes mechanical dehumidification or ventilation may be required to control water vapor condensation that occurs on uninsulated basement walls below grade during humid periods of the summer or during basement laundering activity.

The control of condensation within wall spaces and wall materials is more difficult since condensation here can occur whether the wall has an overall low or high resistance to heat flow. Water vapor will diffuse through materials of building construction at rates that depend upon the water vapor permeability of the materials and the existing vapor pressure differential. Although the passage of water vapor through a material is not harmful in itself, it may become of consequence when, at some point along the water vapor flow path, a temperature level drops below the dew point and condensation occurs.

Excessive levels of condensed water vapor in concrete masonry may cause efflorescence or temporary formation of frost within the wall. In most concrete masonry structures, however, freezing does not constitute a problem due to the daily fluctuations in temperatures. These fluctuations allow frost to thaw and the water is released through the outer wythe as vapor or through weepholes as condensate.

Frost buildup in walls causes the most damage during extended periods of very low temperatures outdoors and relative humidities exceeding about 50% indoors. Fortunately, when outdoor temperatures are very low, high relative humidities indoors are unusual except in laundries, bathing areas, etc.

When the designer is faced with building service conditions conducive to water vapor condensation, he or she usually provides a vapor barrier on or as close as possible to the warm surface of the wall. The vapor barrier is a material, such as plastic film, asphalt-treated paper, or aluminum or copper foil, that will reduce to a minimum the entrance of water vapor into the wall. The small amount of water vapor that does penetrate the vapor barrier will pass through the outer layers of the wall by diffusion.

12.3.6 Acoustics

Masonry walls are well suited to resist the passage of sound or noise from one side to the other. It is an important function of buildings to hold noise to a tolerable level. Noise is commonly defined as sound that is not wanted and this will vary depending on the individual's level of tolerance and on the activity in which he or she is engaged. For noise that cannot be eliminated or reduced, some effort is made to absorb it or to reduce its transmission.

Sound absorption is the amount of airborne sound energy (measured in sabins) absorbed on the wall surface adjacent to the sound. Sound transmission loss is the total amount of airborne sound (measured in decibels) lost as it travels from one side of the wall or floor to the other.

With respect to a building, sound is relatively well transmitted with insufficient absorption through uncaulked openings in walls (for vents and piping) and through doors and windows, particularly open ones. Such transmissions tend to negate building codes written to regulate the amount of noise that must be stopped by walls, floors, and ceilings.

12.3.7 Masonry Chimneys and Fireplaces

Because the fireplace is a dominant and interesting exterior architectural feature as well as a focal point of interior design, fireplaces and chimneys should be aesthetically pleasing as well as functional.

However, many fireplace/chimney combinations that have been constructed mainly for aesthetic and entertainment values are relatively inefficient heaters. Some draw air across a room causing uncomfortable drafts, and others have design problems with respect to smoke clearance. Still others lose so much heat through the chimney that they are actually less than useful as heating devices.

The design and construction of an efficient, functional fireplace require consideration of fireplace location, dimensions, and placement of various component parts. The four basic functions of a fireplace are to (1) assure proper fuel combustion, (2) deliver smoke and other products of combustion up the chimney, (3) radiate the maximum amount of heat, and (4) provide an attractive architectural feature.

Combustion and smoke delivery depend primarily on the shape and dimensions of the combustion chamber, the proper location of the fireplace throat and the smoke shelf, and the ratio of the flue area to the area of the fireplace opening. Heat radiation depends on the dimensions of the combustion chamber. Fire safety depends not only on design, but also on the ability of the masonry units to withstand high temperatures.

Various fireplace and chimney requirements are normally specified by local building codes.

There are several types of fireplaces, but basically design and construction principles are the same. Table 12-4 and Figure 12-17 show these types and standard sizes.

TABLE 12-4 FIREPLACE TYPES AND STANDARD SIZES

Type	Width (w), in. (mm)	Height (h), in. (mm)	Depth (d), in. (mm)	Area of Fireplace Opening in.^{2} (m^2)	Nominal Flue Sizes (Based on 1/10 Area of Fireplace Opening), in. (mm)
Single face	36 (914)	26 (660)	20 (508)	936 (.6)	12 × 16 (286 × 387)
	40 (1,016)	28 (712)	22 (559)	1,120 (.7)	12 × 16 (286 × 387)
	48 (1,219)	32 (813)	25 (635	1,536 (1)	16 × 16 (387 × 387)
	60 (1,524)	32 (813)	25 (635)	1,920 (1.2)	16 × 20 (387 × 489)
Two-face (adjacent)	39 (991)	27 (686)	23 (584)	1,223 (.8)	12 × 16 (286 × 387)
	46 (1,168)	27 (686)	23 (584)	1,388 (.9)	16 × 16 (286 × 387)
	52 (1,321)	30 (762)	27 (686)	1,884 (1.2)	16 × 20 (387 × 489)
	64 (1,626)	30 (762)	27 (686)	2,085 (1.3)	16 × 20 (387 × 489)
Two-face (opposite)	32 (813)	21 (533)	30 (762)	1,344 (.9)	16 × 16 (387 × 387)
	35 (889)	21 (533)	30 (762)	1,470 (.9)	16 × 16 (387 × 387)
	42 (1,067)	21 (533)	30 (762)	1,764 (1.1)	16 × 20 (387 × 489)
	48 (1,219)	21 (533)	34 (864)	2,016 (1.3)	16 × 20 (387 × 489)
Three-face	39 (991)	21 (533)	30 (762)	1,638 (1)	16 × 16 (387 × 387)
	46 (1,168)	21 (533)	30 (762)	1,932 (1.2)	16 × 20 (387 × 489)
	52 (1,321)	21 (533)	34 (864)	2,184 (1.4)	20 × 20 (489 × 489)

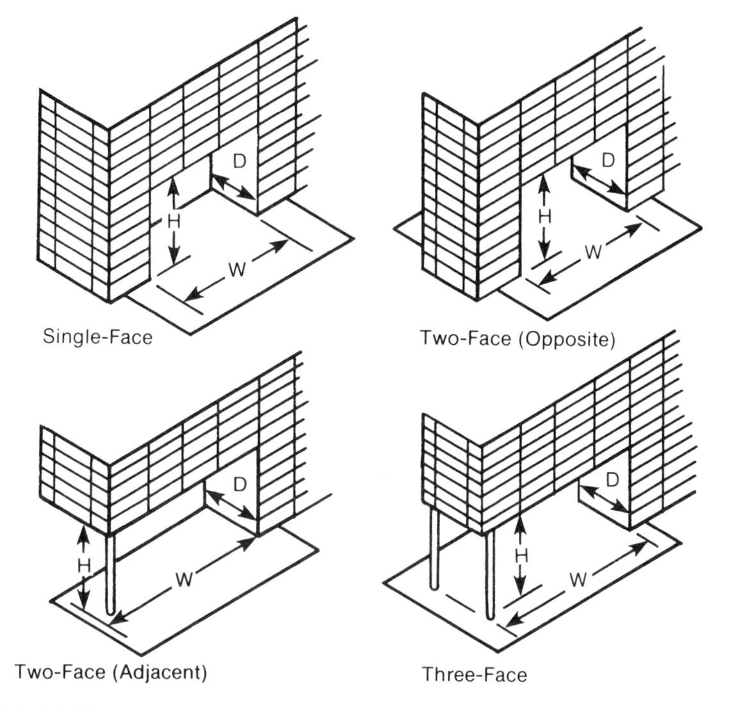

FIGURE 12-17. Types of fireplaces.

The single-face fireplace is the oldest and most common variety, and most standard design information is based on this type. The multiface fireplace, if used properly, is highly effective and attractive, but may cause draft problems.

Fireplace Elements. In the next several paragraphs, the major elements of a fireplace are discussed. These elements are the hearth, lintel, firebox, throat, and smoke chamber.

The fireplace floor is called the hearth. The inner part of the hearth is lined with firebrick. The outer hearth consists of a noncombustible material such as firebrick, concrete brick, concrete block, or just concrete. The outer hearth is supported on concrete.

The horizontal member that supports the front face or breast of the fireplace above the opening is called the lintel. It may be made of reinforced masonry or a steel angle.

The firebox is the combustion chamber where the fire occurs. Its sidewalls are slightly slanted to radiate heat into the room, and its rear wall is curved or inclined to provide an upward draft to the throat.

Unless the firebox is the preformed metal type, it should be lined with firebrick that is at least 2″ (51 mm) thick and laid with thin joints of fireclay mortar. Back and sidewalls, including lining, should be a minimum of 8″ (203 mm) thick to support the chimney weight.

The fireplace is laid out on a concrete slab and its back is constructed approximately 5′ (1.5 m) high before the firebox is constructed and partially backfilled with tempered mortar and brick scraps.

The throat of the fireplace is the slotlike opening directly above the firebox through which flames, smoke, and other combustion products pass into the smoke chamber. The throat must be carefully designed to allow proper draft. It should not be less than 6″ (152 mm), preferably 8″ (203 mm), above the highest point of the fireplace opening.

The inclined back of the firebox extends to the same height as the throat and supports the hinge of a metal damper. The damper extends the full width of the fireplace opening and should open upward and backward.

The smoke chamber compresses and funnels smoke and gases from the fire into the chimney flue. The shape of this chamber should be symmetrical with the center line of the firebox to assure even burning. The back of the chamber is usually vertical, and its walls are inclined upward to meet the bottom of the chimney flue lining. If the solid masonry wall is less than 8″ (203 mm) thick, the smoke chamber should be parged with 3/4″ (19 mm) of fireclay mortar. Metal lining plates are available to give the chamber its proper form, provide smooth surfaces, and simplify construction.

Chimney Elements. The chimney creates a draft and disposes of combustion products. Chimney design and construction must ensure efficient operation and also freedom from fire hazards.

To prevent downward air currents, build the chimney a minimum of 3′ (.9 m) above the flat roof, 2′ (.6 m) above any part of the roof within a 10′ (3 m) radius of the chimney. Increasing the chimney height will improve the draft.

The concrete foundation for a chimney is designed to support the weight of the chimney. Because of the large additional weight, the unit bearing pressure beneath the chimney foundation should be approximately equal to that beneath the house foundation to minimize the possibility of differential settlement.

The footing thickness should not be less than one and one-half times the footing projection. The bottom of the footing should extend below the frostline.

A flue must have the correct area and shape to produce a proper draft. Smoke should be drawn up the flue at a relatively high velocity. Velocity is affected by the flue area, firebox opening, and chimney height.

The total area of the flue opening should be approximately one-tenth of the fireplace opening area. A fireplace or stove chimney may have multiple flues, but each flue must be built as a separate unit.

Flue linings extend from the top of the throat to at least 2″ (51 mm) above the chimney cap. Chimney walls are constructed around the flue lining segments. The space between all masonry joints should be completely filled with mortar.

Modular-size chimney units, whether solid block or one-piece chimney units, can be combined only with modular-size flue lining. Minimum wall thickness measured from the outside of the flue lining should be 4″ (102 mm). Exposed joints inside the flue should be struck smooth.

When a chimney contains more than two flues, they should be separated by 4″ (102 mm) thick masonry bonded into the chimney wall. The tops of the flues should have a height difference of 2″ to 12″ (51 mm to 305 mm) to prevent smoke from pouring from one flue into another.

Chimneys should be built as vertically as possible, but a slope is allowed if the full area of the flue is maintained throughout its length. Slope should not exceed 7″ (178 mm) to the foot (.3 m) or 30°. Where offsets or bends are necessary, miter both ends of the abutting flue liner sections equally to prevent reduction of the flue area.

Details of a masonry fireplace and chimney are shown in Figure 12-18.

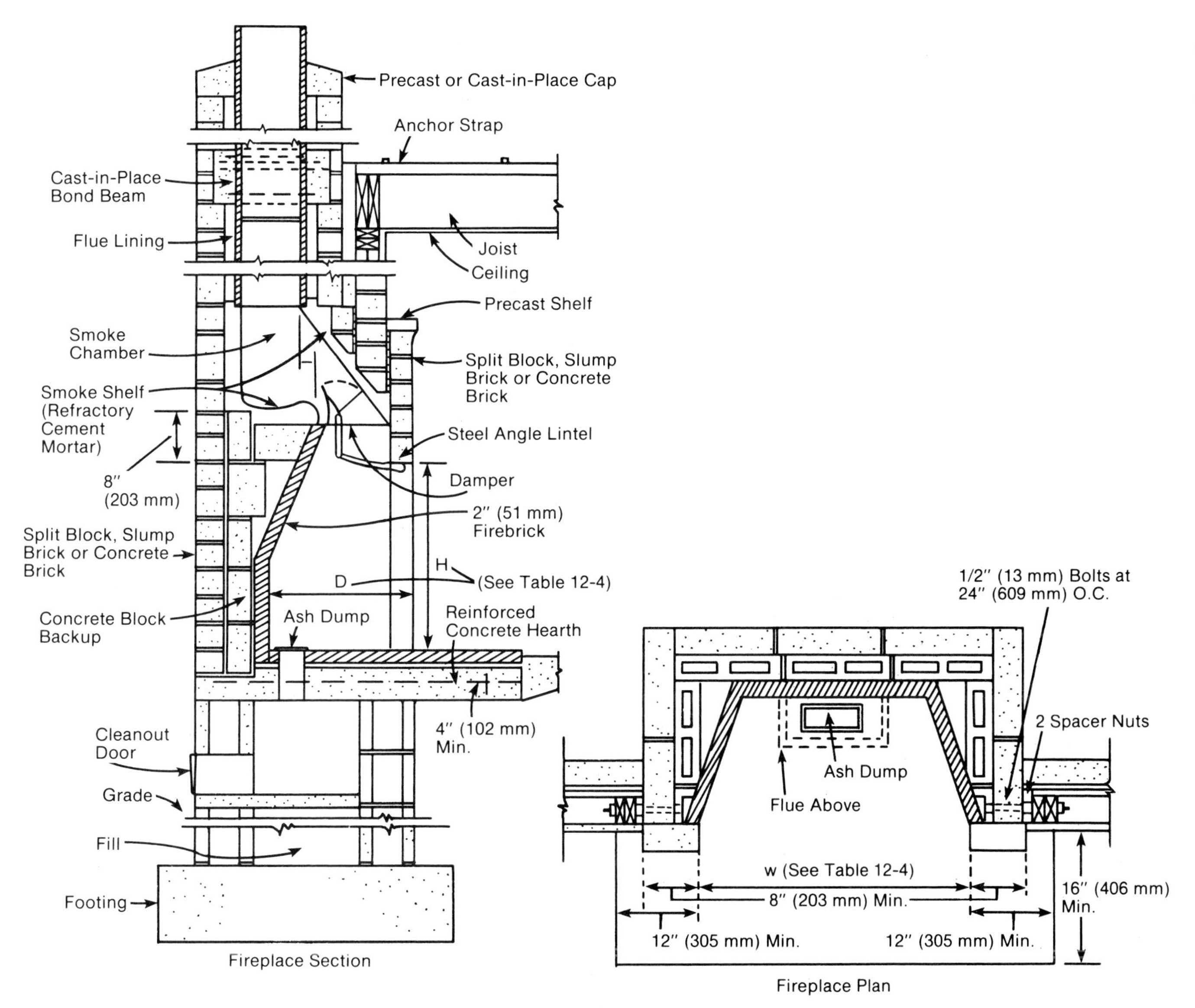

FIGURE 12-18. Details of a masonry fireplace and chimney. **Note:** This does not reflect complete specifications, details, or instructions for fireplace or chimney construction. Consult local and regional codes.

12.3.8 Cold Weather Masonry Construction

Forty degrees Fahrenheit (4°C) is a critical temperature in masonry construction. When the ambient temperature drops below this temperature, mortar heat liberation activity stops. At that time, masonry needs protection, worker comfort conditions need enhancing, and materials handling methods change; thus, construction costs can be expected to increase.

To avoid freezing conditions, the builder needs historic as well as current weather information. The historical data provides averages and extremes of the region; the current data gives predictions of immediate trends. Even if freezing conditions cannot be avoided because of construction schedules, contracts, etc., this weather information is necessary; adequate preparations concerning funds, protective structures, and manpower will be required to maintain good construction practices through the cold weather.

At temperatures below freezing, hydration and strength development are minimal. Fortunately, construction may proceed at temperatures below freezing if the mortar ingredients are heated, and although the strength gain of mortar is less at lower temperatures, final strength may be as high or higher than that of mortar used and cured at more normal temperatures.

Mortar mixed with cold but unfrozen materials possesses plastic properties that are quite different from those at normal temperatures. Water requirements for a given consistency decrease as the temperature drops. More air is entrained with a given amount of air entraining agent. Also initial and final sets are delayed.

During the cold weather, the safe storage of mortar materials can be accomplished by providing an improvised shelter. If the shelter is properly erected, mortar materials may be delivered, stored, protected from the elements, heated, and mixed within it. If shelters are not built, all masonry materials should be covered when the ambient temperature falls below 40° F (4° C).

Temporary enclosures for concrete masonry construction serve several purposes. They help to (1) achieve temperatures high enough to facilitate cement hydration in mortar (and thus obtain the initial strength required for support of superimposed masonry); (2) improve the comfort and efficiency of missions and other craftsmen; and (3) protect materials.

One of the most important practices of concrete masonry work in subfreezing temperatures is to keep the masonry units dry. In addition, these units should be laid only on a sound unfrozen surface and never on a snow or ice covered bed or base; otherwise, no bond will develop and there is a danger of movement when the base thaws. Also, it is considered good practice to heat the surface of an existing masonry course to the same temperature as the masonry units to be added.

During cold weather, mortar should be mixed in smaller quantities than usual to avoid excessive cooling before use. Metal mortarboards with built-in electrical heaters can be used if care is taken to avoid overheating or drying of the mortar. To avoid premature cooling of heated masonry units, only those units which shall be used immediately should be removed from the heat source.

Regardless of temperature, concrete masonry units should never be wetted before being laid down. In cold weather, wetting masonry units only adds to the possibility of freezing any free water on the surfaces, increasing shrinkage and defeating the goal of drawing off water from the mortar to prevent mortar expansion upon freezing.

Upon completion of each section and at the end of every workday, measures should be taken to protect new concrete masonry construction from the weather. In cold weather construction, rapid drying out or early freezing of the mortar must be prevented. Furthermore, the top of the concrete masonry structure should be protected from rain and snow by plastic or canvas tarpaulins which extend at least 2′ (.6 m) down all sides of the structure.

Safety precautions require added emphasis during cold weather construction for a number of reasons. Workers not only direct more of their attention to comfort and out-of-the-ordinary construction problems due to lower temperatures, but also must deal with personal hazards such as uncertain footing on ice and snow and clumsiness due to protective clothing. Extra care must also be taken with flammable building materials and heaters, since the possibility of fire and asphyxiation is increased.

Enclosures for cold weather protection should also be securely guyed or braced if there is danger of snow and ice loads or wind forces. Snow or ice loads can contribute to overloading an enclosure. Wind forces can cause tarpaulins over scaffolds to act like sails. In addition, the combination of wind forces and dead weight of a stack of masonry units can overload the enclosure framework unless it has been carefully sized and installed.

12.3.9 Structural Stresses

Since all structures and structural members can be subjected to various types of loading such as static loads, dynamic loads, and repeated loads, all of which may act over wide temperature ranges, load categories will be discussed first. For our present purposes, we will separate loadings into two broad classifications: static loads and dynamic loads.

The category of static loads can be further broken down into three subdivisions; namely, continuous loads, gradually or slowly applied loads, and repeated gradually applied loads. The term continuous load is used to characterize a load that remains on a member for a long time period. The weight of a structure or a tank subjected to internal pressure for a long period of time is a common example of this type of load. A gradually or slowly applied load is one that slowly builds up to its maximum value and does not cause shock or vibration when it is applied. The last type of static load is the repeated gradually applied load. Loads that are gradually applied but repeated a large number of times are included in this load classification. The repeated gradually applied load is important because it can cause failure under a load that would be safe if the load were applied once or only a few times. Failure of structures due to repeated loads usually occurs catastrophically.

When the application of a load is much more rapid than it is for those loads that are classified as static loads, appreciable shock or vibration can occur. This type of load is known as a dynamic or impact load, and its application can cause internal forces in the structure to momentarily exceed those that occur when the same load is applied gradually. In addition, the deflection of the structure will also momentarily exceed the deflection caused by an equal gradually applied load. As an example, the gradual placing of a weight on the end of a cantilever beam will cause the beam to deflect and gradually come to maximum value. However, if a weight of the same magnitude is dropped on the end of the beam, the maximum deflection will be found to be several times greater than for the same gradually applied load even if the height of drop is small.

All construction materials must resist force. A force is a push or pull that has a value and a direction. The pull of gravity is responsible for most of the force dealt with in construction. However, there are other causes such as the wind. Unit stress, also referred to as stress, is force per unit area over which the force acts. It is obtained by dividing the force by the area on which it acts and is expressed as psi (MPa).

There are three kinds of unit stress—compressive, tensile, and shearing. They depend on the position of the force with respect to the object (Figure 12-19).

Unit stress is determined by dividing the acting force by the original area upon which it acts. This is the area that resists displacement. Tensile and compressive unit stresses act on the cross-sectional area perpendicular to the direction of the force. Shear unit stresses act on the cross-sectional area parallel to the direction of the force. Usually the force is not uniformly distributed across the area; however, in computing tensile and compressive unit stresses, it is assumed to be uniform with satisfactory accuracy. Shearing unit stress acts unequally over an area, and the unit stress at the location of highest units stress must be considered.

A body is in a state of compressive stress when equal and opposite forces are applied along the same line of action and tend to shorten the body (Figure 12-19A). Some examples of structural units under compression are shown in Figure 12-20.

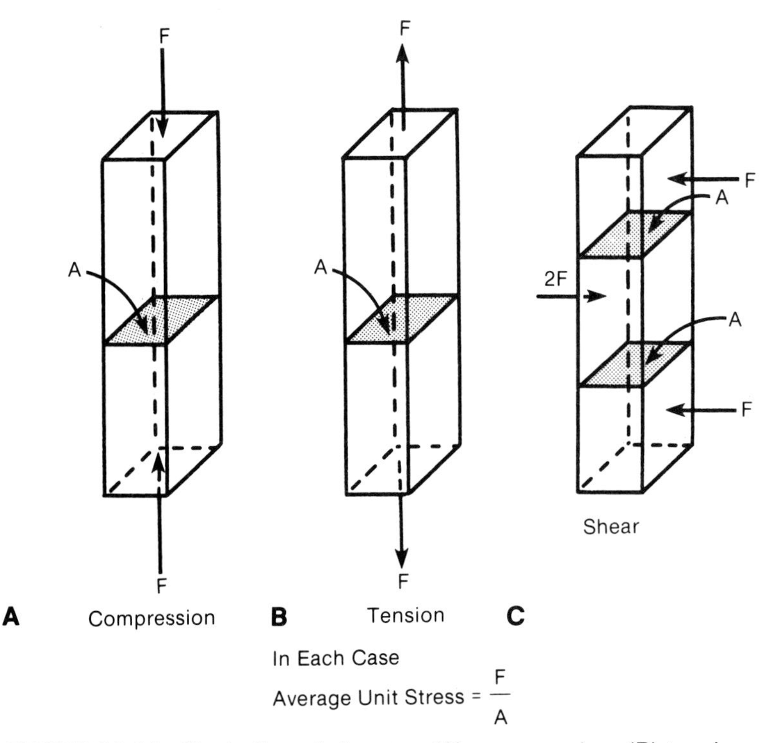

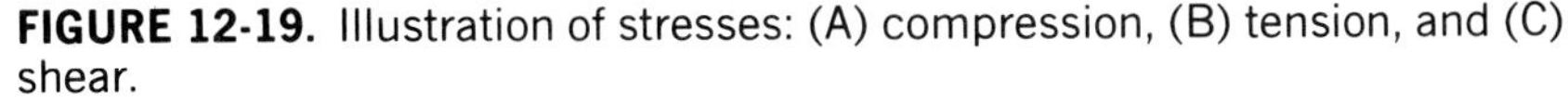

FIGURE 12-19. Illustration of stresses: (A) compression, (B) tension, and (C) shear.

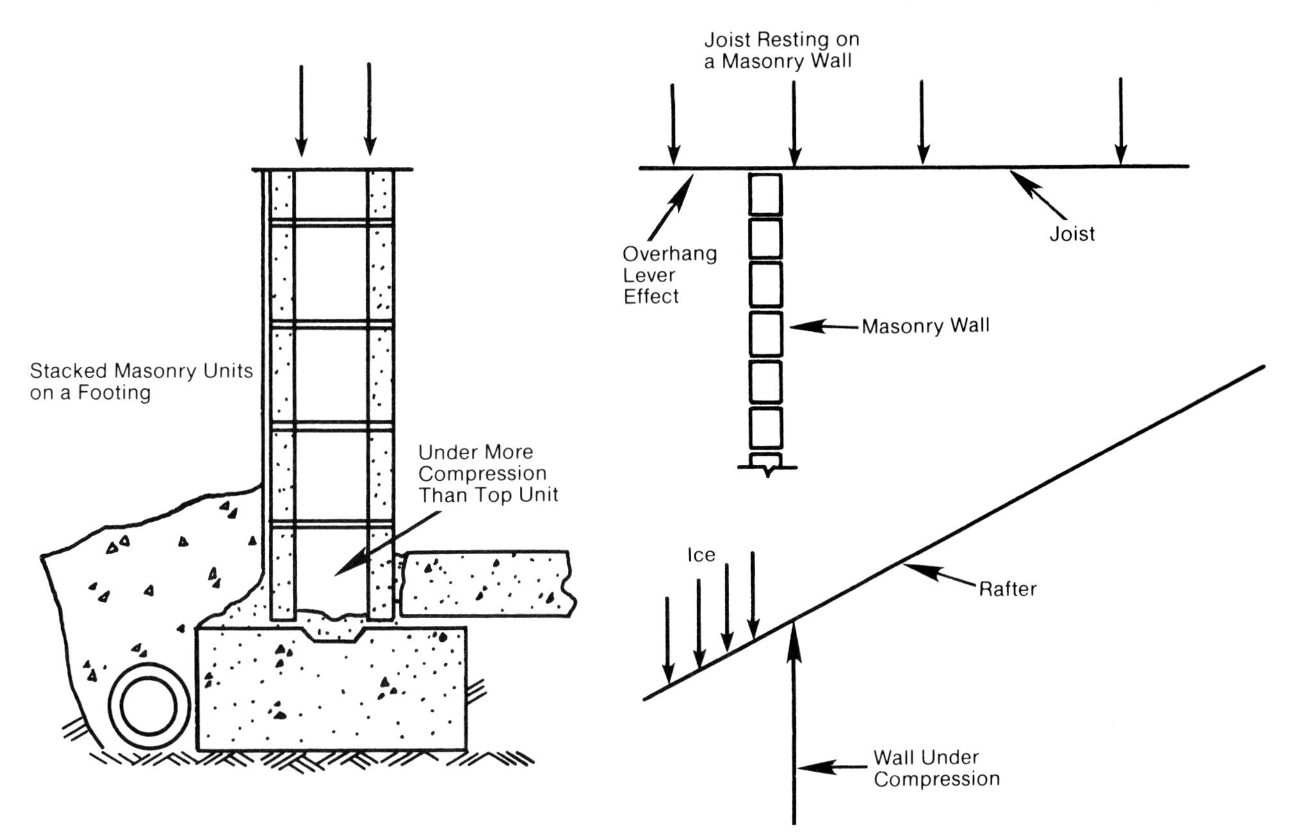

FIGURE 12-20. Examples of structural units under compression.

When masonry units are bonded one to another, they form a building section of certain compressive strength. Unit compression strength for typical units can be measured in a laboratory, and with such information at hand a designer applies the area safety factor when planning for a wall, pilaster, or chimney. A section carries its own weight and possibly the additional loads of roof, snow, equipment, and live weight of persons, animals, or machines moving about. Beams, lintels, and columns or pilasters needing support of a masonry wall or reinforcement may bend under compressive stress (Figure 12-21).

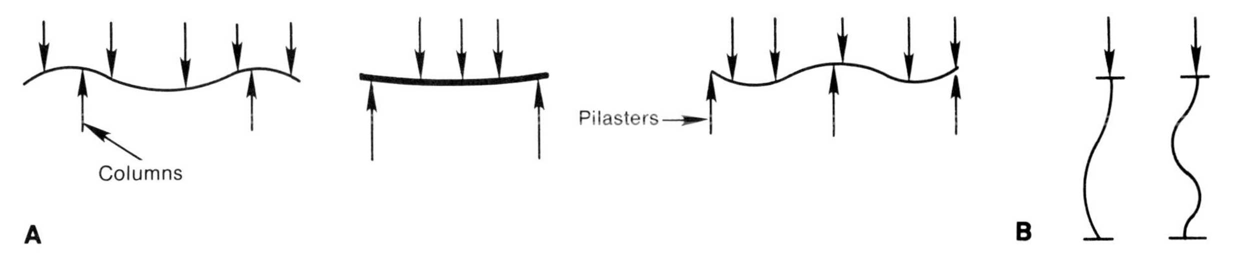

FIGURE 12-21. Bending stresses: (A) beams and lintels deflecting under compression and (B) columns/pilasters needing support of masonry wall and/or reinforcing unit.

When equal and opposite forces are applied along the same line of action so that they tend to elongate the body, the body is said to be in a state of tensile stress (Figure 12-19B). As a wire rope can be under tension from a load, so can other structural members, such as one side of a masonry wall or one edge of a wooden beam resting on a wall (Figure 12-22).

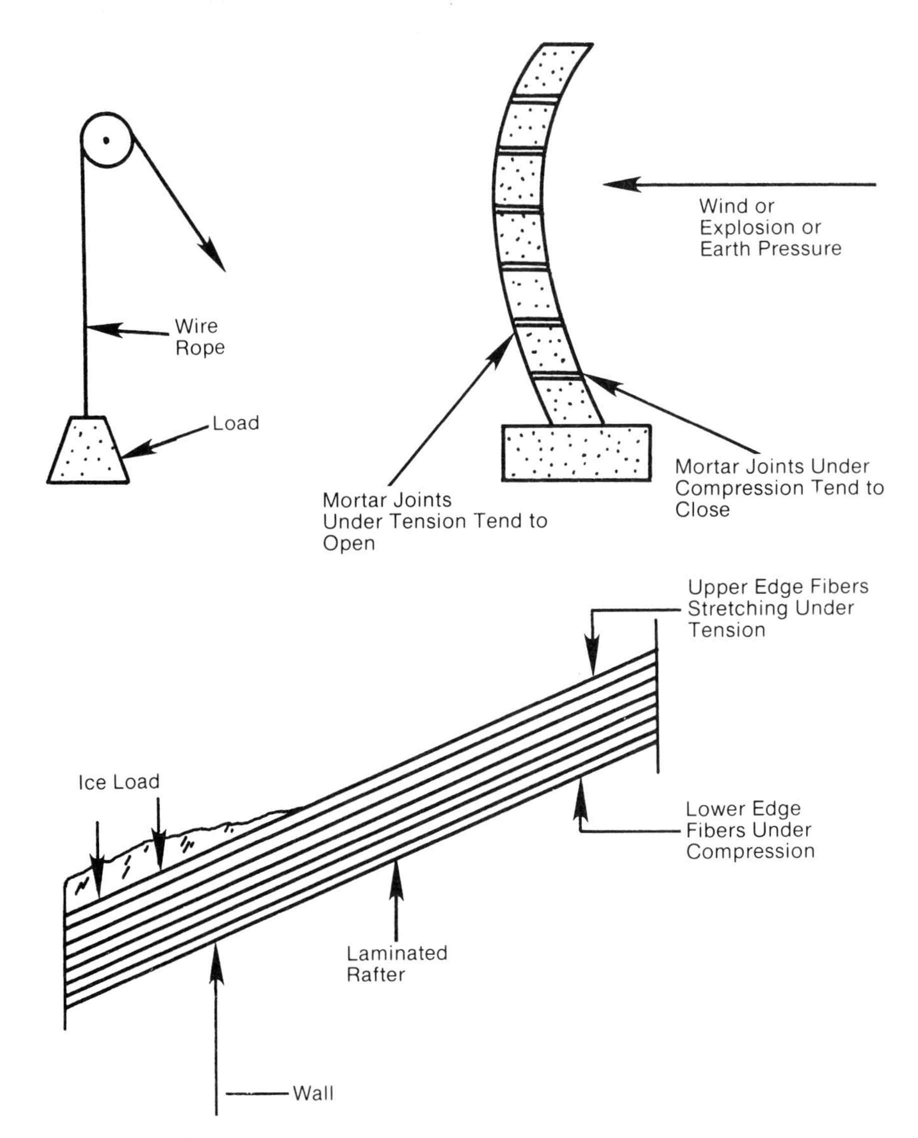

FIGURE 12-22. Tension stresses.

A shearing stress is applied to a body when two parrallel, equal, and opposite forces are applied along different lines of action (Figure 12-19C). A high wind tends to slide a wall over the footing. Such force has a shearing effect on mortar joints and reinforcing rods. The action is similar to cutting a rivet with a chisel or sheet metal with shears (Figure 12-23).

If a building shifts on its foundation, it may twist mortar joints and reinforcing bars, as a wrench turns a nut on a bolt. This is a special type of shearing strain known as torsion.

As one might expect, structural members experience more than one stress simultaneously. Structural stresses may be bending and tension, bending and compression, shear and tension, shear and compression, torsion and compression, or other combinations.

When materials and structural members are loaded and then unloaded, they tend to recover from the stresses, provided they have not reached their elastic limits. If overloaded they tend to deform without recovery.

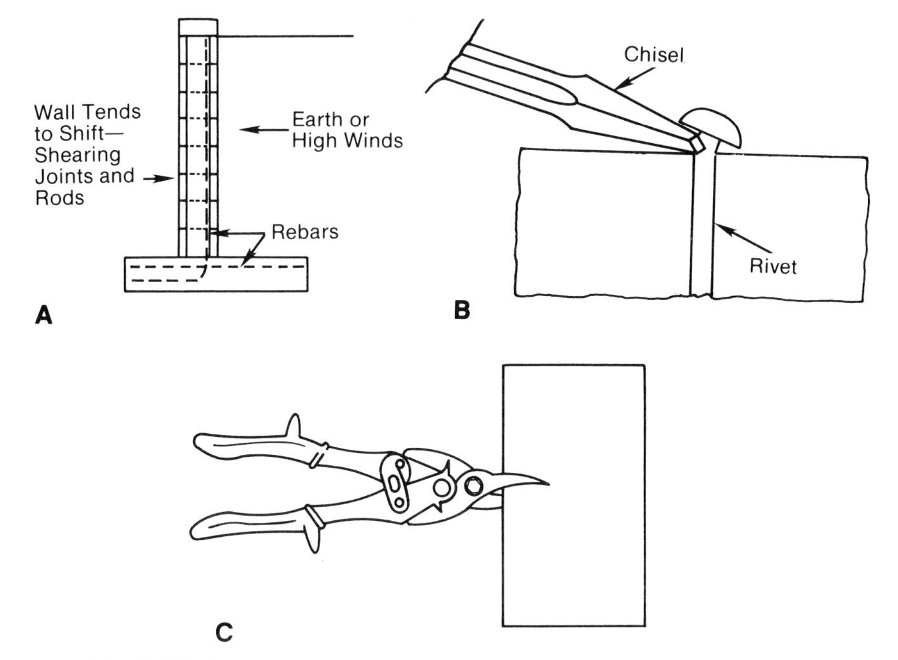

FIGURE 12-23. Shearing stresses: (A) shearing effect of earth pressure or high winds, (B) shearing with a chisel, and (C) metal shears with one blade stationary.

A designer knows generally what conditions a building will experience over its lifetime if he or she knows the conditions the area has previously experienced. Knowing how a building will be used and what loads it will be subjected to, the designer, using safety factors given in a national building code and local regulations, proposes a structure twice, four times, or six times as strong as usually required.

As a building is being created, structural details develop. Stress points are identified and then strengthened or reinforced. To offset natural pressures, the building sections require strong unit members; strong mortar; proper placement of units (bonds), control joints, and reinforcement bars; and wire or wire mesh.

12.4 MASONRY BUILDING DESIGN AS A FIELD OF WORK

Persons wishing to become designers in the field of masonry construction must have an aptitude for mathematics and physics and an understanding of patterns, shapes, textures, and light/color values. They should observe structural testing in a laboratory and see the applications in the field. It is also helpful to read architect's/engineer's plans, data sheets, reference books, and design examples. Design examples are especially useful in obtaining an overview with respect to building design. They should study definitions of terms such as axial, flexural, lateral, live load, dead load, shearing, tensile, bearing, deflection, and compressive strength.

Masonry construction, like any other type of construction, is the product of teamwork with respect to test data, design, construction, materials, and future maintenance requirements.

12.5 RESEARCH

1. Attend a demonstration on how to shape field stone. Make notes about tools and equipment, such as a sand box for supporting the stone being shaped by hand.
2. From a geology department of a college or university, obtain regional data with respect to field stone.
3. Use a library to help define the term acoustics. Identify activities such as noise abatement and record/tape making. Using course outlines from colleges/universities, list courses one might take to help become a professional worker in this field.
4. Critically observe residential zoning with respect to traffic arteries in your area.
5. State how you think engineering traffic problems can be solved while keeping sound levels within healthful bounds.
6. Comment on the interdependence of decision makers in building (place) creation.
7. For building activities which depend mainly on earth materials, describe the formal organization of labor—locally and nationally.
8. Comment on builder ethics with respect to creating unsafe or uncomfortable conditions for unsuspecting buyers or tenants, including low strength concrete, poor chimney design, building below a dam, lack of noise control, known water problems, etc.
9. Obtain charts for your region that describe on a long-term basis soil-water levels, frost depths, temperatures, winds, precipitation, drainage, and soils.
10. From a university civil engineering department, obtain a copy of design notes for a small building.

12.6 REFERENCES

12.6.1 Brick Institute of America. *Recommended Practice for Engineered Brick Masonry*. 1980.

12.6.2 Randall, Frank, A. Jr. and William C. Panarese. *Concrete Masonry Handbook—for Architects, Engineers, Builders*. Portland Cement Association.

13

MATERIALS STORAGE AND CONTROL

Much of the earlier part of this book is spent describing the various materials, tools, and equipment used in the light building construction process. The builder, however, can realize a diminished profit margin if the materials used in or during the construction of a building are ordered, stored, or maintained in a haphazard or inadequate fashion. Materials control involves making sure that the right amount of materials arrive on a jobsite at the proper time during the project. Materials control also requires the keeping of up-to-date inventory counts of work materials at a jobsite, in a workyard or warehouse, or in the workshop.

Proper materials storage ensures materials on a jobsite will not be lost, damaged, or broken before they are used. In this way they can be readily and easily obtained when needed. In addition, materials should be stored so that they do not present a hazard to workers or anyone else inspecting or traversing the jobsite. The probability of theft and vandalism should also be minimized.

Materials losses through improper control or storage can be a serious and expensive form of loss for the builder. For example, ordering materials in excess of what is needed for a particular job is costly, can reduce working capital, and often indirectly leads to losses through waste and improper storage. In other words, jobsite inventory levels should always match the amount of work still to be done. Another common practice of builders who do not control the amount of materials on a job is to add charges for materials losses onto a customer's bill; however, this does nothing to enhance a builder's reputation and in many cases has led to financial and legal battles and an ultimate loss of profit. Equally undesirable for the builder are improper storage and losses due to jobsite waste, breakage, and theft. Not only must these types of losses be replaced, but they also slow down the flow of construction, resulting in increased labor costs.

The builder must gather off-site primary and secondary materials from a variety of places. When they arrive at the site, primary materials are then cut, shaped, mixed, placed, fastened, and so on, as part of the construction process. Secondary materials are used to cover and protect the primary materials. All materials must be stored and handled in ways that will keep an optimum number of individual items from harm and will allow a materials flow during construction of maximum efficiency.

13.1 GENERAL OBJECTIVES

The careful control, handling, and storage of materials is a relatively complex, yet often overlooked, field of work which requires skill and knowledge on the part of those persons in positions of responsibility. Too often construction companies allow for the hidden costs of inadequate control and storage and resultant materials loss. This is a wasteful, negative approach which should have no place in any competitive industry. The bottom line is that wasting material raises costs, lowers cost-effectiveness, and hurts the competitiveness of an individual company.

The contents of this chapter primarily deal with the control and handling of jobsite materials and the on-site receipt, storage, and care of materials from off-site sources. It is only intended as a framework from which each individual student or builder can develop programs for materials control and handling. Specific job conditions and experience will help to guide the builder in these activities. Further reading and job/business experience will be necessary for learning about such additional matters as materials production, materials sources, transportation methods, and the costs involved in materials procurement.

Although the emphasis here is on jobsite materials, general information concerning the do's and do not's of storage will be described also. This will include the storage of common construction materials—lumber, plywood, cement, and so on—at a workyard or in a warehouse, as well as at the jobsite. Other factors considered here are those involving the safety aspect of materials storage, and the increasing problem of theft and vandalism at the jobsite.

13.2 SPECIFIC OBJECTIVES

The specific objectives of this chapter are to enable the student builder to

1. identify and classify building materials;
2. realize why proper care of materials is essential to quality building;
3. understand how a lack of materials control can hurt company profits and how an effective materials control program will not only increase profits as a direct result of less waste and better materials flow, but will also help to control the flow of all other construction work processes;
4. create and use a materials control chart, using such charts as a framework for an overall program of materials control;
5. realize that materials handling and storage are merely physical aspects of an overall program of materials control;
6. demonstrate some degree of awareness regarding the physical and chemical characteristics of materials, not only for the sake of an efficient building process, but to ensure safe working conditions;

7. create a safe working environment by practicing safe storage methods, safe handling of materials, and safe disposal of waste materials;
8. consider the overall economics of materials control and storage;
9. relate materials central location and storage to project planning;
10. realize the potential need for taking jobsite security precautions to prevent losses resulting from theft, vandalism, lawsuits, injuries to bystanders and children, and natural disasters such as fires, storms, and high winds; and
11. be able to device safety formulations for storing work materials, hazardous materials, and fuels on a jobsite.

13.3 IDENTIFICATION AND CLASSIFICATION OF LIGHT CONSTRUCTION BUILDING MATERIALS

A builder must learn to deal with a wide variety of building materials. Some of the more common ones, such as lumber, bricks, concrete, nails, and other fasteners are standard items of the construction industry. The builder may deal with others more rarely depending on such factors as the type of buildings he or she is constructing and where they are being built. The builder learns to work with different materials as he or she learns the techniques involved in using them. To a certain extent, the experience of building is the best guide for learning about the vast array of building materials available on the market. Every builder, nonetheless, should have a good idea of the type of materials he or she will have to work with in the parameters of the light construction field.

This is important for several reasons. A builder should always be aware of the options open for each particular project. This is true of building techniques and methods which are discussed elsewhere in this book, but it is also true of the materials that can be used in every instance. A builder should know building materials so that they can be selected and ordered on the most cost-effective basis. It would be quite impossible for a construction company to develop an effective program of materials control if no one in the company had much general knowledge about construction materials. Also, the concept of safe, efficient materials handling and storage is impossible if a builder does not know what is good or bad for a material, how it is best stored, how it will be affected by different storage environments, what are safe or unsafe storage methods, and so on. Moreover, it is extremely important for everyone involved in construction work to know which materials are hazardous and under what conditions those hazards are increased.

One way to learn more about materials used in light construction is to list, categorize, and classify them. Such a list can also serve as a means of formalizing a company's materials vocabulary. If everyone in a construction company thinks of materials in reference to the same list of classifications, all company employees will speak the same materials language, thereby expediting intercompany materials requests and orders. Carpenter, foreman, purchasing agent, and builder will all refer to the same items in the same way, avoiding confusion and cutting down on incorrect materials orders and waste.

One possible listing and classification of materials used in light building construction is shown in Table 13-1. This list is by no means complete, yet it broadly includes many common building materials. The table is divided into seven sections representing basic materials families including air, water, heating fuels, wood products, earth-derived products, products of high technology chemical industries, and manufactured appliances and fixtures. Some distinguishing features are listed, brief suggestions are given for storage and handling procedures, and primary delivery methods noted. Table 13-1 represents only one way of identifying and classifying materials. A classification system could be much more extensive and detailed. For example, costs could also be included. What is important is that the builder, as well as all other members of a construction team, have sound background knowledge about the building blocks of the profession.

TABLE 13-1 IDENTIFICATION AND CLASSIFICATION OF LIGHT BUILDING CONSTRUCTION MATERIALS

Material	Subcategory	Distinguishing Features	Storage and Handling	Delivery
Air	Clean	Local index		Ventilator fans
	Polluted	Hazardous when under pressure	High pressure lines and air receivers	Compressors
Water	Pure Clean Polluted	Taste Color Hard/soft Vegetable matter	Tanks Well Line/hose Pump Under pressure for fire fighting	Tank trucks Well Municipal system
Heating fuels	Natural gas Producer gas Oil Wood Coal Acetylene	Explosive Toxic Fire hazard/toxic spills Needs protection from weather Toxic gases when slow burning Fire hazard and demurrage costs	Gas lines Tanks Sheds Safe storage	High pressure lines Tank trucks Open trucks
Wood products	Dimension lumber Plywood/ hardboard Cork Shingles/ shakes Floorings Moldings Siding Glued-laminated timbers	Renewable resources Watermarks Molds Warps Twists Stains Fades Easily worked May be prefinished	Easily marked and damaged Needs protection from weather Needs moisture control and air circulation Needs well-located supports	By open or closed trucks Finished pieces may be wrapped (Order only as needed)
Earth-derived products		Heavy Local variations	For building operations (are relatively inactive chemically, except lime, limestone, marble, and terrazzo)	

TABLE 13-1 IDENTIFICATION AND CLASSIFICATION OF LIGHT BUILDING CONSTRUCTION MATERIALS (Continued)

Material	Subcategory	Distinguishing Features	Storage and Handling	Delivery
	Sand Gravel Stone/rock	Usually locally produced	Platforms, bins	Open dump truck
	Marble/terrazzo Slate/tile	Costly, finish materials	Crates indoors	Open truck with hoist
	Brick Terra-cotta Concrete block Structural tile	Many standard sizes, designs, and colors	Pallets outdoors, if fenced	Open truck with hoist
	Lime	Hot or slaked	Hazardous when hot	Bags
	Bitumen		Hazardous when hot	Drums
	Asbestos	May be cancer producing	See national safety regulations	Bags, sheets, boards
	Vermiculite	Fire resistant	Keep dry (holds water)	Bags
	Copper	Sheet, wire	Sort by gauge, expensive, needs safe storage	Open truck
	Aluminum	Siding	Use carefully (easily damaged but durable)	Open truck
	Zinc	Galvanized steel (furnace and duct-work)	Soldering fumes hazardous	Open truck
	Iron/steel (beams, trusses, plates, posts, sheet, rod)	Formed and deformed May be welded on the job Rusts without inhibitors	Off-ground storage	Truck with hoist
	Natural earths (loam, sand, clay, gravel, rock)	Topsoil, fill, backfill, berm material	Outdoor piling Compacting procedures (Find appropriate engineering data)	Onsite backhoe or bulldozer Offsite truck
Products of high technology/chemical industries		May become active chemically Some potential fire hazards Usually nonrenewable	Keep away from exits and in highly regulated safe storage	Closed truck
	Cements: Portland (8 types) Masonry Waterproofing Plastic Expansive Regulated set	40 or 33 kg bags (94 lb./cu. ft. in USA)	Use short-term dry storage/bags on pallets	Covered truck with hoist
	Concrete additives: Air-entraining Water-reducing	Chemical admixtures, expensive	Dry storage	Bags

TABLE 13-1 IDENTIFICATION AND CLASSIFICATION OF LIGHT BUILDING CONSTRUCTION MATERIALS (Continued)

Material	Subcategory	Distinguishing Features	Storage and Handling	Delivery
	Retarding Accelerating Pozzolans Workability agents Bonding agents			
	Plastics: Vinyl (siding, soffits, fascias)	Packaged, May be insulation backed	Safe dry storage	Open truck or specialized truck for applicators
	Glues/adhesives/ sealers/ caulking/fillers	Toxic fumes and fire hazards	Safe dry storage	Closed trucks for tubes, cans, guns
	Machine fuels: Gasoline Coal-oil/solvents Oil Diesel fuel	Flash points, lighter ends volatile and hazardous	See local regulations No smoking Use well-ventilated separate storage	Specialized trucks Approved container
	Paints/finishes: Oil-based Water-based	Colors Fumes Solvents	Safe dry storage Airtight containers	Closed truck
	Cleaners/ solvents: Alcohol-based Petroleum products	Flash points Health hazards	See local regulations No smoking	Safety cans
	Foam insulation	See safety regulations (may not be approved)	Immediate application—no storage	Special truck
	Acids/bases: HCL H_2SO_4 Muriatic Lye Soaps	pH range: acids, water, bases Hazardous at extremes of range	No broken packages or containers	Glass containers for liquids
	Glass: Sheet Plate Blocks	Heavy to handle Easily scratched or shattered	Crates/frames	Special truck
	Building papers/ boards: Wallpapers Sheathing Roofing	Rolls Sheets	Dry safe storage	Truck

TABLE 13-1 IDENTIFICATION AND CLASSIFICATION OF LIGHT BUILDING CONSTRUCTION MATERIALS (Continued)

Material	Subcategory	Distinguishing Features	Storage and Handling	Delivery
	Insulation Hardboard Chipboard			
	Asphalts: Adhesives Waterproofers	Hazardous solvents	Cans	Special truck
	Roofing Paving	Hazardous when hot	Bulk	
	Shingles	Applied when warm	Bundles on pallets	
	Floor coverings: Rolled sheets Rugs	Hazardous solvents and cements	Delivered to place of application	Closed truck
	Carpets Tile	Carpetlaying equipment		
Manufactured appliances and fixtures	Heating/air conditioning	Enameled pieces will chip, spall, scratch	Delivered to points of installation	Special trucks
	Plumbing Electrical Hardware Cabinets/millwork Fixtures	Colors Approval tags Warranties		

13.4 MATERIALS CONTROL

Materials control is an extremely important aspect of the construction business, and one that unfortunately is all too often neglected, especially by small construction companies that lack the middle management personnel who normally specialize in ordering materials and controlling inventories. Materials control simply means being able to order the correct materials in the right quantity when they are needed.

Effective materials control should always be established for a construction job to ensure a smooth operation (workers cannot do their jobs without the right materials) and to keep costs down. Unless adequate and effective materials control methods are used, the waste of unnecessary or excessive labor for materials handling combined with losses in material costs can greatly erode the profits of a project almost without detection. For example, if twice as much lumber is ordered and unloaded at the jobsite than is needed, the cost of ineffective materials control is represented by the cost of the labor and equipment to transport the excess material to the site, to unload it, then to reload it, and to transport it somewhere else. In addition, working capital has been reduced needlessly. This represents the best possible scenario in dealing with unneeded work materials; all too often a significant share of excess material on a jobsite is broken, is stolen, rots, or is otherwise lost before it can be transferred to where it can be properly used.

The job of maintaining proper materials control is really a responsibility for everyone on a job. Even in a well-planned job, individual or ad hoc cases may require jobsite changes in design or technique which then change materials requirements. Every worker should be able to determine how changes in a construction job will change the materials he or she will need to complete his or her particular tasks. Nevertheless, the task of maintaining and monitoring the control methods established in a company ultimately falls on the contractor, foreman, or other management personnel assigned to order, count, and control inventories. Efficient methods with adequate follow-through procedures and monitoring will definitely prevent unnecessary profit depletion.

13.4.1 Factors Preventing Effective Materials Control

Wasted or nonproductive operations involving the handling of materials contribute greatly to inflated labor costs and to materials losses. The following three primary factors can be cited as reasons for ineffective materials control:

1. Not determining exactly when materials will be needed and what materials will be needed for a project. Too often someone in the field scribbles down material needed on an ad hoc basis on a scrap of paper, wasteblock, or something similar. Material order forms are being used more often today, but too frequently the ordering of materials is done in a haphazard manner.
2. Receiving, handling, and storing of unused, excess, or incorrectly ordered materials. It does not require much handling at present labor rates for the cost of such labor to exceed the actual cost of unnecessary material. Moreover, unnecessary receiving and storing costs compound the amount of the loss.
3. Return charges and salvage losses. When unnecessary materials that cannot be used are returned for credit, further handling is required and a restocking charge is assessed in addition to charges for restoring material to a resalable condition. If the material cannot be returned and cannot be used at a later date, it can be either written off entirely or sold for salvage, representing further losses.

13.4.2 Materials Control Programs

Materials control must be a smooth operation which moves materials through the construction process as they are needed. In a sense, an effective materials control program establishes a framework for controlling the rest of the construction process. Four primary considerations must be dealt with for an effective materials control program: information, material, equipment, and labor.

First, adequate information is needed, including drawings, specifications, detailed point-by-point materials lists, etc. On many projects it is advantageous, and sometimes necessary, to prepare layout drawings setting forth the required information necessary for completing the job. When sufficient information is obtained, the materials required must then be ordered. The equipment necessary for working with the material must be at the jobsite and in working order. Then, when the materials are received and the workforce is ready to start, construction/fabrication/installation—in other words, labor—can begin. Although production labor is the factor usually considered the hardest to control on a job (and is the actual money-making activity), it is itself dependent upon the other preceding factors of information, material, and equipment, thus pointing out the overall importance of proper control methods.

A key consideration is for the builder to take the time to collect the right information, instead of having to correct hastily gathered and wrong information during the course of a project. Correcting mistakes always reduces profits. It is poor practice to guess at quantities of bulk materials. Inevitably this practice causes overages or shortages which again cut profit margins. Determine the first time around what is required based on efficient information flow; this will eliminate the need for later corrective efforts.

13.4.3 Materials Control Charts

As required materials for a project are determined, they should be accurately recorded for ordering. By using a simple chart, each component or group of like components can be easily recorded by exact count, location, when required, when ordered, who ordered them, and date received. Making use of such materials control charts greatly expedites inventory monitoring, making it possible to account for each box of nails and virtually every board foot of lumber during the course of each project. Such control, in turn, gives the builder the experience and insight to better estimate needed materials on future jobs.

A simple materials control chart is shown in Figure 13-1. Note that all of the basic information required, listed as follows, is easily recorded on the chart:

1. company name,
2. project name,
3. project address,
4. work category,
5. special instructions,
6. item (description—make and model number),
   (*Note:* This information can be taken from equipment, material, and methods submittals.)
7. location (where the item is to be installed in the project),
8. sizes,
9. date required (determined from the job schedule),
10. date ordered,
11. ordered by,
12. date received, and
13. totals.

The type of chart shown in Figure 13-1 can be used in recording virtually any materials used in a project. The chart can be used by any party involved, including foremen, layout personnel, superintendents, project managers, and purchasing agents. Providing and utilizing the

Materials Control Chart

Company Name ______________________________

Project Name ______________________________

Project Address ______________________________

Work Category ______________________________

Special Instructions ______________________________

Location	Item Sizes											Date Reg'd	Date Ord'd	Ord'd By	Date Rec'd
Totals =															

FIGURE 13-1. A sample materials control chart.

chart will help to develop the habit of using a smooth, efficient method of materials control rather than using hit-or-miss/ad hoc methods. This will help to eliminate waste and increase profits.

13.4.4 Additional Aspects of Materials Control

Another area where materials control can greatly aid the contractor or builder is in regulating and improving the cash flow situation. As materials are received, they must be paid for. Vendor billings are generally sent on a monthly basis. The contractor's payment to suppliers for materials received during the month is usually due and payable by the tenth of the following month. By the same token, work done by the contractor during the course of a month is frequently billable, in accordance with normal progress payment provisions, at the end of each month. Therefore, it may often be advantageous to order materials so that they are received immediately after the first of the month, enabling them to be used at a jobsite during the current month. Payment due the contractor is, therefore, billed and due before or about the same time that the contractor's payment for materials purchased is due to the supplier. If the ordering of materials is controlled in the same sort of fashion, the contractor's cash flow position can be improved. This, in turn, will reduce or eliminate borrowing costs or depletion of working capital and could very well increase profits.

Another area in which materials control is important is if a contractor is paid for material on hand. For example, in most time-and-material contracts and in some fixed-amount contracts, there are provisions which call for the contractor to be paid for materials that are required for the project and are received and stored prior to their installation during the construction process. This type of provision is extremely beneficial to the contractor and, if possible, should be included in every contract because it helps ensure completion of a job without tapping the contractor's own working capital. Under such an arrangement, however, the contractor must maintain a very tight and effective materials control because he or she is effectively responsible for materials that have already been paid for (including the contractor's profit), but which have not yet been installed.

13.5 MATERIALS STORAGE

Proper materials storage simply means placing materials in such a way as to avoid damage or loss before they will actually be used. This includes such considerations as protection of the materials from moisture, heat, and weather. Undesirable chemical reactions must be avoided; therefore, some materials must not be stored together. Materials must be stored in such a way that they will not provide a hazard for workers or passersby. It is also important that materials be stored so that they are easily accessible to the workers who will be using them to construct a building.

Although this chapter deals primarily with jobsite materials storage, practices that are proper and safe on the jobsite are usually equally applicable to storage in a workyard, storage area, warehouse, workshed, or workshop, depending on prevailing conditions.

All construction projects require a coordination of delivery roads to the actual building site and storage areas. The builder, job superintendent, or foreman should decide on the delivery route and storage areas at the beginning of a project to avoid, for example, the possibility of dropping a load of material in front of a delivery route and consequently wasting time and money to remove the material. In addition, there must be access to excavated earth which is to be hauled away, and there must be access to all sides of a foundation where the builder wants to pour concrete directly from a concrete truck. In general, storage facilities should be set wherever they do not impeded construction. As a building rises, a clutter of storage and waste material around the site can be alleviated somewhat by allocating space for storage on the inside.

Careful scheduling is often needed for materials to be on hand when they are required, as discussed in the materials control section. Site planning is also needed for moving materials to points of application or installation during the course of the project.

13.5.1 Safe Storage Practices

As with most construction safety requirements, safe storage involves a certain amount of common sense. Obviously, lumber is not stacked in overly high, unstable piles that could easily fall. Brush should clearly not be burned next to a gasoline storage tank. Suggestions for the safe storage of a wide variety of construction materials are given in Table 13-1.

In addition to common sense and general safety, safe storage practices require procedural rules and regulations. Certain practices are never allowed and others must always be required. An expanded materials identification and classification system could formally describe, in detail, safe storage practices for all materials commonly used by a construction company.

For example, by expanding on some of the data in Table 13-1, one may develop a list of statements that pertain to occupational health and safety. A beginning list would include the following:

1. Piles of dimension lumber may topple if not properly piled or if placed on an uneven foundation.
2. Materials handlers need to know the difference between hot and slaked lime. Under controlled conditions it is well to see the difference.
3. The effect of an acid spill on a marble slab needs to be seen on an experimental basis rather than on the job.
4. Fumes from adhesives and some glues and cements are toxic. Read and heed the labels and always provide proper ventilation.
5. Machine fuels have flash points. Persons handling them should obtain detailed information with respect to the scale of flash points and then establish safe handling limits.
6. Cleaners and solvents are usually toxic and volatile. Learn to be cautious. Be aware of the long-term effects on health. Read the literature.
7. Strong acids and bases are dangerous. Be aware of their affinities for water, including skin moisture or eyes moisture. Learn about the pH scale which indicates strength of acids and bases and their relationships with neutral water.

These statements could in turn be expanded, and specific safety rules set up. Also, it is advisable for persons involved in handling and storing jobsite materials to read local, state, and national occupational health and safety regulations.

Government regulations will not necessarily always provide answers to storage problems; however, they provide a good guide for most storage do's and don'ts. For instance, the United States Department of Labor, Occupational Safety and Health Administration (OSHA) Standards for the construction industry, published in 1983, includes a section specifically regarding materials storage. A sample of this publication is shown in Table 13-2. Again many of these regulations are based on common sense. It makes sense, for example, to remove nails from used lumber before stacking it. At the very least, dangerous protruding nails should be bent and hammered into the lumber, although this could provide problems if the lumber must be cut at some later time.

TABLE 13-2 EXCERPT OF OSHA REGULATION FOR MATERIAL STORAGE

† 1926.250 General requirements for storage.

a. *General.*
   1. All materials stored in tiers shall be stacked, racked, blocked, interlocked, or otherwise secured to prevent sliding, falling or collapse.
   2. Maximum safe load limits of floors within buildings and structures, in pounds per square foot, shall be conspicuously posted in all storage areas, except for floor or slab on grade. Maximum safe loads shall not be exceeded.
   3. Aisles and passageways shall be kept clear to provide for the free and safe movement of material handling equipment or employees. Such areas shall be kept in good repair.
   4. When a difference in road or working level exists, means such as ramps, blocking, or grading shall be used to ensure the safe movement of vehicles between the two levels.

b. *Material storage.*
   1. Material stored inside buildings under construction shall not be placed within 6 feet of any hoistway or inside floor openings, nor within 10 feet of an exterior wall which does not extend above the top of the material stored.
   2. Employees required to work on stored material in silos, hoppers, tanks, and similar storage areas shall be equipped with lifelines and safety belts meeting the requirements of Subpart E of this part.
   3. Noncompatible materials shall be segregated in storage.
   4. Bagged materials shall be stacked by stepping back the layers and crosskeying the bags at least every 10 bags high.
   5. Materials shall not be stored on scaffolds or runways in excess of supplies needed for immediate operations.
   6. Brick stacks shall not be more than 7 feet in height. When a loose brick stack reaches a height of 4 feet, it shall be tapered back 2 inches in every foot of height above the 4-foot level.
   7. When masonry blocks are stacked higher than 6 feet, the stack shall be tapered back one-half block per tier above the 6-foot level.
   8. Lumber:
      i. Used lumber shall have all nails withdrawn before stacking.
      ii. Lumber shall be stacked on level and solidly supported sills.
      iii. Lumber shall be so stacked as to be stable and self-supporting.
      iv. Lumber piles shall not exceed 20 feet in height provided that lumber to be handled manually shall not be stacked more than 16 feet high.
   9. Structural steel, poles, pipe, bar stock, and other cylindrical materials, unless racked, shall be stacked and blocked so as to prevent spreading or tilting.

The same OSHA publication also includes detailed sections on general materials handling, use and disposal, the handling and storage of combustible materials both inside and out, combustible and flammable liquids, and even explosives and blasting agents which may infrequently be used in light building construction. Remember, safe storage practices ultimately benefit everyone associated with a construction company.

13.5.2 Storage Safety Guidelines

The same type of formulations discussed in Chapter 4, regarding the use of tools and machinery, can be devised for setting storage safety guidelines. Concepts and statements concerning storage practices and materials are listed in Table 13-1. From this table, general ideas about safety storage are drawn and a second table of if/then statements of principle can be formed. If certain conditions prevail while storing or handling certain materials, then a safe situation exists. Builders must be aware, however, that new information gathered at the national level, etc., may change the premise of the formulation, in which case similar circumstances might not ensure a safe situation.

Table 13-3A introduces some concepts with respect to the jobsite handling and storage of general light construction materials. Substatements are an expansion of each listed concept. Table 13-3B is a derivative and more general statement about what conditions must prevail to ensure a *safe* handling and storage of general construction materials. Tables 13-4A and B make similar statements about handling and storing hazardous materials at a jobsite. As was stated in Chapter 4, these formulations can be expanded to whatever degree is helpful or useful for improving a company's safety profile. For example, if fuels are introduced as one heading under hazardous materials in Table 13-4A, then that heading can be expanded and subdivided itself if necessary. Table 13-5A breaks down the area of jobsite handling and storage of hydrocarbon fuels into its integral concepts, which are then further defined and subdivided. Finally, this information provides the evidence for forming new statements of principle (Table 13-5B). These formulations can be applied in whatever depth required to any area concerning materials storage.

TABLE 13-3A JOBSITE HANDLING/STORAGE OF GENERAL LIGHT CONSTRUCTION MATERIALS

Concept	Substatements
1. CPM (Critical Path Method)	Scheduling of delivery, handling, and storage
2. Jobsite earth materials	Stockpiling loam, sand, rock
3. Concrete materials	Stockpiling gravel and sand; storage of cement and reinforcing rods; providing water
4. Lumber	Stockpiling, lifting, hoisting, storage of dimension lumber, plywoods, millwork, flooring, shingles, shakes, glued-laminated timbers; regulating moisture control; providing shelter for cabinets and other finishing items

TABLE 13-3A JOBSITE HANDLING/STORAGE OF GENERAL LIGHT CONSTRUCTION MATERIALS (Continued)

Concept	Substatements
5. Sidings	Stockpiling brick, stone, stucco materials (wire, aggregates, cements), storage of gypsum products (boards, tapes, nails, joint fillers), nonwood shingles, siding, glass, paper, plastics, insulation, decorative coatings
6. Electrical	Storage and protection for wire, wiring supplies, equipment, appliances, fixtures
7. Plumbing	Storage and protection for piping, tiles, appliances, fixtures, pumps, tanks
8. Heating/ air-conditioning	Storage and protection for piping, equipment, venting, chimney materials
9. Hardware	Stockpiling and organizing nails, screws, staples, bridging, bolts, hangars, locks, hinges, metal studs
10. Landscaping	Delivery of shrubs, trees, and fencing; watering shrubs prior to planting

TABLE 13-3B SAFE HANDLING/STORAGE OF GENERAL LIGHT CONSTRUCTION MATERIALS

If
1. The materials are received from the supply source (warehouse, external stockpile) according to a CPM-type plan, and
2. The receiver is knowledgeable with respect to the plan and to the characteristics, according to data charts and files, of each material received, and
3. The receiver is trained in the use of materials handling equipment (forklifts, tractors, trucks, cranes, etc.), and
4. The builder has provided adequate and safe storage facilities, and the means of keeping records...
Then
1. The materials can be safely received and stored so that they are available for construction work.

TABLE 13-4A JOBSITE HANDLING/STORAGE OF HAZARDOUS MATERIAL

Concept	Substatement
1. Glues/cements/ adhesives as complex chemicals	Fumes, fire hazards, solvents, effects of freezing
2. Portland cement and affinity for water	Skin/eye burns, moisture control, storage life, use of mortars
3. Lime and affinity for water	Hot lime, slaking, burns
4. Paints, varnishes, lacquers, enamels,	Fumes, fire hazards, spray drift, ventilation, poisonous materials, allergies, general health

TABLE 13-4A JOBSITE HANDLING/STORAGE OF HAZARDOUS MATERIAL (Continued)

Concept	Substatement
solvents, thinners—as chemicals, some volatile	hazards
5. Bitumen as a chemical	Softening temperatures, burn hazard when hot
6. Chemistry/physics of glass	Cutting, shattering hazards; crushing weight
7. Insulations as chemicals	Fire ratings, toxic fumes, irritants
8. Caustic and acid solutions	pH, range, neutralizing methods
9. Plastics as complex chemicals	Classifications, toxic and volatile fumes
10. Fuels as chemicals	Flash points, spillage, cleanup, safe handling and storage
11. Chemical mixing	Hazards
12. Temperature control	Safe limits

TABLE 13-4B SAFE HANDLING/STORAGE OF HAZARDOUS MATERIALS

If
1. The receiver identifies those materials which are hazardous (risky, dangerous) and keeps current data sheets, and
2. The materials are carefully received, stored, and used according to national and local standards, and
3. The materials are used according to manufacturer's specifications and directions, and
4. The users are aware of the personal and group risks associated with the material if manufacturer's instructions and statements of liability are ignored, and
5. The users are trained with respect to fire fighting and the reducing of hazard levels...
Then
1. The general risks associated with jobsite materials are such that a worker may with care, use the materials with some degree of confidence...
But
1. When the level of confidence with respect to a material diminishes as data is gathered nationally and the risk level increases beyond acceptable levels, the handling and storage of same material can no longer be regarded as safe.

TABLE 13-5A JOBSITE HANDLING AND STORAGE OF HYDROCARBON FUELS

Concept	Substatement
1. Flash point	Definition, range, standard method of testing

TABLE 13-5A JOBSITE HANDLING AND STORAGE OF HYDROCARBON FUELS (Continued)

Concept	Substatement
2. Viscosity	Definition, range
3. Vapor Pressure	Measurement, measuring
4. Flammable	Rate of flame spread
5. Flammable liquid	Flash point, vapor pressure
6. Combustion	Oxidation, combustion control, fire fighting
7. Approved container	Testing laboratory, materials, safety can, sizes, dispensing
8. LPG (Liquid Petroleum Gases)	Kinds, characteristics
9. Heating devices	Fuels, ventilation, mounting, clearances, appliances
10. Flammable, combustible liquids	Containers, storage (indoors, outdoors), venting
11. Hazards	Fires, explosions, fumes
12. Safety regulations	Electrical installations, fire wall, ratings, smoking, storage, cleanup.

TABLE 13-5B SAFE JOBSITE HANDLING/STORAGE OF HYDROCARBON FUELS

If
1. The worker is knowledgeable with respect to the characteristics of fuels (heavy ends and light ends), and
2. The jobsite provides for approved storage of fuels and fire prevention capabilities, and
3. The service and refueling areas are designated, properly equipped, and supervised, and
4. The jobsite machines/heating devices are properly maintained and operated...
Then
1. For that worker the jobsite should be relatively safe.

13.5.3 Storage of Common Construction Materials

Although storage principles, such as protecting materials from all types of damage, decay, weather, and so on apply generally, different materials sometimes require specific methods of handling and storage. Also, methods of storage considered suitable for a storage yard or workyard should be duplicated as best as possible at the jobsite. To provide optimum protection, tarpaulins and plastic should always be kept at a jobsite since they are useful for protecting a wide variety of materials from wet, rainy, or cold weather.

Lumber

The objective of lumber storage is to maintain the lumber at or bring it to a moisture content suitable for its end use with a minimum of deterioration. The objective of lumber handling is to load, transport, unload, pile, and unpile lumber economically and without damage. Both of these objectives are obtained easily if good handling and storage practices are followed. Adequate protection of lumber in storage will help prevent attack by fungi, insects, and changes in moisture content that will result in checking, warping, and stain in lumber and make it unsuitable for the intended use.

The storage area should be located where the lumber is easily accessible for use. The ideal location would be on high, level, and well drained ground away from bodies of water or wind-obstructing objects, such as tall trees, and away from operating vehicles or equipment that might damage the lumber. A low level site is likely to be damp and sheltered from the flow of fresh air—conditions that may retard drying and expose the lumber to stain and decay.

The lumber must be stacked on level timber sills with solid supports to prevent direct contact with the ground. All nails protruding from used lumber should be pulled out before it is stacked. The height of the lumber pile should not exceed 16′ (4.9m) if stacked manually and 20′ (6.1m) if stacked mechanically with a forklift or loader; the width should be less than one-fourth the height. Cross strips of wood, called stickers, must be placed in piles that are stacked more than 4′ (1.2m) high. When the lumber is unpiled, each layer must be removed before another is begun.

Cement and Lime

Portland cement is a moisture-sensitive material; if it is kept dry, it will retain its quality indefinitely. Portland cement that has been stored in contact with moisture sets more slowly and has less strength than portland cement that is kept dry. The relative humidity in a warehouse or shed used to store cement should be as low as possible. All cracks and openings in walls and roofs should be closed. Cement bags should not be stored on damp floors but should rest on pallets. Bags should be stacked close together to reduce air circulation, but should never be stacked against outside walls. Those to be stored for long periods should be covered with tarpaulins or other waterproof coverings. Bags of hydrated lime used in mixing of mortar for masonry work should also be kept completely dry and stored in the same manner as concrete.

On smaller jobs where no shed is available, bags should be placed on raised wooden platforms. Waterproof coverings should fit over the pile and extend over the edges of the platform to prevent rain from reaching the cement and the platform. Rain-soaked platforms can damage the bottom bags of cement.

Cement stored for long periods may develop what is called warehouse pack. This can usually be corrected by rolling the bags on the floor. At the time of use, cement should be freeflowing and free of lumps. If lumps do not break up easily, the cement should be tested before it is used in important work. Standard strength tests or loss-on-ignition tests should be made whenever the quality of a cement is doubtful.

Bags of cement or lime should not be piled more than 10 bags high on a pallet except when stored in bins or enclosures built for such purposes. The bags around the outside of the pallet should be placed with the mouths of the bags facing the center. To prevent piled bags from falling outward, the first five tiers of bags each way from any corner must be crosspiled and a setback made commencing with the sixth tier. If necessary to pile above the tenth tier, another setback must be made. The back tier, when not resting against a wall of sufficient strength to withstand the pressure, should be set back one bag every five tiers, the same as the end tiers.

During unpiling, the entire top of the pile should be kept level and the necessary setbacks every five tiers maintained.

Ordinarily, cement does not remain in storage long, but it can be stored for long periods without deterioration. Bulk cement should be stored in weathertight concrete or steel bins or silos. Dry low-pressure aeration or vibration should be used in bins or silos to make the cement flow better and avoid bridging. Due to fluffing of cement, silos hold only about 80% of rated capacity.

Workers who frequently handle cement or lime bags should wear goggles and snug-fitting neckbands and wristbands. They should always practice personal cleanliness and never wear clothing that has become hard and stiff with cement. Such clothing irritates the skin and may cause serious infection. Any susceptibility of a worker's skin to cement and lime burns should be reported to the job foreman, and those workers who are allergic to cement or lime should be transferred to other jobs, as complications from such allergies can be quite serious.

Concrete Aggregates

Concrete aggregates, such as various sized stone, gravel, and sand, should be handled and stored in ways that minimize segregation and prevent contamination by unwanted substances. Store aggregates on a clean, hard surface. Do not store aggregates directly on the ground if it can be avoided. Ground storage wastes material and can result in mud, dirt, and severe moisture contamination. Cover sand and fine aggregate piles to keep rain and moisture out. Do not use the extreme bottom layer of an aggregate pile that has been left uncovered or which has been stored in contact with the ground.

Masonry Units

When delivered to the jobsite, concrete masonry units should conform in dryness with specified limitations for moisture content. To maintain these dryness levels, units should be stockpiled on planks or other supports free from contact with the ground. The units should then be covered with a moistureproof tarpaulin or covering. The top of a concrete structure should be covered with roofing paper or waterproof coverings to prevent rain or snow from entering the units during construction as well.

13.6 JOBSITE SECURITY

Problems involving the theft of and damage to jobsite materials by vandals have greatly increased in recent years and now represent a significant source of loss for many construction companies. The very fact that this form of loss must be considered and possibly prevented creates yet another hidden cost in a construction project and, in a real sense, raises the cost of constructing a building. Nevertheless, the actual losses resulting from theft and vandalism can usually be brought under control through vigilance, care, and proper planning by a prudent builder. This section is not intended to provide a foolproof formula for the prevention of theft and vandalism, an impossible task in any case. It is intended to make the builder more aware of a problem associated with the industry and be more prepared to take preventive measures.

The consideration of theft and vandalism is just one more reason why materials control is so important. The materials stored on a jobsite should never be more than are needed for the job, unless that jobsite is also being used as a storage depot for another job. In regard to theft or damage, following this precept means cutting possible losses: Material which is not needed for a job and is, therefore, not stored at a jobsite cannot be vandalized there. The same is true of equipment, machinery, and tools; avoid keeping anything at a jobsite if it can be kept someplace else which is safer or under greater surveillance.

An obvious preliminary measure to take is for the builder or contractor to make sure that material and machinery located on the jobsite are sufficiently covered by property insurance either through the owner of the building or through the builder's own company.

Although builders will learn how best to deal with security problems by experience—conditions vary from region to region—there are several obvious ways to deal with material which must be stored at a jobsite. For example, the keys should not be left in heavy equipment overnight and such equipment should be locked. Gasoline or diesel fuel tanks should be locked as well to prevent siphoning, a fairly common nuisance. In addition, portable skids or larger tanks for fuel and/or pumps for fueling machinery should be locked when no one from the construction company is around. Either tools should not be kept on the jobsite overnight or they should be locked up in a secured toolbox. Temporary worksheds or work trailers which can be locked at night are often used by construction companies for storing work materials and equipment; however, these are not always feasible or practical on a light building construction jobsite.

Materials such as lumber, bricks, and concrete block should be kept covered overnight, not only as protection from the weather but also to avoid presenting an obtrusive target for thieves and vandals. In general, presenting an overall image of order and unobtrusiveness is probably a good idea. Materials that are stacked should be placed where removing them would be difficult. Some builders set up a temporary lighting system to deter vandals, while others use watchdogs. On jobsites where problems occur, spot checks can be made occasionally, local police can be alerted, and where large amounts of material are stored, a watchman can be stationed. The builder must decide the degree of security necessary in each individual case.

Another area involving security matters on a jobsite is ensuring that bystanders, neighbors, small children, and so on are not injured on the jobsite property. Government regulations do, to a certain extent, stipulate how a construction jobsite must be maintained during non-working hours so as to prevent accidents involving nonconstruction people. The general OSHA rules about materials storage listed earlier in this chapter, for example, are intended to make worksites generally safer places during work hours or off hours. Other regulations governing the use of ladders, scaffolding, shoring, excavations, trenching, and even explosives use are also in part intended to protect nonconstruction personnel. Some of these regulations have already been discussed in Chaper 4; others can be found in the Construction Industry Standards, OSHA 2207, 1983.

Warning or "no trespassing signs" should be placed on a jobsite to keep bystanders away. Stacks or piles of material should always be secured and covered. Open trenches should be fenced off with workhorses, rope, or barricades, and warning signs posted. In general, refuse should be picked up daily and the jobsite put in order so that children who may not read warning signs do not trip over boards, rubble, or other material and do not fall on nails, sharp tools, or jagged workpiece edges. Also, in the event an injury does occur under such circumstances, the builder should be adequately covered by contractor's liability insurance to protect himself or herself from specified claims that may arise out of or result from his or her operations under the building contract (whether such operations are by the builder, by any subcontractor, by anyone directly or indirectly employed by any of them, or anyone for whose acts any of them may be liable).

A builder can only provide so much security on a cost-effective basis to prevent accidents of a secondary nature on the jobsite. Generally speaking, running a smooth, efficient operation is the builder's best bet for avoiding problems of this nature.

Although not exactly a security matter, the builder must take maximum precautions to ensure that fires or other such disasters do not occur at a jobsite. Fuel storage tanks should be located and maintained according to OSHA regulations, and materials should be stored in a way that would minimize damage caused by arson, lightning, a storm, high winds, or some other natural disaster. For example, plywood stacked on the jobsite should be kept off the ground to prevent moisture or water damage, and it should be weighed down by something heavy to prevent uplifting during high winds. Both property insurance and special hazards insurance may be required as protection against losses of this type.

13.7 RESEARCH

1. Create your own identification and classification system for light construction building materials. Create a system that simplifies and defines the most important materials with the least possible verbiage. The system should create a language that will be readily transferable to all other company employees.
2. Diagram a basic materials control program. List ways in which an effective program can increase company profits.

3. What does it mean to say that a contractor is paid for materials on hand? Why is this desirable or undesirable?
4. List ways in which improper materials storage can cost a construction company money.
5. Visit a light building construction site and observe materials storage methods. Do they allow for effective movement of materials? Did you notice any storage practices that seemed unsafe? What were they?
6. Talk to some builders. Ask them if they have problems with vandals and thieves. Are these problems basically only a nuisance or are they costly? Do they disrupt work schedules? How do different builders deal with security problems?
7. Suppose a builder leaves part of a trench or foundation open overnight. Barriers and warning signs are put up around this hole on the jobsite, but a small child, curious to explore, falls into the hole anyway and injures himself or herself. Is the contractor liable for damages even though warning precautions were made? How can a contractor protect himself or herself from problems of this nature?

13.8 REFERENCES

13.8.1 Adrian, James J. *Construction Contracts.* Virginia: Reston Publishing Company, Inc., 1979.

13.8.2 Adrian, James J. *Construction Estimating.* Virginia: Reston Publishing Company, Inc., 1982.

13.8.3 Collier, Keith. *Managing Construction Contracts.* Virginia: Reston Publishing Company, Inc., 1982.

13.8.4 Hardie, Glenn M. *Construction Contracts and Specifications.* Virginia: Reston Publishing Company, Inc., 1981.

13.8.5 OSHA Safety and Health Standards. *Construction Industry.* Washington, D.C.: U.S. Dept. of Labor Occupational Safety and Health Administration, 1983.

14

LIGHT STRUCTURES SYSTEMS

14.1 GENERAL OBJECTIVES

This chapter serves as a necessary prelude to the discussion of framing layouts. In order to choose the most appropriate layout for a given light building project, it is important to understand the various structures systems available. Nine such systems are discussed in this chapter, with emphasis placed on the major differences between the systems.

The many subsystems that comprise a building—including foundations, floors, walls, ceilings, roofs, sheathing, insulation, ventilation, mechanical and electrical services, finishing, and landscaping—are brought together in one structural system. Some of these systems are age-old, while others are the product of recent technological innovations; some are relatively simple in design and construction, while others are more complex; some have become popular lately due to changes in the availability and quality of certain building materials, while others have become less popular. But regardless of their individual differences, all light structures systems have one thing in common: They act as the starting point from which a building project is launched, the concept around which the building takes shape. It is obvious, then, that the importance of choosing an appropriate structural system cannot be overstated. In this chapter, various structures systems are described so that a well-informed decision can be made.

It should be noted that climate is an important consideration when choosing a structural system. While this chapter is limited to structures systems for temperate zones only, climate has a definite influence on systems design. While one part of the country may be well suited to a particular system, that same system may have to undergo extensive changes to be used elsewhere. In other words, extremes in climatic conditions (very high or very low temperatures, high winds, slow water drainage, maritime salt, high humidity, permanent ground frost, etc.) require variations in the structural system being used.

14.2 SPECIFIC OBJECTIVES

On completion of this chapter the reader should

1. have a working knowledge of the following structures systems: balloon, platform (western), post and beam, combined platform and post and beam, energy efficient, rigid frame, arch, A-frame and prefabricated;
2. know the important differences between these structures systems; and
3. be prepared to analyze the advantages and disadvantages of the systems and then choose a specific system for a given light building project.

14.3 BALLOON FRAMING SYSTEM

In the balloon framing system (Figure 14-1), the wall studs originate at the sill plate. Both the perimeter studs and the corner posts rest on sole plates, which are usually bolted to the foundation walls (Figure 14-2).

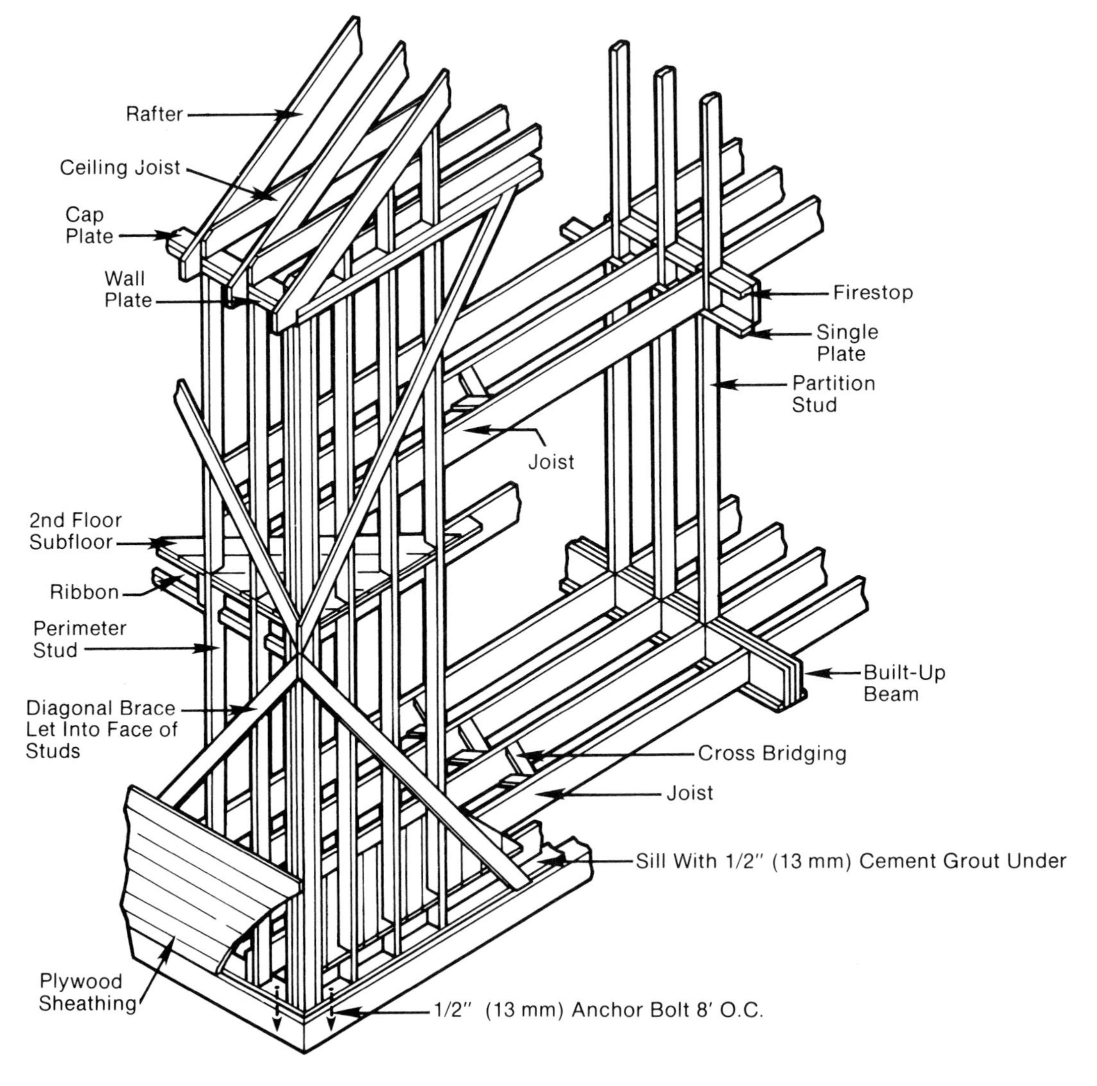

FIGURE 14-1. Balloon framing system.

While these perimeter studs and corner posts can extend more than one story, the partition studs serve only one floor. The studs and posts are topped off by the wall plate and cap plate shown. The first floor joists also rest on the sill plate, while the second-floor joists are secured on a 1″ (19 mm) ribbon set into and flush against the inside edges of the perimeter studs. These second floor joists also form the ceiling joists of the first floor. To reinforce the frame, 1″ × 3″ (19 × 63.5 mm) diagonal bracing is nailed into and flush with the outside edges of the perimeter studs.

Because of the length of the framing members required, balloon framing can be slow, cumbersome, and make the raising of the walls difficult; however, this method is still preferred by some because it reduces the problem of shrinkage while resisting movement between the framing and the masonry veneer.

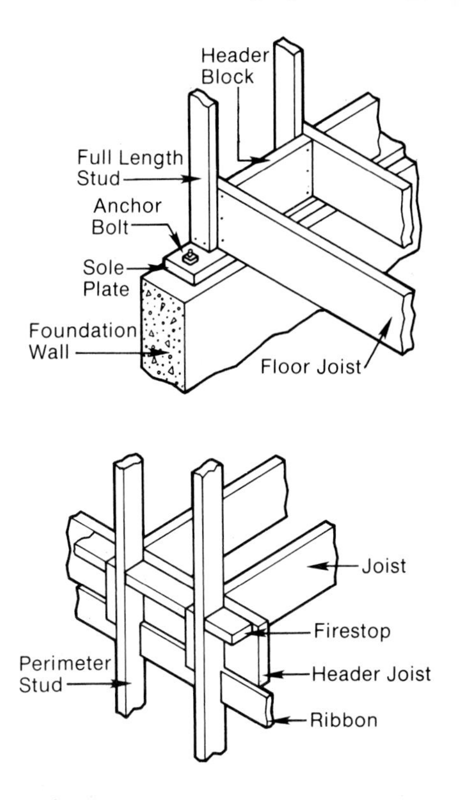

FIGURE 14-2. Sole plate and ribbon details for balloon framing system.

14.4 PLATFORM OR WESTERN FRAMING SYSTEM

The major difference between the platform system (Figure 14-3) and the balloon system is at the floor lines. In the balloon system the wall studs extend from the first floor sill plate to the second floor joists; whereas, in platform framing the walls are complete for each floor. In this system, the studs serve one story only. The studs stand on the sill

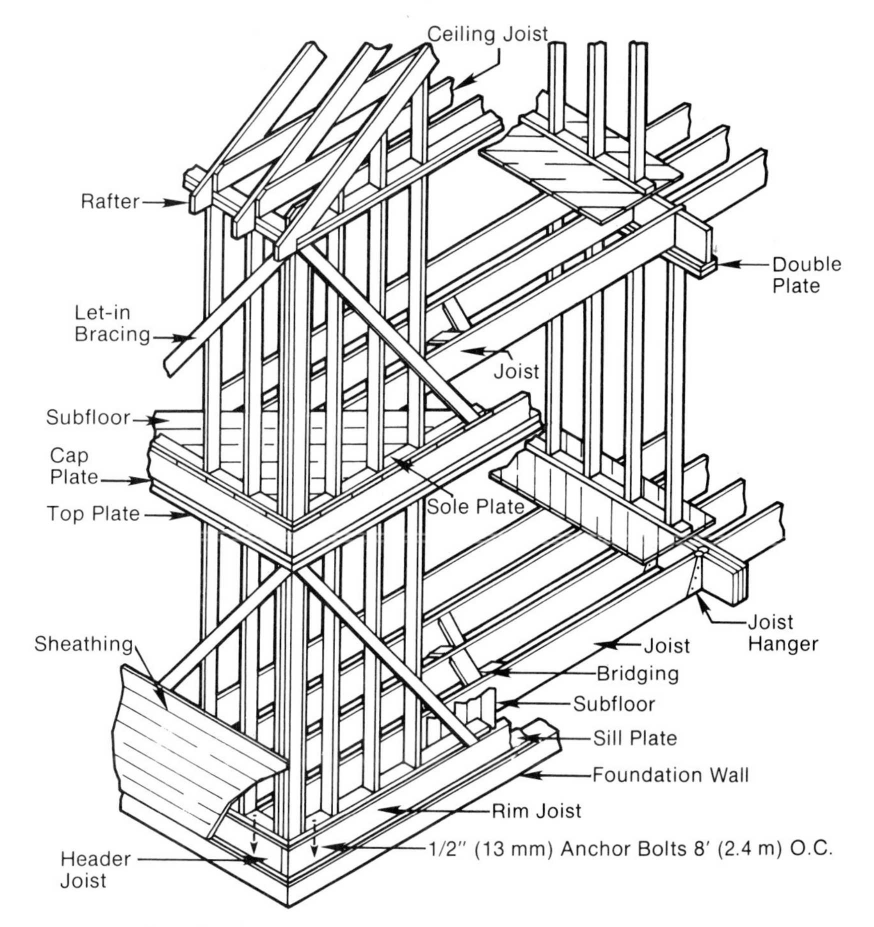

FIGURE 14-3. Platform or western framing system.

plate, over which rest the rim and header joists. A subfloor covers the joists and acts as a platform for the erection of walls and partitions. The wall framing is usually assembled on the floor; the wall is then raised, placed in position, and secured. This same operation is repeated for all subsequent floors, with each story framed and the subfloor placed before the next story is begun.

The platform framing method is much more susceptible to shrinkage than the balloon method due to the many members that are required, but this shrinkage is an acceptable trade-off when compared to the ease with which the walls are raised into place. This system, therefore, is more popular than balloon framing, especially for two- and three-story houses. For a more detailed look at the platform framing system, see Figures 14-4 and 14-5.

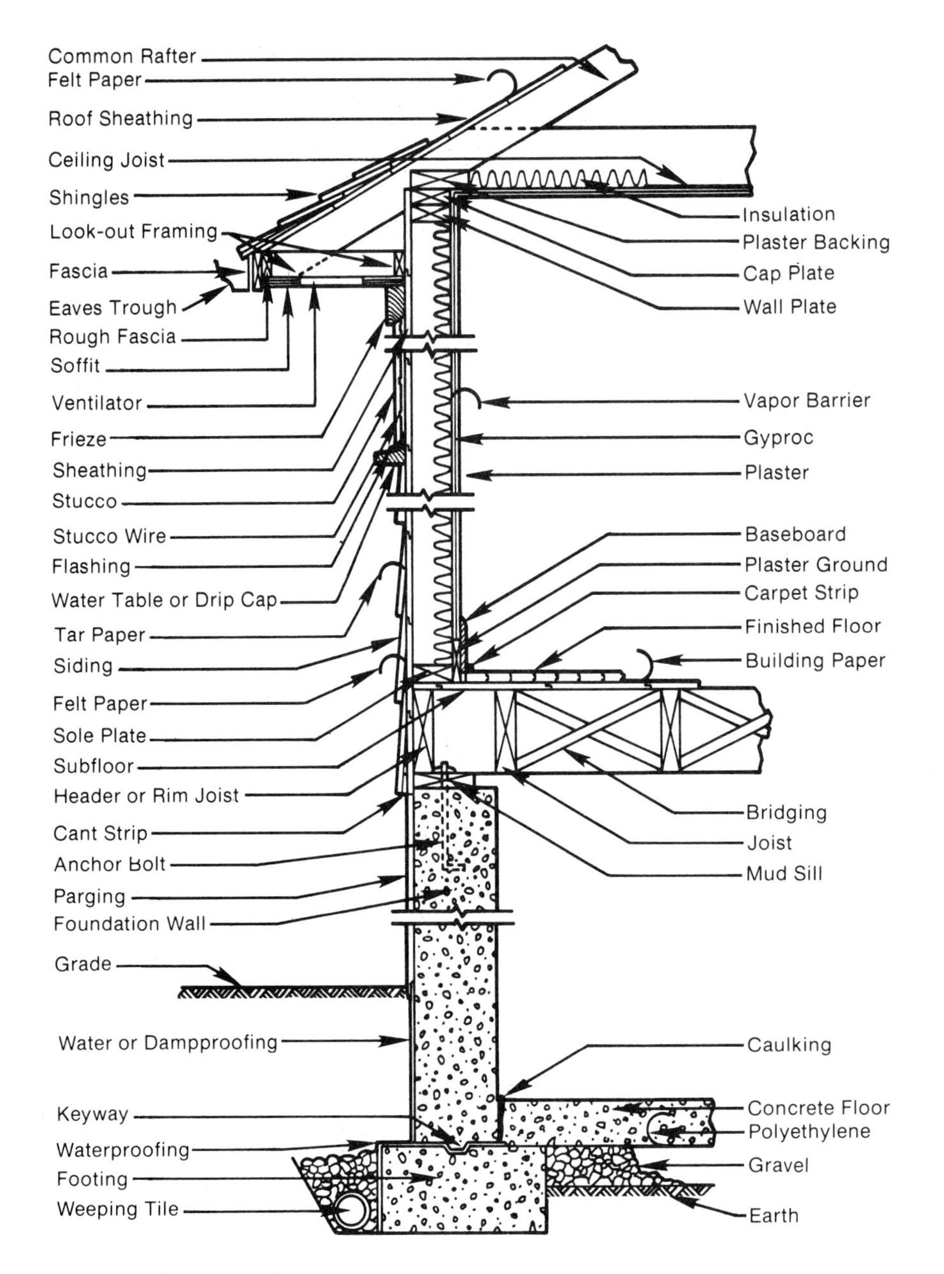

FIGURE 14-4. Typical cross section of platform framing system.

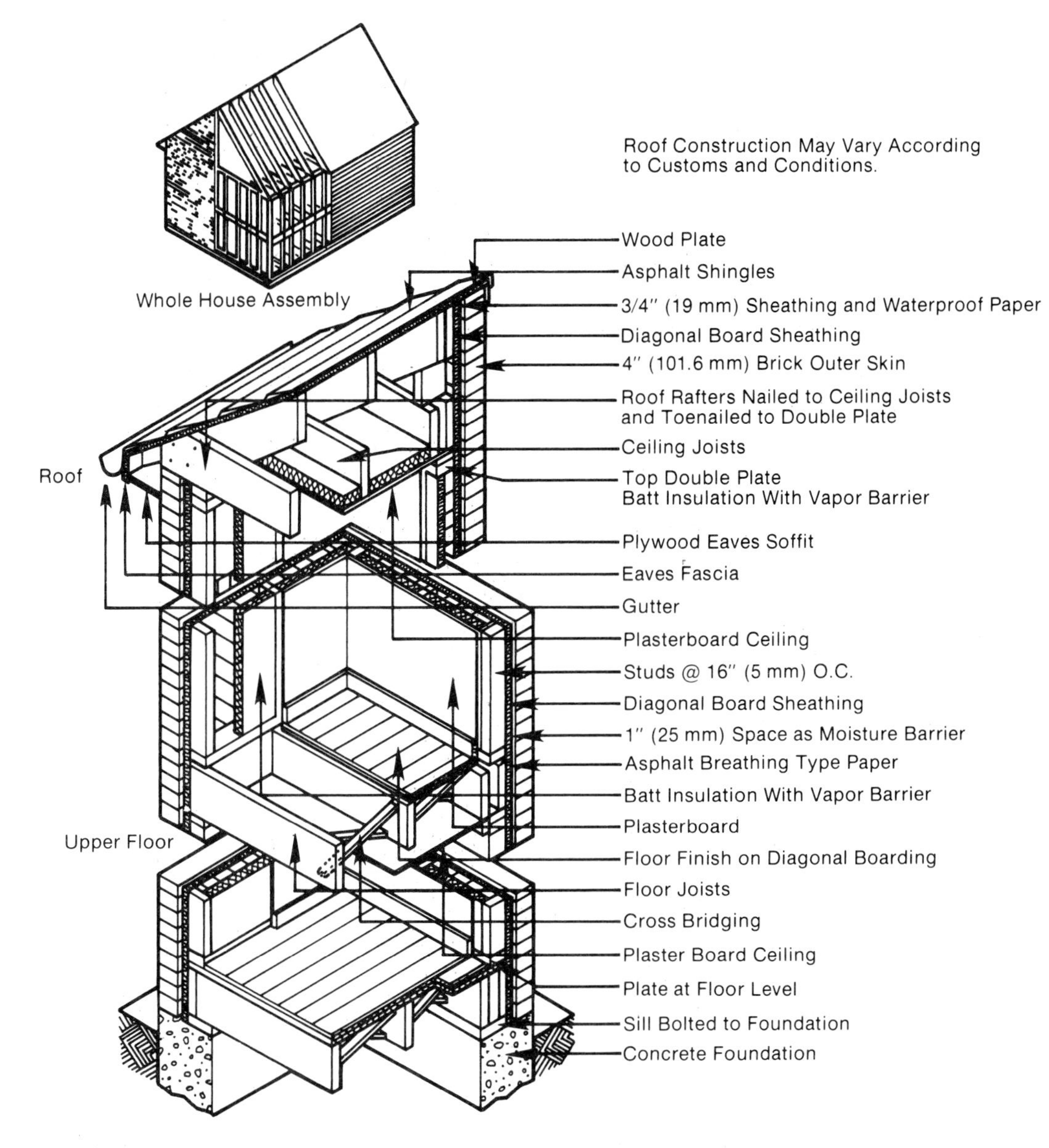

FIGURE 14-5. Brick veneer cross section of platform framing system.

14.5 POST AND BEAM FRAMING SYSTEM

In post and beam framing (Figure 14-6), the system utilizes heavier building materials and fewer members than are used for most construction. The beams, spaced up to 8′ (2.4 m) on center, support the subfloor and roof. Posts support the beams and, for added bracing, supplementary framing members can be placed between the posts. The horizontal spaces between the beams are normally spanned by plank decking, a unique feature of this type of framing (Figure 14-7).

The aesthetic attractiveness of post and beam framing and the creative possibilities it provides are the biggest advantages of this centuries-old system. Instead of being hidden, the posts and beams are kept visible as architectural features. To create a particular effect, the underside of the beams can be finished, and it is also possible to fill in

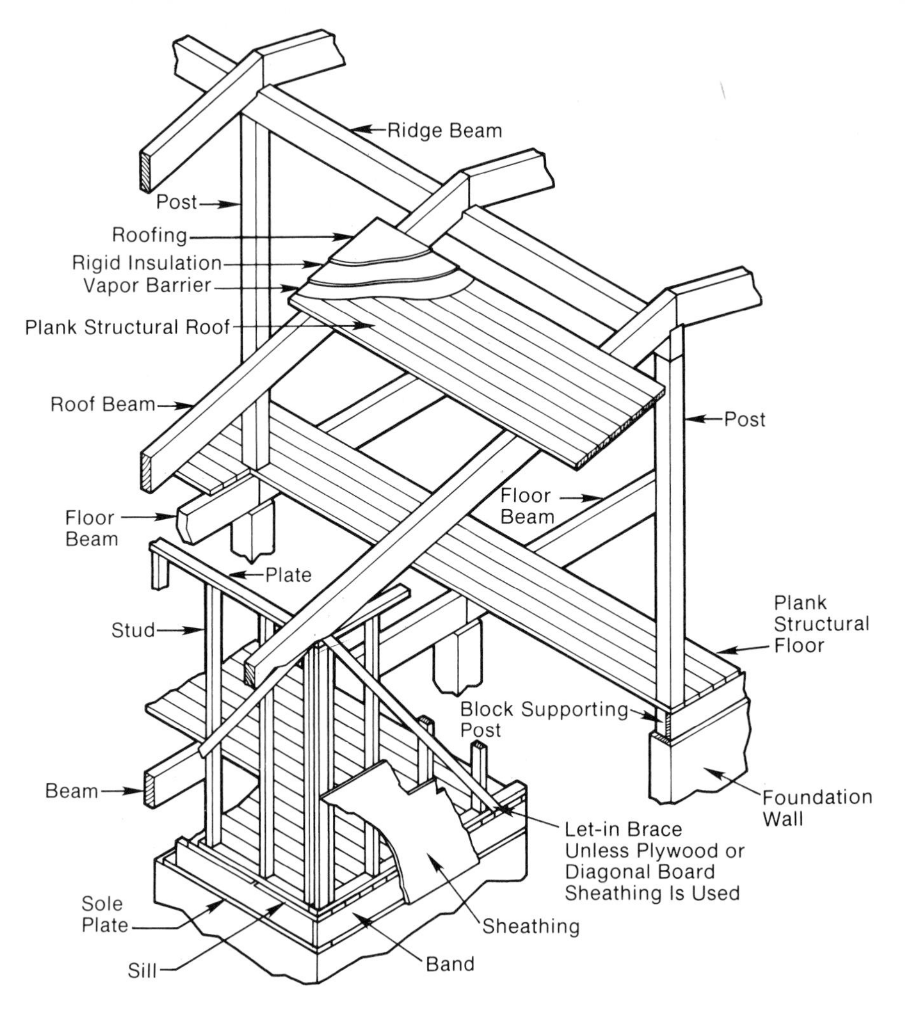

FIGURE 14-6. Post and beam framing system.

FIGURE 14-7. Plank decking for post and beam system.

the spacing between the posts with rustic wall paneling, large panels of glass, or some other decorative covering. This is easily the most visually interesting system; it is also relatively quick and easy to assemble. The biggest disadvantage lies in the fact that supplementary framing is necessary for support and stability underneath and above the partition studs (Figure 14-8). Also, placement of insulation and electrical wiring may present some difficulty in this system.

14.6 COMBINED PLATFORM AND POST AND BEAM FRAMING SYSTEM

This combination system (Figure 14-9) employs the platform framing techniques along with certain post and beam characteristics; namely, exposed beams and plank floors. The spacings between the posts are framed with studs in the conventional fashion, instead of being decorated in the post and beam style. In short, this system combines the sturdiness of the platform system with some of the attractive physical features of the post and beam; however, this system does require longer construction time because of its detailed design.

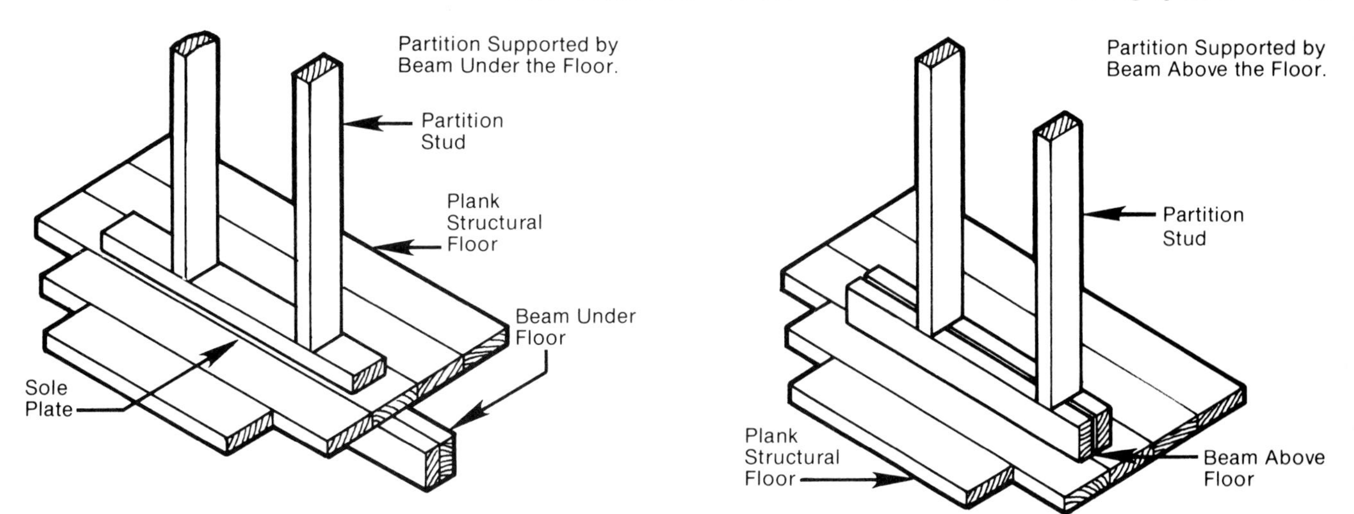

FIGURE 14-8. Partition support for post and beam framing system.

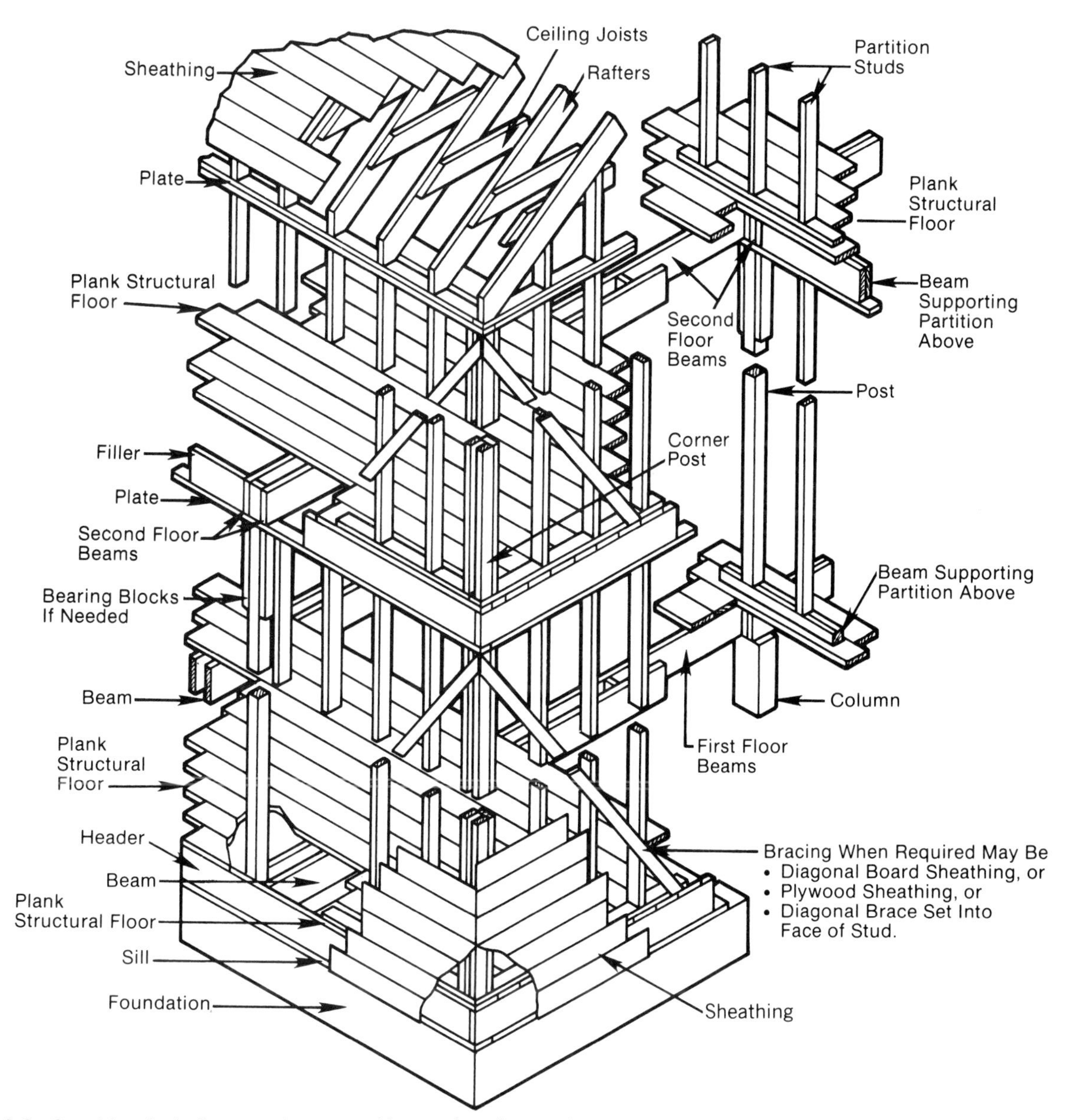

FIGURE 14-9. Combined platform and post and beam framing system.

14.7 ENERGY EFFICIENT SYSTEM

Regardless of the structural system being implemented, energy efficiency is an important consideration during a building project. By improving a building's energy efficiency, energy waste is reduced and the building's comfort level increased. Though this feature is treated here as a separate system, the following energy saving techniques can be used in most light structures systems.

Airtightness is crucial to an energy efficient system, and a vapor barrier can assist greatly in this area. By neither absorbing moisture nor allowing moisture to pass through it, a vapor barrier protects the walls and insulation from harmful water condensation. Sealed 6 mil polyethylene is a very good material for use as a vapor barrier; asphalt-coated or laminated paper, gypsum board, and aluminum foil also work well.

Installing a sufficient amount of insulation is another effective way to cut down on heat loss. Table 14-1 lists the recommended insulation levels for light structures systems. To achieve these levels, double stud wall construction is very helpful (Figure 14-10). In this type of construction, the inner wall serves as the structural unit and the outer wall is the sheathing, with a vapor barrier in between. For comparative purposes, Figure 14-11 illustrates single stud wall construction for an energy efficient system, and Figure 14-12 shows the technique of "superinsulation," the most thorough method of insulating the exterior walls of an existing house.

Use of the pressure treated wood in the building's foundation offers protection against decay caused by weather, fungus, termites, etc. Figure 14-13 illustrates a double stud wall construction on a pressure treated wood foundation.

The following are additional energy saving techniques:

- Design the building with windows on the south side and rooms such as bedrooms, bathrooms, and storage rooms on the north side.
- If feasible, include a solar hot water heater in the building design.
- Place kitchen and bathroom exhaust vents in the walls rather than the ceilings to control air leakage.
- To avoid penetration of cold air from the chimney, "isolate" the furnace room from the rest of the building by insulating it heavily.
- Use a grade beam/piling foundation in areas of permanent ground frost, as shown in Figure 14-14.

TABLE 14-1 RECOMMENDED INSULATION LEVELS FOR AN ENERGY EFFICIENT SYSTEM

	R-Value	RSI (Metric)
Ceilings	50–80	9–14
Above Grade Walls	30–50	5–9
Below Grade Walls	20–30	3.5–6
Floors Over Crawl Spaces	30–40	5–7
Under Concrete Basement Floor	5–10	1–2
Window shutters	10–15	2–3

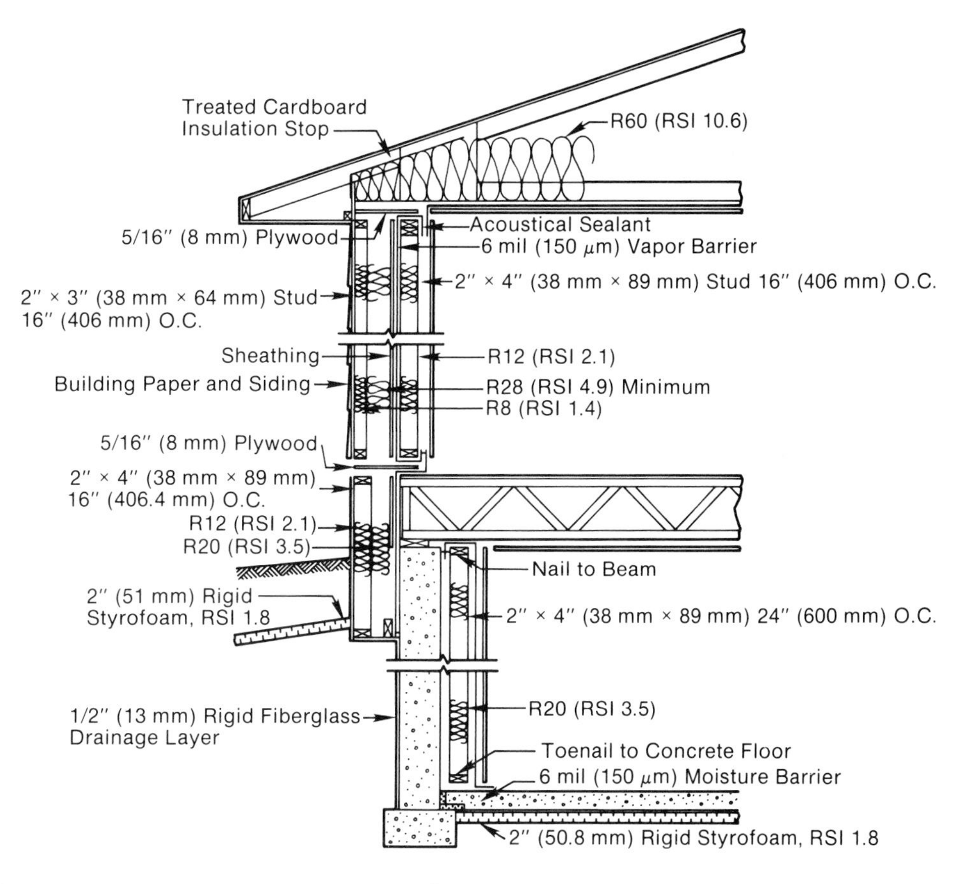

FIGURE 14-10. Double wall construction for an energy efficient system.

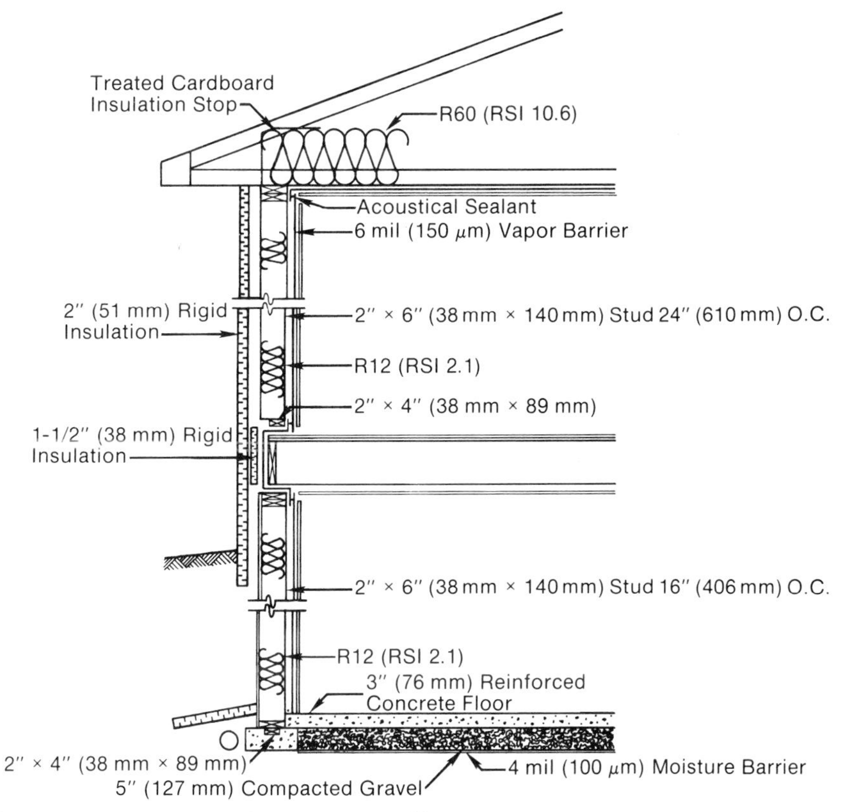

FIGURE 14-11. Single wall construction for an energy efficient system.

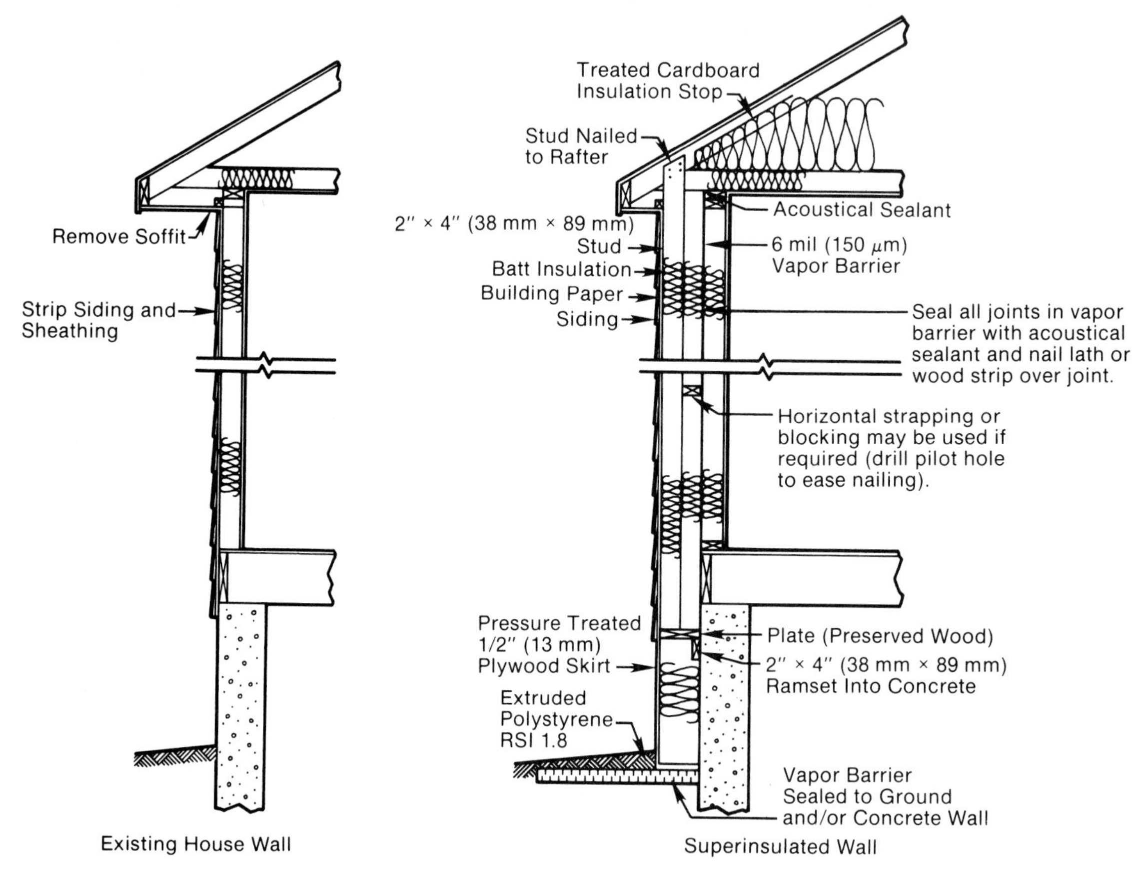

FIGURE 14-12. Techniques for superinsulating exterior walls of an existing house.

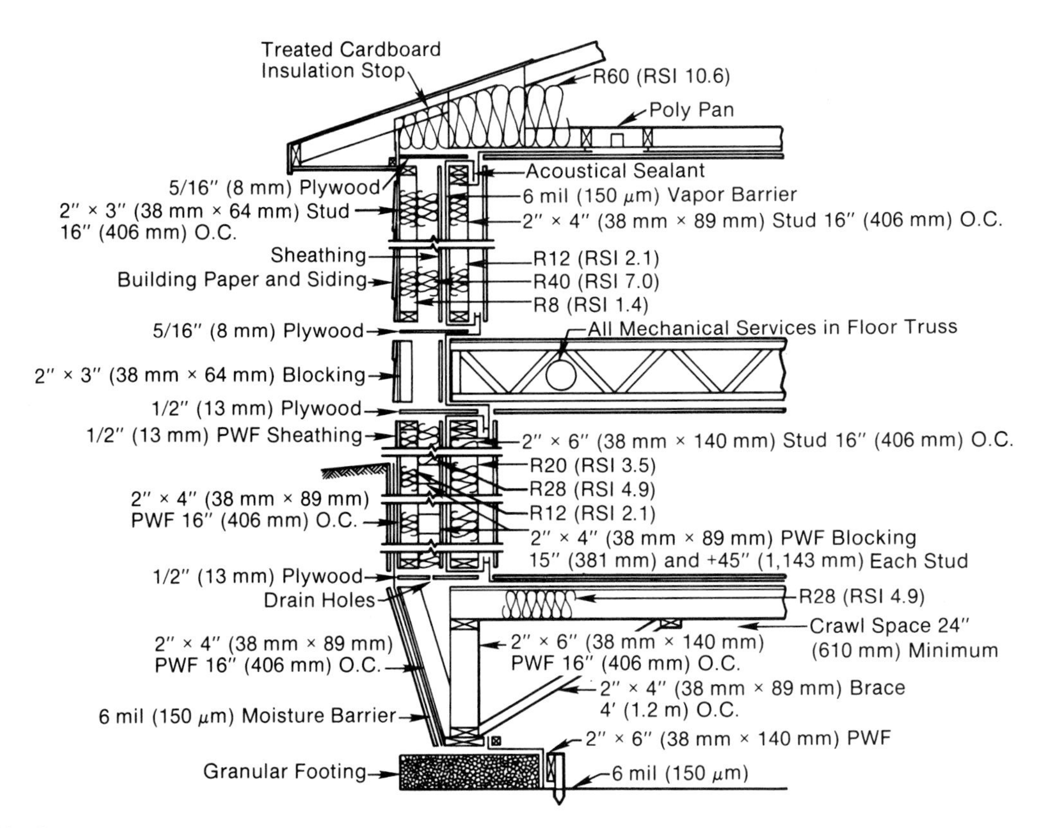

FIGURE 14-13. Double wall construction with pressure treated wood foundation.

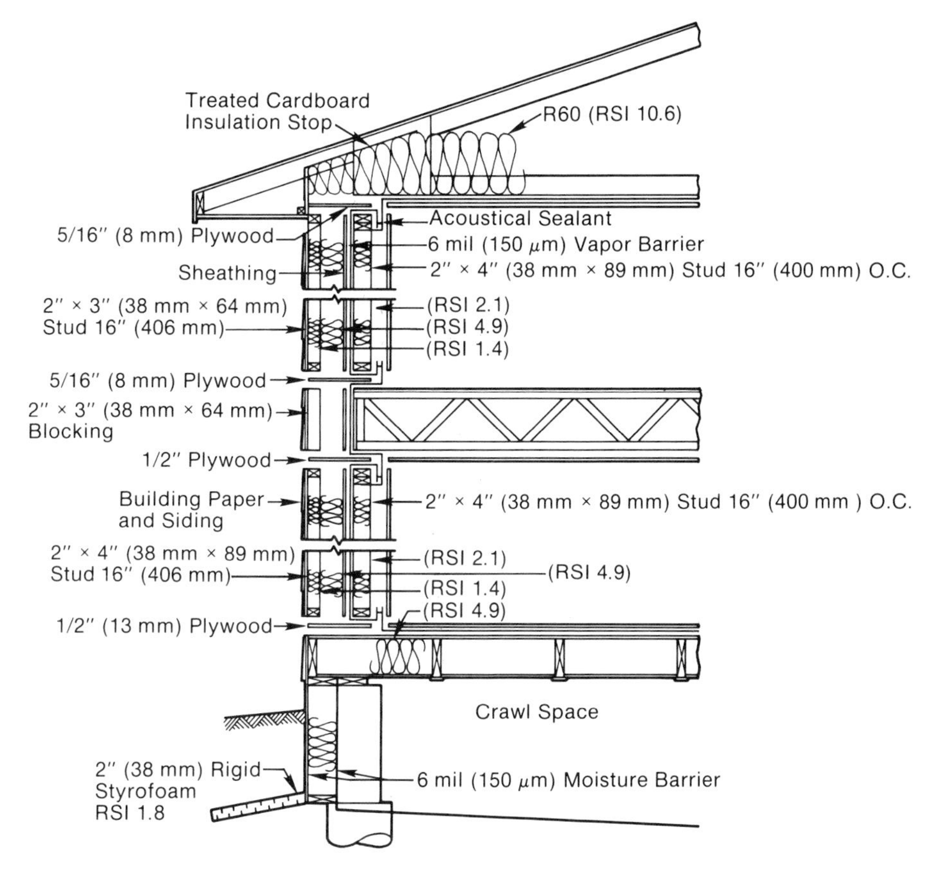

FIGURE 14-14. Grade beam/piling foundation.

14.8 RIGID FRAME SYSTEM

The rigid frame system (Figure 14-15) utilizes a simple one-story unit resting on foundation sole plates. The frame, composed of four pieces of dimensional lumber and three gussetted joints, is assembled in a floor jig. The lower members are anchored to the sole plates, with the side gussets forming window wells. The wall and ceiling areas are sheathed with plywood—which can be overlapped to help drain rainwater—and the end walls may be framed in the conventional manner, with provisions for the desired number of windows and doors. During erection, each frame is raised by itself and spaced on 22" to 24" (approximately 600 mm) centers.

A relatively new concept in light building construction, the rigid frame is appealing for its simplicity, easy construction, and relative inexpensiveness. Originally conceived as a design for farm buildings, it does not offer the architectural features and overall sophistication of many light structures systems, but it is a sturdy and practical alternative.

14.9 ARCH FRAMING SYSTEM

This system features a one- or two-piece arch resting on a concrete sole plate foundation. Two-piece arches are held together with plywood gussets or, if the spans are at least 24′ (7.3 m) long, with collar ties. The

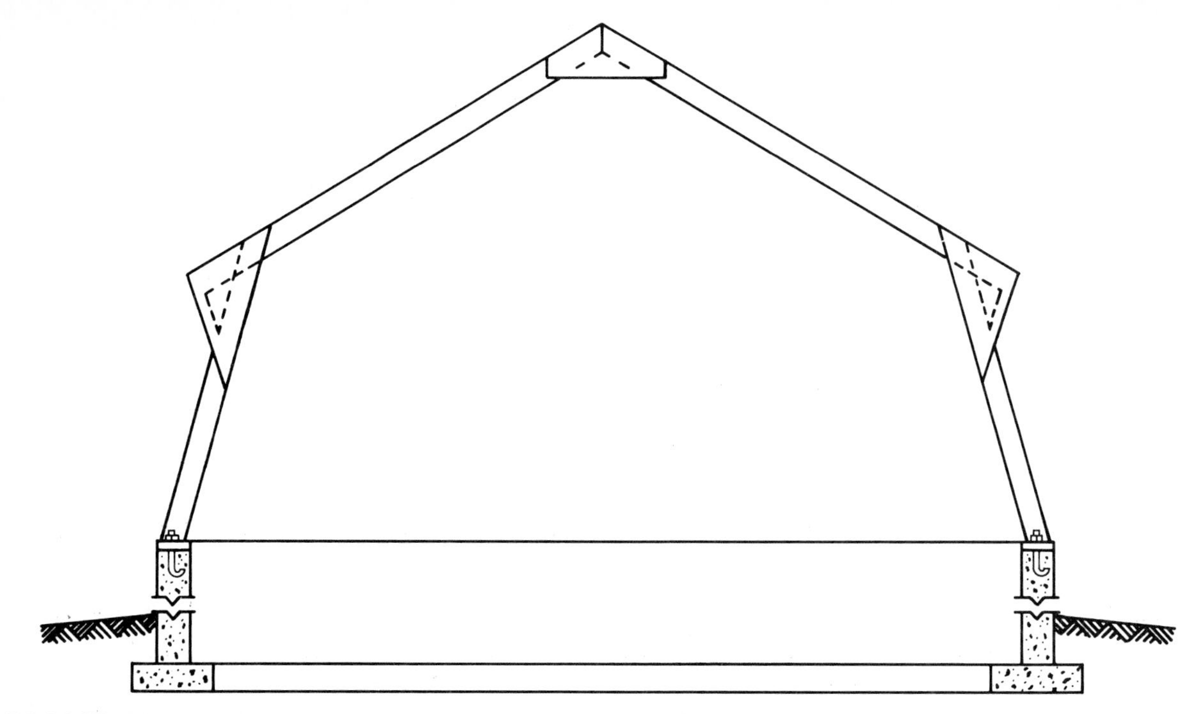

FIGURE 14-15. Rigid frame system.

arch is formed with glue-laminated lumber. Metal fasteners secure the arch to the sole plate. These fasteners are also used to mark the spacings between the members and to help in the raising of the arch to the vertical position (Figure 14-16). As with the rigid frame, the end walls of the arch are framed in the conventional manner to provide for windows and doors. Spacing of the arches depends on the design and span being used, as well as the type of roof deck that is to be applied.

The most popular arch shapes include shed, gothic, semicircular, and parabolic (Figure 14-17). All of these designs are suitable for light building construction projects such as barns, greenhouses, storage sheds, and workshops, as well as residential housing. Though more complex than the rigid frame, the arch system offers a wide range of shapes, sizes, and spans, and thus is becoming increasingly popular.

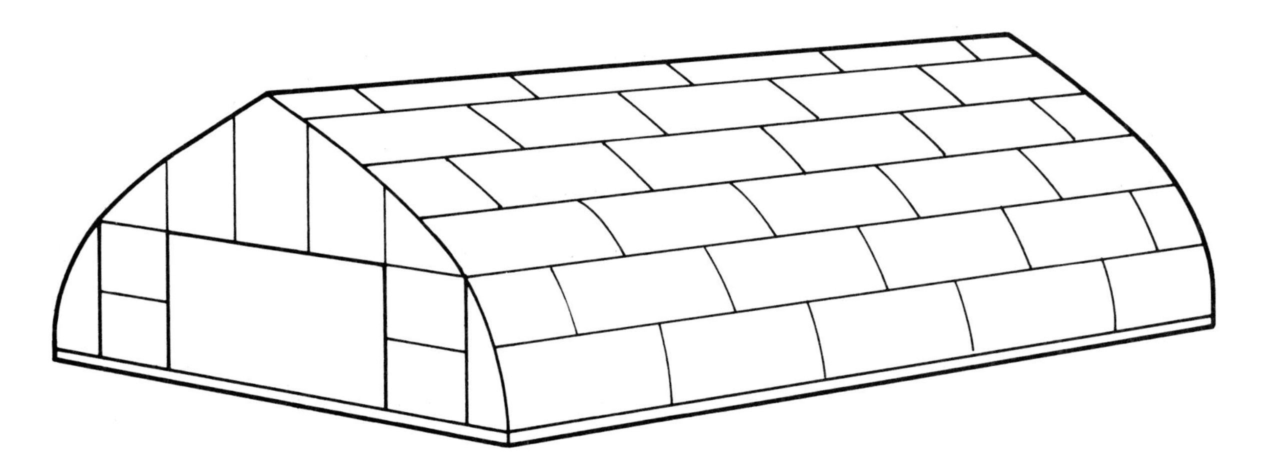

FIGURE 14-16. Arch framing system.

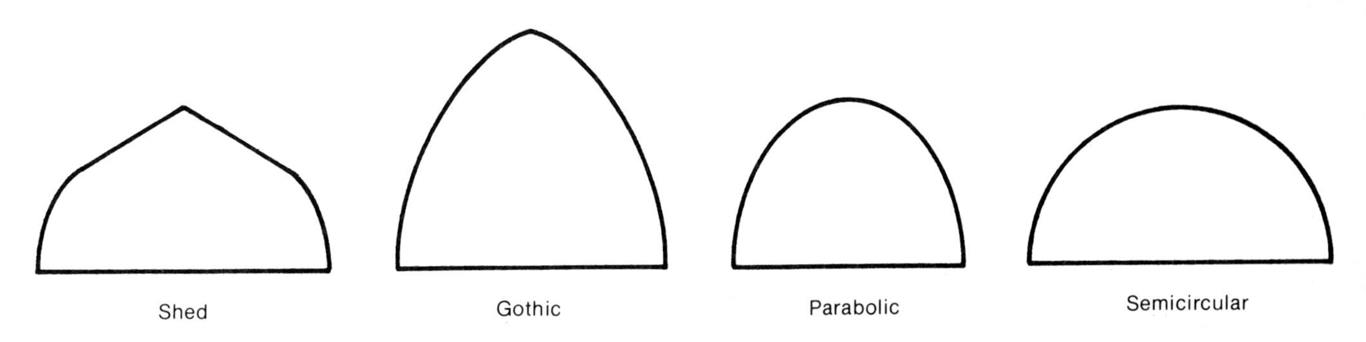

FIGURE 14-17. Arch framing shapes.

14.10 A-FRAME SYSTEM

In the A-frame system, the wall and roof are combined in a single construction (Figure 14-18). Framing members can be dimensional or glue-laminated lumber, and are spaced at 16″, 24″, or 48″ (406.4 mm, 609.6 mm, or 1219.2 mm) intervals on center. The larger members are bolted together at the top of the structure; at the bottom they can either be bolted through the sole plate or secured with metal "shoes." The ends of the building can be framed with conventional framing; if larger areas are needed (for picture windows or sliding glass doors, for example), heavier members should be used.

The A-frame system is often used to construct buildings several stories high, with a set of A-frames resting on sole plates to form a kind of superstructure. This system has long been a popular one in residential construction.

14.11 PREFABRICATED SYSTEMS

Prefabricated buildings are factory-built, portable units without foundations. The subsystems are installed in assembly-line fashion, with the floor, walls, and roof framed and secured with pneumatic fasteners. The major advantage of such a system is its mobility; the various subsystems can be constructed off-site, in an area convenient for the workers, then transported and installed at the selected site. Prefabricated buildings are often used as storage sheds, workshops, and military housing.

14.12 RESEARCH

1. Identify and draw rough sketches of the previously mentioned light structures systems. Make note of the important differences among them.
2. Visit restored and preserved buildings in your area. Make note of the dates the buildings were constructed and the materials that were used.
3. Comment on the changes taking place with respect to light structures systems and suggest why these changes are occurring.
4. Learn more about the history of architecture and suggest how modern design incorporates features from the past. Make a point

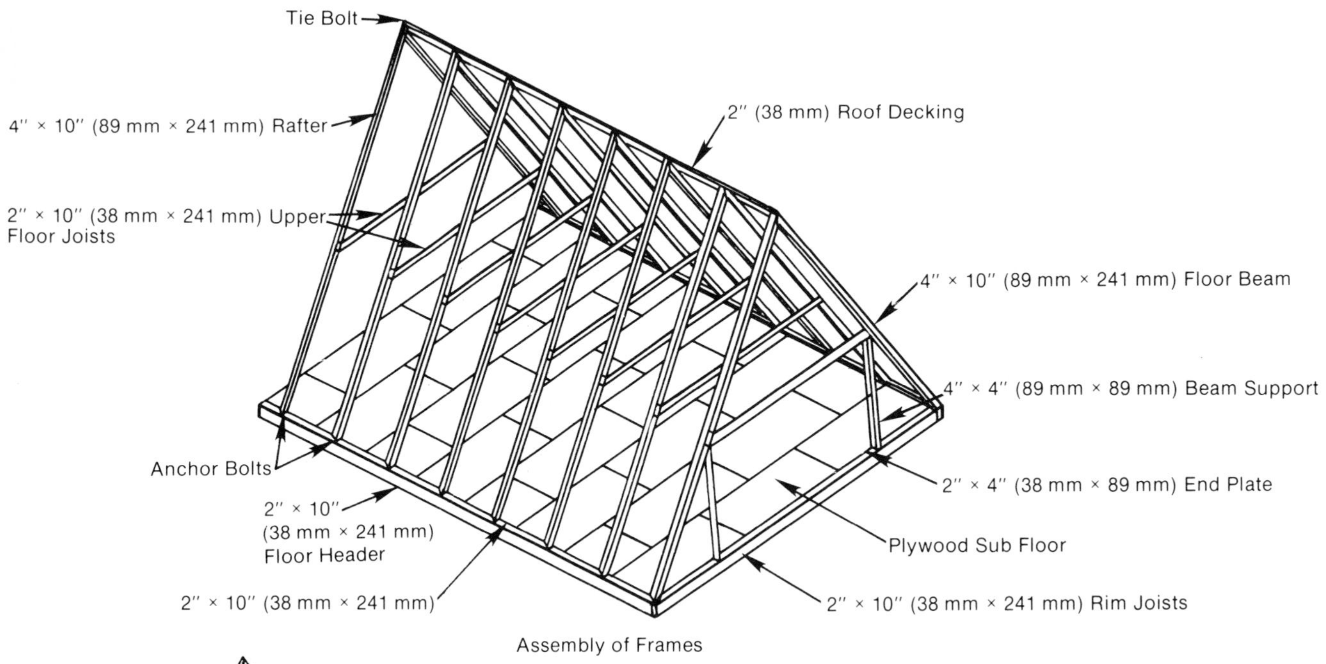

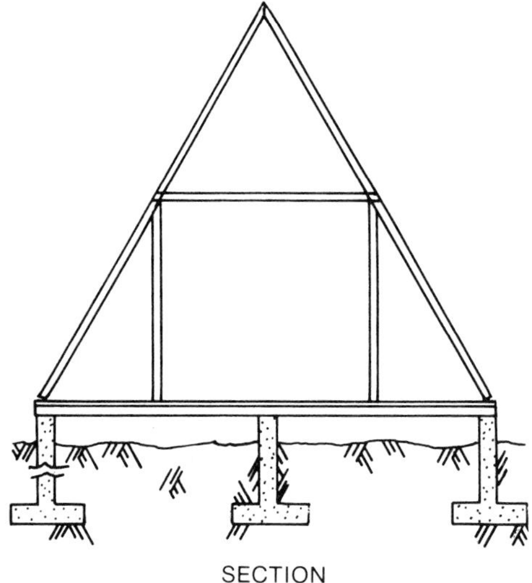

FIGURE 14-18. A-frame system.

of looking at building styles once common in North America—including Georgian, Federal, and Victorian.

14.13 REFERENCES

14.13.1 Ball, John E. *Light Construction Techniques.* Virginia: Reston Publishing Company, 1980.

14.13.2 Scharff, Robert. *The Complete Book of Home Remodeling.* New York: McGraw-Hill, 1975.

14.13.3 Wass, Alzono. *Methods and Materials of Residential Construction.* Virginia: Reston Publishing Company, 1977.

14.13.4 Wass, Alonzo and Gordon Sanders. *Residential Roof Framing.* Virginia: Reston Publishing Company, 1980.

15

FRAMING LAYOUTS

This chapter describes typical layouts for the major components of a light building construction project. The importance of thorough, accurate layouts cannot be overemphasized in any building project. Before attempting any layout work, the reader should fully understand the framing systems discussed in Chapter 14. The builder should also understand how to read and interpret construction drawings and have a thorough knowledge of current local construction codes and practices.

Framing layouts are required for proper footing forms, concrete foundation wall forms, floor joist systems, wall framing systems, ceiling framing, and roof framing. Structural systems develop with the framing layout. As such, materials choices and spacings, locations, openings, the corner assembly, and the assembly unit are all affected by the structural system's influence on the framing layout. Optimum strength and shape of the structure is sought in balancing these factors. As the layout person works, he or she usually thinks in terms of a level, horizontal plane and standard-sized modules of construction, such as 12″, 16″, 24″, 48″ (305 mm, 406 mm, 610 mm, or 1,219 mm).

15.1 GENERAL OBJECTIVES

The chapter is organized to discuss layouts for footing forms, concrete foundation wall forms, floor joist systems, wall framing systems, ceiling framing, and roof framing. It is not a comprehensive treatment of the subject. It does indicate an approach to the learning of framing layouts.

15.2 FOOTING FORMS

Concrete footings provide a strong, level base on which to construct foundation walls and bearing walls (Figure 15-1). The outside dimen-

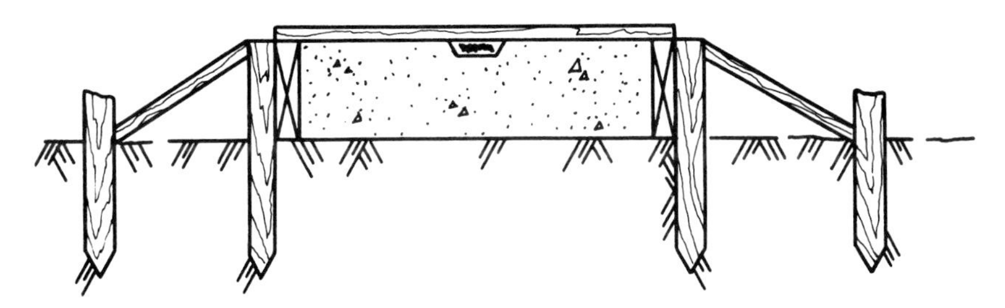

FIGURE 15-1. Cross section of a foundation footing form. Note the support stakes on each side of the form.

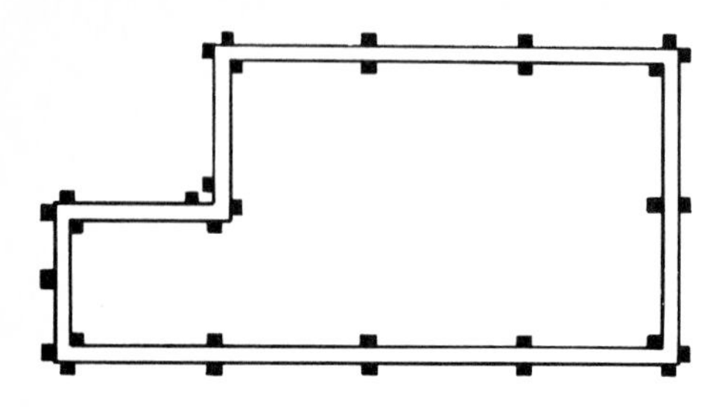

FIGURE 15-2. Simple overhead view of a footing form. The outside dimension of the form determines the finished building dimension.

sion of the footing form determines the outside dimension of the actual footing, the shape of the footing, and the final dimension of the foundation and/or bearing walls (Figure 15-2). Footing forms must provide bracing to hold the wet concrete in place as it sets. They must also have provisions for keying (securing) the footing to the foundation wall. Keying is commonly accomplished through the use of bricks or steel dowels set in the wet concrete or through the use of special keyway forms which provide a sound contact area between footing and wall (Figure 15-3).

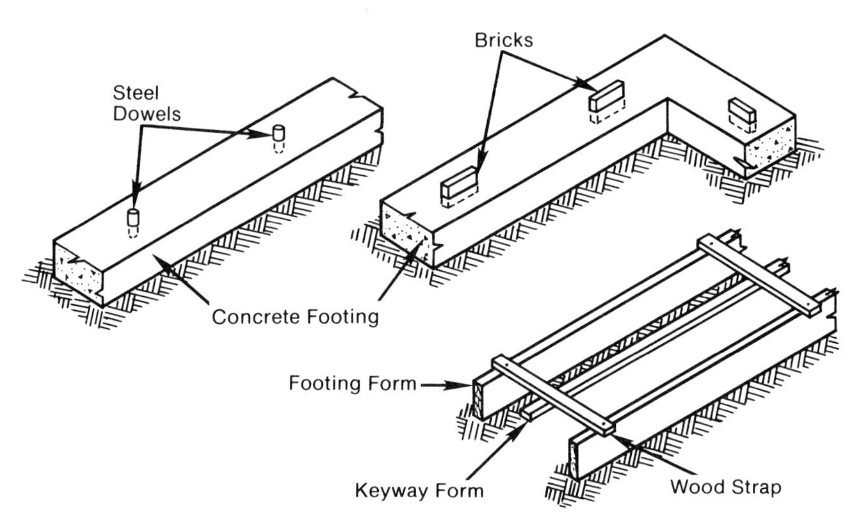

FIGURE 15-3. Methods of keying the footing to the foundation wall.

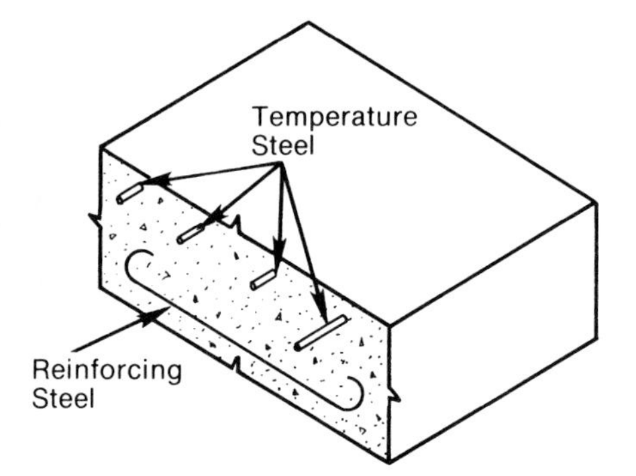

FIGURE 15-4. Reinforcing concrete footings with steel.

Footing forms are usually constructed of boards with a minimum thickness of 1-1/2″ (38 mm). Forms are constructed by placing braced boards in desired positions or by excavating a trench of the required dimensions. Footings must always be placed below the frost line. When reinforcement is needed, steel rods with hooked ends are placed perpendicular to longer straight rods which run in the direction of the footing's length (Figure 15-4). As shown in Figure 15-5, footings should be stepped at the base of the foundation wall to distribute

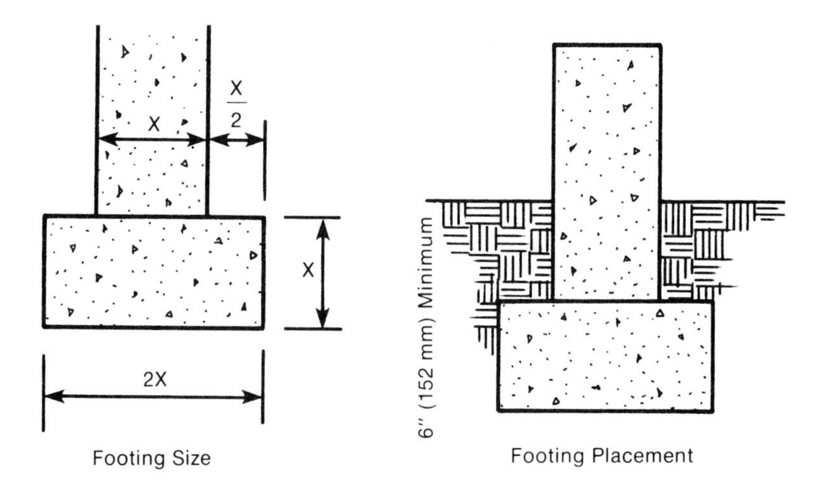

FIGURE 15-5. Footing size in relationship to foundation wall.

building weight. The step projection should equal one-half the width of the foundation wall. When using either board or trench footing forms, care should be taken to ensure that the concrete is sufficiently thick and even at all points.

Because the footing must carry the weight of the building and its foundation, proper footing dimensions are critical. Variations to a basic wood framed structure, such as masonry veneer, require a 2.6″ (66 mm) increase in footing thickness. Each story constructed of masonry materials (in addition to the foundation walls) requires a 5.2″ (132 mm) increase in footing thickness. Footings supporting interior walls must be 4″ (102 mm) thicker for each story of masonry they support.

15.2.1 Footing Form Layout

The footing form layout involves measuring out in two directions from each corner to indicate the width of the footing form and the projection at its base (Figure 15-6). The top of the form should be marked on one external corner stake by nailing a 3/4″ (19 mm) block to the two stakes. A building line is then fastened around the stakes at the expected footing top, made taut, and checked with a line level. This procedure is repeated for each form side. Intermediate stakes should be placed about 40″ (1,016 mm) apart and 3/4″ (19 mm) away from the line.

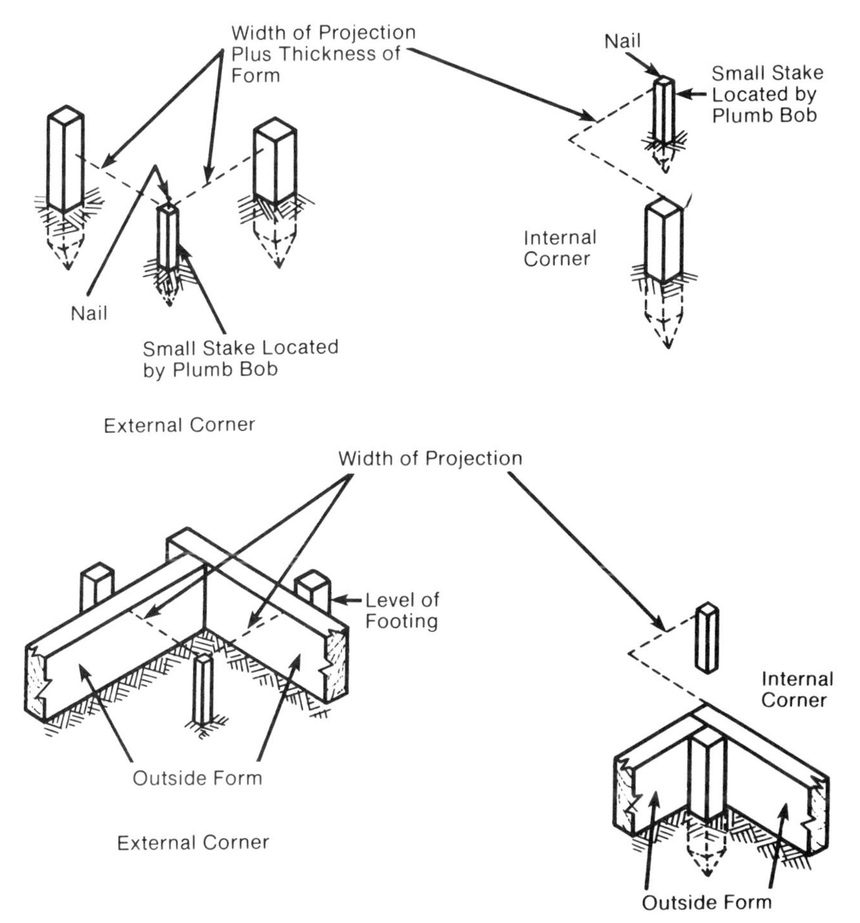

FIGURE 15-6. Footing form corner layout.

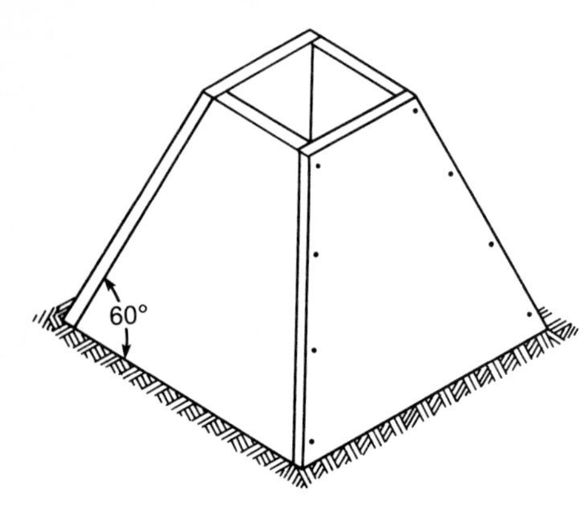

FIGURE 15-7. Tapered post footing form.

Form boards are first attached to external corner stakes and then, after leveling the board, nailed to intermediate stakes. External form boards are butted at corner joints. Inside corners must be lapped to give the footing correct dimensions. Corner braces should be made of wood as thick as that used for forms and positioned with a flat edge flush with the outside face of the form board at a distance equal to the footing thickness.

15.2.2 Post Footings

Girder support posts must have footings so that girder loads are transferred to the basement floor without causing floor damage. This is accomplished by using post footings that have sides sloped at 60° (Figure 15-7). Post footings are typically 10″ × 12″ (254 mm × 305 mm) deep, but depth and area vary with the load.

15.2.3 Bearing Walls

Bearing walls can be used in place of a center girder. A bearing wall typically measures 1-1/2″ × 5-1/2″ (38 mm × 140 mm) and rests on a footing with an elevated center section to protect the wall plate from basement moisture. Figure 15-8 illustrates a bearing wall footing form. Provisions for the bearing wall footing should be made before the basement floor slab is in place as it is positioned beneath the slab (Figure 15-9).

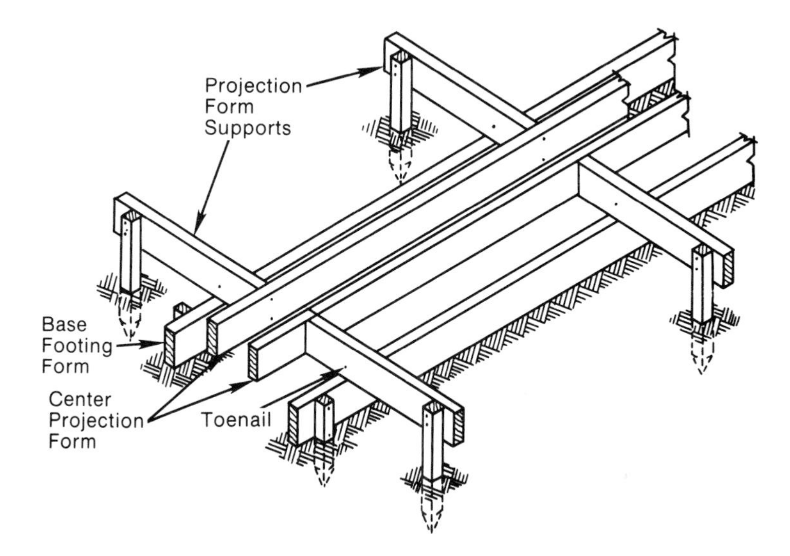

FIGURE 15-8. Bearing wall footing form.

15.2.4 Chimney Footings

Because of the tremendous amount of weight that chimney footings support, special attention should be given to their design. They should be dimensioned 6″ (152 mm) longer and wider than the base of the chimney. Chimney footings must be reinforced transversely and longitudinally with No. 5 reinforcing rods.

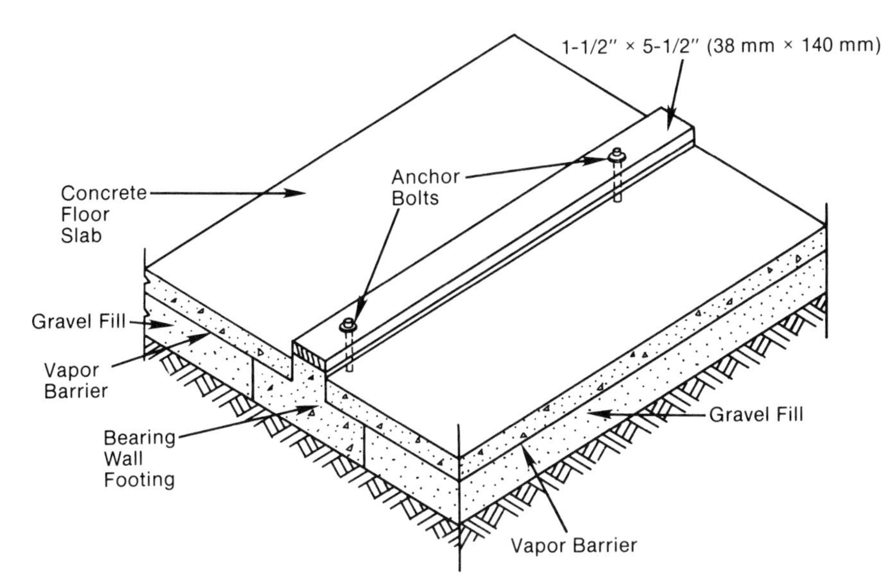

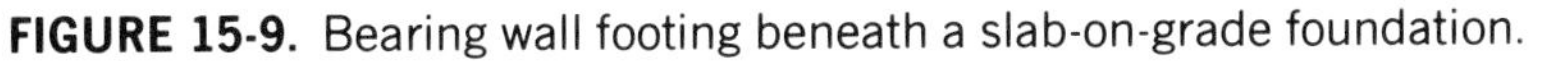

FIGURE 15-9. Bearing wall footing beneath a slab-on-grade foundation.

15.3 CONCRETE FOUNDATION WALL FORMS

When the concrete footings have gained sufficient strength, the foundation wall forms can be erected, tied in place, and braced to receive wet concrete. Fluid concrete must fill the forms to levels shown on the story rod and must flow around windows and door bucks, leaving no voids. Care must be taken so the concrete does not segregate when being placed. Concrete should be vibrated to ensure proper setting (See Chapter 11).

15.3.1 Framed Panel Forms

Many variations of foundation wall form construction exist. One frequently used form style is the type that utilizes reuseable panels joined to form a complete wall (Figure 15-10). The panels, built of 1/2″ to 3/4″ (15 mm × 18 mm) plywood have 1-1/2″ × 3-1/2″ (38 mm × 89 mm) frames. Standard-sized sheets of plywood are used as panels in either a 2′ × 8′ or 4′ × 8′ (600 mm × 2,400 mm or 1,200 mm × 2,400 mm) size. (*Note:* Because of slight dimensional differences, metric and imperial-sized panels cannot be interchanged. Use one or the other on a given job.) The panels are arranged to form a wall by placing them edge to edge and fastening them with nails, bolts or panel ties, and wedges.

15.3.2 Form Ties

Form ties hold the inner and outer form walls in position so that the foundation wall thickness will be uniform. Wire ties (Figure 15-11), bolt ties, bar ties, and rod ties are the most common form tie designs. Bar ties are used where the tie will span the wall thickness plus the sheathing thickness, unlike rod and bolt ties which span the wall thickness, wall sheathing, panel frame, and walers on both walls.

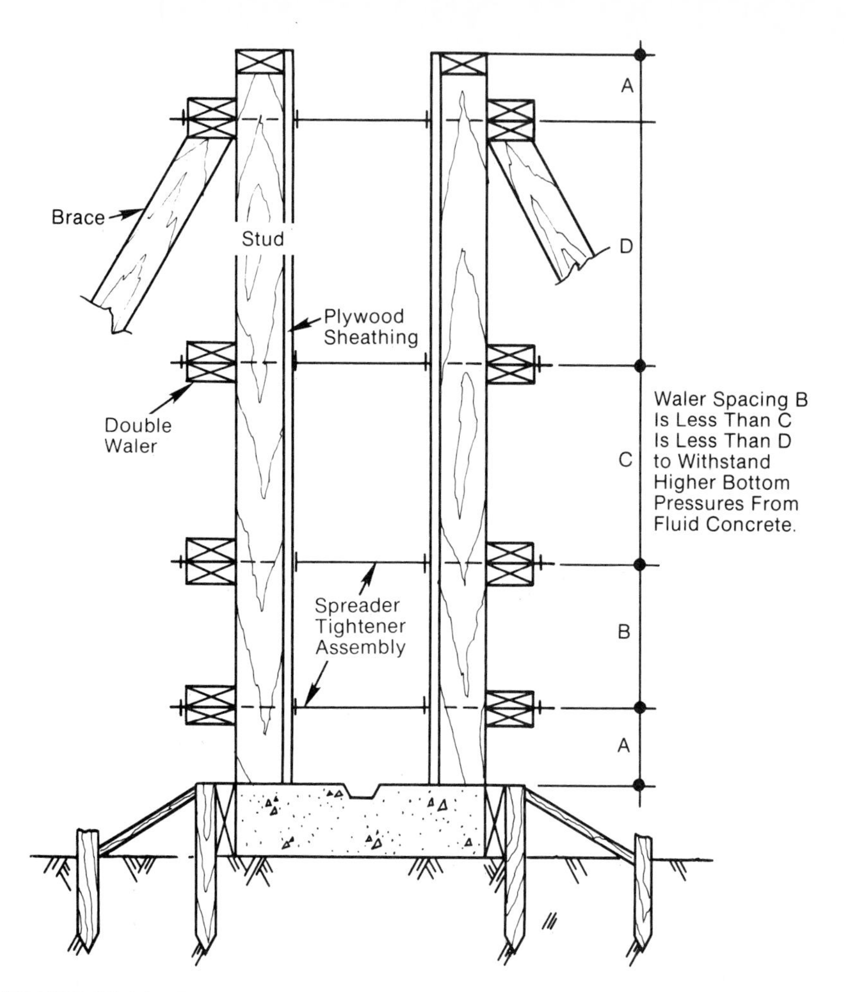

FIGURE 15-10. Cross section of panel-type foundation wall form and wall footing.

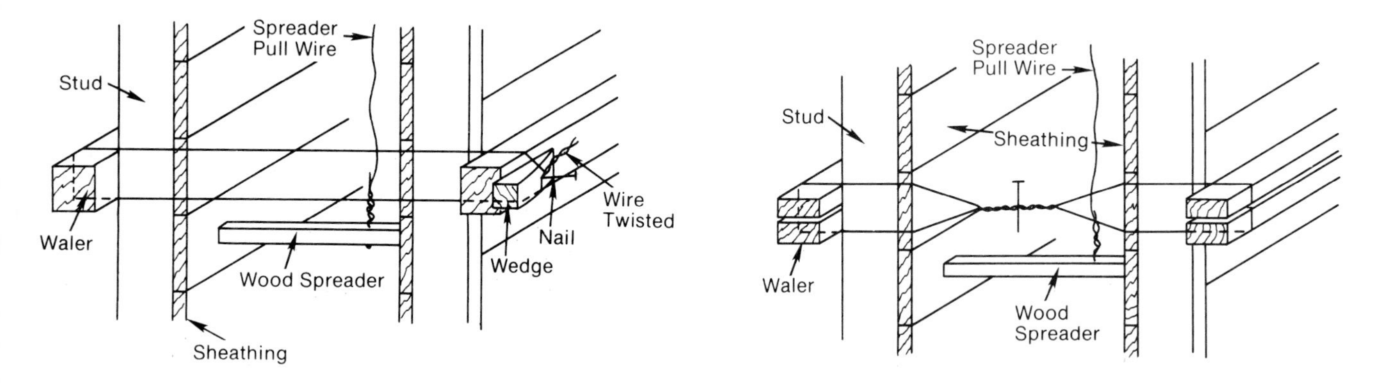

FIGURE 15-11. Examples of common wire ties for wall forms.

15.3.3 Corner Forming

Special attention should be given to the design and construction of foundation form corners. More rigid bracing and a form design that uses standard 2′ or 4′ (.6 m or 1.2 m) panels should be utilized.

To properly construct the forms do the following:

1. Determine the dimensions of the forms and obtain an adequate number of 2′ or 4′ (.6 m or 1.2 m) plywood sections.
2. Prepare any odd sized wall panels for the inside and outside of the form.
3. Construct and brace the form corners.
4. Mark the location for the outside foundation wall with chalk line.
5. Position the outer wall of the form corners. Align the inside sheathing surface with the chalk line and secure them with concrete nails or powder driven pins through the bottom plate or use stakes, braces, and wedges against the outside of the plate.
6. Assemble the other form sections, beginning at the corners and working outward.

15.3.4 Walers

So that wall forms will be adequately rigid and not become deformed under the weight of concrete, horizontal bracing or walers are required. Walers are usually 1-1/2″ × 3-1/2″ or 1-1/2″ × 5-1/2″ (38 mm × 89 mm or 38 mm × 140 mm). Four double walers are needed to brace a 4′ (1.2 m) high, 8′ (2.4 m) thick wall as shown in Figure 15-10. Waler brackets or toenailing secure the walers to the panel studs. Walers should be spaced so that forms can withstand the uneven weight distribution of concrete in the form (Figure 15-10).

15.3.5 Positioning Form Ties

The spacing of form ties will vary with the type of tie used, the form sheathing strength, the spacing and size of studs, waler positions, the temperature of the air and concrete, and the rate at which the concrete fills the form. Table 15-1 is a guide for form tie installation.

TABLE 15-1 GUIDE FOR FORM TIE INSTALLATION

	Horizontal Spacing	Vertical Spacing	Sheathing	Studs	Walers
	32″ (813 mm)	30″ (762 mm)	1/2″ (13 mm) plywood	2″ × 4″ @ 19″ O.C. (38 × 89 @ 483 mm)	Dbl. 2″ × 4″ (38 mm × 89 mm)
			3/4″ (19 mm) plywood	2″ × 4″ @ 19″ O.C. (38 × 89 @ 483 mm)	
2,992 lbs. (1,357 kg)	24″ (610 mm)	30″ (762 mm)	same	2″ × 4″ @ 14″ O.C. (38 × 89 @ 356 mm)	
			same	2″ × 4″ @ 17″ O.C. (38 × 89 @ 432 mm)	Dbl. 2″ × 4″ (38 mm × 89 mm)
			same	2″ × 4″ @ 13″ O.C. (38 × 89 @ 330 mm)	
	24″ (610 mm)	24″ (610 mm)	same	2″ × 4″ @ 16″ O.C. (38 × 89 @ 406 mm)	Dbl. 2″ × 4″ (38 mm × 89 mm)

TABLE 15-1 GUIDE FOR FORM TIE INSTALLATION (Continued)

	Horizontal Spacing	Vertical Spacing	Sheathing	Studs	Walers
	32″ (813 mm)	32″ (813 mm)	1/2″ (13 mm) plywood	2″ × 4″ × 14″ O.C. (38 × 89 @ 356 mm)	3″ × 4″ (38 mm × 89 mm)
			3/4″ (19 mm) plywood	2″ × 4″ × 16″ O.C. (38 × 89 @ 406 mm)	
	32″ (813 mm)	27″ (686 mm)	same	2″ × 4″ × 13″ O.C. (38 × 89 @ 330 mm)	3″ × 4″ (38 mm × 89 mm)
5,869 lbs. (2,662 kg)			same	2″ × 4″ × 16″ O.C. (38 × 89 @ 406 mm)	
	30″ (762 mm)	24″ (610 mm)	same	2″ × 4″ × 12″ O.C. (38 × 89 @ 305 mm)	3″ × 4″ (38 mm × 89 mm)
			same	2″ × 4″ × 14″ O.C. (38 × 89 @ 356 mm)	
	24″ (610 mm)	24″ (610 mm)	same	2″ × 4″ × 12″ O.C. (38 × 89 @ 305 mm)	Dbl. 2″ × 4″ (38 mm × 89 mm)

15.3.6 Form Provisions for Girders, Joists, Plates, and Openings

Before concrete is poured into any set of positioned forms, provisions must be made for positioning of girders, joists, or plates. Several alternative layouts are shown in Figure 15-12.

The joist system shown in Figure 15-13A is positioned on top of the sill. A support girder, at right angles to the joists, is set to carry some of the building load. In this case, the elevation of the top of the sill and the top of the girder would appear at the same mark on the builder's story rod. (*Note:* Crowning of the girder can be done with an adjustable support post.) As shown in Figure 15-13B, this same cast-in type system can be modified to provide for brick veneer.

The joist elevation is determined by the concrete wall form. The elevation is important for under-joist headroom and for stair construction. A finish floor to finish floor measurement is a key measurement for stair design. It should be marked on the story rod in relation to form height and joist position.

The frames and bucks (restraining forms) for all openings that will later appear in the foundation (basement windows, pipe openings, and door openings) must also be placed within the concrete wall forms prior to concrete pouring. Such openings appear as the forms are removed as lined openings. Basement window openings will usually be fitted with frames, keyed to the concrete, and made ready to receive window sash. Installations may be of different materials, usually wood or metal.

Pipe openings for vents or services are designed to provide secure access—secure against water, insects, and air movements. Door openings usually include a rough buck designed to be keyed to the concrete

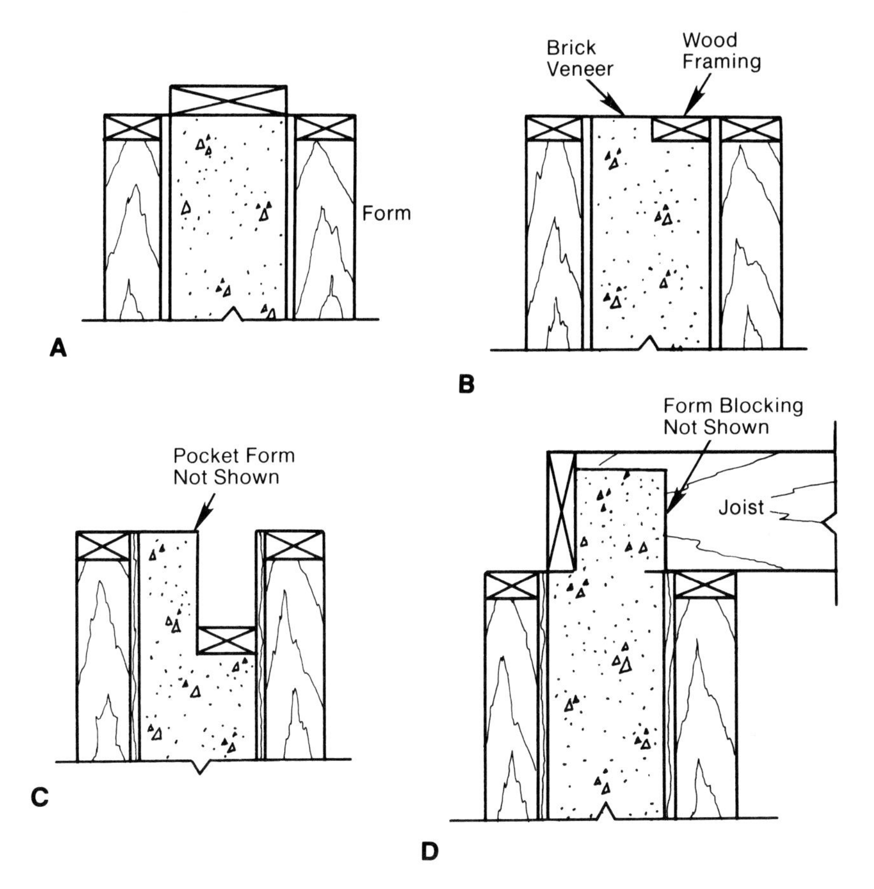

FIGURE 15-12. Several relationships of foundation walls to floor joist systems: (A) sill plate directly on top of foundation wall; (B) sill plate on inside of wall; (C) recessed sill plate for steel floor frame and (D) "cast-in" joist system.

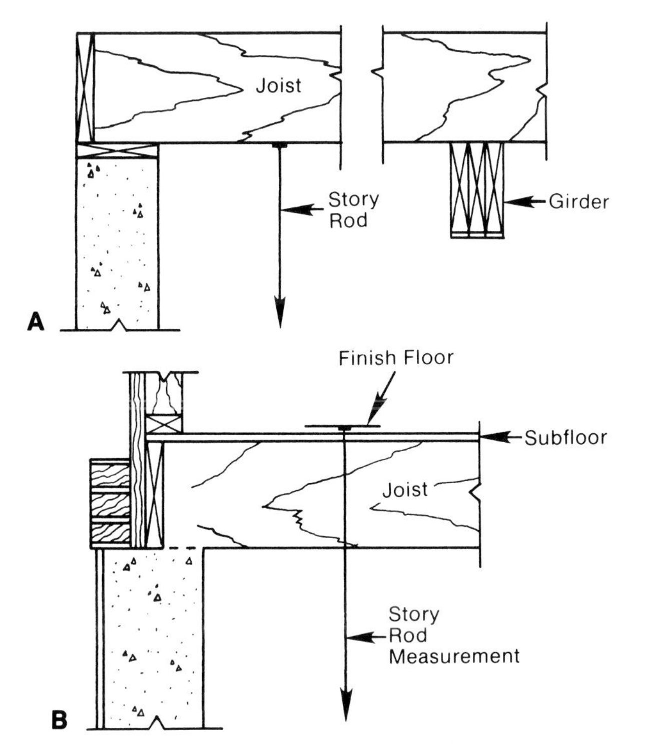

FIGURE 15-13. (A) Joist on top of sill; (B) modifying the "cast-in" joist system to accept a brick veneer.

wall and braced to withstand the pressures of fluid concrete. The key remains in the concrete when the forms and the rough buck are removed (Figure 15-14).

The elevation of a door opening in a basement wall is set in relation to the expected finish grade outside and the landing inside. Flights of stairs may lead from the landing to a basement and to a first floor level. The elevation of the landing provides a uniform unit rise for the stairs.

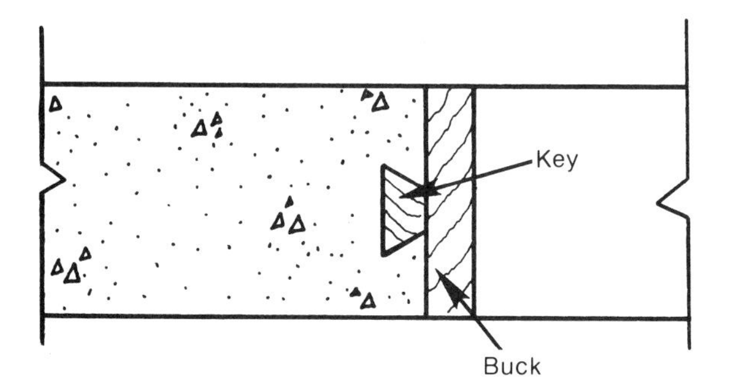

FIGURE 15-14. The buck/key/concrete relationship for keying a door into the foundation wall.

15.3.7 Foundation Anchorage

To prevent movement caused by the pressures of the wind, the superstructure must be firmly attached to the foundation by means of anchor bolts (Figure 15-15). The pressures created by the wind can cause translation, overturning, and rotation. *Translation* is the process by which a building is moved in a lateral direction by strong winds. *Overturning* is the uplifting of light superstructures by wind forces. To avoid overturning, both anchor bolts and framing anchors can be used. The anchors are attached to the individual framing members, usually spaced on 4′ (1.2 m) centers. *Rotation* is the pivoting of a light frame building on its foundation. Rotation sometimes occurs when high winds are not symmetrical. To prevent rotation, the sill plate should be firmly anchored to the foundation and the anchors should possess adequate shearing strength.

The bottom plate of the superstructure is attached to the foundation wall by means of 1/2″ (13 mm) anchor bolts spaced on 4′ (1.2 m) centers. The bolts should be placed so that their threaded ends extend above the top of the sill plate. If the foundation wall is constructed of concrete blocks, the anchor bolts should be placed in a core filled with mortar. To keep the mortar from falling from the core, a metal lath is placed below the hollow core.

15.4 FLOOR JOIST SYSTEMS

The type of wall frame construction used in the building design determines how much of the building load from above is carried on the floor joists. In a box sill system at the building perimeter, the load is transferred to joists, the sill plate, and the foundation wall (Figure 15-16A). For a cast-in type sill configuration (Figure 15-16B), there is no sill plate so the load is transferred directly from the joists to the

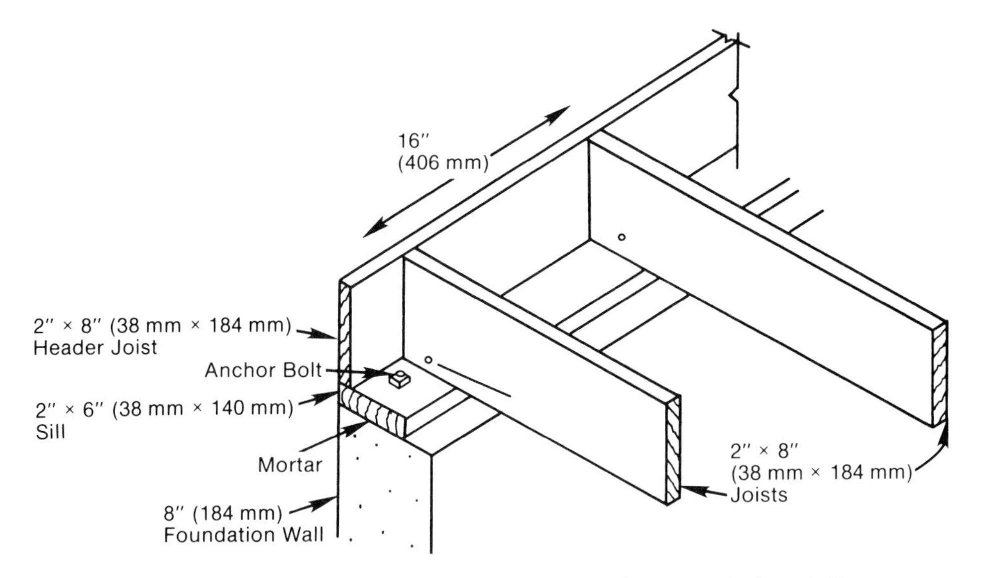

FIGURE 15-15. Attaching the frame superstructure to the concrete foundation wall.

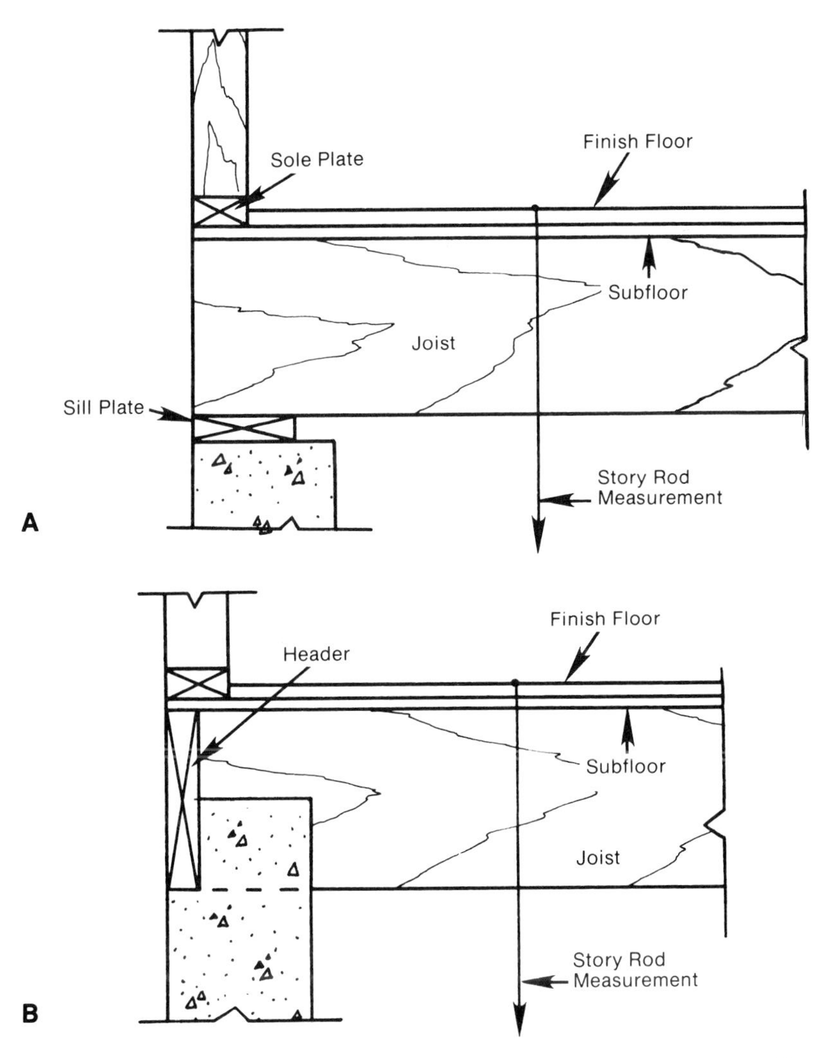

FIGURE 15-16. (A) In a box sill system load is transferred to joists, to sill plate, to foundation wall; (B) in a cast-in joist system load moves directly from joists to foundation wall.

foundation wall. If steel joists rest in foundation wall pockets, the load is on the joists or, if there is a sill plate, on the sill plate (Figures 15-17A and 15-17B).

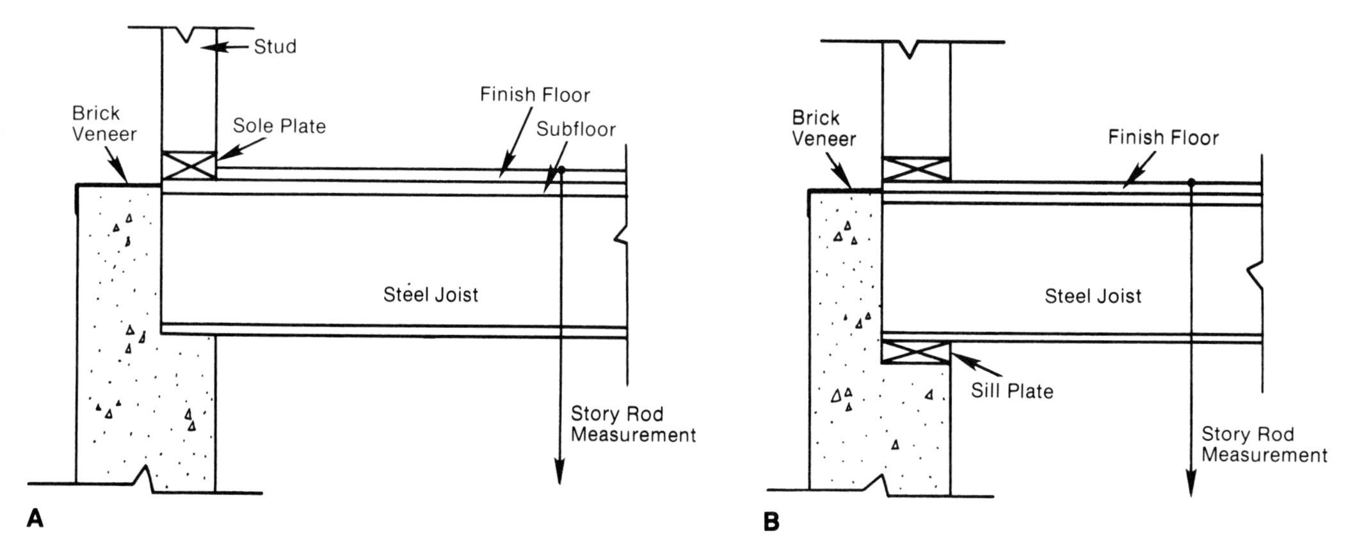

FIGURE 15-17. With steel joist system designs load can be transferred (A) directly to the foundation wall or (B) from sill to wall.

Building load other than that located at the building's perimeter is carried by joist, girder, or post systems as shown in Figure 15-18. If the girder is centered between the foundation walls, it will carry approximately one-half of the building load.

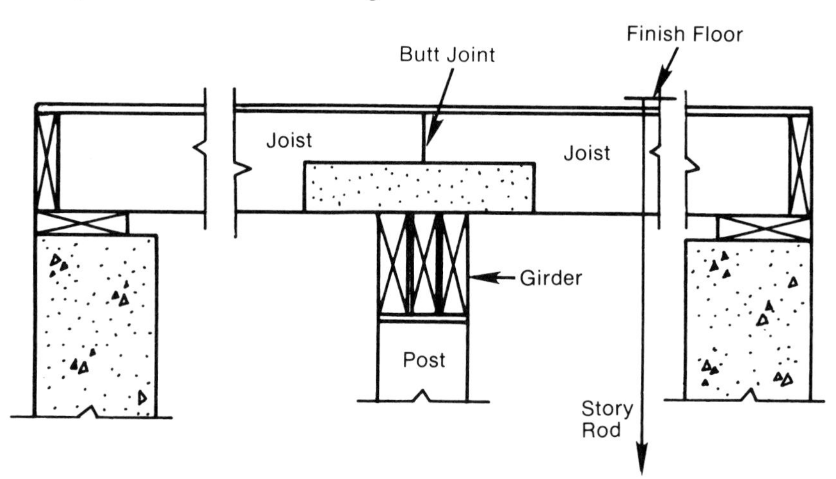

FIGURE 15-18. Details of the joist/girder/post system.

Some builders prefer to use lapped joist joints over the girder (Figure 15-19). The girder may also be located at joist level as illustrated in Figure 15-20. Exact post length is determined by the type of joist girder system used in the floor joist system.

If the joist/foundation wall system is the cast-in type or if pockets are used, the perimeter spacings of joists are determined before concrete is poured. If a sill plate (as different from sole plate) is used, header joists usually carry the layout markings, such as 16″ (406 mm).

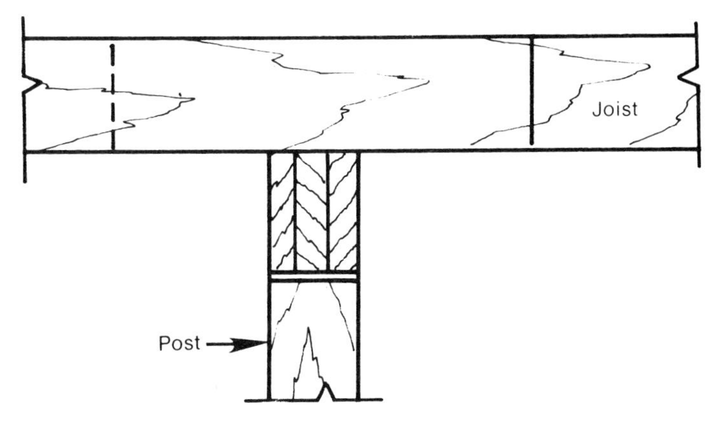

FIGURE 15-19. Joists joined by lap joints over the girder.

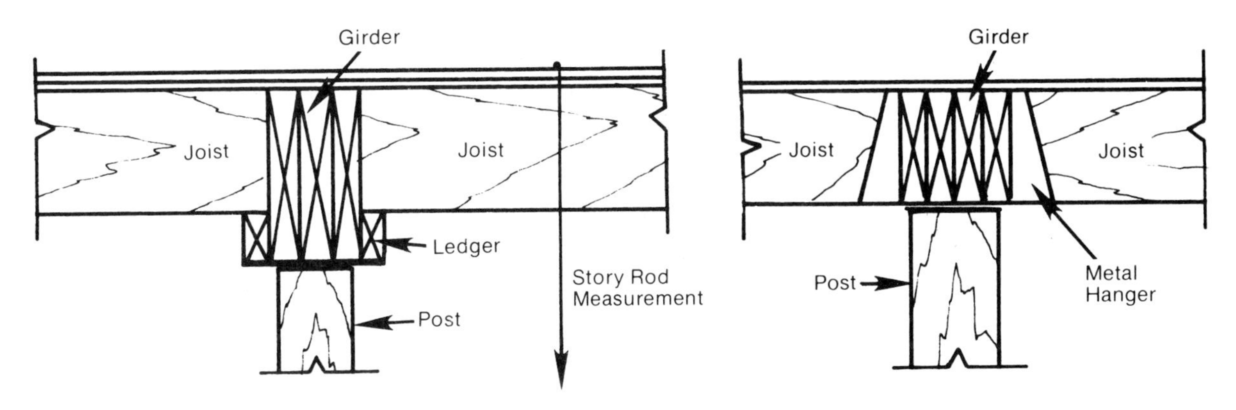

FIGURE 15-20. Examples of girders located at joist level.

Such markings are modular in nature to accommodate standard sizes of subfloor panels, such as 4′ × 8′ (1.2 m × 2.4 m). The initial layout of header joists will be modular to avoid cutting of subfloor panels (Figure 15-21).

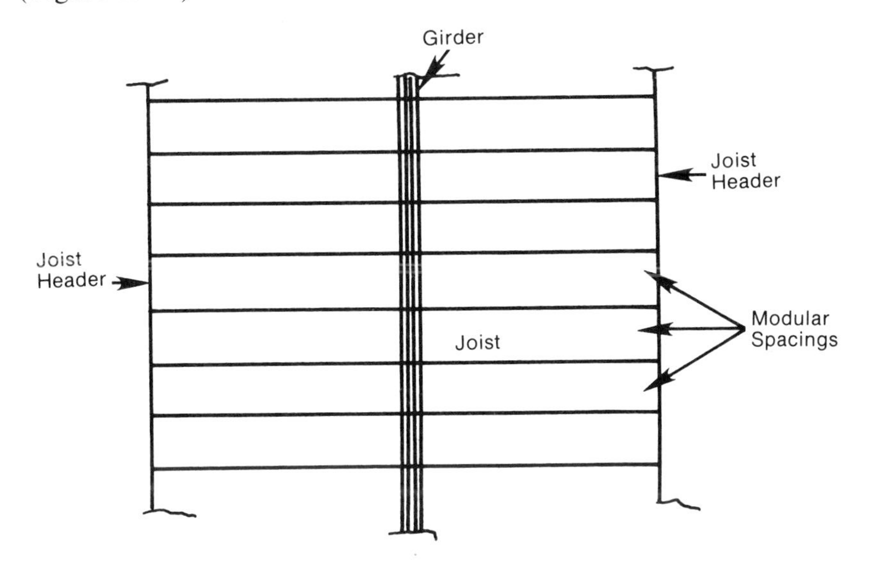

FIGURE 15-21. Joist spacing should accommodate the use of modular sized subfloor materials. This eliminates unnecessary cutting, speeds work, and reduces waste.

15.4.1 Openings in Floor Joist Systems

Joist members must be marked to accommodate openings in the joist frame for stairs, chimneys, fireplaces, etc. In addition, it must be noted where additional joists should be added to give support to load-bearing walls. To account for added loads, joists can be set side by side or spaced with short blocks.

When openings in the joist system must occur, the effective regular joist spacing will be interrupted. To compensate for this loss of joist coverage, additional joist system members, such as trimmers and headers, must be added to the system. As shown in Figure 15-22, trimmers run parallel to the joist while headers run at right angles to it.

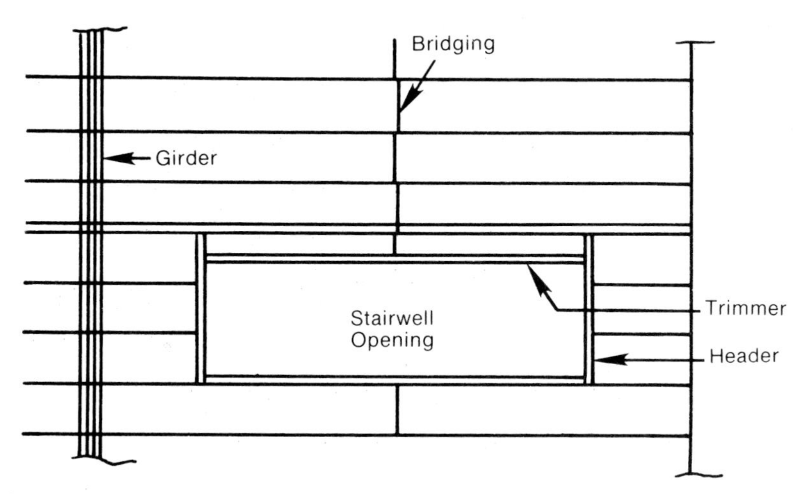

FIGURE 15-22. Trimmers and headers used in framing out openings in the floor frame.

Building codes and regulations determine the required width and headroom of stairs and the clearances which must be placed around chimneys and fireplaces. The openings for these components must be laid out carefully and exactly in accordance with the building plans. Errors are quite expensive to correct.

15.4.2 Joist Bridging

Long joists tend to flex and possibly twist out of shape. To prevent such changes and to provide for better distribution of building loads, bridging is installed. Bridging is of wood or metal and is installed before the subfloor is placed (Figure 15-23).

Nailed bridging is effective in producing a strong and rigid framework. When installed, wooden cross bridging is first cut to fit and then nailed secure at the joist's top edge. Nailing can be started in the lower edge of the bridging pieces, but the bridging can be left hanging and not nailed securely until the building has assumed a considerable load.

Care must be taken to cut bridging to fit. (See Research question 6.) Specifications for bridging can be written out by a knowledgeable layout person so workers on the jobsite can cut all bridging required at one time. Placement of bridging or blocking must also be done with care. Snapping a chalkline across the joists helps assure centered bridging.

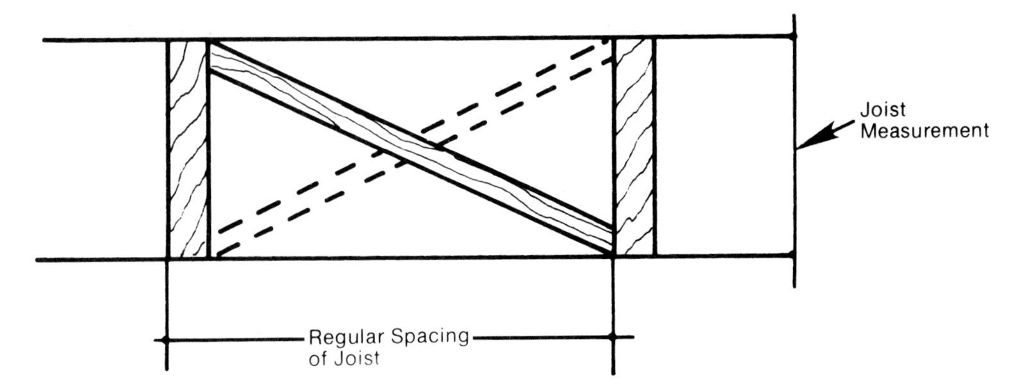

FIGURE 15-23. Supportive bridging between joist members.

The concepts outlined for wooden bridging also pertain to prefabricated metal bridging. Using solid blocking to provide additional support is effective where joist spacing is irregular, such as under bearing walls.

15.5 FRAMING SYSTEMS AND PLATE LAYOUTS

Three major framing systems (western/platform, balloon, and post and beam) were detailed in Chapter 14. Each of these systems contains features that affect how plate layouts are drawn up for the wall framing system. A brief summary of these points is given in the following sections.

15.5.1 Western/Platform

1. Before starting the sole plate layout, the first floor platform is completed with subfloor. Joists, bridging, floor openings, and subfloor are in place and, except for lower ends of bridging, are securely fastened.
2. The perimeter dimensions of the subfloor are the measurements of the perimeter sole plates.
3. After sole plates are marked, the wall frames may be assembled on the platform (or elsewhere) and then raised into place.
4. When perimeter frames and partition frames are in position, platform construction for another story may commence.
5. Sole plate layouts are required for platform framing.

15.5.2 Balloon Framing

1. The perimeter plate layout is of the sill, rather than of the sole.
2. One set of perimeter studs serves more than one story. A ribbon set into the studs carries the joists of the second floor.
3. Partition studs serve one story. (Partition studs and perimeter studs are, therefore, of different lengths.)

15.5.3 Post and Beam Framing

1. There are fewer framing members in post and beam construction than in conventional framing.

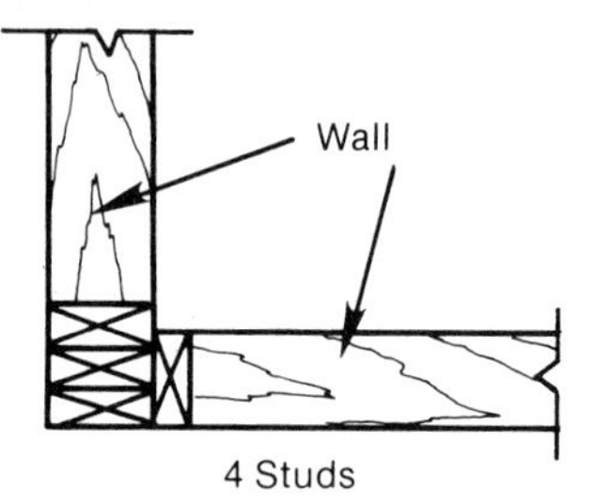

4 Studs

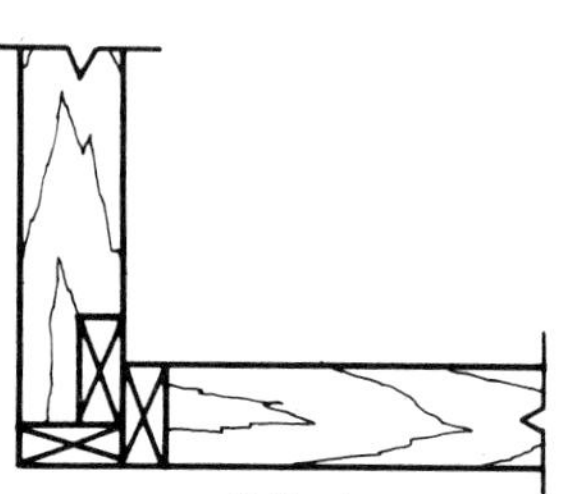
3 Studs

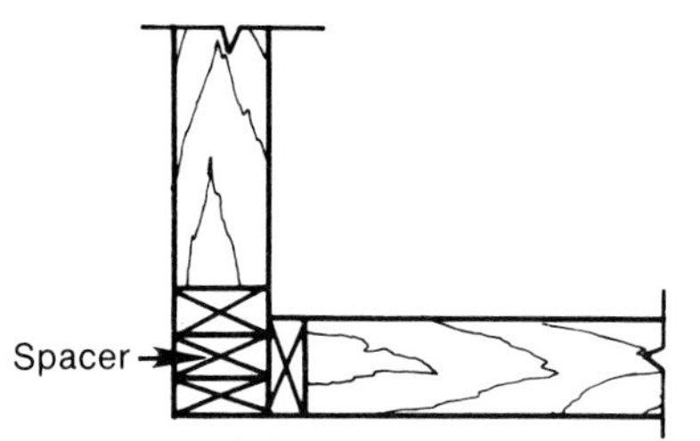

3 Studs and Spacers

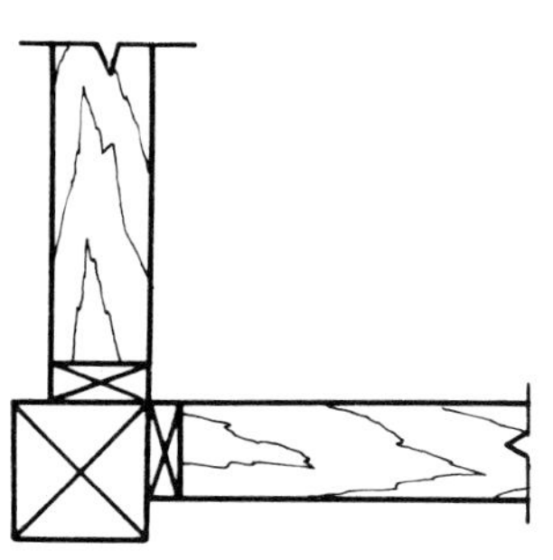
Post and 2 Studs

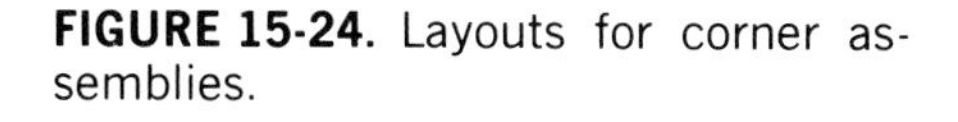
FIGURE 15-24. Layouts for corner assemblies.

2. Partitions do not normally carry direct vertical loads. If they do, however, they should be placed over beams.
3. Post and beam sets are often spaced to take decorative panels of wood or glass or other materials.
4. Properly designed connectors are required to hold the framing elements together.
5. Beams and planks may replace the joists and subfloors of conventional framing.
6. Post and beam framing may be combined with conventional framing.

15.6 PLATE LAYOUTS FOR WALL FRAMING

Marking out the plate layout is the first step in wall framing. Plate layouts should only be undertaken by experienced personnel with a full understanding of the framing system being used on the project. Such layouts can be done on the jobsite or, in the case of prefabricated wall units, at the place of fabrication.

Plate layout involves marking both the sole plate and the top plate. Markings on both plates should match exactly. Both plate members should be marked for the corner assemblies or posts and the position of all regular studs, door/window opening trimmers, partition junctions, and cripple studs of openings.

15.6.1 Corner Assemblies

Corner assemblies provide outside and inside nailing faces for flat walls. As shown in Figure 15-24, corner assemblies can be created from a number of regular studs. In post and beam framing, the post acts as the entire corner assembly.

15.6.2 Stud Spacing for Modular Construction

The spacing of regular studs in the wall framing must allow for modular construction. Because sheathing is available in 4′ (1.2 m) widths, stud spacings of 12″, 16″, or 24″ (305 mm, 406 mm, or 610 mm) are used. Unless modular panels are put between posts, the modular spacing will only apply to outer wall spacing or inner wall spacing, not both (Figure 15-25). As shown, a modular size sheathing sheet would extend from the outside corner to the center of the fourth stud. (The exact same sheet placed on the inside wall would extend beyond the fourth stud to a point where solid nailing would be impossible.) Butt joints for sheathing must be located at studs to provide a solid joint and to prevent lateral movement between the two sheathing pieces. Sheathing can be lapped or unlapped at the corners depending upon the builder's preference. Siding, shingles, or finished brick veneers will cover all corners.

15.6.3 Plate Layout Procedures

Great care should be taken when selecting the stock which will serve as the sole and top plates. Both pieces should be arranged side by side on

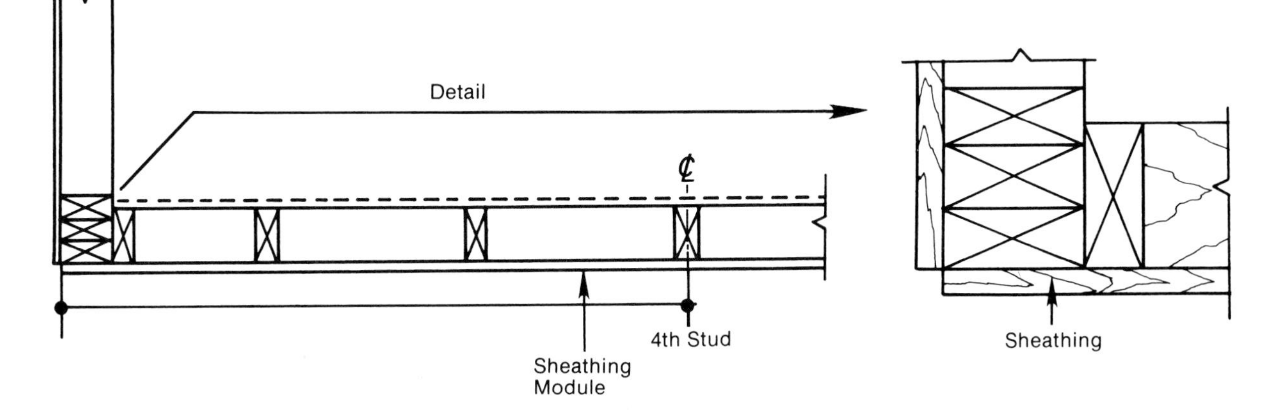

FIGURE 15-25. Proper stud spacing to account for modular sheathing sizes for exterior finishing.

the subfloor. If the size of the building necessitates the use of more than one piece of stock to serve as each plate, butt joints should be used in joining the pieces. Stagger the joints on the top and sole plates to help create a straight and strong wall framework. In conventional framing, a cap plate will also be used to lap the corner joints to help tie side walls to end walls and to strengthen and stiffen the frame.

The sheathing module measurement is often the key measurement for starting the layout. All perimeter regular studs are marked as shown in Figure 15-26A. For lapped corner joints the layout is modified as shown in Figure 15-26B. Once the position of the initial regular stud is marked on the top and sole plates, the layout can continue until places for all regular studs are marked. Keep in mind the following points and considerations:

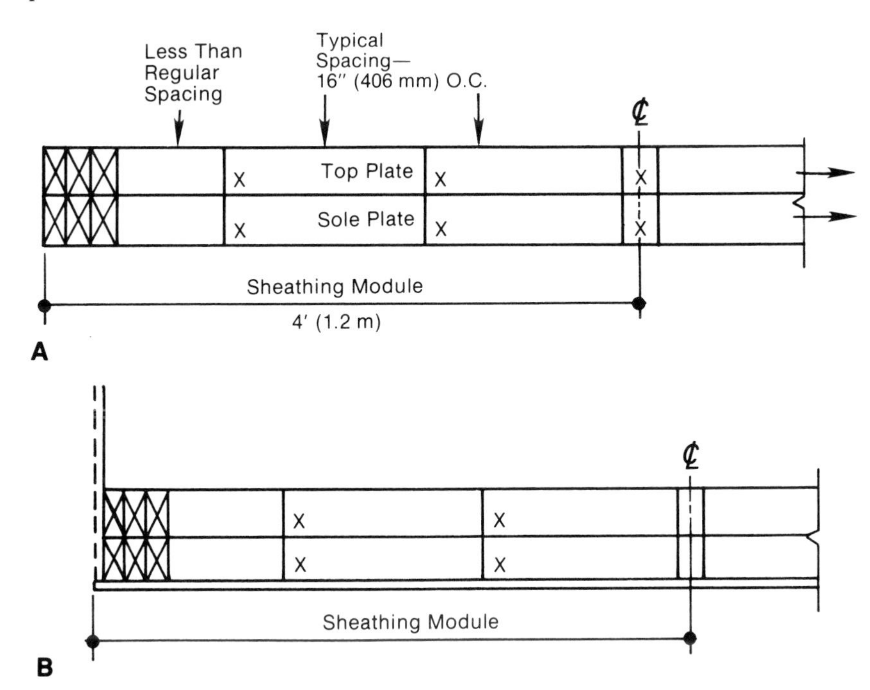

FIGURE 15-26. Starting the plate layout for end-lapping sheathing joints.

1. A left-handed person might work differently and as effectively as a right-handed person.

2. Opposite walls of a rectangular building might be the same, to simplify prefabrication and wall erection. If walls are assembled on the jobsite or on the platform for platform or Western framing, one pair can be tilted and secured in position before the other pair is assembled.
3. For shapes other than rectangular, assembly and tilting sequence need careful planning. It is a matter of jobsite problem solving.

15.6.4 Laying Out Openings in the Wall Frame

Once all places for regular studding have been marked on the top and sole plates, refer to the building plans to locate positions for doors, windows, etc. Openings for doors and windows are located by means of center lines with given distances from building corners (Figure 15-27). The plate layouts for the openings are imposed upon the regular stud layout and must be identifiable. An opening usually requires additional studs, shorter trimmer studs, upper and/or cripple studs, and a lintel. The cripple studs maintain the regular stud spacings whereas trimmers and full-length opening studs are placed to accom-

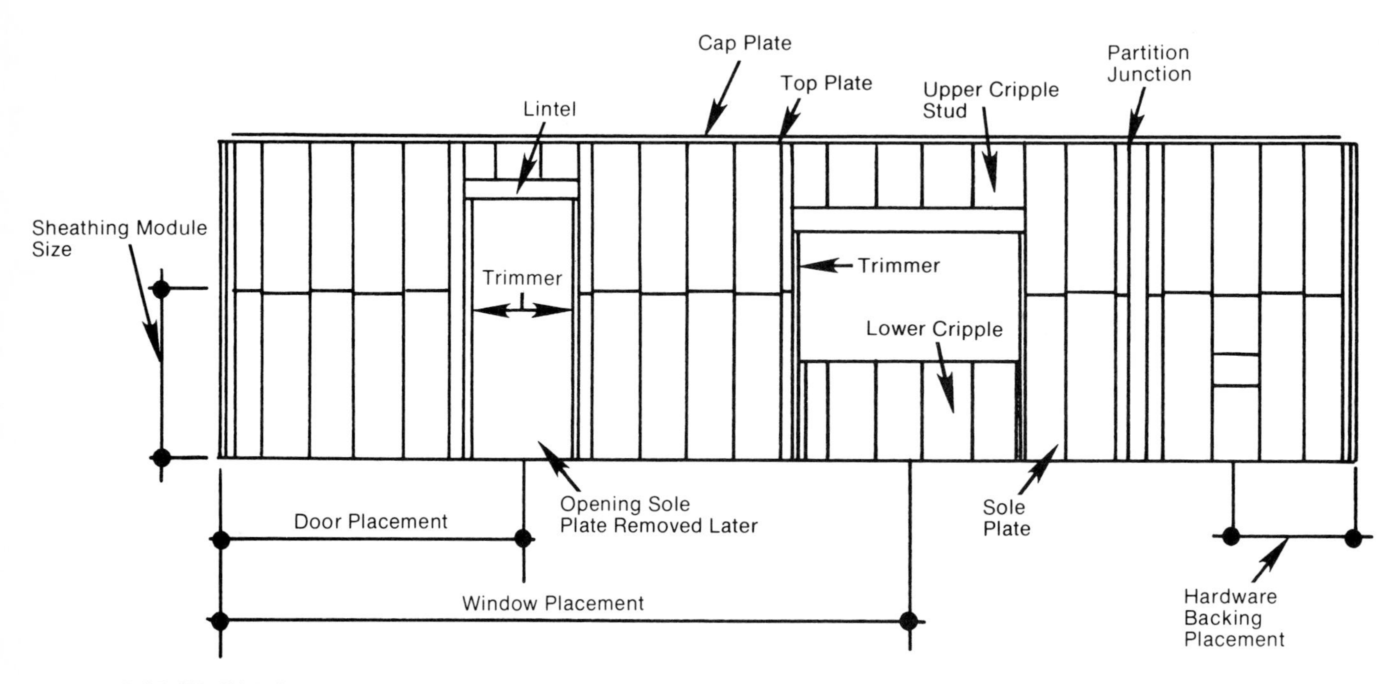

FIGURE 15-27. Wall frame openings for doors and windows.

modate, with recognized manufacturer's clearances, the doorframes or window frames. Fire stops (which serve for nailing of sheathing), wall backing (for hardware installations), and partition junction studs are also located in reference to the building plans. Partition junctions will be above bearing joists in place below the subfloor.

Openings are framed after the doors and windows are known to be on hand and measurements are checked. Recommended clearances between frames and studding are used by the layout person. Consideration is given to spaces for a vapor barrier, plumbing and leveling of the frame, and securing fastenings. Incorrect clearances (too small or too large) may mean wasted installation time later.

15.6.5 Sheathing

When attaching sheathing to the wall frame, care must be taken to produce a wall that is even and true with flush butt sheathing joints (Figure 15-28). Workers must be especially careful not to pull trimmer studs out of position when nailing up sheathing.

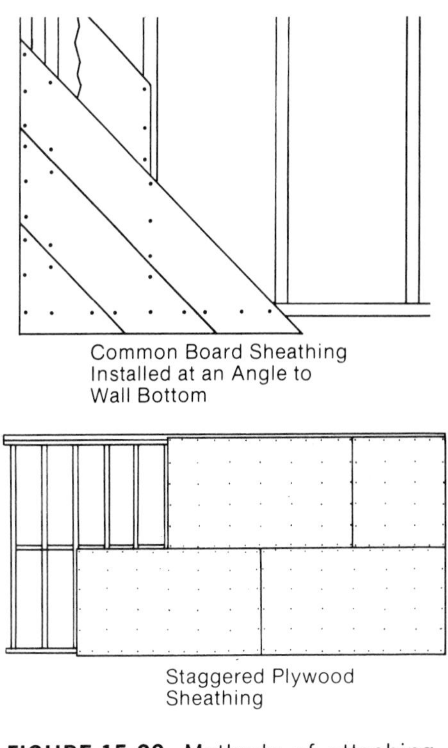

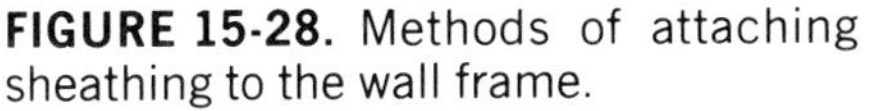

FIGURE 15-28. Methods of attaching sheathing to the wall frame.

15.7 DOUBLE STUD WALL FRAMING SYSTEMS

When a building must be extremely energy efficient, a double stud wall design can be used to provide additional space for insulation. The walls also provide a means of installing the vapor barrier with less risk of puncturing or tearing during the subsequent wiring and plumbing work.

In a double stud wall system, two separate wall frames are used (Figure 15-29). The frames are tied or held together a set distance apart by top and bottom plywood ties which are fastened to the top and sole plate members of both frames (Figure 15-30).

The inner wall in the system is the structural wall. It carries the building load as would a conventional wall frame. It also carries the sheathing, the vapor barrier, some of the insulation, and all of the wiring and plumbing.

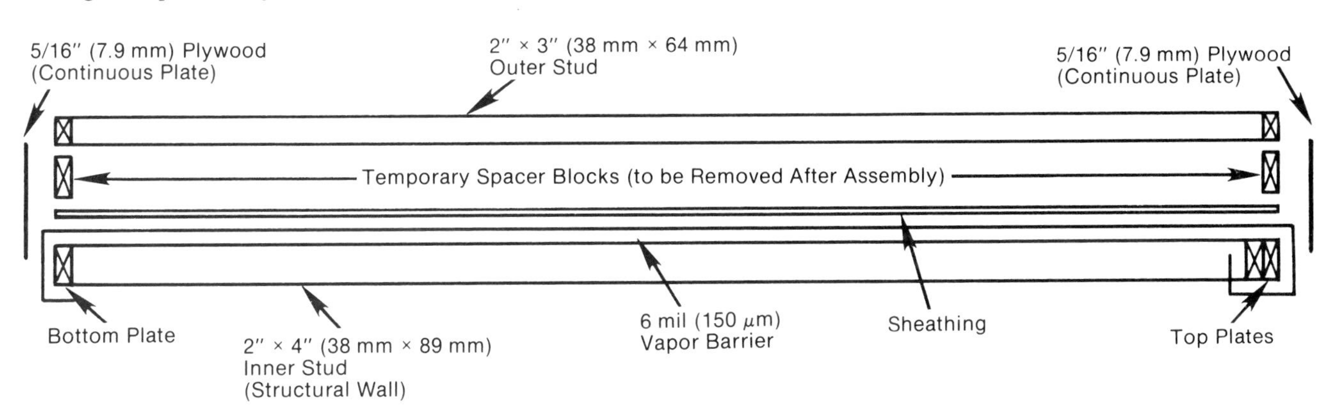

FIGURE 15-29. Double stud wall assembly.

The outer wall is of lighter construction. It serves to carry the exterior siding, shingles, or brick veneer that protects the inner wall and gives space for more insulation.

The width of the plywood tie plates determines the double wall thickness. Temporary spacer blocks are often used during assembly. These are removed when the plywood ties are in place. The basic construction sequence for a double stud wall frame is as follows:

1. Build the inner stud wall (the structural wall for the building) in the conventional manner, but use a double top plate as shown in Figure 15-29.
2. Place 6 mil vapor barrier on top; then nail sheathing in the conventional manner. Take care not to damage the vapor barrier when cutting the sheathing.
3. Build the outer stud wall on top of the inner wall. (To save time mark plates for both the inner and outer walls at the same time.)

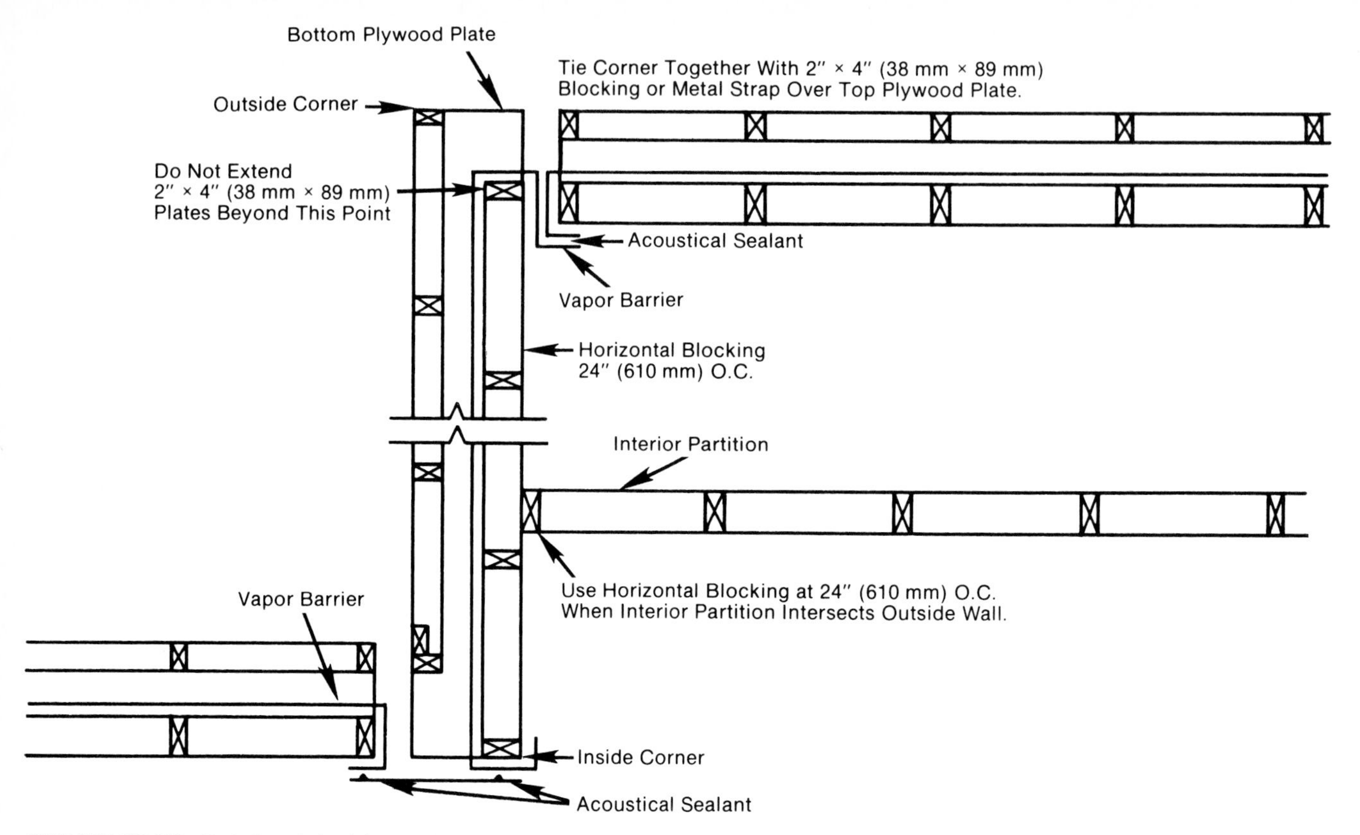

FIGURE 15-30. Details of double stud wall assembly.

4. Using temporary spacer blocks to separate the two walls, nail the walls together using the top and bottom plywood tie plates.
5. Tile the secured wall section up and put it in place. Staple vapor barrier flaps to the inner face of the top and bottom plates to prevent damage. The wall can now be wired and insulated when desired. Since the inner stud is on the inside of the vapor barrier, electrical wires may be run through the studs without penetrating the vapor barrier.

15.8 PLATE LAYOUT OF PARTITION WALLS

Before laying out the plates for partition walls, examine the building floor plan and identify the following:

1. bearing and nonbearing partitions,
2. regular and wide walls (for stacks, sliding doors, and staggered stud walls for sound controlling),
3. longest partitions (to be constructed first),
4. next longest partitions (for developing a construction sequence),
5. locations of openings for ducts, and
6. backings for plumbing fixtures.

Bearing partitions must be supported under the subfloor by properly placed joists or beams, and the superstructure load must be transferred to the footings.

Chalk lines are used to establish partition sole plate locations on the subfloor and to test a finished framed wall for straightness. To produce a properly snapped, straight line, the line must be well chalked, held taut in the proper position, and pulled up vertically from the subfloor when snapping the line. When the wall is ready to be fastened to the subfloor, it should first be checked for straightness. To do this, use three spacer blocks of the same thickness to check the frame against the chalk line snapped on the subfloor (Figure 15-31).

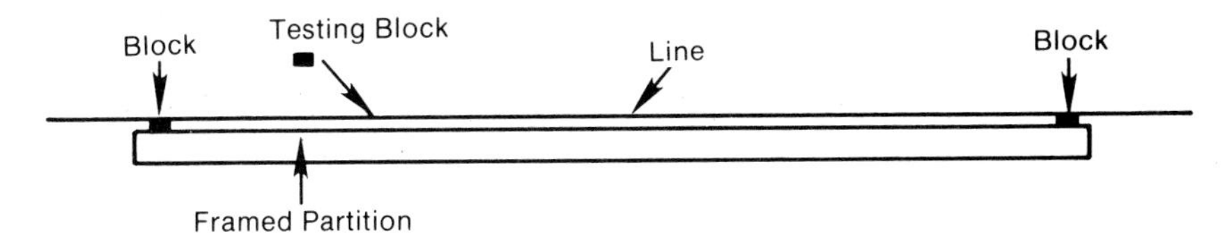

FIGURE 15-31. Using testing blocks and a chalk line to test for partition wall straightness.

As shown in Figure 15-32, the layout for the junction of a partition sole plate and a perimeter sole plate must provide for strong construction and proper nailing or fastening surfaces for interior paneling or finishing materials.

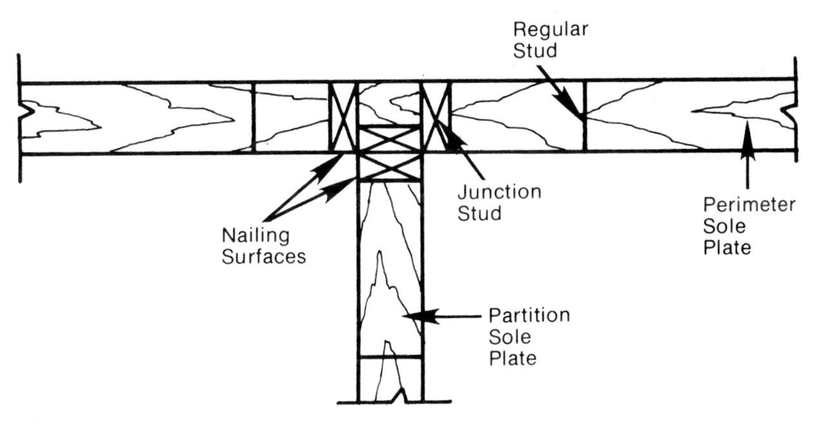

FIGURE 15-32. Junction of partition and perimeter wall.

15.9 CAP PLATES

Cap plates contribute to the load bearing capability of wall frames. They also serve to tie frames together by means of well-fastened lap joints (Figure 15-33). As shown, cap plates can be fastened with nails

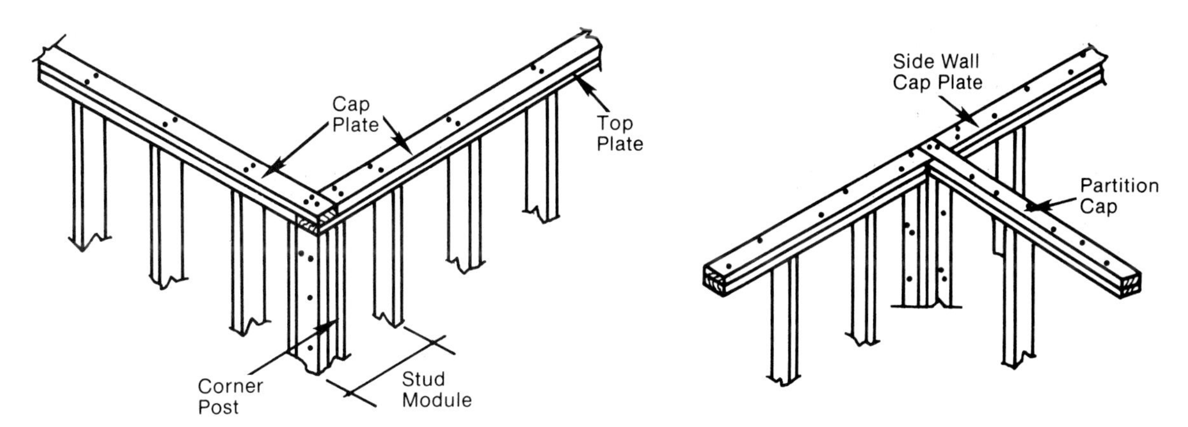

FIGURE 15-33. Lapped joints of cap and top plates in corner and partition wall functions.

although special metal framing anchors or plates can also be used (See Chapter 3).

15.10 FRAME OPENINGS FOR THE HEATING SYSTEM

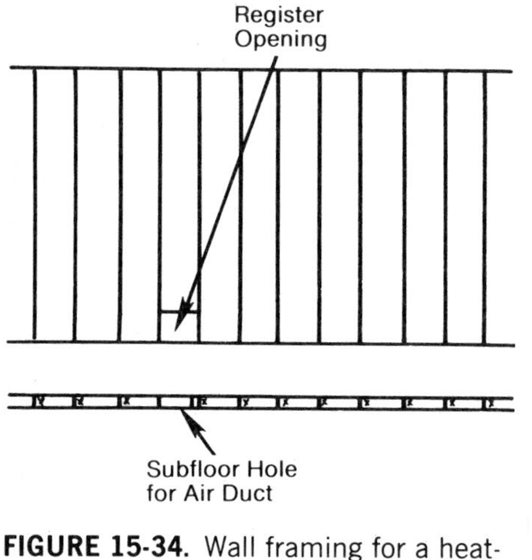

FIGURE 15-34. Wall framing for a heating register.

Before openings are cut for heating ducts and registers, details with respect to the heating system must be known. Delivery of the system elements to the jobsite should have been made. The building plans should clearly show the type of system and installation requirements.

Consider the differences among perimeter hot water, gravity type hot air, forced hot air, electrical, and solar heating (passive and active). All have different piping/ducting requirements.

A forced air system using a natural gas fired furnace usually requires hot air ducts leading to carefully placed registers and at least one cold air return duct. Some installations provide for preheating of fresh air from outside rather than simple circulating of in-building air. Some designers prefer floor registers to wall registers.

Framing for a wall register, as illustrated in Figure 15-34, allows hot air to move to two rooms. It has the disadvantage of being a sound carrier from room to room.

15.11 PLUMBING INSTALLATIONS AND BACKING

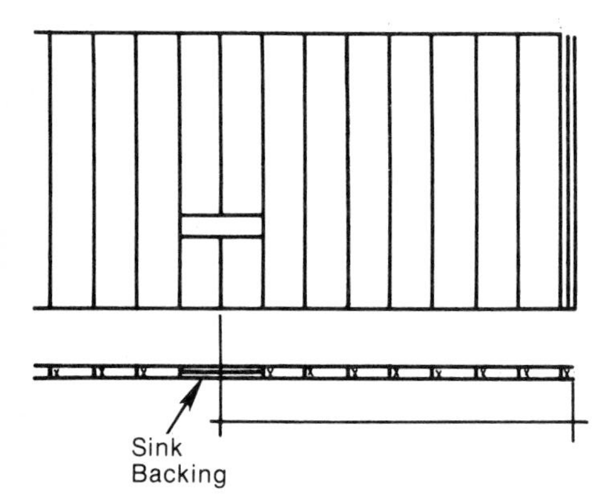

FIGURE 15-35. Plate layout with backing piece for additional sink support.

As with the heating system, plumbing fixtures should also be delivered to the jobsite. All installation instructions given by the manufacturer of the fixtures should be followed. Many plumbing units are now free-standing and do not require wall backing for additional support. However, care must be taken not to miss installing backing where it will be required for sinks, wash basins, towel racks, hanging cabinets, and shelving. Plate layouts may mark the locations. Figure 15-35 is an example of backing built into the wall frame.

Backing provides a fastening surface on a somewhat more flexible basis than standard wall frame members. It can serve new construction and possibly later renovations. Backing installations are easily accomplished during the framing stages but not later in the construction process. Careful planning is needed to assure all backing is installed at the most opportune time.

15.12 CEILING AND ROOF FRAME LAYOUT

Once the framing of each wall section is complete, the frames can be sheathed, tilted into place, and fastened. Fastening systems vary from region to region since they are closely related to the types of weather conditions to which the building will be subjected. When the wall frames are in place, ceiling and roof framing can begin.

In a typical one-story building, the cap plate serves as a nailing surface for the ceiling joists, rafters, and ceiling backing. (Backing is needed to provide a fastening surface for ceiling panels.) In some building designs, trusses may replace rafters and ceiling joists. To create an open ceiling in post and beam framing, posts, ridge beams, and roof beams may be used in place of rafters and joists (or trusses).

Assume that the cap plates are to carry ceiling joists, rafters, and backing and that the ceiling joists are to be placed (insofar as is possible) as lower chords for pairs of rafters. Therefore, for a small gable roof (Figure 15-36)

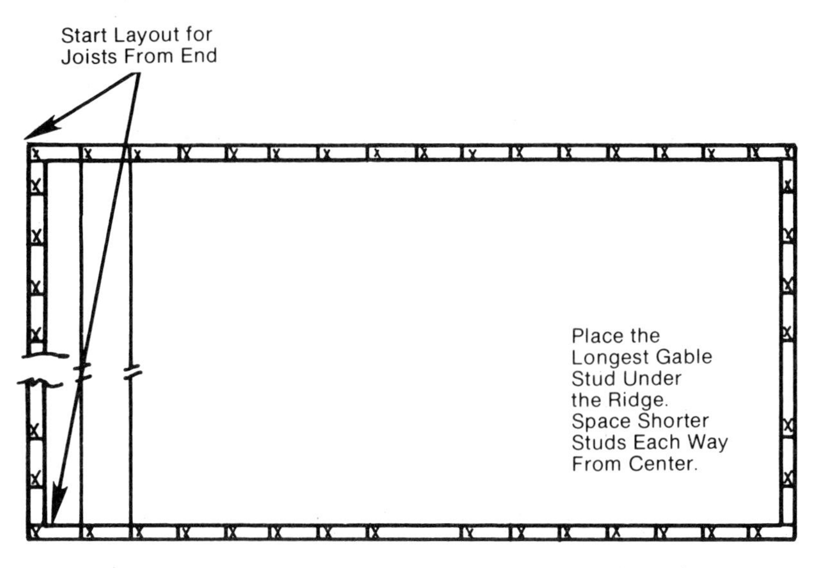

FIGURE 15-36. Typical ceiling joist layout for a small gable roof.

1. the cap plate layout is for two matching sides,
2. the gable stud layout is for two matching ends, and
3. the spacing module will be determined by the
   a. roof design load for snow and wind;
   b. size of sheathing, in terms of length, width, and thickness; and
   c. fastening system.

For a larger gable roof, where small rooms and one larger room are required below, a bearing wall may be used in the small room area and a beam resting on cap plates may be used for the larger room (Figure 15-37). Ceiling joists may

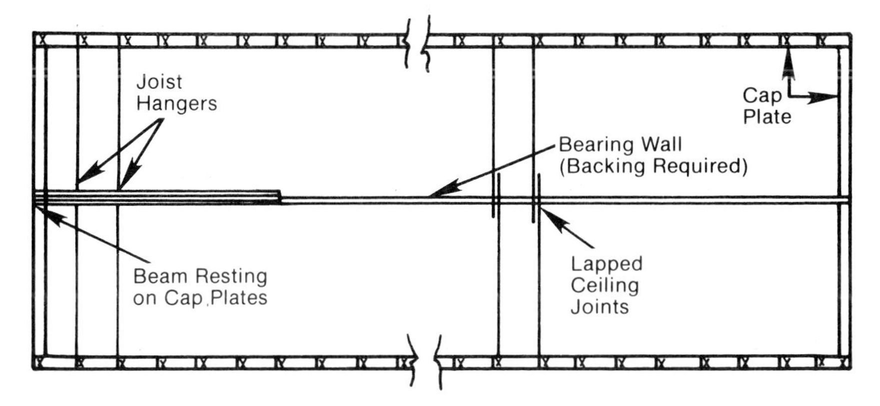

FIGURE 15-37. Typical ceiling joist/cap plate layout for larger gable roofs.

1. lap at bearing walls,
2. butt at the beam (by means of metal hangers),
3. serve as lower chords of trusses, and
4. assume modular spacings such as 16″ or 24″ (406 mm or 610 mm).

Generally roof loads tend to push walls outward where rafters bear on the cap plates. By using ceiling joists as ties, as chords of trusses, or as members of truss/cornice assemblies, the structure can be made secure. Roof framing systems are designed to help solve such problems. Each cap plate layout embodies a solution.

Ceiling joists may run in two directions (Figure 15-38), but

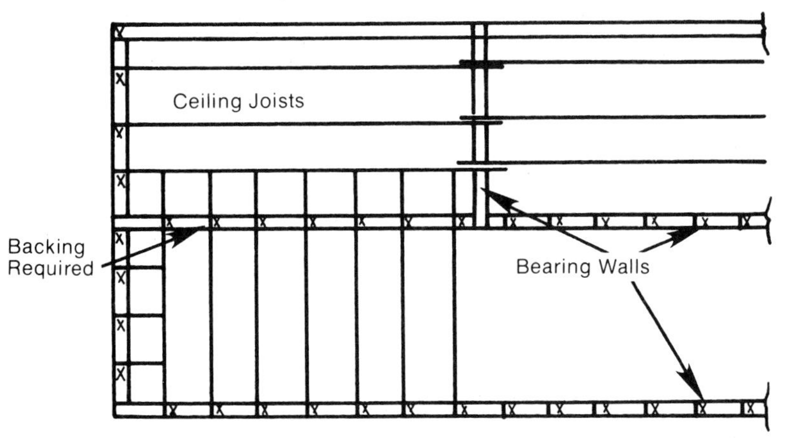

FIGURE 15-38. Ceiling joist layout with joists running in two directions.

1. backing needs careful attention if joists cross a bearing wall,
2. sloped roof construction becomes more difficult, and
3. flat roof construction seems more desirable.

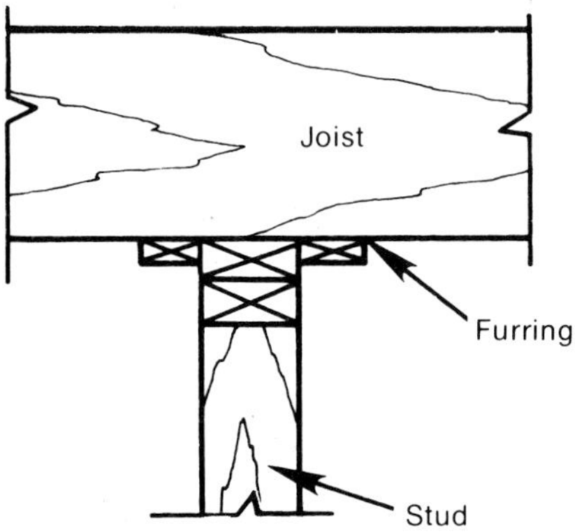

FIGURE 15-39. Furring used as backing beneath ceiling joists.

15.12.1 Furring as Ceiling Backing

Where backing may become a larger than desired task, furring can be installed. After the ceiling joists are placed 1″ × 2″ (19 mm by 38 mm), furring pieces are fastened underneath and at right angles. The pieces serve as ties, as nailing strips for ceiling panels, and as backing between joists at the cap plates (Figure 15-39).

Furring is also used instead of blocking/bridging of ceiling joists. Properly fastened strips will prevent joists from flexing and, to a large extent, from twisting, especially along the bottom edges.

Backing, if cut to fit between joists and to overhang the cap plate, is useful as joist spacers (Figure 15-40).

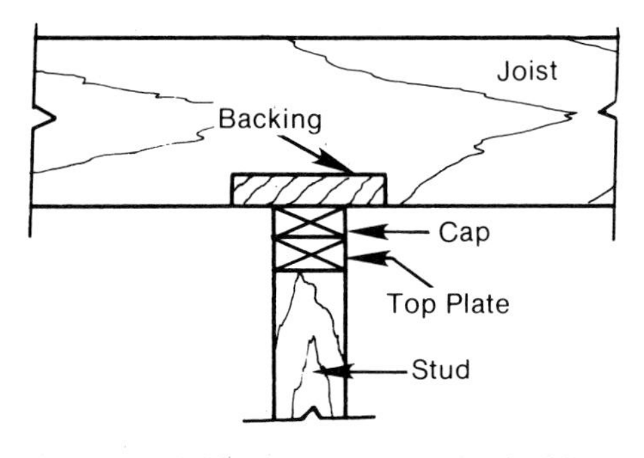

FIGURE 15-40. Alternate method of installing furring backing.

15.12.2 Regular Hip Roof Layout

The following section describes a layout for a typical hip roof design. In this layout the cap plates carry the layout for rafters or trusses, but not studs. Figure 15-41 illustrates the details of this sample layout. The following is a list of factors or data which will affect the layout:

1. Hip rafters cross the corners of the cap plates.
2. Two sets of three common rafters meet hip rafters at the ends of the ridge.
3. Jack rafters are cut in sets. Each set includes left and right subsets.

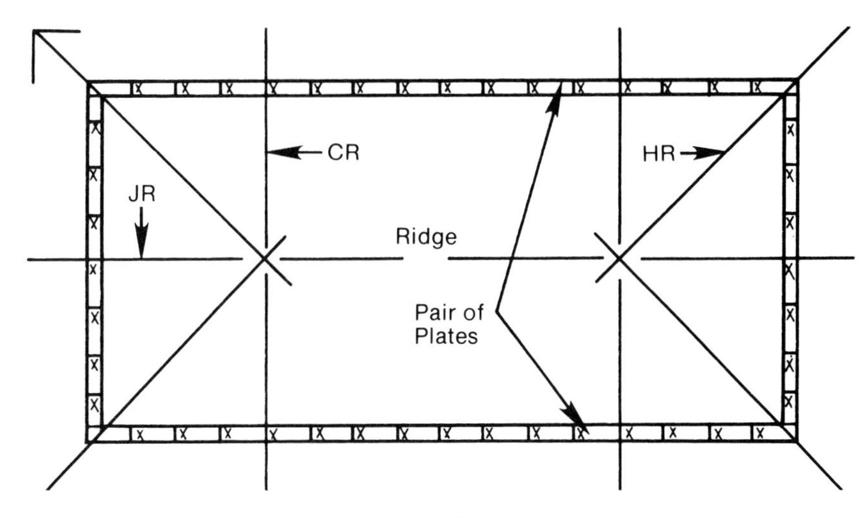

FIGURE 15-41. Layout of simple hip roof.

4. Common rafters, of the same pattern as key common rafters, are spaced on a modular basis to take sheathing.
5. Trusses or partial trusses may be substituted for rafters.
6. Wider spacing of rafters indicates thicker sheathing to carry the loads.
7. The ridge board carries the same CR markings as the cap plates.

These factors, if accepted, become layout constraints.

If spacing is 16″ or 24″ (406 mm or 610 mm) to receive standard-sized sheathing, then it is an on-the-job problem solving situation to do the plate layout to the best advantage. As most rafters overhang the cap plates, cornice details need to be at hand.

A similar listing of data and assumptions can be developed for intersecting roofs of equal slopes (Figure 15-42).

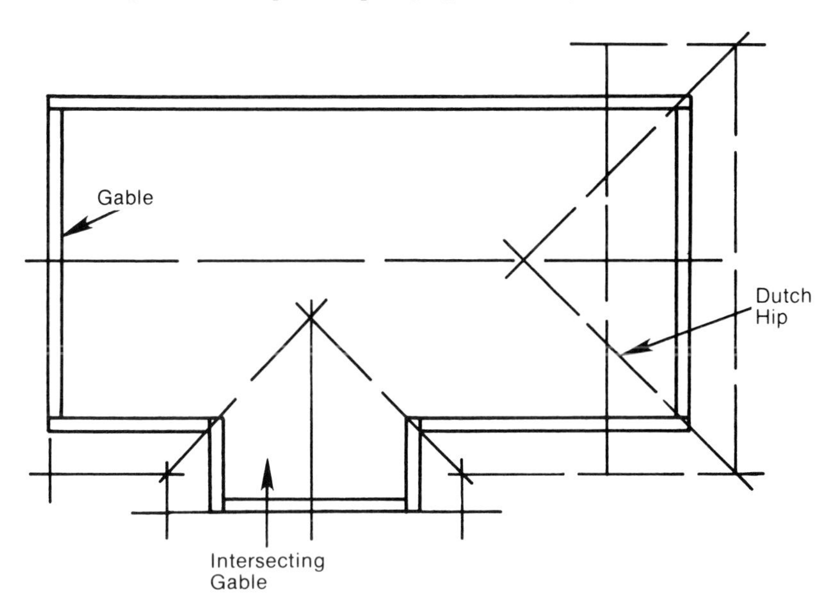

FIGURE 15-42. Layout of cap plates for intersecting roofs of equal slope.

1. The cap plates are in place for a roof of which the major part is gable and dutch hip and the minor part is gable only.

2. The major part will be framed with gable trusses and a set of dutch hip-end trusses.
3. The minor part will be framed with gable trusses and on-the-job framing of the valley (after sheathing of the major part).
4. The spacing of trusses module is 16″ (406 mm), where possible.

For any ceiling or roof layout, dimensions may be assigned to the wall frame and a trial cap plate layout can be made. This trial layout is particularly helpful in estimating the cost of roof framing and sheathing. Practical revisions should be made to sample layouts prior to beginning construction of the ceiling or roof frame.

15.13 RESEARCH

1. With the use of dimensioned sketches, report on the placement of footings for a light construction building. Observe the actual placement, inclusive of how measurements are taken. Compare your notes with those on the project plan. Be specific about the tasks of plan interpretation by the builders.
2. Observe the concrete foundation wall forms for the same building or another light construction building. Dimension your sketches to show sizes and placement of materials. Give technical reasons for the materials selection and placement.
3. For a light building under construction describe the floor joist system by means of sketches. Indicate with arrows how the load above the first subfloor is to be transferred to the footings.
4. For the same building, name and describe the wall framing system. Include subtopics such as perimeter plate layout and partition plate layout.
5. For the same building, describe the cap plate layout for the roof framework. Identify the members used. Write the technical rationale for member spacings.
6. Refer to Figure 15-23. Observe that point A is at the upper edge of one joist and that point B is not at the lower edge of the adjacent joist. Using a framing square, draw up a general rule for the proper cutting of wooden joist bridging.

15.14 REFERENCES

15.14.1 Smith, Ronald C. *Principles and Practices of Light Construction,* Third Edition. New Jersey: Prentice-Hall, Inc., 1980.

15.14.2 *Energy Efficient Housing-A Prairie Approach.* Saskatchewan Mineral Resources Office of Energy Conservation, October, 1980.

16

ROOF FRAMING

This chapter discusses the graphic and framing square methods of roof framing. The graphic method, with its direct approach and limited use of mathematics, has long been utilized by laymen. Though faster, the framing square method is more involved technically and, therefore, is used predominantly by carpenters. By learning the principles and practices of both methods, a full knowledge of roof framing can be gained.

16.1 GENERAL OBJECTIVES

The final step in completing the skeleton of a building is the framing of the roof. In order to complete this task, the builder must: understand framing theory and know how the layout rules were developed, be able to make development drawings, be able to calculate for measurements and cutting angles, and be able to cut the framing members and erect the roof. All of these aspects of roof framing will be covered in this chapter, with emphasis placed on developing roof frames that are both functional and attractive.

By using the graphic and framing square methods, the builder can complete a roof frame for a building regardless of its size, shape, and projected slope. All of the typical framing elements are used, including rafters, sheathing, purlins, and windblocks. In addition, both the British Imperial and Metric systems of measurement are applicable. While there are simpler ways to complete this task, the graphic and framing square methods provide the most thorough understanding of roof framing.

16.2 SPECIFIC OBJECTIVES

On completion of this chapter the reader should

1. have a working knowledge of roof framing from the standpoint of both the graphic and framing square methods,
2. be able to sketch development drawings of buildings,
3. be prepared to explain and demonstrate the use of framing squares and pitch boards, and
4. be able to layout and cut members for a given roof framing job.

16.3 TYPES OF ROOFS

A roof frame can take one of several of the following common designs:

- *Shed*—This roof slopes in one direction only, with the slope extending the width of the building. It is the easiest roof to build (Figure 16-1).
- *Gable*—This roof has two slopes, much like having two shed roofs placed together (Figure 16-2).
- *Hip*—This roof has four slopes. If it is a square hip, the slopes terminate at a point; if it is a rectangular hip, the slopes terminate in a ridge (Figure 16-3).
- *Gambrel*—This roof is basically a modification of the gable, with more than one slope on each side (Figure 16-4).
- *Intersecting*—This roof is formed by the meeting of two sloped roofs of one type or another (Figure 16-5).

16.4 ROOF TERMINOLOGY

In order to be able to understand and solve the problems involved in roof framing, familiarity with the common terminology is a prerequisite. A plan view of a roof frame (Figure 16-6) provides an excellent look at the different types of rafters.

Hip rafters extend from the corners of the wall plates to the ridge at a 45° angle to the plates; they form the intersection of two adjacent roof surfaces. Common rafters run at right angles to the plates and connect with the ridge at their top end. Valley rafters occur where two roofs intersect.

Jack rafters run parallel to the common rafters, but are shorter; they are named according to their position. Hip jacks extend from the wall plates to the hip rafters and are normally laid out in pairs. Valley jacks extend from the valley rafters to either the ridge or the wall plates. Cripple jacks extend from the hip rafters to the valley rafters or from one valley rafter to another; these rafters contact neither the ridge nor the plates.

The part of a rafter with a plumb and level cut that is secured to the wall plates is known as the birdsmouth. For measurement purposes,

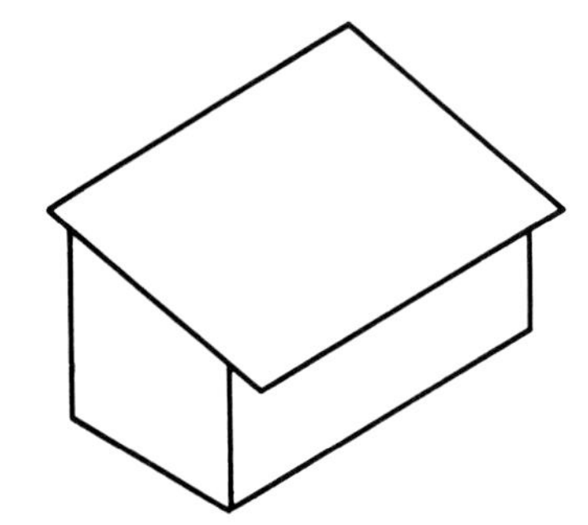

FIGURE 16-1. Shed roof.

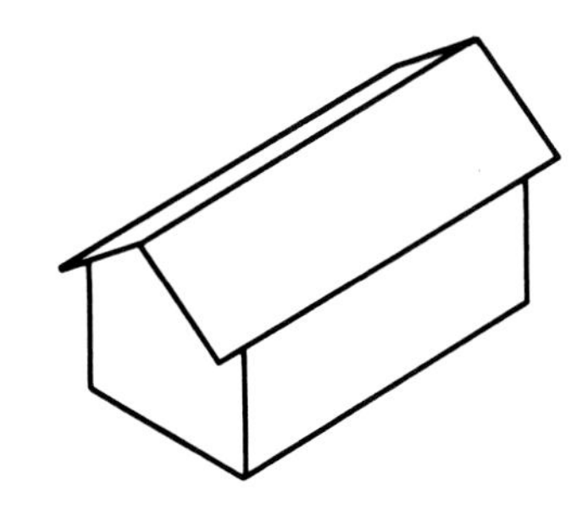

FIGURE 16-2. Gable roof.

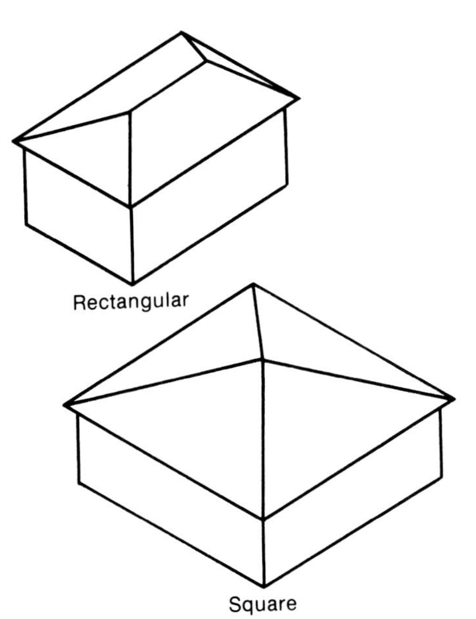

FIGURE 16-3. Two types of hip roof.

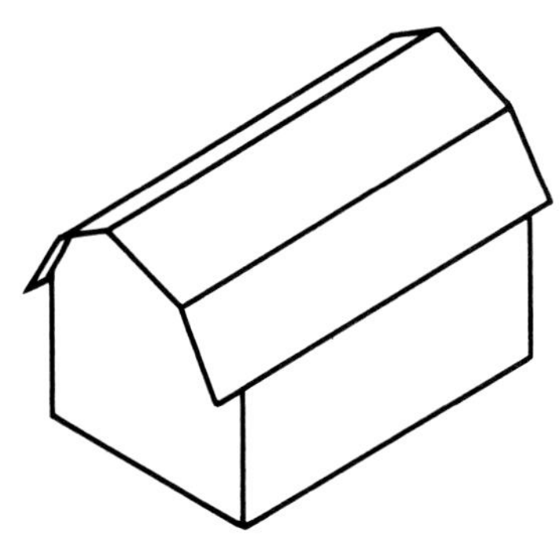

FIGURE 16-4. Gambrel roof.

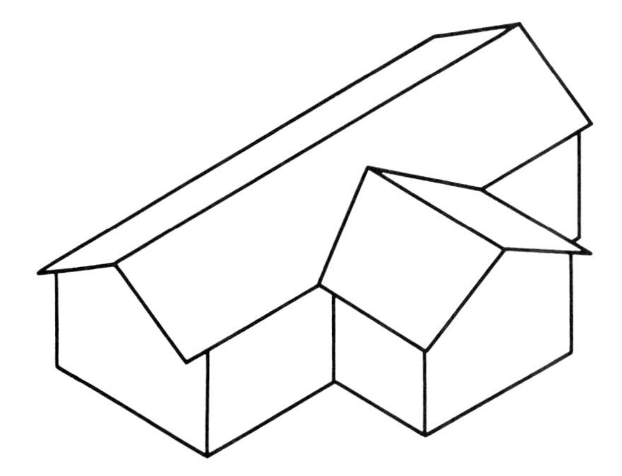

FIGURE 16-5. Intersecting roof.

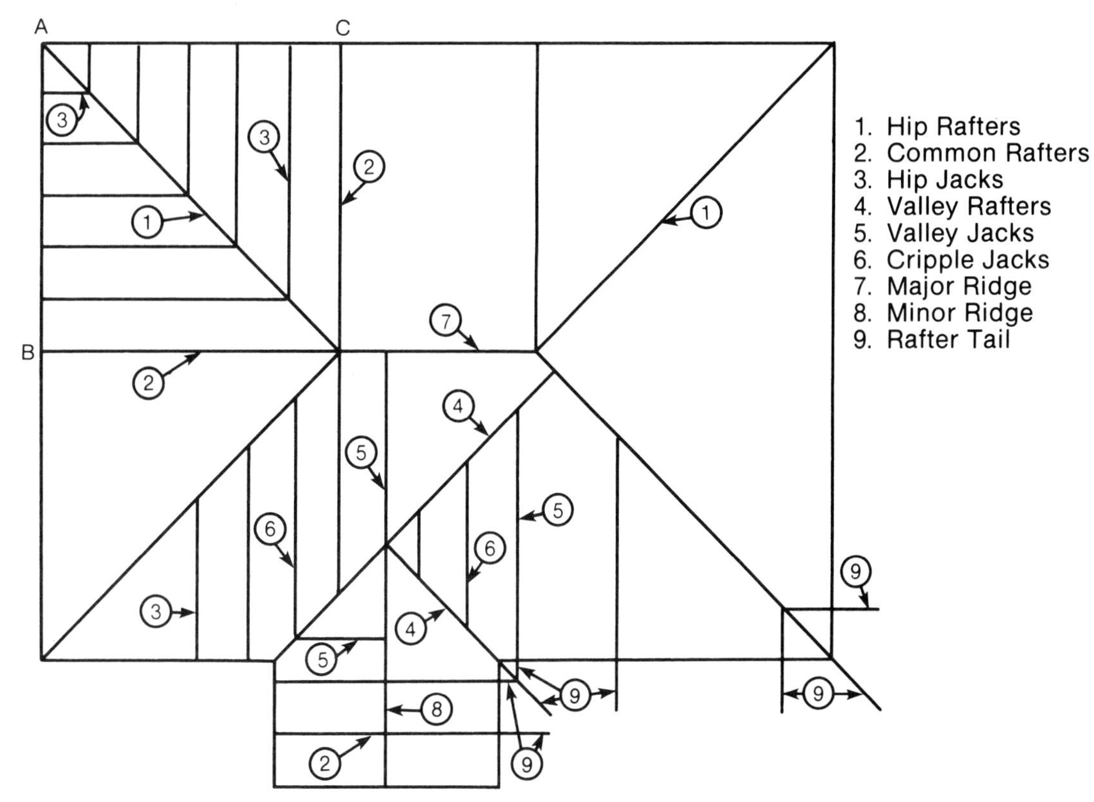

FIGURE 16-6. Plan view of a roof.

any vertical line on a secured rafter is called a plumb line and any horizontal line is a level line. The part of a rafter that extends beyond the wall plates is known as the rafter tail.

The following definitions apply to regular, or equal-sloped, roofs (Figure 16-7):

- *Span*—The dimension of a building bridged by a pair of common rafters; considered to be the width of the building.
- *Run*—The distance covered by one common rafter; half the span.
- *Rise*—The height of the roof as measured from the tops of the wall plates to the ridge.
- *Pitch*—The slope of the roof; the ratio of the rise to the span.
- *Line length*—The length of a rafter as measured on a line through the outside edge of the building frame.

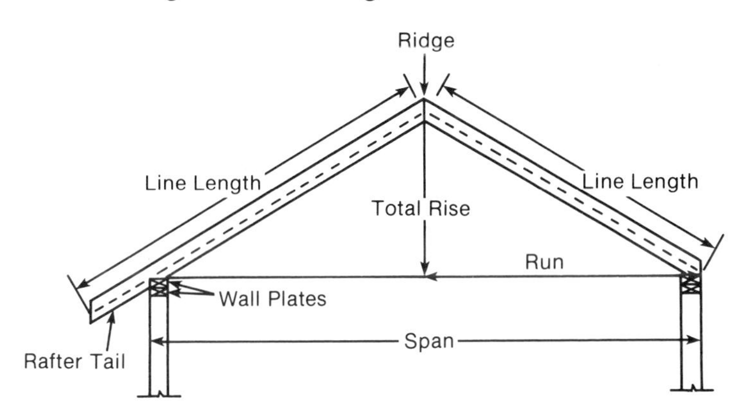

FIGURE 16-7. Equal-sloped roof.

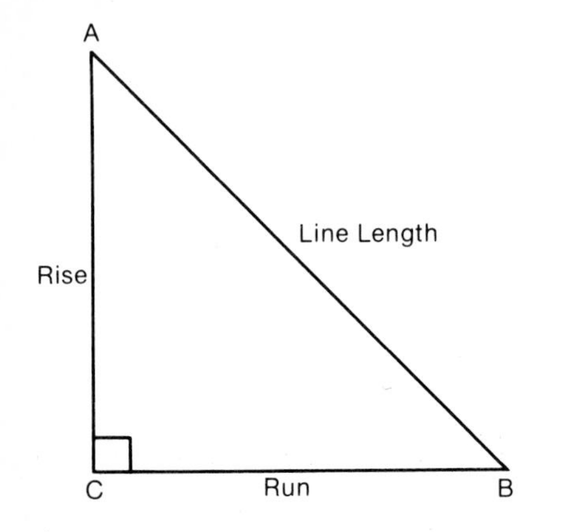

FIGURE 16-8. Roof framing as applied to right-angle triangulation.

16.5 ROOF FRAMING PRINCIPLES

The basis of roof framing is right-angle triangulation. Thus, a logical method of study of roof framing is to analyze it in terms of right triangles which describe, on a scaled basis, the rise, run, and line length of each framing member.

If two sides of a right triangle are known, the third can be found using the Pythagorean theorem: The area of the square formed by the side of the hypotenuse of a right triangle is equal to the sum of the areas of the squares formed by the other two sides of the triangle. The following formulas can be deduced from the theorem:

$$\text{Hypotenuse} = \sqrt{\text{Base}^2 + \text{Altitude}^2}$$
$$\text{Base} = \sqrt{\text{Hypotenuse}^2 - \text{Altitude}^2}$$
$$\text{Altitude} = \sqrt{\text{Hypotenuse}^2 - \text{Base}^2}$$

In roof framing, the base of the right triangle is called the *run*, the altitude is called the *rise*, and the hypotenuse is called the *line length* (Figure 16-8). For example, if the run of a given building is four units and the rise of the roof is three units, the line length of the roof rafters can be determined by using the formula for calculating the hypotenuse of a right triangle (Figure 16-9).

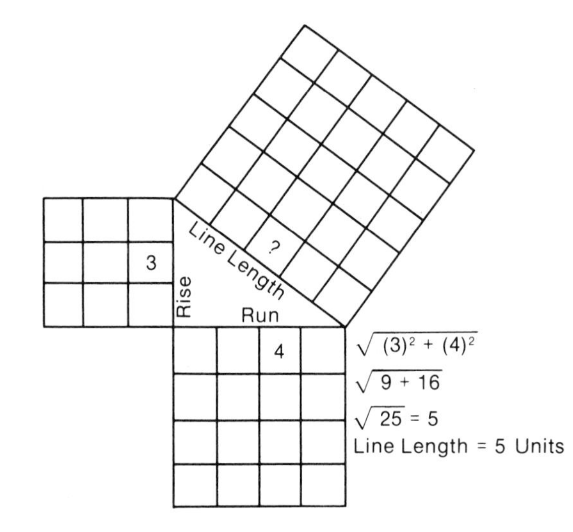

FIGURE 16-9. Calculating line length of a common rafter.

In order to make rafter layout simpler, run, rise, and line length are broken down into units. The basic one is the unit run, a standard length of 10″ (250 mm). The unit rise is the vertical distance which a rafter rises for one unit of run, while the length of the rafter resulting from one unit of run is the unit line length.

16.5.1 Span

To determine the span of a building, divide the total rise by the pitch, as shown:

Rise = 6′ Pitch = 1/4

$$6 \div 1/4 = \frac{6 \times 4}{1} = 24$$

Span = 24′

16.5.2 Rise

To determine the rise of a roof, multiply the span by the pitch, as shown:

Span = 24′ Pitch = 1/4

$$24 \times 1/4 = \frac{24 \times 1}{4} = 6$$

Rise = 6′

16.5.3 Pitch

To determine the pitch, or slope, of a roof, divide the total rise by the span, as shown:

Rise = 6′ Span = 24′

$$6 \div 24 = \frac{6}{24} = 1/4$$

Pitch = 1/4

16.6 PLANS, SPECIFICATIONS, AND DEVELOPMENT DRAWINGS

To be able to construct even a relatively simple shed roof, a builder must know the intentions of the designer or architect. Plans are the drawings of the intended structure; specifications are the written details about the intended structure. When read together, they give the full intentions of the designer or architect.

Figure 16-10 shows the plan and elevations of a shed. The specifications are as follows:

- Plan view—rectangular, 12′ (3.7 m) × 20′ (6.1 m)
- Wall height—8′ (2.4 m)
- Pitch—1/3
- Overhang—12″ (305 mm)
- Rafters 16″ (406 mm) on center (O.C.)
- Rafters made of 2″ × 4″ (38 mm × 89 mm)
- Rafters at top to be secured to a piece of 1″ × 6″ (25 mm × 140 mm) nailed to the existing wall.
- Exposed rafter tail to be plumb and level cut.

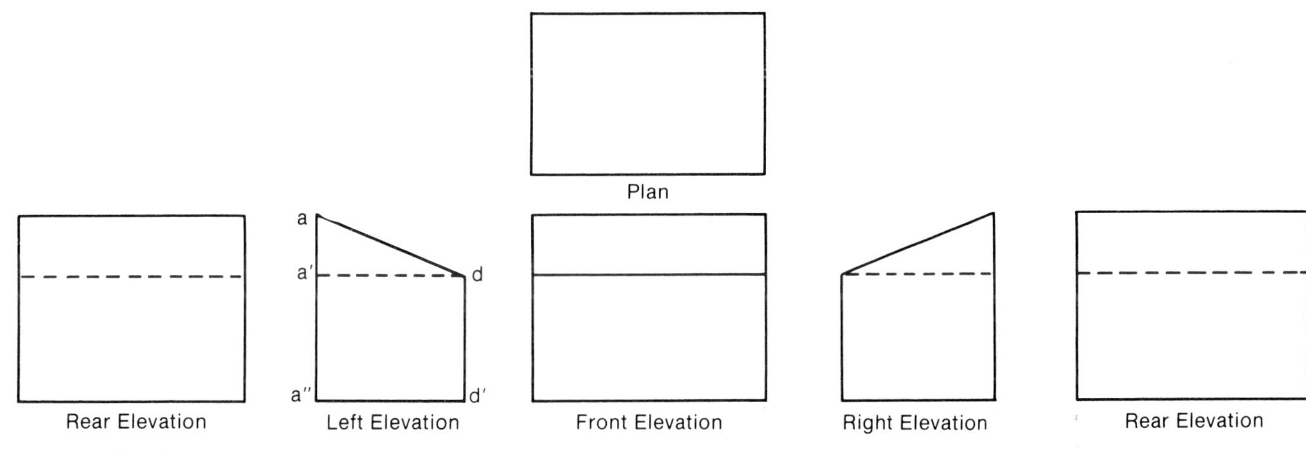

FIGURE 16-10. Views of a shed.

16.6.1 Development Drawing

Figure 16-11 illustrates how a development drawing is made from the existing plans and specifications. Keep in mind that accuracy is very important when making development drawings. For this particular shed construction, the process is as follows:

1. On building paper draw to the scale of 1/4″ to 1′ (6.4 mm to 305 mm) a plan view and an attached end view of the roof.
2. It can be seen that a′-b-c-d is the plan of the building. In addition, a′-d is the 12′ (3.7 m) run over which the common rafter a-d passes; a-a′ is the 8′ (2.4 m) rise; and a-d is the line length of the common rafter.
3. Because a-a′-d is a right triangle, the line length of the common rafter can be determined using the formula: Hypotenuse = $\sqrt{\text{Base}^2 + \text{Altitude}^2}$. This calculates the line length to be 14′5″ (4.4 m).
4. Upon further examination, it can be seen that a′-b measures 5″ (127 mm) and represents 20′ (6.1 m), the actual length of the building; a′-d measures 3″ (76 mm) and represents 12′ (3.7 m), the actual width of the building; a-a′ measures 2″ (51 mm) and represents 8′ (2.4 m), the actual rise of the roof; and a-d measures 3.6″ (91 mm) and represents 14′5″ (4.4 m), the line length of the common rafter.
5. By utilizing the a-a′-d triangle, the common rafters can now be measured, cut, and laid out (Figure 16-12).

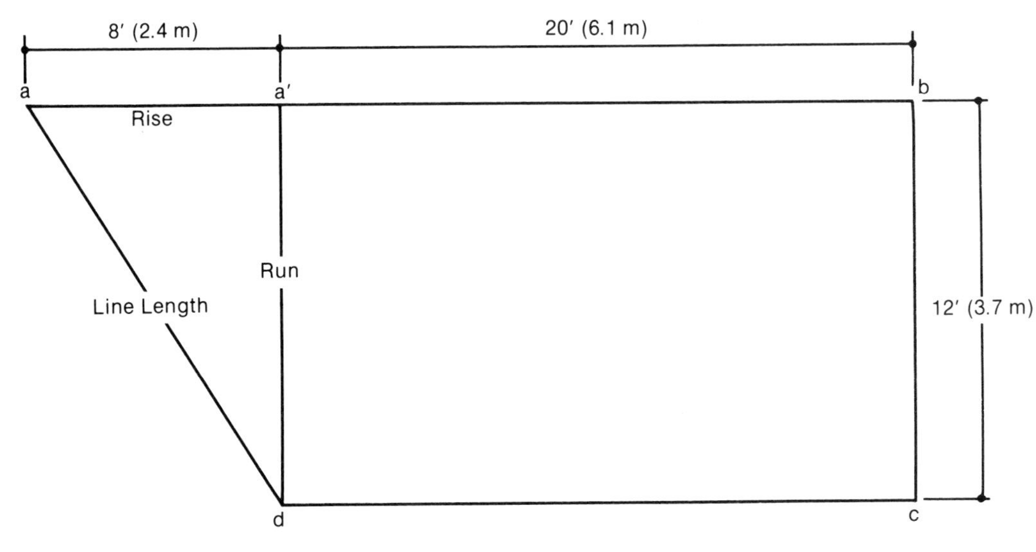

FIGURE 16-11. Development drawing of a shed.

16.7 LAYOUT OF THE COMMON RAFTER

In this section, layout of common rafters will be explained from both the graphic and framing square perspectives. In this way, each system can be learned and subsequent rafters (hip, jack, etc.) can be laid out using the method with which the builder is most comfortable.

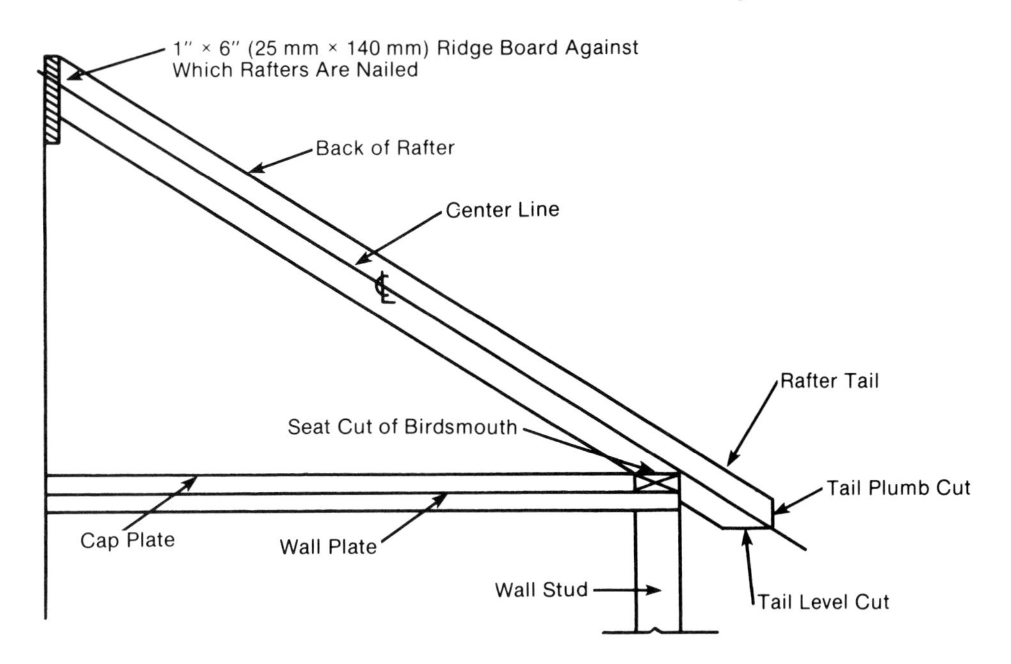

FIGURE 16-12. Graphic method illustrating common rafter layout.

16.7.1 Graphic Method

The first step is to obtain a piece of 1/8″ (3 mm) hardened composition board. Cut it to the dimensions of the development drawing discussed earlier, but substitute the 1/4″ to 1′ scale with one of 1″ to 1′ (25.4 mm to 305 mm). Mark on it the run, rise, and line length. This instrument, called a pitch board, is crucial to the graphic system (Figure 16-13).

Select a straight, clean piece of 2″ × 4″ (38 mm × 89 mm) of suitable length and place it on a pair of sawhorses. Examine the stock for the crown side by holding it with the wide face flat in the palm of the hand and sighting along the narrow face. Place the stock on the sawhorses with the crown side away from you.

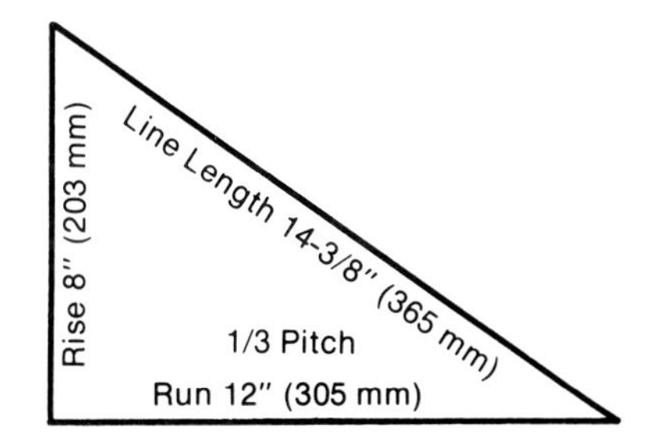

FIGURE 16-13. Pitch board.

Center Line. To find the center line of a piece of 2″ × 4″ (38 mm × 89 mm), place a tape square across the 3-5/8″ (92 mm) face of the stock. Traverse the tape until the 4″ (102 mm) mark may be read on the top edge of the stock. Spot the center; then set an adjustable square to the center spot and draw in a center line the full length of the 2″ × 4″ (38 mm × 89 mm), as seen in Figure 16-14.

When the common rafter is raised into place, the center line is the line length of the common rafter; the horizontal distance over which it passes is the run; and the height that it is raised over the plate is the rise. The theoretical center on this common rafter is at the center of the end section of the 2″ × 4″ (38 mm × 89 mm), as seen in Figure 16-15.

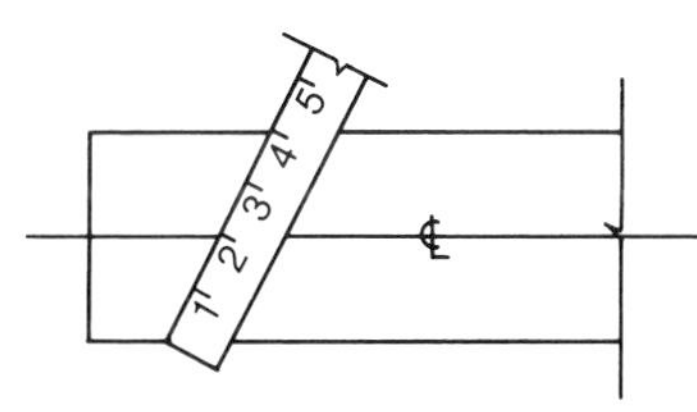

FIGURE 16-14. Locating the center line.

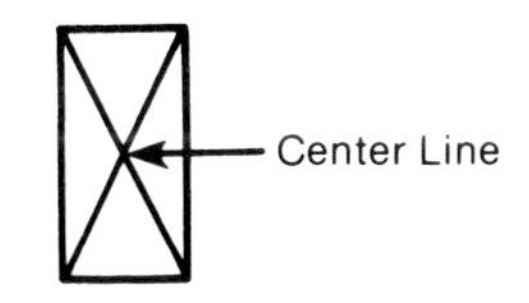

FIGURE 16-15. Theoretical center of common rafter.

Plumb Cut. Place the rafter stock on the sawhorses with the crown side away from you. Lay the hypotenuse of the pitch board, with the plumb cut to the left, on the back of the rafter (Figure 16-16). Draw in the plumb line. *Note:* In Figure 16-17, the numbered lines give the sequence in which the common rafter cutting lines are drawn. This plumb line, as seen in the illustration, is the first line drawn. It is important to remember that this same sequence is used for the framing square layout method.

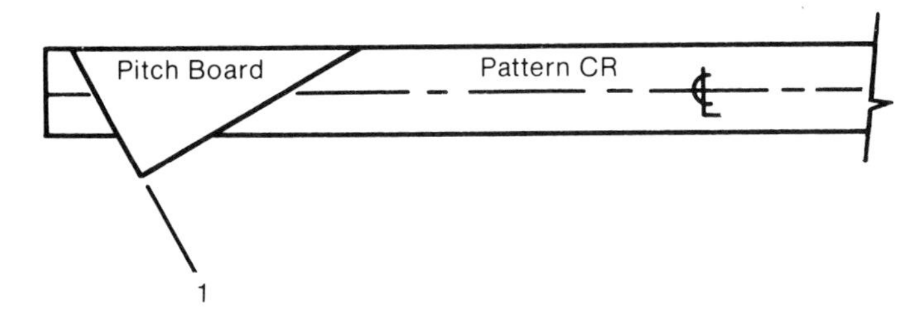

FIGURE 16-16. Locating the first plumb line.

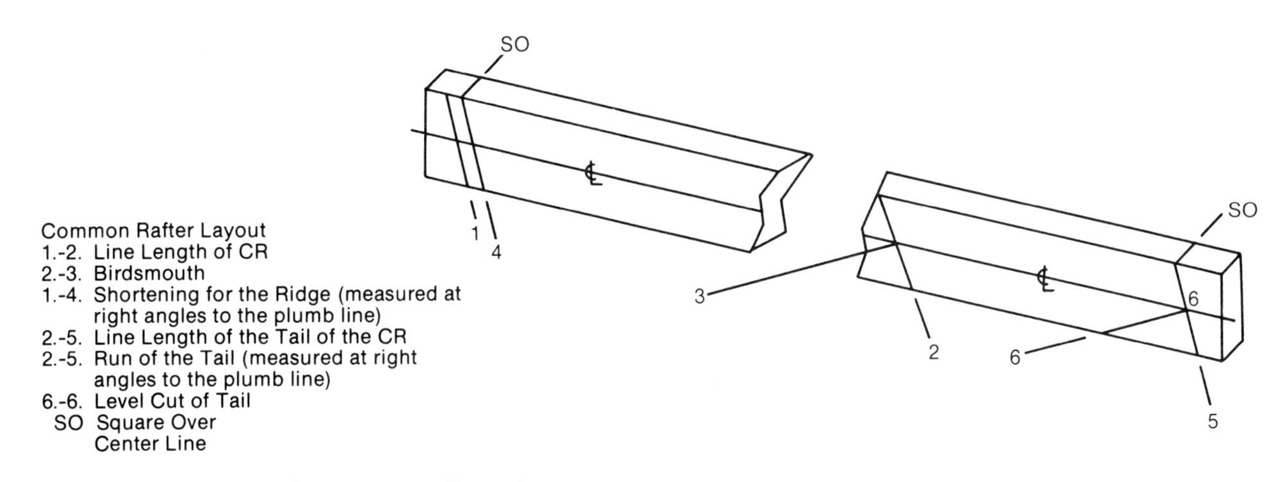

FIGURE 16-17. Line sequence for common rafter cuts.

Line Length. Measure the line length of the common rafter (a-d) on the development drawing (Figure 16-11). Check the line length of the common rafter to a tolerance of 1/8″ (3.2 mm) by mathematics. This will be, when laid out, the pattern rafter from which all other similar rafters are to be cut; therefore, accurate measurement is crucial. In this case, the line length is 14′5″ (4.4 m).

From the plumb line, measure 14′5″ (4.4 m) on the top edge of the face of the rafter. Place the hypotenuse of the pitch board to the back of the rafter with the plumb cut of the pitch board to the left. Draw in another plumb line at the 14′5″ (4.4 m) mark, as seen in Figure 16-18. Observe that the line length is the same if measured on the back of the rafter, the center line, or the bottom of the rafter.

Stepping Off. As a further check, place the hypotenuse of the pitch board along the top edge of the face of the rafter and step it off twelve times. The first application of the pitch board is at Line 1 (the first plumb line) and the twelfth application is at Line 2 (the second plumb line), 14′5″ (4.4 m) from Line 1. The twelfth step should reach within

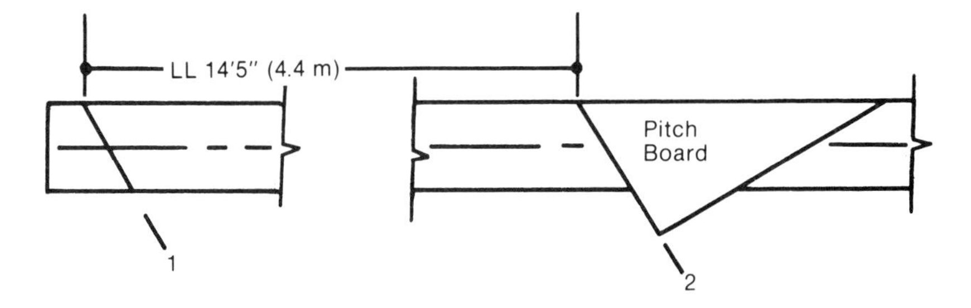

FIGURE 16-18. Locating the second plumb line.

3/8" (10 mm) of the mathematical line length mark. While this practice is very effective in helping avoid error, it should be emphasized that stepping off is only a further check and *should not* be used to obtain the true line length.

Seat Cut. To draw in the level cut for the birdsmouth, first place the pitch board on the back of the rafter with the plumb cut to the left. Adjust the hypotenuse of the center line of the rafter. The run of the pitch board must intersect Line 2 (Figure 16-19). Draw in the seat cut line; this is Line 3.

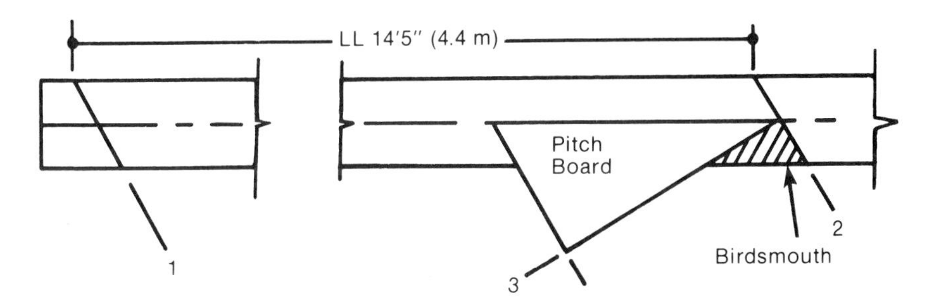

FIGURE 16-19. Locating the seat cut line.

The shaded area formed by Lines 2 and 3 meeting at the center line is the birdsmouth. When the rafter is in position, the birdsmouth fits over the wall plate. Lines 2 and 3 should now be squared under the bottom of the rafter.

Shortening. Referring to the specifications, it can be determined that the top rafters are to be secured to a 1" × 6" (19 mm × 140 mm) ridge board nailed to the side of the building. Since the actual size of this piece is 3/4" × 5-1/2", the common rafter will have to be shortened by the thickness of the board. To obtain the actual height at which to nail the ridge board, the detail should be drawn full size. The theoretical height is governed by the center line of the common rafter (Figure 16-12).

The shortening is measured at right angles to Line 1. Place the pitch board hypotenuse to the back of the rafter at Line 1, and slide it back until it assumes a position 3/4" (19 mm) back to Line 1. (This must be measured at right angles.) Draw in the shortening line; this is Line 4 (Figure 16-20).

Tail Length. The tail length of a rafter is the additional length required for it to pass over the horizontal distance from the birdsmouth to the

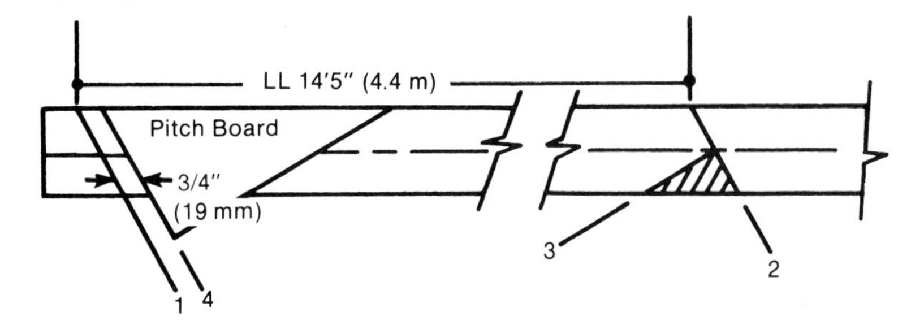

FIGURE 16-20. Locating the shortening line.

measuring point at the tail end of the rafter. It is described in the specifications as the overhang; in this particular case it measures 12″ (305 mm).

To lay out the tail, place the pitch board on the tail with the plumb cut to the left. The hypotenuse must be to the back of the rafter, as seen in Figure 16-21; the plumb line must be aligned to the plumb line of the birdsmouth, Line 2.

Measure the mark 12″ (305 mm) along the run of the pitch board, i.e., the base of the triangle. Place the pitch board with the hypotenuse on the bottom edge of the rafter, plumb cut to the right. Slip the pitch board until the plumb cut intersects the 12″ (305 mm) mark on the top edge of the rafter as seen in Figure 16-22. Draw in the tail end cut line; this is Line 5.

The plumb and level tail cut, as seen in Figure 16-23, is drawn by placing the pitch board with the hypotenuse on the center line. The pitch board plumb cut is to the left. Slip the pitch board and draw in the level cut on the run; this is Line 6. While there are variations of the tail end cut for common rafters, in each case the tail line length must first be determined before the shape of the tail is drawn.

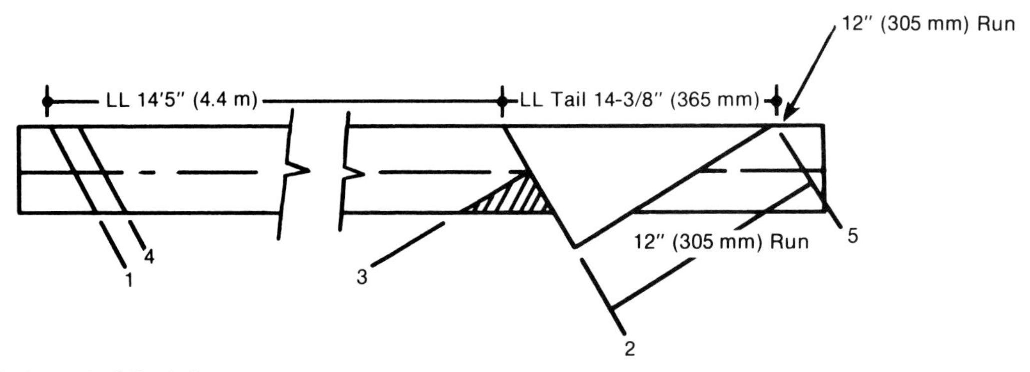

FIGURE 16-21. Layout of the tail.

16.7.2 Framing Square Method

The precision-made framing square, with its rafter tables and fine graduations, is of comparatively recent origin. Available in both British Imperial and Metric system styles, its job is identical to that of a pitch board. (There are more than one metric squares in use; in this book, the more common Frederickson square is used.) A typical

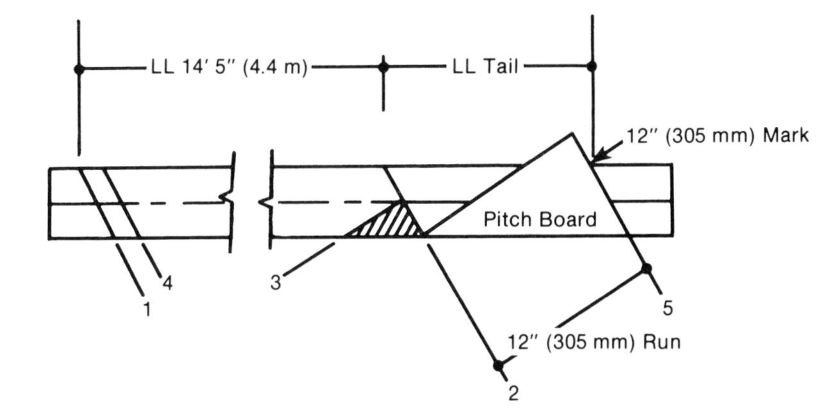

FIGURE 16-22. Drawing the tail end cut.

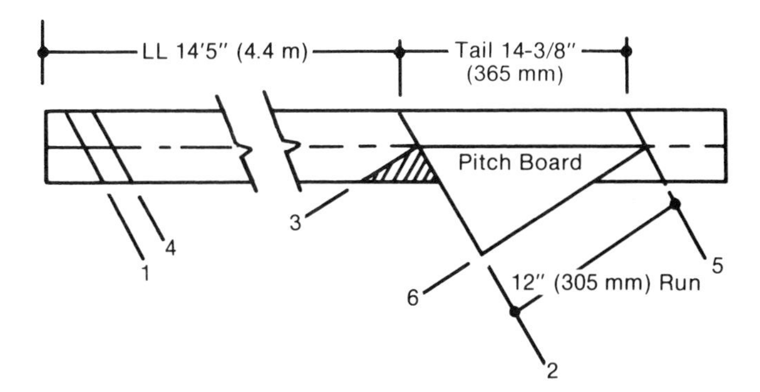

FIGURE 16-23. Drawing the level cut.

framing square is two sides of a right triangle, with the rafter tables on the face side. *Note:* To avoid confusion, dimensional conversions are not given in discussions pertaining to metric or Fredrickson squares.

Traditional Framing Square Dimensions

The body of the traditional, or imperial, framing square is 2″ wide and 24″ long. The tongue is 1-1/2″ wide and 16″ long. There are eight edges on the imperial framing square, each divided into inches and fractions of an inch.

Figure 16-24 illustrates, in broken form, a traditional framing square. It reads as follows:

- First line—length of common rafter per foot run
- Second line—length of hip or valley per foot run
- Third line—difference in length of jacks 16″ centers
- Fourth line—difference in length of jacks 24″ centers
- Fifth line—side cut of jack rafter use
- Sixth line—side cut of hip or valley use

It should be noted that the length of rafter per foot run refers to the number of feet of run in a common rafter. The length of hip or valley per foot run also refers to the number of feet or run in the common rafter.

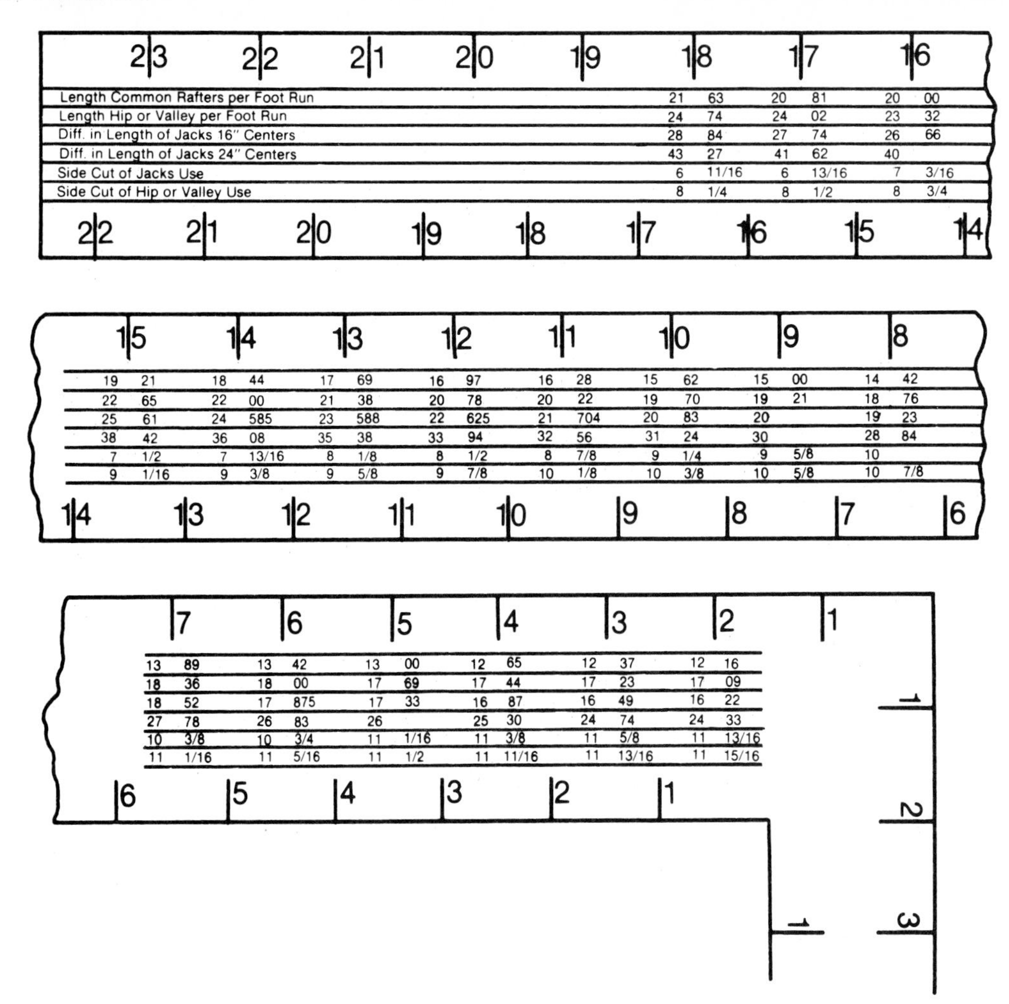

FIGURE 16-24. Traditional framing square.

Relative Pitches. Figure 16-25 shows graphically the relative pitches as used on a traditional framing square. They are based on the unit of 12″ of run and a 24″ span. Two common pitches are already drawn in: 6″ rise per foot run (for a 1/4 pitch) and 8″ rise per foot run (for a 1/3 pitch). These were calculated using the previously mentioned formula:

$$\text{Pitch} = \frac{\text{Rise}}{\text{Span}}$$

Rise per Foot Run When Span and Rise Are Given. The first consideration in understanding the rafter tables is the ability to find the number of inches of rise to every single foot of run in the common rafter. For instance, assume that a gable roof is to be erected on a building with a span of 36′ and a rise of 9′. To find the number of inches of rise for every single foot of run of the common rafter, simply divide the total number of inches of rise in the roof by the actual number of feet of run of the common rafter. Since it is known that the run is half the span, or 18′ in this case, the calculation is simply: $\frac{9 \times 12}{18} = \frac{108}{18} =$ 6″ rise per foot run.

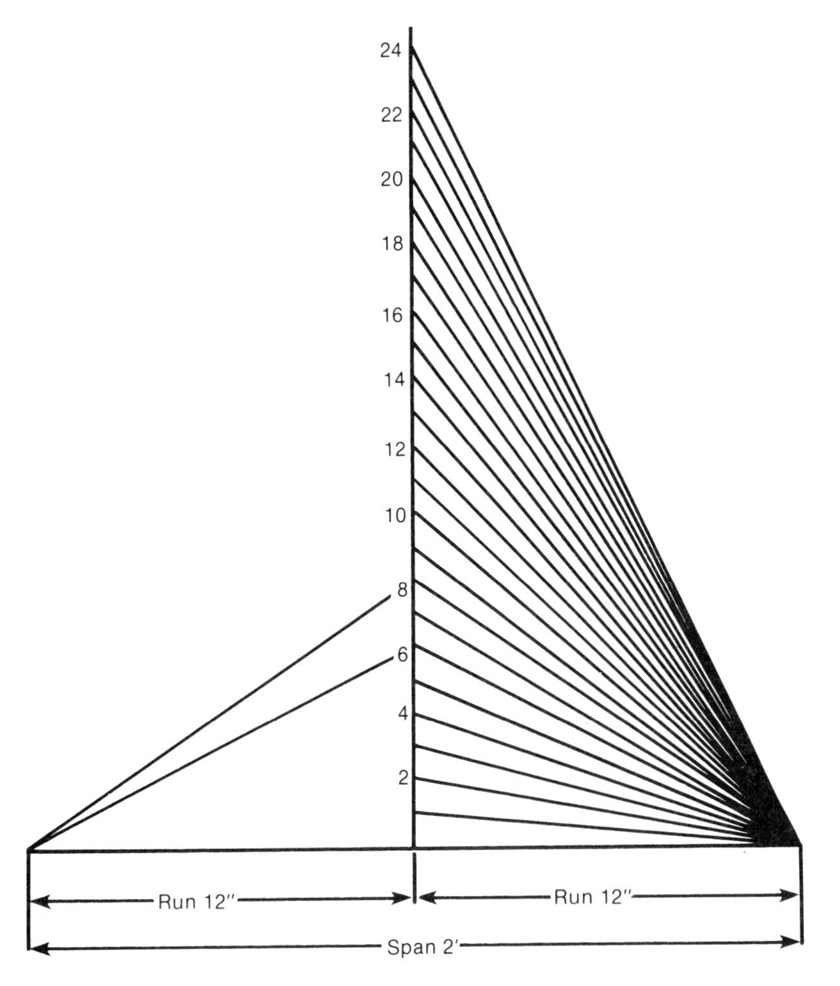

FIGURE 16-25. Relative pitches on a framing square.

Rise per Foot Run When Pitch Is Given. In this example, assume that a 1/3 pitch roof is to be erected on a building with a span of 36'. To find the rise of the roof, multiply the pitch by the span: $1/3 \times 36' = 12'$. To find the number of inches of rise for every single foot of run of the common rafter, use the same formula as above: $\frac{12 \times 12}{18} = \frac{144}{18} = 8''$ rise per foot run. Most architects show the rise per foot run on drawings as shown in Figure 16-26.

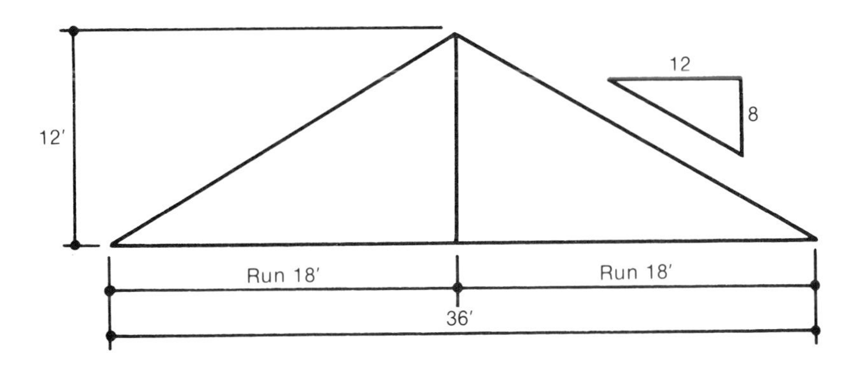

FIGURE 16-26. An 8" rise per foot run.

Line Length per Foot Run. To determine the line length per foot run for a 1/3 pitch roof, measure with a steel tape measure across the framing square from the 12″ mark on the body to the 8″ mark on the tongue. Read 14-3/8″ to the nearest 1/8″ (Figure 16-27). Read on the rafter tables, under the 8″ mark on the first line, 14.42″. It will be seen that 14.42″ is 14-3/8″ to the nearest 1/8″. Thus, the line length per foot run for a 1/3 pitch roof is 14.42″.

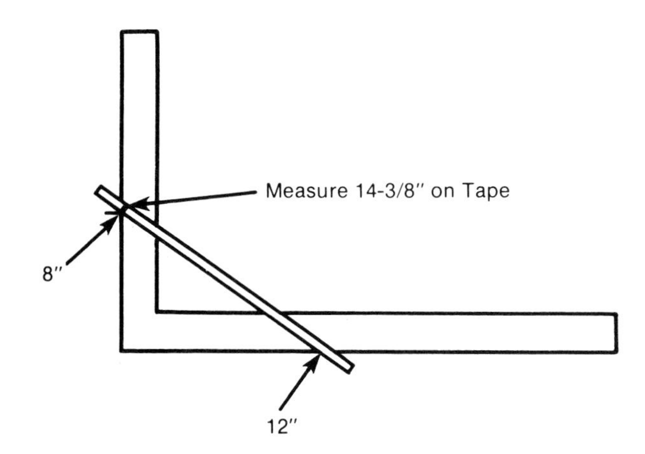

FIGURE 16-27. Measuring for line length per foot.

Line Length of Common Rafter. For this example, assume that a 1/4 pitch gable roof is to be erected on a building with a 36′ span. First, find the rise per foot run. Using the perviously mentioned formula, this calculates to 6″.

Read on the first line of the rafter tables under the 6″ mark. The figure 13.42″ is the determined hypotenuse of a right triangle with a base (run) of 12″ and a height (rise) of 6″. The reading on the framing square for a 1/4 pitch roof is 13.42; the length of the common rafter for every foot of run of the common rafter is 13.42″.

To find the line length of the common rafter for this roof, multiply the length of the common rafter per foot run (13.42″) by the actual number of feet of run in the common rafter (18′): 13.42 × 18 = 241.46″. Reduce this to feet and inches, and express the result to the nearest 1/8″. (All rafters should be laid out to a tolerance of ±1/8″.) Thus: 12 241.56 = 20′ and 1.56″ = 20′ + 1″ + .56″.

To change .56″ to eighths of an inch:

$$.56 = \frac{?}{8} \qquad .56 \times 8 = 4.48 \qquad .56 = \frac{4.48}{8} \qquad .56 = \frac{4''}{8}$$

Thus, the line length of the common rafter would be 20′ 1-4/8″.

Specifications for Layout of a Common Rafter With a Framing Square. Assume that a 1/3 pitch gable roof with a 1″ × 6″ ridge is to be erected on a building with a 36′ span and a 16′ overhang. It should be noted that all positions of the framing square for obtaining plumb and level cuts, as well as the line numbering, are the same as in the graphic method. Also, stair gauges are indispensable when rafter framing with a square, and it is recommended that several be purchased before beginning.

Select a straight, clean piece of 2″ × 4″ of suitable length and place it on a pair of sawhorses. Examine the stock for the crown side by holding it with the wide face flat in the palm of the hand and sighting along the narrow face. Place the stock on the sawhorses with the crown side away from you. To find the center line, follow the same procedure used with the graphic method.

Plumb Cut. Set the stair gauges for a 1/3 pitch roof by attaching one gauge to the 8″ mark on the tongue and the other on the 12″ mark on the body. Be sure that both gauges are placed on the outside edges of the square. Place the square with the plumb cut to the left and the gauges snug to the back of the rafter stock as shown in Figure 16-28. Draw the plumb line (Figure 16-29); this is Line 1.

Line Length. Just to the right of the plumb cut (and actually on the common rafter) work out the line length of the common rafter. Multiply 14.42 (as read in the rafter table) by the actual number of feet of run in the common rafter: 14.42 × 18 = 21′ 7-5/8″. With a steel tape measure, mark off 21′ 7-5/8″ on the back of the rafter stock.

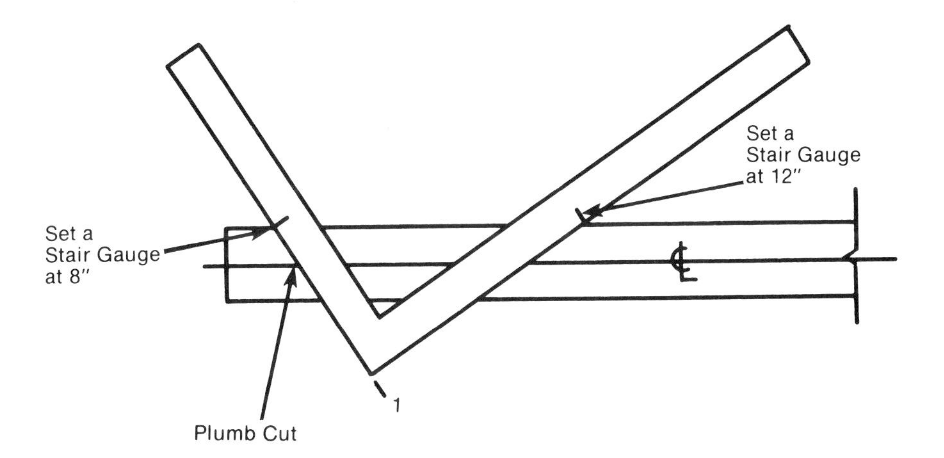

FIGURE 16-28. Placing the framing square for a plumb cut line.

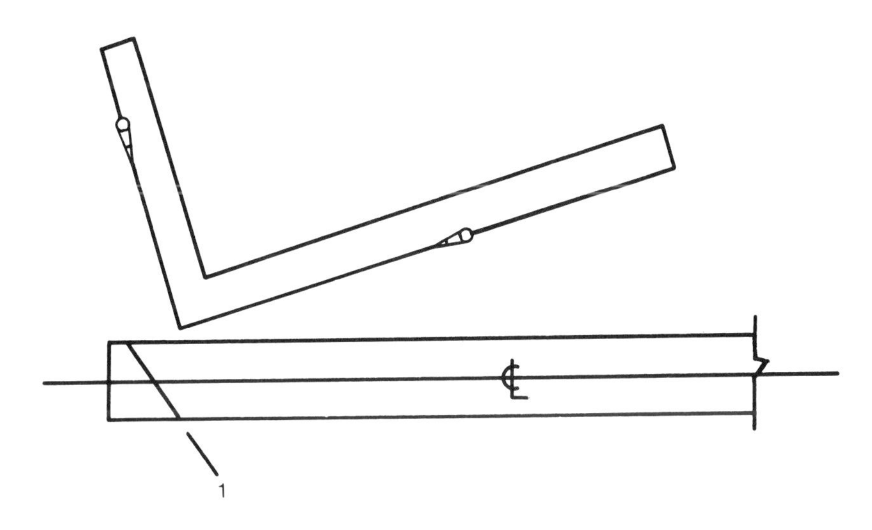

FIGURE 16-29. Drawing the first plumb line.

Stepping Off. Stepping off is also an effective method of checking line length with the framing method. Place the stair gauges for the pitch of the roof and set the square at Line 1 as seen in Figure 16-28. Mark a short line on the run of the square to intersect with the top edge of the rafter stock. Slip the square along the rafter stock until the plumb marking line intersects with the run line already drawn. Repeat this operation until the number of steps taken is the number of feet of run in the common rafter; in this case, 18 times for a run of 18′.

The last application at the run end of the framing square should reach to within 3/8″ of the line length as determined from the rafter table. When it checks to such a tolerance, the line length can be accepted. As with the graphic method, however, it must be emphasized that stepping off is only to be used a check and *should never* be used to obtain the true line length. With the plumb cut of the framing square to the left, draw in another plumb cut (Line 2) at the 21′ 7-5/8″ mark, as shown in Figure 16-30.

Birdsmouth. Slip the framing square to the left until the level cut may be drawn to intersect with Line 2 at the center line, as shown in Figure 16-31. The meeting of Lines 2 and 3 lays out the birdsmouth. Both lines should be squared under the bottom edge of the rafter. Once the birdsmouth is removed, the level and plumb cuts fit snug to the wall plate.

Shortening. As with the graphic method, some shortening of the 1″ × 6″ ridge board is necessary. In the framing square method, the amount of shortening necessary on each common rafter is half the

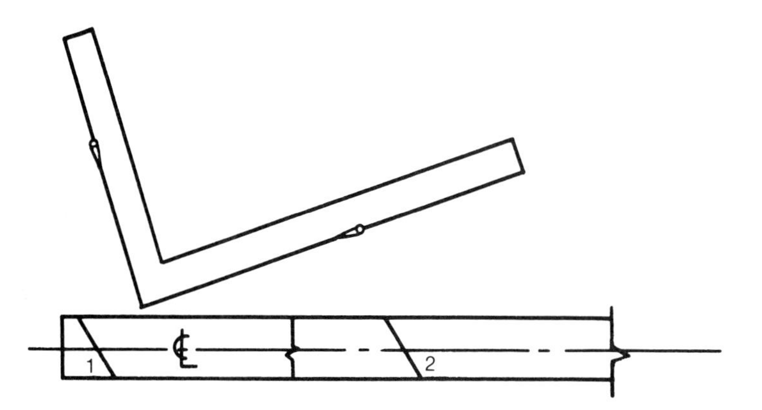

FIGURE 16-30. Drawing the second plumb line.

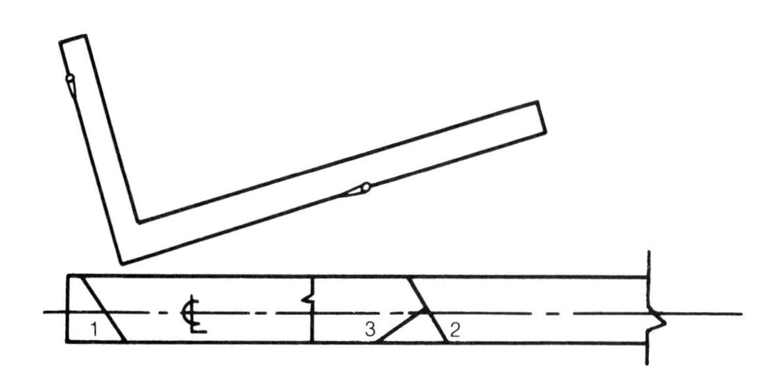

FIGURE 16-31. Drawing the level cut line.

thickness of the ridge board, measured at right angles to the plumb cut. Place the square with the plumb marking line to the left on Line 1. Slip the square so that it is 3/8″ back at right angles to Line 1, as shown in Figure 16-32. (The tape is seen placed in the position to check the shortening.) Draw in the shortening line (Line 4). Square over Line 4 to the back of the rafter; this is the cutting line.

Tail Length. In this case, the tail length, or overhang, is 16″ and the pitch is 1/3. On the rafter, just to the right of Line 2, make the calculations for the line length of the tail for the overhang. Reading on the framing square for a 1/3 pitch roof, it can be seen that the rise per foot of run is 8″. On the first line of the rafter tables, 14.42 is read under 8″.

To find the line length of the tail of the common rafter, multiply the length of the rafter per foot run (14.42) by the actual number of feet of run in the tail (this is 16″, or 1-1/3′ of run): 14.42 × 1-1/3 = 19-1/4″. Thus, the line length of the tail is 19-1/4″. Use the tape to check the measurement, as seen in Figure 16-33. Draw in Line 5. *Note:* When the rafter stock is of just sufficient length for the common rafter, it is necessary to turn the framing square over to draw Line 5. It should be turned over so that the heel is away from the carpenter; the plumb side should be to the right and the stair gauges should fit snug against the bottom of the stock.

To further check the line length of the tail by stepping off, place the framing square, with the plumb marking side to the left, on Line 2.

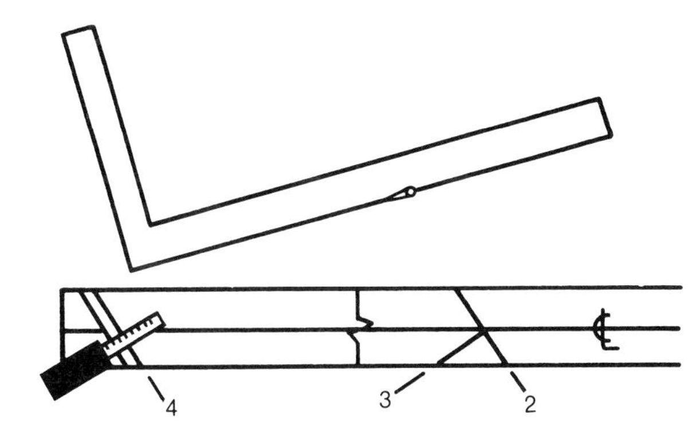

FIGURE 16-32. Locating the shortening line.

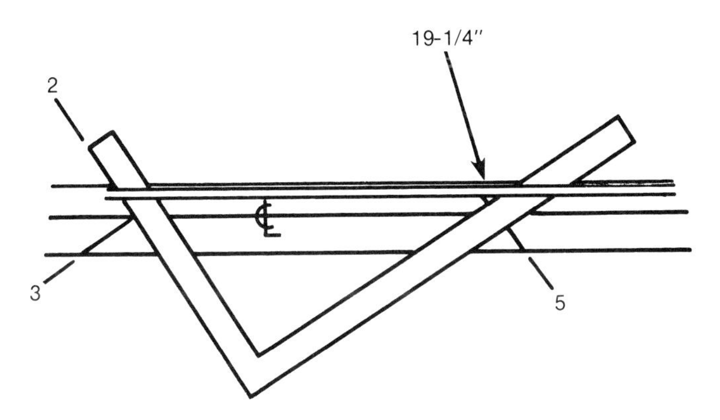

FIGURE 16-33. Checking the line length of the tail with the tape.

Measure 1″ on the run, slip the square to the right, and measure an additional 4″ (Figure 16-34). The 12″ and the 4″ together equal 16″, the given overhang measured on the run.

Level Cut of the Tail. Where a plumb and level cut is to be made at the end of a common rafter, place the square with the plumb side to the left. Adjust its position until it intersects Line 5 at the center line. Draw in the level cut; this is Line 6 (Figure 16-35).

Actual Cutting Lines. The actual cutting lines of the common rafter are as follows:

- Line 4—the shortening cut
- Line 3 and the lower part of Line 2—the birdsmouth
- Lines 5 and 6—the plumb and level tailcuts

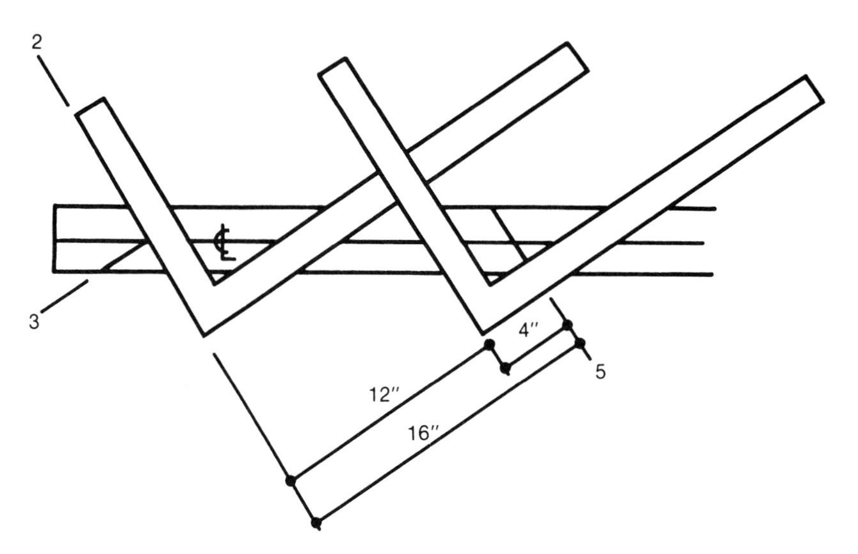

FIGURE 16-34. Checking the line length of the tail by stepping off.

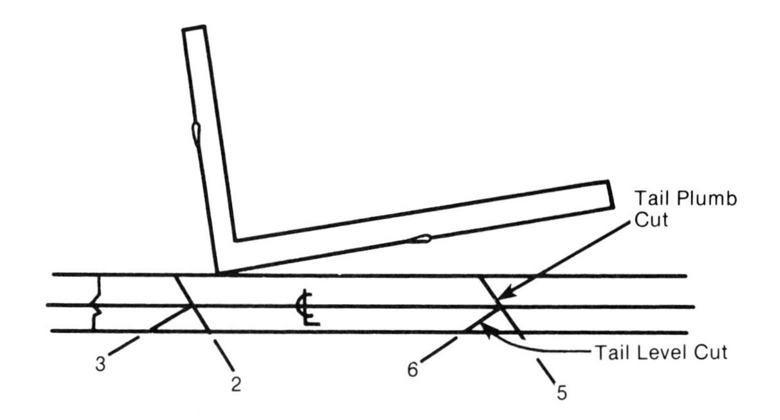

FIGURE 16-35. Drawing the level cut of the tail.

Frederickson Metric Framing Square Dimensions.

The body of Frederickson metric framing square is 50 mm wide and 600 mm long. The tongue is 38 mm wide and 400 mm long. Like the traditional framing square, it has eight edges, each graduated in millimeters. The Frederickson framing square is based on a unit run of 250 mm.

Figure 16-36 illustrates, in broken form, the Frederickson metric framing square. It reads as follows:

- First line—length of common rafter per unit run of 250 mm
- Second line—length of hip rafter per unit run of 353.5 mm
- Third line—difference in jack rafters at 400 mm O.C.
- Fourth line—side cut of hip or valley rafters
- Fifth line—side cut of jack rafters
- Sixth line—angle of common rafter at the plate

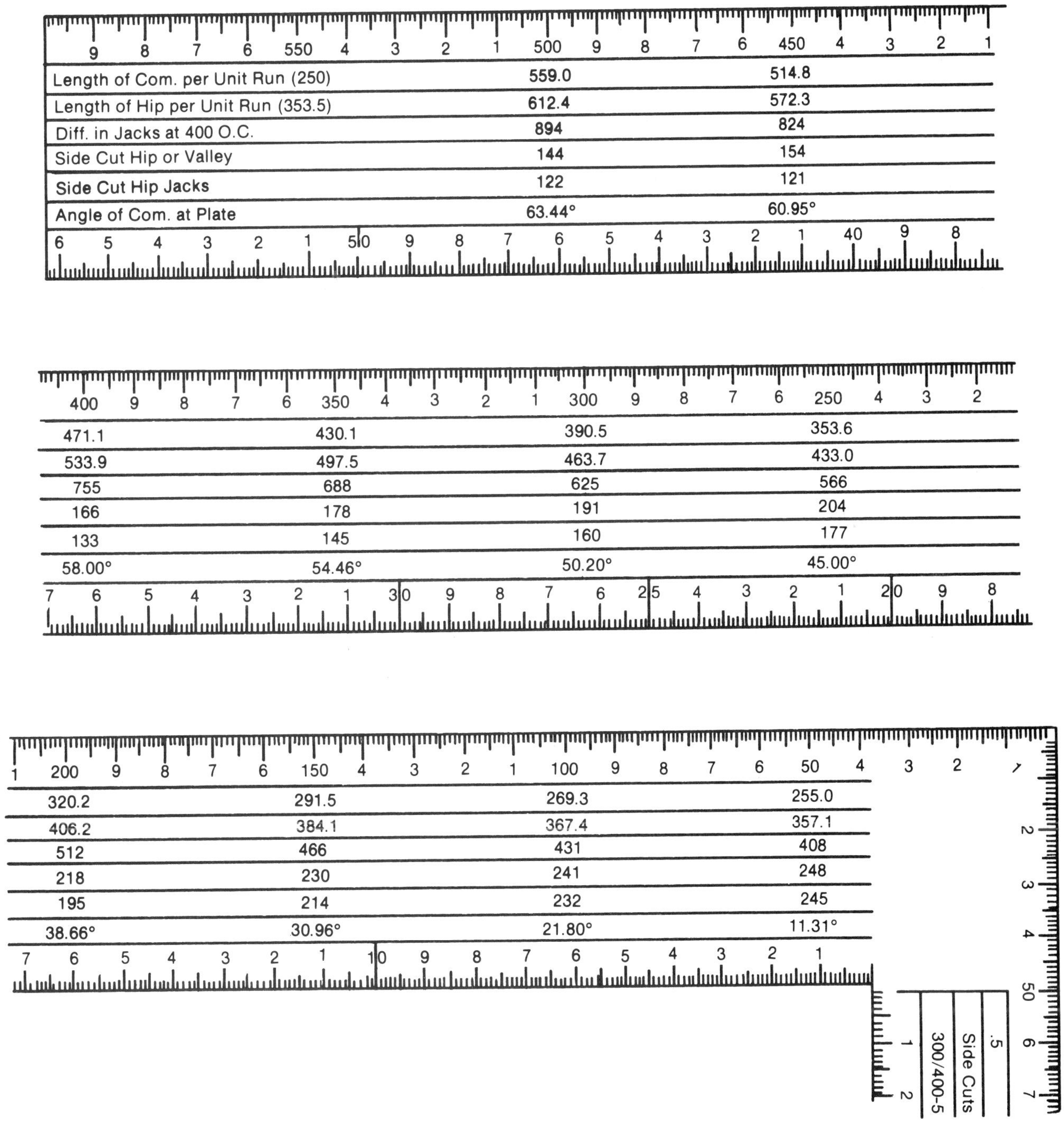

	500	450	400	350	300	250	200	150	100	50
Length of Com. per Unit Run (250)	559.0	514.8	471.1	430.1	390.5	353.6	320.2	291.5	269.3	255.0
Length of Hip per Unit Run (353.5)	612.4	572.3	533.9	497.5	463.7	433.0	406.2	384.1	367.4	357.1
Diff. in Jacks at 400 O.C.	894	824	755	688	625	566	512	466	431	408
Side Cut Hip or Valley	144	154	166	178	191	204	218	230	241	248
Side Cut Hip Jacks	122	121	133	145	160	177	195	214	232	245
Angle of Com. at Plate	63.44°	60.95°	58.00°	54.46°	50.20°	45.00°	38.66°	30.96°	21.80°	11.31°

FIGURE 16-36. Frederickson metric framing square.

Relative Pitches and Slopes. Figure 16-37 graphically shows some of the relative pitches and one slope triangle common to a metric framing square. The rafter tables are considered to be developed on the basis of a 250 mm run, which may or may not imply a 500 mm span. Two pitches are already drawn in: 250 mm rise for a run of 250 mm for 1/2 (0.5) pitch and 150 mm rise for a run of 250 mm for 3/10 (0.3) pitch. For the latter, a slope triangle was also drawn.

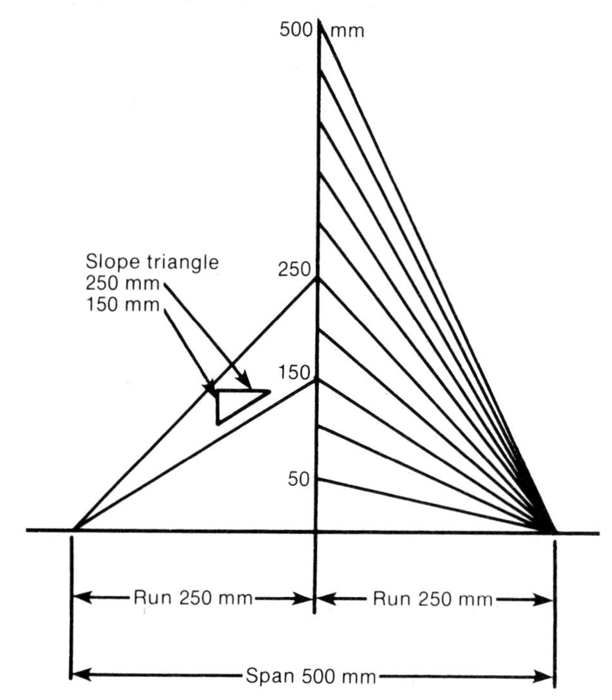

FIGURE 16-37. Relative pitches on a metric framing square with a 250 mm unit run.

Slope Triangles. The pitch of a roof may be indicated on a drawing as a slope triangle. The two sides of the triangle represent the unit rise and the unit run and are dimensioned to help avoid errors in communication between the architect or planner and the builder. In Figure 16-37 (and in more detail in Figure 16-38) the pitch 3/10 (0.3) is represented by the slope triangle dimensioned with a 150 mm rise and a 250 mm run for an implied unit span measurement of 500 mm. Slope triangles are not necessarily to scale and are mainly used to indicate the slope (pitch) of a roof without reference to the span.

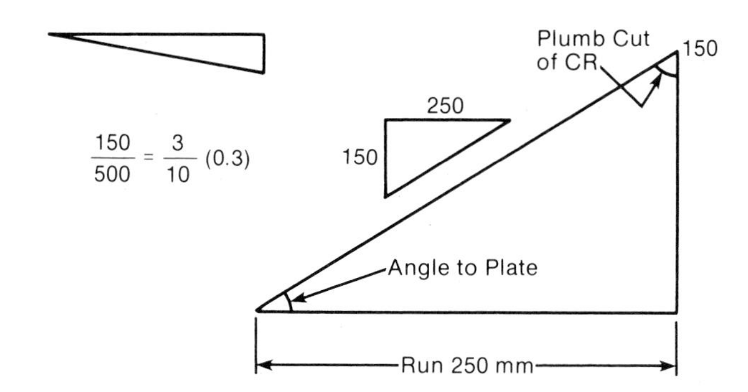

FIGURE 16-38. Dimensioning a slope triangle.

Table 16-1 relates selected roof pitches to the Frederickson metric framing square settings.

TABLE 16-1 RELATIVE PITCHES* FOR UNIT RUN OF 250 MM (Frederickson Metric Framing Square)

		Framing Square Setting	
Rise over Span	**Pitch**	**Rise**	**Run**
500/500	1.0 (Full)	500	250
450/500	0.9	450	250
400/500	0.8	400	250
350/500	0.7	350	250
300/500	0.6	300	250
250/500	0.5	250	250
200/500	0.4	200	250
150/500	0.3	150	250
100/500	0.2	100	250
50/500	0.1	50	250

*Only 10 pitches are shown in the table. There are many others. One, for example, is 125/500 or 1/4 pitch.

Readling Line Lengths of Common Rafters Using the Fredrickson Framing System. Figure 16-39 indicates three readings of line length for a common unit run of 250 mm and three rise measurements: 200 mm, 250 mm, and 500 mm. The first line of the rafter table (length of common rafter per unit run of 250 mm) indicates these line lengths: (1) 320.2 mm, (2) 353.6 mm, and (3) 559.0 mm. In addition, the angle to the plate for a common rafter, the plumb cut angle of the same common rafter, and one slope triangle are also provided.

Table 16-2 is a summary of the Frederickson metric framing square based on a unit run of 250 mm. It differs from Table 16-1 in that additional data is given in the columns designated "Angle to Plate" and "Slope Triangle."

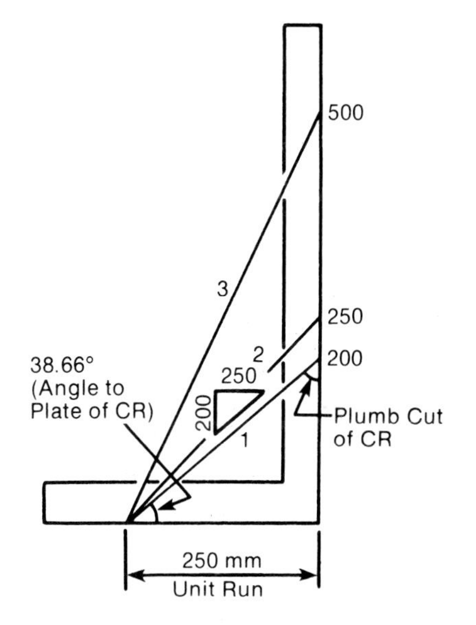

FIGURE 16-39. Line lengths and slope triangle for a unit run of 250 mm.

TABLE 16-2 ROOF FRAMING DATA WITH RESPECT TO THE FREDERICKSON METRIC FRAMING SQUARE

Unit Run of 250 mm						
	Slope Triangle (Run, Rise)					
Angle to Plate (°)	**Rise Run**	**Unit Run**	**Span**	**Rise**	**Pitch**
63.44	500 in 250	250	500	500	1.0
60.95	450 in 250	250	500	450	0.9
58.00	400 in 250	250	500	400	0.8
54.46	350 in 250	250	500	350	0.7
50.20	300 in 250	250	500	300	0.6
45.00	250 in 250	250	500	250	0.5
38.66	200 in 250	250	500	200	0.4
30.96	150 in 250	250	500	150	0.3
21.80	100 in 250	250	500	100	0.2
11.31	50 in 250	250	500	50	0.1

16.8 GABLE STUDS

The line length of the longest gable stud, when centrally placed, is equal to the rise of the roof. The remaining gable studs are erected in pairs on either side of the longest and are spaced at regular intervals along the plate. The line length is also equal to the pitch multiplied by the span. For example, if a 1/4 pitch roof has a 36′ span, the line length of the longest gable stud is 9′.

16.8.1 Rise per Foot Run

To find the rise per foot run of a gable stud, multiply the pitch by 24″. This is the unit of span and also the length of the blade of the framing square. For example, the rise per foot run for a 1/4 pitch roof is 1/4 of 24″, or 6″. *Note:* Although the traditional framing square method is used in this and all remaining sections of this chapter, the choice of which layout method to use is entirely up to the individual.

16.8.2 Differences in Gable Stud Lengths

To find the difference in lengths of the gable studs for any regular gable roof, multiply the pitch by twice the spacing of the gable studs. Table 16-3 illustrates this rule with common roof pitches.

TABLE 16-3 DIFFERENCES IN LENGTHS OF GABLE STUDS

Roof Pitch	Gable Studs Spacings	Twice Gable Stud Spacings	Differences in Lengths of Successive Gable Studs
1/4	12″	24″ 1/4 of 24″	6″
1/4	16″	32″ 1/4 of 32″	8″
1/3	16″	32″	10-2/3″
3/8	16″	32″	12″

The difference in the lengths of the gable studs for a 1/3 pitch roof where the studs are placed 12″ O.C. is 8″. The difference in the length from the wall plate to the center line of the gable studs, for every inch of run on the plate for a 1/3 pitch roof, is 2/3″; that is, 8″ ÷ 12″. It follows, then, that the difference in the lengths of the gable studs for a 1/3 pitch roof where the studs are placed 16″ O.C. is 16 times 2/3″, or 10-2/3″. It should be noted that the pitch does not change when the O.C. measurement changes.

16.8.3 Layout of Gable Studs

The layout of gable studs is usually made on the wall plate from the center gable stud. The position of the longest gable stud should be located first and the remaining studs then laid out on either side of the center gable stud.

Layout of Gable Studs at 16″ O.C. for a 1/3 Pitch Roof. Figure 16-40 shows a center line drawing of the common rafters and gable studs for a

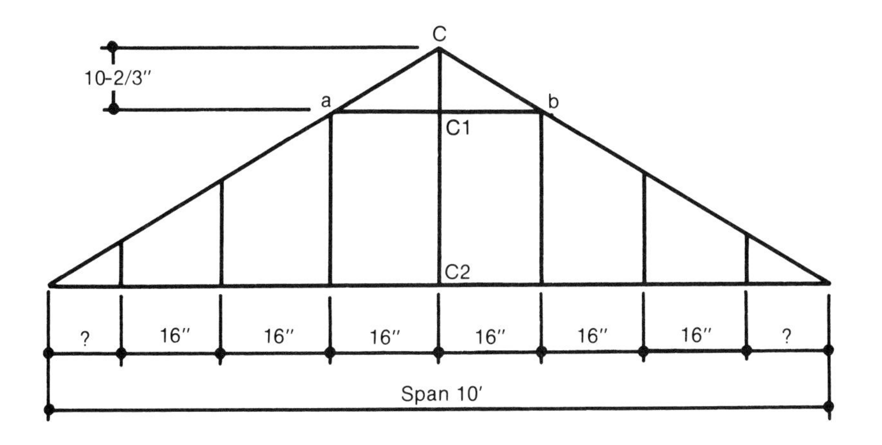

FIGURE 16-40. Common rafters and gable studs for a 1/3 pitch roof.

1/3 pitch roof. The distance a-b is the center line-to-center line distance between a pair of studs. In this case, it is 32″ and is twice the spacing or run distance.

C-C2 represents the line length of the longest gable stud. C-C1 shows the rise of the center gable stud over a-b, a distance of 10-2/3″. With the roof having a 1/3 pitch and using the "pitch times span equals rise" formula, it can be determined that 1/3 × 32″ (a-b) is 10-2/3″. The latter measurement is the common difference in the length of the gable studs for a 1/3 pitch roof with studs placed at 16″ O.C.

The line length of the longest gable stud is equal to the rise of the roof from the wall plate to the center line of the common rafters. It can be seen extending from Line 1 to Line 2 in Figure 16-41. Both the line length and the shortening cut should be laid out on the edge of the gable stud stock. The shortening is equal to the vertical measurement of the birdsmouth from the bottom edge of the common rafter to the level cut. This can be seen extending from Line 1 to Line 3 in Figure 16-41 and is equal to the measurement of a-b at the birdsmouth.

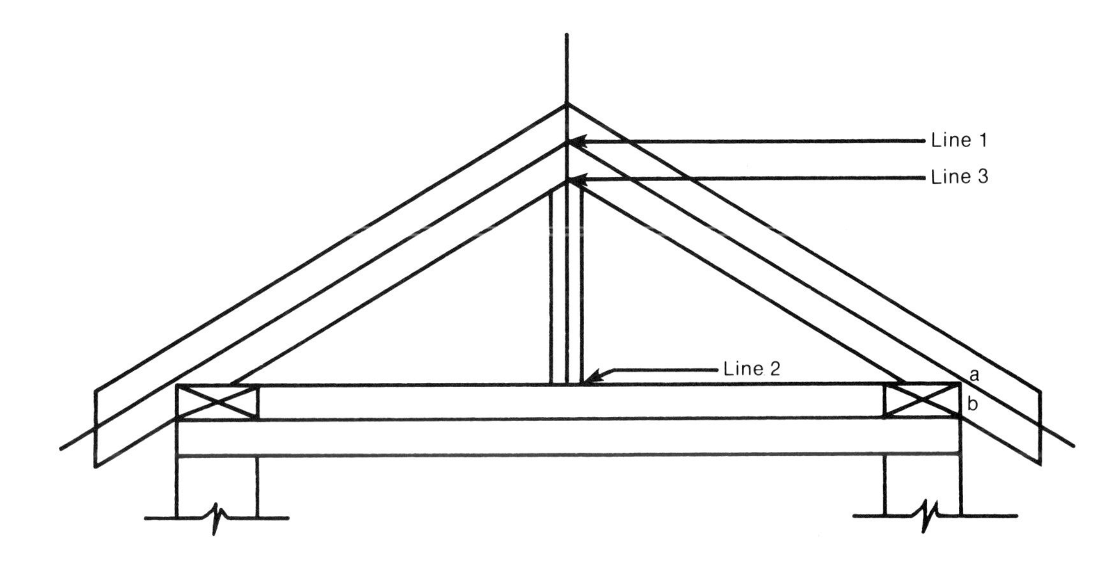

FIGURE 16-41. Shortening the longest gable stud.

Layout of a Pattern Gable Stud for a 1/3 Pitch Roof. Select a straight, clean piece of 2″ × 4″ of convenient length. Place it on a pair of sawhorses with a narrow face uppermost. Use the previously mentioned procedure to find the center line, and draw it approximately 8″ × 10″ long from the left-hand end as seen in Figure 16-42.

Mark off the line length of the longest gable stud (Lines 1 and 2). Mark off from Line 1 a measurement equal to the vertical distance between Lines 1 and 3 (Figure 16-43). Place the stair gauges at the 8″ mark on the tongue of the framing square and at the 12″ mark on the body (identical to the common rafter setting). Set the framing square with the plumb cut toward the top of the gable stud and the stair gauges fitting snugly along the bottom edge of the stud stock. Draw in the angles for the rake cuts as shown in Figure 16-44.

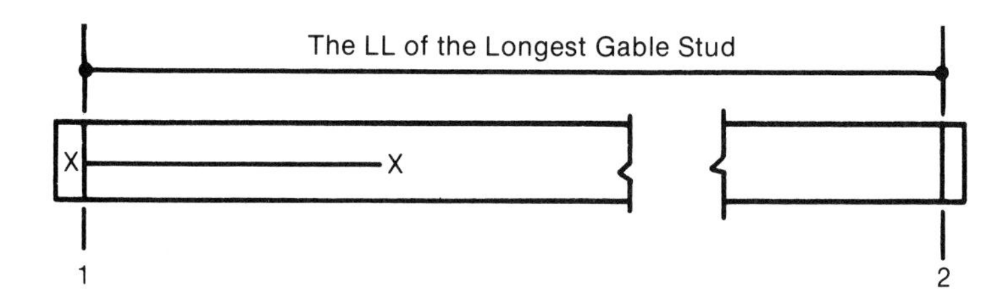

FIGURE 16-42. Locating the center line on the longest gable stud.

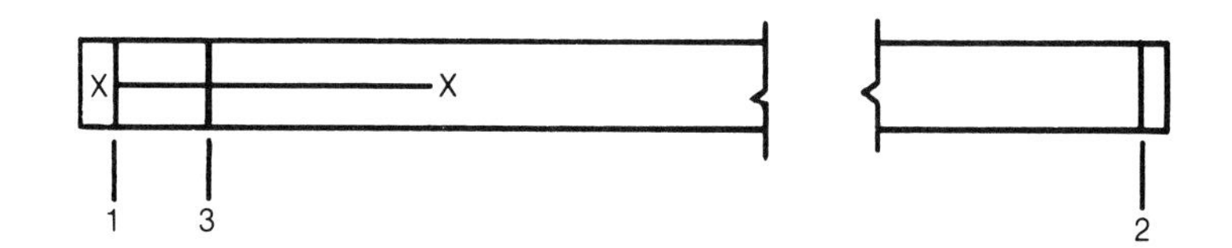

FIGURE 16-43. Drawing shortening line on gable stud.

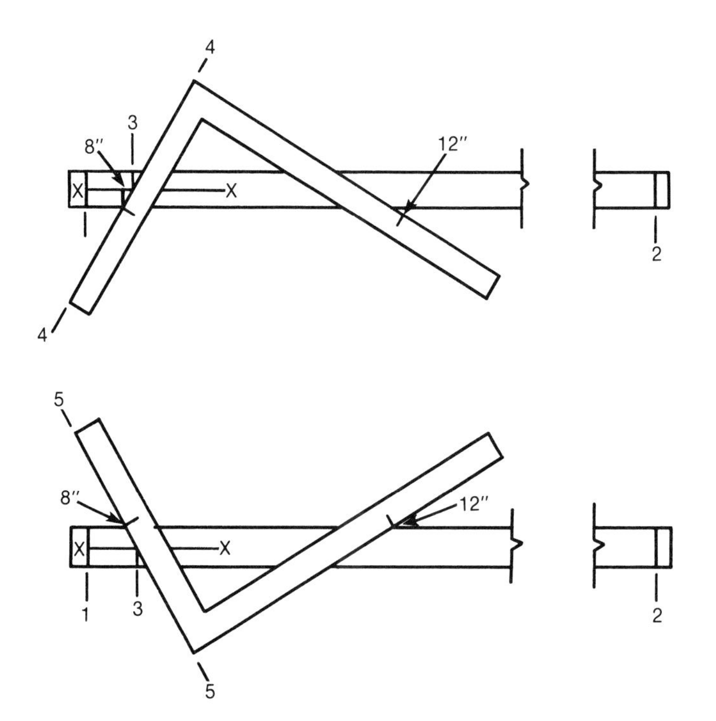

FIGURE 16-44. Making rake cuts on pattern stud.

Layout of the Gable Studs From the Pattern Stud. ,From the outside edge of the gable stock at Line 4, the layout of the remaining gable studs can be completed. The common difference in the length of each succeeding gable stud is determined from the pitch of the roof and the spacing of the studs; as was pointed out earlier, in this case it is 10-2/3″ (see Lines 5-6 and 6-7 in Figure 16-45). Measurement 2-6 is the line length of the second longest pair of studs and measurement 2-7 is the line length of the third longest pair.

Pieces of stock may now be squared off in readiness for the layout from the pattern stud (Figure 16-46). Set the stair gauges for the pitch of the roof and mark off the cutting lines for each identical pair of studs.

It is important that the cutting lines be drawn in from the outside edge of the stud. The rake cut should always be toward the peak of the roof (Figure 16-47).

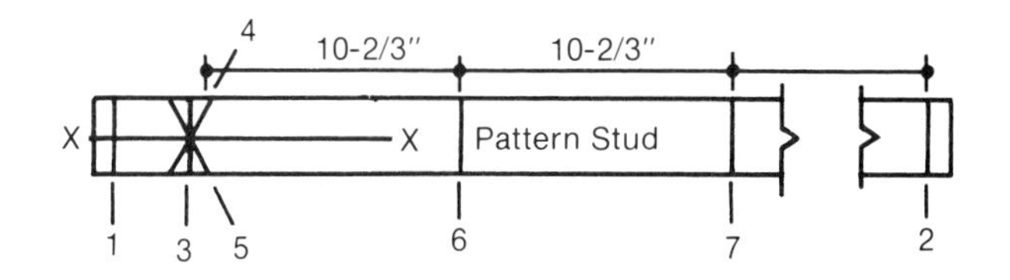

FIGURE 16-45. Marking lengths of additional studs.

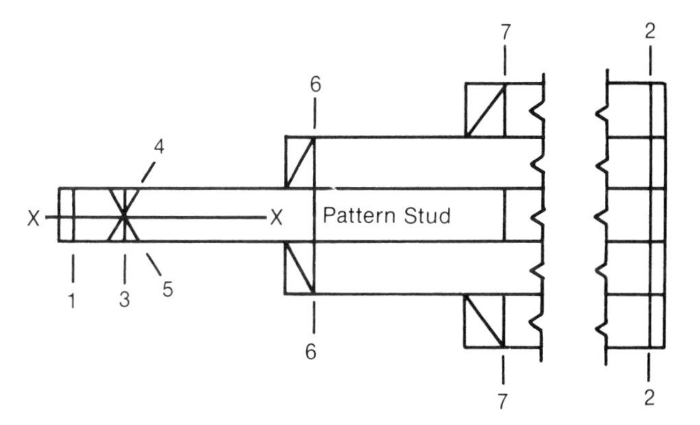

FIGURE 16-46. Using pattern stud to square additional studs.

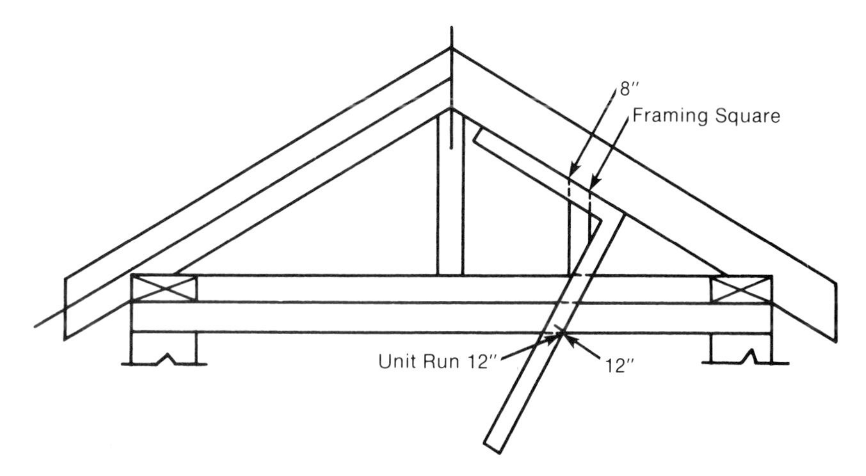

FIGURE 16-47. Making rake cuts on additional studs.

16.9 CEILING JOISTS AND COLLAR TIES

Ceiling joists are used to form a truss effect (Figure 16-48). By securing them to opposite rafters, the desired effect is achieved, with each member subject to longitudinal stress only.

After selecting a sound piece of stock on which to lay out the pattern ceiling joist, place it on a pair of sawhorses with the crown side facing away. Set the stair gauges on the framing square to the pitch of the roof (identical to the common rafter setting).

16.9.1 Rake Cut of the Ceiling Joists

The rake cut of the ceiling joists is the angle cut at the end of each joist. This is made to match the angle of the pitch of the roof. Place the framing square with the stair gauges set snugly along the bottom edge of the ceiling joist. The 12″ setting should be at the left-hand end (Figure 16-49).

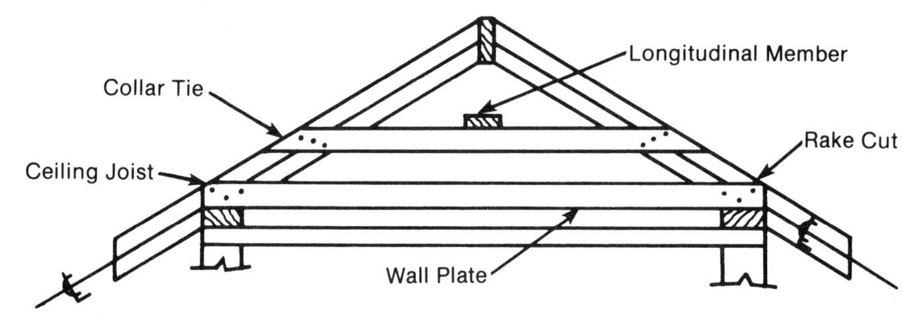

FIGURE 16-48. Truss effect of ceiling joists and collar ties.

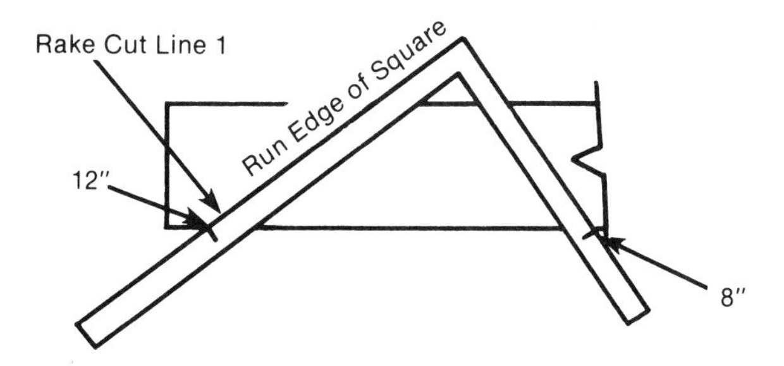

FIGURE 16-49. Making rake cut on ceiling joist.

16.9.2 Plumb Cut of the Ceiling Joists

Ceiling joists should be cut plumb at the wall plate end. Set a carpenter's combination square equal to the vertical measurement a-b on Line 2 above the birdsmouth of the common rafter. With the carpenter's square setting a-b of the common rafter, draw a plumb line on the joist to intersect with the Line 1 rake cut (Figure 16-50). When the joist is in position, Line 2 should be plumb over the outside edge of the wall plate. The rake cut should be flush with the back of the common rafter (Figure 16-51).

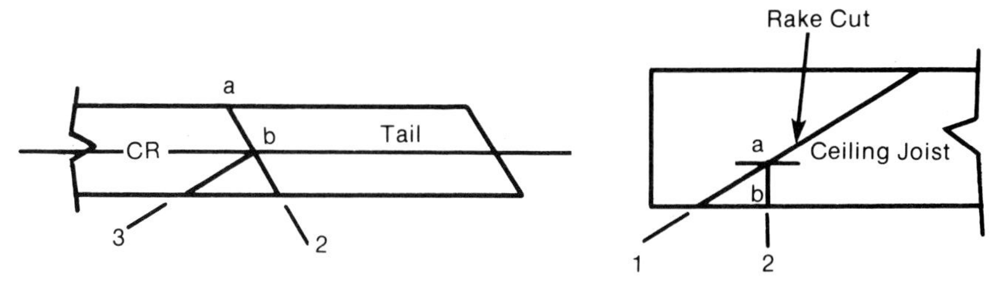

FIGURE 16-50. Using a common rafter to draw a plumb line.

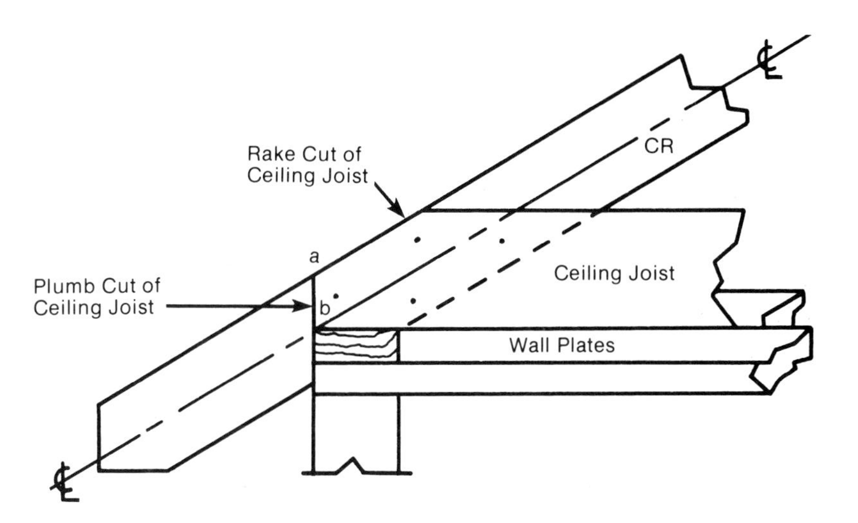

FIGURE 16-51. Ceiling joist in position at the common rafter.

16.9.3 Ceiling Joist Template

A template should be made of the wall plate end of the ceiling joist. It can be made from a piece of material approximately 18″ in length and exactly the same width as the ceiling joist. Where portable hand saws are being used, the template may be made into a jig. If a stationary power saw is used, the angle of the saw may be set by using the jig or the framing square.

16.9.4 Layout of the Collar Ties

Many building authorities state that, for residential construction, the collar ties (where there is no attic) must be placed in the upper third of the line length of the common rafter. The length of the collar ties can be determined by reading the plans and specifications. They are laid out in a manner similar to that of ceiling joists; a template, framing square, or pitch board may be used.

The actual nailing position of the collar ties should be marked on the pattern common rafter. The nailing position should be transferred to all the common rafters during the layout of these members. Also, it is very important to stress that all lines and cuts on roof framing members should be made on the ground; this will help guarantee exactness of layout. A longitudinal member is usually secured at right angles to the center of the collar ties (Figure 16-48). Read the drawings and specifications for the dimensions of this member and mark the rafter spacings on it before it is passed up for nailing.

16.10 LAYOUT OF HIP RAFTERS

This section deals with the layout of hip rafters for a regular hip roof. A regular hip roof has an equal pitch from each wall plate toward the ridge. Figure 16-52 shows an outline of a hip roof; the specifications are as follows:

- Plan view—rectangular, 28′ × 40′
- Pitch—1/3
- Overhang—16″
- Rafters 16″ on center (O.C.)
- Common rafters made of 2″ × 4″
- Hip rafters made of 2″ × 6″
- Ridge made of 1″ × 6″

The rise of a hip rafter is equal to the rise of the common rafter against which it fits. The run of a hip rafter is equal to the diagonal of a square whose sides are equal to the run of the common rafters (Figure 16-52). Most importantly, the run of a hip rafter tail is equal to the diagonal of a square whose sides are equal to the run of the common rafter tail.

Figure 16-52 reveals a great deal about the hip roof by showing the outline of the wall plates (A, B, C, and D); the runs of the six common rafters (g-1, f-f′, f′-f2, f′-j, l-k, and l-e); the runs of the four hip rafters (A-I, D-I, B-f′, and C-f′); and the ridge (l-f′).

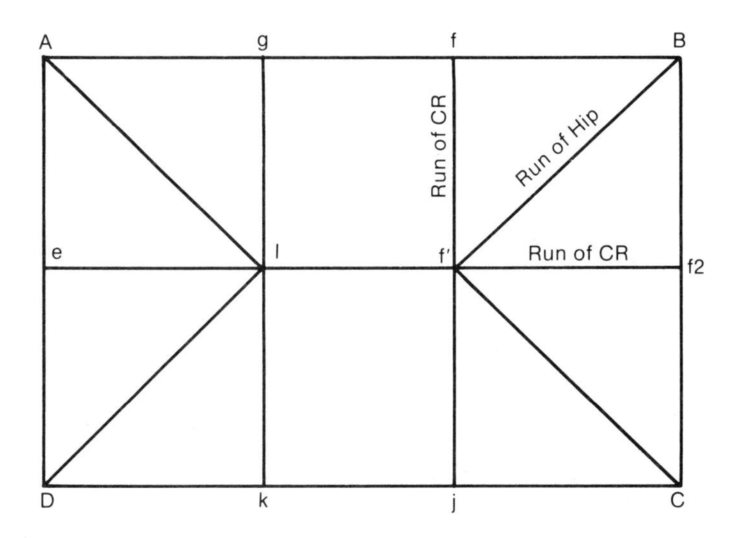

FIGURE 16-52. Outline of hip roof.

16.10.1 Layout of the Hip Rafter

Hip rafter stock is usually of wider material than common rafter stock. While 2″ × 4″s are used for common and jack rafters, 2″ × 6″s are best for hip rafters. Select a sound piece of stock and place it on a pair of sawhorses with the crown side facing away.

For a 1/3 pitch roof, the stair gauges are set on the framing square at 8″ and 12″ for the common rafter. Because the rise of a hip rafter is equal to the rise of the common rafter against which it fits, one gauge is likewise set at 8″ for the rise of the hip rafter.

As was mentioned previously, the run of a hip rafter is equal to the diagonal of a square whose sides are equal to the run of the common rafter. The run of the common rafter is 12″. The diagonal of a square with 12″ sides is 17″, so the other stair gauge is set at 17″ for the run of the hip rafter (Figure 16-53). Use the framing square to draw in the first plumb line (Line 1) on the hip rafter as shown in Figure 16-54.

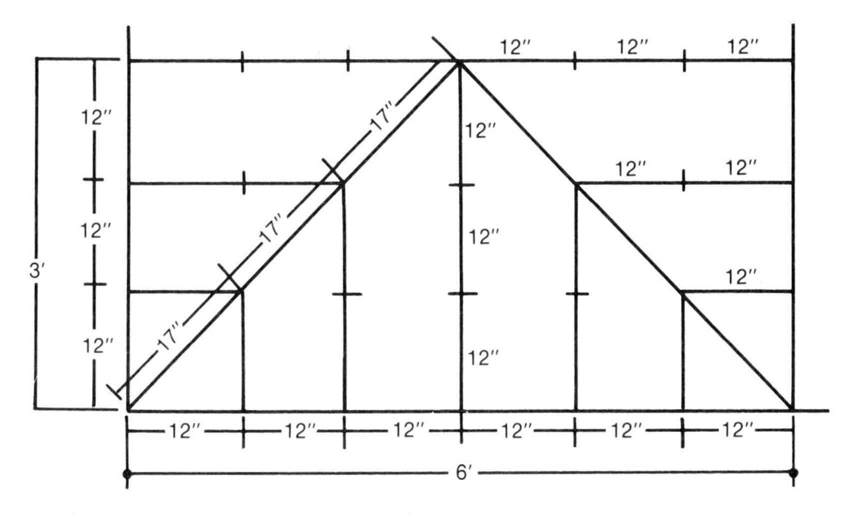

FIGURE 16-53. Measuring the run of a hip rafter.

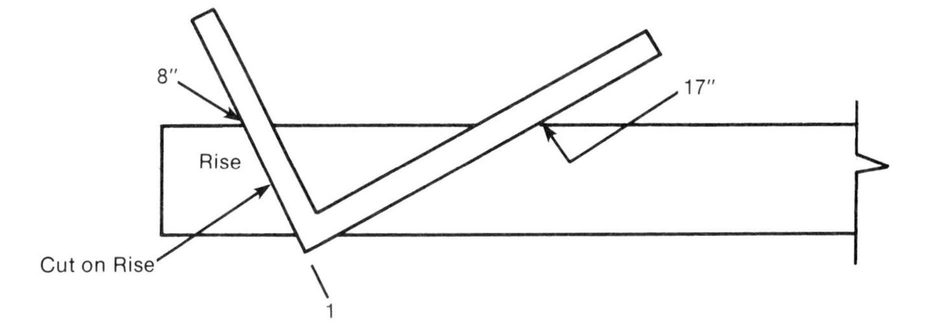

FIGURE 16-54. Drawing the first plumb line.

16.10.2 Line Length of the Hip Rafter

Figure 16-55 shows the sequence in which the cutting lines are made on a hip rafter. This sequence remains the same regardless of which rafter layout method is being used; also, all positions of the framing square for obtaining plumb and level cutting lines are in exactly the same relative positions as those for pitch boards. All line length calculations should be made directly on the face of the rafter stock, a bit to the right of Line 1 for the line length of the hip rafter and a bit to the right of the birdsmouth for the line length of the hip rafter overhang.

Read on the framing square rafter tables under the 8″ mark and on the second line. The figure 18.76″ is the determined hypotenuse of a right triangle with a base (run) of 17″ and a height (rise) of 8″. Multiply 18.76 by the number of feet of run in the common rafter, 14: 18.76 × 14 = 21′ 10-5/8″. This is the line length of the hip rafter.

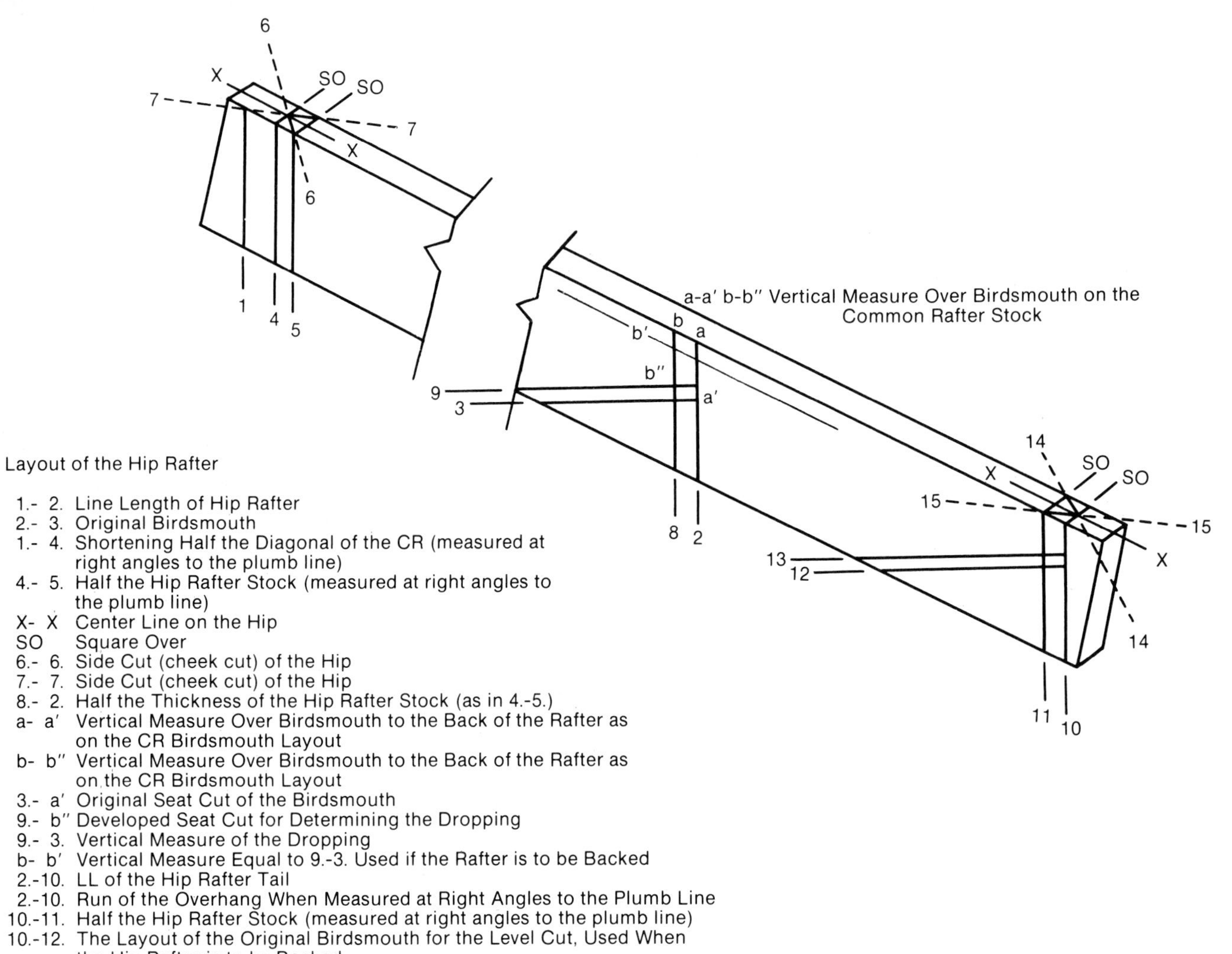

Layout of the Hip Rafter

1.- 2. Line Length of Hip Rafter
2.- 3. Original Birdsmouth
1.- 4. Shortening Half the Diagonal of the CR (measured at right angles to the plumb line)
4.- 5. Half the Hip Rafter Stock (measured at right angles to the plumb line)
X- X Center Line on the Hip
SO Square Over
6.- 6. Side Cut (cheek cut) of the Hip
7.- 7. Side Cut (cheek cut) of the Hip
8.- 2. Half the Thickness of the Hip Rafter Stock (as in 4.-5.)
a- a' Vertical Measure Over Birdsmouth to the Back of the Rafter as on the CR Birdsmouth Layout
b- b" Vertical Measure Over Birdsmouth to the Back of the Rafter as on the CR Birdsmouth Layout
3.- a' Original Seat Cut of the Birdsmouth
9.- b" Developed Seat Cut for Determining the Dropping
9.- 3. Vertical Measure of the Dropping
b- b' Vertical Measure Equal to 9.-3. Used if the Rafter is to be Backed
2.-10. LL of the Hip Rafter Tail
2.-10. Run of the Overhang When Measured at Right Angles to the Plumb Line
10.-11. Half the Hip Rafter Stock (measured at right angles to the plumb line)
10.-12. The Layout of the Original Birdsmouth for the Level Cut, Used When the Hip Rafter is to be Backed
10.-13. The Layout of the Level Cut Used When the Hip Rafter is to be Dropped
X- X Center Line on the Back of the Hip
14.-14. Side Cut (cheek cut) for the Tail of the Hip Rafter
15.-15. Side Cut (cheek cut) for the Tail of the Hip Rafter

FIGURE 16-55. Line sequence for hip rafter cuts.

Measure with a tape measure from Line 1 on the back of the rafter a distance of 21′ 10-5/8″, then draw Line 2 as shown in Figure 16-56. To check the length, step off the framing square 14 times; it should check to a tolerance of ±3/8″.

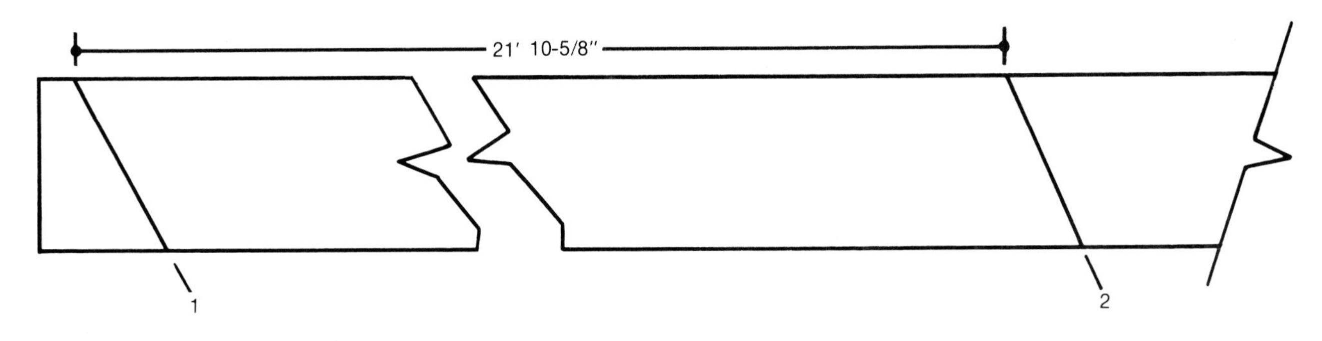

FIGURE 16-56. Drawing Line 2.

16.10.3 Hip Rafter Original Level Line at the Birdsmouth

Figure 16-57, the pattern common rafter, shows Line 2 at a-a′ to be the perpendicular height of the rafter from the seat cut, Line 3, to the back of the rafter. Figure 16-58, the pattern hip rafter, shows Line 2 at a-a′ to be the amount of rafter above the plate as measured from the common rafter and reproduced on the hip rafter. Since 2″ × 6″s are being used for the hip rafters, a larger birdsmouth must be removed. Place the framing square in position to draw the level line, Line 3 (Figure 16-59).

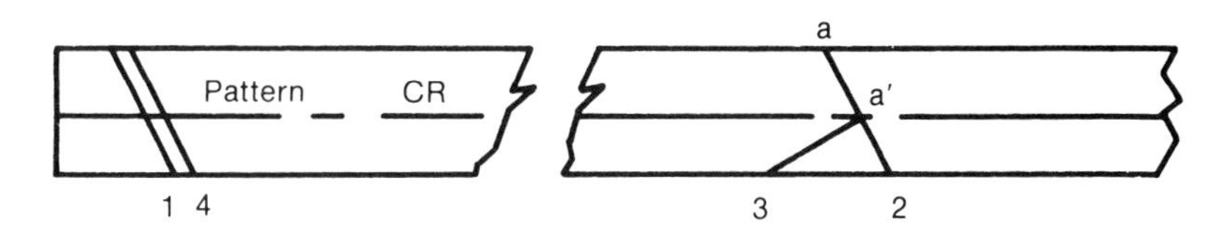

FIGURE 16-57. Height of common rafter from seat cut.

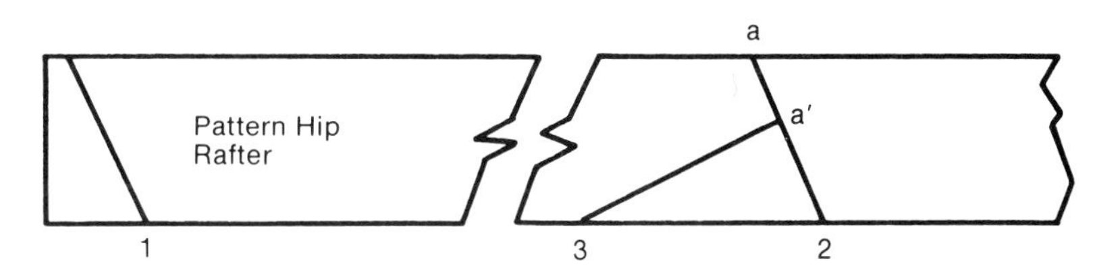

FIGURE 16-58. Measuring the amount of pattern hip rafter above plate.

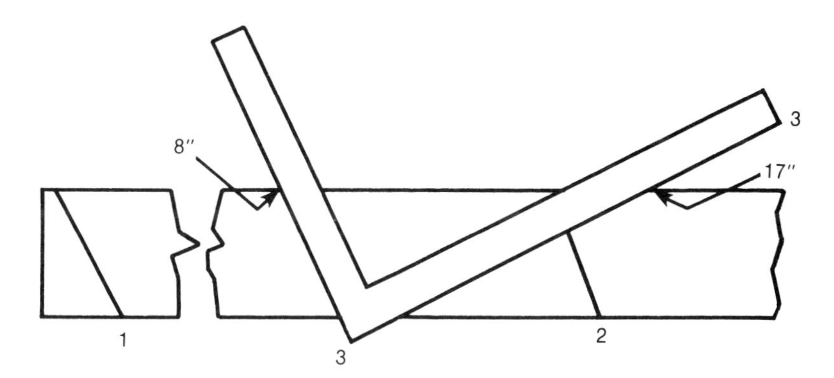

FIGURE 16-59. Drawing the level line.

By observing Figure 16-60, it can be determined that the run of the shortening of the hip rafter is equal to half the diagonal thickness of the common rafter as at Lines 1 and 4. To shorten, measure back, at right angles to Line 1, half the diagonal thickness of the common rafter; in the case of a 2″ × 4″, the distance is 1-1/8″. Draw in Line 4 measured at right angles to Line 1 (Figure 16-61).

Upon inspection, it can also be determined that the run of the side cuts of the hip rafter is equal to half the thickness of the hip rafter stock. To make a side, or cheek, cut on a hip rafter, measure back at right angles from Line 4 half the thickness of the rafter stock. In the case of a 2″ × 6″, half the thickness is 13/16″. Draw in Line 5 as shown in Figure 16-62. The side cut is completed by drawing Lines 6 and 7 (Figure 16-63).

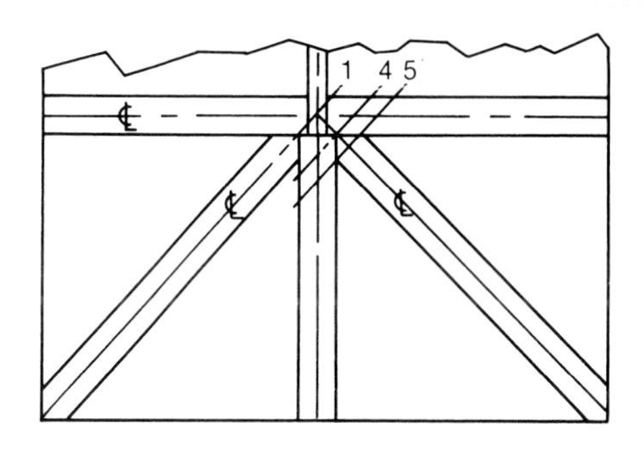

FIGURE 16-60. Amount of necessary hip rafter shortening.

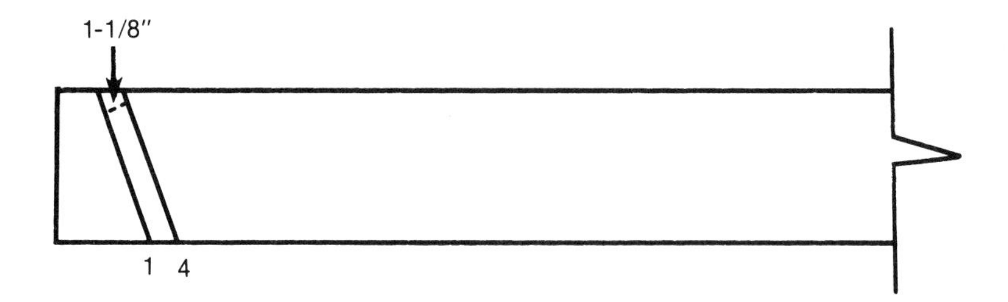

FIGURE 16-61. Drawing Line 4.

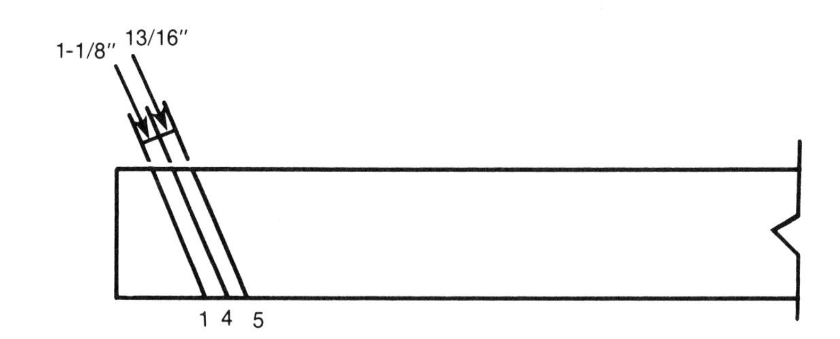

FIGURE 16-62. Drawing Line 5.

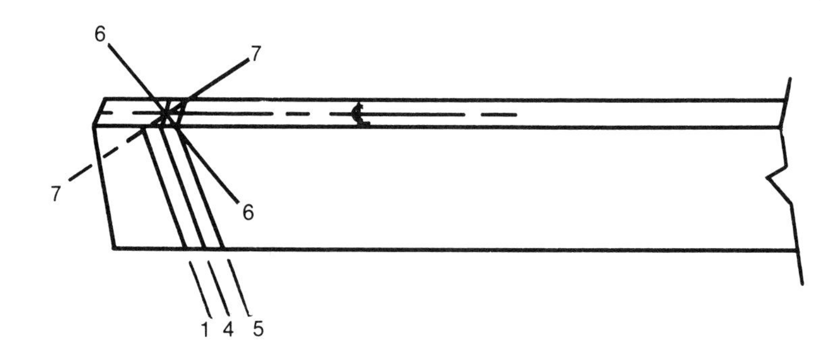

FIGURE 16-63. Completing the side cut by drawing Lines 6 and 7.

16.10.4 Backing of the Hip Rafter

The sheathing of adjacent sides of a roof meets at a point above the center of the back of the hip rafters; for this reason, it is good practice to have the edges of the hip rafters beveled to receive the sheathing. A close look at Figure 16-64 reveals that a perpendicular line through the center of the back of the hip on Line 2 at point a would intersect with the inside intersection of the wall plates. This perpendicular height corresponds to the perpendicular height of the common rafter from the seat cut Line 2 of the common rafter.

The length of the run between Lines 2 and 8 on the plan is equal to half the thickness of the hip rafter stock. (It is also equal to the run between Lines 4 and 5.) The length of the run can be seen between Lines 2 and 8 in Figure 16-65. On the pattern hip rafter, draw Line 8 forward from Line 2, a distance of 13/16", as seen in Figure 16-66.

The edges of the hip rafter are vertical over the wall plates, as seen at b and d on Line 8 in Figure 16-64. Points b and d are the end views of plumb lines from the wall plate to the top edges of the hip rafter. Close

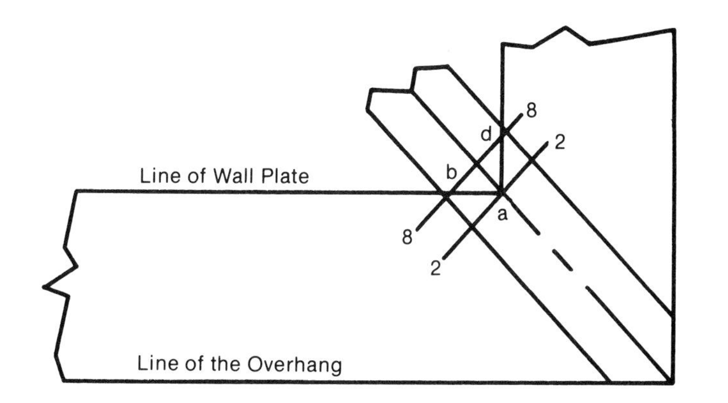

FIGURE 16-64. Perpendicular line intersecting hip rafter and wall plate.

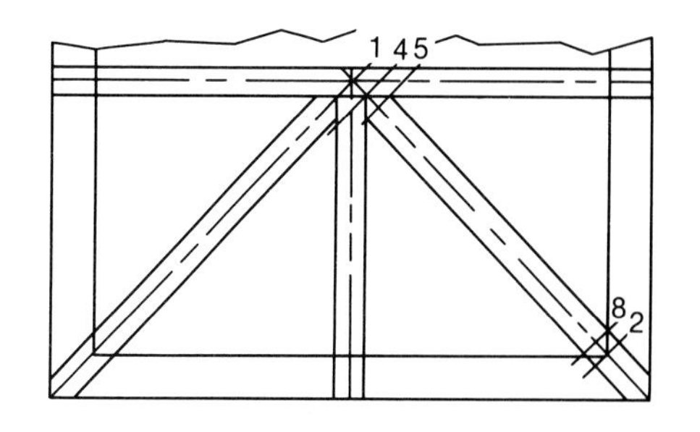

FIGURE 16-65. Plan view of length of run between Lines 2 and 8.

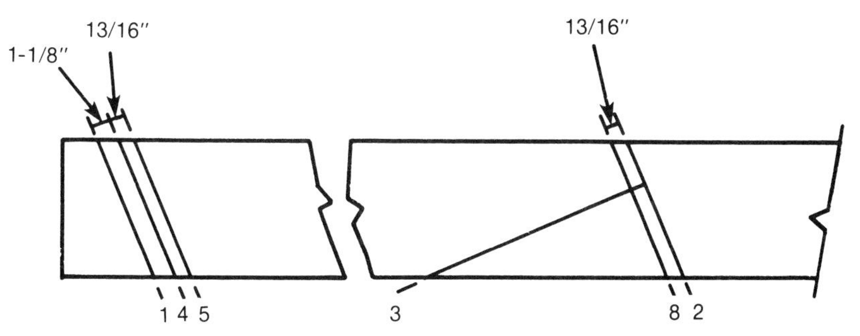

FIGURE 16-66. Drawing Line 8.

inspection of Figure 16-64 reveals that b and d on Line 8 and point a on Line 2 are on the same level plane as corresponding points on the jack and common rafters.

The height of the center of the hip rafter above the wall plate was located at a-a′ on Line 2 of the pattern hip rafter layout. This is the same height as is measured from the back of the common rafter on Line 2 of the common rafter layout. The measurement from the wall plates to the outside edges of the hip rafter, where the hip rafter passes directly over the wall plate, is a-a′ and must be measured from the back of the rafter on Line 8.

Measure down from the back of the hip rafter on Line 8 the same distance as on Line 2 at a-a′. Spot b′ on Line 8 as shown in Figure 16-67. Place the framing square with the run to intersect Line 8 at b′, then draw a level line (Line 9) as shown in Figure 16-68.

The measurement between b′ and b_2 on Line 8 is the amount of bevel required. From the back of the rafter on Line 8, use a carpenter's square to measure a distance equal to b′-b_2 as at c-c (Figure 16-69).

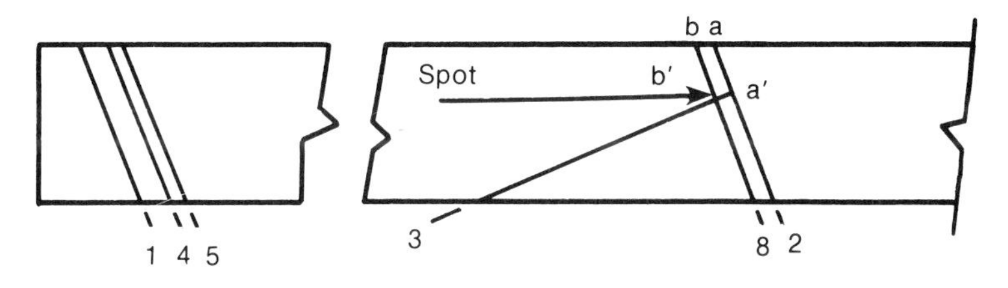

FIGURE 16-67. Locating b′ on Line 8.

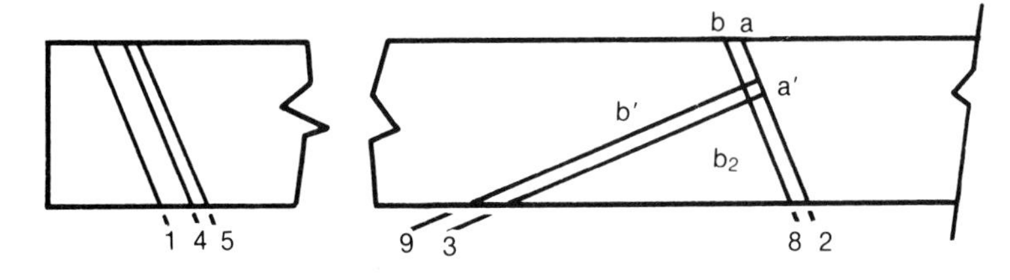

FIGURE 16-68. Drawing Line 9.

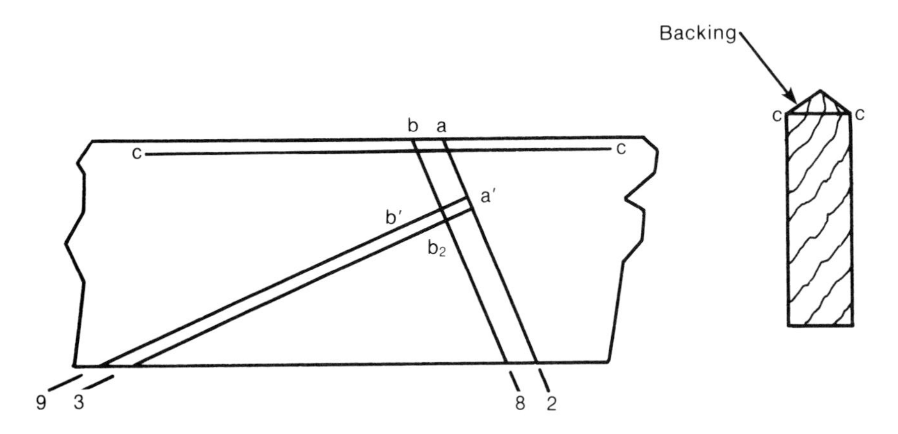

FIGURE 16-69. Amount of beveling on hip rafter.

Draw a center line down the back of the rafter; the amount to be beveled is shown in detail in Figure 16-69. Note: b'-b_2 equals b-c.

16.10.5 Dropping the Hip Rafter

As an alternative to backing, a hip rafter may also be dropped. The amount that it is dropped is equal to the amount that it would otherwise be backed. Thus, it follows that if the rafter is to be dropped, the seat cut must be made on Line 9 (Figure 16-68). On the other hand, if it is to be backed, the seat cut should be on Line 3. Remember that a hip rafter *must* be either backed or dropped to ensure that its edges will line up with the main plane of the roof.

16.10.6 Line Length of the Tail of the Hip Rafter

The line length of the tail of the hip rafter is marked off from the plumb line (Line 2) of the original birdsmouth. Make the calculations on the face of the rafter, a bit to the right of Line 2. Read on the rafter tables of the framing square under the 8″ mark on the second line. The figure 18.76 is the length of the hip rafter per foot run. Multiply 18.76″ by the number of feet of run in the common rafter tail: 18.76 × 1-1/3 × 25″. Measure 25″ along the back of the hip rafter from Line 2 and draw the plumb line, Line 10; this is the line length of the tail from Line 2 (Figure 16-70). Check the line length with the step-off method to ensure accuracy.

16.10.7 Tail End Side Cuts of the Hip Rafter

If the eaves are to be boxed in, the tail end side cuts will be the reverse of the side cuts at the ridge end of the hip rafter. From Line 10 of the plan

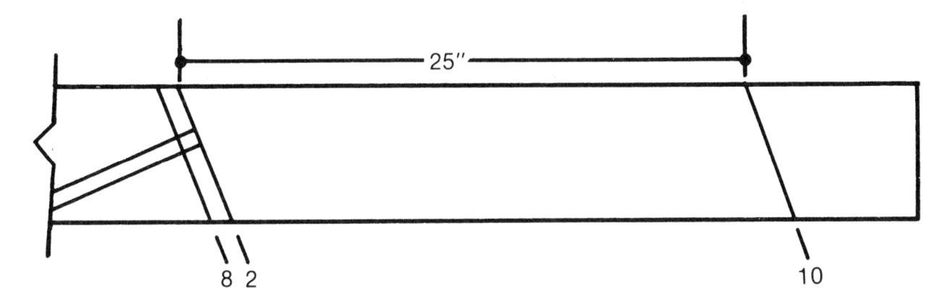

FIGURE 16-70. Drawing Line 10.

view, mark off half the thickness of the rafter stock, as seen at Line 11 in Figure 16-71. On the hip rafter, measure back (at right angles from Line 10 towards the birdsmouth) half the thickness of the rafter stock, or 13/16". Draw Line 11 (Figure 16-72).

Draw a center line on the back of the rafter as seen at x-x. From Line 11 draw Lines 14 and 15 to intersect at the center line on Line 10. From Line 10 draw two level lines in the same relative positions as Lines 3 and 9 from the back of the rafter at the birdsmouth; these will be Lines 12 and 13. The final cutting level line at the end of the rafter will correspond to the cutting line at the birdsmouth. If the rafter is to be dropped, cut on Line 9 at the birdsmouth and on Line 13 at the tail; if it is to be backed, cut on Line 3 and Line 12. *Note:* It is common practice to allow the hip rafter tail end to run out past the tail ends of the common rafters and then to mark the actual length with a straight edge lined up with the tail ends of the common rafters, from both sides of the roof.

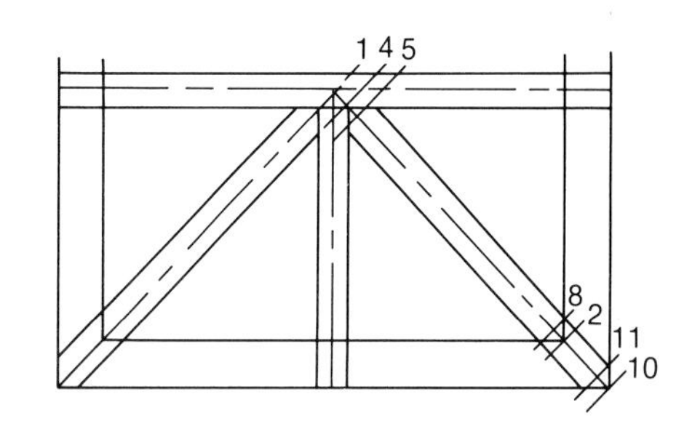

FIGURE 16-71. Plan view of hip rafter tail end side cuts.

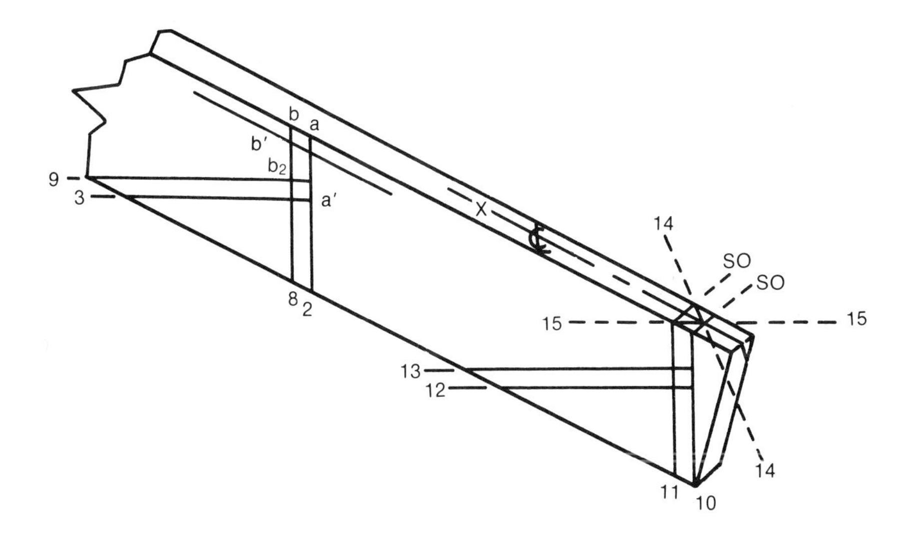

FIGURE 16-72. Final hip rafter cuttting lines.

16.11 LAYOUT OF JACK RAFTERS

This section deals with the layout of jack rafters for a regular hip roof. With the exception of the side cuts, all other cuts of jack rafters are identical with those of common rafters for the same regular hip roof. The runs of pairs of jack rafters and the wall plate lines form perfect squares on a plan view. A simple rule to remember is this: The run of any jack rafter on a regular hip roof is equal to its distance from the outside corner of the wall plates (Figure 16-73).

16.11.1 Common Differences in Lengths of Jack Rafters

The difference in the lengths of succeeding jack rafters on a regular hip roof is equal to the number of feet and portions of feet in the center-to-center spacings of the jack rafters multiplied by the line length of the common rafter per foot of run. For example: For a 1/3 pitch roof with jack rafters placed at 14″ O.C., the common difference in the runs of the jack rafters is 1-1/6′. The line length per foot run of the common rafter, as read from the rafter tables, is 14.42. The common difference in the lengths of the jack rafters is 14.42 × 1-1/6, which calculates to 16.83 or 16-7/8″ to the nearest 1/8″. While this is an odd O.C. placing of jack rafters and is not indicated on the rafter tables, it shows that any problem may be solved in this manner.

The jack rafters are laid out on the pattern common rafter. This is then used for the layout of the jack rafters for the entire roof. The example used is for a regular hip roof of 1/3 pitch with rafters placed at 16″. Figure 16-73 shows the plan view of the jack rafters (not to scale).

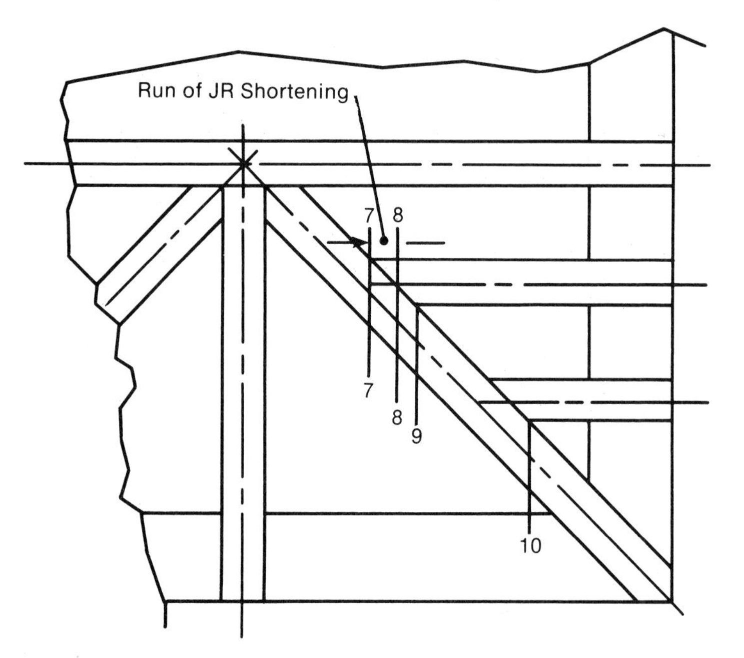

FIGURE 16-73. Plan view of jack rafters.

16.11.2 Line Length of the Longest Jack Rafter

The line length of the longest jack rafter is 19-1/4″ shorter than the line length of the pattern common rafter. It is measured from its center line intersection with the center line of the hip rafter through to the birdsmouth (Figure 16-73). From Line 1 on the face of the common rafter, mark off 19-1/4″. This is the difference in the line lengths of the common rafter and the jack rafter. Draw the plumb line, Line 7, but do not square this line over (Figure 16-74).

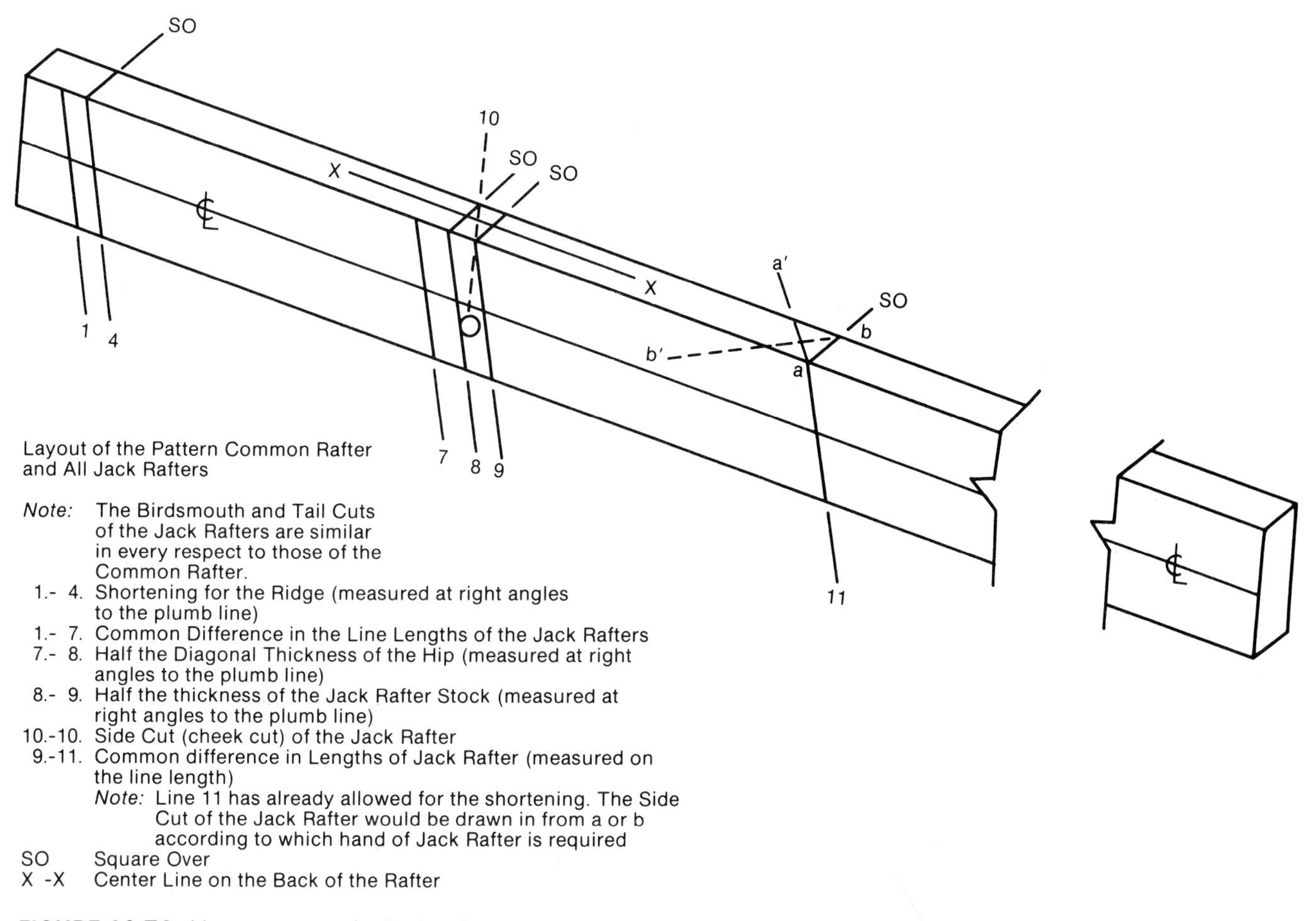

FIGURE 16-74. Line sequence for jack rafter cut.

16.11.3 Shortening of the Longest Jack Rafter

Because the jack rafters meet the hip rafters at a 45° angle, they must be shortened half the diagonal thickness on the run of the hip rafter stock. From Line 7 measure at right angles half the diagonal thickness of the hip rafter stock and draw another plumb line, Line 8. Square this line over (Figure 16-75).

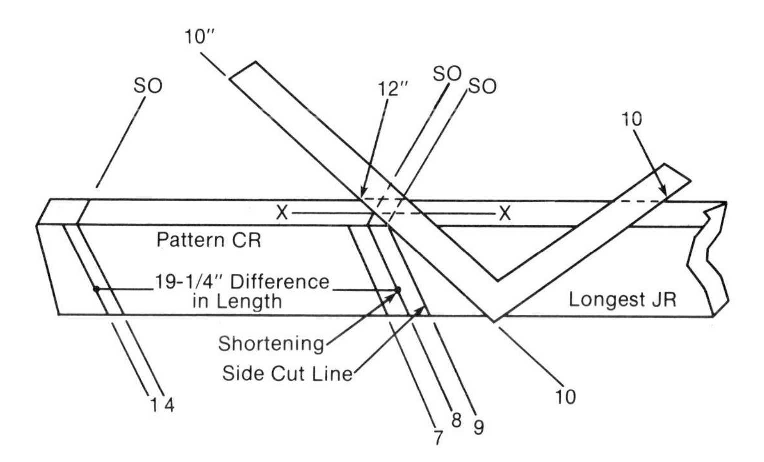

FIGURE 16-75. Layout of jack rafter on pattern common rafter.

16.11.4 Side Cut of Jack Rafters

Read on the fifth line of the framing square rafter tables, under the 8″ mark. Set one stair gauge on the outside edge of the tongue of the square at the 10″ mark, the figure read from the table. Set the other stair gauge on the outside edge of the blade on the square at the 12″ mark. Place the square on the back of the rafter with the 10″ setting towards the ridge end as seen in Figure 16-75.

Draw the side cut line, Line 10. All other side cut lines are drawn with the framing square. Once the longest jack rafter has been laid off, the second and succeeding jack rafters can be laid out (also on the pattern common rafter). From the squared-over Line 9 on the back of the rafter, mark off the remaining differences in lengths of the jack rafters. Square over the spaces of 19-1/4″ on the back of the rafter as shown in Figure 16-76.

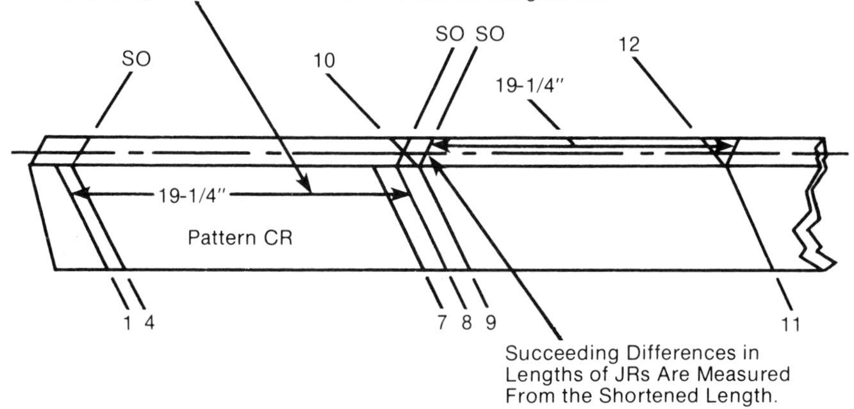

FIGURE 16-76. Marking differences in the line lengths of the remaining jack rafters.

16.11.5 Layout of Pairs of Jack Rafters From the Pattern Common Rafter

From the squared-over Line 9 on the back of the pattern common rafter, mark off all the remaining common differences of 19-1/4″ for the line lengths of the jack rafters. Pieces of stock of suitable length may now be placed in pairs on either side of the common rafter, as seen in Figure 16-77.

Square over four pairs of jack rafters from the pattern common rafter. Each pair will be 19-1/4″ different in length from the next pair. With stair gauges set on the framing square at 10″ and 12″, draw in the bevel side cut lines on the backs of the jack rafters as seen in Figure 16-77. Reset the stair gauges to 8″ and 12″ for the plumb line of a 1/3 pitch roof. Draw in the plumb cutting lines on all the pairs of jack rafters. The completed pairs of jack rafters are now ready for nailing into position.

16.11.6 Additional Rafters

For added architectural effect and functional advantage, architects and designers often draw roofs with an intersecting portion of either

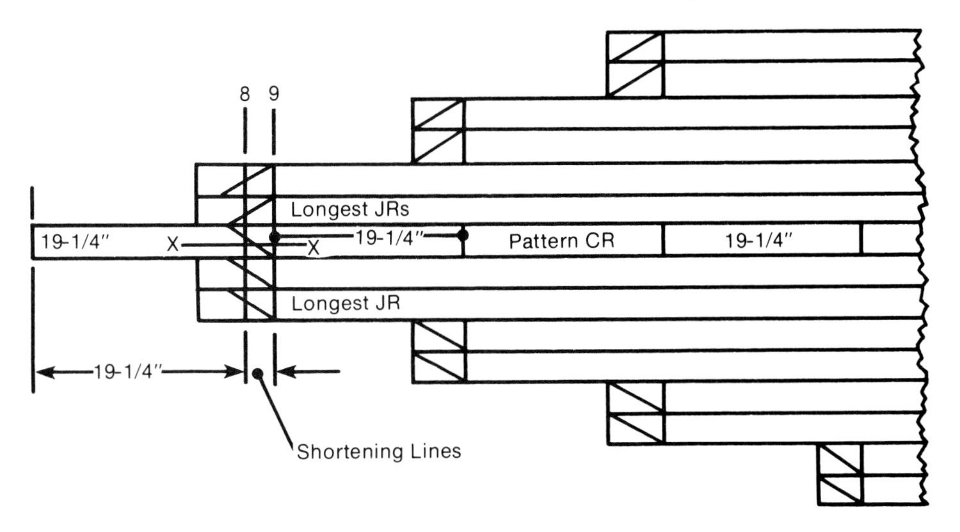

FIGURE 16-77. Drawing bevel side cut lines on additional jack rafters.

equal or unequal pitch. The part of the roof covering the greatest area is called the major roof, and the remaining part is the minor roof. The scaled drawings, layout, and erection of the common rafters, hip rafters, and jack rafters for the major roof are identical to those of ordinary roofs (hip, gable, etc.); in addition, the common rafters for the minor roof are the same as those of the major roof.

There are additional rafters, however, that must be laid out and erected for intersecting roofs. Valley rafters, whose run and rise measurements are identical to those of the hip rafters, intersect both roofs at the center lines of their ridges and also at their wall plates. These rafters are handled in much the same manner as hip rafters. *Note:* One important difference is that, unlike the seat cut of the hip rafter, there is only one level line at the birdsmouth. Valley jack rafters (built to fit against the ridge and the valley rafter) and valley-cripple jack rafters (built to run from valley rafter to supporting valley rafter) are also part of such building projects. Both the graphic and framing square methods of rafter framing are still used for these rafters, but the work involved is much more complicated and time-consuming. Unquestionably, the builder should be sufficiently experienced in the more basic roof framing projects before taking on projects with intersecting roofs.

16.12 SHEATHING, PURLIN, AND WINDBLOCK CUTS

Roof sheathing gives rigidity and provides insulation to the structure; it also forms the base for the roof covering. Roof sheathing may be shiplap, tongue and groove, common boards, plywood, or composition boards. While shiplap or common boards must be applied solid if composition roofing is to be used, the boards may be spaced if wood shingles or tile roofing is specified (Figure 16-78).

The only specifications required by a builder to lay out a regular hip roof sheathing cut are the span and the pitch. In this section, the span will be 28′ and the pitch 1/3.

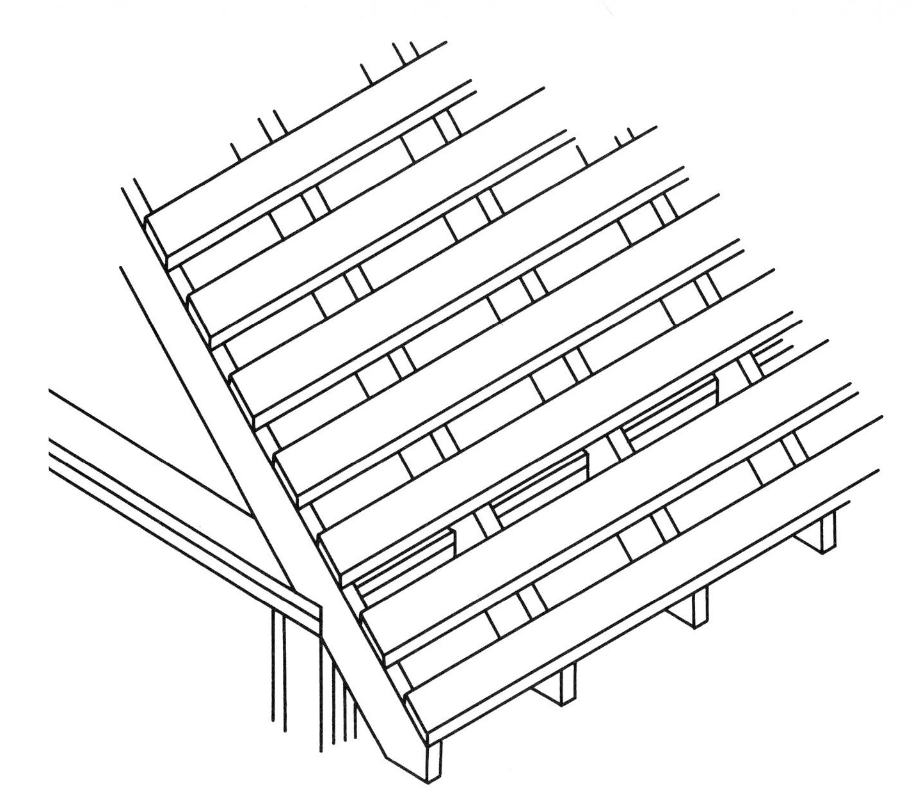

FIGURE 16-78. Spaced roof sheathing.

16.12.1 Angle of the Sheathing Face Cut

Figure 16-79 shows a regular hip roof with one portion projected to a horizontal plane. The right triangle a-b-c has a base equal to the run, a height equal to the line length of the common rafter, and a hypotenuse equal to the line length of the hip rafter. This triangle is shown in a horizontal plane at a-b-c′. The face angle of the sheathing is shown on both triangles at b. The sheathing face cut is the same angle as the side cut of the purlin and the windblock (Figure 16-80).

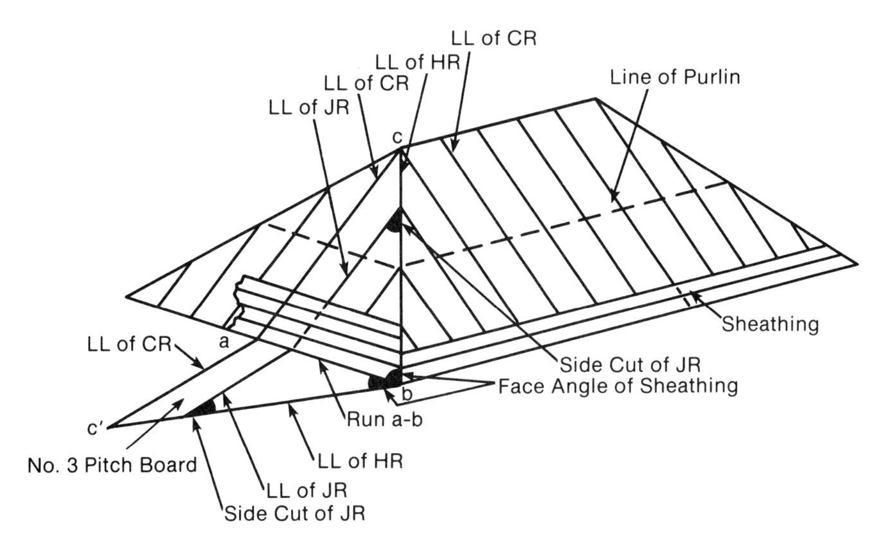

FIGURE 16-79. Hip roof showing angle of sheathing.

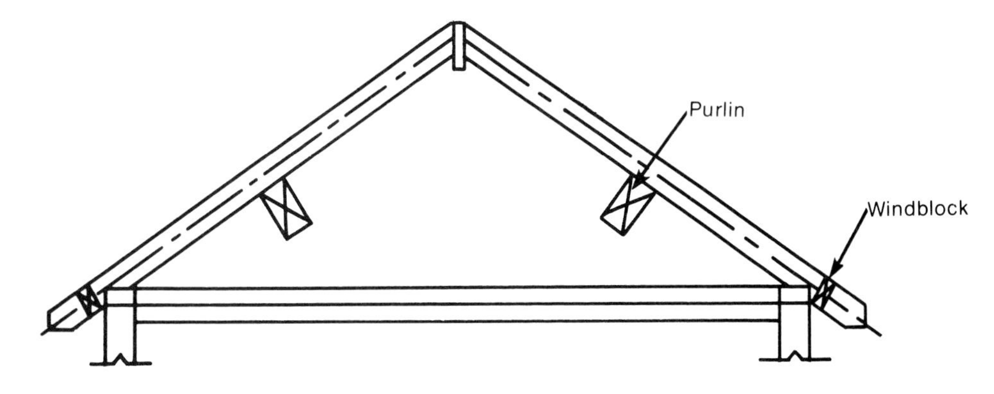

FIGURE 16-80. Purlin and windblock.

16.12.2 Sheathing Face Cut

Lay the framing square on a piece of sheathing stock with the stair gauges set at 14″ for the run of 16-5/6″ for the line length of the common rafter (as read from the development drawing). Draw a line along the blade of the square (Figure 16-81).

Slip the framing square along the drawn line until the 12″ mark can be read at the bottom edge of the stock. Notice that, with the framing square in this position, although 12″ is read on the blade, 10″ is read on the tongue. Mark off the sheathing cut on the 10″ side. For a regular hip roof, stair gauges are always set with one on 12″ for the layout of the sheathing face cut, the purlin side cut, and the windblock side cut.

Consult the rafter tables for a 1/3 pitch roof. On the fifth line under the 8″ mark, the 10 refers to the 10″ in the previous paragraph. Set one gauge to 12″ and the other to 10″, and cut on the 10″ side. Use the same figures on the framing square for the face cut of the sheathing as are used for the side cut of the jack rafters (Figure 16-82). The framing square is based on a unit of 12″. In this case, the unit is related to the line length of the common rafter which gives a relative figure of 10″ on the run side.

When the framing square is being used to find the face cut of the sheathing for any regular hip roof, the only information needed is the pitch. For example, say the pitch of a roof is 1/4. What figures should be used on the framing square for the sheathing face cut angle? The rise

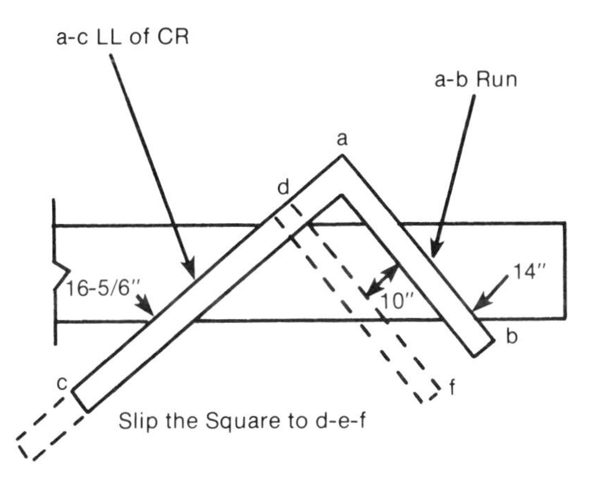

FIGURE 16-81. Slipping the framing square.

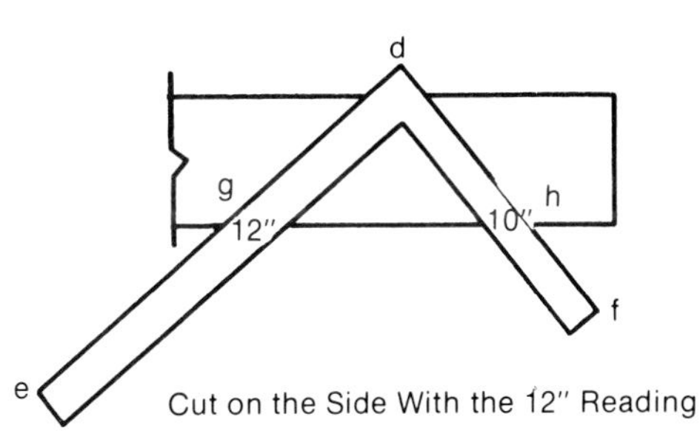

FIGURE 16-82. Locating the sheathing face cut.

per foot run is calculated to be 6″. Read on the fifth line of the rafter tables under 6″: Use 10-3/4. Set the stair gauges to 12″ and 10-3/4″ and mark and cut on 10-3/4″.

On many roofs the sheathing is nailed into place with the ends left running untrimmed past the hips. A chalk line is then struck from the ridge to the tail, over the center line of the hip. A hand power saw is used to cut through the sheathing from the ridge to the tail. *Note:* This method is not practical for the valley sheathing.

16.12.3 Purlin and Windblock Cuts for a Regular Hip Roof

The miter cuts are identical for both the purlin and the windblock. The purlin is very important because it gives support to the rafters and, providing that it is mitered and fitted snugly at the hips, gives strength to the entire roof. The windblock, as its name implies, is secured in place to close the open top of the wall frame to the underneath side of the roof sheathing.

The surface of the purlin that can be seen in Figure 16-83 are at a-b and a-c in section. In the center of the drawing a single line plan view of

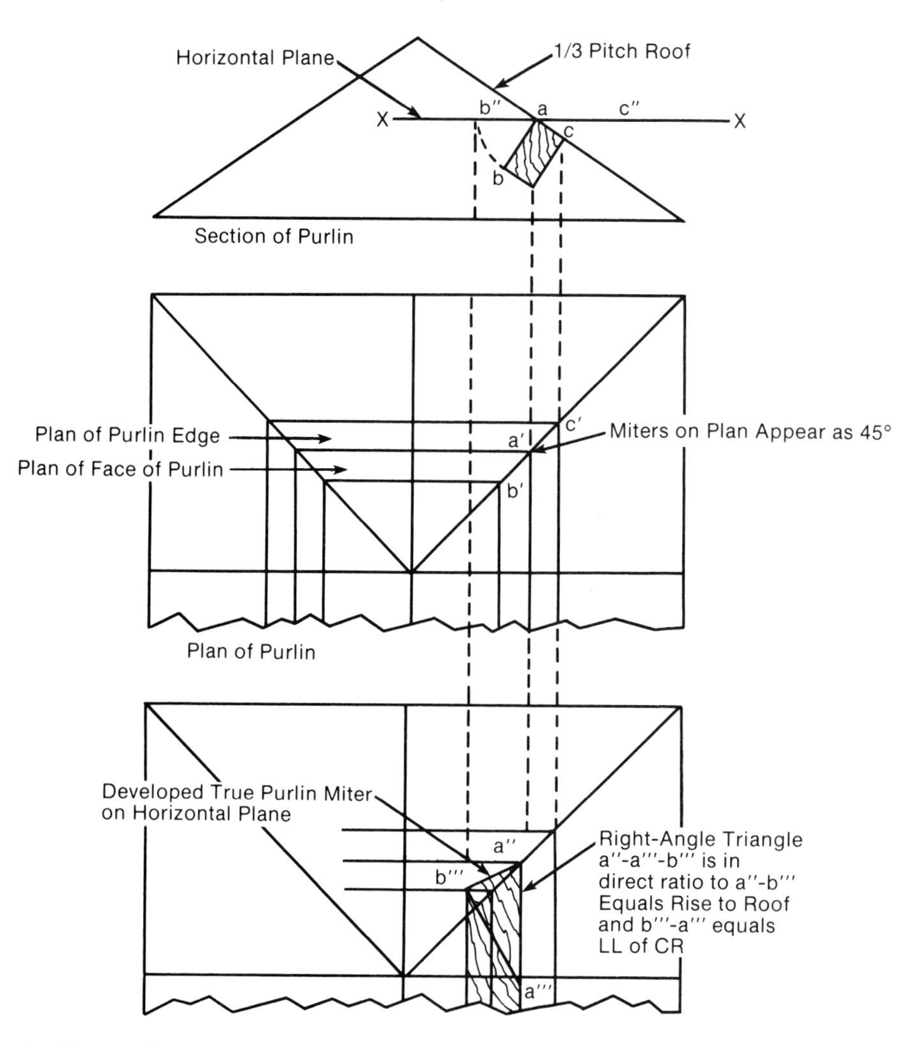

FIGURE 16-83. Miter cut of the purlin.

the miters of the purlin is shown at a′-b′-c′. On the plan they appear as at 45°. The side cut angles of the purlin and windblock are the same angle as the sheathing cut.

In the bottom of Figure 16-83 a single line plan view of a portion of the hip roof is shown, with a developed true miter of the purlin face cut shown at a″-a‴-b‴. To establish the true miter cut of the purlin, the face and the edge must be projected to a horizontal plane. The face a-b of the purlin in section is assumed to have revolved about the point to a horizontal plane x-x at a-b″. The true face width is now projected to the bottom portion of the drawing at b″-b‴.

The shaded portion of the purlin on the plan is assumed to have revolved about a″-a‴. By studying the right triangle a″-a‴-b‴, it is observed that it is in direct ratio to a right triangle with a base equal to the rise of the roof and a height equal to the line length of the common rafter. This can be verified by taking a piece of building paper or composition board and making a full-size layout of a section of a purlin for a 1/3 pitch roof. Develop the face angle of the purlin; check by placing a framing square with the 8″ mark of the tongue on a″ and the 14.42″ (14-3/8″) mark of the blade at a‴. Draw and cut the miter on the 8″ side.

For all purlin face cuts, set one stair gauge to the rise per foot run and the other to the line length of the common rafter per foot run, in this case 8″ and 14-3/8″. Always cut on the rise side (8″) as seen in Figure 16-84. When fully sheathed, a regular hip roof is like a closed inverted hopper (Figure 16-85).

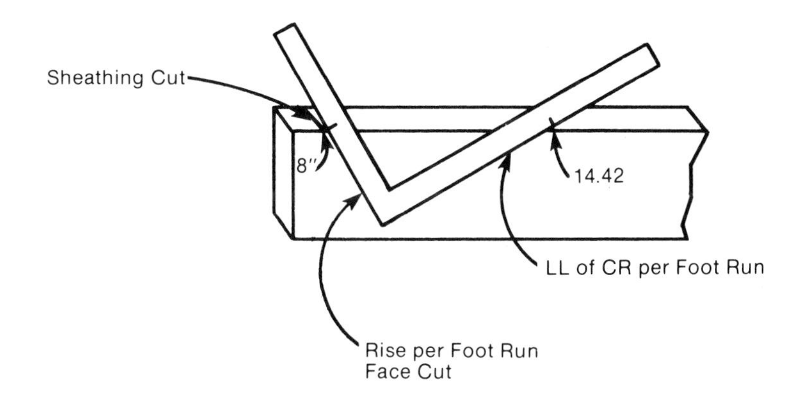

FIGURE 16-84. Making the purlin face cut.

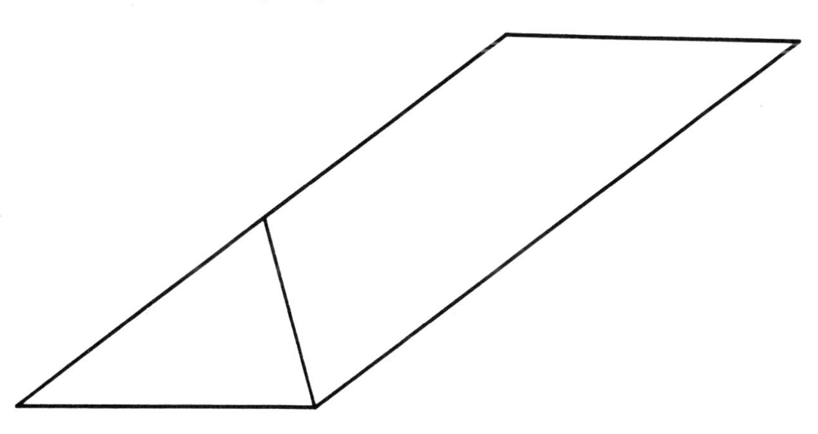

FIGURE 16-85. Closed inverted hopper.

16.13 RESEARCH

1. Identify development drawings of the various types of roofs.
2. Learn the basic terminology involved in roof framing. Know the roof framing principles and how they lend themselves to a particular roof framing project.
3. Read and study actual plans and specifications from a roof framing project and compare them with those in this chapter.
4. Learn more about roof framing theory and practice by visiting building sites in your area and noting the different styles used.

16.14 REFERENCES

16.4.1 Ball, John. *Light Construction Techniques.* Virginia: Reston Publishing Company, 1980.

16.4.2 Wass, Alonzo. *Methods and Materials of Residential Construction.* Virginia: Reston Publishing Company, 1977.

16.4.3 Wass, Alonzo, and Gordon Sanders. *Residential Roof Framing.* Virginia: Reston Publishing Company, 1980.

17

CORNICE LAYOUT WITH THE STORY ROD

This chapter illustrates how story rod measurements relate to cornice layouts for the purpose of solving light building construction problems. The specific example used here is of a full-scale cornice layout for a one-car garage. While this chapter will explain how to make and use the story rod, the actual construction details are omitted; they will be dealt with in the next chapter.

17.1 GENERAL OBJECTIVES

A story rod is usually made from a clean, straight piece of wood about 1″ × 1″ (19 mm × 19 mm). It is carefully marked to record a specific set of vertical measurements. Provided that the measurements are within a tolerance of 1/64″ to 1/32″ (.4 mm to .8 mm), the degree of accuracy is satisfactory for most stock cutting, including stud and trimmer lengths, fitting allowances, cabinet parts, and jig constructions.

Prior to marking off measurements on a story rod, the builder must have a general building plan and specifications. Story rod development is essentially a problem solving technique used to answer the question "What are the measurements?" rather than "How are the construction elements brought together?" It is very helpful in checking the practicability of building designs and specifications. Thus, the purpose of this chapter is to teach the builder—using the data from a garage building project—how the story rod and cornice layouts are used to determine the set of vertical measurements required for cutting members.

17.2 SPECIFIC OBJECTIVES

On completion of this chapter the reader should

1. be able to make a story rod for a given light building construction project,
2. know how to use trial layouts to solve construction measurement problems, and
3. be able to relate story rod measurements to a cornice layout for the purpose of solving a given construction problem.

17.3 ASSESSING THE PROJECT

In this example, a one-car garage is to be built to match the design of an already existing house. The house has a gable roof, as seen in Figure 17-1. The following details reveal more about the project:

- The slope of the garage will be identical to the slope of the house; the slope must be measured accurately.
- The frame of the garage will rest on a flat concrete slab (Figure 17-2).
- The garage will feature a metal overhead door, two conventional doors, and two windows.
- The header trim for the conventional doors and windows will be set at the same level as the trim for the overhead door.
- The eaves will be boxed.

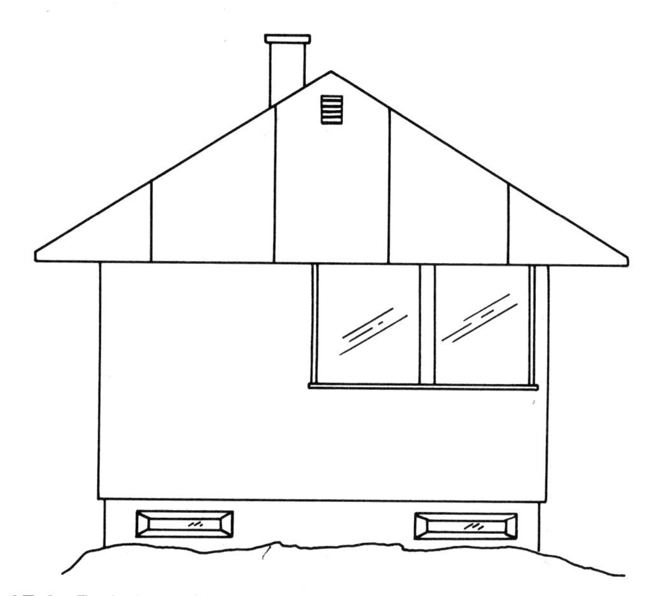

FIGURE 17-1. End view of house with gable roof (not drawn to scale).

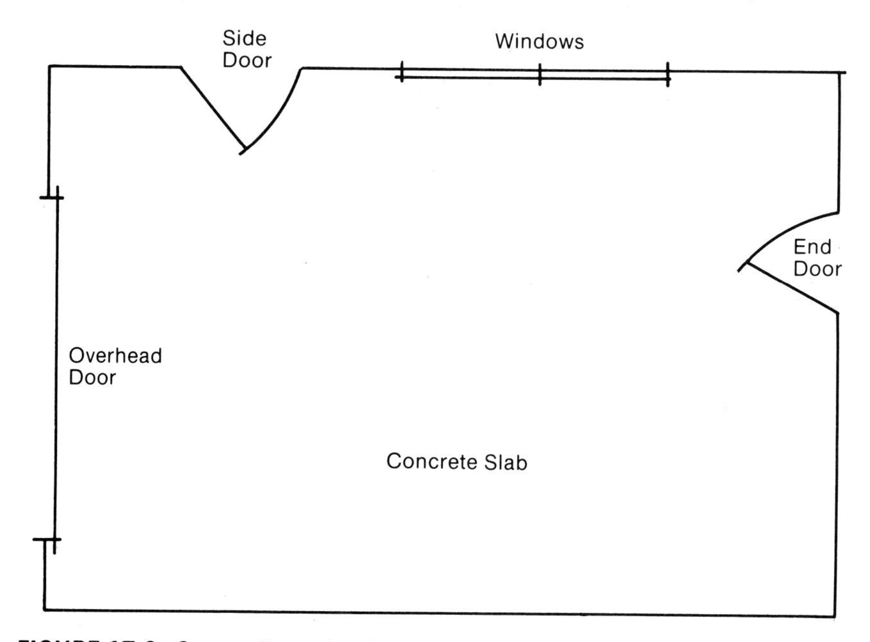

FIGURE 17-2. Garage floor plan (not drawn to scale).

In order to be able to cut the studs to the exact lengths which will allow the garage to be constructed in accordance with the design specifications, certain steps must be followed. These include:

1. measurement to obtain the slope of the roof of the house,
2. use of the framing square as a shop drawing tool,
3. the cornice design,
4. the story rod, and
5. trial cornice layout and story rod development.

Each step will now be examined in detail.

17.4 SLOPE MEASUREMENT HOUSE ROOF

To determine the slope of the roof of the house, place a framing square on the roof (Figure 17-3). Set a spirit level at the 12″ mark on the blade of the square. When the bubble in the level is centered, read the mark on the tongue of the square. In this case, it reads 8″ as shown in Figure 17-3. The slope of the roof, therefore, is 8″ in 12″. This can also be expressed on a drawing as a slope triangle, as shown in Figure 17-4. The pitch, using the previously mentioned formula, is $\frac{8}{2 \times 12} = \frac{8}{24} = \frac{1}{3}$.

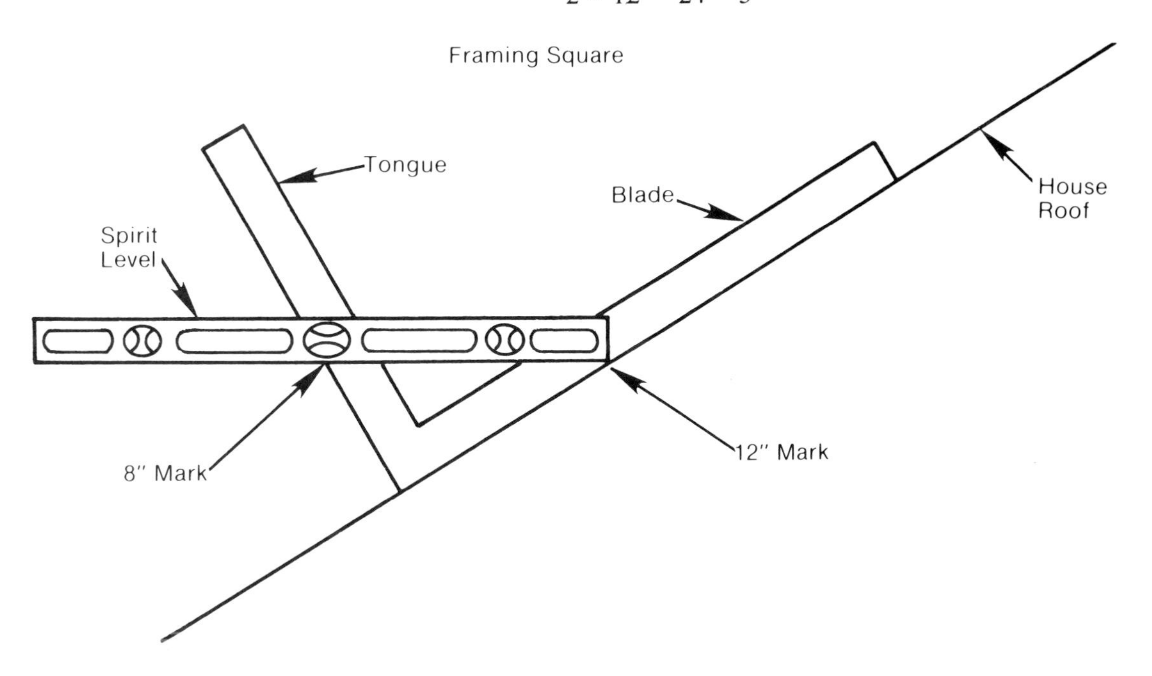

FIGURE 17-3. Using a framing square to obtain the roof slope measurement.

To go from the measurement of the roof to a workable statement of slope, it is necessary to further develop the relationship between the framing square and spirit level measurements. Consider the two right triangles ABC and DEF (See Figure 17-5). Sides AB and ED measure 8″; sides BC and EF measure 12″. Triangle ABC represents the measurement on the roof with the framing square and the spirit level; whereas, triangle DEF is the slope triangle. Side ED represents the unit rise and side EF represents the unit run of the common rafter.

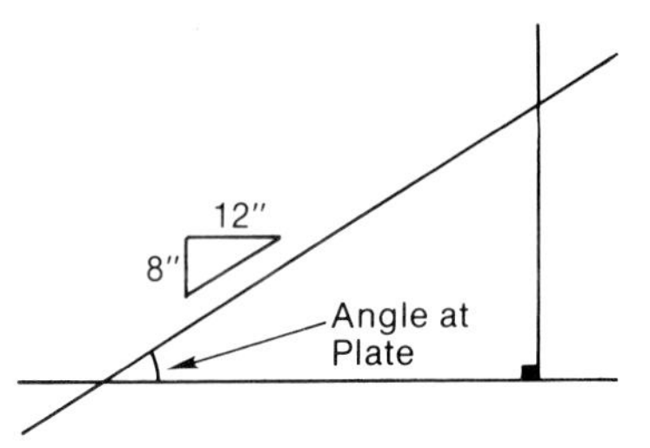

FIGURE 17-4. Slope triangle for 8″ in 12″ slope.

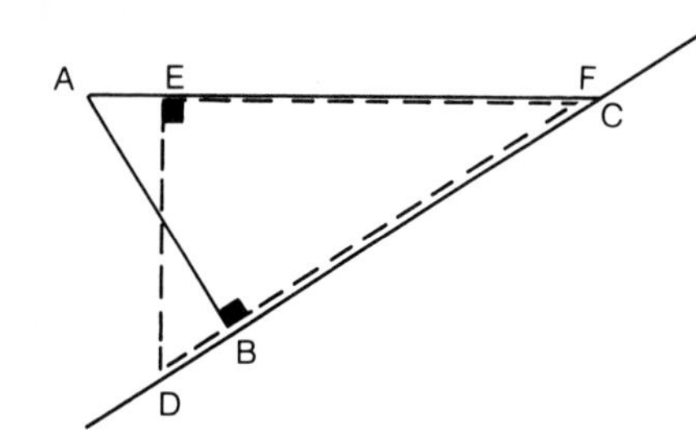

FIGURE 17-5. Expanding the framing square and spirit level measurements.

17.5 THE FRAMING SQUARE AS A SHOP DRAWING TOOL

Before preparing a full-scale drawing of the garage cornice layout, the garage roof line must be drawn. On building paper draw a straight line 30″ (750 mm) long to represent a "level" line. Set the framing square to the line and mark at 8″ and 12″ as shown in Figure 17-6. Run a roof line through the two points. Draw a slope triangle (not to scale) and dimension it without any inch or millimeter designation.

A roof line can also be drawn by placing the framing square on the drawing with the heel upward (Figure 17-7). A roof line can be drawn in this method (which is easier than the first) for any given slope or pitch.

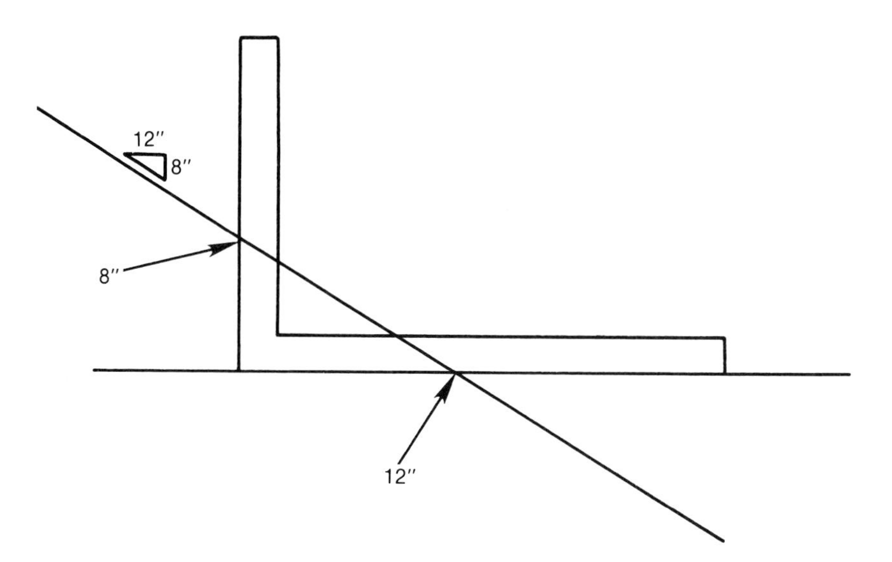

FIGURE 17-6. Drawing a roof line (method 1).

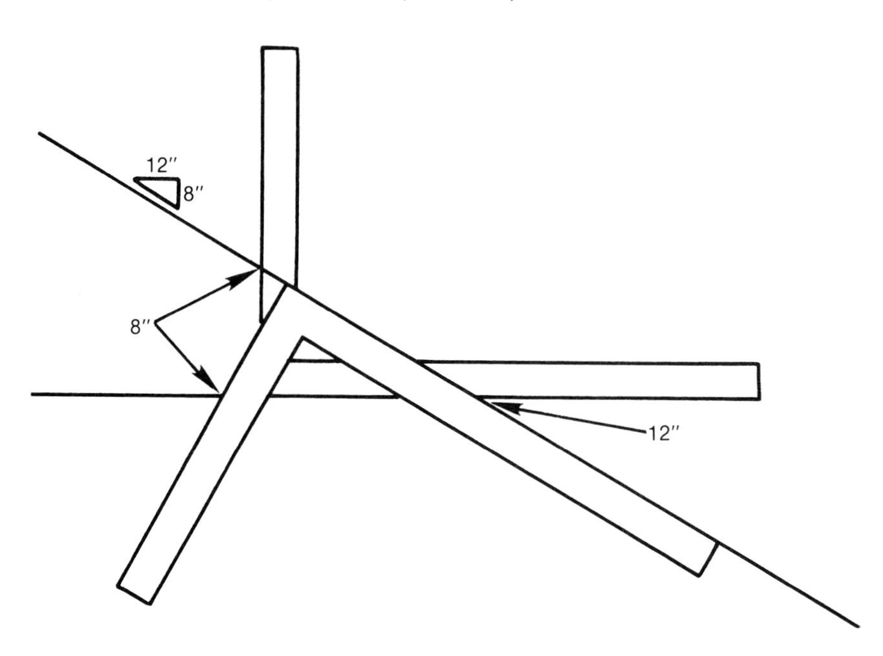

FIGURE 17-7. Drawing a roof line (method 2).

17.6 THE CORNICE DESIGN

The construction materials used to build the house included a common rafter (birdsmouth and tail), wall plates and studs, wall sheathing, two fascias, roof sheathing, a lookout, a quarter round, a soffit, and building paper. In order for the garage to match the house design, the same elements must be used, though the sizes of the stock must be determined. Knowing that the garage is to be a small, one-car unit, a materials list can be started:

- Studs and wall plates—2″ × 4″ (38 mm × 89mm)
- Rafters—2″ × 4″ (38 mm × 89 mm)
- Wall sheathing (plywood)—5/16″ (8 mm)
- Lookout parts—1″ × 4″ (19 mm × 89 mm)
- Roof sheathing (plywood)—3/8″ (10 mm)
- Rough fascia—1″ × 4″ (19 mm × 89 mm)
- Soffit (plywood)—5/16″ (8 mm); width to be determined from shop drawing
- Finish fascia—1″ × 4″ (19 mm × 89 mm)
- Quarter round—cut to fit

The soffit measurement is critical. If the plywood sheets are cut accurately and with as little waste as possible, a 4′ × 8′ (1.2 m × 2.4 m) sheet can produce four strips 12″ (305 mm) wide or six strips 8″ (203 mm) wide. Remember that the garage cornice design should be in keeping with the existing house cornice.

Trial cornice and story rod layouts may be done to accurately and quickly determine the soffit stock width. A rough sketch is very helpful in detailing the locations of critical measurements (Figure 17-8). As

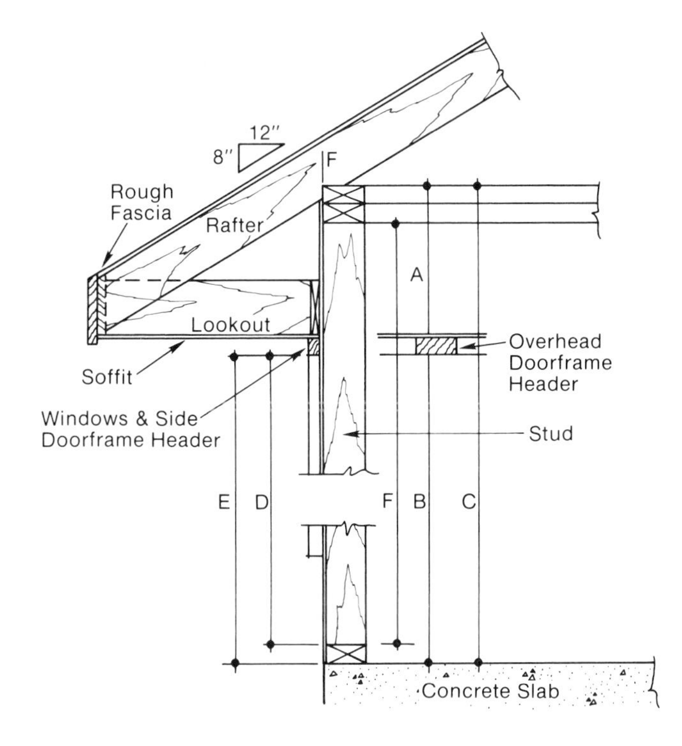

FIGURE 17-8. Critical measurements for story rod development (not drawn to scale).

can be observed, the slope of the roof is known but the rafter overhang must be determined. As the rafter overhang increases, so does the width of the soffit and the measurement from the top of the wall plate to the bottom of the soffit (A). In this particular design, the soffit level will control the placement of the frame headers for the windows, the conventional doors, and the overhead door.

The conventional doors should be standard sizes, such as 2′8″ × 6′8″ × 1-5/8″ (813 mm × 2,032 mm × 41 mm), to avoid lost time due to cutting and fitting. The overhead door must be installed according to the manufacturer's instructions. In this area, the finish opening size (B) is critical. This measurement is common to all doors and windows and must be seen that way on the story rod. While details of the gable end construction are not given here, it should be noted that a lookout ladder will carry a soffit. This soffit would probably be on the same level as the side soffit but have less width. Assuming that the thickness of the overhead door frame header is 1-5/8″ (41 mm), the story rod layout can begin.

17.7 THE STORY ROD

Select a straight piece of stock wood approximately 1″ × 1″ (19 mm × 19 mm). Its length should be greater than measurement C in Figure 17-8. The rod will stand on the concrete slab when exact height measurements are marked on it. Each of the four faces is used for different sets of corresponding measurements, all read from the concrete floor (Figure 17-9).

Face # 1 (C) is used for the following measurements: concrete floor, overhead door opening height, overhead doorframe header, and vertical measurement of the soffit and lookout. *Note:* More details may be added if the stud length and other construction details are known.

Face # 2 (D/E) is used for the following measurements: concrete floor, sole plate, conventional door and sill, door frame header, vertical measurement of the soffit and lookout, cripple and trimmer studs, and the conventional door lintel.

Face # 3 is used for the following measurements: concrete floor, sole plate, window cripple stud, window rough sill and header, window lintel, window and frame, studs, and wall plates.

Face # 4 is used for collecting length measurements for studs such as cripple studs, trimmer studs, etc. Additional measurements can be taken if joint details for the doors and window frames and framing details above the doors and windows are known.

Always remember that the common measuring point when using a story rod is the concrete floor. Face # 1 does not show the sole plate; whereas, Face # 2 and Face # 3 do show it. (A decision must be made with respect to leaving the sole plate under the door sill or cutting it away. The builder must decide whether to use measurement D or E and whether to change the size of the conventional door.)

Tools used to mark on the story rod include a hard pencil and a combination square and tape. Carefully mark all known measurements on the rod; corresponding marks can be transferred from one face to another using the combination square and pencil. Stud length and soffit width must be determined after the trial cornice layouts.

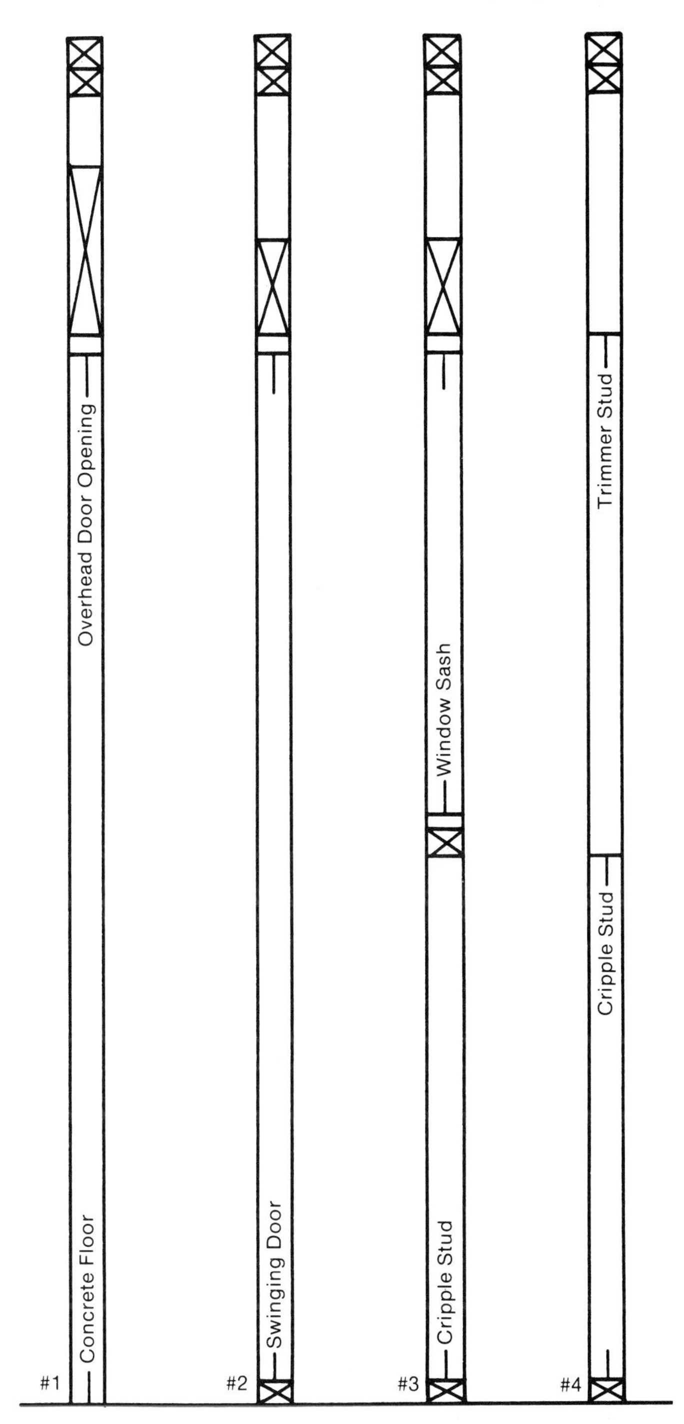

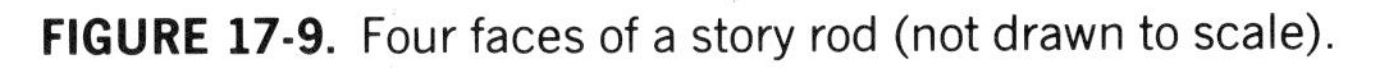

FIGURE 17-9. Four faces of a story rod (not drawn to scale).

17.8 TRIAL CORNICE LAYOUT AND STORY ROD DEVELOPMENT

On building paper make a trial cornice layout as shown in Figure 17-10. This can also be done on a sheet of plywood. From a "level" reference line, use the framing square to draw the roof line at a slope of 6″ in 12″. Draw the rafter stock with its actual width of 3-5/8″ (92 mm) and with a length suitable for cornice construction.

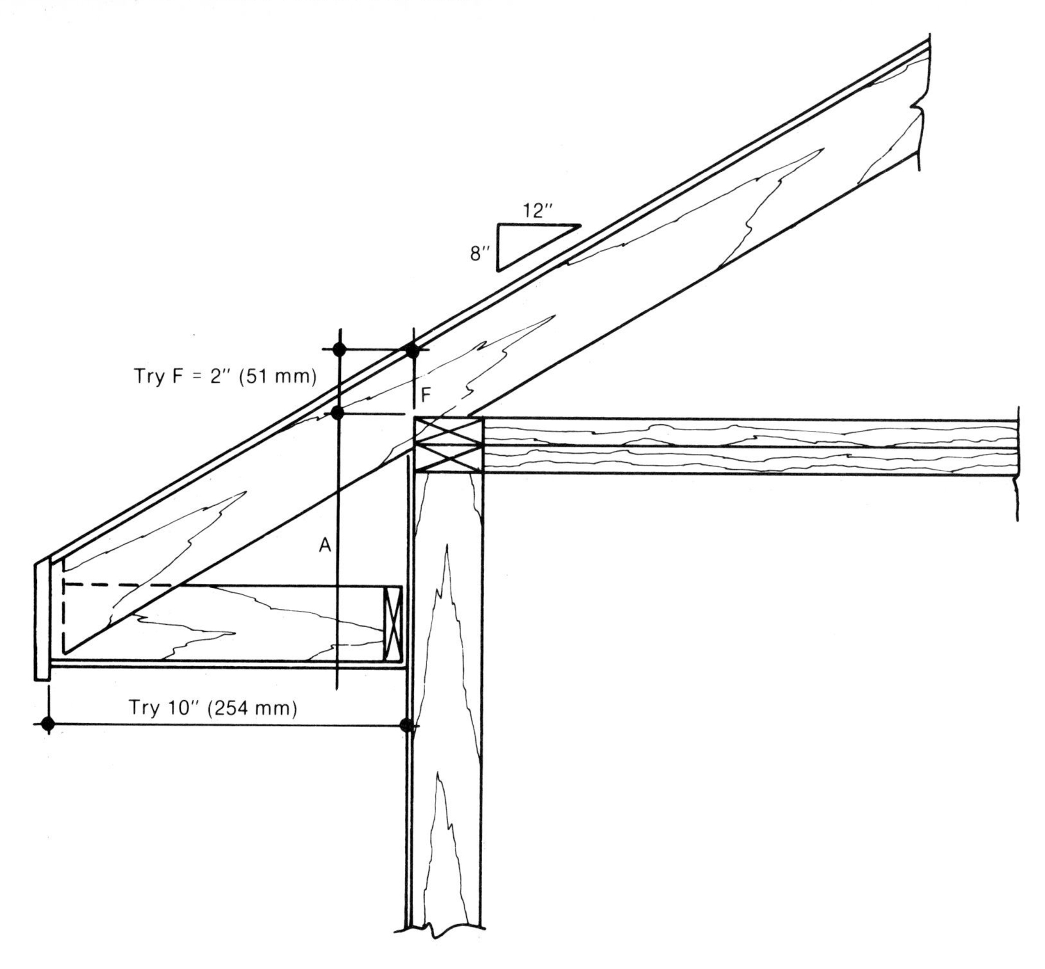

FIGURE 17-10. Trial cornice layout (not drawn to scale).

After establishing a position for the rafter birdsmouth, draw the wall plates, studs, and wall sheathing. In Figure 17-8, note that the measurement rafter above the plate (F) may vary from "all of the rafter" to "part of the rafter," depending on the depth of the birdsmouth. If all of the rafter is above the plate, a truss type construction would be necessary to secure the rafter in place. On the other hand, a birdsmouth cut too deep would create a weak rafter. Thus, the builder has the options open to him or her with respect to measurement F; a good length to try is 2" (51 mm).

Next, draw the wall sheathing and the trial lookout ladder. In Figure 17-8 the soffit covers the lookout ledger, the lookout, and the rough fascia. In this design, the finish fascia has not been grooved to receive the soffit (Figure 17-11). For the soffit width, try one-fourth the width of a sheet of plywood, less saw cut.

Connect the rafter lines to the lookout assembly. Draw the soffit in place, not as in the exploded view seen in Figure 17-11. Add the critical measurement A (from the underside of the soffit to the top of the wall plate) to Face # 1 and Face # 2 of the story rod. If the measurement A fits (for example, if the header trim for each door just touches the soffit), then the design is complete and the stud length can now be found on the story rod. If the measurement A does not fit, some of the design measurements must be altered.

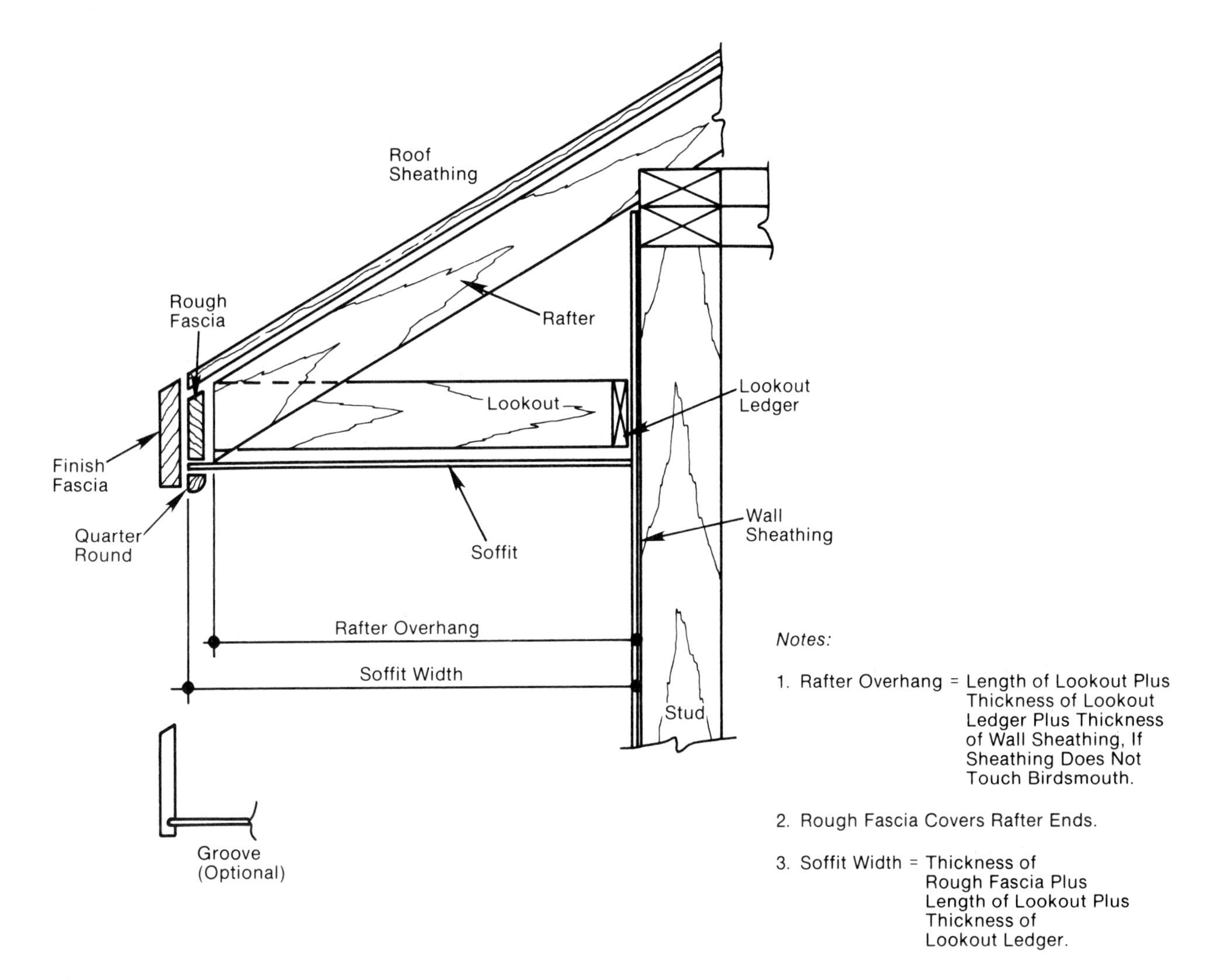

FIGURE 17-11. Partially exploded view of cornice.

Design changes can be made in a number of places, including the depth of the birdsmouth; the soffit width; the use of the sole plate under the sills of the two conventional doors; and in the height of the doors. Show all the necessary marks on Face # 3.

Complete the cornice layout to show only the necessary measurements. Date the drawing and label it "Do not destroy." Keep it for future reference. Check the story rod as shown in Figure 17-9 for accuracy; be sure to leave only useful marks. Label the faces by name and number. It is imperative that no frame cutting begin until the measurements have been checked for accuracy. Store the story rod in a safe place until it is time to prepare the cutting list.

17.9 RESEARCH

1. Sketch the floor plan and one wall section of an existing attached or detached one-car garage. Prepare a cornice layout and develop a story rod. What is the stud length? What is the length of the cripple studs?

2. For your home, make a sketch of the vertical cross section of one wall, including a door. After deciding where a story rod would stand and from where the measurements would be taken, prepare a story rod to show all vertical measurements. Assuming wood framing, what is the stud length? What is the length of an upper cripple stud?
3. For your home, make a sketch of a first floor-to-basement vertical section through the stairwell. From the measurements taken after sketching, prepare a story rod to show a set of vertical measurements. Observe carefully where the stairs rest. What is the finish floor-to-finish floor measurement? What is the unit rise?

17.10 REFERENCES

17.10.1 Maguire, Byron. *Carpentry for Residential Construction.* Virginia: Reston Publishing Company, 1975.

17.10.2 Wass, Alonzo, and Gordon Sanders. *Materials and Procedures for Residential Construction.* Virginia: Reston Publishing Company, 1981.

17.10.3 Wass, Alonzo, and Gordon Sanders. *Residential Roof Framing.* Virginia: Reston Publishing Company, 1981.

18

ROOF CONSTRUCTION

With roof framing, cornice layout, and story rod development having been discussed previously, this chapter deals with the next step in the roofing process: construction. Because this topic lends itself particularly well to a graphic treatment, extensive use is made of plan views and elevation drawings. This chapter builds on the introduction to light structures systems presented in Chapter 14 and the roofing information in Chapters 16 and 17.

18.1 GENERAL OBJECTIVES

This chapter discusses the construction of a variety of roofs, from flat to arch. In dealing with each type of roof, it draws heavily on the information presented in the preceding chapters; for this reason, a good working knowledge of roof framing principles and terminology is essential. The use of trusses in roof systems is emphasized, including the different design effects that can be achieved.

In roof construction, it is very important to follow a prescribed sequence of steps, from plate layout to lifting the trusses into place to sheathing. Working in a planned, orderly fashion not only produces the best results, but it also helps avoid the costly and time-consuming mistakes that can occur during such a job. This chapter attempts to present roof construction in such a manner, with each plate layout reflecting precise measurements and workmanship. Components can be cut and assembled either on a subfloor or off-site. Again, an understanding of roof framing techniques is mandatory to a successful roof construction project.

18.2 SPECIFIC OBJECTIVES

On completion of this chapter the reader should

1. be able to identify roof components and their functions,
2. be able to recognize roof component assemblies and critical component measurements, including those of the cornice, and
3. understand the methods of truss, post and beam, flat roof, and arch construction.

18.3 ROOF PLANS WITH PLATES AND CORNICES

Figure 18-1 shows the plan views and front elevations of nine different buildings with roofs. Framing members are shown in solid lines and plates as broken lines. Except where noted, the roofs are regular, or equal-sloped. Notice that regular roof members have 90° and/or 45° relationships in plan and slopes as indicated. Irregular roofs are intersections of roofs of different slopes; they may meet at angles other than 90°. It should also be pointed out that the frame for a flat roof can accommodate a gable or hip roof as well.

The objective of ceiling joist/rafter truss layout is twofold. First, it maintains regular spacing of the framing members (with the use of plywood sheets and standard lumber lengths) for modular construction. Second, it achieves the desired roof strength for the expected loads and stresses.

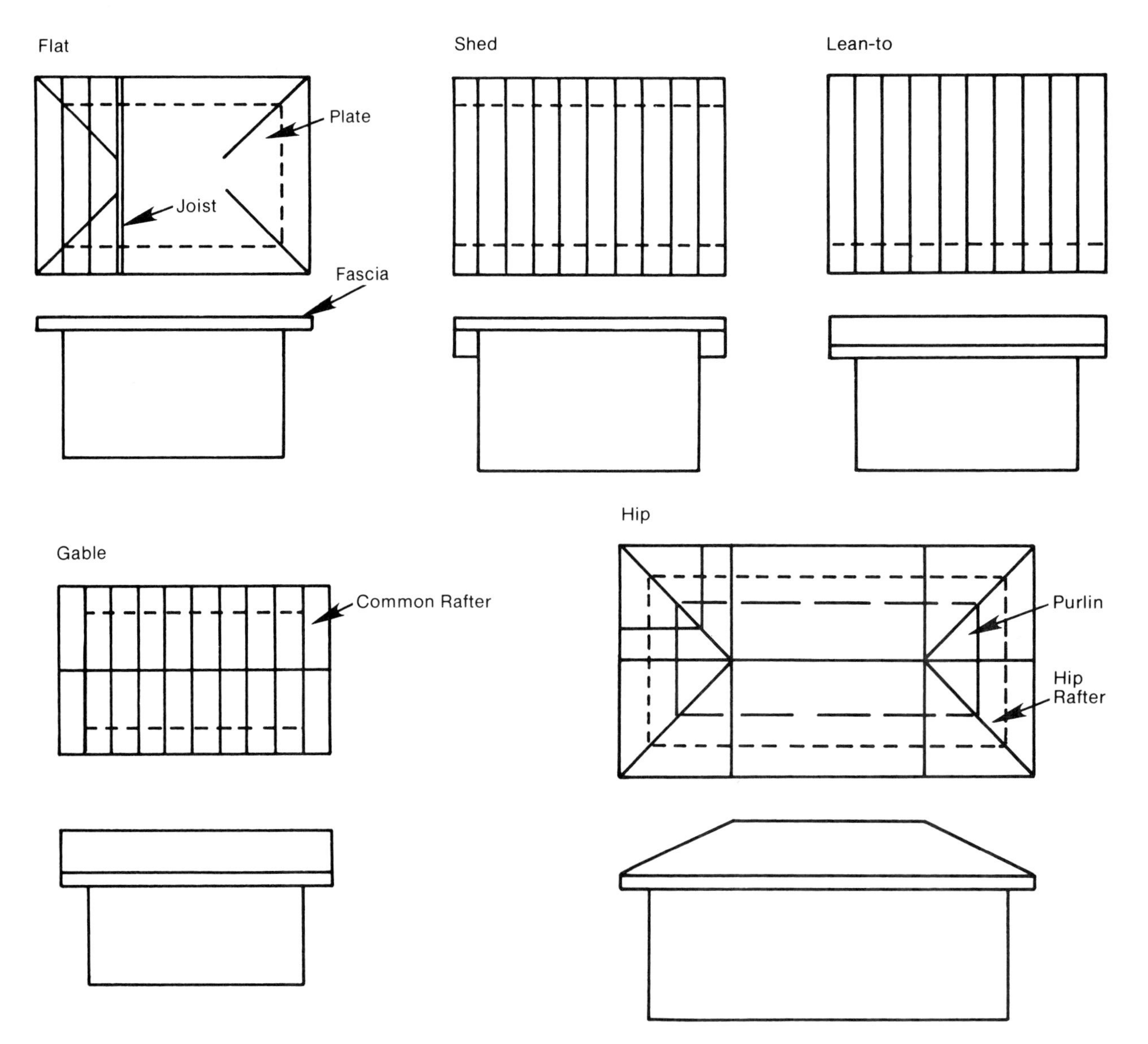

FIGURE 18-1. Roof plans with plates and cornices.

Equal Slope Intersecting

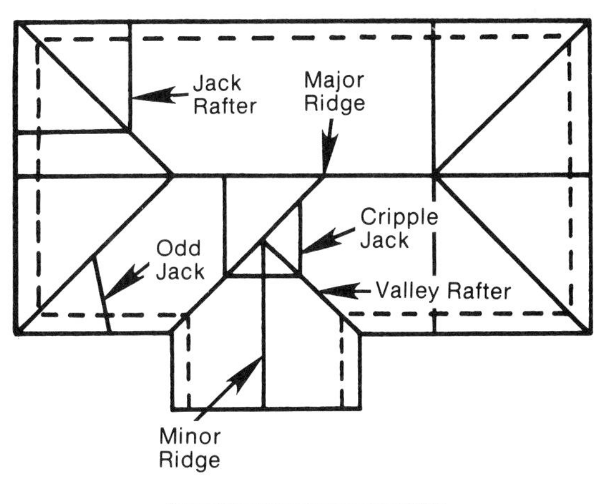

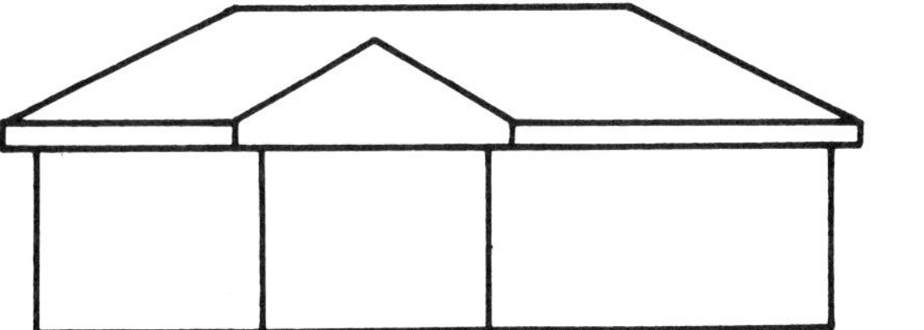

Unequal Slope Intersecting

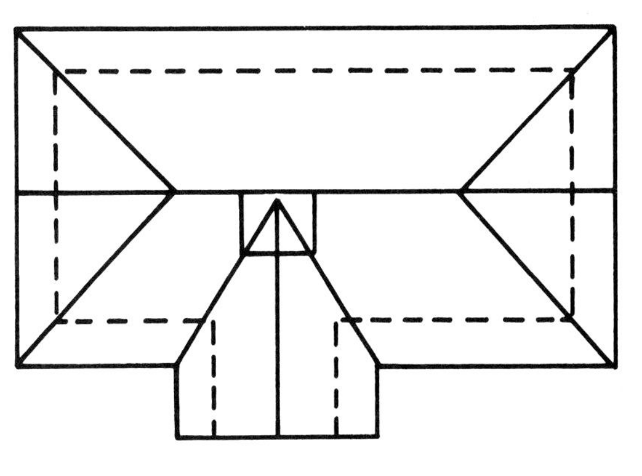

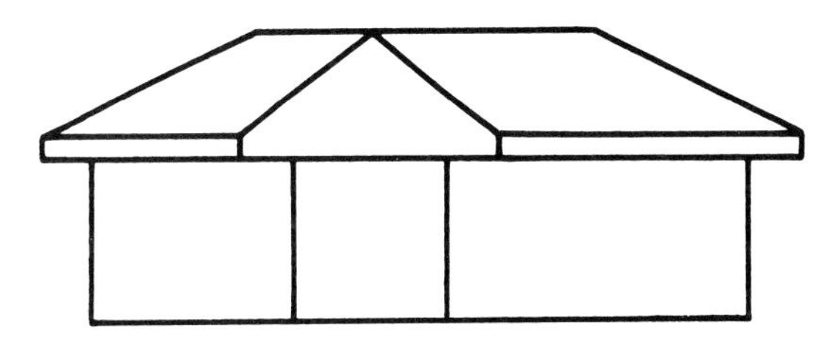

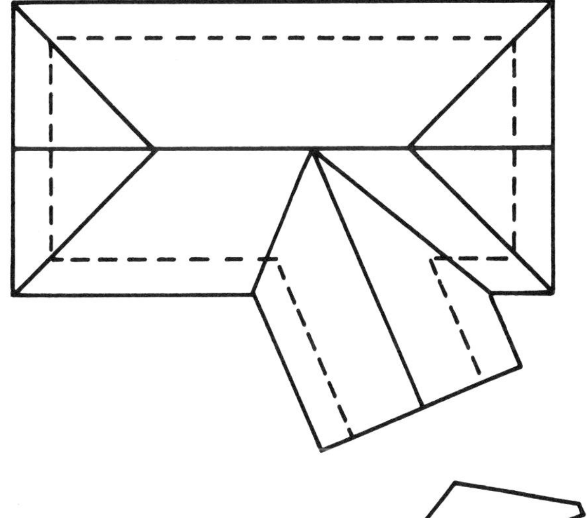

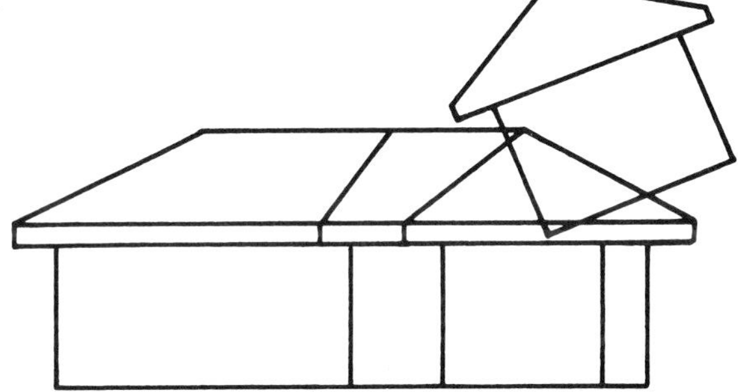

Octagonal Roof Plan

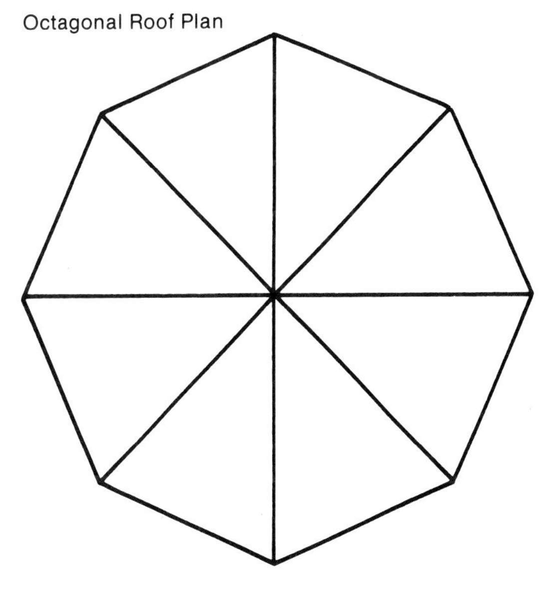

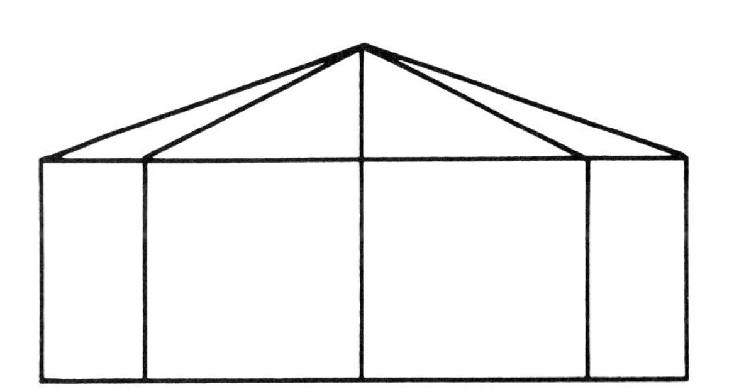

FIGURE 18-1. Continued.

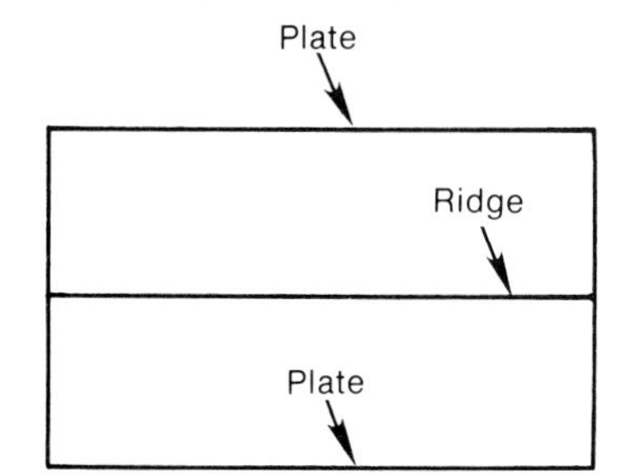

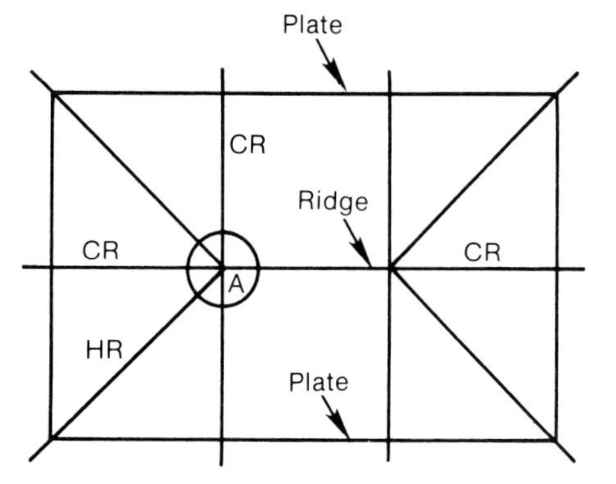

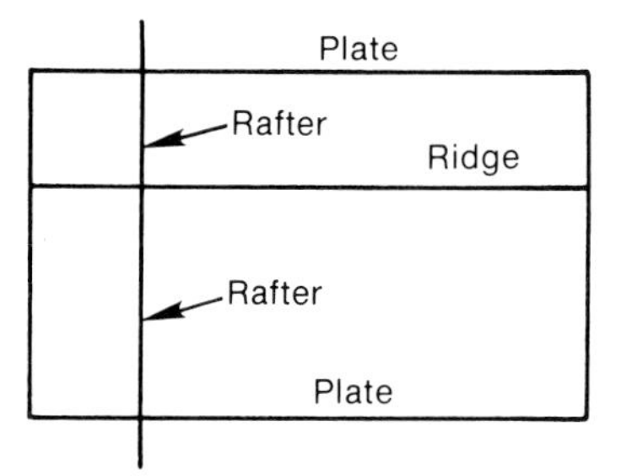

FIGURE 18-2. Plates and ridge boards for gable and hip roofs.

18.4 PLATES AND RIDGE BOARDS

Assuming stock sizes for the rafters and ridges, the plate layouts shown in Figure 18-2 were designed to maximize the fastening of components on the roof and on the ceiling below. Placement of the members is described with exactness; decisions concerning particular construction methods can be left to the discretion of the builder. Designers and builders should be aware of the increased costs associated with building unequal sloped, intersecting roofs.

18.5 TRUSSES

Roof trusses are used to cut down on both time and materials needed for a roof construction job. Because no load bearing partitions are needed, large open spans are possible. Truss work requires a certain degree of engineering data, including the span and slope of the roof. Accuracy in measuring is crucial; the lumber and plywood bracing plates must be cut to just the right size if the building is to withstand high winds and other adverse conditions.

Figure 18-3 shows how rafters are used to form trusses in a gable roof. The simplest truss design consists of two rafters and a ceiling joist, though a chord may also be added. A tie adds rigidity and helps maintain spacing during erection. A variation is the purlin assembly, consisting of a purlin, purlin stud, and strongback. For a gable end truss design, all but the longest stud are cut in pairs and the common rafters are notched where they meet the wall plate.

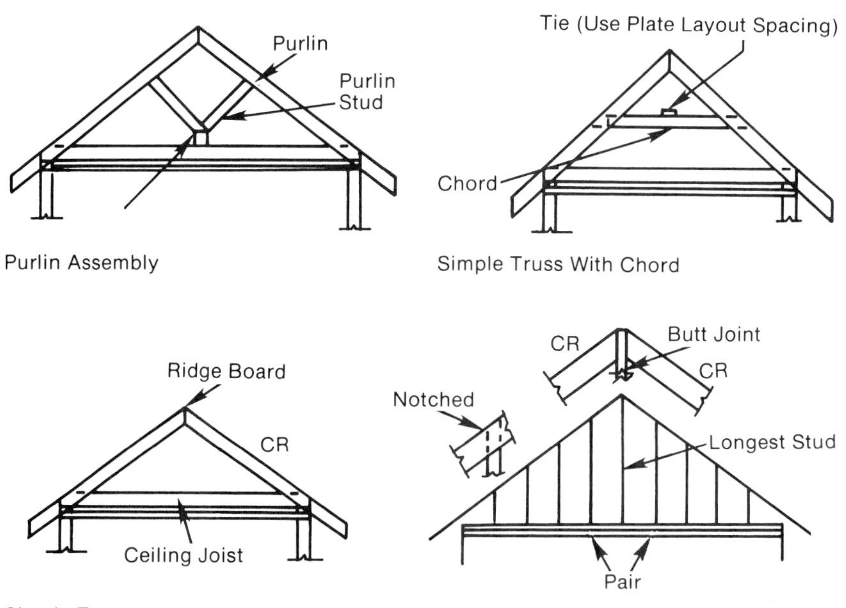

FIGURE 18-3. Constructing a truss for a gable roof.

The W-style truss is a commonly used design in light building construction (Figure 18-4). It utilizes a bevel-heel gusset, a peak gusset, and upper and lower chord intermediate gussets. The lower chord must also be spliced as shown. Other common truss designs include the

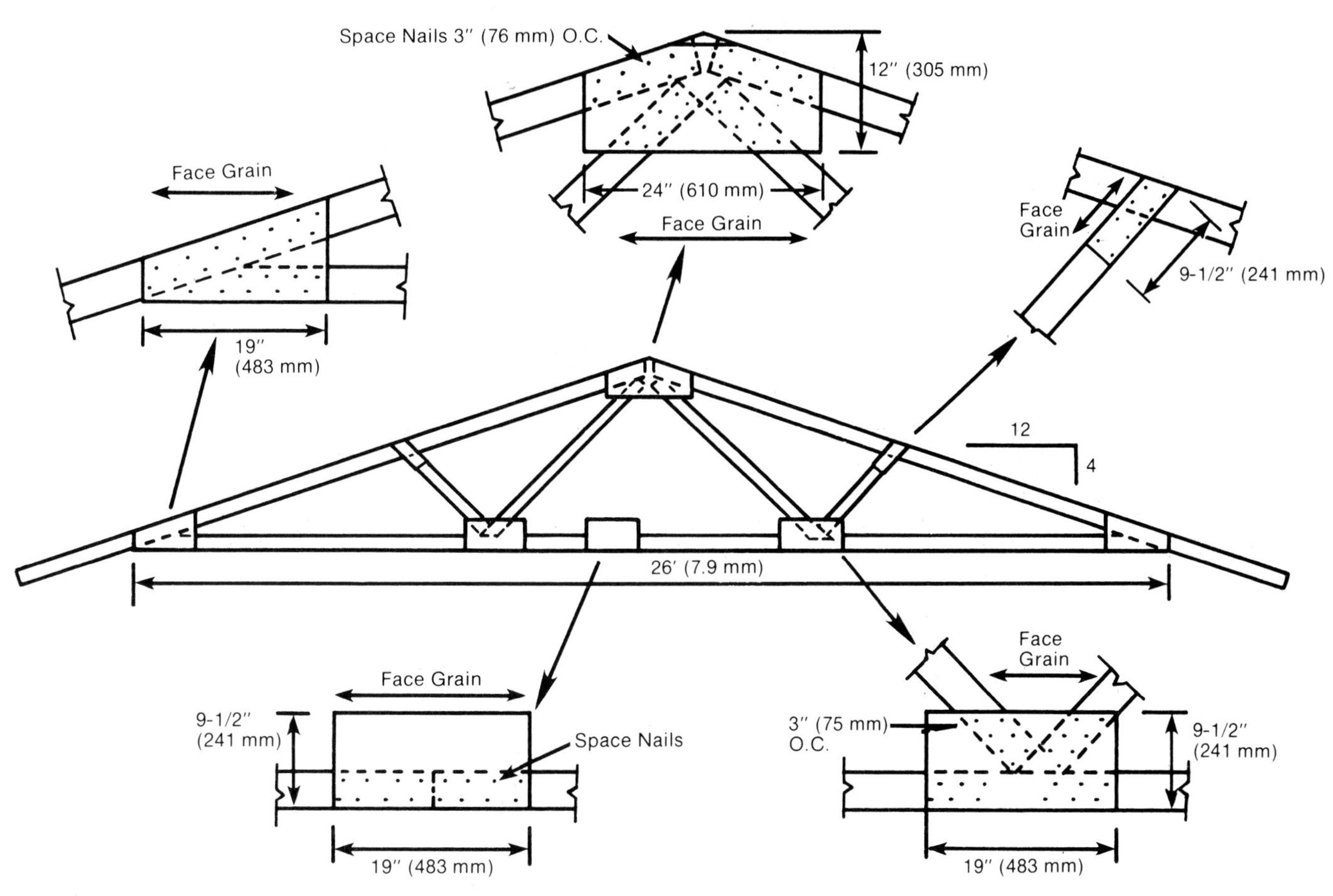

FIGURE 18-4. Constructing a 26′ (7.9 m) W-style truss.

step-down, hip louver, girder and valley, and flat roof systems. The step-down design (Figure 18-5) provides a ceiling framework different from the conventional joist style. It utilizes step-down trusses and hip end trusses. The hip louver design (Figure 18-6) adds an extra common truss and utilizes hip and hip end trusses to create an interesting effect.

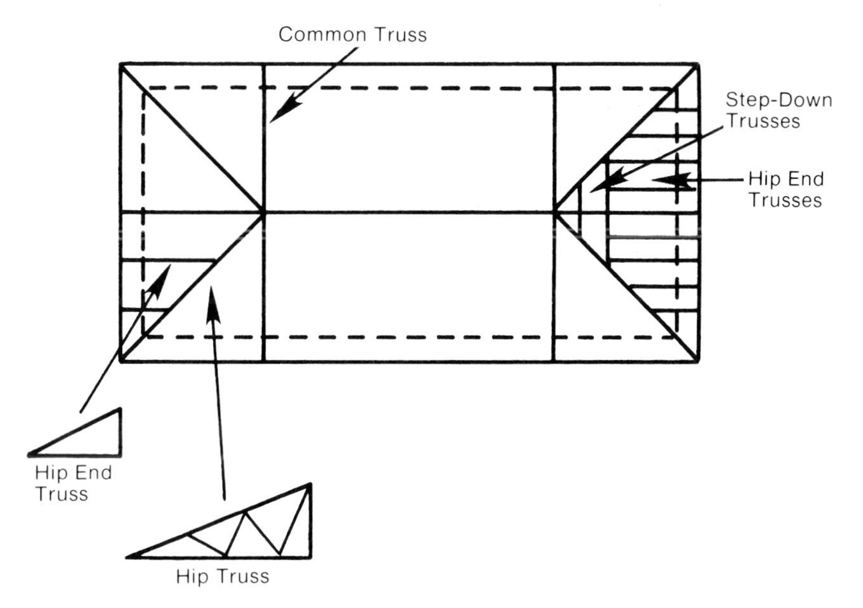

FIGURE 18-5. Step-down truss system.

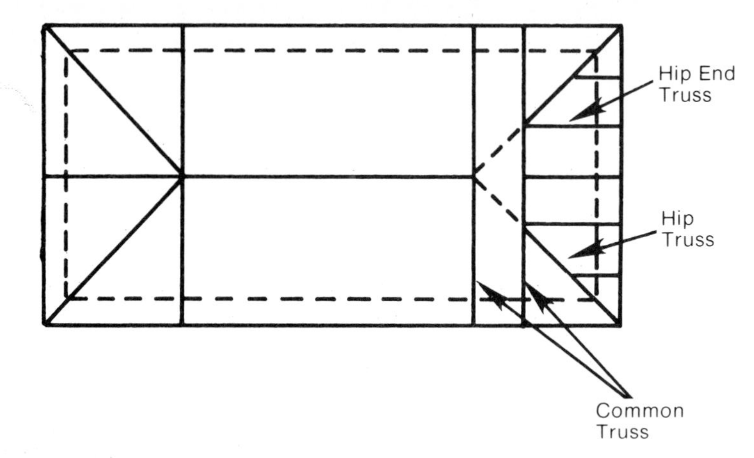

FIGURE 18-6. Hip louver truss system.

If the minor roof is part of a building extension, it can be constructed without opening the major roof by using the girder and valley truss system, as shown in Figure 18-7. Valley trusses fastened to the sheathing, a girder truss at the wall plate line, and common trusses on the building extension create the desired roof framework.

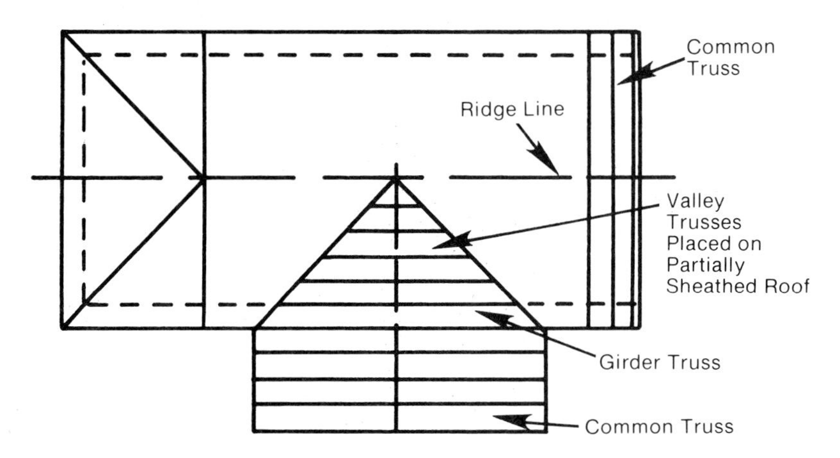

FIGURE 18-7. Girder and valley truss system.

Flat roof systems such as the Pratt and Warren (Figure 18-8) can be used for both floor and roof construction. Special connector plates are used to fasten the members into a truss. The trusses can then be ganged, or fastened into groups, for hoisting and securing. Where gang-nailing is done, the nails are pressed into the members to create an extra measure of rigidity.

Assembly hardware is now available on the market and automated set-ups are available for gang-nailing of trusses. The designs for the latter are determined from input data to a computer. With computer software available for each type of truss, it is possible to obtain detailed data for construction. Specific data is fed into the computer regarding the specific type of truss being used. Information such as span, slope, type and species of wood, and type of fasteners being used would be fed in, and output data would include length of members, positions of splices, position of webs, and angles of cuts.

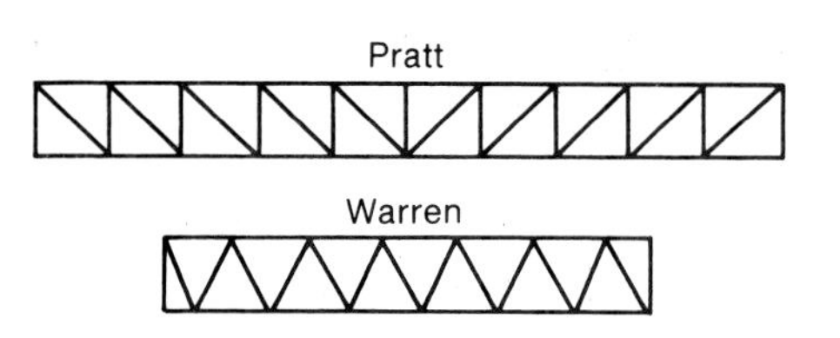

FIGURE 18-8. Two flat roof truss systems.

When the design of a truss is known, a jig can be assembled to allow for rapid positioning and fastening of the members. Cutting angles are taken from a pattern truss, and a cutting list is drawn up to build a set of trusses. A typical truss assembly jig is shown in Figure 18-9. While suitable for use in any type of construction, trusses are best in buildings where the design is rectangular.

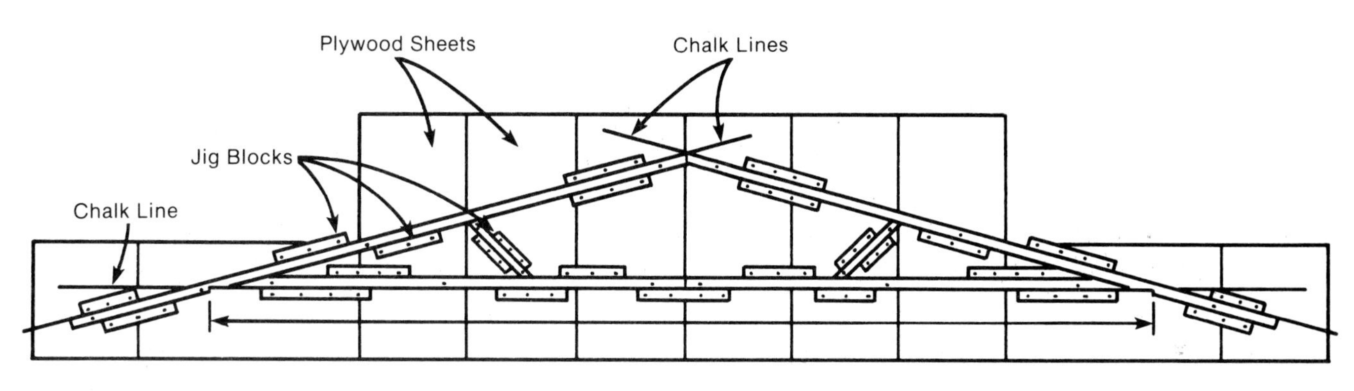

FIGURE 18-9. Truss assembly jig.

18.6 CORNICES

While a cornice undoubtedly adds to the appearance of a structure, its main purpose is to protect the building from the elements. Figure 18-10 shows the cornice for a flat roof; because of its simplicity, there are no rafters and, therefore, the overhang is the end of the roof joist. In a more conventional design, the ceiling joist acts as a platform framework for the cornice (Figure 18-11). Note that the overhang includes rough fascia.

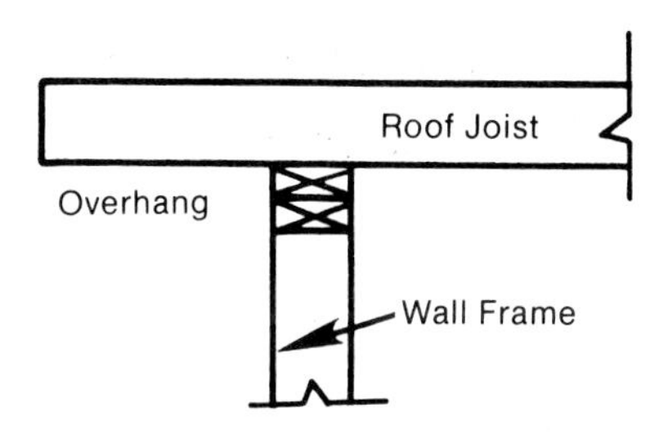

FIGURE 18-10. Flat roof cornice.

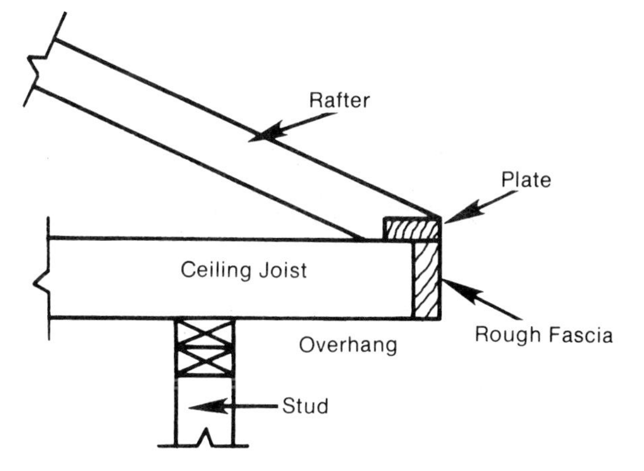

FIGURE 18-11. Cornice with ceiling joist as platform.

In Figure 18-12, the cornice is open and includes windblocks. The windblocks have hopper cuts where they touch the hip rafters. In this design, a ceiling joist or truss chord may be added if desired. Figure 18-13 shows a boxed-in, or closed, cornice. This design features additional nailing surfaces for the soffit known as lookouts. The lookouts are face-nailed to the rafter extensions.

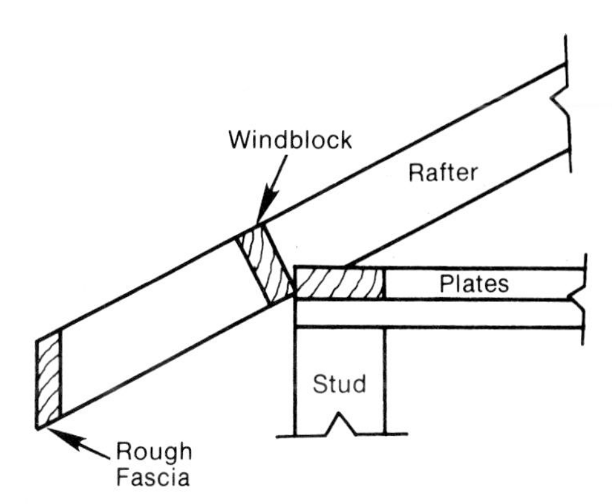

FIGURE 18-12. Open cornice.

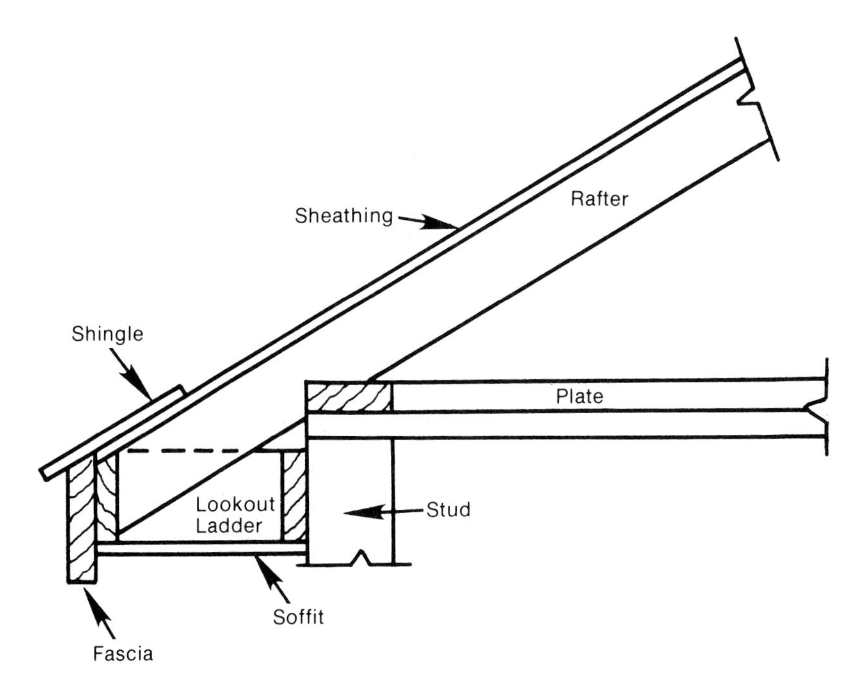

FIGURE 18-13. Boxed-in cornice.

When the cornice is built on a structure with a brick veneer exterior, it can be a truss or a rafter/ceiling joist assembly (Figure 18-14). With the post and beam system, a framing anchor is used for added support (Figure 18-15).

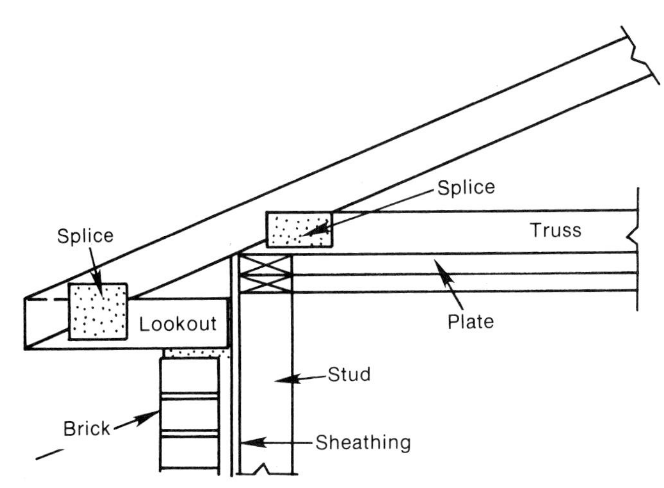

FIGURE 18-14. Cornice with brick veneer.

18.7 GABLE ROOF CONSTRUCTION

Erecting the rafters is a job requiring a hammer, nails, bracing, level, plumb line, and, if possible, a few helpers. With the rafters already cut and measured, they can be erected fairly easily by following an orderly plan. Following is the procedure for constructing a gable roof; while individual details will differ from roof to roof, the general procedure is the same and can be used for any type of roof.

Begin by nailing some 2″ × 4″ (38 mm × 89 mm) at the gable ends in such a position where they can be easily shifted to support the end

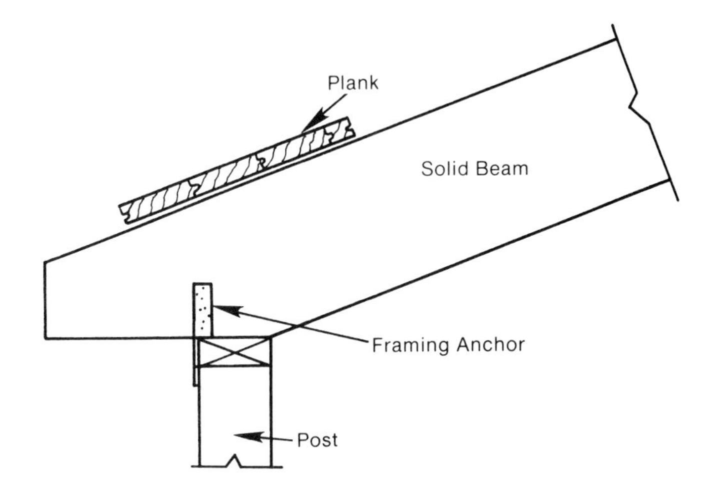

FIGURE 18-15. Cornice with post and beam.

rafters. These props must be long enough to reach from the top plate up to the bottom of the ridge board. Cut and double toenail to the top plates directly under the spot for the first rafter pair at the ridge board. If the ridge board is to be a single piece unit, do the same thing at the opposite end of the building.

If, on the other hand, the ridge board is to be spliced, set a second brace about 1′ (.3m) from where the short ridge board will end. The ridge board at the tops of these braces can be tacked with 10d nails partially driven; this will hold them in place until at least one pair of rafters is raised. The braces can be kept near plumb by setting in diagonal braces. While exact plumb isn't essential, the whole construction job will be easier if the braces are reasonably close to plumb when the rafters start to go up.

Lift the end rafter pairs (or the last short ridge board rafter pair and one end pair) onto the working surface along with the ridge board. If the ceiling is framed but not floored, lay sheets of plywood down as a safety measure. Next, erect the first set of rafters. Move the ridge board into place, and erect a supporting set of rafters (Figure 18-16). Toenail the first set to the ridge board at the correct marks; use 8d nails, two to each side. Toenail the first set of rafters to the top plates with 10d nails; then nail the rafters to the ceiling joists using at least four 16d nails per

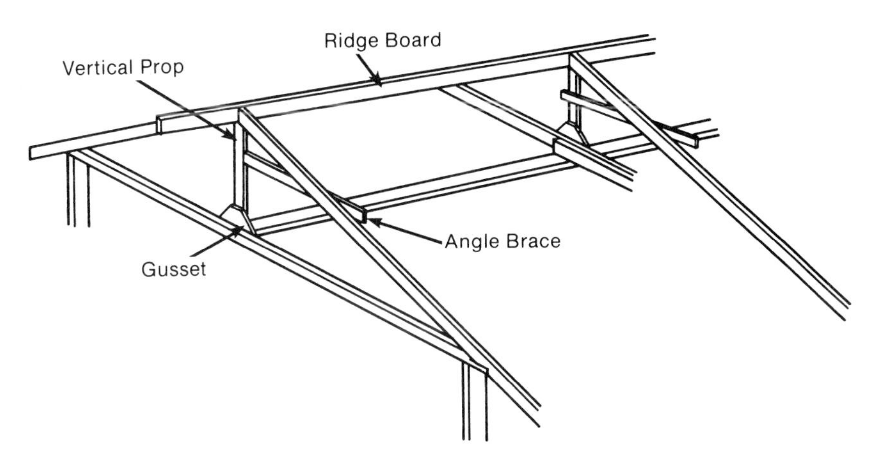

FIGURE 18-16. Erecting the first set of rafters.

side. If the span is over 24′ (7.3 m) or if the building is located in a high wind area, a support strap can be added at this point.

With the second set of rafters toenailed to the ridge board, check the board levelness and drop a plumb line to make sure it is over the center of the house. Continue with the ridge board if it is in two pieces, setting in another pair of rafters at the inside end, and moving to the final set of gable end rafters. Check this section for levelness and centering over the house. Nail at the ridge board, top plates, and ceiling joists.

Once the full ridge board is in place, continue with placing and nailing the rafters. Do a pair at a time, checking for levelness and centering about every third pair. If all the cuts were made correctly and the original pattern rafter was laid out properly, the roof will practically pull itself into alignment with no extra work.

Though not always essential, collar beams do add ridigity, cut the span width, and provide nailers for second floor or attic ceilings. Usually 1″ × 6″s (19 mm × 140 mm) are used, placed on each (Figure 18-17). Collar beams should be kept in the upper third of the rafter length, with each end of the collar board getting four 8d nails. Once the collar beams are in place, all temporary props can be taken down.

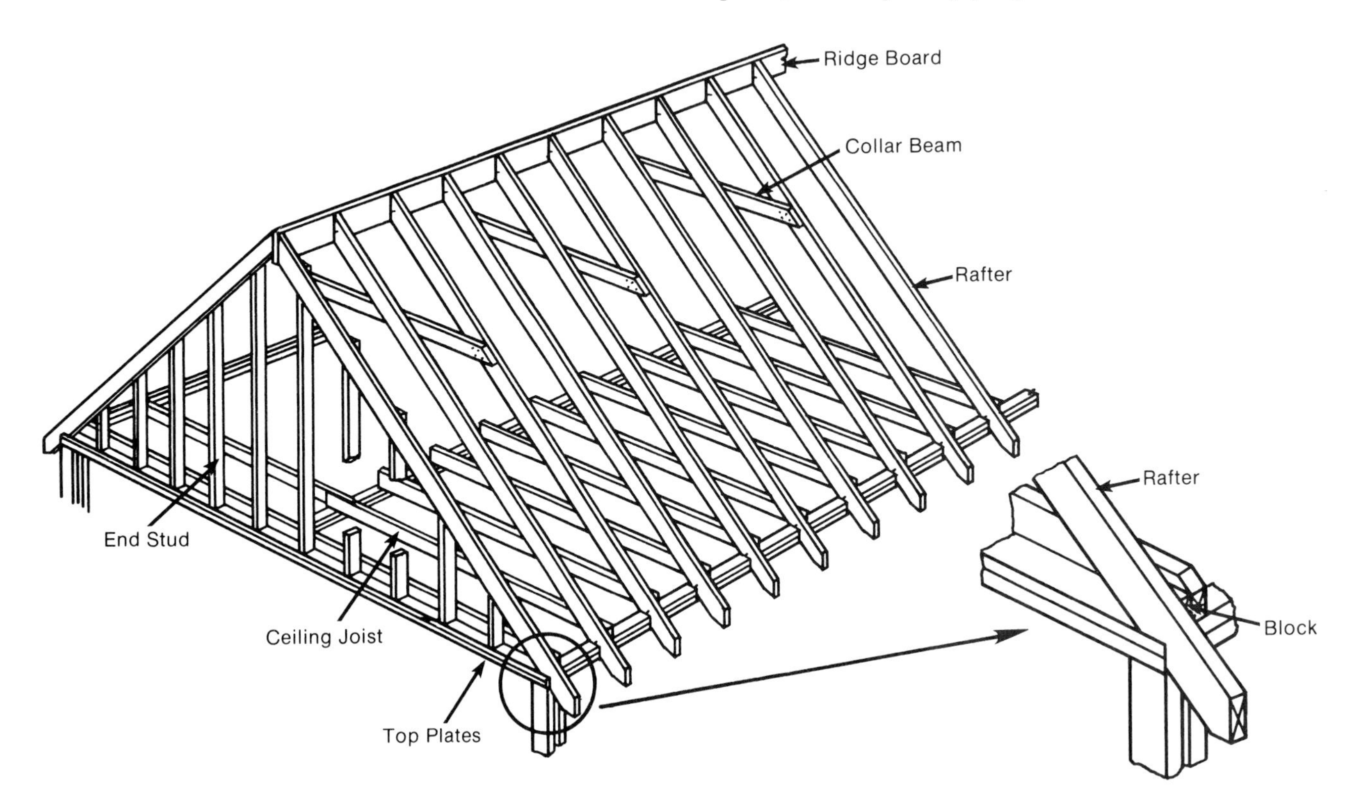

FIGURE 18-17. Adding collar beams to roof framing.

Gable ends require framing so that the siding can be continued. Framing starts just to either side of the ridge board, with 2″ × 4″s (38 mm × 89 mm) notched to fit the particular angles; these are also known as cripple studs. To start, drop a plumb line down from the center line of the ridge board, and mark on the top plate. Now square a line from this and measure it and divide that in half. Use half this measurement on each side of the plumb line to show where the first two cripple studs will go. After those are measured, notched, and nailed, measure on 16″ (406 mm) centers for the remainder, cutting and notching to fit as you go along.

If the overhang at the ends requires a floating rafter, it should be cut now. The floating rafter will nail to the end of the ridge board, so it should not be cut half the distance on each side to allow for the ridge board. It should also not have a birdsmouth cut, but it is like any other rafter in all other respects. The bottom of the floating rafter is nailed to the fascia, which is installed first. The fascia should be of sufficient stock size to cover the rafter ends and must be nailed with galvanized or aluminum nails.

Before moving on to finishing the roof, it might be a good idea to add a shed dormer. Generally, the shed dormer runs from the ridge board right out to the edge of the house, and it is a very efficient way to add extra space to an attic story. Double trimmer rafters are used at the outside edges of a dormer to carry any extra loads, with nailer strips on each side for the roof sheathing. The slope of the rafters on the dormer is determined by the style of the original house and the expected loadings, as are the rafter sizes. Studs from the outside of the rafters drop down to the top plates of the floor below, with bottom plates, also in use with the dormer. It, too, has its own doubled top plate on which its rafters are nailed; the end walls are framed with notched cripple studs like those used for gable end framing.

18.8 POST AND BEAM CONSTRUCTION

Post and beam construction, explained fully in Chapter 14, offers some distinctive construction features for the builder. When combined with conventional plank framing, the differences in design are many and varied. Among them are the following:

- Posts and beams are larger and spaced farther apart than studs, joists, and rafters of conventional framing (Figure 18-18).
- Partitions may be placed wherever they are needed.
- The surface finishes may be exposed and finished to match interior finishes.
- A reduction in building height is possible, thus leading to fewer materials being used.
- The relatively heavy wooden posts and beams have high levels of thermal performance.
- The ceiling beams may form the lower chord of a truss to absorb the horizontal thrust from the roof load. The spaced beams can carry electrical wiring.
- The ridge beam can carry roof beams for open beam construction to absorb the horizontal thrust from the roof load.
- When roof beams are not used, a ridge beam and top wall plate may carry planks as rafters to make a solid deck.

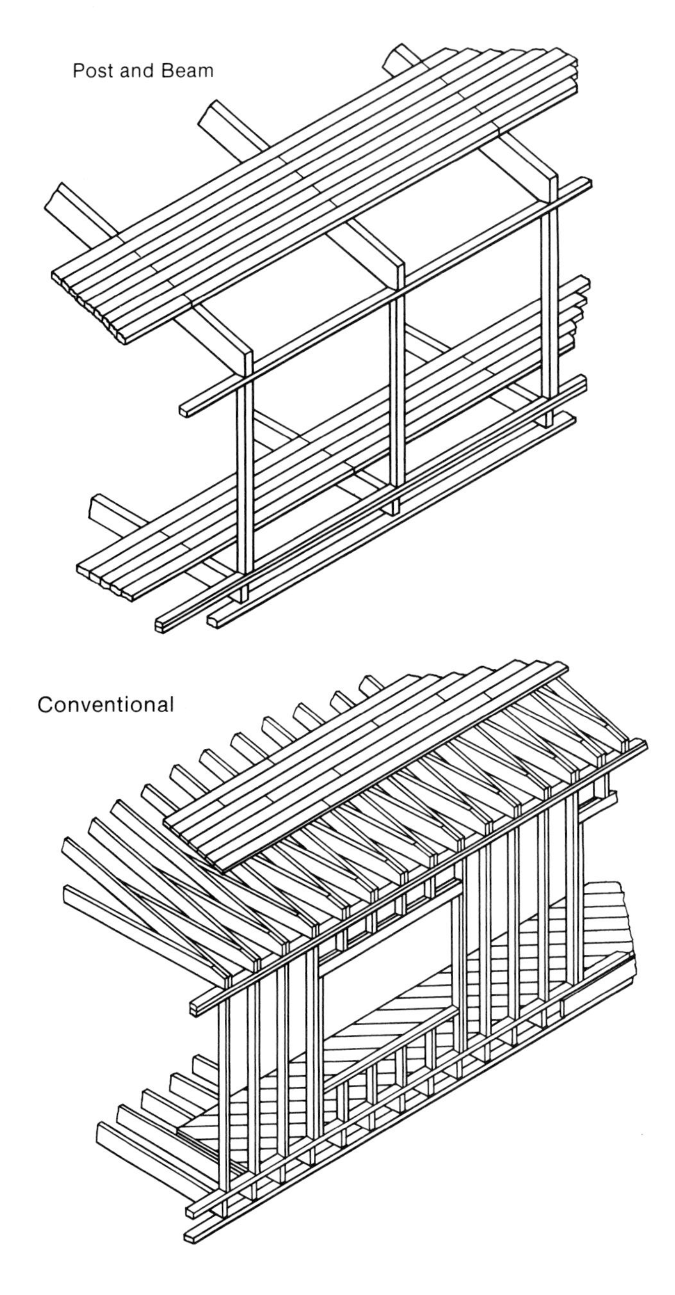

FIGURE 18-18. Post and beam spacing vs. conventional spacing.

18.9 ARCHES

Some facets of light building construction are accomplished using foundation arches. Arches are designed and finished according to the requirements of span, stress expectations, service conditions, and appearance. Table 18-1 lists common arches, their spans, and applications. Metal fasteners are used in the construction of all arches because of their durability and strength.

TABLE 18-1 COMMON ARCHES

Type of Arch	Typical Spans	Common Applications
Circular	From 60′ to 200′ (18.3 m × 61 m) with rise from 1/5 to 1/2 of L.	Roof support for industrial, commercial, and recreational buildings such as warehouses, shopping centers, arenas, lodges, etc.
Parabolic	From 50′ to 100′ (15.2 m × 30.5 m) with rise at least equal to L.	Roof support for religious buildings or entranceways to recreational buildings.
Segmental (Gothic)	From 30′ to 100′ (9.1 m × 30.5 m).	Roof support for agricultural and industrial buildings, especially for storage.
Tudor	From 30′ to 80′ (9.1 m × 24.4 m) with roof pitch of 3:12 to 16:12.	Roof support for religious and recreational buildings such as churches and auditoriums.
Three-Hinged Frame	From 50′ to 200′ (15.2 m × 61.0 m) with pitch about 4:12.	Roof support for industrial and recreational buildings such as warehouses, factories, arenas, etc.
A-Frame	From 30′ to 140′ (9.1 m × 42.7 m).	Roof support for religious and industrial buildings.

18.10 RESEARCH

1. Using sketches, describe the construction of a roof with trusses.
2. Using a wood truss handbook, make a list of design data required for a particular truss.
3. Study the design and construction of buildings in your area with arch-style foundations. Make note of their spans.

18.11 REFERENCES

18.11.1 Capotosto, John. *Step-by-Step Home Carpentry*. Virginia: Reston Publishing Company, 1977.

18.11.2 Wass, Alonzo. *Methods and Materials of Residential Construction.* Virginia: Reston Publishing Company, 1977.

18.11.3 Wass, Alonzo, and Gordon Sanders. *Residential Roof Framing*. Virginia: Reston Publishing Company, 1980.

19

SUBFLOORS, SHEATHING, SIDING, AND FINISH FLOORS

This chapter deals with two aspects of light building construction. The framing process is completed with the placement of the subfloors and sheathing. Meanwhile, the addition of siding and finish floors begins the job of finishing the building. The necessary continuity of any construction project is evident; even while one phase is being completed, the next phase is underway.

19.1 GENERAL OBJECTIVES

As was pointed out in Chapter 14, a building is the sum of many different subsystems. In looking at subfloors, sheathing, siding, and finish floors as such, their importance becomes obvious. Siding, for instance, definitely enhances the appearance of a building, but that is only one of its functions. More importantly, it protects and helps make functional other subsystems, including electrical, mechanical, insulation, painting, and furnishing. So it is with the other subsystems being dealt with in this chapter; each one, while important in its own right, is a prerequisite to the installation and operation of other subsystems. For this reason, the principles of construction presented in this chapter are crucial as a foundation upon which future subsystems will be built.

19.2 SPECIFIC OBJECTIVES

On completion of this chapter the reader should

1. understand the concept of subsystem checklists,
2. know the principles of construction with respect to subfloors, sheathing, siding and finish floors, and
3. have a better understanding of building construction objectives.

19.3 SUBFLOORS

A subfloor is the final step in completing a floor frame. Materials that can be used for subflooring include plywood, shiplap boards, center match boards, common boards, and particleboard; they are fastened with nails, screws, staples, or adhesives. The placement of a subfloor provides a walking and working surface and helps bring construction to the finish floor elevation shown on the story rod.

Large unsanded sheets are commonly used to create subfloors because they encourage modular construction and help produce strong and rigid floor systems through diaphragm action. When boards are used, they are often applied diagonally across the joists for added support (Figure 19-1). Common boards in particular should be covered with a floor underlay of particleboard or plywood panels for the purpose of leveling, covering the spaces between the boards, and bringing the assembly to the required elevation.

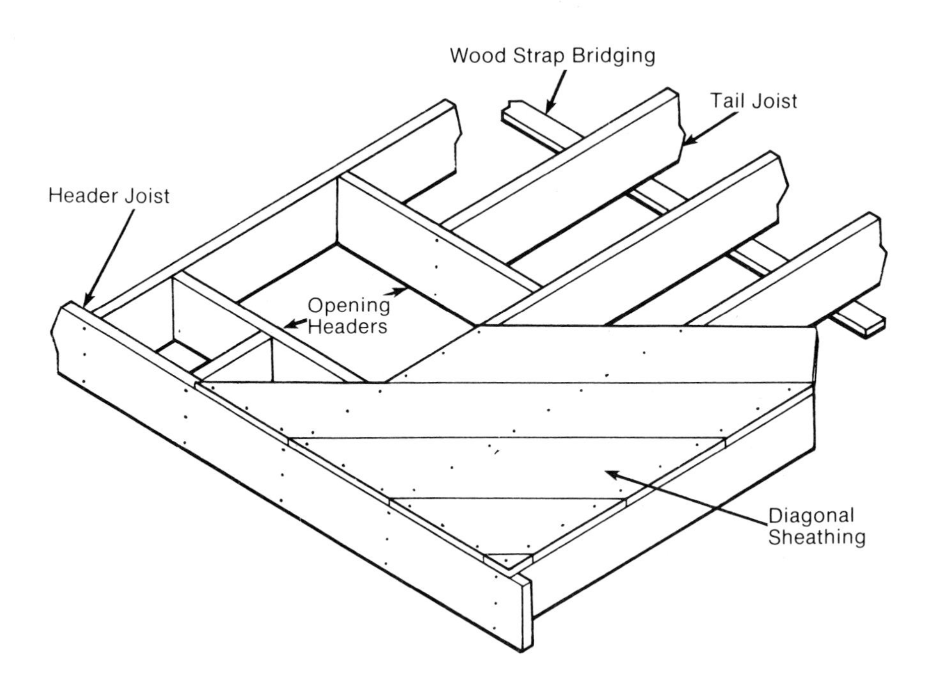

FIGURE 19-1. Diagonal subflooring.

Tongue-and-groove plywood panels reduce the need for blocking at the panel joints because the interlocking ensures that loads are transmitted evenly across the joints. A placed plywood subfloor provides a smooth base for finish floors such as rubber, linoleum, plastic tile, and carpet. Field-glued plywood subfloors, with proper framing underneath, increase floor rigidity. Adhesives are effective in bonding plywood to untreated joists, and a glued assembly also helps eliminate common floor squeaks. Special uses for plywood subflooring and underlayment are shown in Figure 19-2.

19.4 SHEATHING

Sheathing covers a framework and helps secure it. It provides a degree of insulation and helps prevent air infiltration while holding the framing members flush together. A stress-neutralizing effect takes place when sheathing is fastened to a frame, creating a uniform wall; a similar effect occurs when fastening subfloors to floor framing and roof sheathing to rafters or trusses. In addition, a sheathed building is

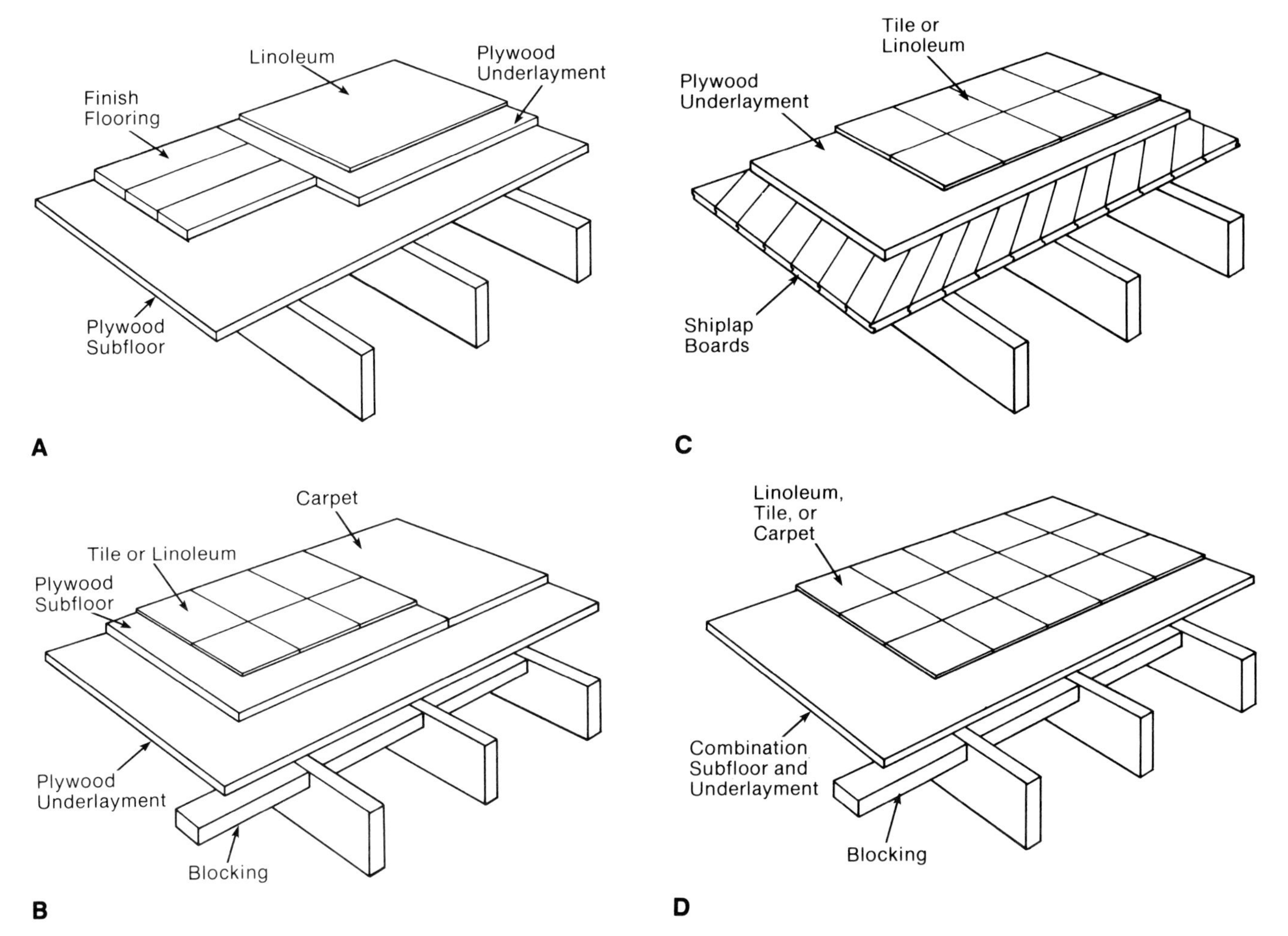

FIGURE 19-2. (A) Plywood underlayment used to compensate for difference in thickness between linoleum and wood flooring. (B) Plywood underlayment used to compensate for difference in thickness between carpet and tile. (C) Plywood underlayment over board subflooring. (D) Plywood combined subfloor and underlayment.

ready for an outer covering of siding, shingles, stucco, bricks, or any other material that may be specified on the building plans.

Sheathing panels can be made of plywood, fiberboard, gypsum board, or styrofoam, with plywood the most widely used. The larger the panel, the fewer the joints and the greater the resistance to air movement. Panel size must relate to the construction industry's recognized spacings for studs, rafters, and trusses. The spacings are 12″ (305 mm), 16″ (406 mm), 24″ (610 mm), and 32″ (813 mm); the standard panel size is 4′ × 8′ (1.2 m × 2.4 m). Panel length also depends on the particular spacing, though this relationship is lost when boards are applied diagonally.

Wall sheathing panels must be applied according to manufacturer's specifications. For example, plywood panels are usually spaced about 1/8″ (3 mm) apart to allow for changes in the moisture content. For energy efficient double-stud wall construction, the sheathing is placed on the inner wall. The outer wall need not be sheathed, as its function is to carry insulation and provide protection.

Roof sheathing panels must be strong enough to carry loads such as workers, tiles, shingles, asphalt, gravel, and shakes, as well as to withstand the ravages of nature. Plywood panels provide a suitable deck for fastening the outer covering. To start plywood sheathing after a single fascia has been fitted, place the sheathing over the edge of the fascia board, leaving a 1/16″ (2 mm) gap between the panel edges (Figure 19-3). (In wet service conditions, increase the gap to 1/8″ [3 mm] to be safe.) This method will prevent leakage into the soffit in the event water should back up at the eave due to ice, snow, or general debris in the eaves trough. Tables 19-1 and 19-2 give plywood roof sheathing data for both the British Imperial and Metric applications, respectively.

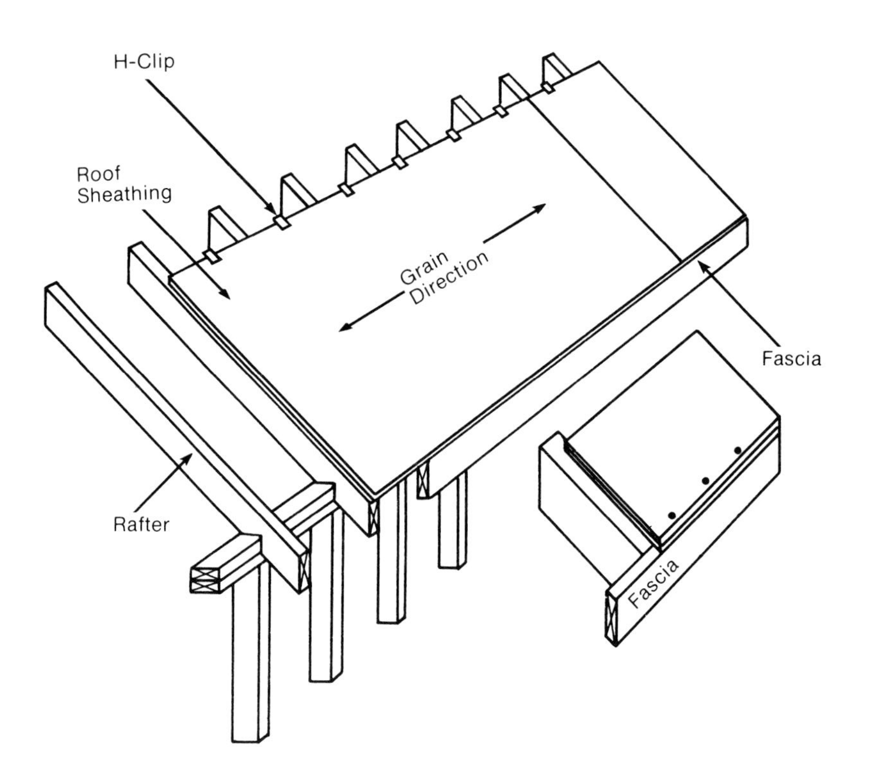

FIGURE 19-3. Plywood roof sheathing.

TABLE 19-1 PLYWOOD ROOF SHEATHING DATA (IMPERIAL)

	Framing	Plywood Thickness (in.)					Remarks
		5/16	3/8	1/2	5/8	1/4	
Spacing of Supports (in.)	Panel edges supported by 2″ × 4″ headers fitted between rafters or other framing members	16	24	24 32	42 36	48 40	
	Panel edges supported to prevent differential deflection; H-clips, T&G plywood, Spline, etc.	16	24	24 32	40 36	46 40	
	Panel edges not supported	12	16	24	32	Not recommended	

TABLE 19-1 PLYWOOD ROOF SHEATHING DATA (IMPERIAL) (Continued)

	Framing	Plywood Thickness (in.)					Remarks
		5/16	3/8	1/2	5/8	1/4	
Nail Length (in.)		1-1/2	1-1/2	1-3/4	2	2-1/4	Panels shall be fastened to supports with common nails at intervals not exceeding 6″ along all edges supported on framing nor 12″ along intermediate bearings, except that when primary supports are spaced 36″ or greater on center, nails shall be spaced at intervals not exceeding 6″ along all supports

TABLE 19-2 PLYWOOD ROOF SHEATHING DATA (METRIC)

Method of Support	Plywood Species Designation	Spacing of Supports (mm)					
		300	400	480	600	800	1,200
		Plywood Thickness (mm)					
Panel edges supported by blocking	DFP	7.5	7.5	9.5	9.5	12.5	18.5
	CSP	7.5	7.5	9.5	9.5	15.5	20.5
Panel edges supported by tongues and grooves or H-clips	DFP	7.5	7.5	9.5	9.5	12.5	18.5
	CSP	7.5	7.5	9.5	9.5	15.5	20.5
Panel edges not supported	DFP	7.5	9.5	12.5	12.5	15.5	NR
	CSP	7.5	9.5	12.5	12.5	15.5	NR

Maximum Nail Spacing	Nail Type and Length				
	Type	Plywood Thickness (mm)			
		7.5, 9.5	12.5, 15.5	18.5	20.5
150 mm on center along edges and 300 mm on center along intermediate supports except that when primary supports are spaced greater than 800 mm on center, nails shall be spaced at 150 mm intervals along all supports	**Common**				
	in.	1-1/2	1-3/4	2	2-1/4
	mm	38	44	51	57
	Annularly Grooved				
	in.	1-1/4	1-1/2	1-3/4	2
	mm	32	38	44	51

Boards may also be used for roof sheathing. If shingles or shakes are to be applied, the boards can cover the entire roof, much like plywood, or those above the cornice can be spaced to conserve materials. The spacing is determined by the length of the shingles or shakes being used.

19.5 ROOF COVERINGS

The application of a roof covering is normally begun (and often completed) before the exterior wall covering is installed. Following are a few examples of typical skeleton specifications for roof coverings:

- Solid or spaced sheathing; cedar wood shingles (available in various species, grades, and for use in different climates; fire retardant brands are also available); galvanized steel or aluminum flashing (copper is not compatible with cedar); nails made from corrosion-proof stainless steel, hot-dipped galvanized steel, or aluminum (two per shingle); and no shingle stain, for a gray, weathered appearance. Figure 19-4 shows sections of roof with shingles.

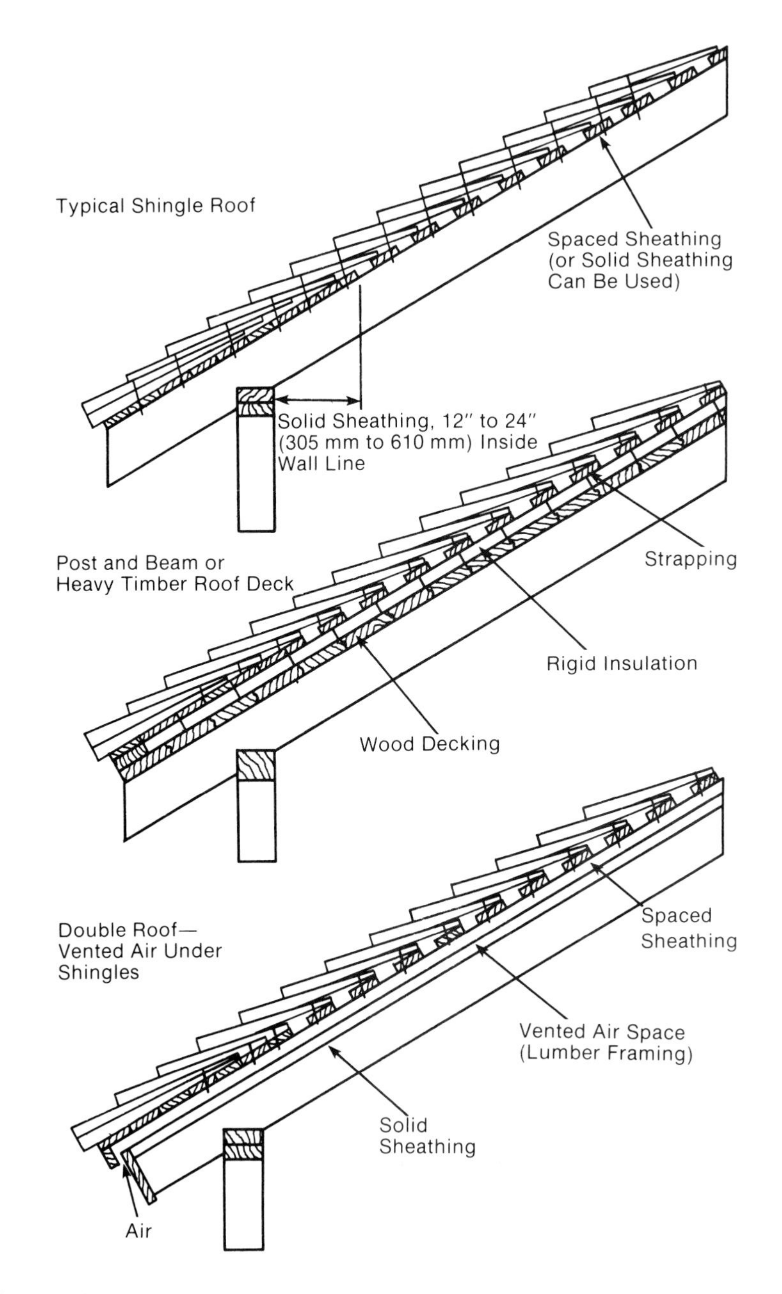

FIGURE 19-4. Typical shingle application.

- Solid or spaced sheathing; cedar shakes (preferably fire retardant); roofing felt (weight per unit area); and wood shingle nails long enough to penetrate the sheathing (two per shake). This application is shown in Figure 19-5.
- Solid sheathing; asphalt shingles (triple-tab); wood starter shingles; roofing nails (13 gauge/1-1/2″); and asphalt roofing gum to fasten and secure the tabs.
- Solid sheathing; 2″ (51 mm) furring strips; Spanish style clay tile; and copper nails.
- Sheathing paper; flashing; 1″ (25 mm) roofing nails; two lapped layers of asphalt-impregnated felt paper; three additional layers of felt paper, mopped with hot tar or asphalt; and a cover coat of hot tar with pea gravel rolled into it.

Keep in mind that this is by no means a comprehensive list. Specifications for additional subsystems can be written up using slate, tile, corrugated metal, etc. Whatever subsystem is used, the known weather patterns for the area should always be considered when covering a roof, as harsh conditions can cause failure and the need for costly repairs. Whenever possible, roof coverings should be fire retardant; most such materials are laboratory-tested and rated for their fireproof qualities. Some building specifications may indicate a required fireproof rating, as well as the expected life of the roof in terms of years.

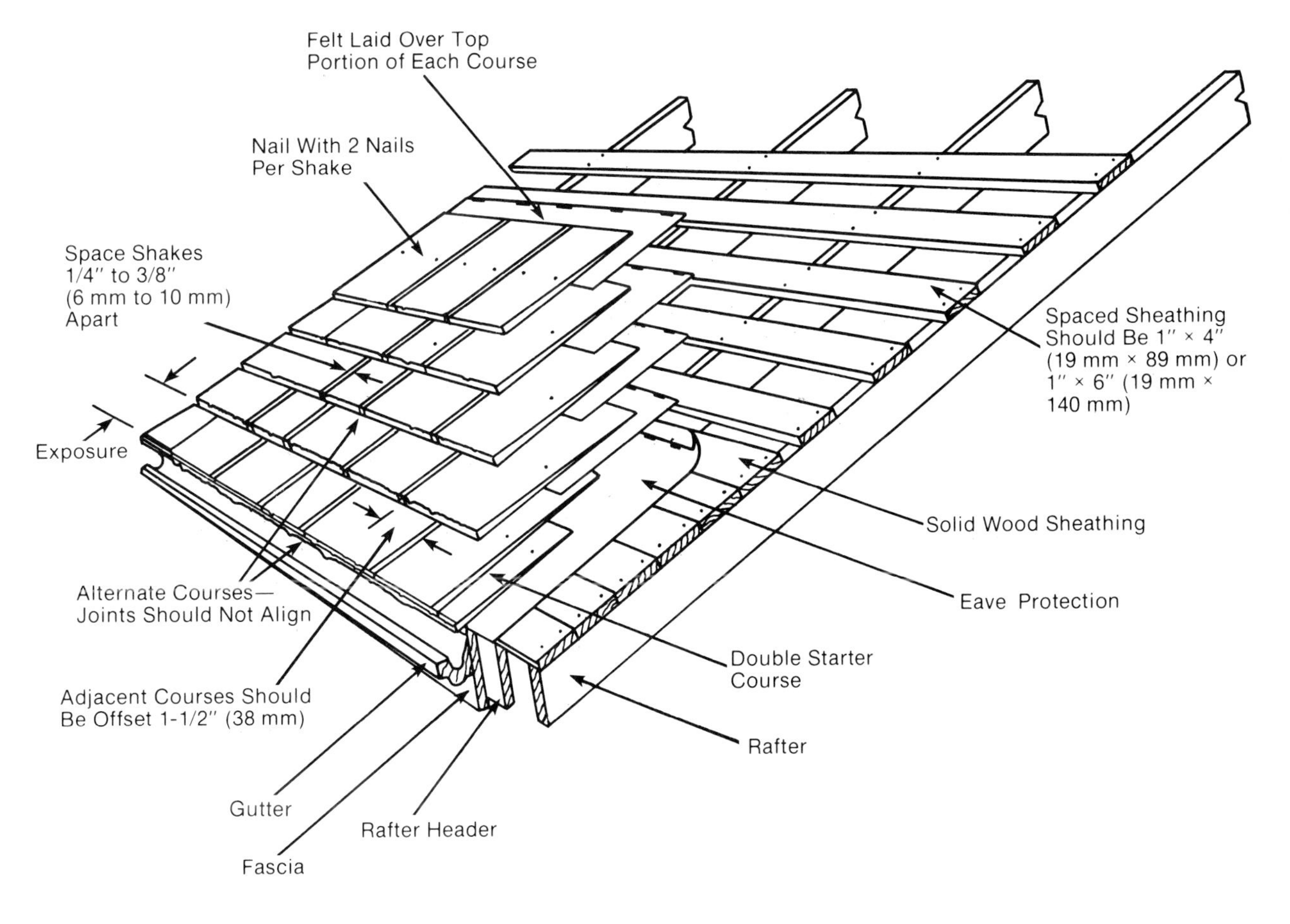

FIGURE 19-5. Application of cedar shakes.

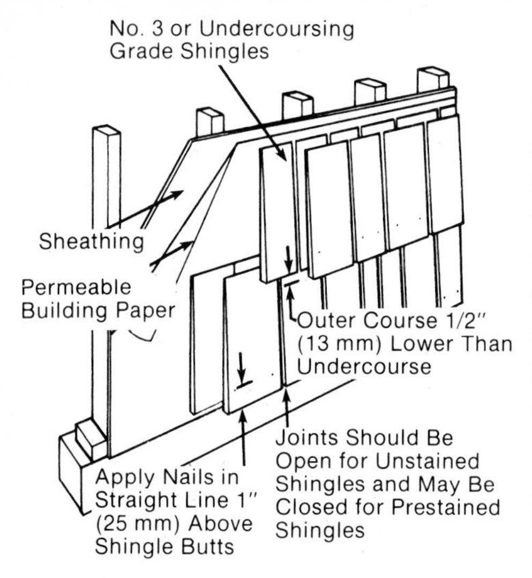

FIGURE 19-6. Side wall shingle application.

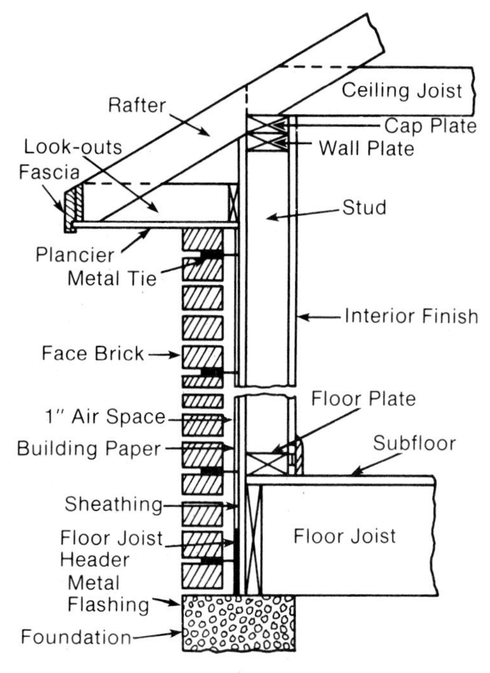

FIGURE 19-7. Brick veneer wall covering.

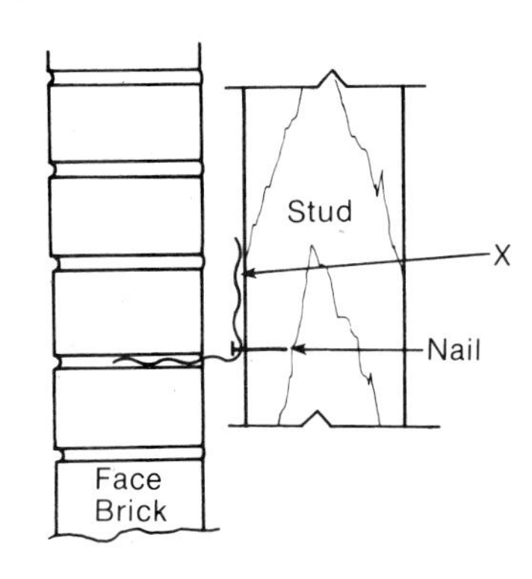

FIGURE 19-8. Brick veneer tie.

19.6 WALL COVERINGS

Beside the obvious functions like privacy and protection from the elements, walls accommodate service piping, electrical wiring, heating/air conditioning ducts, insulation, and, in some recent designs, vapor/air seals. Wall covering design determines how a wall will withstand weathering, how it will resist destruction by fire, and how much time and money will need to be spent in the future for maintenance and repairs. As with roof coverings, all elements of an area's climate (temperature changes, humidity, precipitation, etc.) must be considered when designing a wall covering subsystem.

A sidewall composed of plywood sheathing, permeable building paper, and double coursed shingles is excellent for shedding water (Figure 19-6). This design also dries out quickly to prevent decay and organism growth. Where bricks are economical to use, brick veneer is a common wall covering. A typical wall section might include sheathing, building paper, an air space, metal ties, and face bricks (Figure 19-7).

In a brick veneer wall covering design, the sheathing and building paper control air movement. The air space has an insulating effect and provides a moisture barrier; any seepage through the face brick drains away through the weep holes at the base of the wall. The brick, set in mortar, sheds water and provides protection. The foundation wall, naturally, must be heavier and stronger than the one needed for the shingled sidewall seen in Figure 19-6. Brick veneer (and all veneers, for that matter) carries its own loads, but none of the load from the roof.

Corrosion-resistant ties are used to secure a brick veneer wall covering. Figure 19-8 shows the location of the nails at the end of the tie. Note the spot marked with an "X"; if the nail is placed there, the tie would become dysfunctional, thus adversely affecting the entire design.

Figure 19-9 illustrates the combining of two subsystems in one wall. The two, stucco and siding, are separated by a drip cap so they function independently; the entire wall is protected from above by the cornice. Materials in the stucco portion include sheathing, tar paper, stucco wire, and two or three coats of stucco. The stucco is met at the top of the wall by the frieze and at the bottom by the flashing. The flashing, in turn, rests on the drip cap, which is grooved underneath to allow drops of water to fall free rather than forming rivulets down the wall.

Stucco is carried on wire mesh which, because of its configuration, sets out from the walls and becomes embedded in the first coat. Without tar paper or an equivalent paper between this "scratch coat" and the dry wood sheathing, moisture would be drawn into the wood from the stucco, and the stucco would lose its strength and eventually crumble.

Materials in the siding portion of the wall include sheathing, tar paper, felt paper, and beveled siding (or vinyl, metal aluminum siding, steel, or hardpressed fiberboard). The tar paper is water repellent and also controls air movement. The felt paper acts as a cushion to protect the soft wood. While wood with a low rate of expansion and contraction should be used to prevent much wall movement, the nailing should be done so as to allow some movement. The lapping of the siding is functional if the upper part is free to move.

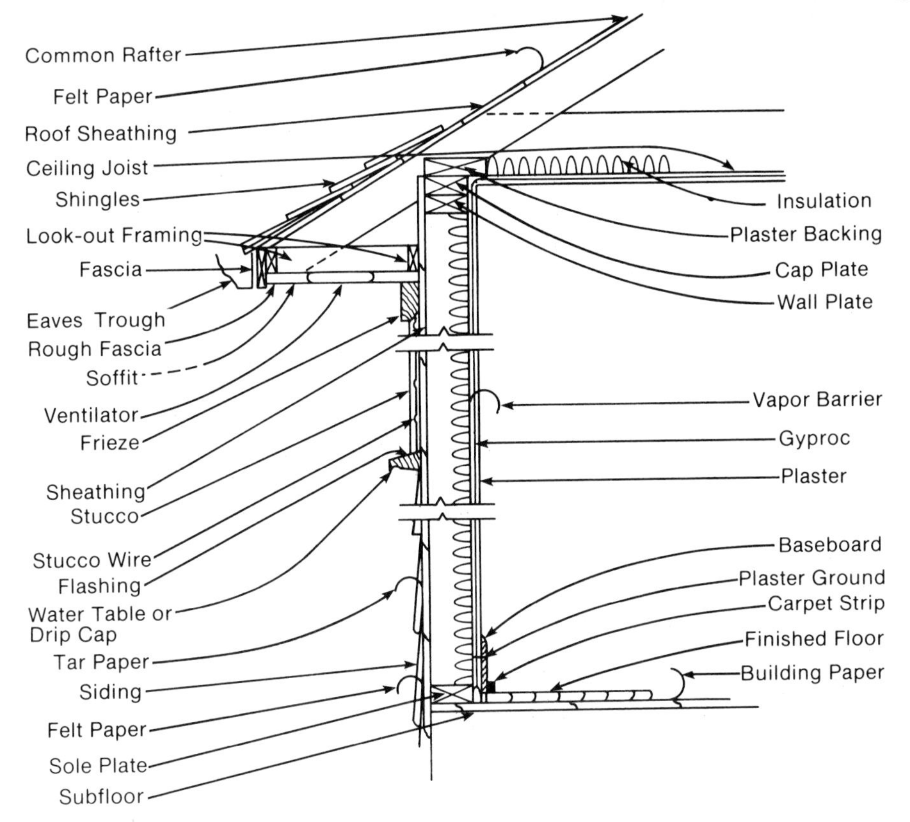

FIGURE 19-9. Combination stucco and siding subsystems.

Corrosion-resistant nails are driven into the studs above the lap. Siding butt joints, fitted and sealed, are then staggered vertically to prevent water seepage and to control the joints. Do not use nails that corrode, as they have a tendency to stain and lose their effectiveness as fasteners.

All wall covering materials must be kept away from the ground below and the roof above to protect them from damaging moisture. Minimum distances of 8″ (203 mm) from the ground and 2″ (51 mm) from the roof should be observed. These spaces should be wide enough to prevent any capillary action, since constantly damp woods are prone to decay.

In any building installation, the drawings and written specifications for a given subsystem often require special skills in order to coordinate the job. For example, when writing the specifications for tie nails, precise data with respect to the shape, gauge, length, material, and coating of the nails would be given. The person writing the specifications must be familiar with the wall assembly and the materials being used. In turn, the specifications for the tie nails would closely follow those of the tie, for the sake of compatibility. Finally, the builder must be certain to use the specified nails and ties. Working closely together in this manner leads to better quality work.

19.7 BUILDING PAPERS AND VAPOR BARRIERS

The designer or builder must decide whether to use a vapor barrier or a water barrier. Keep in mind that while a vapor barrier is also a water barrier, a water barrier may or may not be a vapor barrier. Asphalt-impregnated and coated papers are impervious to water, but not to water vapor; polyethylene sheets are, to a high degree, impervious to both.

The decision of which to use is made easier with knowledge of the various papers and vapor barriers. Table 19-3 lists properties and uses for the most common ones.

19.8 CONSTRUCTION CHECKLISTS

After the roof covering has been installed and the exterior walls have been at least sheathed, the next step is formulating a checklist to ensure

TABLE 19-3 BUILDING PAPER AND VAPOR BARRIER CHARACTERISTICS

Type	Properties	Application/Uses	Principle
Sheathing	1. Plain, made from pulp or kraft pulp, 36″ (914 mm) roll, 400′ (122 m) per roll.	Stapled.	Controls movement of air and dust.
	2. Asphalt-impregnated and coated papers of various weights, 4 to 10 pounds per 100 ft^2 (1.8 to 4.5 kg per 9 m^2).	Stapled.	Impervious to water, but not to water vapor.
	3. Polyethylene, usually 6 mil (150 micrometers).	Sealed joints with nonhardening sealer; stapled.	Prevents both movement of air and diffusion of vapor.
Roofing	1. Rolled heavy mineralized surfaced paper, 18″ and 36″ (457 mm and 914 mm) rolls, 45 to 120 pounds 100 ft^2 (20 to 54 kg per 9 m^2).	Galvanized roofing nails; lapped joints sealed with asphalt.	Easy application on sloping roofs.
	2. Built-up roof felts, 36″ (914 mm) rolls, 3 to 20 pounds per 100 ft^2 (1 to 9 kg per 9 m^2).	Layered, cushioned asphalt with sand/pebble topping.	Bonds for 10 to 20 years on flat roofs.
Wallpapers	Decorative, self-adhering rolls, patterned for butt joints.	Interior wall coverings.	Replaceable interior design materials.
Cushioning	Wood fiber insulating paper.	Cushioning under linoleum and slate roofs.	Leveller between subfloor and finish floor, and between sheathing and slate.
Concrete form	Spiral tubes of kraft paper.	Piers, reinforced concrete columns, duct or core forms.	Strong enough to contain fluid concrete; in the case of duct forms, can withstand outside pressure.

TABLE 19-3 BUILDING PAPER AND VAPOR BARRIER CHARACTERISTICS (Continued)

Type	Properties	Application/Uses	Principle
Envelope	Plain or asphalted paper covers for insulating materials or building panels such as gypsum board or straw board.	Nailed or cemented.	Paper contains the mineral or straw, making it manageable.

that the building is in readiness to receive interior covering. Municipal inspectors will use their own list to see that building standards are being met with respect to the strength of the structure, plot plan, elevations, fire protection, electrical safety, heating, air-conditioning, plumbing, and site drainage. Table 19-4 illustrates a typical construction checklist.

TABLE 19-4 CONSTRUCTION CHECKLIST PRIOR TO PLACEMENT OF INTERIOR COVERING

System	Subsystem	Typical Checks	Reasons
Heating	Chimney(s)	Design/Construction Foundation pads Lining Caps Clearances Height above roof Spark control	Fire regulations: compatibility of fuels with chimney materials
	Fireplace	Design/Construction Footings Clearances Air intake Smoke control Clean-out Flashings	Efficiency, fire-box safety, and fire-fighting potential
	Furnace	Design Size Piping Shut-down capability Approvals Footings Elevation Installation Testing Ductwork Clearances Services Venting Air supply Preheating	Avoid dangerous over-sizing; balance output to needs (Automatic shut-down necessary)
	Ductwork	Secure installation Air circulation Venting Blockages	Blockages reduce effective sizes; need positive linkages, with adequate clearances, to prevent fires

TABLE 19-4 CONSTRUCTION CHECKLIST PRIOR TO PLACEMENT OF INTERIOR COVERING (Continued)

System	Subsystem	Typical Checks	Reasons
Electrical	Panel	Sizing Type Materials and hardware	Adequacy in terms of voltage, amperage, and circuits; automatic shut-down in case of surges or high demand; long-term safety
	Wiring	Plan vs. actual owner's wiring plan	Grounding; safe installations of appliances; shut-down safety; quick locating of service line, if underground
	Meter	Location Readability Seal Mast	Safety and service charges
	Air-conditioning	Humidity control Temperature control	Effectiveness, safety, and noise level
Communications	Telephone	Wiring pattern Jacks	Flexibility
	Doorbell Television/radio	Wiring pattern Power supply Cable supply Dish	Testing of circuits Adequacy of services
	Computer	Power supply Telephone links	Testing the concept
Plumbing (roughed-in)	Supply piping: main to service pipe	Potable water	Health and system life
	Drainage piping traps Storm sewer Waste sewer	Actual installations vs. local regulations	Safety
	Venting	Stacks inspection Number of fixture equivalents	Pest control, water and sewer gas seals, and adequacy of the system
	Hot water system	Piping Controls Storage volume Adherence to code Warranty/guarantee	Adequacy of the system and safety
	Piping insulation	Materials vs. specifications	Energy efficiency
Building Insulation	Roof/Ceiling	R-values	Energy efficiency
	Outside and inside walls	Type of material	Health; vapor barrier relationship; noise control

TABLE 19-4 CONSTRUCTION CHECKLIST PRIOR TO PLACEMENT OF INTERIOR COVERING (Continued)

System	Subsystem	Typical Checks	Reasons
Vapor/Air barrier		Material type and thickness Seals Placement	Life of material; control of air/moisture movements Prevention of in-wall and in-ceiling condensation
Stairs		Construction vs. building plans	Ease of stair headroom; safety regulations
Blocking/Backing		Nailing/fastening surfaces	Fixtures and panel edges need support
Windows/ Exterior doors		Actual vs. specifications	Conformity to building design and security of unfinished building

19.9 COVERING CEILINGS AND INTERIOR WALLS

Once the builder and building authority checklists have been completed and clearance given, interior covering can begin. First, inspect the blocking and lower edge line of the ceiling joists to determine whether the lower edges need trimming. Also, decide if a strongback should be installed across the upper edges of the joists to bring them into line and if strapping is required on the lower edges for a leveling and nailing surface.

Ceiling treatments are in keeping with the type of wall frame being used. The two principal finishes are plaster and drywall. These are commonly found in balloon and platform framing systems, while solid wood is used with post and beam.

Plastered ceilings and walls are made with up to three coats of mixture on a base of gypsum or metal lath. Before placement of the mixture, screeds are installed to control the thickness of the plaster at openings, inside corners, and exterior corners.

The base for plaster must provide a good bond and be secured to the frame. Care should be taken to stagger the joints, to use gypsum-lathing nails, and to allow the corners freedom of movement by using reinforcements instead of nails (Figure 19-10). For optimum drying, building operations should be shut down and the temperature kept between 42° F and 68° F (6° C and 20° C). Good ventilation is also very helpful. Table 19-5 describes various interior plaster coats.

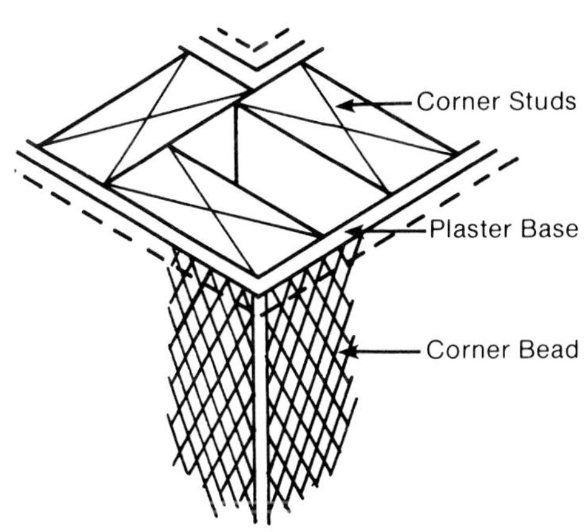

FIGURE 19-10. Reinforcing plaster.

TABLE 19-5 INTERIOR PLASTERING

Coat	Material	Application
First (scratch)	1 part gypsum plaster and 2 parts sand, by weight	While still soft, it is scratched to ensure bond of second coat

TABLE 19-5 INTERIOR PLASTERING (Continued)

Coat	Material	Application
Second (brown or leveling)	1 part gypsum plaster and 3 parts sand, by weight	
Third (putty-coat)	1 part gypsum plaster and 3 parts lime, by volume	Troweled to a smooth, hard finish

Note that for a two-coat application, the scratch and brown coats are combined and then the putty coat is applied. The final thickness should be approximately 0.5″ (13 mm).

Drywall describes any finish that requires little or no water for application. Drywall finish includes gypsum board, plywood, and fiberboard, all of which need well prepared nailing/fastening surfaces. A strongback is useful in leveling and aligning the ceiling joists (Figure 19-11).

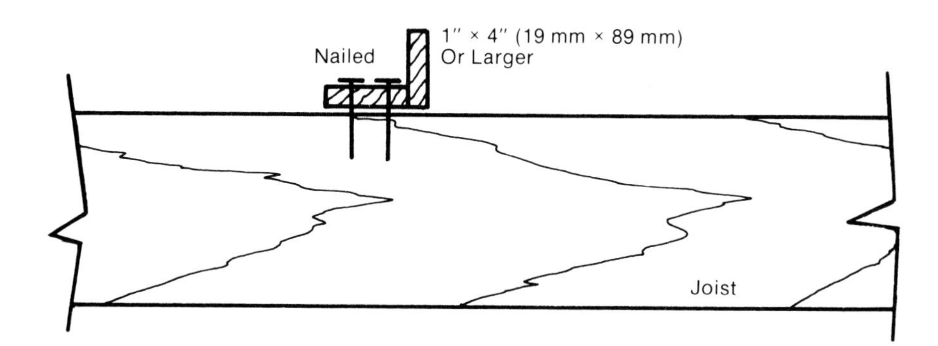

FIGURE 19-11. Joist strongback.

When the ceilings are covered first, the wall boards and panels support the edges and allow movement without secure fastening at the edges. The movement takes place in the framework more than in the covering.

Ceilings may be covered with one or two layers of gypsum board. If two are used, the first is properly nailed in place and the second, with panels placed at right angles, is cemented to the first. Temporary bracing and double-headed nails hold the panel until the cement sets. Later, the nails are pulled, the holes filled, and the joints taped and finished.

Walls may also be covered with one or two layers of gypsum board (three in some cases, where the wall is meant to be nearly self-supporting on a light frame). The first layer is applied horizontally. The longer the sheet the better, to reduce the number and lengths of joints. Nailing patterns are important to secure the panels to the framework and to prevent nail popping. Care must be taken to ensure framework backing for nailing, as well as openings for outlet boxes.

The second layer of gypsum board is applied vertically, with double-headed nails to hold it in place. Joint cement is available in powder form and is mixed with water to attain a soft putty consistency. It can be applied with hand tools or mechanical applicators. To finish a gypsum board, set a nail with a crowned hammer, cement and tape the joint, and then tape at the inside corners (Figure 19-12).

Plywoods, hard-pressed fiberboards, and fiberboards are also applied vertically. These may be prefinished or unfinished. If the finished wall is to be clean and unmarked, special attention must be given to joint treatment.

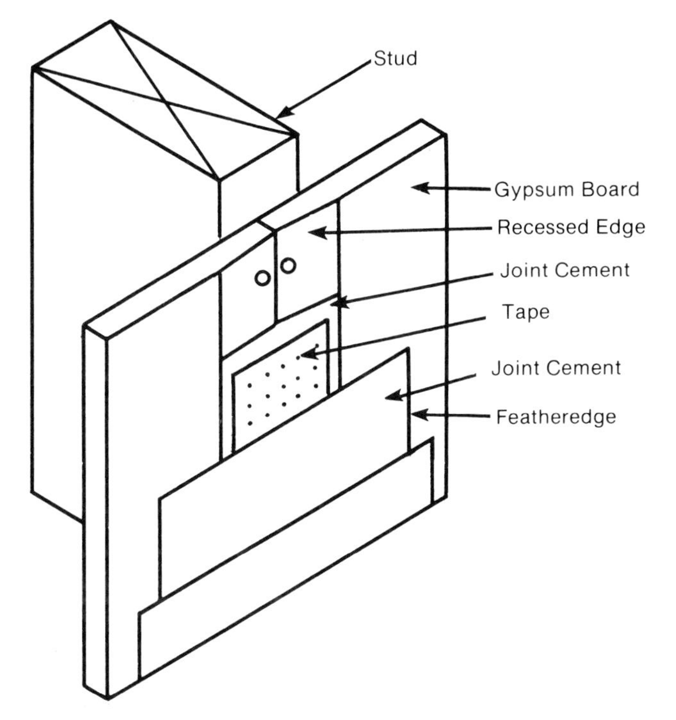

FIGURE 19-12. Finishing gypsum board.

19.10 FINISH FLOORS

Any material used as the final wearing surface on a floor is called the "finish" flooring. Of the many such materials on the market, each has specific advantages for a particular use. Two essential properties of any kind of finish flooring are durablity and ease of handling; fire resistance, propensity to produce toxic fumes when burning or smoldering, and cost of material and application are other important considerations.

Hardwoods such as birch, maple, beech, and oak are used in a variety of widths and thicknesses as strip flooring, and some species are also available in parquet form. Vertical grain strips of soft woods such as fir or hemlock are sometimes used as well. Wood finish flooring is widely used in living and dining rooms, bedrooms, corridors, and general-purpose rooms such as family rooms and dens. Other materials suitable for finish flooring include resilient flooring in tile or sheet form, terrazzo, or ceramic tile. These materials provide water resistance and are used in bathrooms, kitchens, public entrance halls, and general storage areas. Carpets may also be used as finish flooring, except where water resistance is required.

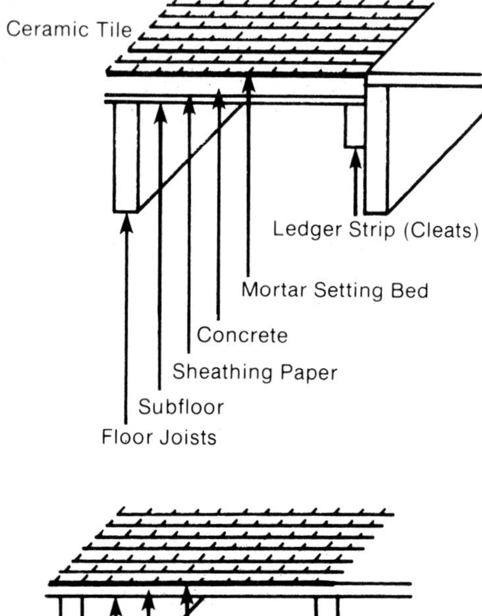

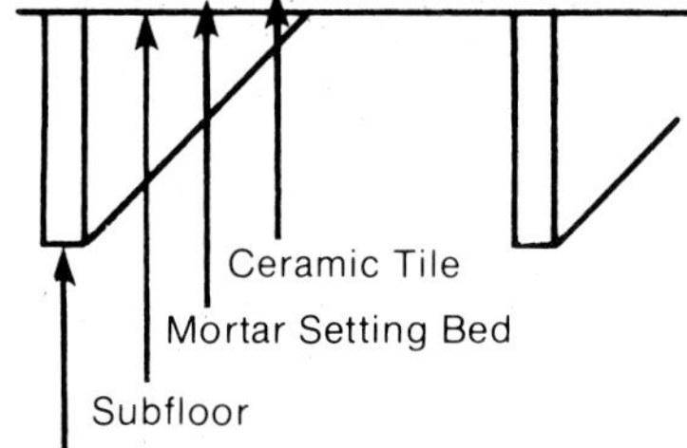

FIGURE 19-13. Two methods of installing a ceramic tile floor.

Tongue-and-groove strip flooring is hollow-backed, with the top face wider than the bottom. The strips are driven together by hammer blows against a fitted short end of the same stock, and the faces come together. Humidity control is imperative when storing, handling, and placing the strips. The strips are secured to the subfloor by driving nails at 45° angles, then set without marking the face edges.

A finish floor of carpet or resilient flooring requires an underlay to provide unbroken, cushioned support. Underlay materials (plywood panels, hard-pressed fiberboard or wood particleboard) are securely nailed or stapled and the joints filled and sanded. Resilient flooring and carpets should be laid after most other building tasks have been performed. Traffic in the area should be kept to a minimum, so as not to interfere with the job.

Procedures for ceramic tile are different, whether it's used as a covering for bathrooms or vestibules, or on fireplace hearths. Ceramic tile may be installed on a concrete slab floor, applied to a mortar base supported on a subfloor, or attached by a special adhesive to a panel-type underlay such as plywood or hardpressed fiberboard. Figure 19-13 illustrates two methods of installing a ceramic tile floor.

For further information on finish flooring, see Chapter 23.

19.11 RESEARCH

1. Collect and file materials data with respect to subfloors, wall and roof sheathing, roof and exterior wall coverings, interior ceilings and walls, and finish floors.
2. Write general specifications for light construction subsystems. Choose one for roofs, one for exterior walls, one for interior walls, and one for finish floors.
3. Observe several construction checklists, noting the significant differences between them. Decide which offers the most organized approach and be prepared to adopt it.

19.12 REFERENCES

19.21.1 Crowdis, David and Kay. *Designing and Building Your Home.* Virginia: Reston Publishing Company, 1981.

19.12.2 Maguire, Byron. *Carpentry: Framing and Finishing.* Virginia: Reston Publishing Company, 1979.

19.12.3 *Plywood Sheathing.* Council of Forest Industries of British Columbia, Vancouver, British Columbia, 1979.

19.12.4 Wass, Alonzo. *Data Book for Residential Contractors and Estimators.* Virginia: Reston Publishing Company, 1979.

20

SOUND BARRIERS, AIR/VAPOR BARRIERS, VENTILATION, AND THERMAL INSULATION

20.1 GENERAL OBJECTIVES

The technologies associated with sound barriers, air/vapor barriers, ventilation, and thermal insulation have been developing quickly in the past few years. It is the general purpose of this chapter to encourage recognition of these technologies and their relationships to the light building construction industry.

20.2 SPECIFIC OBJECTIVES

After reading this chapter, the student should be able to

1. relate technical terms and light construction practices with respect to creating comfortable living and working environments.
2. indicate how measurements are used to define terms such as noise, sound absorption, sound isolation, heat loss (transfer), and thermal resistance.

20.3 SOUND AND NOISE BARRIERS

Sound may be defined as wave motion that reaches one's ear as minute fluctuations in air pressure. Sound waves travel through gas (air), water, and solids at different rates. In air at 70°F, (21.1°C) audio sounds travel at an approximate 1,130′ (344.4 m)/sec. The following list represents some of the different rates of travel for solids: wood, 11,000′ (3,352.8 m)/sec; glass and steel, 16,000′ (4,876.8 m)/sec; and quartz, 18,000′(5,486.4m)/sec. From this it can be seen that wood will damp sounds better than glass, steel, or quartz, which act more like diaphragms. The denser and stiffer the solid material, the faster the sound will travel through it. Any interruption in the path of the sound waves or any coating of a smooth, nonporous substance will appreciably reduce the level of energy induced in the material.

Based on this brief look at sound, it seems logical to consider that soft, fibrous materials will not readily transmit sound energy, nor will painted or glazed surfaces. Soft materials will absorb large percentages of the sound energy, while painted or glazed surfaces will reflect most of the sound energy.

The amplitude or strength of sound energy is frequently measured in decibels. The decibel (dB) is a unit used to express the ratio between the amounts of power existing at two points. It may be expressed as negative dB or positive dB, where dB is the point of hearing threshold. Negative dB levels are not audible to the human ear. The average dB ratings of several common sounds are listed in Table 20-1.

TABLE 20-1 SOUNDS AND dB RATING/LEVEL

Sound	Decibel (dB)
Jet engine noise (close up)	160
Threshold of pain	130
Threshold of feeling	120
Thunder	120
Wood saw at 3′ (0.9 m)	100
Noisy office	80
Automobiles on highway	60
Conversation	60
Average office	45-60
Quiet radio, average residence	35-40
Whisper	18-20
Leaves rustling	10
Normal breathing	10
Reference	0

Noise may be defined as any unwanted or noticeably unpleasant sound. At high levels, noise may pose health hazards.

Sound control on an effective basis is an activity with which the builder/designer must cope. With sounds originating both inside and outside a building, it is necessary for the builder/designer to choose a site carefully, to improve it, and to create a building which in itself is acoustically treated.

20.3.1 Sites

Whenever possible, building sites should be located far from potential sources of noise such as airfields, industrial plants, railroads, and traffic arteries. Buildings should be placed to take advantage of rolling terrain and stands of trees.

Rolling terrain or berms can be used as noise and wind barriers. Combinations of trees, low foliage, and ground cover can be used as buffers to absorb and reflect unwanted sounds. Generally, in order to properly diminish the intensity of normal traffic noises, these buffers should be from 500′ to 1,000′ (152.4 m to 304.8 m) deep (Figure 20-1). When preparing the site, consider how screen walls, fences, and other buildings may be used to improve berm effects. Put planning high on the priority list.

Contour maps help with site selection and building orientation. On larger sites (acreages and farms), the general layout of buildings is

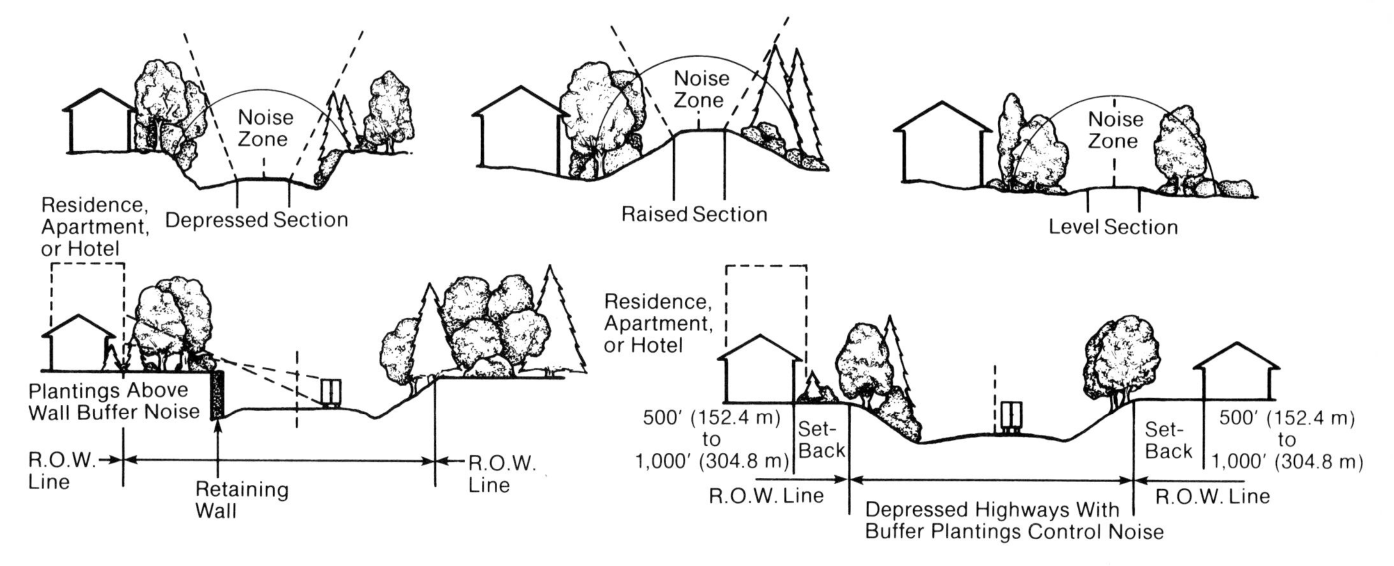

FIGURE 20-1. Buffers may be employed to help diminish unwanted noise.

important in relation to ground configuration and to prevailing winds. Pumps and generators need to be housed away from living quarters to avoid the hearing and feeling of vibrations from them.

The shape of a building in relation to the ground around it, as shown in Figure 20-2, is important in sound control. If the building is higher, the berm is less effective.

Pioneers in a new land often knew where to locate log houses or sod houses in relation to a river bank. Knowing of the prevailing winds, they would place the building on the down-wind side and far enough away from the bank to allow high winds to deflect over (Figure 20-3).

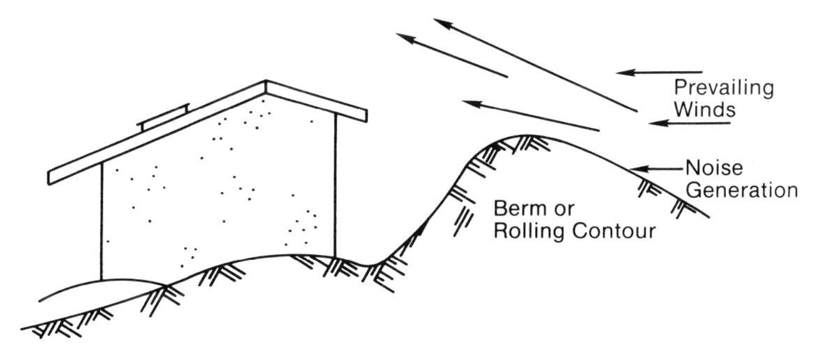

FIGURE 20-2. Building shape vs. ground contour.

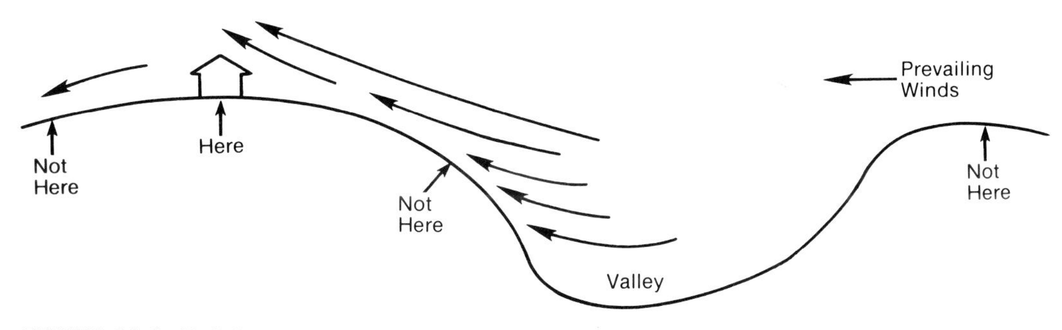

FIGURE 20-3. Building site near a river valley.

20.3.2 Building Layouts

Sound control begins with site selection and continues with building layout. Before the construction of a new building can begin, the design for that new building must be approved. In order to obtain approval, the designer must meet certain sound control requirements. There are several helpful "rules" that simplify the problem of achieving adequate noise insulation. A few such rules follow:

- In manufacturing plants, locate the offices as far as possible from sources of intense noise.
- In multiple dwelling structures, arrange layouts so that the most critical rooms (bedrooms, living rooms) are protected from adjoining apartments by a buffer zone of noncritical sound areas such as bathrooms, kitchens, closets, and hallways. The next best arrangement is to place quiet rooms such as bedrooms on the two sides of a party wall.
- Keep windows and doors away from the noisy side of a building if possible. For example, the berm-side wall of the structure in Figure 20-2 would be solid and well insulated; there would be no doors or windows.
- In shops and small manufacturing plants, dampen vibrations from machines by using special foundations or mountings.

20.3.3 Sound Absorption

Different materials absorb and reflect sound waves in varying amounts. For this reason a variety of building materials have been tested for their ability to absorb sounds at different frequencies. Table 20-2 provides the coefficients for many of these materials. The noise reduction coefficients (NRC) given in this table represent only one of the three common ways used to define and evaluate sound absorption. The sound absorption coefficient (SAC) and sabin are also used.

TABLE 20-2 NOISE REDUCTION COEFFICIENTS OF MATERIALS (NRC)

Material	NRC
Brick wall, unpainted	0.05
Brick wall, painted	0.02 (1/6 dB)
Concrete	0.02
Lightweight blocks, medium surface texture	0.45 (4 dB)
Heavyweight blocks, medium surface texture	0.27
Plaster, gypsum, or lime	0.04
Plaster, acoustical	0.21–0.75
Glass	0.02
Marble or glazed tile	0.01
Wood paneling	0.06
Acoustical tile	0.55–0.85 (5–6 dB)
Carpeting	0.45–0.75
Linoleum, asphalt, rubber, or cork tile	0.01–0.05

The noise reduction coefficient (NRC) is a rating used to measure the ability of materials such as acoustical tile to absorb sound. Part of this absorbed sound is "killed" within the material, while some of it is passed through to the other side of the material. The NRC is a rating that averages the absorption characteristics of the middle frequencies, rounded off to the nearest 5%. Theoretically, these ratings can range from 0 to 1; a perfectly reflective material would have a rating of 0 and a perfectly absorbing material would have a rating of 1. An NRC rating of 0.25 indicates that an average of 25% of the sound that strikes the material would be absorbed and 75% would be reflected. A difference of 0.10 in the NRC is seldom detectable in a completed installation.

The sound absorption coefficient (SAC) is a rating for sound absorption that represents the peak performance of the material. It is not representative of the overall sound absorbing characteristics of the material, but it simply represents its highest rating which may be at any of the frequencies tested. The SAC rating ranges from 0 to 1, similar to the NRC.

The sabin, sometimes called a square-foot unit of absorption, is the equivalent of 1 ft^2 of material having an absorption coefficient of 1. A totally absorbing material having an area of s square feet is said to have an absorption of s sabins. Similarly, a totally absorbing material having an area of s′ square meters is said to have an absorption of s′ metric sabins (1 sabin = 0.093 metric sabin).

20.3.4 Sound Isolation

The transmission of airborne sound between rooms depends on the materials and methods used in the construction. Sound transmission through walls, floors, and ceilings depends on the mass (or unit weight) of the materials and their inelasticity.

Sound transmission loss (STL) is the amount of airborne sound, the amount of sound energy in decibels, that is lost as the sound passes through a wall or ceiling for a given frequency. The average STL for a group of frequencies is the rating by class and is called the sound transmission class (STC). The higher the numbers, the better the sound barriers. Values of 45 to 55 are acceptable as good barriers. It is important to understand, however, that any opening in the assembly nullifies the rating since sound will readily pass. Figure 20-4 shows a typical situation. Airborne sounds originating from machines, for example, have an 80 dB level. An office wall is built with an STC 55

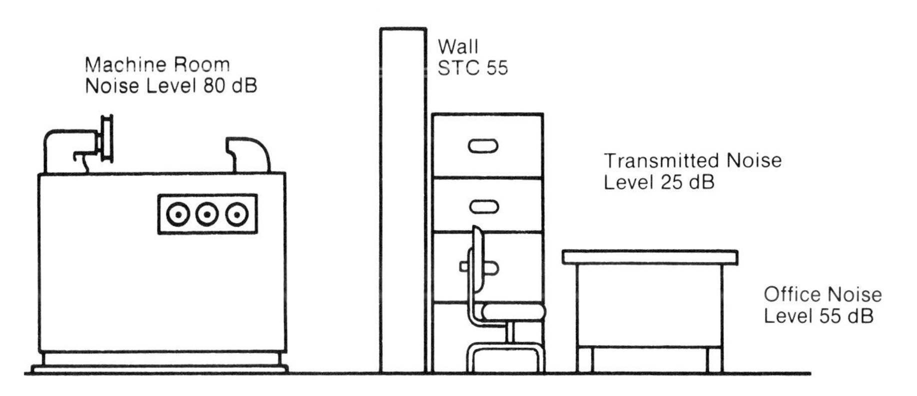

FIGURE 20-4. Sound transmission class for an office wall.

rating. The machine noise level in the office is STC 25, which is about the same level as normal speech. Notice that the transmitted noise level is below the average decibel level for an office.

Floors also transmit sound energy; however, the sound from floors is generally impact noise. Therefore, tests and test data are available on different impact noises. Rated by individual frequencies, these are made from standardized impacts made on top of the floor or floor covering. The rating by class for these noises may be either impact noise isolation (INR) or impact sound isolation (IIC).

The greater the positive value of the INR, the more resistant is the floor to impact noise transfer. For example, an INR of -2 is better than one of -17, and one of +5 INR is a further improvement in resistance to impact noise transfer. An INR rating of +5 indicates the assembly averages 5 dB better than the standard. The standard is based on average background noises that might exist in typical moderately quiet suburban apartments. Where higher background noises are found, as in urban areas, a minus INR rating (to about -5) may be used without any detrimental effects. In areas that are quieter, an INR rating of +5 to +10 would be used since not much background noise is available to mask the sound.

The IIC rating is expressed in decibels and more closely relates to the STC rating for airborne sound transmission than does the INR rating. The IIC rating of any given assembly is usually about 51 dB higher than its INR rating; that is, an assembly with an IIC rating of 61 is comparable to an INR rating of +10. However, deviations in this can occur and individual test data should be confirmed.

20.3.5 Selecting Walls, Floors, and Ceilings

Partitions and ceilings of many different configurations have been tested for sound absorption and have been assigned STC ratings. American Society for Testing and Materials, U.S. Gypsum Corp., Underwriter Labs, National Bureau of Standards, Western Wood Products Assoc., and others have extensive data available upon request that gives designers, as well as students, reliable STC ratings. Partitions designed with wood or steel studs, single- and double-stud design, drywall, resilient channels, glass or wool insulation batts, sound-deadening board, and acoustical caulking are used in combinations to achieve STC values ranging from a poor low of 32 to a very good 60. Ceilings also made of wood joists or bar (steel) joists and sheathing plywood, drywall, suspended ceilings, glass or wool insulation batts, resilient channels, and, where needed, acoustical caulking also produce 40 to 60 STC ratings. Examples of walls and ceilings are shown in Tables 20-3 through 20-6.

20.3.6 Choosing Construction Subsystems

Not all rooms require soundproof walls or partitions. In fact, there are many techniques that may be employed to accomplish sound control goals.

Vinyl floor tiles reduce impact noise levels as well as airborne sounds. If the tile were installed over a 4″ (102 mm) slab of concrete, for

TABLE 20-3 SOUND INSULATION OF SINGLE WALLS

Wall Detail	Description	STC Rating
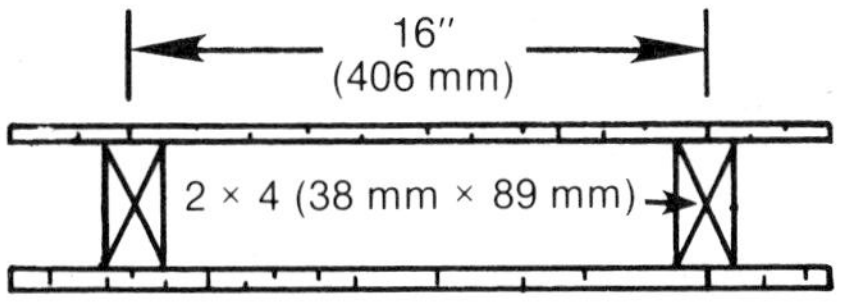 16″ (406 mm); 2 × 4 (38 mm × 89 mm)	1/2″ (13 mm) Gypsum wallboard 5/8″ (16 mm) Gypsum wallboard	32 37
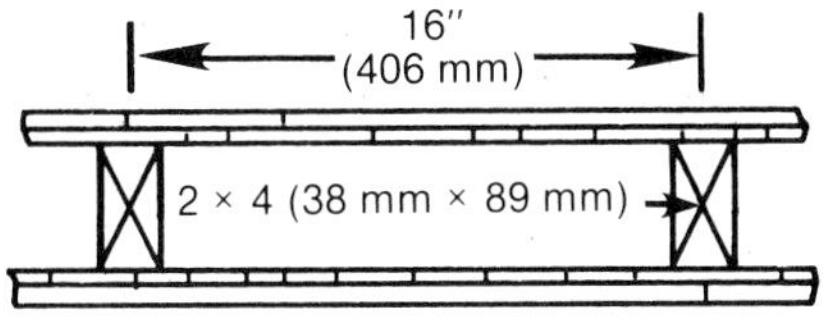 16″ (406 mm); 2 × 4 (38 mm × 89 mm)	3/8″ (10 mm) Gypsum lath (nailed) plus 1/2″ (13 mm) gypsum plaster with whitecoat finish (each side)	39
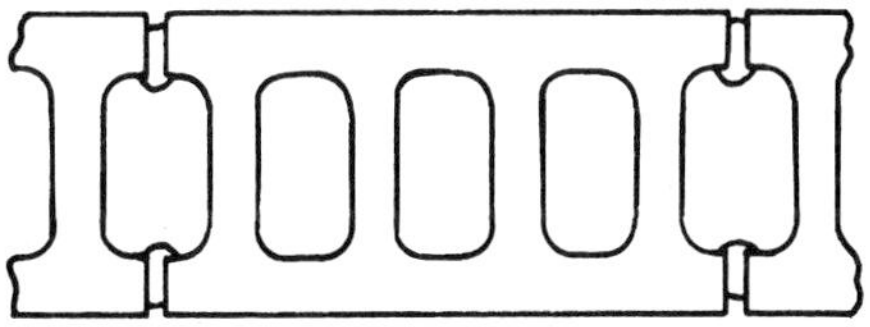	8″ (203 mm) Concrete block	45
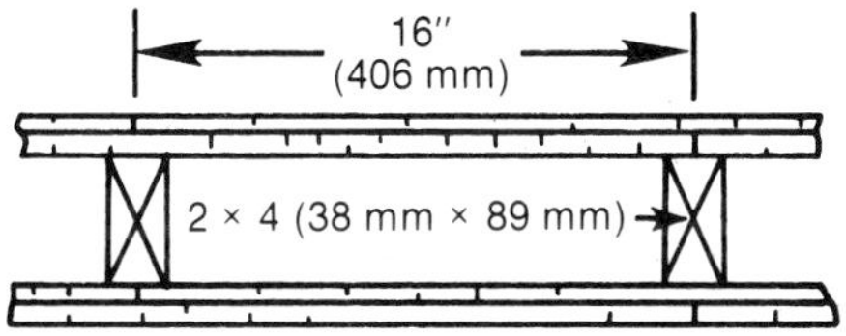 16″ (406 mm); 2 × 4 (38 mm × 89 mm)	1/2″ (13 mm) Sound deadening board (nailed) 1/2″ (13 mm) Gypsum wallboard (laminated each side)	46
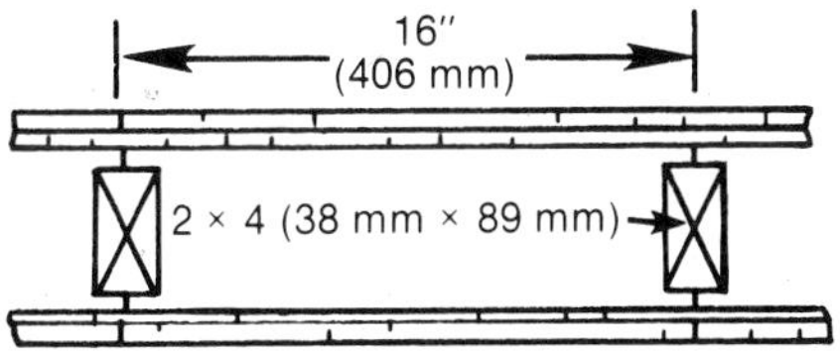 16″ (406 mm); 2 × 4 (38 mm × 89 mm)	Resilient clips to 3/8″ (10 mm) gypsum backer board 1/2″ (13 mm) Fiberboard (laminated each side)	52

TABLE 20-4 SOUND INSULATION OF DOUBLE WALLS

Wall Detail	Description	STC Rating
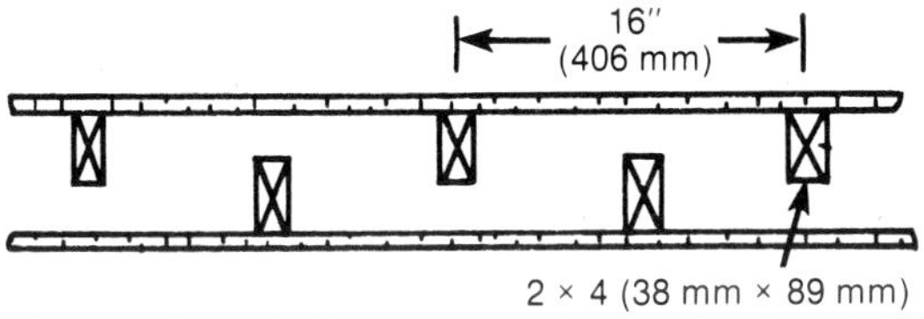 16″ (406 mm); 2 × 4 (38 mm × 89 mm)	1/2″ (13 mm) Gypsum wallboard	45
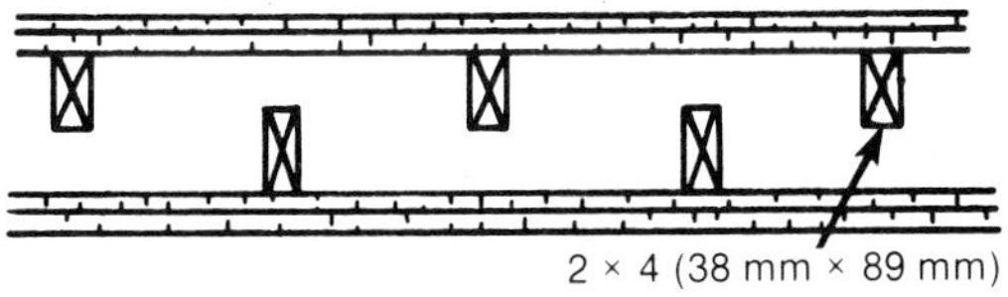 2 × 4 (38 mm × 89 mm)	5/8″ (16 mm) Gypsum wallboard (double layer each side)	45

TABLE 20-4 SOUND INSULATION OF DOUBLE WALLS (Continued)

Wall Detail	Description	STC Rating
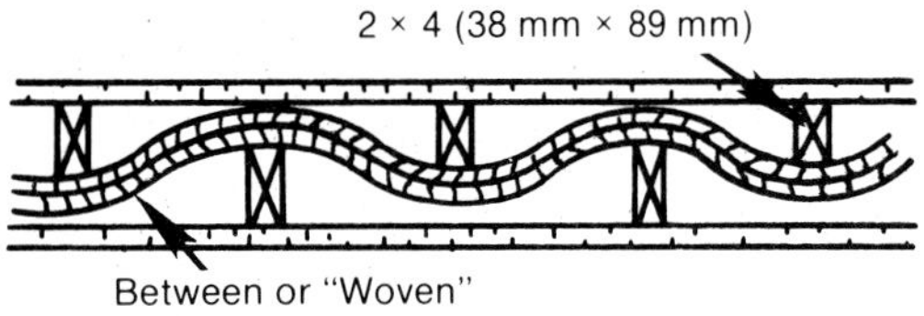 2 × 4 (38 mm × 89 mm); Between or "Woven"	1/2″ (13 mm) Gypsum wallboard 1-1/2″ (38 mm) Fibrous insulation	49
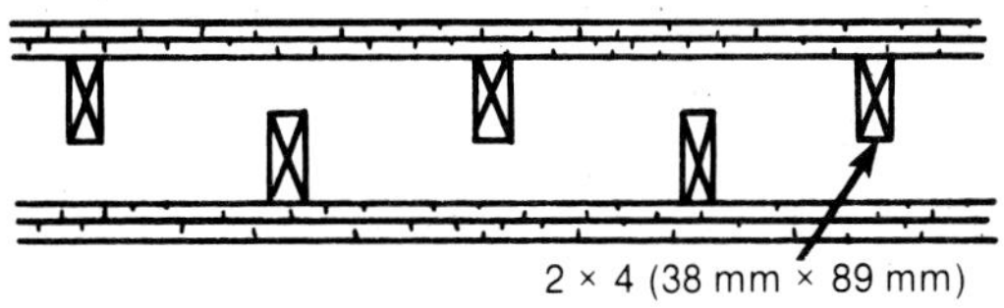 2 × 4 (38 mm × 89 mm)	1/2″ (13 mm) Sound deadening board (nailed) 1/2″ Gypsum wallboard (laminated)	50

TABLE 20-5 RELATIVE IMPACT AND SOUND TRANSFER IN FLOOR-CEILING COMBINATIONS FOR 2″ × 8″ (38 MM × 184 MM) JOISTS

Detail	Description	Estimated Values	
		STC Rating	Approx. INR
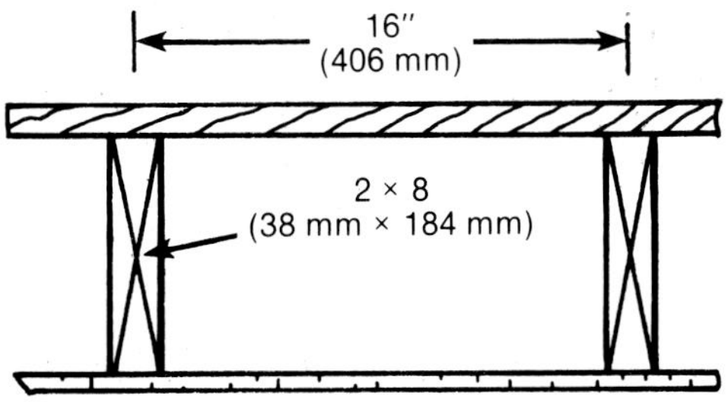 16″ (406 mm); 2 × 8 (38 mm × 184 mm)	Floor 7/8″ (22 mm) T & G flooring Ceiling 3/8″ (10 mm) gypsum board	30	−18
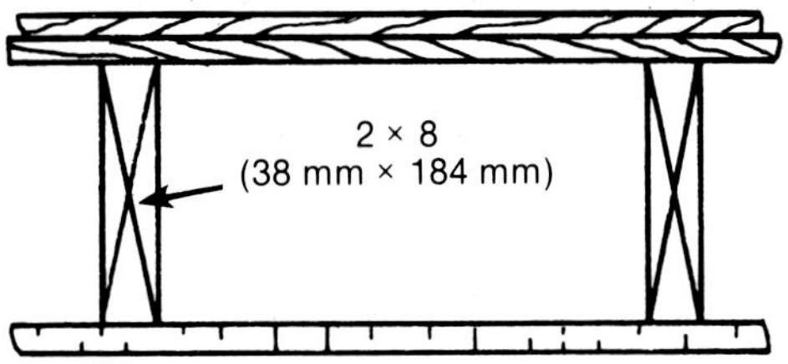 2 × 8 (38 mm × 184 mm)	Floor 3/4″ (19 mm) subfloor 3/4″ (19 mm) finish floor Ceiling 3/4″ (19 mm) fiberboard	42	−12

TABLE 20-5 RELATIVE IMPACT AND SOUND TRANSFER IN FLOOR-CEILING COMBINATIONS FOR 2″ × 8″ (38 MM × 184 MM) JOISTS (Continued)

Detail	Description	Estimated Values	
		STC Rating	Approx. INR
2 × 8 (38 mm × 184 mm)	Floor 3/4″ (19 mm) subfloor 3/4″ (19 mm) finish floor Ceiling 1/2″ (13 mm) fiberboard lath 1/2″ (13 mm) gypsum plaster 3/4″ (19 mm) fiberboard	45	–4

TABLE 20-6 RELATIVE IMPACT AND SOUND TRANSFER IN FLOOR-CEILING COMBINATIONS FOR 2″ × 10″ (38 MM × 235 MM) JOISTS

Detail	Description	Estimated Values	
		STC Rating	Approx. INR
16″ (406 mm) 2 × 10 (38 mm × 235 mm)	Floor 3/4″ (19 mm) subfloor (building paper) 3/4″ (19 mm) finish floor Ceiling Gypsum lath and spring clips 1/2″ (13 mm) gypsum plaster	52	–2
2 × 10 (38 mm × 235 mm)	Floor 5/8″ (16 mm) plywood subfloor 1/2″ (13 mm) underlayment 1/8″ (3 mm) vinyl-asbestos tile Ceiling 1/2″ (13 mm) gypsum wallboard	31	–17

TABLE 20-6 RELATIVE IMPACT AND SOUND TRANSFER IN FLOOR-CEILING COMBINATIONS FOR 2" × 10" (38 MM × 235 MM) JOISTS (Continued)

Detail	Description	Estimated Values	
		STC Rating	Approx. INR
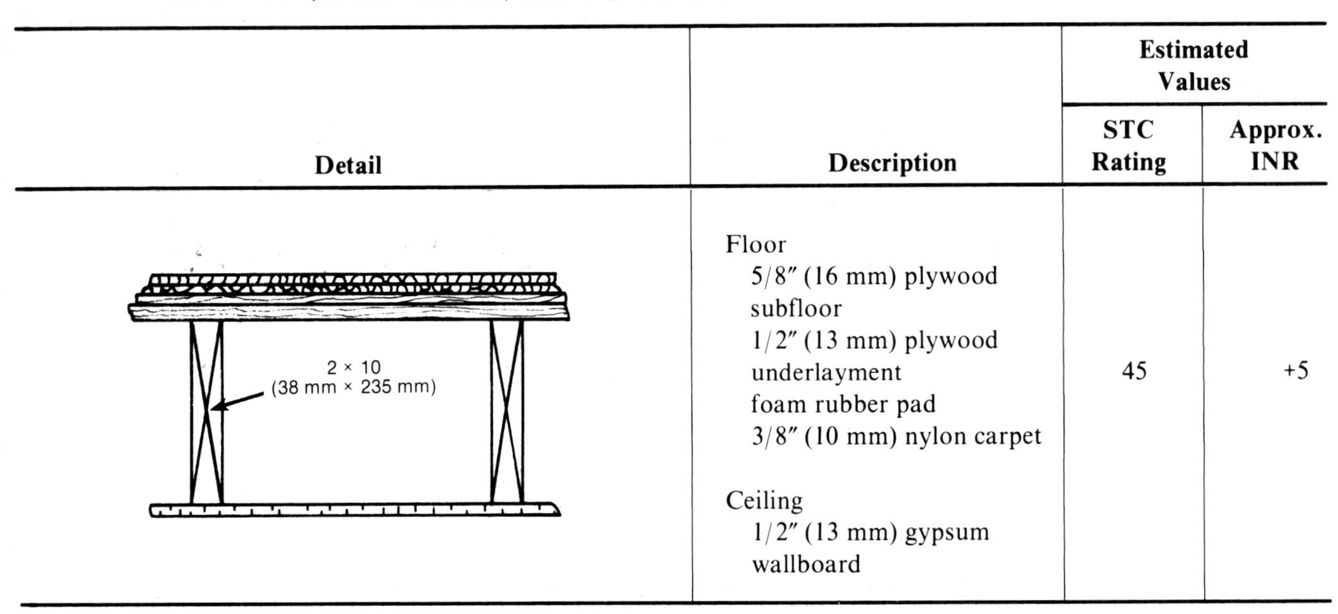	Floor 5/8" (16 mm) plywood subfloor 1/2" (13 mm) plywood underlayment foam rubber pad 3/8" (10 mm) nylon carpet Ceiling 1/2" (13 mm) gypsum wallboard	45	+5

example, the INR or IIC improvement would be 4 dB (from 25 dB to 29 dB). The NRC improvement over a bare concrete floor is 0.06 (from 0.02 to 0.08).

Acoustical tiles, which can be easily installed as shown in Figure 20-5, will provide NRC ratings of 0.55 to 0.85, depending upon thickness of the tile and design of the surface texture. Ceilings of this type are used quite frequently in public places and office buildings; in most instances, they account for the largest number of absorbed units of sound. These tiles measure 3/16" to 1-1/4" (5 mm to 32 mm) thick and are available in sizes from 12 in.^{2} (7,742 mm^2) to 2' × 4' (0.6 m × 1.2 m). Edges and surface textures of the tiles vary greatly.

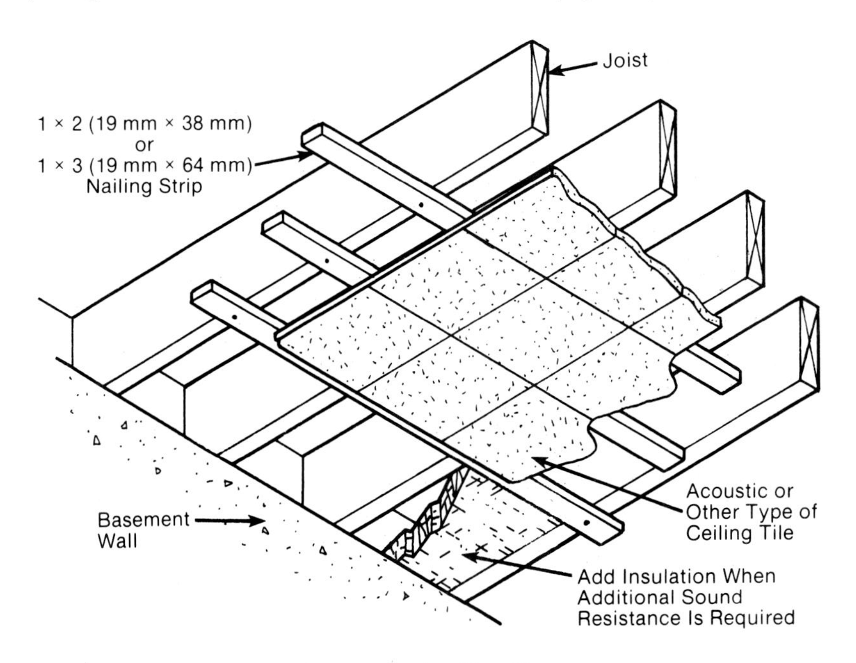

FIGURE 20-5. Installation of acoustical ceiling tile.

Carpeting is also an excellent acoustical material. It is effective on both airborne (NRC) and impact (INR and IIC) noises and sounds. NRCs of airborne sounds may be reduced by 0.45 to 0.70 with the use of carpeting, depending upon the type used.

Acoustical caulking, used to seal off cracks and holes around windows, pipes, outlets, and so on, guarantees that airborne sounds will be attenuated by the wall.

In general, the provision of effective sound barriers is expensive. In large communities it might be better (more cost effective) to eliminate or dampen the cause of the noise.

20.4 VAPOR/AIR BARRIERS

Most building materials are permeable to water vapor. This presents problems because considerable water vapor is generated in a house from cooking, dishwashing, laundering, bathing, humidifiers, and other sources. In cold climates during cold weather, this vapor may pass through wall and ceiling materials and condense in the wall or attic space; subsequently, in severe cases it may damage the exterior paint and interior finish or even result in decay in structural members. Since wetting of the structure, cladding, and insulation is obviously undesirable, a method for containing the water vapor within the dwelling must be employed. This is the function of a component which has traditionally been called a *vapor barrier.*

Water vapor is driven through a building shell in two ways. The first occurs because in the winter there is more water vapor in the air inside the house than there is in the air outside the house. This difference in vapor pressure tends to force the water vapor to diffuse through the materials which constitute the shell. Unlike most building materials, the materials classified as vapor barriers have very low permeability to the passage of water vapor and, thus, are very resistant to this diffusion.

The second manner in which water vapor is forced through the building shell is due to air movement. There are often differences in air pressure from the inside to the outside of a house. When the air pressure inside is greater than that outside (due to stack effect, the operation of fans, or the action of the wind), air tends to flow outward through any holes or cracks in the building envelope, carrying with it the water vapor it has absorbed. It has only recently been recognized that this air movement has far more potential for the unwanted transmission of water vapor than does diffusion. As a result, building scientists today tend to talk about "air barriers" rather than "vapor barriers."

Among the effective vapor barrier materials are asphalt laminated papers, aluminum foil, and plastic films. Most blanket and batt insulations are provided with a vapor barrier on one side and some have paper-backed aluminum foil. Foil-backed gypsum lath or gypsum boards are also available and serve as excellent vapor barriers.

Some types of flexible blanket and batt insulations have a barrier material on one side. Such flexible insulations should be attached with the tabs at their sides fastened on the inside (narrow) edges of the studs, and the blanket should be cut long enough so that the cover sheet can lap over the face of the soleplate at the bottom and over the plate at the top of the stud space. However, such a method of attachment is not the common practice of most installers. When a positive seal is desired, wall-height rolls of plastic-film vapor barriers should be applied over studs, plates, and window and door headers. This system, called enveloping, is used over insulation having no vapor barrier or to ensure excellent protection when used over any type of insulation. The barrier should be fitted tightly around outlet boxes and sealed if necessary. A ribbon of sealing compound around an outlet or switch box will minimize vapor loss at this area. Cold-air returns in outside walls should consist of metal ducts to prevent vapor loss and subsequent paint problems. Paint coatings cannot be considered a substitute for the membrane types of vapor barriers, but they do provide some protection for houses where other types of vapor barriers were not installed during construction.

Of the various types of paint, one coat of aluminum primer followed by two decorative coats of flat wall or lead and oil paint is quite effective. For rough plaster or for buildings in very cold climates, two coats of the aluminum primer may be necessary. A primer and sealer of the pigmented type, followed by decorative finish coats or two coats of rubber-base paint, are also effective in retarding vapor transmission.

20.5 VENTILATION

Condensation of moisture vapor may occur in attic spaces, under flat roofs, and in crawl spaces during cold weather. Even where vapor barriers are used, some vapor will probably work into these spaces around pipes and other inadequately protected areas and some even through the vapor barrier itself. Although the amount might be unimportant if equally distributed, it may be sufficiently concentrated in some cold spot to cause damage. The most practical method of removing the moisture employs the use of ventilation.

A warm attic that is inadequately ventilated and insulated may cause formation of ice dams at the cornice. During cold weather after a heavy snowfall, heat causes the snow next to the roof to melt. Water running down the roof freezes on the colder surface of the cornice, often forming an ice dam at the gutter which may cause water to back up at the eaves and into the wall and ceiling. Similar dams often form in roof valleys. Ventilation thus provides part of the answer to the problems. With a well-insulated ceiling and adequate ventilation, attic temperatures are low and melting of snow over the attic space will be greatly reduced.

In hot weather, ventilation of attic and roof space offers an effective means of removing hot air and, thereby, materially lowering the temperature in these spaces. Insulation should be used between ceiling joists below the attic or roof space to further retard heat flow into the rooms below and materially improve comfort conditions.

Types of ventilators and minimum recommended sizes have been generally established for various types of roofs. The minimum net area for attic or roof-space ventilators is based on the projected ceiling area of the rooms below (Figure 20-6). The ratio of ventilator openings as shown are net areas, and the actual area must be increased to allow for any restrictions such as louvers and wire cloth or screen. The screen area should be double the specified net area shown in Figures 20-6 to 20-8.

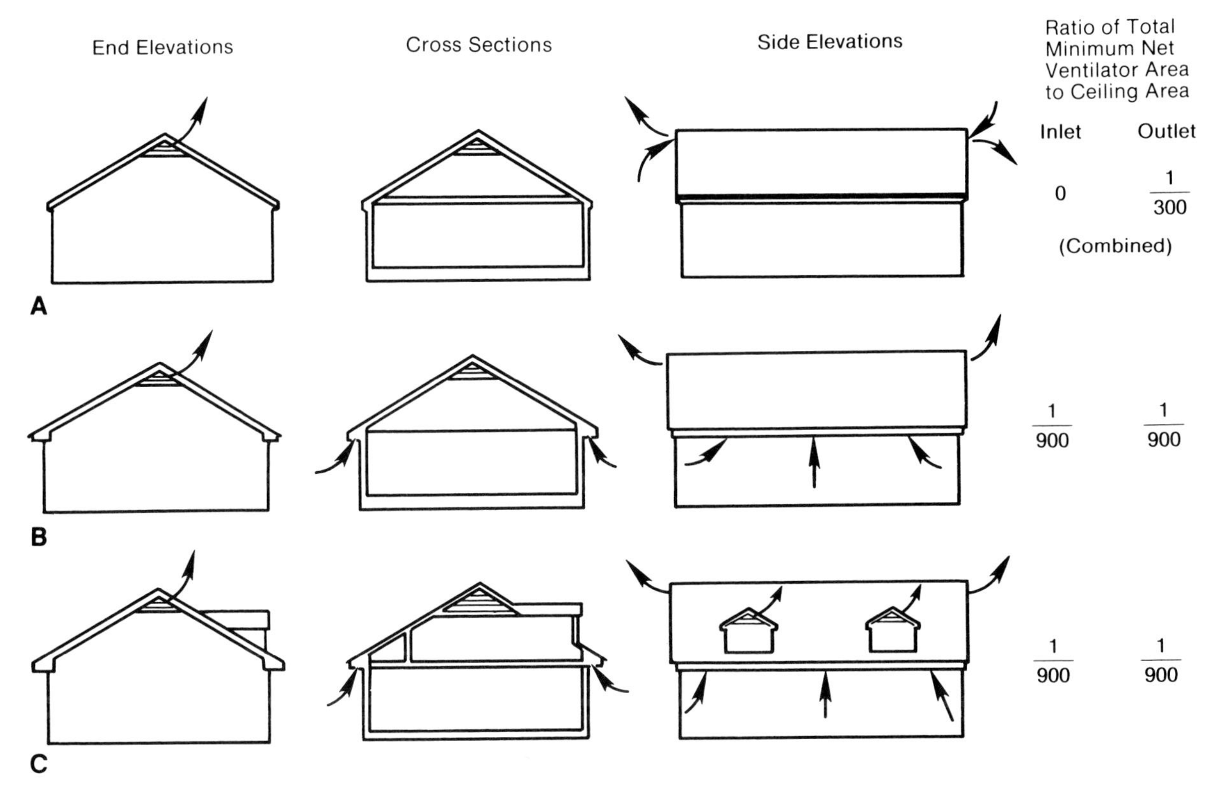

FIGURE 20-6. Ventilating areas of gable roofs: (A) louvers in end walls; (B) louvers in end walls with additional openings in soffit area; (C) louvers at end walls with additional openings at eaves and dormers. Cross section of C shows free opening for air movement between roof boards and ceiling insulation of attic room.

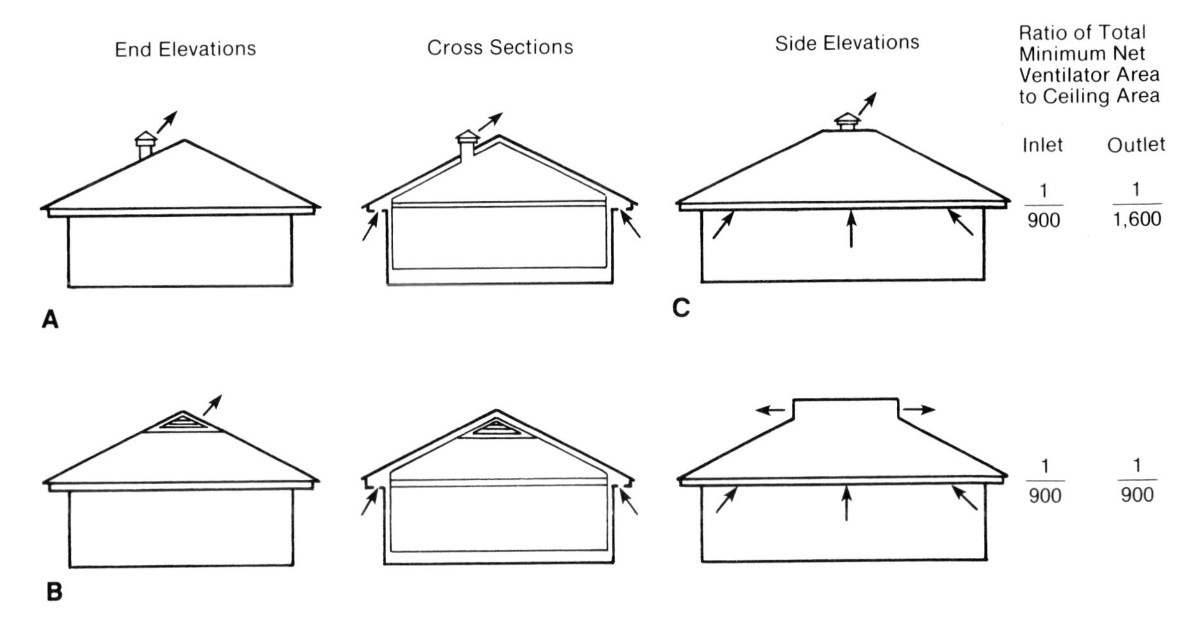

FIGURE 20-7. Ventilating areas of hip roofs: (A) inlet openings beneath eaves and outlet vent near peak; (B) inlet openings beneath eaves and ridge outlets.

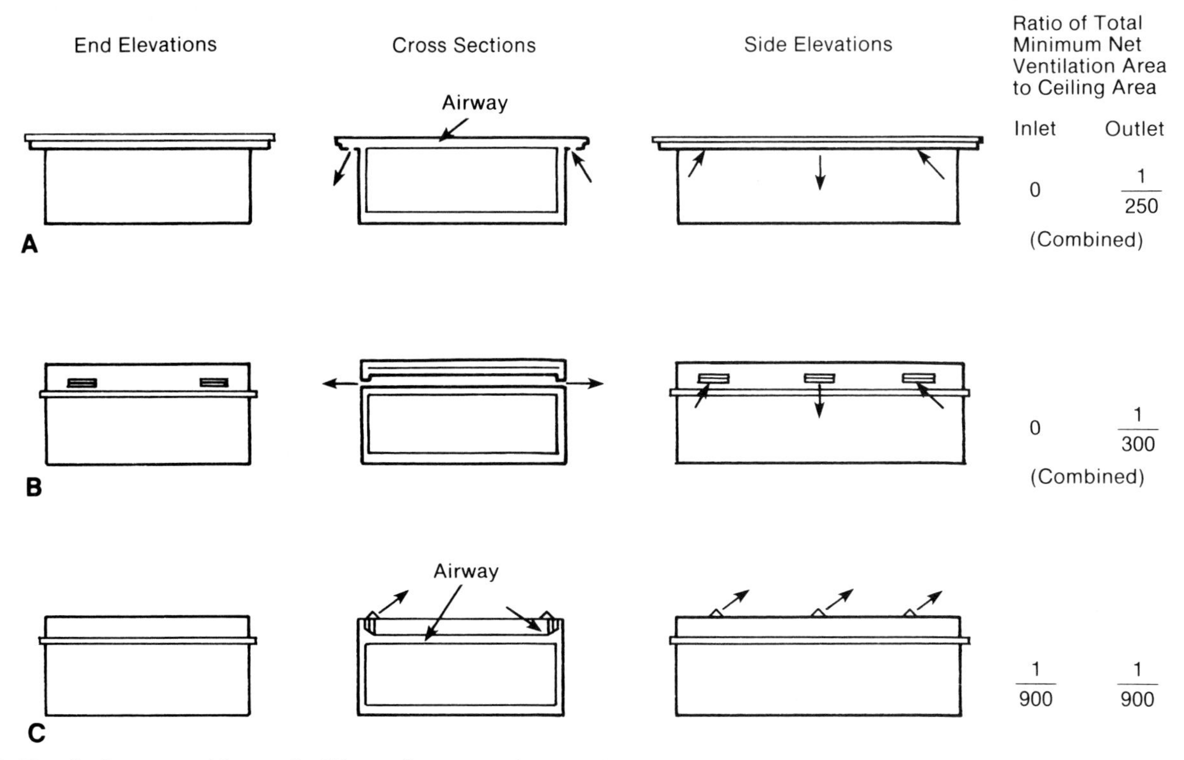

FIGURE 20-8. Ventilating area of flat roofs: (A) ventilator openings under overhanging eaves where ceiling and roof joists are combined; (B) for roof with a parapet where roof and ceiling joists are separate; (C) for roof with a parapet where roof and ceiling joists are combined.

To obtain extra area of screen without adding to the area of the vent, use a frame of required size to hold the screen away from the ventilator opening. Use as coarse a screen as conditions permit, not smaller than No. 16, because lint and dirt tend to clog fine mesh screens.

Louvered openings are generally provided in the end walls of gable roofs and should be as close to the ridge as possible (Figure 20-6A). The net area for the openings should be 1/300 of the ceiling area (Figure 20-6A). For example, where the ceiling area equals 1,200 ft^2 (111.5 m^2), the minimum total net area of the ventilators should be 4 ft^2 (.4 m^2).

As previously explained, more positive air movement can be obtained if additional openings are provided in the soffit area. The minimum ventilation areas for this method are shown in Figure 20-6B.

Where there are rooms in the attic with sloping ceilings under the roof, the insulation should follow the roof slope and be so placed that there is a free opening of at least 1-1/2″ (38 mm) between the roof boards and insulation for air movement (Figure 20-6C).

Hip roofs should have air-inlet openings in the soffit area of the eaves and outlet openings at or near the peak. For minimum net areas of openings, see Figure 20-7A. The most efficient type of inlet opening is the continuous slot, which should provide a free opening of not less than 3/4″ (19 mm). The air-outlet opening near the peak can be a globe-type metal ventilator or several smaller roof ventilators located near the ridge. They can be located below the peak on the rear slope of the roof so that they will not be visible from the front of the house. Gabled extensions of a hip roof are sometimes used to provide efficient outlet ventilators (Figure 20-7B).

A greater ratio of ventilating area is required in some types of flat roofs than in pitched roofs because the air movement is less positive and is dependent upon wind. It is important that there be a clear open space above the ceiling insulation and below the roof sheathing for free air movement from inlet to outlet openings. Solid blocking should not be used for bridging or for bracing over bearing partitions if its use prevents air circulation.

Perhaps the most common type of flat or low-pitched roof is one in which the rafters extend beyond the wall, forming an overhang (Figure 20-8A). When soffits are used, this area can contain the combined inlet-outlet ventilators, preferably a continuous slot. When single ventilators are used, they should be distributed evenly along the overhang.

A parapet-type wall and flat roof combination may be constructed with the ceiling joists separate from the roof joists or combined. When members are separate, the space between can be used for an airway (Figure 20-8B). Inlet and outlet vents are then located as shown, or a series of outlet stack vents can be used along the centerline of the roof in combination with the inlet vents. When ceiling joists and flat rafters are served by one member in parapet construction, vents may be located as shown in Figure 20-8C. Wall inlet ventilators combined with center stack outlet vents are another form of variable in this type of roof.

Various styles of gable-end ventilators are available ready for installation. Many are made with metal louvers and frames, while others may be made of wood to fit the house design more closely. However, the most important factors are to have sufficient net ventilating area and to locate ventilators as close to the ridge as possible without affecting house appearance.

One of the types commonly used fits the slope of the roof and is located near the ridge (Figure 20-9A). It can be made of wood or metal. When made of metal, it is often adjustable to conform to the roof slope. A wood ventilator of this type is enclosed in a frame and placed in the rough opening much as a window frame (Figure 20-9B). Other forms of gable-end ventilators which might be used are shown in Figures 20-9C, D, and E.

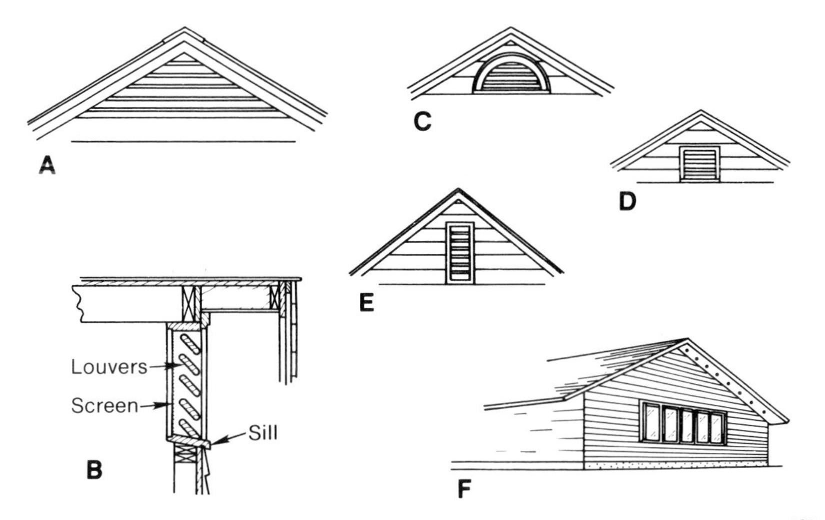

FIGURE 20-9. Outlet ventilators: (A) triangular; (B) typical cross section; (C) half-circle; (D) square; (E) vertical; and (F) soffit.

A system of attic ventilation which can be used on houses with a wide roof overhang at the gable end consists of a series of small vents or a continuous slot located on the underside of the soffit areas (Figure 20-9F). Several large openings located near the ridge might also be used. This system is especially desirable on low-pitched roofs where standard wall ventilators may not be suitable.

Small, well-distributed ventilators or a continuous slot in the soffit provide inlet ventilation. These small louvered and screened vents can be obtained in most lumberyards or hardware stores and are simple to install.

Only small sections need to be cut out of the soffit; these can be sawed out before the soffit is applied. It is more desirable to use a number of smaller well-distributed ventilators than several large ones (Figure 20-10A). Any blocking which might be required between rafters at the wall line should be installed so as to provide an airway into the attic area.

A continuous screened slot, which is often desirable, should be located near the outer edge of the soffit near the fascia (Figure 20-10B). Locating the slot in this area will minimize the chance of snow entering. This type may also be used on the extension of flat roofs.

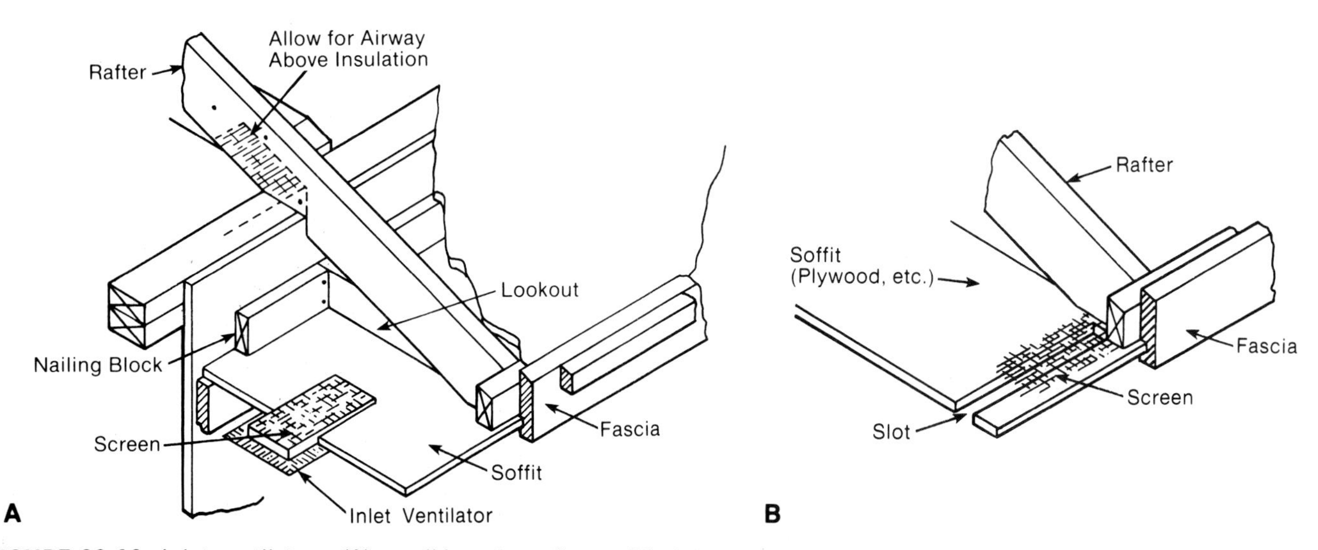

FIGURE 20-10. Inlet ventilators: (A) small insert ventilator; (B) slot ventilator.

The crawl space below the floor of a basementless house and under porches should be ventilated and protected from ground moisture by the use of a soil cover as illustrated in Figure 20-11. The soil cover should be a vapor barrier with a perm value of less than 1.0. This includes such barrier materials as plastic films, roll roofing, and asphalt laminated paper. Such protection will minimize the effect of ground moisture on the wood framing members. High moisture content and humidity encourage staining and decay of untreated members.

Where there is a partial basement open to a crawl-space area, no wall vents are required if there is some type of operable window. However, the use of a soil cover in the crawl space is still important. For crawl spaces with no basement area, provide at least four foundation-wall vents near corners of the building. The total free (net) area of the

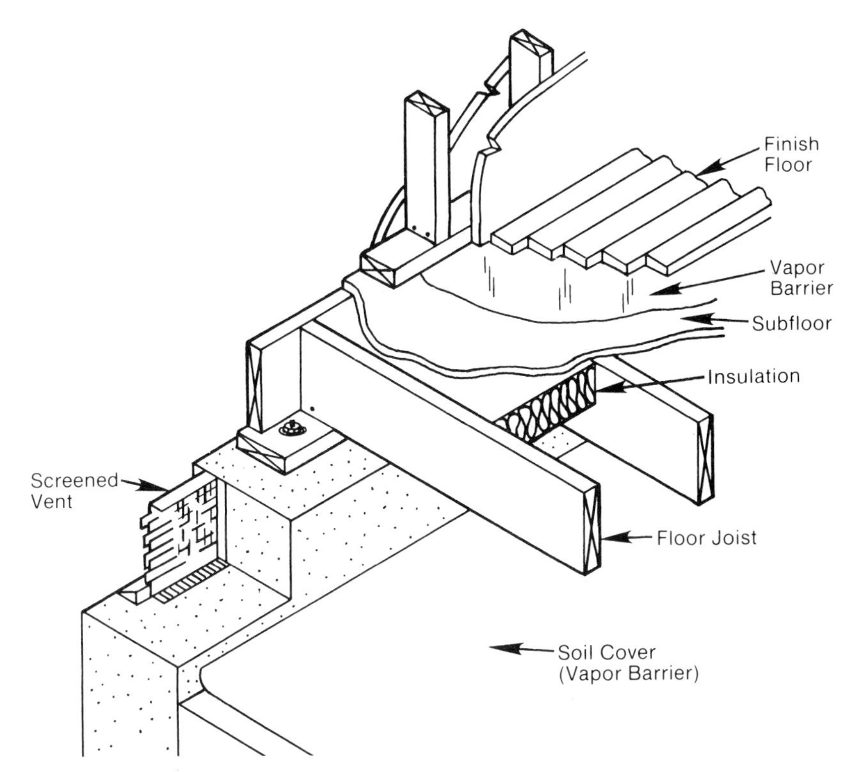

FIGURE 20-11. Crawl-space ventilator and soil cover.

ventilators should be equal to 1/160 of the ground area of when no soil cover is used. Thus, for a ground area of 1,200 ft^2 (111.5 m^2), a total net ventilating area of about 8 ft^2 (0.7 m^2) is required, or 2 ft^2 (0.2 m^2) for each of four ventilators. More smaller ventilators having the same net ratio are satisfactory.

When a vapor barrier ground cover is used, the required ventilating area is greatly reduced. The net ventilating area required with a ground cover is 1/1,600 of the ground area, or for the 1,200 ft^2 (111.5 m^2) house, an area of 0.75 ft^2 (.07 m^2). This should be divided between two small ventilators located on opposite sides of the crawl space. Vents should be covered (Figure 20-11) with a corrosion-resistant screen of No. 8 mesh.

The use of a ground cover is normally recommended under all conditions. It not only protects wood framing members from ground moisture but also allows the use of small, inconspicuous ventilators.

20.6 INSULATION

Thermal insulation plays an important role in the economics and comfort of a residence. A properly insulated building means lower costs for fuel in the winter and for electrical energy in the summer. The purpose of such insulation is to minimize the inflow of heat through building walls in the summer and its outflow during the winter. The greater the difference between inside and outside temperatures, the faster the flow of heat. Insulation effectively slows down this transfer of heat and maintains a uniformly comfortable temperature throughout the building.

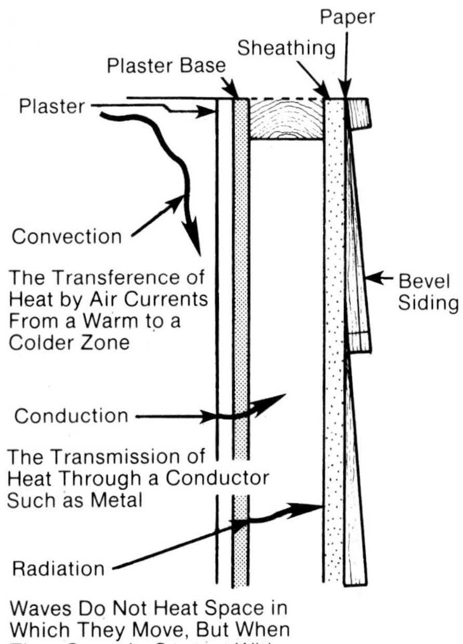

FIGURE 20-12. Three ways in which heat is transferred from one mass to another or to the surrounding atmosphere.

The transfer of heat takes place in one of or in any combination of three ways: conduction, convection, or radiation.

Conduction occurs when heat travels through any solid body from one interacting molecule to another. Different materials conduct heat at different rates. For example, a piece of steel conducts heat much faster than a piece of wood. Generally, the denser the material, the faster the flow of heat.

Convection is the movement of heat through air. The density of air decreases as its temperature rises; this, in turn, causes heavier colder air to move in and replace the warmer air. As the cycle continues, whether upward or downward, these convection currents carry heat from one area to another.

Radiation is a process in which heat is transmitted from one body to another by electromagnetic waves which are emitted and travel through either the air or a vacuum. The radiated heat does not heat the air it moves in, but heat is generated when it strikes a receptive surface. The flow of such radiated heat is from the warmer surface to the cooler one. The three methods of heat transfer are illustrated in Figure 20-12.

While all materials have some insulating qualities, there are certain ones that work better than others. Materials that have many small air spaces or pockets are good insulators since air is a poor conductor if confined to a small area that restricts movement. Such restriction eliminates the loss of heat through convection because the barrier of air cells or other insulating material effectively obstructs the flow of heat. Obviously, this means that less heat passes through the insulation and a useful reduction in heat loss results.

Terms used in describing insulation are called heat-loss coefficients. They include Btu, k, C, U, and R.

Btu or British thermal unit is the amount of heat needed to raise the temperature of 1 pound of water by 1°F.

Thermal conductivity or k is the amount of heat (Btu's) that passes through 1 ft^2 of homogenous material 1″ thick in 1 hour when there is an initial 1°F difference between the two surfaces. The lower the k value, the higher the insulating value.

C or thermal conductance, representing the conductance or transmitting capacity of a material regardless of its thickness, shows the amount of heat in Btu's that will pass through 1 ft^2 of material in 1 hour with a 1°F difference between its two surfaces. The lower the C, the higher the insulating value.

U represents the heat loss through a building section in Btu's per hour per square foot per degree-Fahrenheit difference between inside and outside temperatures. The lower the U, the higher the insulating value.

R is the measured resistance of a material to the flow of heat. All building components of a wall, roof, or floor have an R value. The higher the R value, the more efficient the insulation. The R values of various wall components are listed in Table 20-7.

The U value of a construction section is the reciprocal of the R values of the various components. Expressed mathematically,

$$U = \frac{1}{R_t} = \frac{1}{R_1 + R_2 + R_3 + Rn}$$

TABLE 20-7 R VALUES FOR BUILDING MATERIAL COMPONENTS

Building Material Components	R Value
Outside air film	0.17
Wood siding	0.85
1/2″ (13 mm) sheathing	1.32
Air space	1.01
Interior finish (1/2″ [13 mm] drywall)	0.45
Inside air film	0.68
Total resistance (R_t)	4.48

Study of this formula clearly shows that adding insulation to a wall makes it possible to reduce the U value or heat loss. For example, the U value of a typical wood frame wall without insulation is 0.22. It is possible to reduce this to 0.10 or 0.07 if one of the following materials is used:

U value, without insulation = 0.22
U value, with R7 (2-1/4″ [57 mm]) insulation = 0.10
U value, with R11 (3-1/2″ [89 mm]) insulation = 0.07

In order to lower the U value, the R value (thickness) must be increased. The R value is always expressed as a numeral, and each numeral equals a resistance unit. An insulating material with an R11 factor insulates better than one with an R7 rating. In other words, regardless of the material or its thickness, all products with the same R factor have equal insulating value.

Thermal resistance in the metric system is described by means of RSI values. A constant ratio exists between the English R system and the metric RSI systems of thermal resistance: RSI = 0.176 R. (See Table 5-10 for more direct conversions.)

20.6.1 Types of Insulation

Insulating materials are manufactured in a variety of forms and generally fall into one of the following structural groups: flexible, rigid, reflective, or loose fill.

Flexible insulation is made up in blankets and batts. The blankets are actually fabricated in roll form and in lengths up to several hundred feet. The usual thickness is from 1″ to 3″ (25 mm to 76 mm) and the widths fit standard stud and joist spacing. Insulation blanket consists of fluffy mats of mineral or vegetable fibers such as rock, slag, glass wool, wood fiber, and cotton. The organic fibers are treated to make them resistant to fire, vermin, and decay.

Blanket insulation generally has paper covers on one or both sides and is provided with side tabs for fastening to studs and joists. One covering usually serves as a vapor barrier; the other simply encloses the mat. Aluminum foil, asphalt, and plastic are also used as vapor barrier coverings.

Batt insulation is also formed of fluffy fibrous materials in thicknesses up to 6″ (152 mm) or more. Widths are 15″ (38 mm) and 23″ (584 mm) and lengths are either 24″ (610 mm) or 48″ (1,219 mm). Batts are available with and without covers.

Rigid insulation is manufactured from wood, cane, and other fibrous vegetable materials. These are formed into large lightweight boards which combine tensile strength with thermal and acoustical insulating properties.

The material is available in a wide variety of sizes, from 12" (305 mm) squares to boards measuring 4' (1.2 m) wide and 12' (3.7 m) long with 1/2" to 1" (13 mm to 25 mm) thicknesses.

Where the main purpose is structural, these insulating boards are fabricated as building boards, roof decking, sheathing, and wallboard. Their insulating qualities are then secondary. Such insulating sheathing is made in 1/2" (13 mm) and 25/32" (20 mm) thicknesses. The sheets are 2' × 8' (.6 m × 2.4 m) for horizontal application and either 4' × 8' (1.2 m × 2.4 m) or 4' × 12' (1.2 m × 3.7 m) for vertical use.

Other materials used for rigid insulation include polystyrene, polyurethane, cork, and mineral wool. These are nonstructural boards.

Reflective insulations have high reflective properties. Aluminum foil, sheet metal, and reflective coatings are the most common materials used for this purpose. To function properly the insulation must have at least a 3/4" (19 mm) air space facing the reflective surface. In ceiling applications, the air space should be 1-1/2" (38 mm) or greater.

The most efficient of the reflective insulations is the accordion type which is made up of several spaced layers of foil. As it is applied, the material is stretched out to form the necessary air space.

Loose-fill insulation consists of bulk packed materials made of rock wool, wood fibers, shredded bark, cork, sawdust, vermiculite, and other granular products.

Poured or blown into wall or ceiling spaces, it is ideally suited for use in existing structures where the installation of other forms of insulation is not practical.

Foamed-in-place insulation is also used extensively in existing structures. Two chemicals are mixed together to produce a urethane foam with a k factor of 0.12. Special applicators with far-reaching nozzles are used to blow it into place.

Other types of insulation include corrugated paper in multiple layers, sprayed insulations, and lightweight aggregates like vermiculite and perlite mixed with plaster.

20.6.2 Insulation of New Constructions

Any room that is to be cooled in the summer or heated in the winter should be properly insulated. Provide walls, floors, and ceilings with a blanket of insulation. This retains heat in the winter and keeps it out during the warm weather. Place such insulation as close as possible to the rooms used as living quarters. Thus, in an unused attic, insulate the attic floor, not the roof over it. Also insulate the floors over unheated basements, crawl spaces, and porches.

If basement areas are to be finished as living rooms, use thermal insulation for the walls. Also put acoustical insulation on the ceiling to reduce the passage of sound from the basement to other parts of the house. This is especially important if the basement is to be used as a playroom for children. Acoustical insulation adds to fuel savings as well. Figure 20-13 shows the placement of insulation in various types of structures.

When homes are built without basements, many builders install a crawl space under the floor. This is better than placing the slab directly on the ground and it provides access to plumbing, heating, and other services if the need arises.

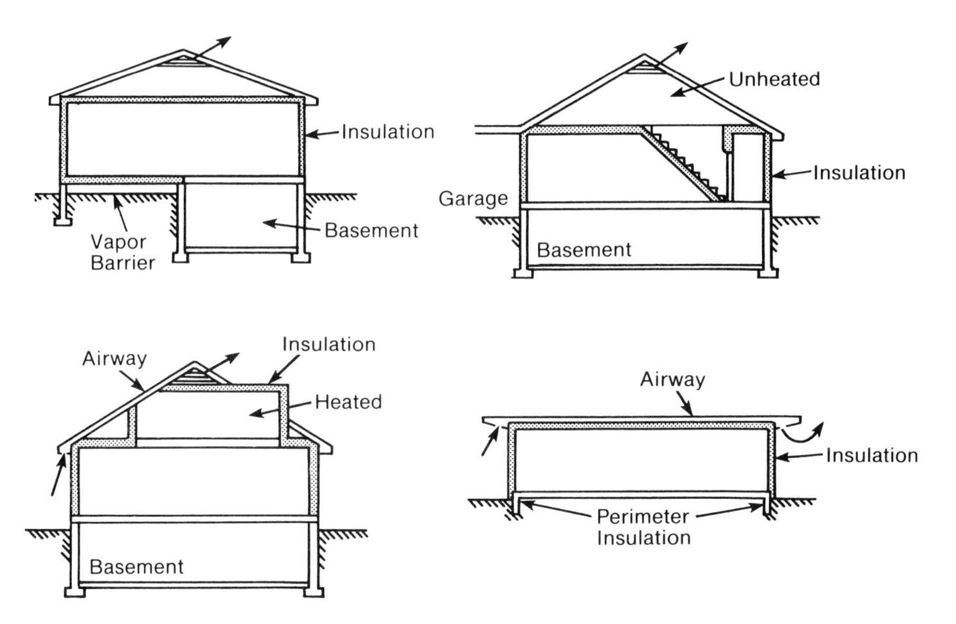

FIGURE 20-13. Diagram showing areas for insulation in various structures.

The floor above the crawl space should be insulated and the area should be ventilated as well. If the crawl space is unvented, insulate the foundation walls as shown in Figure 20-14. Over a vented crawl space insulate the floors by placing the insulation between the joists. Position it with the vapor barrier side facing toward the room (Figure 20-15). Wire mesh, laced wires, or special bowed wires may be used. If preferred, use polyethylene on the ground.

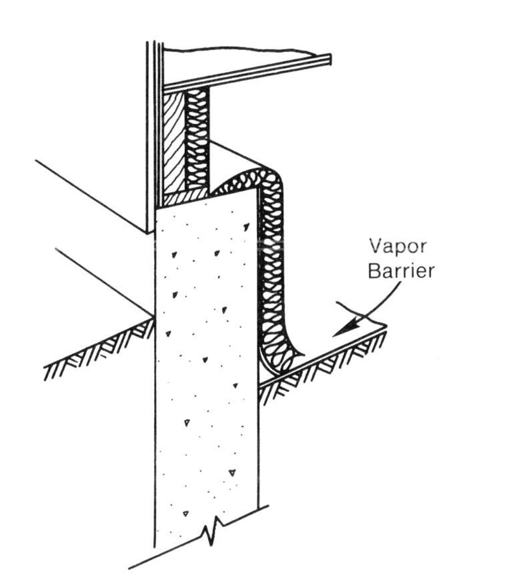

FIGURE 20-14. Foundation wall insulation for unvented crawl space.

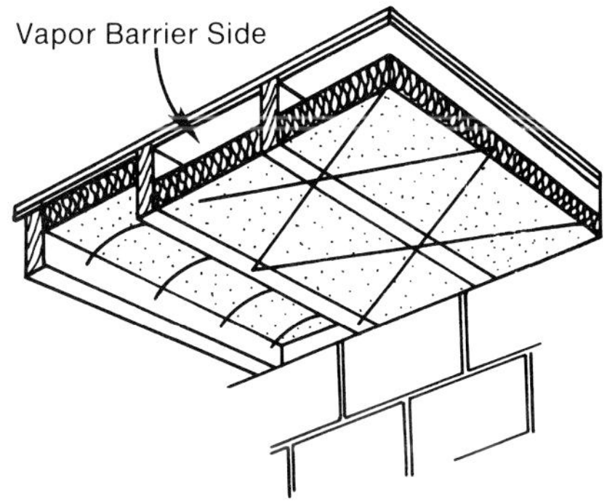

Vapor Barrier Side

FIGURE 20-15. Insulation over vented crawl space is placed between floor joists with vapor barrier side facing toward room above.

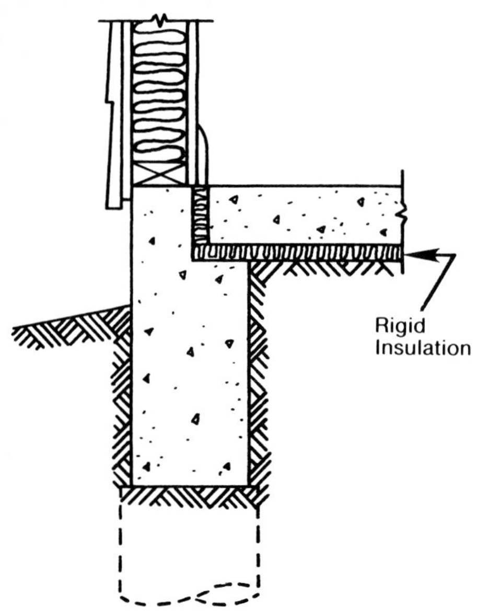

FIGURE 20-16. Application of blanket insulation at floor overhang.

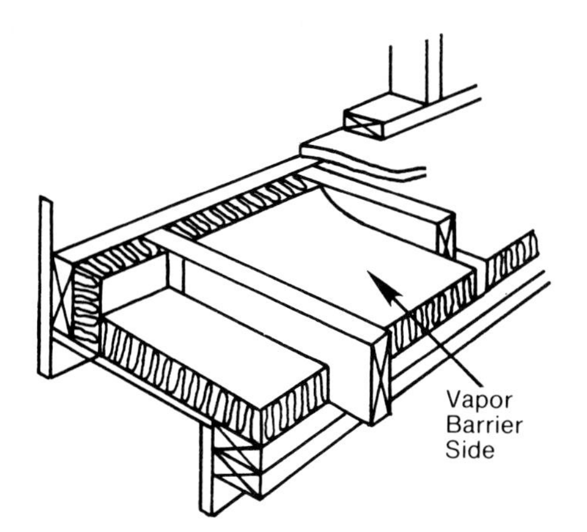

FIGURE 20-17. Insulation around perimeter of concrete slab.

Use a sill sealer in basement and unvented crawl space construction to prevent air infiltration between the sill and foundation.

Cantilevered floors can be insulated as illustrated in Figure 20-16. Cut the blanket to fit or fold it as shown.

Slab-on-ground construction may require insulation. If so, lay it only where needed, around the perimeter, since little heat is lost toward the center of the floor. Figure 20-17 shows how rigid insulation is used between slab and foundation wall.

Install either batt or blanket insulation in open spaces in walls, floors, ceilings, and roofs. Do not install either where compressive loads are present. Squeezing or compressing any flexible insulation causes it to lose its effectiveness.

Batts or blankets can easily be cut with a handsaw or a sharp knife. Most workers, however, prefer a knife since it is less likely to scatter particles which can be very irritating to the skin (especially when fiberglass insulation is used).

Make lengths slightly oversize so that a stapling flange can be formed at each end. Do this by removing some of the insulation after cutting.

Most workers find it easiest to cut blanket insulation on the floor. The usual procedure is to mark the required length on the floor, roll out the material, and then cut on the mark. Compress the material with a metal straightedge and cut with a sharp knife.

Ceiling insulation may be installed from below by stapling it to the ceiling joists as shown in Figure 20-18 except when unfaced insulation is used. In that case, pressure-fit the blanket between the joists. In both cases, the insulation must extend across the top plate, and it may be necessary to fill in any gap left at the inside edge of the plate.

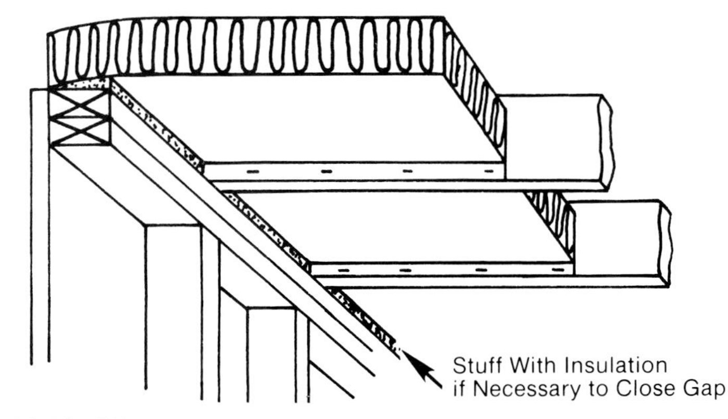

FIGURE 20-18. Efficient method of applying ceiling insulation by stapling it to ceiling joists.

If the specifications do not permit face-stapling of the flange, staple it to the sides. Staples should be placed every 6″ to 12″ (152 mm to 305 mm).

Masonry walls require furring strips when flexible insulation is used. The furring may be 1″ × 2″ or 2″ × 2″ (19 mm × 38 mm or 38 mm × 38 mm) stock. Spacing is either 16″ or 24″ (406 mm or 610 mm) OC, depending on the thickness and type of wall finish. Use either pressure-fit, batt, or blanket type insulation. For the 1″ × 2″ (19 mm × 38 mm) furring, a 1″ (25.4 mm) pressure-fit insulation without vapor barrier is generally used. Add either a separate 2 mil polyethylene or foil-backed gypsum board as the vapor barrier.

With 2″ × 2″ (38 mm × 38 mm) furring, use an R7 insulation with vapor barrier (Figure 20-19). Because of this insulation's shallow depth, the flanges must be stapled to the face of the furring.

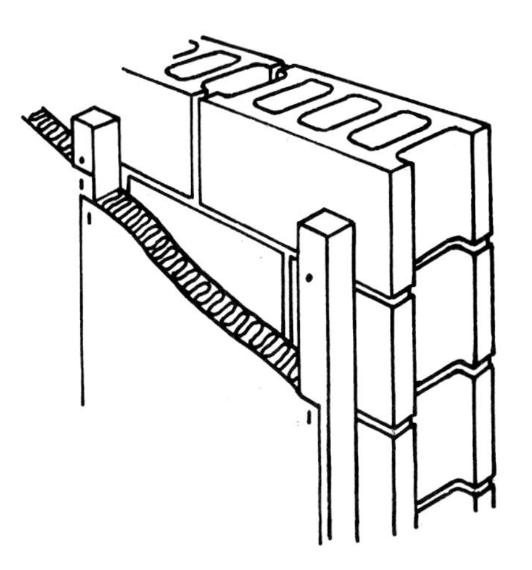

FIGURE 20-19. Insulation fastens to furring on masonry wall.

The procedure for installing batt or blanket insulation in frame walls is similar to that for ceilings. Place insulation between studs and staple the flange to the side or face. If foil-faced insulation is used, do not let the reflective surface touch the back of the drywall or paneling. In fact, inset stapling is required for this work.

Start at the top of the wall, working down toward the plate. Remove about 1-1/2″ (38 mm) of insulation at the top and bottom of each piece to form a stapling flange. See Figure 20-20 for two ways to terminate this insulation at the top and bottom plates when the inset method is used.

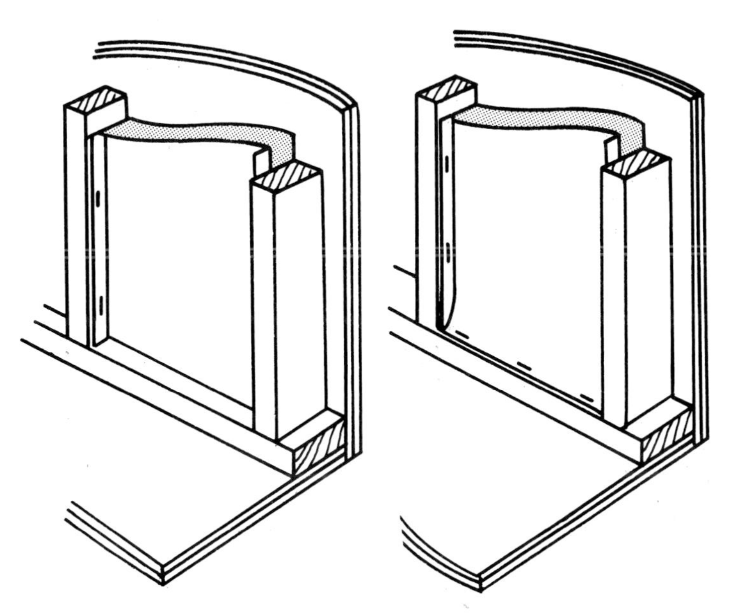

FIGURE 20-20. Two ways of terminating ends of insulation.

Use great care when installing insulation around plumbing and other outlets. If the facing is accidently torn, be sure to repair it with tape. Stuff small spaces around doors and windows with pieces of insulation. To keep them effective, do not pack the pieces tightly. Cover these spaces with paper or plastic vapor barriers.

Fill insulation is commonly used in ceiling areas and is poured or blown into place. A vapor barrier should be used on the warm side (the bottom, in case of ceiling joists) before insulation is placed. A leveling board will give a constant insulation thickness (Figure 20-21). Ceiling insulation 6″ (152 mm) or more thick greatly reduces heat loss in the winter and also provides summertime protection.

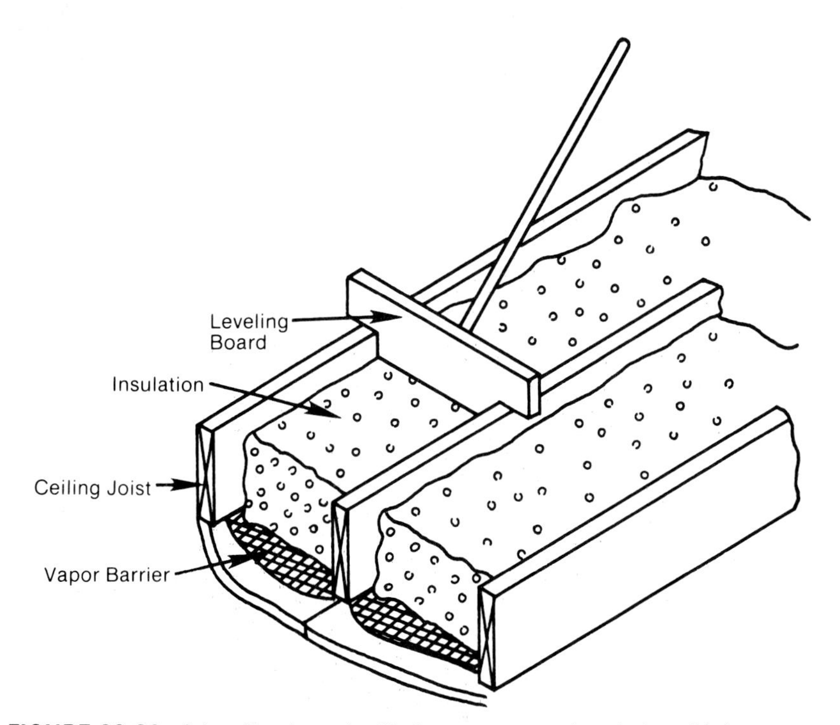

FIGURE 20-21. A leveling board will give a constant insulation thickness.

Reflective insulation is installed in the same manner as batt or blanket insulation. It is attached to either the side or the face of the framing. Be careful to maintain an air space between its reflective surface and the wall covering. Normally this is 3/4″ (19 mm) for walls and at least 1-1/2″ (38 mm) for floors and ceilings.

Expanded polystyrene foam plastics (types of rigid insulation) have excellent thermal qualities, are waterproof, and do not require the use of vapor barriers. Lightweight and easy to handle on the job, these materials are available in convenient 2′ × 8′ (.6 m × 2.4 m) sizes and in a variety of thicknesses. They are also available in 4′ × 8′ (1.2 m × 2.4 m) sheets and are easily cut with a knife or saw.

Furring is not required when rigid insulation is used to insulate masonry walls, such as those in a basement. The slab may be installed with a mastic. Use mastic also to join gypsum or paneling directly to the insulation. Special mastics are made to different specifications; be sure to employ the right one for the job.

Rigid plastic foam is excellent for insulating exposed beam roof decks. Special high-density foam with high water resistance is especially suitable. An application over 2 × 6 (38 mm × 140 mm) decking is shown in Figure 20-22.

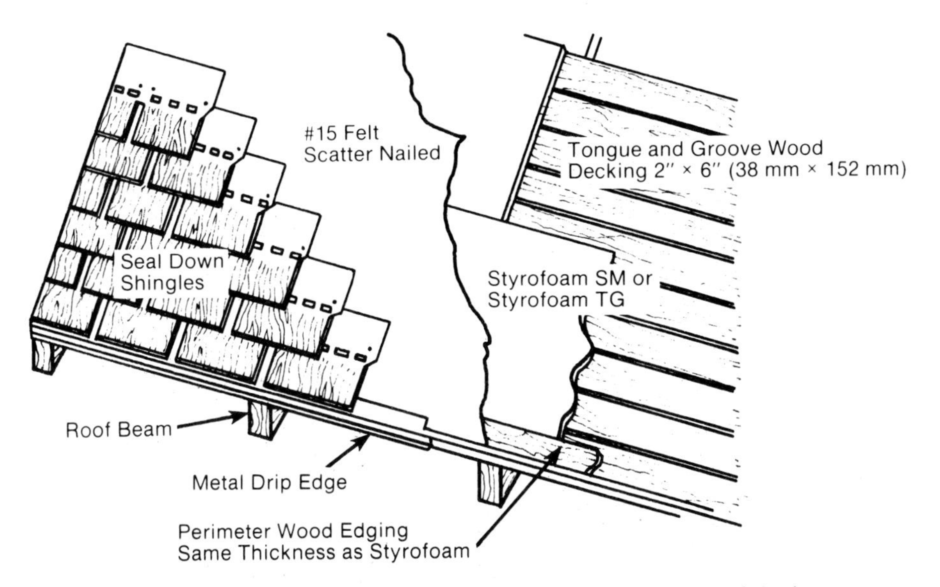

FIGURE 20-22. Details for application of rigid plastic foam on roof decks.

20.6.3 Insulation of Existing Constructions

The approach to old construction is quite different from new. Most owners would set priorities and then approve work orders on the basis of payback years. They might ask the question: If I insulate now how many years will it take to break even? If the answer is a few months, the work would probably commence. If the payback period is 20 years, one might move the item down the priority list.

If the objective is mainly to reduce energy costs rather than to increase the comfort level, there might be other options such as thermostat setback, a better schedule for furnace filter cleaning, and replacing 100 watt bulbs with 60 watt bulbs.

Additional insulation above truss/rafter type roof ceilings is usually not too expensive to do. Putting it into the walls is not as easy. Small amounts can be added to the exteriors of walls, but large amounts are much more difficult.

To add insulation into walls that are surfaced on both sides, access into the stud cavities must be attained. This can be done from inside or outside the house. Insulating a wall from the inside involves major work and is only practical if you are completely remodeling. The wall's surface (drywall or paneling) can be removed so insulation batts or blankets can be installed between the wall studs as though it were an unfinished wall.

Most contractors, however, cut holes through either the interior or exterior surfaces and blow in or foam in insulation instead of removing entire sections of wood-frame wall surfaces.

The holes are usually bored from the exterior if the siding is clapboard or shingle. If the exterior walls are of masonry, metal siding, or a material that is especially difficult to patch, the holes are cut into the interior walls.

This is not a minor job. A 1-1/2″ to 2″ (38 mm to 51 mm) hole must be bored into each cavity between each pair of wall studs—both at the top end of the cavity and below each fireblock or obstruction. For a multistory house, this must be repeated on each story.

Foamed-in insulation totally fills cavities, creates its own vapor barrier, and has high R values per inch of thickness (Figure 20-23). Be

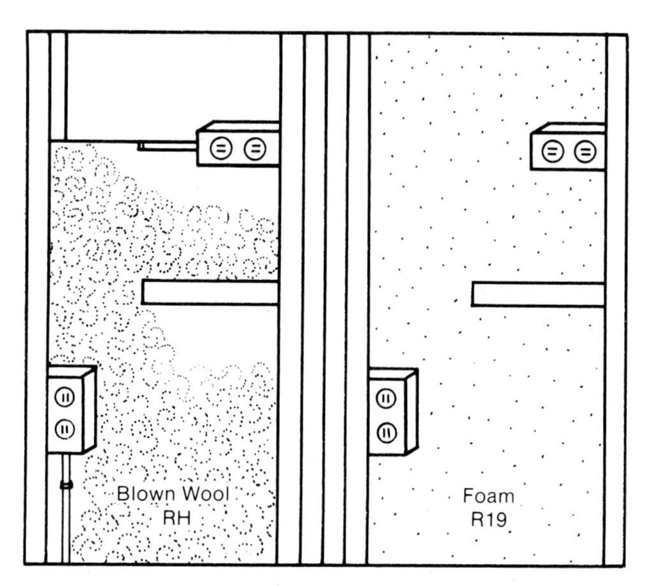

FIGURE 20-23. Foamed-in vs. blown-in insulation.

aware that some foams are flammable or toxic when ignited; building codes must be strictly followed.

Blown-in fibers often catch on obstructions such as nails, pipes, and electrical outlets. This results in only partially filled cavities. Blown-in fibers do not automatically provide vapor barriers and often settle with time. To create a vapor barrier when using them, interior wall surfaces can be painted with two coats of glossy oil-based enamel or a pigmented wall sealer, followed by one coat of alkyd.

20.7 LIMITS

Building technology is advancing to the point where one can live and work in comfort. Some are wondering if we can afford the levels to which we aspire. Maybe it depends on one's personal priorities. Some will want the best and others will set limits.

The designer/builder needs to know what limits an owner or potential owner is actually willing to set. An owner usually puts a general dollar value limit on a project. The designer must then interpret this amount in terms of the degree of comfort desired and practical economics. He or she must also make the project safe for future years.

Specifications need to be written in explicit terms with respect to expected sound levels, efficiency of the air/vapor seal, and temperature controls. Test data must be applied to the construction of a building and interpreted to the persons doing the construction.

With respect to practical economics, everyone is interested in break-even points for costs against returns on investment. For this reason, one might ask, "How much ceiling insulation is needed for this area? Do I need an R value of 60 or will R28 or even R12 insulation suffice?" Other questions might include: "Is double-stud wall construction warranted for this area? Should construction workers be expected to actually create an air/vapor barrier without punctures?"

The answers to these questions will reflect current and expected costs of energy and the need for energy conservation.

Long-term safety features and comfort features are integral to building design. Fresh air control, humidity control during winter and summer, and temperature control during winter and summer are three such features. Building materials must add to comfort rather than detract (through use of irritants and toxics). Fire and fume risks must be lessened rather than enhanced.

Putting construction subsystems together in a building in a given geographical area is a demanding vocation.

20.8 RESEARCH

1. Help organize a project to measure sound levels at a particular place, such as a public building, a dwelling, a sidewalk, or a building site. Limit activities to a small area. Make sketches and record levels measured in decibels.
2. Locate a berm. Make sketches of its section and then take sound level readings at selected positions on the berm or adjacent to it. Describe conditions and explain differences in the readings, if any.
3. Visit a structures laboratory at a university. Observe measurement of sound absorption and of sound transmission class (STC).
4. Visit a light construction building site to observe the placement of an air/vapor barrier. Make notes and sketches on how the envelope is created.
5. At the same site make sketches of how roofs are vented. Compare total vent areas with volume to be vented.
6. At the same site make sketches and a list giving R (RSI) values of the different insulations used in the building. Comment on the expected effectiveness of the insulation.
7. Study earth sheltering techniques used in your area. Before organizing the study, obtain technical data sheets on the subject. If you get them from material suppliers, be sure to not limit yourself to one. Be wary of biased statements.
8. Obtain permission to observe retrofitting of a light construction building. Report on the selection of options to make an older building more energy efficient. Use technical terms and measurements.

20.9 REFERENCES

20.9.1 Alberta Energy and Natural Resources Energy Conservation Board. *Alberta Energy Savers.*

20.9.2 Canadian Mortgage and Housing Corporation. *Canadian Wood-Frame House Construction.* 1979.

20.9.3 Canadian Wood Council. *Acoustics.* 1978.

20.9.4 Editors of Sunset Books and Sunset Magazine. *Do-It-Yourself Insulation & Weatherstripping.* Menlo Park: Lane Publishing Co., 1980.

20.9.5 National Concrete Masonry Association. *NCMA TEK* (Numbers 18, 119, 112, 14, 12, 38-14, 67 of years 1969 to 1981).

20.9.6 Saskatchewan Mineral Resources Office of Energy Conservation. *Energy Efficient Housing—A Prairie Approach.* 1980.

20.9.7 U.S. Department of Agriculture. *Agriculture Handbook No. 73.*

20.9.8 U.S. Naval Education and Training. *Builder 3 & 2 RTM.*

21
STAIR BUILDING

Stair building is a fairly complex and time-consuming job. It is often considered a trade in itself, so it's not unusual to have main stairways millmade and delivered to the jobsite ready for installation. Most service stairs, however, are constructed on the job; thus, a basic knowledge of stair building is necessary for most aspects of light construction. This chapter will discuss the design and construction of the most commonly used types of stairs.

21.1 GENERAL OBJECTIVES

Stairs are mainly functional, but they can also add to the appearance of a home. When correctly designed and built, stairs have the proper incline, tread size, and riser height; safety and comfort must also be given strong consideration. This chapter will concentrate on the most practical methods of building stairways, all of which meet design and safety standards. Emphasis is placed on learning the terms that apply to stair construction, as well as the mathematical formulas that are needed to determine critical measurements. As with most aspects of light construction, stair building utilizes knowledge learned earlier, i.e., use of the story rod and framing square.

21.2 SPECIFIC OBJECTIVES

Upon completing this chapter, the reader should

1. understand the basic terminology used in stair building,
2. be able to lay out a stairway of predetermined dimension on graph paper, and
3. be able to build a stairway at a jobsite.

21.3 STAIR TYPES AND LANDINGS

Stairs can be either closed (with walls on both sides) or open (with a wall on just one side or no wall on either side). The simplest type is the single flight straight stair (Figure 21-1). This conventional stairway

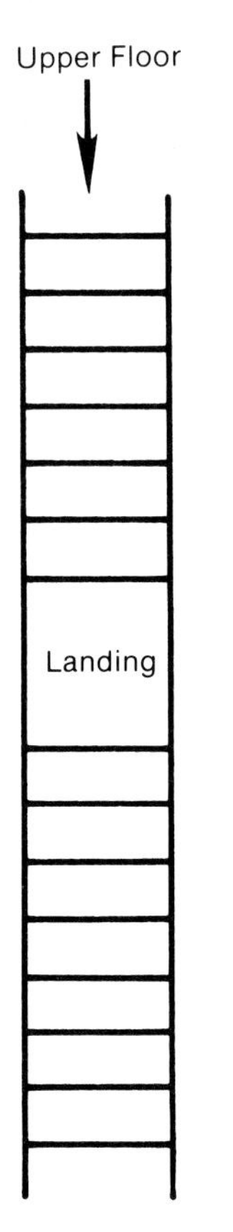

FIGURE 21-2. Long stairway with a landing.

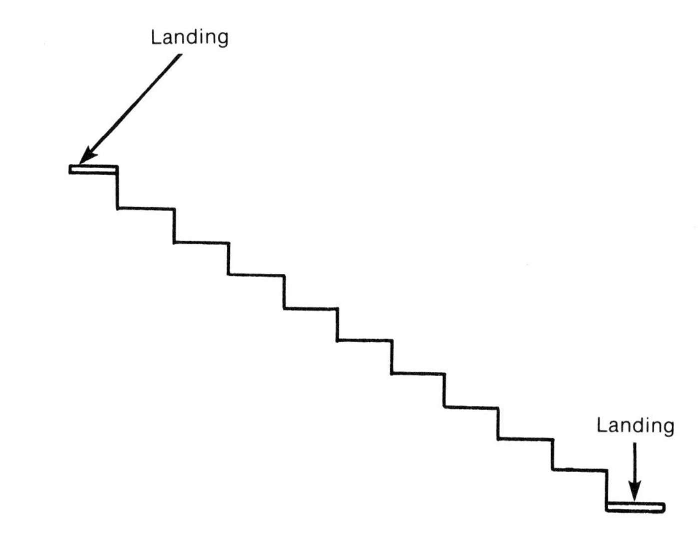

FIGURE 21-1. Conventional straight stairs.

leads from one floor to the next and generally is made up of 13 or 14 steps without any landings. In a small house, the straight stairway can be somewhat troublesome, as it requires a long hallway.

If a flight of stairs is longer than 15 steps, it should be interrupted with a landing (Figure 21-2). By providing a pause, stairways with landings are less tiring. A landing also estimates the potentially frightening effect that long stairways can have. Landings can be used to change the direction of stairways as well. An L-shaped stairway has two flights placed at right angles to each other with a landing in between (Figure 21-3).

When space is limited, a narrow U-shaped stairway is a good idea. As seen in Figure 21-4, this stair is laid out so that the lower level will extend up to a landing at some intermediate point and then double back on itself. When space permits, a wide U-shaped design can be used. This type of stairway utilizes two landings with a short flight of steps in between (Figure 21-5).

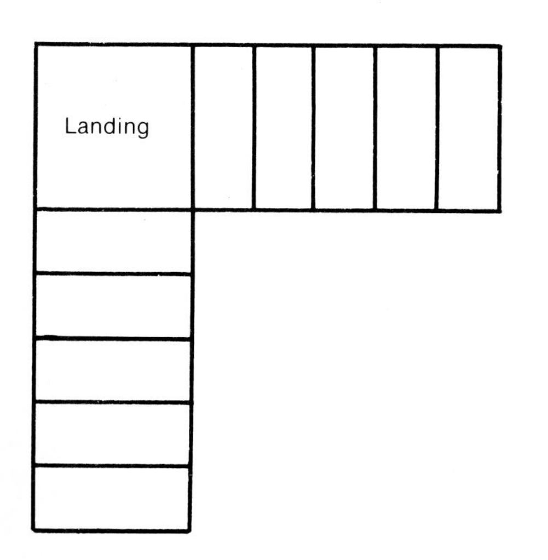

FIGURE 21-3. L-shaped stairs.

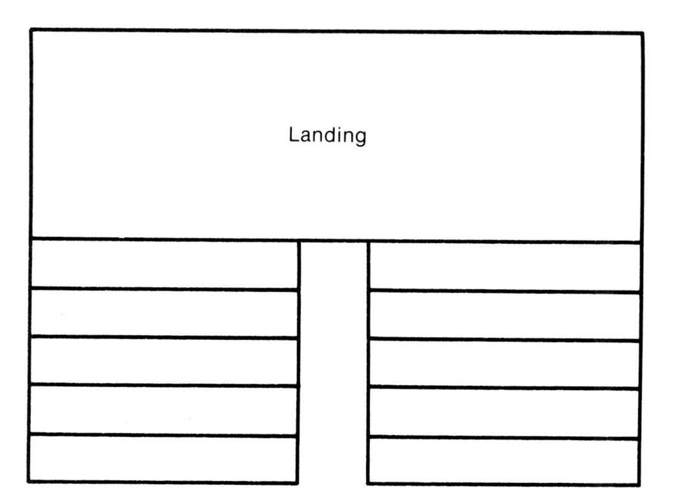

FIGURE 21-4. Narrow U-shaped stairs.

FIGURE 21-5. Wide U-shaped stairs.

21.4 STAIR TERMS AND PARTS

The following terms and parts are common in stair construction:

- **Riser:** the vertical member of the step. Without one, the step is described as an open riser.
- **Tread:** the horizontal member of the step.
- **Nosing:** the front edge of the tread that projects beyond the riser.
- **Stringer:** an inclined member that supports the risers and treads.
- **Nosing:** the projection of the tread beyond the face of the riser.
- **Unit rise:** the vertical distance from the top of one tread to the top of the next.
- **Total rise:** the sum of all the risers.
- **Unit run:** the horizontal distance from one riser to the next.
- **Total run:** the horizontal distance covered by a flight of stairs, including any intermediate landings.
- **Headroom:** the open space between the front edge of the tread and the ceiling above the stairway.
- **Carriage:** rough timbers cut to fit below the risers and treads. Mainly used for support, they are also known as rough stringers.
- **Winder:** a step whose tread is narrower at one end than the other. Winders bring about a change in the direction of a stairway and are used where space is limited (Figure 21-6).

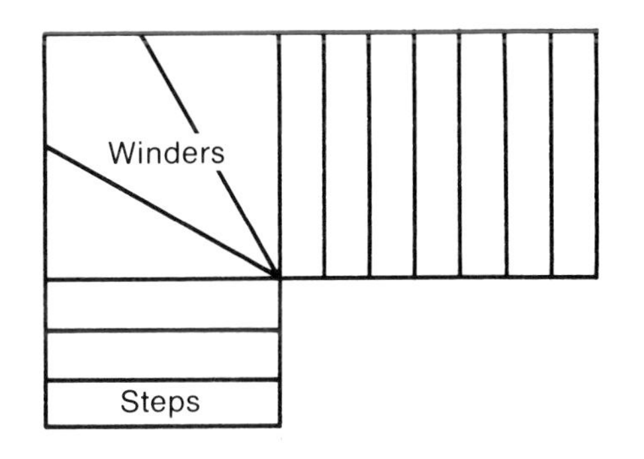

FIGURE 21-6. Stairway with winders.

Figure 21-7 illustrates the layout of a common stairway. Main stairways should be wide enough to permit two people to pass comfortably. The accepted minimum width for main stairways is 3′ (.9 m), though 3′6″ (1.1 m) is preferred. The accepted minimum width for service stairways is 2′6″ (.8 m). It is important to give consideration to the passage of furniture when planning stairs. Remember that stairs open on one side offer the best opportunity for moving large pieces of furniture; those with a closed or narrow U-shaped design present the most difficulty in this respect.

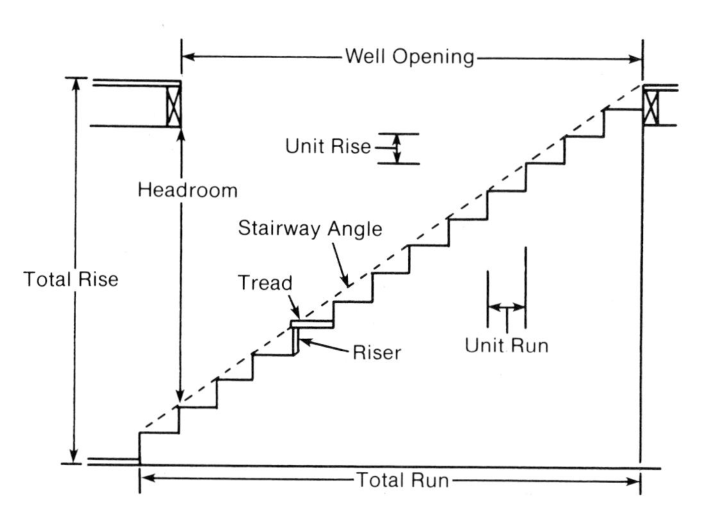

FIGURE 21-7. Common stair layout.

21.5 STAIRWAY SPECIFICATIONS

In order to make stairways as safe as possible, architects have developed rules for stairway designs that are incorporated into building codes. Naturally, there will be instances when a builder must stray from predetermined specifications due to a change in plans. For the most part, however, the following measurements apply to all light building construction projects.

21.5.1 Rail Heights

The most comfortable handrail height is 34″ (864 mm) on the stairs and 30″ (762 mm) on the landing. Wherever possible, the handrail should be continuous from floor to floor. It should also be smooth and free from splinters and sharp edges. A handrail is required on both sides of an open stairway; on a closed stairway, a handrail on one side is sufficient.

21.5.2 Headroom

Stairway headroom should be sufficient to permit a person to ascend or descend with ample clearance between his head and the soffit or ceiling above. A minimum headroom distance of 6′6″ (2 m) is required, though a distance of 7′4″ (2.2 m) to 7′7″ (2.3 m) is best.

21.5.3 Rise and Run

Stairs that are too steep or too shallow are both uncomfortable and dangerous. Those with too short a riser and too wide a tread force the user to lean backward while climbing the stairs; too steep an incline forces the user to lean too far forward. These tendencies are reversed when descending such stairs. To achieve the correct incline, a riser-to-tread ratio has been established. The following rules must be adhered to when designing stairs:

1. The sum of two risers and one tread should equal 24″ to 25″.
2. The sum of one riser and one tread should equal 17″ to 18″.
3. The tread width multiplied by the unit rise should equal 72 to 75 sq. in.

Using these rules, it can be determined, for example, that a step with an 8″ unit rise and a 9″ tread will meet the guidelines:

1. In Rule 1, 8 + 8 + 9 = 25.
2. In Rule 2, 8 + 9 = 17.
3. In Rule 3, 9 × 8 = 72.

All three results are within the guidelines, so the user will be able to ascend and descend the stairs safely and without undue strain. It should be noted that main stair risers are normally kept between 7″ and 7-5/8″. An exception is made for basement stairs, which often have lower ceiling heights. In these cases, 8″ risers are permitted. As the riser height increases, the tread width decreases, and vice versa.

21.6 STAIRWAY FRAMING

The openings for stairways are framed during the floor framing stage of a building project. The long dimension of the opening can run either parallel to or at right angles to the joists, though construction is much easier when the two are parallel (Figure 21-8). If the span is greater

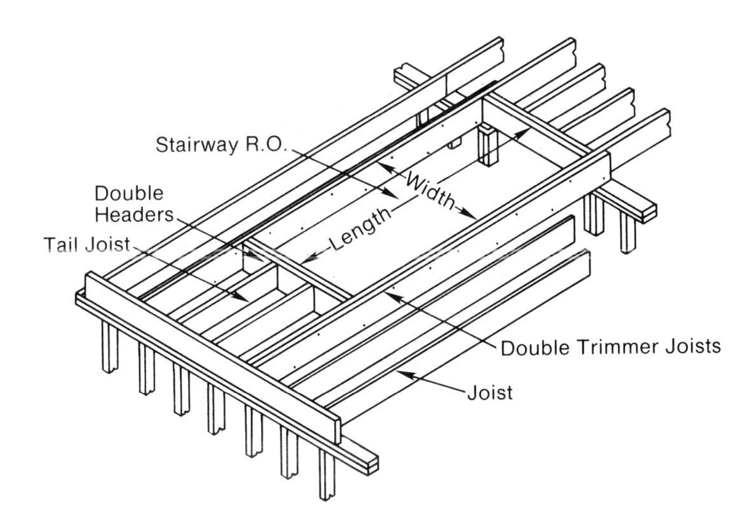

FIGURE 21-8. Framing for rough opening for stairway parallel to joists.

than 4′ (1.2 m) the headers and trimmers must be doubled. The length of headers should be limited to 10′ (3 m) unless secured by a wall; headers more than 6′ (1.8 m) long must be secured with framing anchors unless they are supported by a beam, post, or partition.

The rough opening for basement stairs is usually 9′ (2.7 m) long by 2′8″ (.8 m) wide. The minimum rough opening for main stairs is 10′ (3 m) long by 3′ (.9 m) wide. In Figure 21-9, the stringers have been installed for two flights, one above the other. Figure 21-10 shows the rough framing for a long L-shaped stairway with the landing and one stringer supported by a corner post.

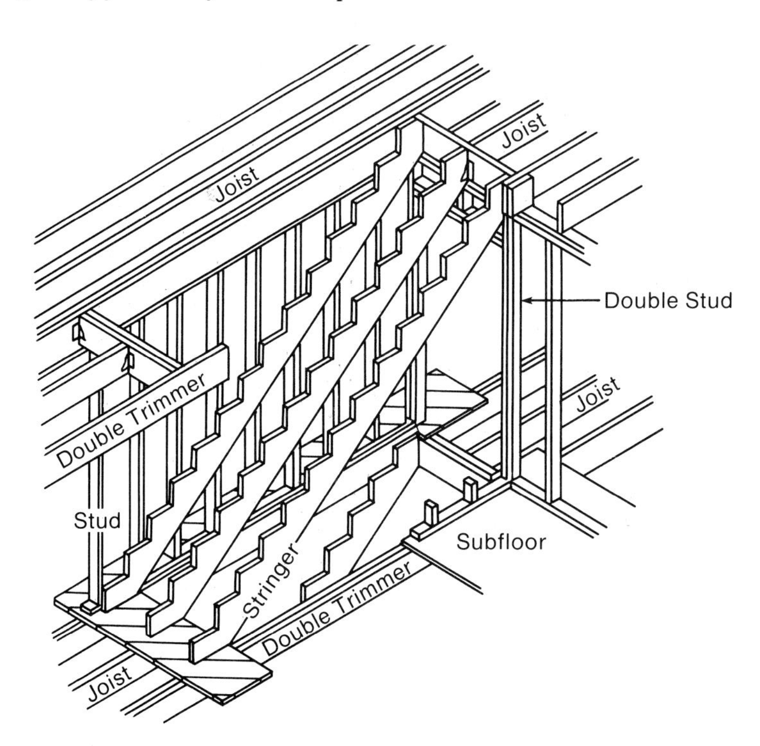

FIGURE 21-9. Rough framing details for two-flight stairs.

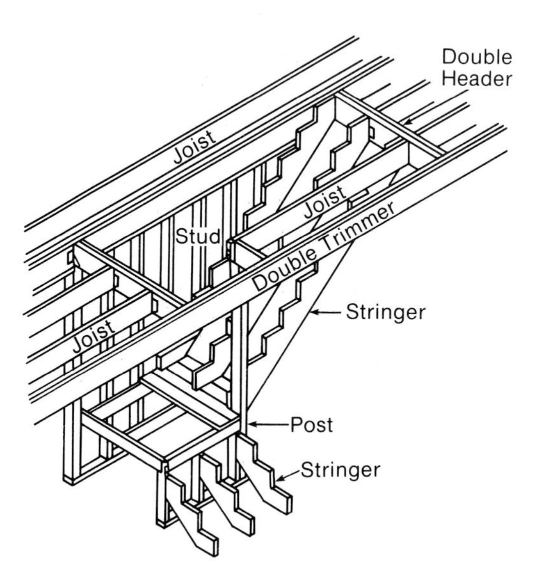

FIGURE 21-10. Rough framing detail for L-shaped stairs.

21.6.1 Stringers

It is very important that stringers are well-made, as they must support the load on the stairs. They must also be carefully laid out and cut. The best lumber to use is 2″ × 10″ (38 mm × 241 mm) or 2″ × 12″ (38 mm × 292 mm) for common stringers; housed stringers are made from 1-1/16″ (27 mm) or thicker material. At least 3-1/2″ (89 mm) of stock must remain on the underside of the stringer after the notches are cut. In addition, a 4″ (102 mm) bearing is required where the stringer bears against the header. Three stringers will be needed if the treads are 1-1/16″ (27 mm) thick and more than 2′6″ (.8 m) wide or if the treads are 1-5/8″ (41 mm) thick and more than 3′ (.9 m) wide.

21.7 CALCULATIONS FOR RISER HEIGHT AND TREAD WIDTH

The tread and riser sizes, the angle of slope, and the rough opening dimensions are all interrelated. Most plans will specify the opening sizes, and some will include the number of treads and risers as well. If the opening size is known, the tread and riser sizes can be calculated using the stair ratio principle. If, on the other hand, the opening size is not known, it can be determined after the tread and riser sizes have been calculated.

A scale drawing on graph paper is helpful for calculating and checking the stair layout and the length of the stairwell opening. The drawing should include the total rise (floor-to-floor height), two diagonals representing the slope limits, and the steps (Figure 21-11). The steps are determined by using the stair ratio. The tread/riser combination is determined as follows.

Find the number and size of the risers by dividing the total rise in inches by 7. For example, if the total rise is 8′5″, or 101 inches, divide 101 by 7. This equals 14.428, and since many risers are needed, there is a choice of using 14 or 15. If 14 is decided upon, divide 101 by 14 to determine the unit riser (in this case, 7.214, or approximately 7-3/16″).

Refer to Rule 2, which states that the sum of one tread and one riser must equal 17 or 18. If 17 is used, then 17 minus 7-3/16″ equals 9-13/16″, which is the tread (less nosing). If 18 is used, then 18 minus

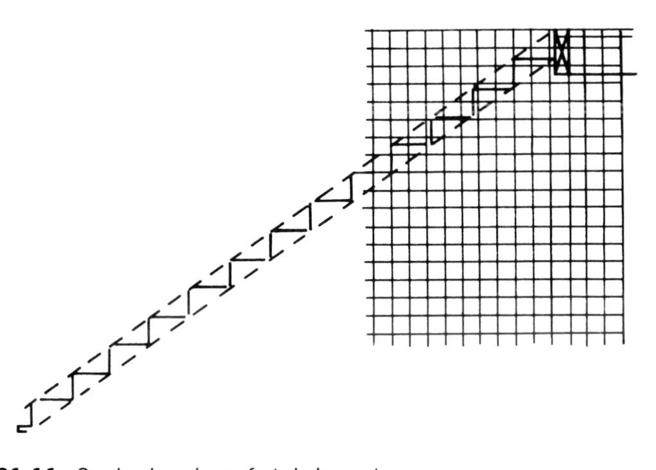

FIGURE 21-11. Scale drawing of stair layout.

7-3/16″ equals 10-13/16″. Take the average between the two and make it 10-5/16″. Thus, the riser height is 7-3/16″ and the tread width (less nosing) is 10-5/16″. Both fall within the range of desired angles, as shown in Table 21-1.

The number of treads in a stairway is always one less than the number of risers. In this example, since the number of risers is 14, the number of treads is 13. To find the total run, multiply the number of treads by the tread width: 10-5/16 × 13 = 134.06″. This calculates to a total run of 11′2″ (fractions are dropped).

TABLE 21-1 PREFERRED ANGLES FOR STAIRS WITH TREAD + RISER = 17-1/2″

Angle with Horizontal				
Degrees	**Minutes**	**Riser in Inches**	**Tread in Inches**	
22	00	5	12-1/2	
23	14	5-1/4	12-1/4	
24	38	5-1/2	12	
26	00	5-3/4	11-3/4	
27	33	6	11-1/2	
29	03	6-1/4	11-1/4	
30	35	6-1/2	11	Preferred
32	08	6-3/4	10-3/4	Preferred
33	41	7	10-1/2	Preferred
35	16	7-1/4	10-1/4	Preferred
36	52	7-1/2	10	
38	29	7-3/4	9-3/4	
40	08	8	9-1/2	
41	44	8-1/4	9-1/4	
43	22	8-1/2	9	
45	00	8-3/4	8-3/4	
46	38	9	8-1/2	
48	16	9-1/4	8-1/4	
49	54	9-1/2	8	

21.8 STAIR LAYOUT

Using the figures obtained above, the profile of the stairs can now be laid out on the scale drawing. From this layout, the stairwell opening size can be determined; the slope angle and headroom clearance can also be checked before cutting and laying out the stringers.

The length of the stringers can be determined from the layout and then checked with a story rod. Place the rod vertically on the finished floor below and mark the top of the finished floor above. Set a divider to the calculated riser height and step off the number of risers (14 in this example). If the last step-off does not fall on the mark, adjust and step off again. Repeat until the 14 divisions are equal. This will give the actual riser height, which can be used on the tongue of the framing square; the tread dimension is used on the blade of the square. The stringer can now be laid out. Use stair gauges or blocks clamped to the square to ensure accuracy (Figure 21-12).

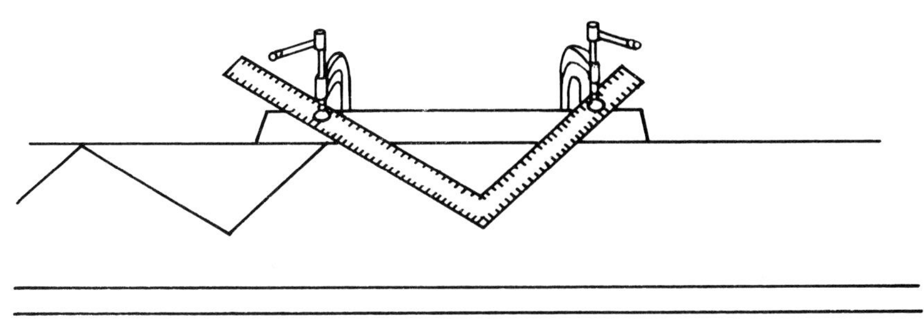

FIGURE 21-12. Layout of stringers with the aid of a framing square.

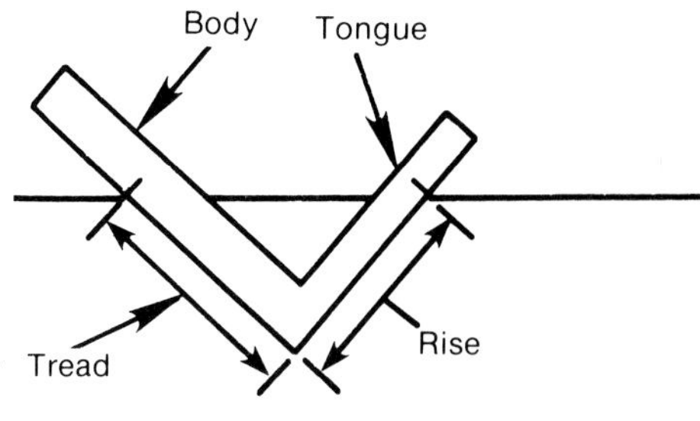

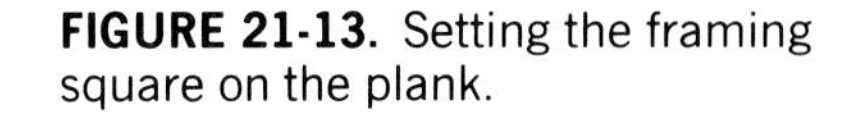

FIGURE 21-13. Setting the framing square on the plank.

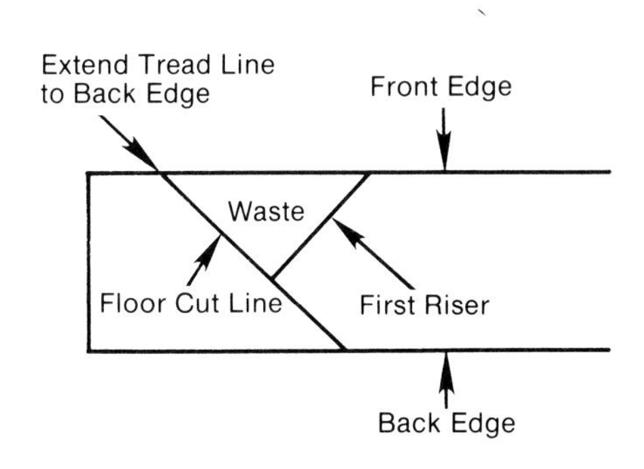

FIGURE 21-14. Marking the floor cut line.

Select a sound 2″ × 12″ (38 mm × 292 mm) board about 2′ (51 mm) longer than the stringer. Lay it on a pair of sawhorses and check it for any cracks or other defects. Place the framing square on one end of the plank and draw a line along the body and tongue (Figure 21-13). Now remove the square and extend the body or tread line to the back edge of the plank to represent the floor cut (Figure 21-14). Mark off the floor cut line and then move the square along the edge until the required number of steps have been drawn. When the last riser has been drawn, extend the line to the back edge of the plank (Figure 21-15). This part of the stringer will bear against the header; for extra strength, notch it to fit around the header joist.

The stringer must be dropped at the bottom to allow for the tread thickness. (This was not included in the layout.) To do this, draw a line equal to the tread thickness and parallel to the floor line (Figure 21-16). Keep in mind that the stringers must be cut carefully, and make all outboard cuts first. During the second or notching cut, the base is well supported by the remaining stringer stock. A handsaw is best for completing the cut.

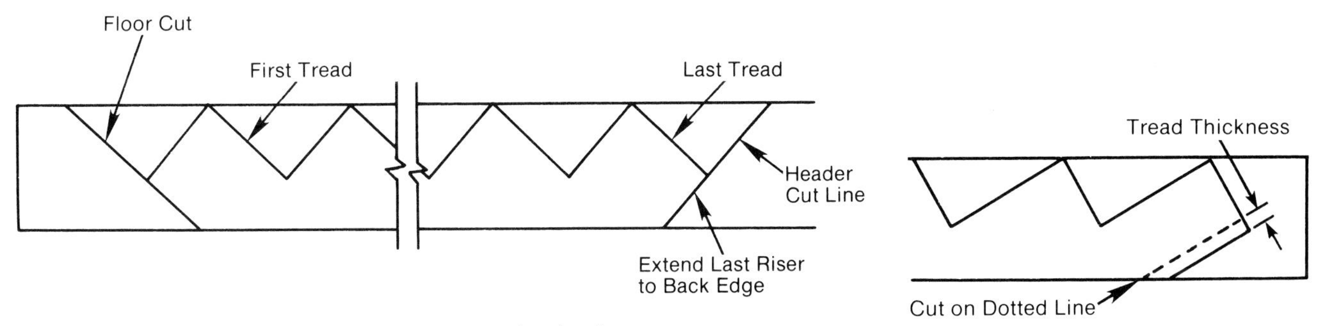

FIGURE 21-15. Extending the last riser to the back edge.

FIGURE 21-16. Dropping the stringer.

21.9 TREAD AND RISER DETAILS

The treads for main stairways are usually 1-1/16″ or 1-1/8″ (27 mm or 29 mm) thick. While both hardwoods and softwoods can be used, softwoods must have a vertical grain or, if flat-grained, they must be covered with a suitable floor finish material.

Risers are normally made of 3/4″ (19 mm) stock of the same species as the treads. Less expensive lumber may be used if they're going to be covered, and plywood can be used when the stairs are to be carpeted. Treads and risers are fastened with finishing nails at the edges. Drive

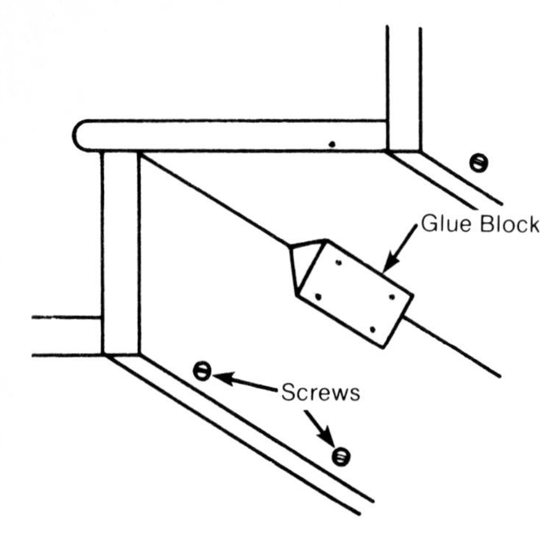

FIGURE 21-17. Preferred method of fastening treads and risers.

nails or screws into the treads at the rear through the lower part of the riser, using glue blocks at the intersection of the rear surface of the tread and riser (Figure 21-17).

21.10 STRINGER DETAILS

Open riser stairs are often used for basement stairways. When installing stairs, either the dado or cleat method can be used (Figure 21-18). The depth of the dado should be one-third the stringer thickness. When cutting a dado, use the tread line on the stringer for the top edge of the dado cut. Fasten the stair by driving nails through the stringer into the ends of the threads; use glue for an improved joint.

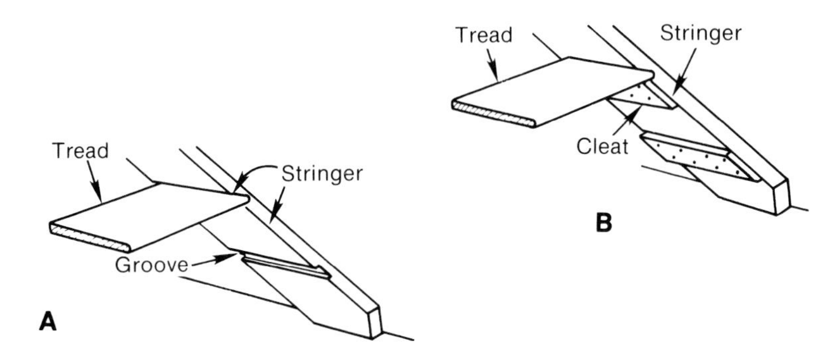

FIGURE 21-18. Two methods of installing basement stairs: (A) dado and (B) cleat.

For a cleated stringer, draw a line below and parallel to the tread mark on the stringer. The space between the two lines should be equal to the tread thickness. After making certain that the cleat is long enough to adequately support the tread, bevel all exposed edges and fasten the cleat to the stringers by driving nails through the stringers into the ends of the treads. In the event that the side walls or floor are masonry, wood blocks or some other suitable masonry anchor will be necessary to secure the stringers.

21.10.1 Finish Stringers

Either a full finish stringer or a notched finish stringer may be used at the sides of the stairway. When full stringers are used, fasten them directly to the side walls. The rough stringer is then fastened to the finish stringer and the treads and risers are cut to fit between the finish stringers (Figure 21-19). When using a notched stringer, the rough stringer is fastened directly to the wall. The finish stringer is then notched to match the steps of the rough stringer (Figure 21-20) and the treads and risers are cut and installed. A low-cost alternative to these methods is to use metal stringers, which are easily secured with screws.

21.10.2 Housed Stringers

Special stair templates are necessary when constructing housed stringers. The templates guide the router for making the grooves that receive the treads, risers, and wedges. When one of the walls of the stairway is open, a mitered stringer is used. Figure 21-21 shows details of both housed and mitered stringers.

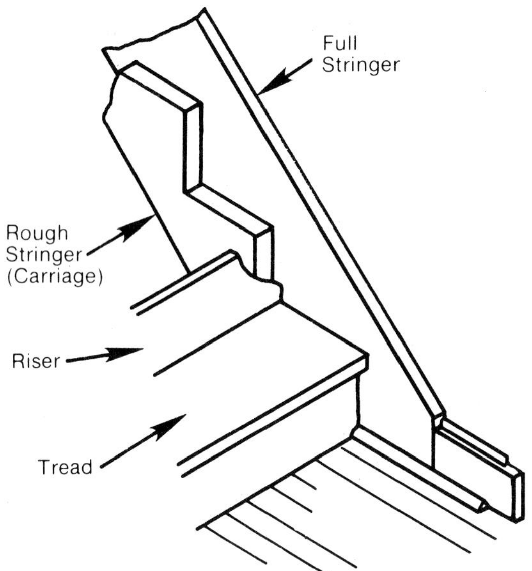

FIGURE 21-19. Enclosed stairs with full stringer.

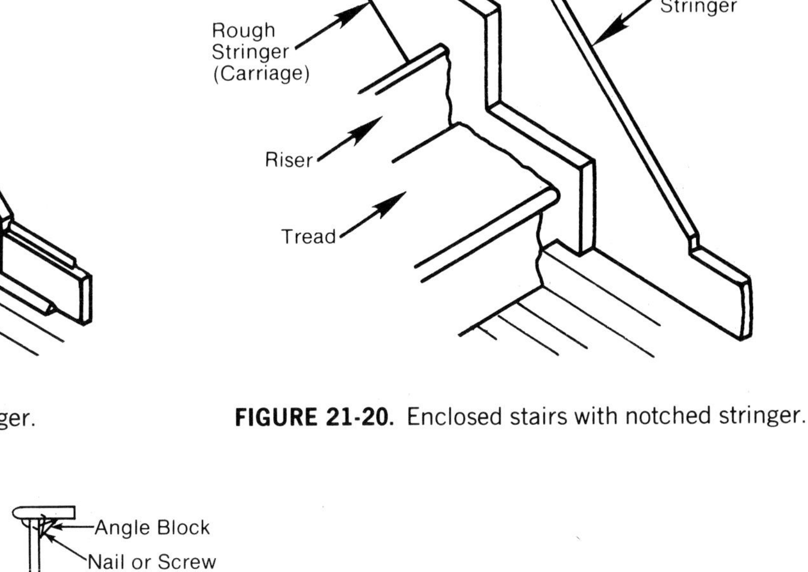

FIGURE 21-20. Enclosed stairs with notched stringer.

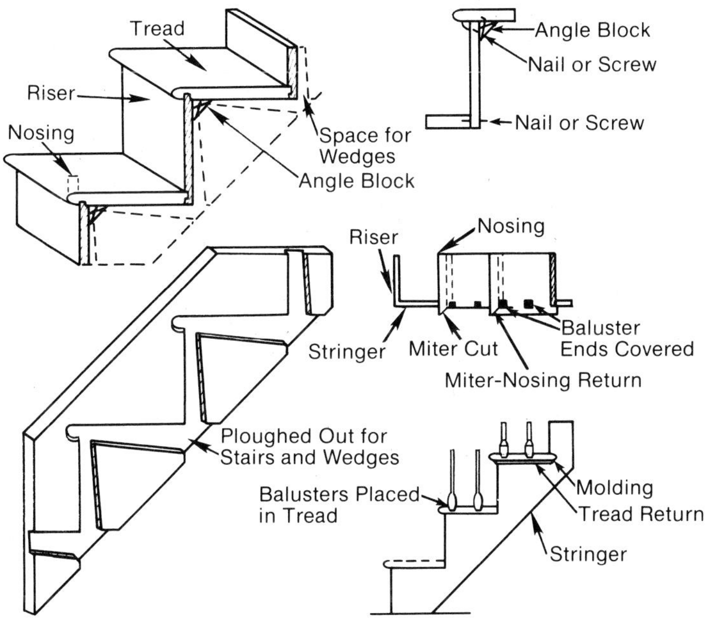

FIGURE 21-21. Details of housed and mitered stringers.

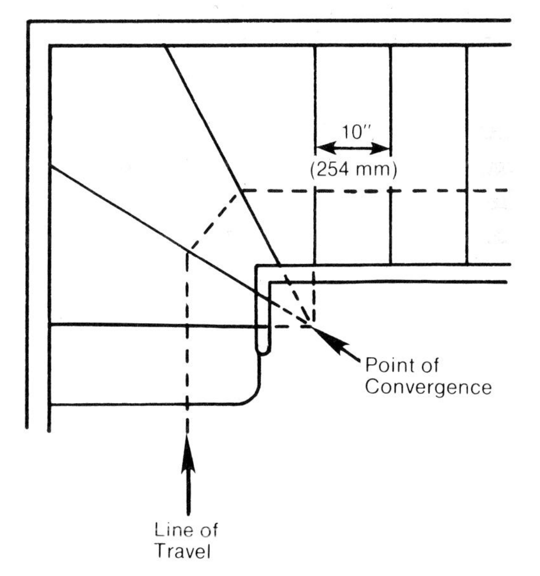

FIGURE 21-22. Winder details.

21.11 WINDERS, BALUSTERS, NEWELS, AND RAILS

Though convenient in buildings where space is at a premium, winders are, nevertheless, dangerous because of their tapered treads. A degree of safety can be introduced by bringing the center of convergence to a point out from the corner (Figure 21-22). This converging point should be located to provide a tread width equal to that of the regular treads along the line of travel. This imaginary line is about 18" (457 mm) from the inside stair edge.

Winders can be calculated mathematically, though a full-size plan view layout showing positions, lengths, and angle cuts is preferable. The riser heights must be the same as those of straight stairs. Spiral stairways are an acceptable form of winder and are used mainly for decorative effect.

The open section of stairs must be enclosed with a handrail parallel to the slope of the stair. The handrail is fastened to the newel post and is supported by balusters. The balusters may be square or turned spindles which are fastened between the handrail and tread or between the handrail and a lower rail. This entire assembly—including the newels, balusters, and rails—is known as the balustrade (Figure 21-23). All of these components are millmade and then installed by the builder. The rails are fastened to the walls with special brackets which are screwed directly to the wall frame. Do not use nails, as they have a tendency to loosen. Spacing for the brackets should not exceed 10′ (3 m).

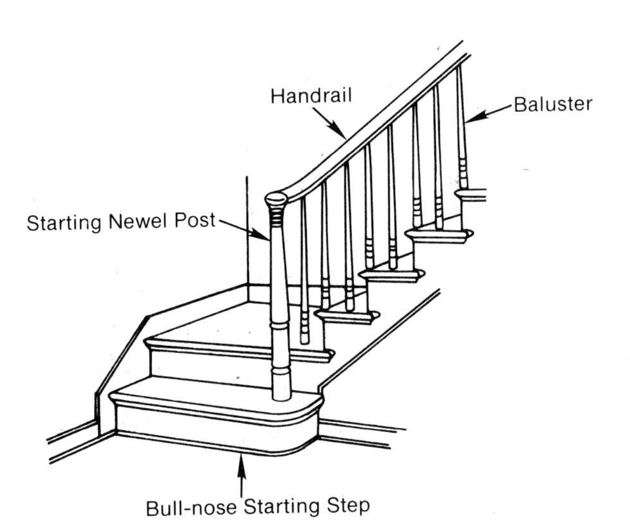

FIGURE 21-23. Parts of a balustrade.

21.12 RESEARCH

1. Take measurements of the stairs in your home, noting the tread and riser sizes, the number of steps, and the floor-to-floor height. Compare your measurements with those of other students at class.
2. Identify and draw rough sketches of the various types of stairs and comment on the advantages and disadvantages of each.
3. Examine the stairways in older homes in your area, noting the changes that have occurred over the years in stairway design and construction.

21.13 REFERENCES

21.13.1 Capotosto, John, *Basic Carpentry*. Virginia: Reston Publishing Co., 1980.

21.13.2 Reed, Mortimer. *Residential Carpentry*. New York: John Wiley and Sons, Inc., 1980.

22

CABINET LAYOUT BOARD AND CUTTING LIST

Cabinet building is common to the light construction field. Most small buildings contain cabinets of various designs and sizes. These designs vary as needs arise and new materials appear on the market.

Builders must concern themselves with cabinet construction, whether the cabinet is built totally on the jobsite or prefabricated by others and simply installed at the site. A pleasing cabinet design is composed of materials sized to specific dimensions and selected hardware combined by a set of construction details. Cabinet materials, hardware, and construction details are brought together by the cabinet designers and craftspersons who must be able to communicate clearly with one another.

The layout board is one means of communication which links the elements of cabinet construction so that the designer and craftspersons can work toward the same final objective. Actual measurements are clearly set out on the layout board which becomes a full-scale measurement data bank for a particular cabinet. The board is different from a story rod or shop drawing in that it carries both vertical and horizontal measurements. Using a hard lead shop pencil and a combination square, only absolutely necessary marks are made on the board.

The layout board is usually a piece of light-colored plywood longer than the proposed cabinet. If the builder's shop is equipped with a power planer, a white spruce board of suitable width, about 10″ to 12″ (254 mm-305 mm), can be utilized.

In using the general principle of the cabinet layout board, each builder will develop a set of markings to suit his or her particular style. If the builder is consistent in marking off critical measurements, few errors will be made in creating a cutting list of materials for the project. Builders who can successfully mark out a layout board and draw up a cutting list from this data will fully understand what the finished cabinet will look like, how it will be built, and how it will be installed. Items such as material selection, composition, and measurement are also important to the builder in estimating project cost.

22.1 SPECIFIC OBJECTIVES

After reading this chapter and conducting the suggested research, the reader should understand all phases of cabinet construction and installation. Specifically, the reader should be able to

1. prepare a layout board for a small cabinet project by working from a given sketch,
2. create a cutting list for that project, and
3. interpret a given cabinet sketch/layout board when preparing a cutting list.

22.2 WORKING THROUGH A SAMPLE PROBLEM

The simplest means of understanding the concept of a cabinet layout board is to work through a sample cabinet problem. In this problem the builder is given a sketch of kitchenette cabinets (Figure 22-1). The builder is to construct and install the cabinets. The following assumptions also hold true:

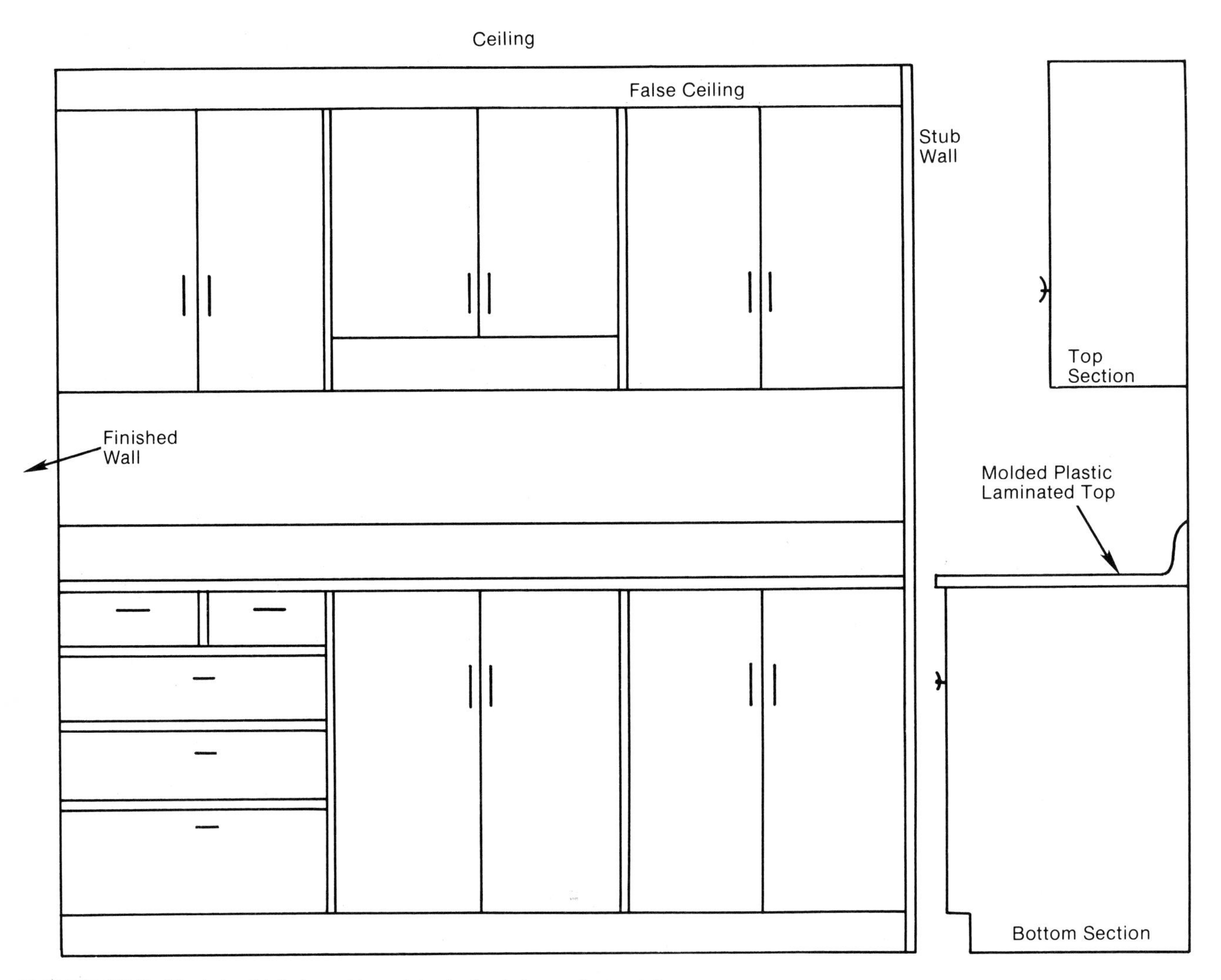

FIGURE 22-1. Sketch of kitchenette cabinets (not drawn to scale).

1. Overall measurements for the cabinets will be taken on-site by the builder (carpenter or cabinetmaker).
2. Although the sketch will control the general design, the builder will make final decisions with respect to construction methods. (The builder will take into consideration the squareness of the corner, the flatness of walls and floor, and the roughed-in services: electrical and plumbing.)
3. One side of the cabinets will be built against a finished wall; the other side will be exposed at a stub wall.
4. The top section will be separate from the bottom section, but critical vertical lines will correspond. Space between the sections will make provision for kitchen electrical appliances, but not a stove.
5. The cabinets will be of plywood construction. Matching wood trim is optional.
6. Hardware will be selected by the builder.
7. The cabinets will be constructed, painted, and installed by the builder.
8. The layout board and cutting list will become the property of the builder.

22.3 MEASURING

Prior to taking any measurements, the cabinetmaker should make a thorough inspection of the installation site. The builder should be equipped with a framing square, measuring tape, level, and edge-grain straightedge so that accurate measurements of the cabinet installation area can be made. The edge grain straightedge should be 6′ (1.8 m) in length and equipped with a handle so that the level and straightedge may become one tool as illustrated in Figure 22-2.

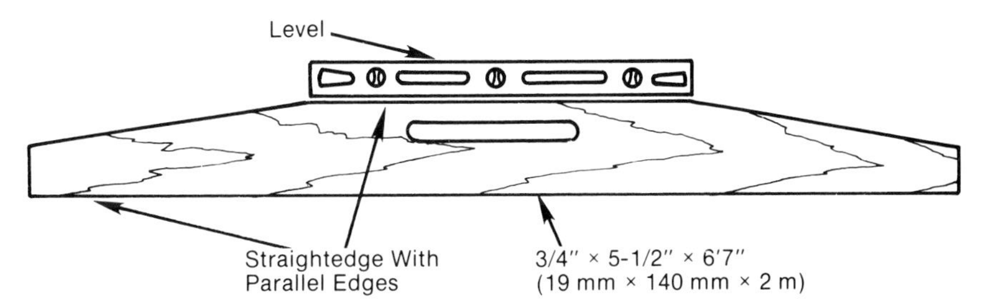

FIGURE 22-2. Level and straightedge combined for use with one hand.

The level/straightedge is used to check for flatness of walls and floors and whether or not they are plumb and level. The framing square is used to check the squareness of adjoining walls and ceiling. The location of service outlets must also be noted, and construction details must be adjusted to allow for any irregularities in the floor, wall, and ceiling.

22.4 CABINET LAYOUT BOARD

To prepare to record sets of measurements, the layout board is divided into four, five, or six divisions as required for the job (Figure 22-3). In

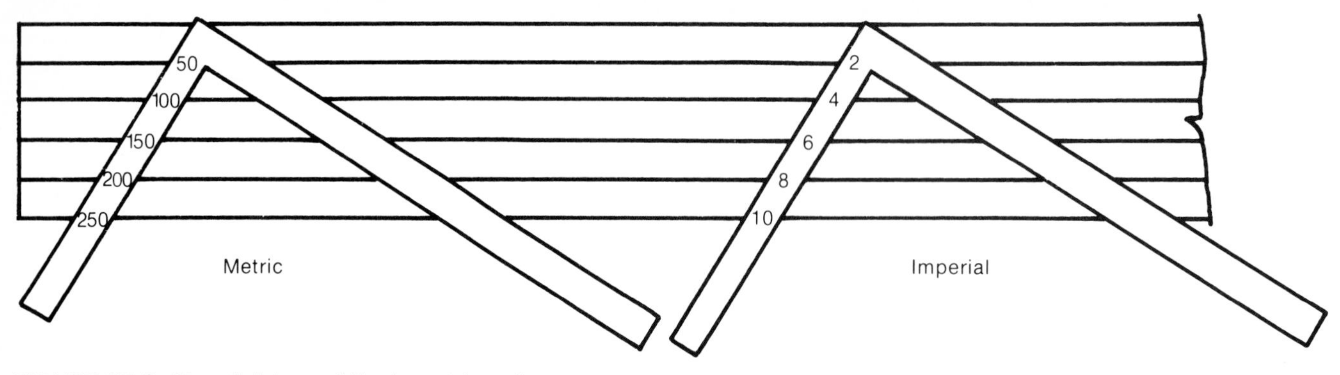

FIGURE 22-3. Five divisions of the layout board.

this sample problem, five divisions are used and numbered as shown in Figure 22-4. Explanations of the types of measurements recorded in each division are given below.

FIGURE 22-4. Number the divisions on the layout board (not drawn to scale).

22.4.1 Division #1: Length of Lower Cabinets

Use division #1 to record horizontal measurements for the lower cabinets. Design the cabinets to go against the finished wall if the two walls and floor are respectively plumb and level, corners are square, and walls and floor flat (not warped). Mark the left end as a measuring point for length of cabinet top. Use a short pencil mark as shown in Figure 22-5.

Hold the board against the wall in the approximate final position of the cabinet top. Mark at the right end of the board the position of the stud wall. This mark establishes the measurement limit. However, for easy installation it might be better to measure back 1-1/4″ (32 mm) from the stud wall for the countertop measuring point. See Figure 22-5. Print "Length of Top."

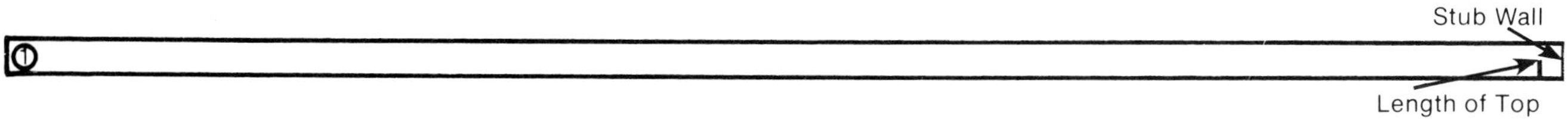

①
Stub Wall
Length of Top

FIGURE 22-5. Length measuring points for lower cabinet (not drawn to scale).

22.4.2 Division #2: Length of Upper Cabinets

Use division #2 for the length of the upper cabinets. Transfer the overall measuring points for division #1. There is no need for printed designations as they are already on division #1 (Figure 22-6).

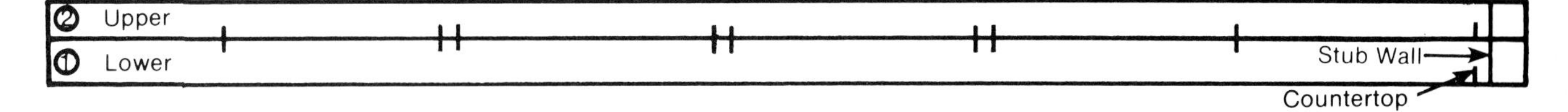

FIGURE 22-6. Transfer of measurements from division #1 to #2 for length measurement of upper cabinet (not drawn to scale.) **Note:** drawer fronts, partitions and doors added.

22.4.3 Division #5: Floor to Ceiling Measurements

Division #5 is used to show actual floor to ceiling measurements. If it is not possible to stand the board on end to mark the height, measure with a tape. Measure, mark, and measure again to check. (*Note:* It is not sufficient to only record in writing a measurement of 7′6″ or 7′9″ (2.3 m or 2.4 m). The actual measurement must be on the board. Someone else might misread 7′6″ as 7′9″ or might measure 2.2 m instead of 2.3 m. See Figure 22-7 for marking division #5.

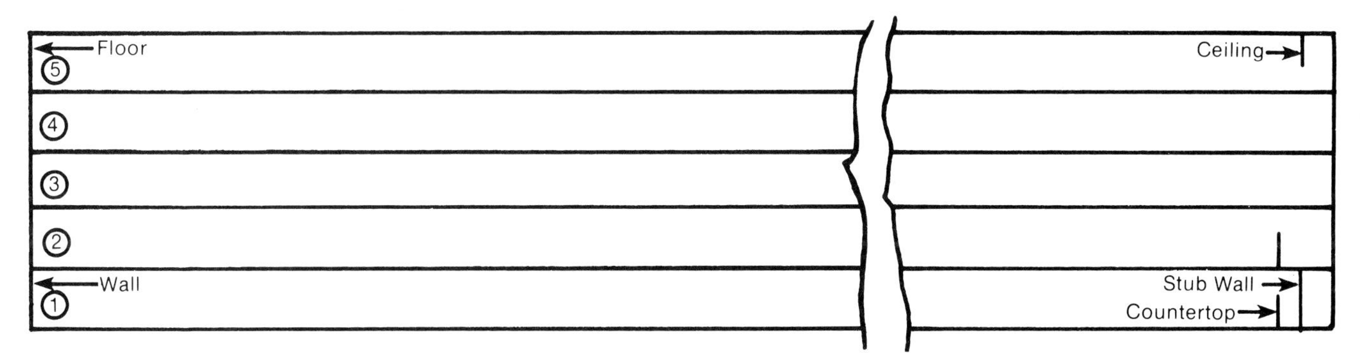

FIGURE 22-7. Floor to ceiling measurement (not to scale).

22.4.4 Division #5: Optimum Space Between Two Cabinets

Division #5 is also used to mark the top of the lower cabinets and the bottom of the upper cabinets. The top of the kitchenette counter is about 36″ (914 mm) high and the space between the two cabinets is 15″ to 16″ (381 mm to 406 mm)—24″ (610 mm) if there is a stove. Adjust the height of the upper cabinets to obtain the desired space between cabinets (Figure 22-8).

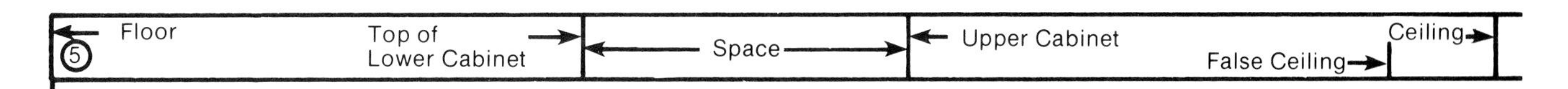

FIGURE 22-8. Top of lower cabinet and bottom of upper cabinet (not drawn to scale).

Mark the depth of the false ceiling, keeping in mind the appearance of the finished cabinets and the use of plywood to make cabinet doors. In division #5 limit the number of marks to major measuring points. Leave details for another division to avoid confusion with respect to marks.

22.4.5 Division #3: Depth of Countertop and Upper Cabinet

The depth of countertop is usually one-half the width of a sheet of plywood or chipboard: one-half of 4′ (1.2 m) less a saw kerf. The depth of the upper cabinet shelf may be one-quarter of a sheet of plywood—one-quarter of 4′ (1.2 m) less a saw kerf.

These are basic no-waste measurements for cutting plywood of chipboard panels. If a solid wood face trim were to be added, the overall measurements would change by 3/4″ (19 mm) (Figure 22-9). Position these overall measurements against the installation place to be sure that a window, door, or service outlet will not interfere.

Consider once more the construction of the countertop. If a plastic laminate is to be cut and used to cover the top and to face the edge, width of sheet may determine the depth of countertop, as in Figure 22-10.

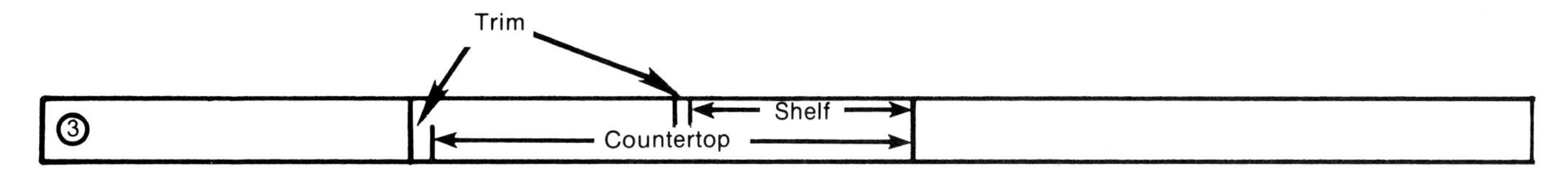

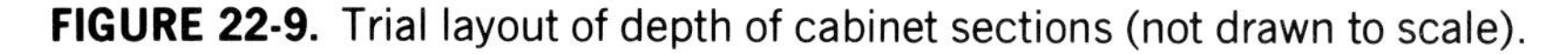

FIGURE 22-9. Trial layout of depth of cabinet sections (not drawn to scale).

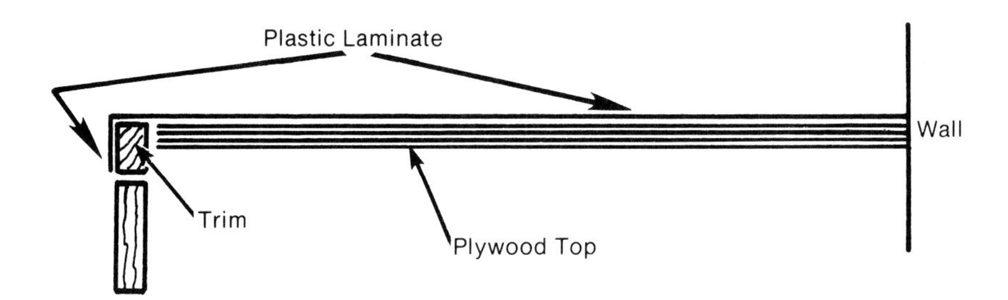

FIGURE 22-10. Use of plastic laminate on countertop (not drawn to scale).

As sawing may leave a broken edge, the top sheet of plastic laminate must be trimmed after being cemented to the plywood/chipboard top. An examination of the sawn edge will determine the amount of trimming, say 1/64″ (.4 m), probably with a special router bit. The trial layout shown in Figure 22-9 may be modified, as in Figure 22-11. The width of plywood top may have decreased about 3/4″ (19 mm), as compared with the construction method shown in Figure 22-9.

If the countertop is to overhang the front, say 1″ (25 mm), the sketch and layout board critical measurements must be specific (Figure 22-12). A set of critical overall measurements are now on the board (Figure 22-13).

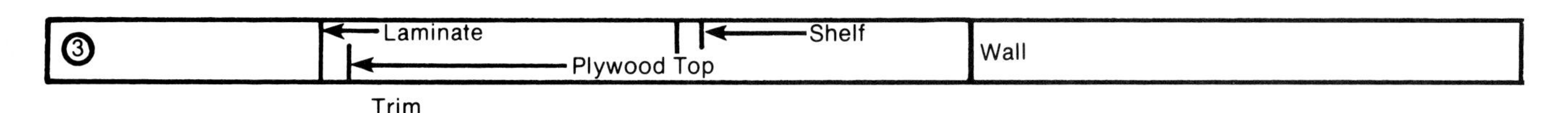

FIGURE 22-11. Division #3 use of plastic laminate on countertop (not drawn to scale).

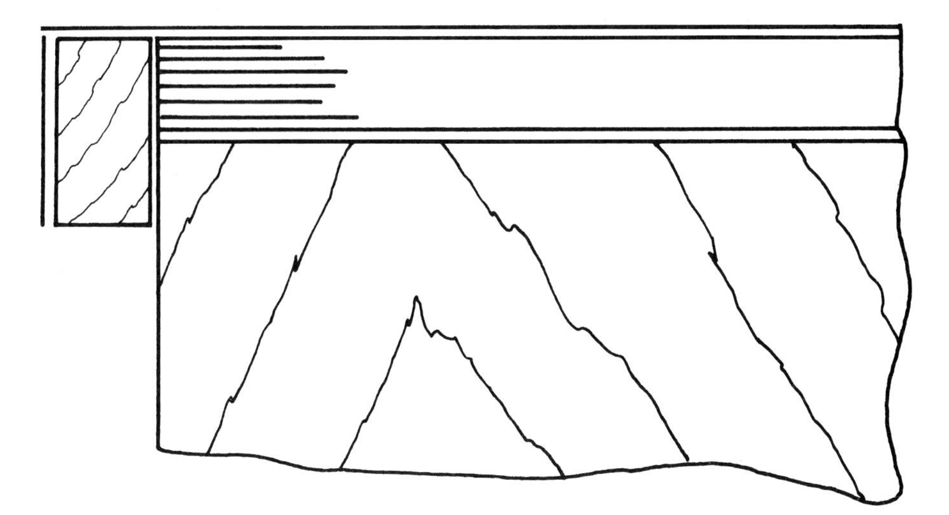

FIGURE 22-12. Countertop with overhang (not shown on layout board).

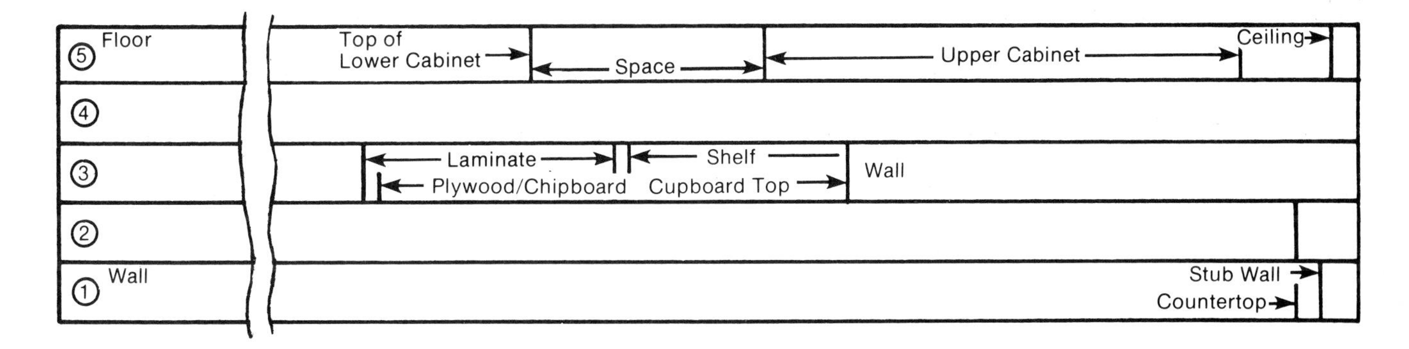

FIGURE 22-13. Layout board with critical overall measurements (not drawn to scale).

Countertops may be of prefabricated molded plastic laminate selected by the architect or the potential owner. If this is the case, the cabinetmaker must have the measurement details. Such countertops should be in stock, on hand, and available when needed (Figure 22-14).

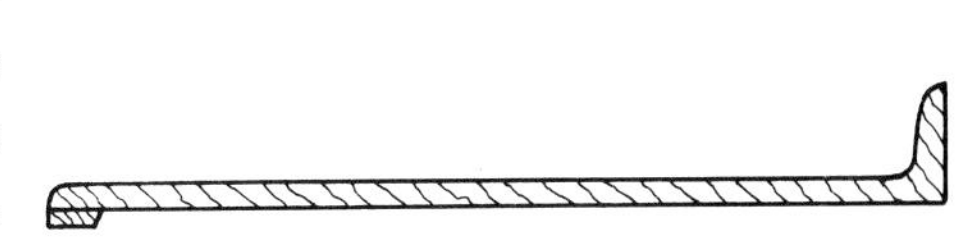

FIGURE 22-14. Prefabricated countertop.

If a prefabricated countertop is to be used, the measurements are critical as far as designing the remaining parts of the cabinet is concerned. The same may be said for cabinet hardware and wood stock. All items must be in the shop, on-site, or available from a known supply before the layout board and cutting list are approved.

Should the builder, on speculation, prepare the board and list, with implications of a completely designed cabinet, and then discover after the project is partially finished that some items are not available, much time and money are wasted, and the project may appear poorly finished and designed. If specified hardware or materials are not available, a compromise design can develop before construction commences. Someone must assume responsibility. The decision-making process must be understood by all concerned.

At this point in the layout process, having overall measurements clearly marked on the board and the installation place inspected, the builder can proceed either on the jobsite or in the off-site shop. Construction details and a cutting list are needed before cutting the stock.

22.5 CONSTRUCTION DETAILS

Return to division #1 to mark measurements for trim or facing strips, drawer fronts and doors, framing details (rabbets, dadoes, glue blocks, etc.), sink opening, top, and backboard (if any, and if a molded one). Depths of rabbets and dadoes may be shown on the board with broken lines, as in Figure 22-15. Flush doors and drawer fronts with clearances can also be shown.

If dadoes are shown, the assumption is that framing to carry drawers will be constructed and fitted. These may be expensive operations unless the general scale of building is such that a shop setup is available.

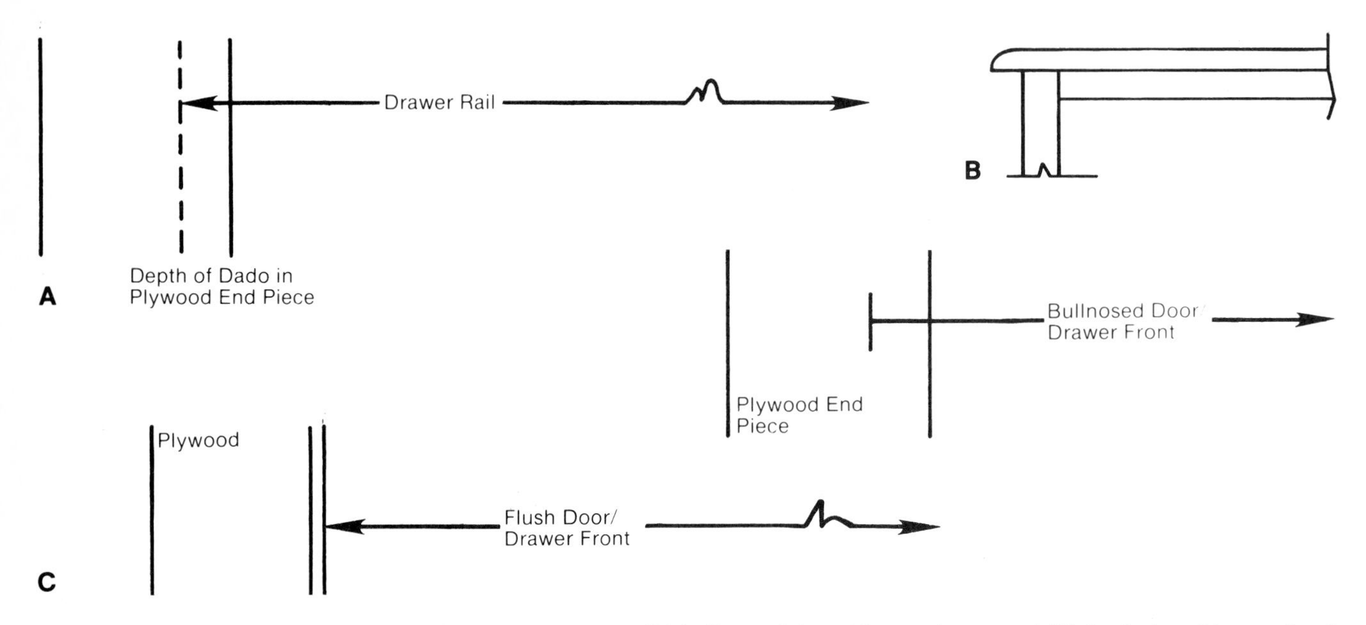

FIGURE 22-15. (A) Depth of dado in plywood end piece, (B) bullnosed door/drawer front, and (C) flush door/drawer front.

22.5.1 Division #2: Length of Upper Cabinet

Mark measurements for the following: trim or facing strips to divide the cabinet length into the same three major sections as in division #1 (vertical lines must correspond); framing details, to indicate rabbets and dadoes; and overall lengths of shelves. (Decisions must also be made with respect to the number of cabinet subsections, the assembly of parts, and installation.)

22.5.2 Divisions #3 and #4: Depths of Cabinets

With reference to the construction method shown in Figure 22-12, Figure 22-16 indicates useful depth measurements; shelf width; flush door thickness; and position for the upper section, laminate width, trim thickness for overhang, widths of top, shelving partitions and ends, kick space, lengths of breadboard (if any), and drawer frames. If space allows, drawer details may show overall length, stops, thickness of front with plow for bottom, back thickness, and length of bottom. Division #4 can well be used for this data, as in Figure 22-17. Sink and top data may be added.

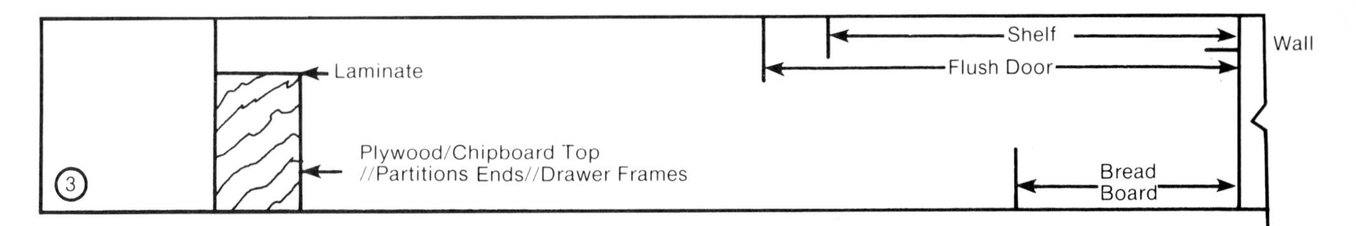

FIGURE 22-16. Depths of cabinets (not drawn to scale).

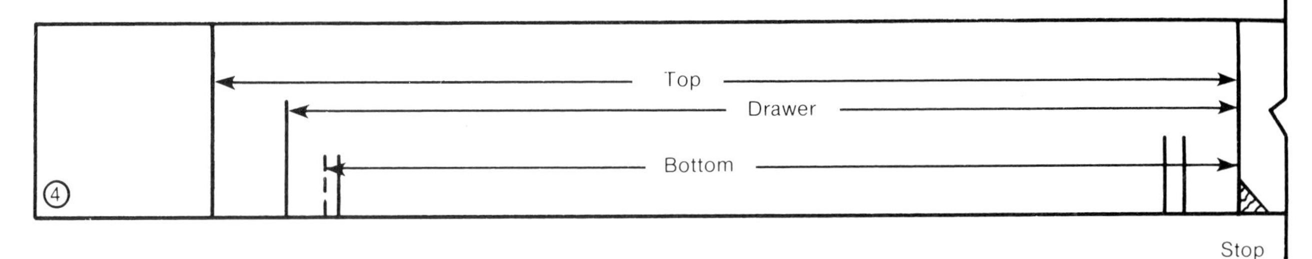

FIGURE 22-17. Division #4 drawer length measurements.

In Figure 22-18 detail measurements of height are added to division #5 and division #4. The latter division is used as it is convenient to separate dado measurements from other marks for drawers and shelves. Part of division #4 is borrowed for that purpose. (Possibly six rather than five divisions of the layout board might be better for such layouts, in order to reserve one division for dado positioning.)

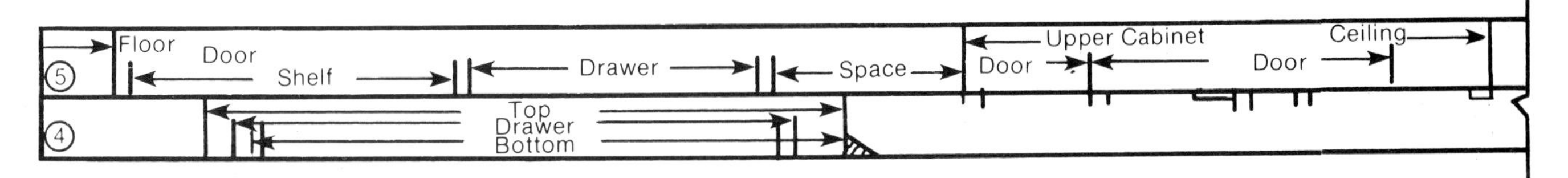

FIGURE 22-18. Divisions #4 and #5 heights of cabinets.

Overall measurements and sets of detailed measurements are now on the board (Figure 22-19). Probably the number of marks is sufficient for taking off a cutting list, except that a few decisions about construction have not been made: base framing; miter joint between base kickplate and the plywood right end; the bottom shelf (floor) of the bottom section; the hinge hardware to allow for flush doors; the guides for drawers; and sink cutout.

Because the purpose of this chapter is to present a *measuring technique* for cabinet construction, assembly procedures will not be discussed. The overriding building principle seems to be "If you can draw it, you can build it." For the cabinetmaker/carpenter, this drawing may be a rough sketch and a layout board layout.

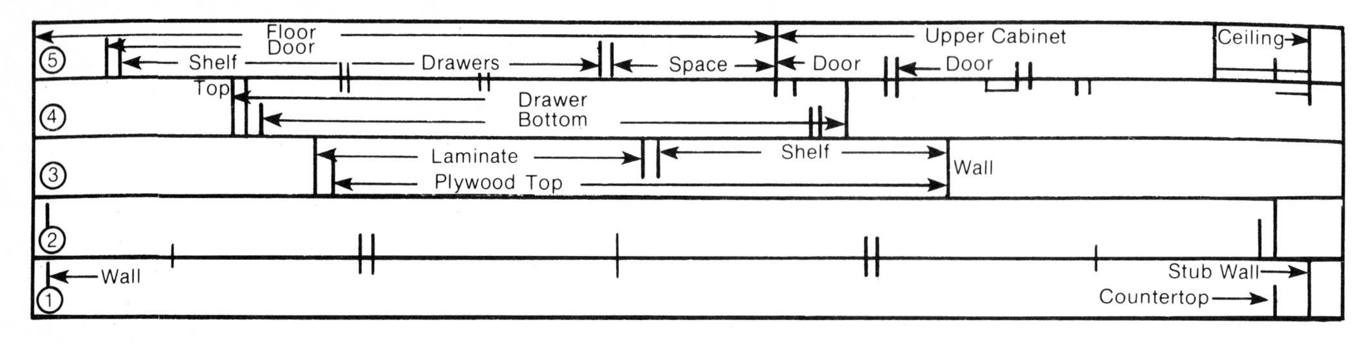

FIGURE 22-19. A layout board for a kitchen cabinet.

22.6 BOARD CUTTING LIST

The list should be put into some order so that it reflects the actual construction sequence. Usually the base is first and the drawers and doors are last. A tabular format for the list would allow for additional data, such as material, labor, and overhead costs. Costs for power, shop time, transportation, installation, and finishing may be categorized as overhead. A break-even point should be observed with respect to the amount and nature of detail data recorded (Table 22-1).

With actual measurements taken from the layout board and recorded on the cutting list, a breaking-out phase can proceed. Attention must be given to proper selection of materials: clear, undamaged, suitable grain, matched pieces, and optimum cutting to minimize

TABLE 22-1 CUTTING LIST FOR KITCHENETTE CABINETS

						Material		Labor	
	Size			Number of		Unit Cost	Cost	Rate	Hours Amount
Part	T	W	L	Pieces	Material	$	$	$	$
				Bottom Section					
Base	21/32″ (17 mm)	3-15/16″ (100 mm)	83″ (2,100 mm)	2	Sanded poplar				
	21/32″ (17 mm)	3-15/16″ (100 mm)	19-11/16″ (500 mm)	4	Plywood				
Ends/ partitions	21/32″ (17 mm)	*	22-27/32″ (580 mm)	4	Plywood				
Top	21/32″ (17 mm)	*	*	1	Plywood				
Shelves	21/32″ (17 mm)	*	*	*	Plywood				
Countertop	*	*	*	*	Plastic laminate				
Drawer fronts	21/32″ (17 mm)	*	*	2	Sanded poplar plywood				
	21/32″ (17 mm)	*	*	1	Sanded poplar plywood				
	21/32″ (17 mm)	*	*	1	Sanded poplar plywood				
	21/32″ (17 mm)	*	*	1	Sanded poplar plywood				

TABLE 22-1 CUTTING LIST FOR KITCHENETTE CABINETS (Continued)

Part	Size T	Size W	Size L	Number of Pieces	Material	Material: Unit Cost $	Material: Cost $	Labor: Rate $	Labor: Hours Amount $
Bottom Section									
Sides	7/16″ (11 mm)	*	*	4	Sanded poplar plywood				
	7/16″ (11 mm)	*	*	4	Sanded poplar plywood				
	7/16″ (11 mm)	*	*	2	Sanded poplar plywood				
Backs	5/16″ (8 mm)	*	*	2	Sanded poplar plywood				
	5/16″ (8 mm)	*	*	2	Sanded poplar plywood				
	5/16″ (8 mm)	*	*	1	Sanded poplar plywood				
Bottom	1/4″ (6 mm)	*	*	2	Sanded poplar plywood				
	1/4″ (6 mm)	*	*	2	Sanded poplar plywood				
	1/4″ (6 mm)	*	*	1	Sanded poplar plywood				
Frames etc.									
Doors	21/32″ (17 mm)	*	*	4	Sanded poplar plywood				
Top Section									
False ceiling	21/32″ (17 mm)	5-29/32″ (150 mm)	83″ (2,100 mm)	1	Sanded poplar plywood				
Ends/ partitions	21/32″ (17 mm)	11-5/8″ (295 mm)	*	4	Sanded poplar plywood				
Shelves	21/32″ (17 mm)	10-7/8″ (278 mm)	*	1	Sanded poplar plywood				
	21/32″ (17 mm)	10-7/8″ (278 mm)	*	4	Sanded poplar plywood				
Wall strap	21/32″ (17 mm)	2-9/16″ (65 mm)	81″ (2,057 mm)	2	Sanded poplar plywood				
Doors	21/32″ (17 mm)	*	*	4	Sanded poplar plywood				
		*	*	2	Sanded poplar plywood				

*Data to be confirmed and/or read from the layout board.

waste. A painted surface direction of grain is of lesser consequence, but for transparent or translucent finishes on expensive woods it is most important (Figure 22-20).

For vertical panels on a kitchenette cabinet, examples A, B, and C of Figure 22-20 would not be mixed. B and C would probably not be used. Disagreement with respect to direction of grain and matching of grain can be costly for the builder.

It is unlikely that a no-waste situation will happen for a particular cabinet, but it should be an objective. Plan the cutting with care. See Figure 22-21 for a breaking-out plan for one sheet of plywood.

Without obtaining a current cost-per-unit-length of cabinet from industry, it would be difficult without collected data on the part of the builder to provide a customer with a cost estimate for a cabinet. For one's own building the materials cost may be the major item. For a project builder materials, labor, and overhead must be known.

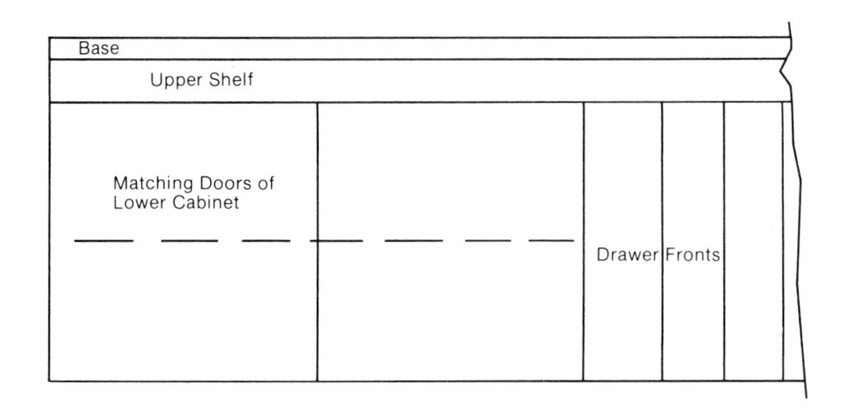

FIGURE 22-21. "Breaking out" plan for one sheet of plywood.

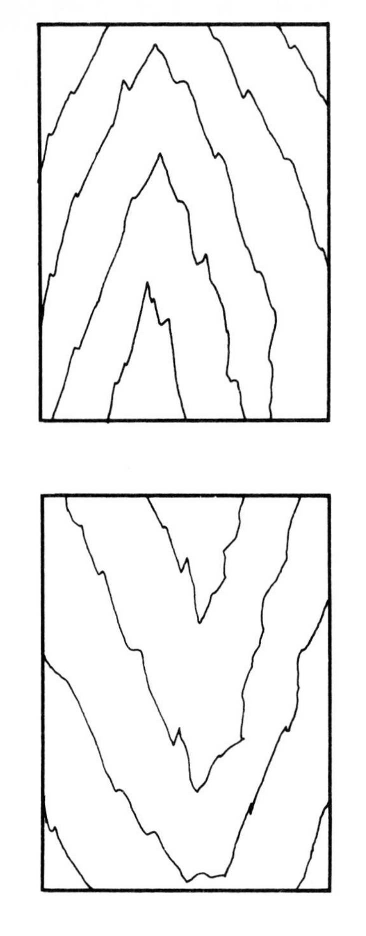

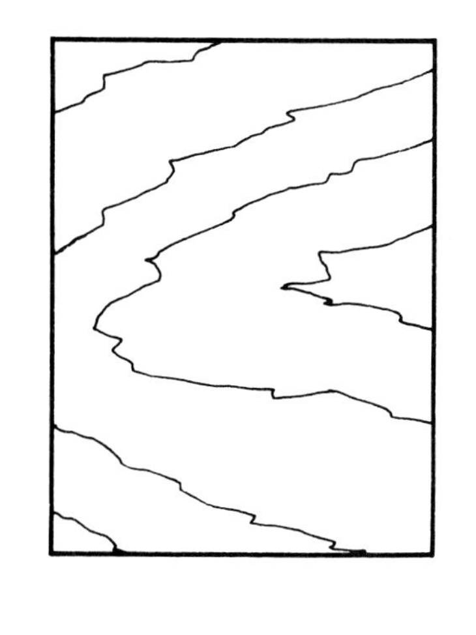

FIGURE 22-20. Wood grain.

22.7 RESEARCH

1. This chapter described the development and use of a layout board to create a cutting list of materials. How would the cutting list be used to prepare a materials list?
2. What format would you recommend for a materials list?
3. Prepare a materials list including hardware for the kitchenette cabinets partially described in this chapter.
4. Obtain pricing from local supplies for those materials.
5. Calculate the per unit length cost of materials.
6. Estimate the time it would take for you to plan, build, finish, and install the set of kitchenette cabinets. What hourly rate would you expect?
7. What would be a reasonable overhead expense charge for the building and installation of these cabinets?
8. After adding the estimated costs of planning, building, finishing and installation, calculate a per unit length cost. How does it compare with the current rate in your area?
9. Examine the construction of a small portable piece of furniture. Sketch the joint details used and then develop a layout board which would carry all the measurements. Explain your layout to another person.

22.8 REFERENCES

22.8.1 Smith, Ronald C. *Principles and Practices of Light Construction.* New Jersey: Prentice-Hall Inc., 1980.

23

BUILDING FINISHING

Every builder is relieved when a structure is in place with siding and roofing applied. Even with temporary windows and doors and without stucco or brick facing, the shape of the building is evident in its rough setting, and the building can be made secure. Once secure, finishing work can proceed in good or bad weather. Services can be roughed-in and inspected, insulation and the air/vapor barrier placed, windows and doors installed, and interior finishing materials applied.

23.1 SPECIFIC OBJECTIVES

The objectives of this chapter are to

1. reinforce the concept that a building is a system composed of many integrated subsystems,
2. illustrate how doors and windows may effectively contribute to building functions,
3. emphasize the point that selections and applications of finishing materials contribute to building design,
4. reinforce the idea that workers in subsystems expect to have their activities coordinated, and
5. convince the reader that interior and exterior designs of buildings need coordination.

23.2 THE DEVELOPMENT OF THE BUILDING AS A SYSTEM

At this point in the building project all the building parts or subsystems are coming together to create a building system. Decisions made earlier in the project are beginning to become evident in the applications. The foundation, enclosed framework, and exterior sheathing and siding provide shelter for electrical, plumbing, heating and air-conditioning subsystems. Provision is being made for sound control, energy conservation, and acceptable humidity levels. Insulation is in place and interior walls are covered but probably not decorated.

Before beginning the finishing work, the builder should assess the progress of the project to assure that everything is in order. The following points should be evident.

The site is generally clean and tidy. The services are roughed-in and inspected. Electrical service entrance and panel are installed as is the wiring to all the receptacles for outlets, controls, and fixtures. Few, if any, electrical and plumbing fixtures have been installed except perhaps in the basement, if some services were required for the on-going work activities. If there is a basement, the concrete floor has been placed to cover the building drainage system. The heating/air-conditioning service may be functioning to control interior temperature levels.

The walls, ceilings, and subfloors are in place and clean. Depending on specific construction schedules, the door and window openings are there but frames, doors, and windows probably have not been installed. The finish floor is yet to be placed.

The subsystems in the building have been developed by persons, probably teams of persons, skilled and knowledgeable in their respective activities. The critical path chart indicates when they were on-site and how many teams were there at any given time.

23.3 DOORS

Dwellings have exterior and interior doors. Frames for exterior doors are complete in that a sill is normally included. Interior doors usually have headers and side jambs and no sills. Sliding doors and folding doors have, in addition to headers and side jambs, tracks or guiding hardware.

Garages, shops, and light industrial/commercial buildings have regular exterior exit doors and larger, heavier doors which allow the movement of equipment and freight. These doors may be overhead swing-up or roll-up or sliding. Each one requires a different building space for opening and closing. In addition to regular door control hardware, some doors may be remotely actuated.

23.3.1 The Hand of Hinged Doors

Before installing door jambs, particularly rabbeted ones, check the plans to see which way the door is to swing and on which jamb the hinges are to be set. In other words, observe the *hand* of the door. The possibilities are swing-in, swing-out, left, and right. The selection of the correct combination is most important with respect to the function of a building.

To study the hand of doors, a set of rules recognized in industry must be considered. One must first face the door from the outside, even for interior doors. One is *outside* when on

1. the street side of an exterior door,
2. the corridor side for doors leading from corridors to rooms,
3. the room side for doors from rooms to closets, and
4. the stop side (the side from which the hinge butts cannot be seen) for doors between rooms.

From a position outside a door the following conditions hold true.

1. If the hinges are on your right, it is a right-hand door.
2. If the hinges are on your left, it is a left-hand door.

3. If the door swings away from you, it is a regular door.
4. If the door swings toward you, it is a reverse door.

Doors should be installed in a consistent manner throughout the building. Rabbeted jambs randomly installed on a set of corridor doors would create confusion. Also, the header casings would be located at different levels.

23.3.2 Door Openings

Door openings must be wide enough to accommodate the door, the side jamb, the clearance space (at each side) of the side jamb, and the clearance space (at each side) between the jamb and the trimmer stud to allow plumbing of the jambs (Figure 23-1). Finish sizes of doors and jambs must be known before the layout is done on the sill plate for stud layout. Care must be taken when measuring the effective jamb thickness of rabbeted frames (Figure 23-2).

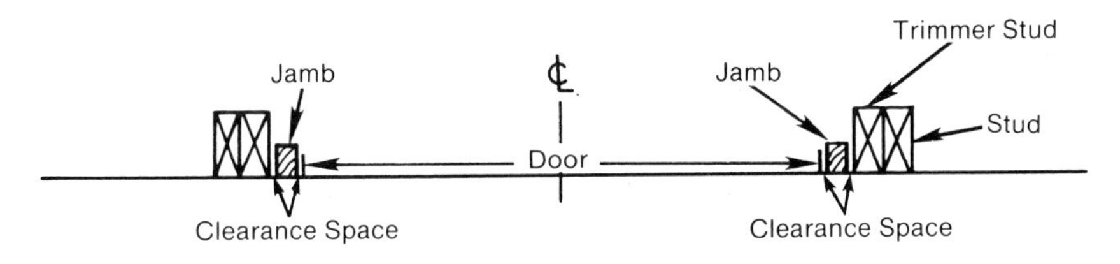

FIGURE 23-1. Sill plate layout for a door opening.

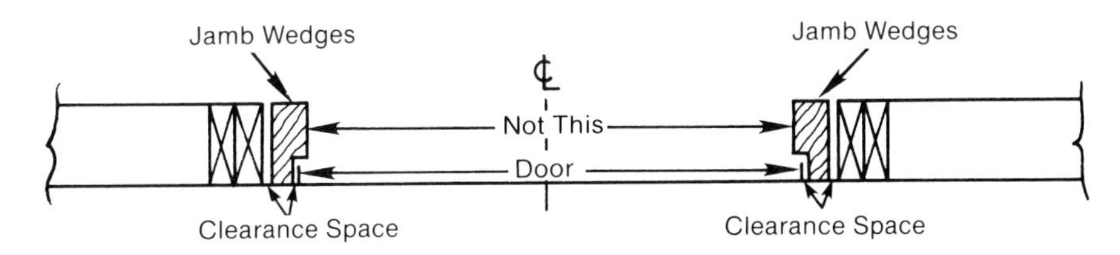

FIGURE 23-2. Measuring the effective jamb thickness.

Heights of door openings may be ticked-off as on a story rod. It is imperative that the person doing the layout and the person interpreting the layout follow the same set of assumptions. The following assumptions refer to the installed interior doorframe illustrated in Figure 23-3.

1. The sole plate has been removed from the opening far enough back to allow the trimmer stud to run to the subfloor.
2. The doorframe is rabbeted. (*Note:* Jambs for interior doors are often not rabbeted. Instead, a simple jamb with nailed-on stops is used.)
3. The side jambs rest on the subfloor.
4. Provisions are made for a carpet to serve as a finish floor. The height allowance is 3/4″ (19 mm).
5. Clearance between the jamb and door is set at 1/16″ (1.6 mm) at the top, 3/32″ (2 mm) at the lock side, and 1/32″ (1 mm) at the hinge side. The bevel on the lock edge is set at 5°.
6. Clearance between the jamb and framing is set at 3/8″ (9 mm) at the top between the jamb header and framing header and 3/8″ (9 mm) at sides between jambs and trimmer studs for double shingle wedges.

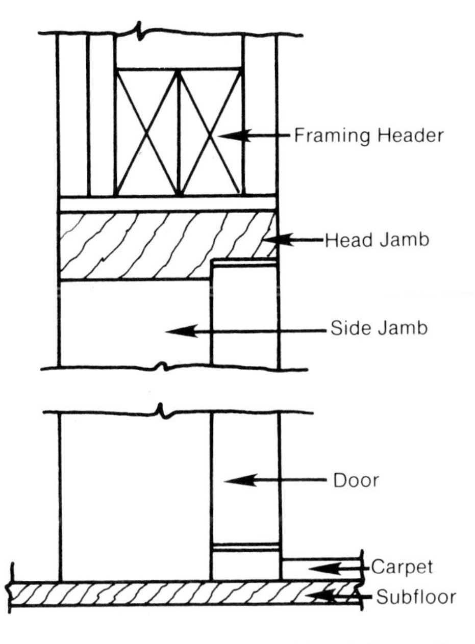

FIGURE 23-3. Rabbeted jamb for an interior door.

7. A story rod standing on a 3/4″ (19 mm) piece of wood will reach above the cap plate and accommodate without problems the door (without cutting), doorframe framing headers, and top/cap plates. This method can be used as a quick check of the doorframe opening.

Figure 23-3 and similar drawings of other doors are helpful but lack definitive information for installers. Doors of standard sizes need not be cut if overall planning is done. Figure 23-3 does not show door height, door opening height, or framed wall height. The drawing is done out of context and creates problems for persons learning framing/installation.

To be certain of member relationships, it is useful for a learner to draw a complete section full-scale, as for a cabinet drawing/story rod (Figures 23-4 and 23-5). The story rod can be drawn correctly only by persons who are sure of actual measurements and member relationships. The recording method is simple and direct and capable of being double-checked before cutting expensive materials of wood or metal. This layout principle is also applicable to any door or window installation.

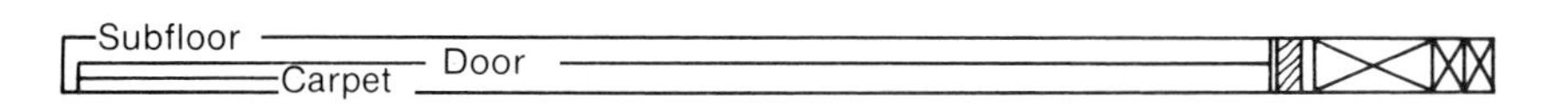

FIGURE 23-4. Sectional view of an interior door/frame/wall arrangement.

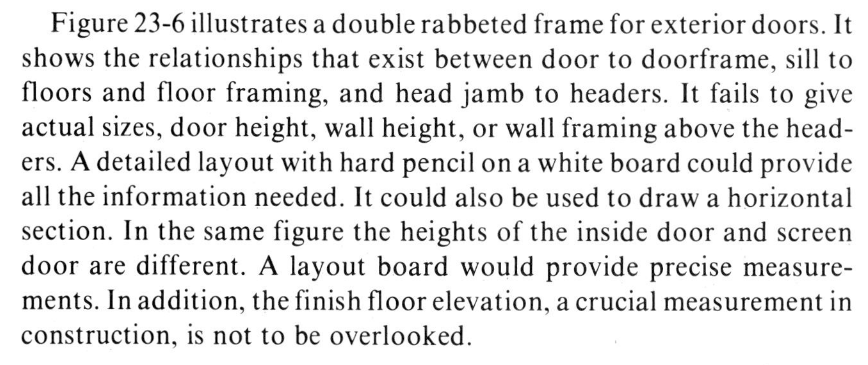

FIGURE 23-5. Figure 23-4 as represented on a simple story rod.

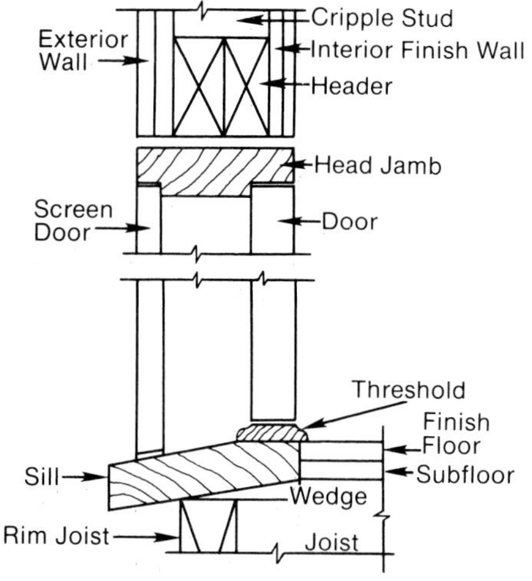

FIGURE 23-6. Double rabbeted frame for an exterior door.

Figure 23-6 illustrates a double rabbeted frame for exterior doors. It shows the relationships that exist between door to doorframe, sill to floors and floor framing, and head jamb to headers. It fails to give actual sizes, door height, wall height, or wall framing above the headers. A detailed layout with hard pencil on a white board could provide all the information needed. It could also be used to draw a horizontal section. In the same figure the heights of the inside door and screen door are different. A layout board would provide precise measurements. In addition, the finish floor elevation, a crucial measurement in construction, is not to be overlooked.

23.3.3 Door Hardware

Hinges of the butt design, rather than the surface design which can easily be removed, are usually used for security reasons. Butt hinges are of different sizes and weights, and two or three are fitted to a door. They are set into matching gains in the door and jamb. The pitch/slope of the gain is important to the way the door swings. The optimum placement of hinges has, by convention, been "7 and 11." The rule has been 7″ down from the top of the door and 11″ up from the bottom of the door. Metric measurements are becoming 178 mm and 279 mm. A third hinge would be centered.

Cylinder locks set 36″ (914 mm) above the subfloor are common. There are many different designs and keying systems. As security becomes a priority, locks become sophisticated.

When a completed building is a commerical or public building, keying levels are important. The levels generally follow the organizational pattern of persons who use the building. A set of keys allows the holder of a master key access to all rooms. A submaster would allow limited area access, and a low level key would permit entry to one room only. Keying/lock systems contribute to vocations for responsible persons working in the security field.

23.4 WINDOWS

For energy efficient buildings, windows that are designed to open must also fit tightly to prevent air leaks. Preferred designs in this respect are the casement, awning, and hopper units roughly described in Figure 23-7. Sliding windows tend to be poor performers since they must be loose to slide. Live loads in buildings and changes in temperature levels also cause movements and expansion or contraction of metal trim and gaskets which break the seal. Design of fixed windows requires careful consideration. Table 23-1 indicates insulation values for four designs.

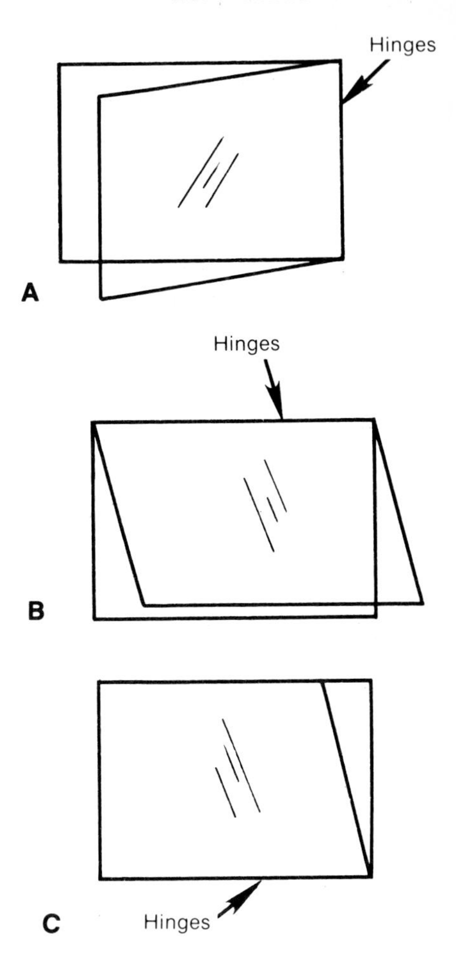

FIGURE 23-7. Types of window designs: (A) casement, (B) awning, and (C) hopper.

TABLE 23-1 INSULATION VALUES FOR FOUR FIXED WINDOW DESIGNS

Frame	Window Description	R-value	RSI Value
Wooden	Single glass	1.8	0.317
Wooden	Double glass	2.6	0.458
Wooden	Triple glass	3.7	0.652
Wooden	Double glass with insulated shutter	12.7	2.237

The placement of windows in relation to the orientation of the building needs consideration, especially in colder climates. In northern areas an unbroken insulated north-facing wall helps create an energy efficient building.

To keep building lines simple, designers may place window and door headers at the same level. The design also simplifies construction. To accomplish this on the job, story rods are compared. Lengths of cripple and trimmer studs can then be identified. From such information setting sticks can be cut to length to ensure proper header framing and elevations of window frames. Such sticks standing on a subfloor can prop a frame in temporary position for final plumbing and leveling.

Figure 23-8A illustrates a board which compares the story rod layout for a door and a window set in the same wall and at the same header level. (Trimmer stud lengths could be added.) A typical window

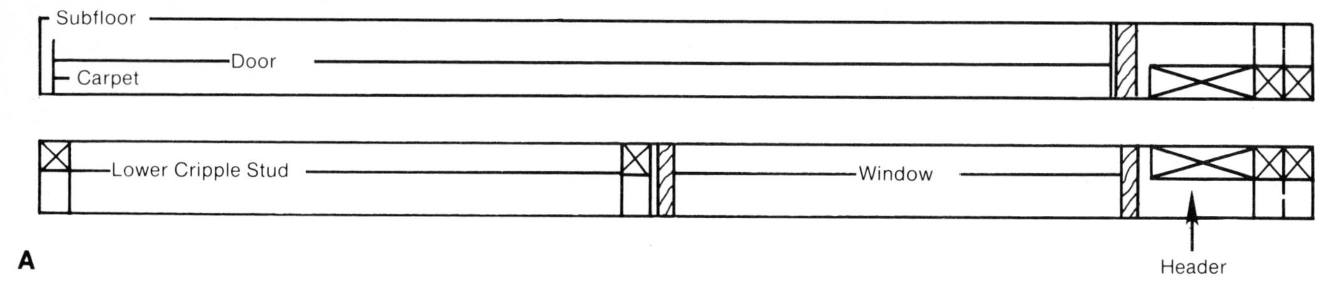

FIGURE 23-8. (A) Story rods for a door and window set in the same wall at the same header level and (B) sectional view of the window installation. Note how the angle window sill is not represented on the story rod.

installation is given in Figure 23-8B and shows a sloping sill as opposed to the sill shown on the story rod.

Whether a builder is making windows or ordering them for installation, care must be taken in stating measurements. Figure 23-9 illustrates this need and the consequences of misinterpretation. If the long side measures X and the short side Y, then pane A measures X by Y. Pane B measures Y by X. In other words the horizontal measurement is given first and the height second. The same measurement rule applies generally to windows. It would be a costly mistake to order incorrectly or to make a set from an incorrect order. Window openings in walls are not changed to accommodate such errors.

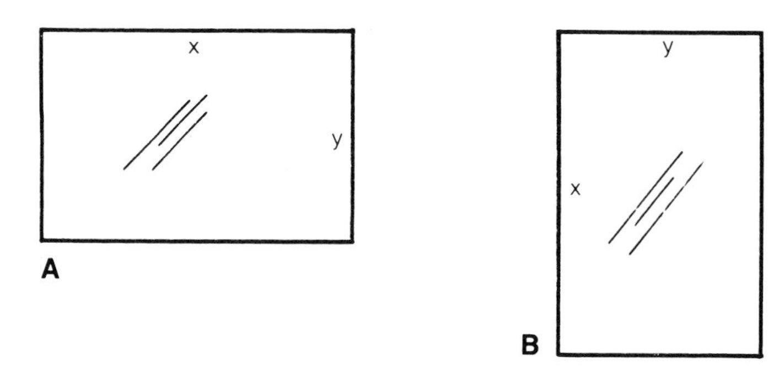

FIGURE 23-9. These two identically sized panes of glass can be used in windows of totally different sizes, depending upon which dimension is specified first. When ordering, the horizontal measurement should always be given first, the vertical measurement second.

23.5 FLOORS

A high quality wooden subfloor can be made with plywood if the joints are tongue-and-groove and the assembly of plywood and joists is fastened with elastomeric glue and annular grooved nails. With a good nailing pattern such a subfloor becomes a composite stiff T-beam unit which helps prevent deflection between joints.

Before installing the finish floor, the subfloor needs careful cleaning and checking. Loose nails, dust, and pieces of gypsum or concrete must be removed. All butt joints require fastening with blocking underneath to prevent movement between panels or boards. Floor squeaks must be avoided, even when temperature and humidity levels change. Should the subfloor be seen to be less than minimum thickness, it may be necessary to apply an additional underlay at some additional expense to the owner. Table 23-2 provides typical data with respect to minimum thicknesses of subfloors.

TABLE 23-2 MINIMUM THICKNESS OF SUBFLOORS

Maximum Joist Spacing		Minimum Plywood Thickness		Minimum Particleboard Thickness	
Inches	mm	Inches	mm	Inches	mm
16	406	5/8	16	3/4	19
20	508	3/4	19	1	25
24	610	3/4	19	1	25

Further examination of the subfloor is necessary to determine whether all openings have been cut and whether any changes must be made. Nothing of this sort should happen after the finish floor is placed. Openings for air ducts, utility lines, and bathroom and kitchen fixtures need positive identification and size checks. All roughed-in services must be ready for installation of fixtures and appliances.

Should the floor be of concrete, the surface should be checked for spalls and rough surfaces. Patching may be necessary. The surfaces should be clean and free of oil, grease, and dust. Clean surfaces assure good bonding of the finish floor materials. Additional openings are not easily made in concrete floors. All checking should have been done before the pouring. Each subsystem should have been traced to assure continuity.

23.5.1 Finish Flooring

Any material used as a final wearing surface is the finish flooring. The material is chosen for its aesthetic and utilitarian appeal, its durability, its ease of cleaning, and its fire hazard rating.

Woods are used in strip or parquet form (Figure 23-10). Color, grain, and ability to positively respond to nontoxic finishes are some selection criteria. Care of the materials while on the jobsite includes moisture control and prevention of marking and staining by water, oil and grease. Accidental staining of woods may necessitate removing and discarding expensive pieces. Thus, the critical path of construction activities would put any damp work (concrete, brick, or plaster) well ahead of wood floor laying. Laying patterns, nailing methods, sanding operations, and finishing are important details.

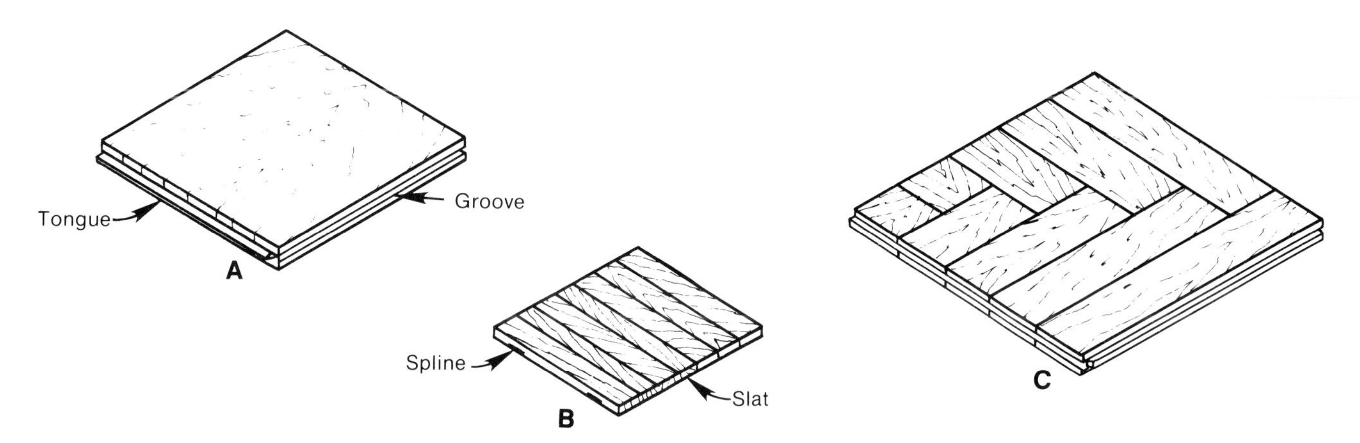

FIGURE 23-10. Types of wood flooring: (A) tongue-and-groove blocks, (B) splined square edged, and (C) parquet flooring.

Parquet flooring, a wood-strip square tile flooring, is usually placed in a checkerboard pattern according to manufacturer's instructions. With two edges grooved and the other two tongued, parquet flooring is effective as a finish flooring. Objectives when installing strip or parquet flooring are to match color and grain, have no hammer marks or stains, and have no nails showing. Nailing is at 45° through a pilot hole drilled in the tongue (Fig. 23-11A). Face nailing at the perimeter of rooms may be covered with base and shoe moldings (Figure 23-11B).

Resilient flooring in tile or sheet form for kitchens, bathrooms, workrooms, craftrooms, and laundry rooms requires an underlay over uneven subfloors. Plywood, particleboard or hard-pressed fiberboard nailed every 6″ (152 mm) at the edges and every 8″ (203 mm) within, provides an unbroken surface if the joints are filled with nonshrinking filler compound and sanded flush. Resilient flooring (linoleum, asphalt tile, vinyl, and rubber) is cemented to underlay with waterproof adhesive. Manufacturer's instructions usually include rolling both ways and then sealing the surface with a recommended floor wax. If placed directly on a concrete slab supported by the ground, resilient flooring is fastened with a waterproof adhesive according to manufacturer's instructions.

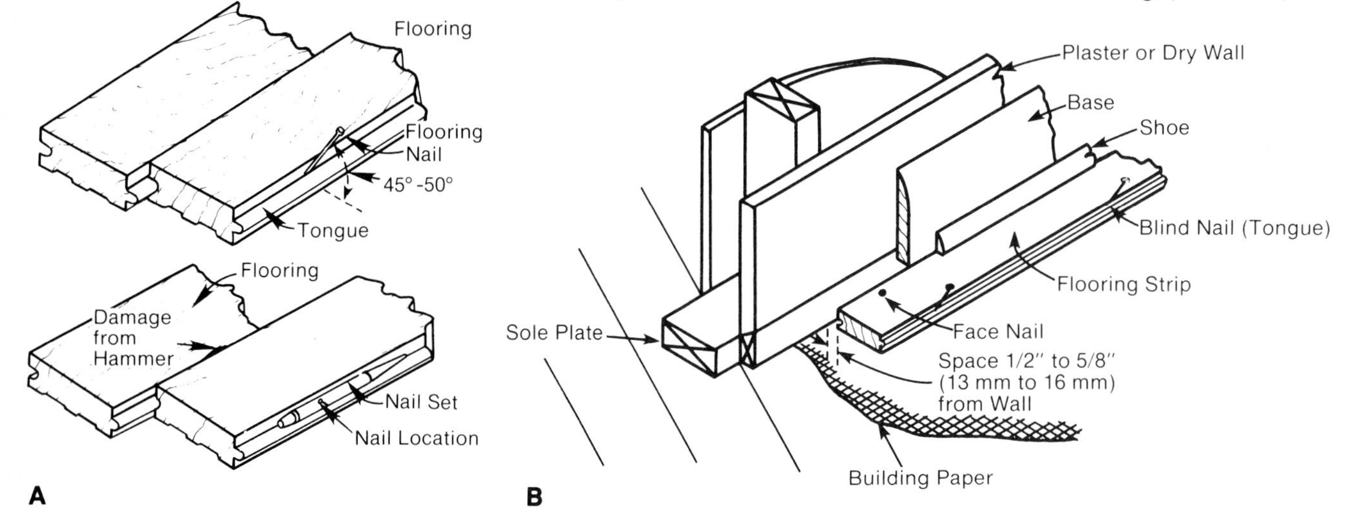

FIGURE 23-11. (A) Nailing and setting the nail in place. (B) Face nailed edge pieces covered with base and shoe molding.

Carpeting can also be used as finish flooring. Carpeting, however is not recommended where there is the possibility of it getting wet, such as in a bathroom. Kitchens generally are not carpeted because of the likelihood of frequent staining from spills. Where there is risk of water damage or staining, synthetic fiber types are used. In other building areas synthetic or natural fibers provide for sound absorption and for feelings of warmth. Carpet colors and textures are important to the decorator who might work from such data to schedules for interior and exterior work, to draperies, and to fixtures and furnishings. In most cases, carpetlayers handle all phases of carpet installation. Care should be taken to make sure that the subfloor is suitably flat and is free of imperfections so that the carpet will stay secure and wear evenly.

Some wood or tile floor materials can be installed without special techniques or equipment, giving excellent results even when the installer has minimal flooring experience. Careful estimating and strict adherance to manufacturer's installation instructions will assure proper results at minimum cost.

23.5.2 Estimating Materials

After selecting the desired carpet material, resilient flooring, wood-strip flooring, or parquet flooring, an estimate must be made of the quantity required to cover the given area. Accuracy is essential to avoid waste.

Carpets and resilient flooring are ordered in standard widths so that they can be installed with a minimum of cutting and sizing work. Attention must be given to sheen and direction of fibers and the matching of carpet patterns. Those performing the installation must know how the estimate was made to avoid costly mistakes. In the same regard, persons making the materials estimate must have knowledge of carpet laying techniques. Standard estimating forms and diagrams aid in clear communication between personnel.

To insure a pleasing wood floor or tile design, materials should be inspected with regard to grain and patterns before installation begins. Working up a small sample layout is also an excellent idea. Off-square rooms can cause installation problems, so careful sizing of border tiles is important. In many rooms with suspended tile ceilings and floors, it is important to balance the ceiling and floor patterns. As shown in Figure 23-12, establishing center lines to work from helps assure overall balance in design.

Cut pieces of wood or ceramic tile should be kept for use at room edges and around problem areas. Of course, grain or pattern matching of cut pieces is essential.

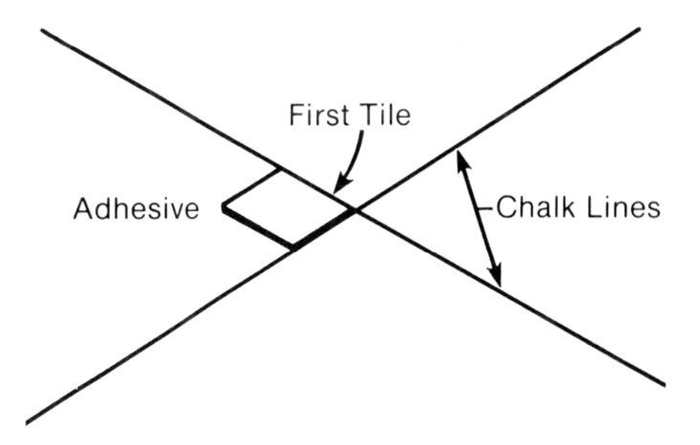

FIGURE 23-12. Center lines can be used to assure an even tile layout.

23.5.3 Baseboards

Design problems exist where finish flooring meets walls and, therefore, questions arise. Does carpet and resilient flooring stop at the walls without a baseboard? If there is to be a baseboard, will the material be the same as the flooring or will it be rubber, ceramic, or wood? Is the baseboard to be like the wall or the floor? What is the purpose of a baseboard? Does the floor seem to go up the wall, does the wall just meet the floor, or is the baseboard a dividing line between floor and wall?

Baseboard is expensive to install and to finish but is effective for covering gaps between floor and wall and for protecting walls, particularly at the corners. Installation needs planning and knowledge of the cutting techniques.

A useful technique for cutting and coping inside joints at 90° angles or other angles is to butt one piece to the wall and then miter-cut and cope on the miter line to make the second piece fit over the first (Figure 23-13). To do this

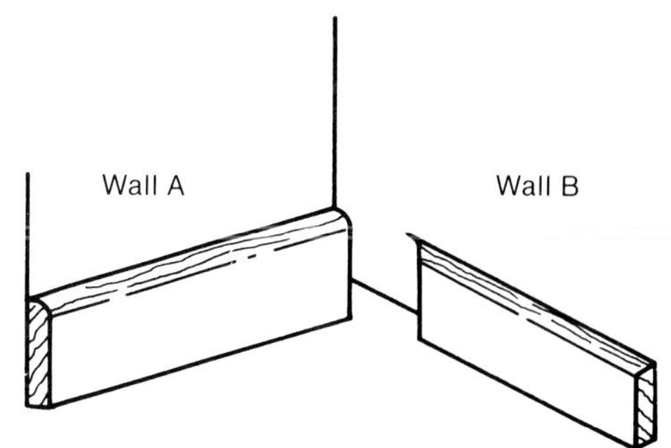

FIGURE 23-13. Fitting an inside corner molding baseboard joint.

1. butt piece #1 against wall B,
2. cut piece #2 on a miter of 45°, and
3. using a coping saw, follow this miter line and cut away excess material within the miter to match the shape of piece #1. Piece #2 should fit over piece #1 without an open or false joint.

This technique can be used to form the inside corners for baseboard of any shape, although accurately shaping the coping cut may be more difficult on smaller pieces. Outside corners should always be mitered to form a flush joint.

23.6 PLUMBING, HEATING, AND WIRING

Each of these services requires a distribution system which must be accommodated by the structure. Without careful planning, conflicts arise among workers as to where and how the systems will run. Since they are discrete systems but interdependent, the planners must know and appreciate the objectives of each and how all of them are to be integrated within a given building.

Furthermore, it benefits the planner to work with those subcontractors who can install compatible systems and who individually appreciate the limitations of a particular structural design. It is better to work with problem solvers than with those whose main objectives are to install and go—leaving joists and studs improperly notched or drilled, in a weakened structure, or leaving other systems working under disadvantaged conditions.

Each person, the planner and the subcontractor, must accept and improve upon the local codes. Such behaviors tend to bring benefits to a community in the form of better living conditions in better buildings.

23.6.1 Plumbing

The following questions must be answered before the framing for the plumbing system is put into place.

1. Will the walls be thick enough to carry piping? Is double wall framing needed?
2. What materials (copper, plastic or cast iron) will be used in vent pipes? What will be the outside diameter (O.D.) sizes?
3. Will water lines be kept away from cold outside walls?
4. Will gravity be allowed to help with drainage? Do any of the drainage lines run up hill? Can they be repaired if necessary?
5. Is the venting from each fixture secured?
6. Are there provision for water supply shutdown—generally and at each fixture—so that repairs/maintenance can be done?
7. Are there provision for cleanout and for easy maintenance?
8. Will the piping be permanently enclosed or will there be access to it for maintenance or freeze-up repair?
9. If notching and drilling of framing members will be done, what are the limits? How are weakened members to be strengthened to bring the structure back to design level?
10. Will the floor framing allow piping installations and yet carry building loads (Figure 23-14)?
11. Fixtures and cabinets are obtainable in different colors and materials. What level of sophistication is expected by the owner (whether individual or corporate)?

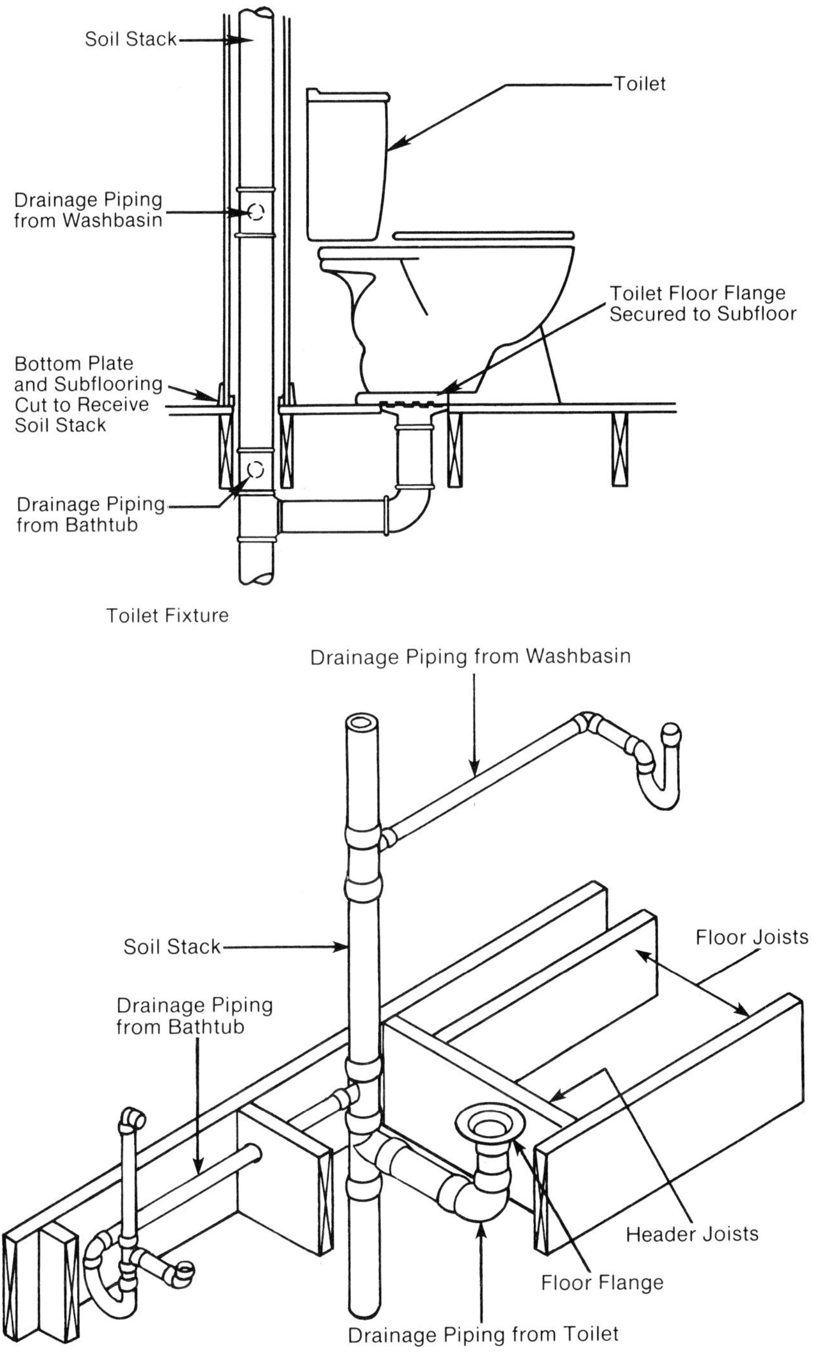

FIGURE 23-14. Typical example of framing provisions made for a bathroom drainage system.

23.6.2 Heating

A building heating system must be sized to match expected conditions. Heating needs for an energy efficient building are much less than for a traditional one. The insulation values for all parts of a building must be known.

In colder climates auxiliary heating units—units using different fuels or different controls—may be installed. Such subsystems may be quite different from the main one. For example, many furnaces using natural gas as fuel require a low voltage electrical supply to function, i.e., to operate controls, and a full electrical supply to operate a fan motor which moves the hot air through the piping. Other natural gas heaters which are space heaters without a fan or blower do not require electrical power and would, therefore, function when the electrical power supply is shut down. Two such furnaces complement one another to offer a high degree of probability that one or the other would provide heat to the building.

Other dual arrangements might be

1. natural gas furnace and wood burning fireplace,
2. propane fuel and wood burning fireplace,
3. oil burning furnace and wood burning fireplace,
4. electric baseboard heating and wood burning fireplace, or
5. natural gas or propane or oil burning furnace and some electric baseboard heating.

Dual or triple discreet systems mean careful planning. They are expensive in terms of time and the cost of installations. Costs and benefits need assessment since multiple systems require more building space. Building codes will probably indicate that only one type of fuel, except coal or wood, may be burned in a given building.

Propane and natural gas furnaces are different from one another in terms of supply and controls. If one is in a building, the other is probably not. Extensive electrical baseboard heating would probably not be found in a building heated by natural gas. Some electrical heating, however, might complement propane, natural gas, or solar heating.

As a planner one must be actively aware of the economics of building heating and know the relative costs of fuels, equipment, installations, and space requirements. Safety factors must be known and judgments made on a practical basis as well as for aesthetic reasons.

A heating system has to be compatible with the framing/structural system. One-story and multiple-story buildings present different heat distribution problems for warm air systems. Hot water systems may more easily be installed while recognizing that supervision needs may increase in colder climates where a shutdown would cause breaking pipes and expensive repairs. Area space heaters help overcome the need for long piping runs and offer area controls, but common areas of the same building require attention.

Heating equipment location is a design problem. Natural gas, propane and oil burning furnaces may stand on a floor or may be suspended from a ceiling. Coal and wood burning fireplaces and stoves or heaters are built in or free-standing. Electric baseboard heaters form decorative baseboards and are relatively easy to install because they are surface mounted without framework requirements except for concealed wiring.

Figure 23-15 shows a typical heating layout with respect to an oil burning furnace in a one-story building.

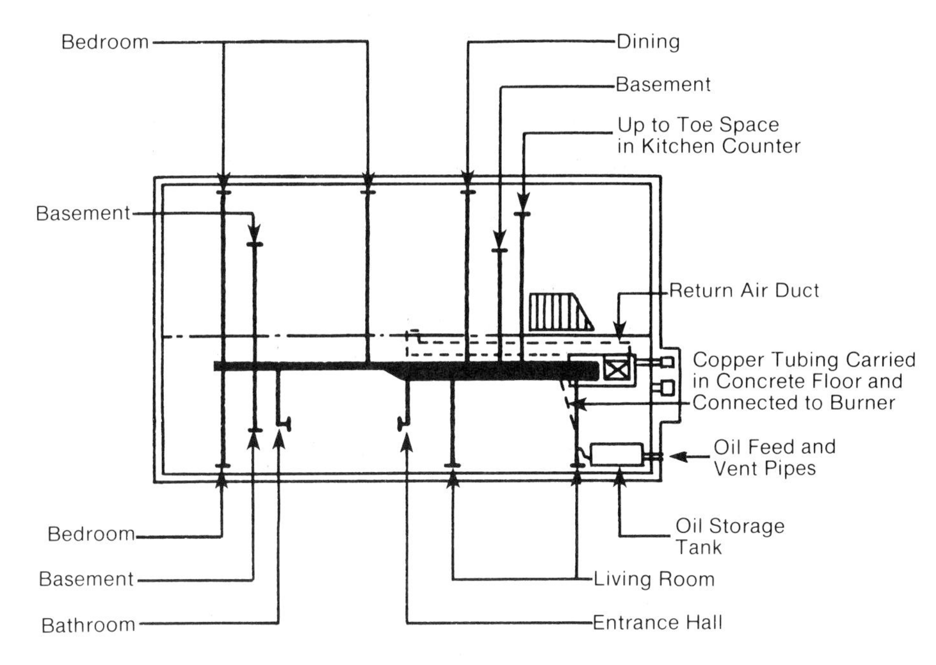

FIGURE 23-15. Basement plan of a typical heating layout.

Figure 23-16 describes the construction and insulation of a furnace room in a basement of an energy efficient building.

Furnace/heating unit designs have changed remarkably over the past 25 years, with trends towards automatic shutdowns in case of malfunction, simplified controls, and smaller sizes. Mechanical features such as fan shaft bearings and drive belts continue to wear out and to cause noise problems.

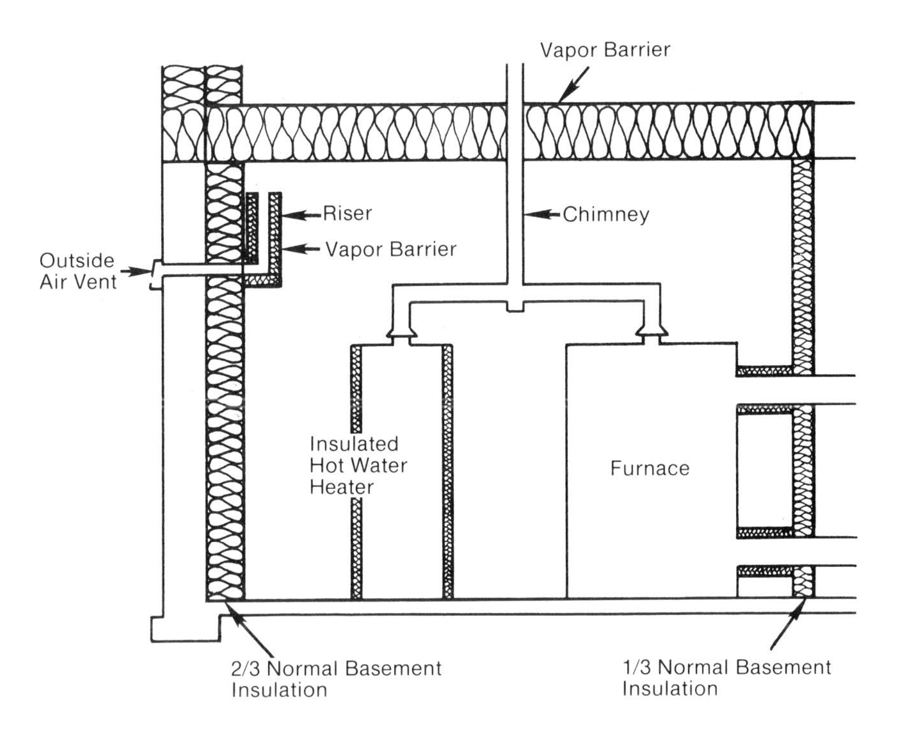

FIGURE 23-16. A furnace room layout which isolates the fuel burning equipment from the remainder of the building.

23.6.3 Wiring

Electrical power is transmitted over long distances at high voltages and for local consumption is transformed more than once to about 120 to 240 volts. Power electricians can safely bring the power lines to a building. Construction electricians create the wiring harness for a building—a wiring system. The power supply authority, private or public, wants to know on an on-going basis how much power is being delivered. The building owner pays accordingly. Measurements recognized by both parties are read on a meter at the service entrance. The power electrician and the construction electrician, therefore, have professional interests in the design of the service entrance. Figure 23-17 illustrates a typical arrangement.

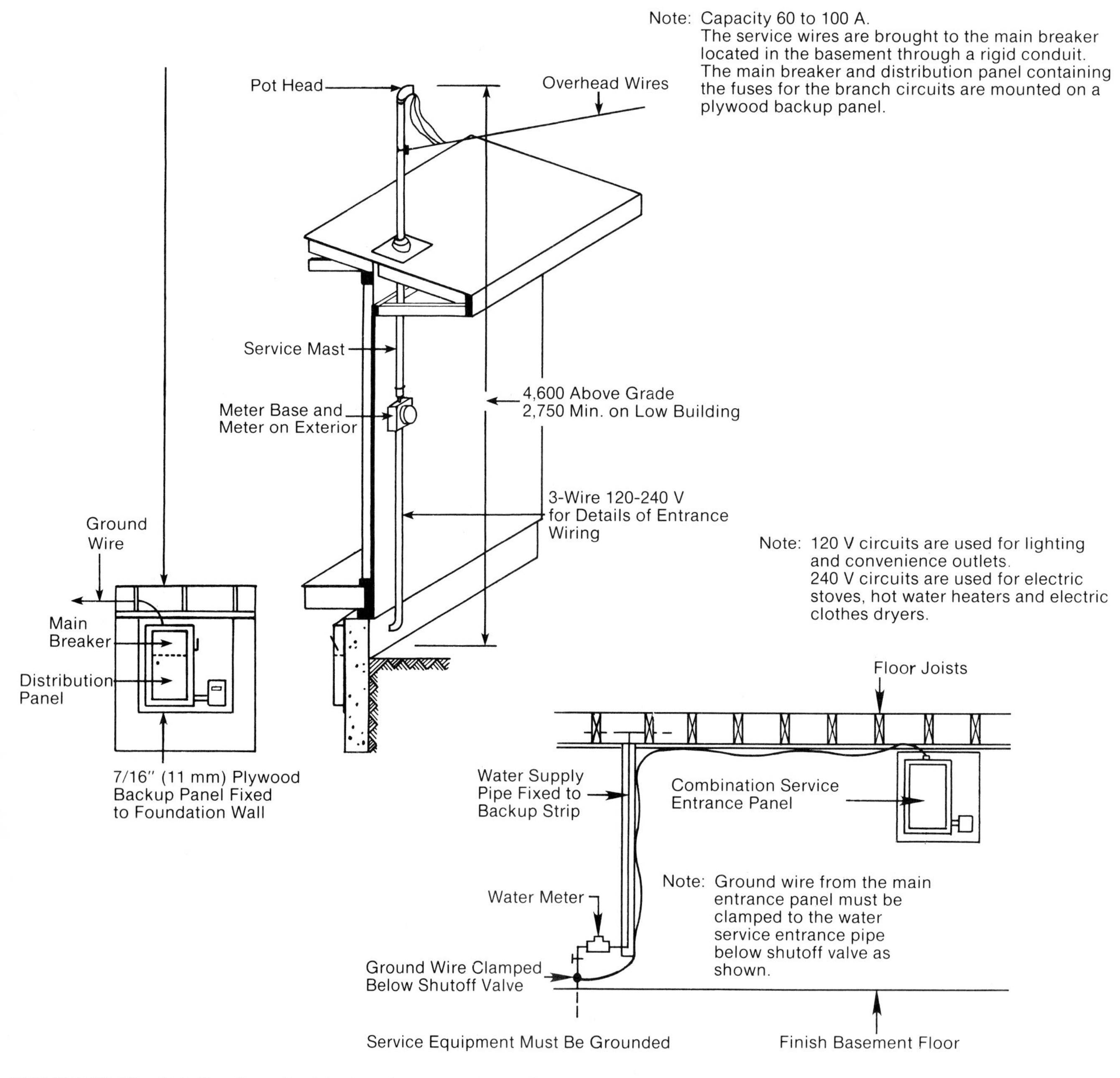

FIGURE 23-17. Details of an electrical systems service entrance.

The construction electrician establishes electrical controls with the installation of a main breaker and a distribution panel. A surge of electrical power, from a lightning stroke for example, will trip the main breaker and give protection to the building system. Attempted overloading of a circuit from within the building is prevented by breakers or fuses in the distribution panel. Distribution panels are large enough to accommodate the required circuits and a limited number of additional ones. In other words, the electrical system needs to be in balance with the rest of the building project. Figure 23-17 also illustrates the relationship of the power supply to the meter, the main breaker, the distribution panel, and the ground wire clamped to the water supply pipe.

The design and installation of the entire wiring system are usually regulated by state or provincial codes. System inspections at the roughing-in and finish levels of installation ensure compliance with respect to regulations.

Installations may be technically correct and may or may not be in worker-like condition. Visual standards are set by the designer, the builder, and/or the owner and can help make or break a project. To set standards one must have definite ideas or sets of criteria for measurement such as boxes, receptacles, and plates being set plumb and level and at consistently convenient positions. Circuits will have to be uniformly loaded to prevent easy and annoying shutdown. Fixtures must be properly and cleanly mounted and installed to function. Work done through other building activities such as framing, painting, or decorating will be aided rather than adversely affected.

There are times when installation options are offered and the owner makes decisions. At one time, for example, aluminum wiring could be substituted for copper. Subsequent experience seems to indicate that aluminum wiring is less than adequate because the oxidation of the wires at joints creates overheating. It is expensive to change a wiring harness at any time, whether at the roughing-in stage or after finishing.

Wiring for multiple control switches, dimmer switches, telephone jacks, controls for use by handicapped persons, security systems, and television/sound system/computer requirements all need assessment and then installation at the roughing-in stage, not later.

23.7 FINISHING HARDWARE AND INSTALLATION TOOLS

Finishing hardware includes nails, screws, and cabinet and door hardware. When nails are used, they are usually hidden from view by setting or by covering, such as with a shoe mold. Set nails in wood are covered with an opaque material such as putty, which is made to match the surface finish. Exposed screws are of decorative design, of one material, and of one head pattern. Door hardware may, of necessity, have oval and flathead screws or bolts, but generally it is a nuisance to have a variety of patterns. Exposed hardware on doors and cabinets is a dominant decorative feature to be properly fitted and left in a clean state (not partially painted and scratched with screw heads burred).

Installation of hardware is done by responsible persons who appreciate properly matching doors, straight and level door lines, functioning doors and drawers, and generally a simplicity of placement.

Doors must open, close, and lock. Exterior doors must additionally create a seal from a proper fit. Cabinet doors must swing without interfering with one another, and drawers must glide smoothly without wearing and without touching the framework. Well-made drawers are often slightly tapered to improve the fit of the drawer in its opening.

Hardware must be of sufficient size and strength to function safely. Fasteners and the materials in which they are set need to be compatible over a long-term, not a short-term, period. Door hardware is sized to match the weight and thickness of the door and jamb.

Bathroom hardware such as towel racks and grab bars should be securely installed in relation to the framework, rather than with temporary fasteners in gyproc. False settings are hazardous to persons who unknowingly will use them and trust them only to have them fail.

When hardware is fastened to concrete or concrete blocks, spalling of the concrete needs to be avoided. Spalling produces an unsightly wall and diminishes the holding power a fastener might have. Plywood backing boards are easily pinned to concrete with power actuated tools. Fixtures can then be hung from the plywood.

23.8 PREPARATION FOR FINISHING

Knowledge of materials and some good planning will help avoid finishing problems. Nails and screws which will rust should be used in dry places, not where there is moisture. Gyproc needs a well-framed wall or ceiling on which to be nailed to prevent nail popping and then a second sheet without nails (cemented to the first) to present an unbroken surface for final finishing.

Finish wood usually has planer marks removed before sanding with the grain. Stair newels, handrails, and guards need to be sliver-free in readiness for final finishing. Wood floors, especially strip flooring, requires operation of sanding machines to produce a flat surface, unbroken by sander gouges. First cuts with a power sander are across the grain or at 45° to the grain of the wood, and successive cuts are with the grain, using finer and finer sanding paper. Edge sanding needs to be done with extreme care.

When all sanding is done, whether on-the-job or in-shop, dust needs to be removed so that applied wood finishes will be smooth to the touch and visually pleasing. Dust in a finish produces a rough surface.

Before finishing commences, the building needs to be clean: sanding dust, gyproc dust, and street dust removed; papers and other waste cleared; and all flammable materials taken out of the building.

Prepared surfaces need to be finished immediately before they can become soiled again. Conditions of humidity and temperature need strict control. Wood should be seasoned (not just surface tested) to a maximum moisture content of 12% to 14%. Testing meters can indicate the measurements directly.

Humidity of the building interior as well as the wood itself needs attention. Condensation on interior walls can ruin interior finishes with water runs. Without a vapor barrier in walls and ceilings, water and/or ice can build up within walls and ceilings and cause staining of finished surfaces. A building without a vapor barrier is better not built!

Before actually starting a finishing procedure application, equipment and finish materials are assembled and cleanup methods determined. Water-based finishes are cleaned away with water, and oil-based finishes are removed from equipment with the recommended solvent. Obey regulations with respect to disposal of cleanup materials.

Ventilation is required to protect the health of the applicator. Everyday exposures to volatile finish materials are often hazardous. Respirators may be needed. Consult local health authorities and safety councils. Read the material labels carefully. Have fire fighting equipment at hand.

To observe how strict finishing rules may be, visit an automotive spray booth or a furniture finishing establishment. Notice how overspray is handled. Ask how direction of airflow and a high pressure condition within a booth keep unwanted dust from entering.

Be sure that the specified finish material is actually on hand in sufficient quantities to complete the job. Check on the solvent, color, drying time, and work to be done between coats. Read the warranty label.

23.9 INTERIOR FINISHING

Unlike exterior finishes, most interior finishes are not exposed to large or sudden temperature and humidity fluctuations. They are sheltered from wind, dust, hail, snow, rain, and ice. They do not need to withstand as many spills, scuffs, and scratches. However, floors, room doors, and cabinet doors are vulnerable, and over time other areas accumulate residue from fumes, smoke, and dust.

The finishes protect the base materials and are usually renewable. On wood surfaces they add depth, warmth, and character with a clear finish or slightly pigmented finish. They add color with paints or stains. On gyproc or other panel boards the finishes may provide color and texture and design. Paints and papers produce pleasing surfaces to complement finishes in other parts of a room. On concrete they add color, seal pores, and somewhat change the acoustics of the room. Most finishes make materials cleanable or possibly washable.

23.9.1 Composition of Finishes for Woods

Pigmented finishes, or paints, are composed of two basic materials—the pigment and the vehicle. Clear finishes usually contain only the vehicle. Prime pigments are very fine powders that provide color and hide the underlying surface. The vehicle, or liquid portion, generally consists of two parts, one nonvolatile and the other volatile. The nonvolatile vehicle forms the film and is also called the binder because it binds or holds the pigment to the surface. The volatile part, called the solvent, dissolves the film-forming material and is used to adjust the viscosity of the mixture so that it can be easily applied (Figure 23-18).

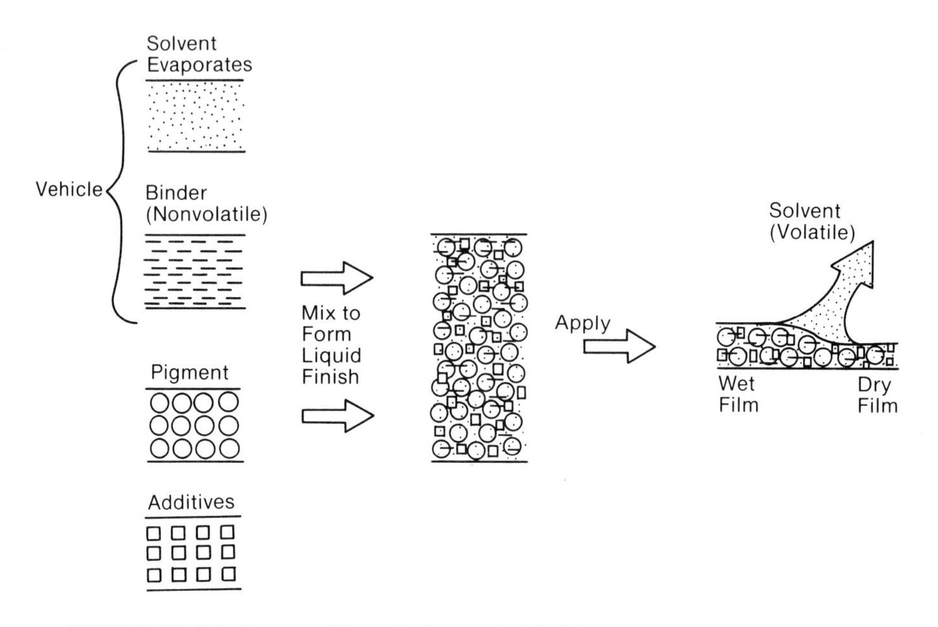

FIGURE 23-18. Composition of pigmented finishes.

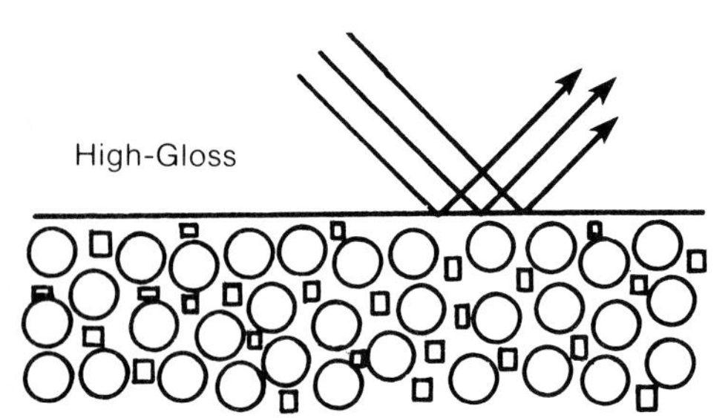

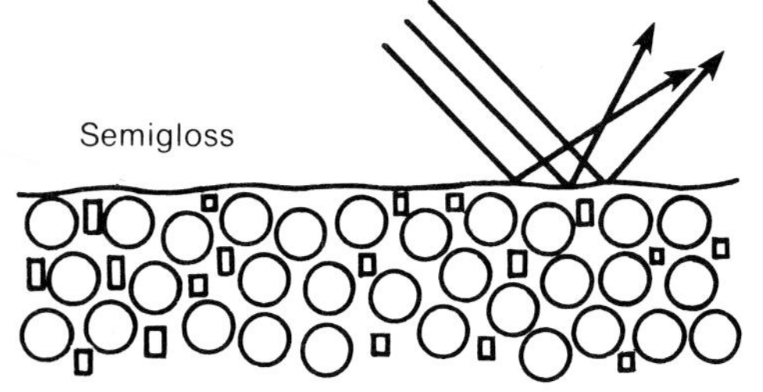

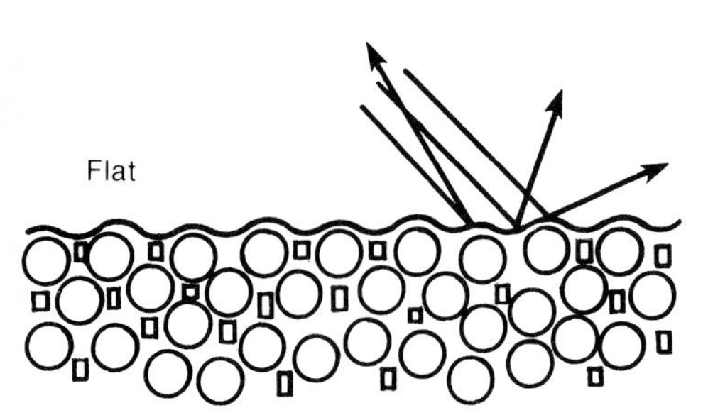

FIGURE 23-19. Effect of pigment content on finish gloss.

Pigment concentration is varied to produce the desired degree of gloss. A low pigment-to-vehicle ratio produces a high gloss surface (Figure 23-19). As the proportion of pigment increases, the gloss decreases. The three basic categories are high-gloss, semi-gloss, and flat, but manufacturers use their own terms to describe various gloss ranges (e.g., eggshell).

The two categories of finishes are *solvent-based* materials that contain or are soluble in organic solvents and *water-based* materials that dissolve in or are dispersed in water.

Solvent-based Finishes. Oils, oil paints, varnishes, enamels, and lacquers are generally classed as solvent-based finishes. The simplest types are drying oils and their pigmented version, paints. Linseed oil is the most common of this type and, if unpigmented, usually requires no solvent. Solvent may be added to increase penetration. Except for a few special cases, oils and oil paints have been replaced over the last 30 years for most interior and exterior uses.

Varnishes are clear finishes made from a combination of drying oils and synthetic resin cooked together (earlier varnishes, made from oil and natural resin, were called oleoresinous). Solvent is added to produce the desired viscosity, and the film dries by reaction with moisture in the air (polymerization).

Enamels are pigments dispersed in varnish or resin. Resins developed for varnishes and enamels include alkyds, epoxies, polyurethanes, and acrylics. Lacquers are binders dissolved in organic solvents and drying occurs solely by evaporation of the solvent. Shellac is a solution of natural resin in alcohol.

Water-based Finishes. Latex or emulsion finishes are the dispersion type. The binders are either soluble in or are dispersed in water. Currently, there are three types of latex binders in use. Styrene-butadiene (SB) was used in the first latex paints and is still present in some interior finishes. The other two are polyvinyl acetate (PVA) and acrylic, both of which are used on interior and exterior surfaces.

23.9.2 Notes on Interior Wood Finishes

1. Durable polyurethane varnishes are used on hardwood floors.
2. Alkyd and latex enamels, used on interior woodwork, can be cleaned.
3. Washable paint, latex and alkyd enamels, is used on baseboards, trim, and doors which tend to become scuffed.
4. Clear finishes and pigmented finishes are used on cabinets to enhance grain or to introduce color to the decor.
5. Hardwoods and softwoods have different finishing characteristics.
6. Before commencing a finishing procedure, one needs to know the characteristics of the wood product to be finished and the methods of application.
7. Before finishing softwoods, inspections need to be made with respect to resin or oil pockets, the seasoning of the wood, and the moisture content.
8. Moisture content limits for finishing are 7% to 15% with alkyds used on woods up to 12% and latex up to 15%.
9. Knots and pitch pockets and streaks need covering with a sealer or shellac.
10. Hardwoods may have open or close grain. Fillers may be applied to open grain woods to fill the pores for a smooth finish or to change or enhance the color.
11. Tannin, a water-soluble substance in deep colored woods such as cedar, redwood, mahogany, and walnut, may bleed. A stain-control primer prevents bleeding into the final finish by keeping moisture from getting into the wood.
12. Semitransparent pigmented stains are applied to woods to give protection and to emphasize grain and surface figures.
13. Wood surfaces need preparation before finishing. When pigmented stains and clear finishes are applied, one can see through to the wood surface and any marks thereon. Planer marks are particularly noticeable.
14. Clear finishes, which require intensive labor to apply, are expensive. However, if fine woods are to be seen and are to be protected, proper finishing procedures need to be known before commencing.
15. Pigmented stains are classified by the solvent in which they are dispersed: (mainly) water, alcohol, and oil.

23.10 EXTERIOR FINISHES

Exterior materials include brick, tile, concrete, concrete block, wood, vinyl, stucco, metals, and stone. Materials are selected and treated to contribute to an exterior design goal which carries lines around the structure in a manner which unifies it and brings it all together. Continuity of material line or color is an obvious means of doing this.

1. A building form which is a bit heavy and awkward will appear more slick and slim if a blended material treatment is applied.
2. If a building form is well proportioned but appears to lack excitement, an unusual form in a different treatment of the same material introduces variety.

3. The use of materials taken from the natural environment (like wood or stone) cannot fail to harmonize with the site.
4. Design in material selection is one aspect, but the inherent qualities of materials, like durability and maintenance, are equally important.
5. Although many materials are prefinished, there are some which require on-the-job finishing. Such is often the case for woods used as siding, fascia boards, window frames, shutters, and soffits.
6. Latex-based paints allow materials underneath to breathe. This prevents blistering and alligatoring.
7. Semitransparent and opaque stains are easily applied to wood. Preservatives and color may be added to the stain, and, if the color is not too intense, restaining is required less often.
8. Cedars and redwoods are durable, even without finishes.
9. When natural materials are used in the building design, color schemes seem to develop. Mellow colors of stone and brick combine with natural wood tones. Wood as siding and shingles blend with the natural surroundings.
10. If painting is to be done, check samples in bright and shaded areas as for a camera setting. Consider three major types of color schemes: monochromatic, related, and contrasting. (But first, examine the colors on the site.)
    - Monochromatic: a neutral brown or white tone with one color (yellow, green or orange).
    - Related: a neutral brown or white with two or more colors (yellow, orange, or other).
    - Contrasting (to be used with great care).
11. Try to create a unified image. Use some major material as a starting point. If the building is a house, consider also the visual impact of drapes that can be seen from outside.
12. Pigment volume concentration is crucial to decorative waterproofing of concrete masonry walls. A high concentration, without sufficient binder, is less durable. Lower concentrations prevent water absorption, helping masonry withstand exposure to water or high humidity. The PVC needs to be stated.
13. Materials must be functionally and aesthetically satisfying.

23.11 RESEARCH

1. Sketch the floor plan of the place in which you live. Show the doors and directions of swing. For each door mark with an X the outside observer position. Designate the hand of each door.
2. Sketch the floor plan of a motel corridor. Include side room doors, fire doors, and exit doors. Designate the hand of each door. Give reasons for the design.
3. Using a white board and sharp pencil, draw full-scale the section of an exterior doorframe, doors, and framing. Put only necessary marks and explanatory notes on the board. Note the length of the cripple stud.
4. Write specifications for a reasonable home lock-up system for exterior doors.

5. Visit a door-hardware wholesaler. Make notes on three typical keying systems for buildings of different sizes.
6. Use a white board and sharp pencil to explain how exterior window and door trim headers may come to the same elevation on a building. Start with standard window sizes. Show complete wall framing in elevation. Identify the lengths of cripple studs.
7. With baseboard materials, demonstrate the creation of an inside corner joint by coping on the miter line and by scribing.
8. Conduct a limited survey of heating systems used in your area. Include such topics as fuel, fuel supply, automatic shutdown features, furnace space requirement, metering, annual cost of fuel, annual maintenance costs, installation cost, and in-building heat distribution.
9. Visit an automotive spray booth. Describe the surface preparation, taping, dust control, air movements, application of paint, and safety standards. Compare the finishing conditions with those found in a typical building, for example a house or a small office building.
10. Paint typical color scheme tabs for building exteriors. Include monochromatic, related, and contrasting schemes. Indicate in color how they might or might not blend with natural surroundings for a unified image.
11. Examine the doors in a typical office building, including those in the rest rooms. Do they fit properly? Note whether or not the swing of the doors hinders movement. Mentally reverse the swing. Would worse problems ensue if the doors were installed in this fashion?
12. Compare two doors, one with butt-type hinges and the other with surface-type hinges. Determine why butt hinges are preferred in terms of security.
13. Check the fit of the different windows in the building in which you live, feeling for air leaks. Note where and why the leaks occur. If you were to redo the installation, how would you improve on it?
14. Visit a local flooring materials dealer and compare the prices of parquet flooring, vinyl flooring, and carpeting. Which is the least costly per square yard? Which is the most costly? What positive qualities are lost within each of the three groupings as the price goes down? Make a list beneath each type heading of the applications for which that type of flooring would be most practical.
15. Suppose you were going to cover the floor of the room in which you are presently seated with vinyl tile. Draw up a sample layout to estimate how many cartons of tiles you would need. Take into consideration waste from cutting openings and matching patterns.
16. Find out how much it costs the owner of the building in which you live to heat that structure yearly. Determine whether savings would result if the building was converted to a different fuel type. Keep in mind costs for any extensive renovations that might have to be made.
17. Familiarize yourself with local electrical codes. Even if you will not be doing any wiring, it is a good idea to be familiar with the codes that govern such installations. This will help you set visual design

standards and advise the owner on any installation options that arise.

23.12 REFERENCES

23.12.1 Ball, John E. *Light Construction Techniques: from foundation to finish.* Virginia: Reston Publishing Company, Inc., 1980.

23.12.2 Capotosto, John. *Residential Carpentry.* Virginia: Reston Publishing Company, Inc., 1979.

23.12.3 *Carpenter.* Washington, D.C.: Headquarters, Dept. of the Army, 1984.

23.12.4 Crowdis, Kay, and Dave G. Crowdis. *Designing and Building Your Home.* Virginia: Reston Publishing Company, Inc., 1981.

23.12.5 Duncan, S. Blackwell. *Guide to the 1981 National Electrical Code.* Virginia: Reston Publishing Company, Inc., 1982.

23.12.6 Jones, Peter. *Homeowner's Guide to Plumbing, Heating, Wiring, and Air Conditioning.* Virginia: Reston Publishing Company, Inc., 1980.

23.12.7 Philbin, Thomas. *Basic Plumbing.* Virginia: Reston Publishing Company, Inc., 1977.

23.12.8 Powell, Evan, and Ernest V. Heyn. *Popular Science Book of Home Heating (and Cooling).* Virginia: Reston Publishing Company, Inc., 1983.

23.12.9 Self, Charles R., Jr. *Building Your Own Home.* Virginia: Reston Publishing Company, Inc., 1978.

23.12.10 Sullivan, James A. *Plumbing: Installation and Design.* Virginia: Reston Publishing Company, Inc., 1980.

23.12.11 Traister, John. *Electrical Inspection Guidebook.* Virginia: Reston Publishing Company, Inc., 1979.

24 LANDSCAPING

In terms of the physical process of constructing a building, landscaping often represents the final step. In other words, landscaping is the final subsystem to be integrated into the total building system. The landscaping process generally includes two primary objectives: cleanup and aesthetics.

Builders frequently use this end phase to perform a thorough cleanup operation on the building site. All scraps, refuse, and waste should be removed. Any leftover building materials (lumber, bricks, nails, cement, etc.) should be picked up and transported to another jobsite or back to a storage yard. The undesirable side effects of the initial excavating work on the lot should be taken care of at this time. All too often, scraps, assorted debris, and leftover building supplies are left scattered around on the building site when a job is done. Not only is this practice sloppy, wasteful, and potentially hazardous (nails in scraps of wood can be stepped on, for example), but it reflects badly on the builder and the workers.

Second and more basic to the concept of landscaping is the need for the builder to more fully integrate the newly constructed building into its immediate surroundings and into its local environment. This includes both cosmetic and practical features ranging from planting of shrubbery and laying of flower beds, to sodding, terracing, and all-around drainage control.

24.1 GENERAL OBJECTIVES

When the building is operating as a functional system, with all subsystems coordinated, attention must be given to its setting. The backfilling must be done and all disturbed earth must be compacted and allowed to settle. After compaction and settling, finish grading is begun. Surface drainage patterns will often have become more visible during the course of the building process and must be fully dealt with at the end of the job. This can be done concurrently with and as part of the grading work.

To ensure a proper and lasting grading job, the builder should be equipped with knowledge about the local vegetation (what grows best in what circumstances), the microclimate, and the effects of gradings and plantings.

24.2 SPECIFIC OBJECTIVES

The material in this chapter is designed to

1. encourage study of the effects of backfilling and compaction and the relationship between these factors and drainage,
2. present reasons for using site analysis data for estimating what the microclimate will be for a new structure,
3. show the need for ascribing limits to the different site slopes,
4. advance the idea that a study of natural site vegetation will provide sound clues with respect to new plantings,
5. emphasize the need to integrate a site and the building or buildings,
6. encourage planning of a site on the basis of public and private areas, and
7. introduce the reader to the area of landscaping and encourage an in-depth study of the means and methods of the landscaping process.

24.3 BACKFILLING

The use of various soils and backfilling have already been discussed to some degree in Chapter 10. However, when the building has been erected (or at least is nearing completion) and the time comes to reexamine the site in anticipation of landscaping activities, it is important for the builder to be able to observe and test the results of any backfilling and compaction done earlier, including the resulting settlement. Problems found at this point can then be rectified during the landscaping phase of the building operation and before the building is turned over to the owner. For example, has severe overcompaction cracked a foundation wall? Is the compacted soil following the contours as planned? Does rainwater drain away from the building?

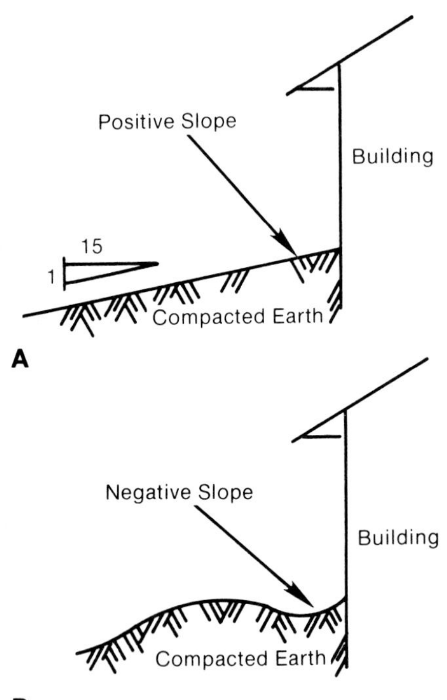

FIGURE 24-1. (A) Achieving a positive slope away from a building when backfilling and grading cause rainwater and runoff to drain away from the building; (B) a negative slope will often cause drainage problems.

The builder should almost always try to achieve a positive slope away from a building on all sides to expedite proper drainage, as shown in Figure 24-1A. Obviously, some cases, such as a building on a steep hillside, will present problems; nevertheless, a positive slope or the best alternative—a neutral or parallel slope—should always be attained to whatever degree is possible. What is to be avoided is a negative slope, as shown in Figure 24-1B, which will cause water to sit and stagnate around a building or will force runoff back towards the foundation.

Backfilling around a building without any further compacting offers an assortment of risks. For example, if the backfill eventually settles, as shown in Figure 24-1B, the owner will be left with problems to cope with for years to come. By the time such a settling problem becomes pronounced enough to demand correction, the owner may have already added plantings and put in new flower beds and other site features which he or she does not want to dig up. Unfortunately, many owners simply live with drainage problems, neglecting to correct the moisture and water buildup at the foundation.

The grading must be done, and done properly, before the more cosmetic aspects of landscaping work commence. Needless to say, the builder cannot be held accountable for earth movements years after a job is completed. This is a matter out of the builder's hands. What the builder is responsible for is correcting the disruptions in the earth caused by the original excavation and for ensuring that backfill is compacted, settled, refilled, and graded before the owner takes possession of the building. Seeding, landscaping, or other site work can then be undertaken by the owner with relative impunity from the kind of disruptions of the earth that would require major surgery on the site.

If the building footings or foundation were set too low or too high, other problems are created. If too low, then window wells, extra weeping tile, and gravel may be needed. In more severe cases, a restructuring of the site may be necessary to carry away the runoff. Shallow depressions, called swales, may have to be made to direct water around and away from a building. If the footings or foundation were set too high, the doorsills will be too far from the ground and additional fill must be placed around the building and compacted.

This is a good example of the type of work a builder must do to achieve a finished, quality building project; yet, such jobs require extra material and labor and represent additional costs for the builder. The best way to avoid such extra work and costs is to be aware of and to begin the landscaping process (the landscaping subsystem) while the rest of the building is being done. Major backfilling and compacting problems can be worked out, sometimes even done, before the final landscaping is even started. In this way the final phase of the building process can be used to make corrections and to add the final cosmetic touches.

Elevations and levels are established before construction even commences. They need repeated checking, using the story rod technique, as construction proceeds. This prevents any unpleasant surprises for the builder when the final landscaping is ready to be done, and it also avoids later problems for the owner.

24.4 FINISH GRADING

Prior to adding topsoil and shaping the overall site, reference should be made to the initial site analysis data: geology, soil, water table, topography, slopes, and natural vegetation. The estimate regarding the microclimate of the site should be reviewed, including the direction of prevailing winds, the sun pattern across the site, rainfall, snowfall, and temperature range (with attention given to the extremes). Such information will help determine the placement of plantings, berms, other buildings, fences, screens, and the use of natural topography and vegetation. The following three statements can serve as a very general guide for finish grading around a building site.

1. The degree of slope and the orientation of the slope or slopes to the sun affect decisions about erosion control, roadways, walkways, plantings, and passive energy.
2. If one is sensitive to or wishes to retain some of the existing

landform characteristics, as in the case of a view, topographic features may determine the organization of the site. A further example of topographic influence is placement of drainage. Natural drainage flowing away from the site is preferred. For water accumulating above the site, some correction, such as a swale or an interception and redirection, may be necessary.

3. Limits for site slopes may be 4% to 10% for roads, 2% to 6% for adequate drainage, 0% to 25% for allowing machine cutting of lawns, and up to 25% for soil erosion control.

In general, the final sloping and grading is done in relation to the general site plan and the expected microclimate.

24.5 SITE VEGETATION

Attention given to the natural vegetation on the site will yield clues as to the nature of the soil and of the microclimate existing before the site was disturbed. Conditions that caused water or wind erosion and plant growth characteristics will probably still be present after a building has been constructed. The builder should consider the microclimate which enabled specific kinds of vegetation (pine, spruce, willow, palm, and oak trees, cactus, and so on) to flourish. A pine tree would indicate a particular soil and water supply, direction of wind and its velocity, and amount of sunlight. Attention should also be given to valley versus ridge climates, frost pockets, and proximity to a body of water.

Microclimate information will suggest the possible need for natural or other windbreaks, berms, erosion control, and the appropriate placement of other buildings on the site. It will also suggest the placement of individual or group plantings and the types of plants to use in relation to all of these possibilities. Site vegetation can serve a wide variety of practical as well as aesthetic purposes. For example, trees, shrubbery, and hedges have frequently been used as an effective means of erosion control.

Increasing knowledge about the beneficial effects of passive solar energy—the southerly orientation of a building and its windows to take advantage of the free heat and light of the sun—provides another area of landscaping where site vegetation can play an important part. In the summer, greenery keeps itself and surrounding areas cooler because of respiration and shading. The more leaf surface there is around a building, the more natural cooling will be provided. Deciduous trees make excellent selective sunscreens because they grow leaves and throw their biggest shadow when it is needed the most—during the time of summer heat. In the winter they shed their leaves, blocking little of the southerly sunlight that can be used to help keep a building warm.

Research on passive solar heating has shown that a judicious planting of deciduous trees on the south, southeast, and southwest sides of a building will allow the maximum amount of sunlight into a glassed-in part of a building in the winter, while providing considerable supplementary natural cooling in the summer (Figure 24-2).

Actually, it is best not to have deciduous trees directly in front of south-facing windows. If they are too close, their trunks and branches,

FIGURE 24-2. Landscaping for heat control: (A) deciduous trees on the south side of a building help to shade the building and keep it cool in the summer; (B) in the winter the same trees lose their leaves and allow sunlight to enter the house through south-facing glass, creating a supplementary form of heating.

even without leaves, can block out too much sunlight in the winter. However, if they are too far away from the windows or other building glass, the sharply-angled summer sun will slant around them, making the trees of little value for shading. Perhaps the best alternative is to plant deciduous trees for shading on the southeast and southwest to combine shading with solar access. The west side of a building probably benefits most from summer shading because of the long hours of afternoon sun there, and in very warm climates even the eastern, early morning side of a building can benefit from some shading.

Hedges and shrubs planted close to a house also help with natural summer cooling while providing an aesthetic feature to a site. Using overhead trellises, arbors, or any type of climbing plant or vine are other options for providing temporary natural summer shading for the hot spots on a building. Unless an open, unobstructed southerly view is not wanted by the owner, evergreen trees are best not planted on the south side of a building because they will blot out the sunlight in the winter.

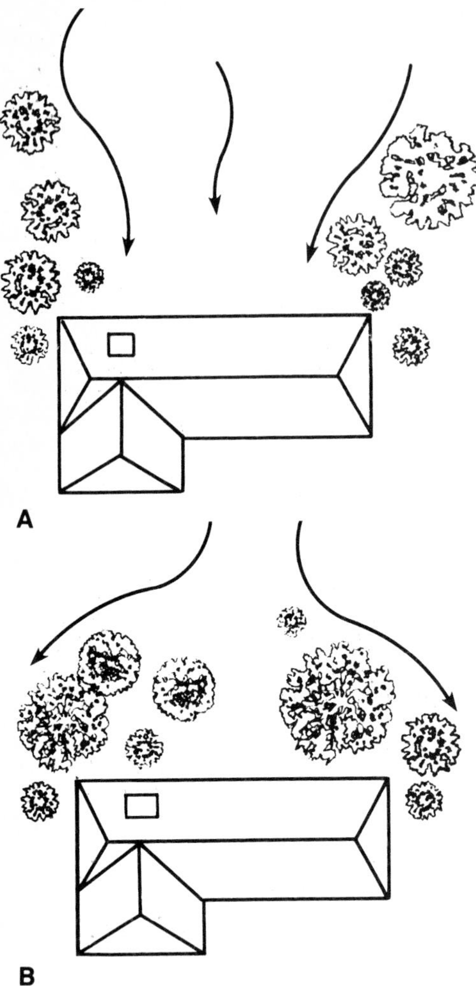

FIGURE 24-3. (A) Poorly designed landscaping can actually channel cold winter winds to the north side of a building, helping to raise heating costs; (B) trees and shrubs properly placed can deflect rather than channel a cold airflow.

Aside from shading a single building, the builder or owner may, for a variety of reasons, decide to shade other parts of the site. For example, shading may be a desirable feature for lowering the temperature in some area on the site to make it more hospitable for outdoor use in the summertime. The general effects of shade from plantings, buildings, or other man-made structures can be estimated for a particular situation. Site testings from different seasons and drawings made from estimates can show the amount and angles of shading for objects of definite heights and sizes.

Remember, though, that these ideas are only meant as suggestions for combining aesthetic and utilitarian values. A building may be put up in the middle of an evergreen forest; the owner may not want to cut any more trees down than are absolutely necessary, regardless of solar considerations. Contrarily, some owners may decide to eliminate trees on the site altogether in favor of hedges, formal flower beds, rock gardens, or wide expanses of lawn. Some methods and types of planting will work better on a certain site, in a known climate, and with a specific topography, as determined by reference to microclimate information. However, the landscaping must be in accordance with what the owner wishes and has agreed to finance.

In addition to shading, trees and other vegetation can also be valuably used as windbreaks by themselves or in combination with berms, fences, and walls. Protecting a building from cold northerly winds in the winter is an often overlooked form of energy efficiency which can help to keep heating bills down. As shown in Figure 24-3, it is better to deflect rather than to channel a cold airflow. Evergreen trees planted on the north, northwest, and northeast sides of a building are particularly suited for this purpose as they do not lose their foliage in winter. Evergreens planted close to a north wall also create a pocket of still air which helps to insulate a building and cuts down on air infiltration. Dense plantings, however, tend to create a zone of turbulence which may strike a building during high winds; whereas, a partially penetrable windbreak frequently provides better protection with streams of air flowing one over the other.

Also, the placement of trees by size, short to tall, needs to be considered in relation to the direction of expected prevailing winds. (The prevailing winter winds in different parts of the United States range from northeast to northwest. They are even more diverse in Canada. An evergreen wind barrier should be planted or preserved—if trees are being cut down or thinned out—with this in mind.) Always research information on local conditions. Figure 24-4 shows three ways in which plantings may be used to create windbreaks: on a berm, combining small and large trees, and using a snow fence with spaced plantings and a berm.

In summary, the spacing and selection of plants is important if the energy efficiency of a building is to be improved by the site vegetation. This is equally true of plants used for erosion control. The builder must know what plants can be used for different purposes or be able to work with someone who does.

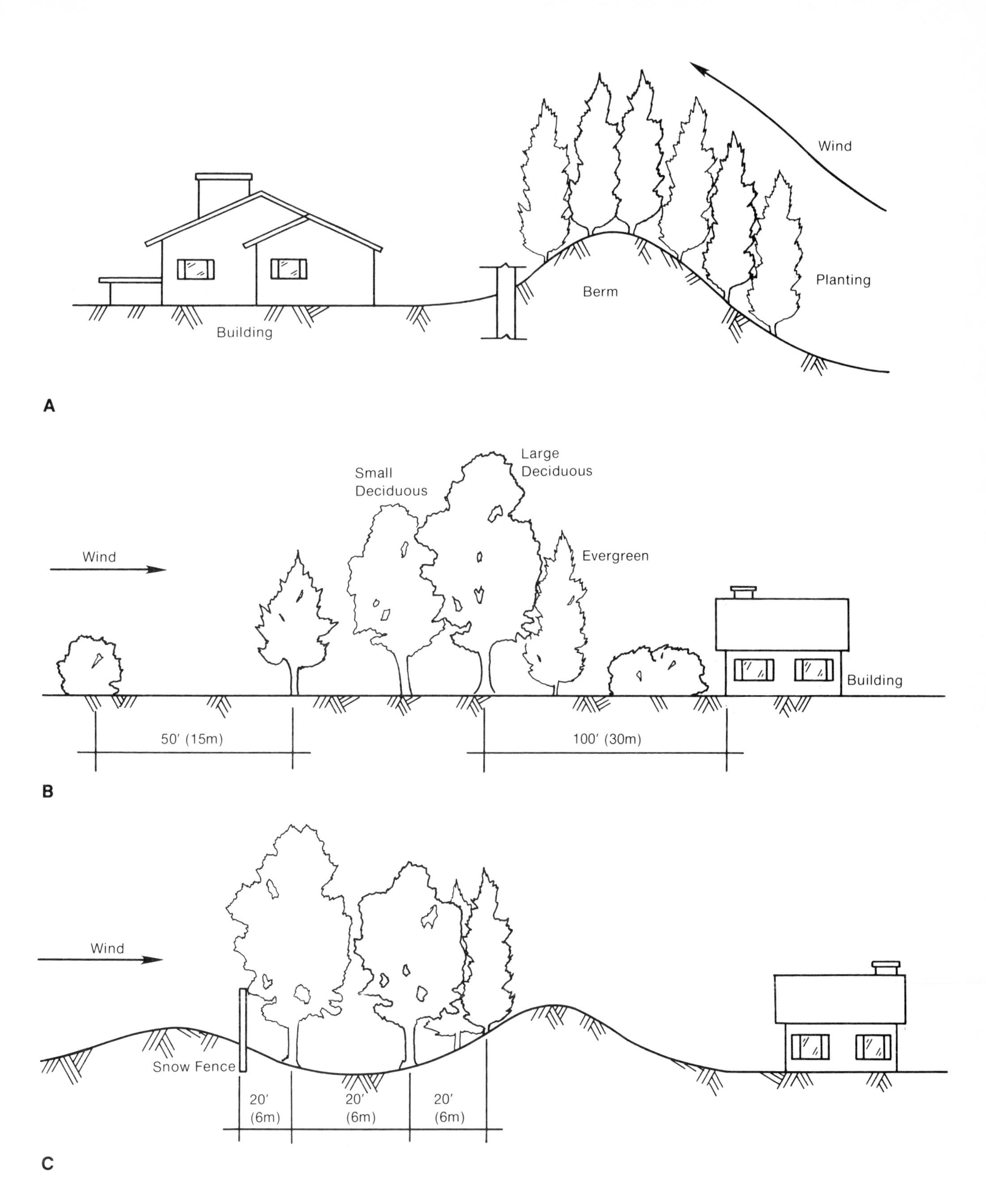

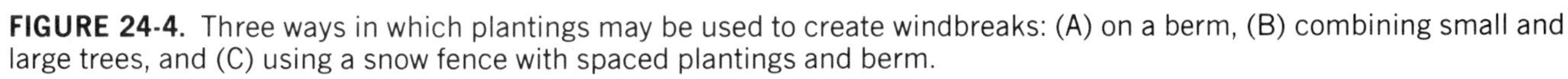
FIGURE 24-4. Three ways in which plantings may be used to create windbreaks: (A) on a berm, (B) combining small and large trees, and (C) using a snow fence with spaced plantings and berm.

24.6 INTEGRATION OF SITE AND BUILDING

Too often a new building on a light construction building site doesn't blend in with its surroundings. Perhaps the brand new building is surrounded by a wide expanse of freshly seeded lawn with one or two newly planted ornamental trees forlornly occupying opposite corners of the site. Even worse is that frequently seen sea of mud, or its counterpart, the dustbowl, lacking even the beginnings of a lawn. The inference is often made that the contractor got in and out of these jobs as quickly as possible—not the best way to build a reputation as a solid, competent builder. Furthermore, such a building job, no matter how well the building itself is constructed, is not environmentally sound.

Building contractors should, and more frequently are, considering the development and integration of the entire building site as part of the light construction job. People are realizing that a building cannot simply be plopped onto a site without some regard for how that building is going to affect the site, as well as the rest of the local environment. A badly graded site that continually runs off into a small brook, polluting it, can irreparably damage the checks and balances of this environment. How the building process on each and every building site affects its setting, visually and environmentally, has, in recent years, increasingly become a responsibility of the builder.

It is also important for the owner and builder to consider the effects of the local environment on the building. Not only is this important in the sense of using sound and appropriate construction techniques, but it is also important in the sense of adapting the building and the site to the special features that determine the character of the local setting and environment. In addition, the site itself should accent the building design and help to define the style of the building. If the building design suggests a formal life-style, then the site should do likewise.

This is possible by clearly defining spaces and emphasizing formality with respect to fencing, plantings, color, roads, walkways, gardens, and secondary buildings. If the expected life-style is more casual or informal, the plantings can be more natural—an English-style rock garden rather than a formal, geometric, French-style garden, for example. Colors can be earthier—fences, for instance, can be stained rather than painted. Even security measures can be integrated into the overall style of a building and site. A building site, when finished, should be apparent to the casual observer, but not necessarily emphasized.

The components of the site should appear unified and attractive. Uninteresting or unattractive parts of a site can be altered, camouflaged, or concealed. In general, a building site should make a statement about the owner's tastes and style without wreaking environmental or aesthetic havoc on its surroundings.

Plants are useful in defining spaces and may be used in conjunction with landforms, buildings, and architectural features such as walls, fences, driveways, and outdoor living areas. Perhaps more than any other type of site ornamentation, plants are best for achieving integration between a building and its own site, and often, depending on the setting, for integrating a building and its site into the local environment. This frequently even holds true where greenery is not especially abundant, as in urban areas. Care must be taken when establishing plantings to space them as if they were fully grown—or to plant additional temporary ones to be removed later. A ten-year plan is advised.

24.7 PUBLIC AND PRIVATE USE

The site plan reflects the degree of expected use of a building by the public. A commercial building, for example, is used differently than a housing unit. Some buildings are specifically designed to combine public and private use; a building with both a doctor's office and his or her residence or an apartment building with a series of small shops are examples. The zoning designation reflects the expected amount and kind of usage. Check zoning and research the intended use of the entire site to fully integrate all of its features.

A public area generally includes an entrance road leading to a parking sector on the site. Municipal regulations usually prescribe the setback of buildings from secondary and main roads and, therefore, help to determine the minimum length of the entrance road. Other factors determining the proper setback distance of a building are safety, noise, dust, and security (including precautions against vandalism and theft). Parking areas should indicate who is to park where, an example of delineating a limited amount of site space to a certain number of necessary functions. Visitors should know where space has been set aside for them to park. Boundaries should be clearly marked and controlled.

A building may have the entry at grade level. Such an entrance without steps gives a building a low profile, and, in certain cases, a comfortable, attractive appearance. An entrance that will receive a lot of public use might be oriented away from prevailing winds—especially in a cold, northerly area—and sheltered by plantings, a screen, or other buildings. At night, proper outdoor lighting should mark the way to the door. An entrance needing steps, particularly one which will bear a lot of public use, presents certain design problems that require careful attention. The problem is often a space problem with the steps' space being out of proportion to the building's space. Occasionally a small public building will be seen with huge steps and an enormously wide entrance which are largely disproportionate. Even more frequently the steps' space is too small and the steps are too short. Some building projects are left with such a design fault for the life of the building.

The private areas of a site might include a garden or gardens, flower beds, recreation areas, and safe playing areas for youngsters. Outdoor living areas will frequently require both sunny and shady places with provisions for easy supervision. Decks and patios, on the other hand, become extensions of indoor living areas and tend to architecturally link the indoor and outdoor spaces. Provisions should be made for a service area—garbage collection and so on—on virtually all building sites regardless of whether public or private access is dominant.

To service a building and to accommodate visitors, walkways need to be wide enough to move furniture and service equipment and to allow two persons to walk side by side. Wide, smooth walkways are especially important where public access exists. Walkways are placed to encourage, not discourage, use; they should seem to provide the best path to a destination.

When walkways and driveways are finished, they establish elevations and contour restraints for other areas such as seeded lawns, sodded lawns, storage areas, parking areas, gardens, and sites for additional buildings. Their placement, therefore, should be carefully planned beforehand, and these passageways should be as fully integrated with the building and the rest of the site as possible. Driveways and walkways are also perhaps the best means available for defining public and private areas and for limiting access to certain areas.

Fences and screens are often used to mark public and private areas and to communicate the tone of the expected life-styles of the occupants. When this is a matter of importance, the fence and screens require designer attention and careful building. Sometimes a building design suggests one style and a fence another. Such designs are not compatible and create wrong impressions. The creation of fences and screens, virtually a specialty job for designers and builders, will complete the intended landscaping goal.

24.8 RESEARCH

1. Visit the office of an architect to read a site plan. Describe the features.
2. Determine from the site plan the site slope limits for roads, driveways, drainage/soil erosion prevention, and walkways.
3. Describe the building entrance and its orientation. Note the orientation of all the windows, sliding doors, and other glazing. Is the building's orientation sound from a passive solar point of view?
4. Identify all plantings in the site plan. Write brief notes on expected growth characteristics, shading effects, and wind breaking effects.
5. Review the initial site analysis data on geology, soil, water table, topography, slopes, and natural light. From this information, determine which factors, (for example, dry soil and shade from adjacent buildings) would affect plantings; then develop ways to combat the problems.
6. Evaluate the intended integration of indoor and outdoor design features. Refer to building sketch plans for the same project.
7. Measure the effect of shade from the roof of a local building or from a wall. Use two thermometers. Report on the findings and suggest some implications.
8. For an existing building, measure the on-ground shade pattern over a period of one year. Suggest some implications for the site planning of a new building in the same locality.
9. Consult locally prepared booklets and information sheets from the fields of horticulture, silviculture, and agriculture. Consider the relationship between these fields of study or enterprise and the landscaping responsibilities of a light construction worker.

10. Read local college course outlines with respect to landscaping for your area. Obtain reading lists.
11. Become familiar with local vegetation, noting where the different types are found. Read up on the various plant growing seasons and the care required.
12. If possible, visit a construction site where backfilling and compacting has just taken place. Notice the slope of the ground near the building, and try to predict if any problems will arise. Visit the site again on a rainy day or just after a heavy downpour. Have any of your predictions held true?
13. Review common landscaping measures that can be employed to combat erosion. What types of vegetation work best for this purpose? Which of these types are suitable for planting in your area?
14. Assume that the building in which you are living is heated by a passive solar system. Determine the south face; then draw up a landscaping plan that would be complementary to solar heating. To do this, take into consideration how close the trees must be to the house to provide adequate shade in the summer, yet not block out too much sun during the winter.
15. Make a list of common shrubbery types. Note their final growing heights for reference when landscaping around window areas.
16. Using the solar landscaping plan of your building you made earlier, add special landscaping features for wind control. Consult your local weather service for the direction of prevailing winds in your area.
17. Walk down a street in the residential part of your city or town. How well do the homes blend in with one another and with the area in general? What common elements have been employed to integrate sites and buildings? Do any homes "stick out?" Try to determine what keeps them from fitting in with the rest.
18. Study a zoning map of your area. Which areas are zoned for commercial use? Which are for residential buildings only? What are the rules concerning dwellings that combine public and private use?
19. Look up municipal regulations regarding the setback of buildings from the street as well as rules regarding on-street parking.
20. Review federal regulations that deal with building access; pay particular attention to sections dealing with accessibility for the handicapped.

24.9 REFERENCES

24.9.1 *Earth Manual.* Washington, D.C.: U.S. Dept. of the Interior, 1974.

24.9.2 Hannebaum, Leroy G. *Landscape Operations: Management, Methods and Materials.* Virginia: Reston Publishing Company, Inc., 1980.

24.9.3 Herubin, Charles A. *Principles of Surveying.* Virginia: Reston Publishing Company, Inc., 1978.

24.9.4 Meyers, L. Donald, and Richard Demske. *Guide for Outdoor Building and Maintenance.* Virginia: Reston Publishing Company, Inc., 1977.

24.9.5 *Solar Dwelling Design Concepts.* Washington, D.C.: U.S. Dept. of Housing and Urban Development, 1976.

24.9.6 Turgeon, A. J. *Turfgrass Management.* Virginia: Reston Publishing Company, Inc., 1980.

24.9.7 Untermann, Richard K. *Principles and Practices of Grading, Drainage and Road Alignment: An Ecologic Approach.* Virginia: Reston Publishing Company, Inc., 1978.

24.9.8 Waschek, Carmen, and Brownlee Waschek. *Inflation Fighter's Big Book: beat the high cost of operating your home.* Virginia: Reston Publishing Company, Inc., 1979.

24.9.9 Waschek, Carmen, and Brownlee Waschek. *Your Guide to Good Shelter.* Virginia: Reston Publishing Company, Inc., 1978.

24.9.10 Wass, Alonzo. *Methods and Materials of Residential Construction.* Virginia: Reston Publishing Company, Inc., 1977.

24.9.11 Wass, Alonzo, and Gordon A. Sanders. *Materials and Procedures for Residential Construction.* Virginia: Reston Publishing Company, Inc., 1981.

24.9.12 Wicks, Harry. *How to Plan, Buy or Build Your Leisure Home.* Virginia: Reston Publishing Company, Inc., 1976.

25

BEING CREATIVE IN THE LIGHT BUILDING CONSTRUCTION INDUSTRY

Some people are willing to work within a construction job—to perform tasks well, to improve skill levels, and to accept changes in materials and handling techniques as they come—to work within an activity. Others wish to take responsibility to change or to go beyond just one operation. Such people are approaching industry from a different perspective by becoming involved with many, or possibly all, of the activities associated with a building project.

To work with the whole project and to influence design and site development is to be creative within the light construction industry. The purpose of this chapter is to help a person shift from one activity to multiple activity interests in the light building construction industry.

25.1 SPECIFIC OBJECTIVES

On reading the chapter and conducting the suggested research the reader should better be able to

1. gather and interpret physical environment data which influences building design in a region,
2. gather and interpret social environment data which indicates something of building needs,
3. gather and interpret business environment data which provides direction with respect to building design thrusts,
4. identify means of increasing his or her knowledge of industry, and
5. prepare himself or herself for taking increasing responsibilities within and for the industry—to creatively prepare for changing roles.

25.2 PHYSICAL ENVIRONMENT

The data found in Tables 25-1 through 25-4 helps describe a particular urban construction environment, that of Edmonton, Canada. After studying these tables one might ask questions about weather patterns of the area, particularly about the extremes with respect to temperature, wind, rainfall, snow load, and solar energy. Additional data may be needed to answer the questions in depth, especially about the velocity of wind gusts. Answers to questions may help a builder decide when to commence construction and when to have it closed in. By planning to accept the extremes data, precautions may be taken to allow construction activities to proceed without too much disruption.

TABLE 25-1 TEMPERATURE[1]

1. Mean temperature 1931-1960

January	2° F	(−16.7° C)
April	38° F	(+3.3° C)
July	62° F	(+16.6° C)
October	40° F	(+4.4° C)

2. Degree-days[2/3] above 42° F (5.6° C), the approximate measure of plant growth: 2,400.
3. Degree-days below 65° F (18.3° C), to estimate fuel consumption for the heating period: 11,000.
4. Temperature range (Mean annual): 55° F (12.8° C)
5. Warmest month (Average maximum July): 70° F (21.1° C)
6. Coldest month (Average minimum January): −5° F (−20.6° C)
7. First fall frost (Average date 1951-1964): after September 15.
8. Last spring frost (1951-1964): May 15-31.
9. Frost free period (1951-1964): 105 days.
10. Winter days with maximum above 40° F (4.4° C) (Mean number 1955-1964): 30 days.
11. Summer days with maximum above 80° F (26.7° C) (Mean number 1955-1964): 15 days.
12. Very cold winter (December 1949-February 1950): 0° F (−17.8° C) to 10° F (−12.2° C).
13. Very warm winter (December 1930-February 1931): 20° F (−6.7° C) to 30° F (−1.1° C).
14. Hot spell July 12, 1964: 91° F (32.8° C) down to July 13, 1964: 73° F (22.8° C).
15. Cold spell December 13, 1964 (11 a.m.): −18° F (−27° C) to December 14, 1964 (11 a.m.): −24° F (−31° C).

[1]Reference: Atlas of Alberta (1969).

[2]Degree-day: One degree-day represents one day of departure of the mean daily temperature from a given standard. For instance, if a day has a mean temperature of 50°, this represents eight degree-days above 42° and 15 degree-days below 65°. Degree-days are calculated over a season.

[3]Data for Edmonton, Alberta, Canada.

Based on the information in Tables 25-1 through 25-4, the following questions should be considered:

1. If frost-free ground is desired, when should excavations be started?
2. Which months are more likely to be wet and, thereby, create problems in placing footings and foundation walls?
3. For a concrete block wall, how much bracing is needed to safely withstand wind?

TABLE 25-2 WIND[1/2] PATTERNS

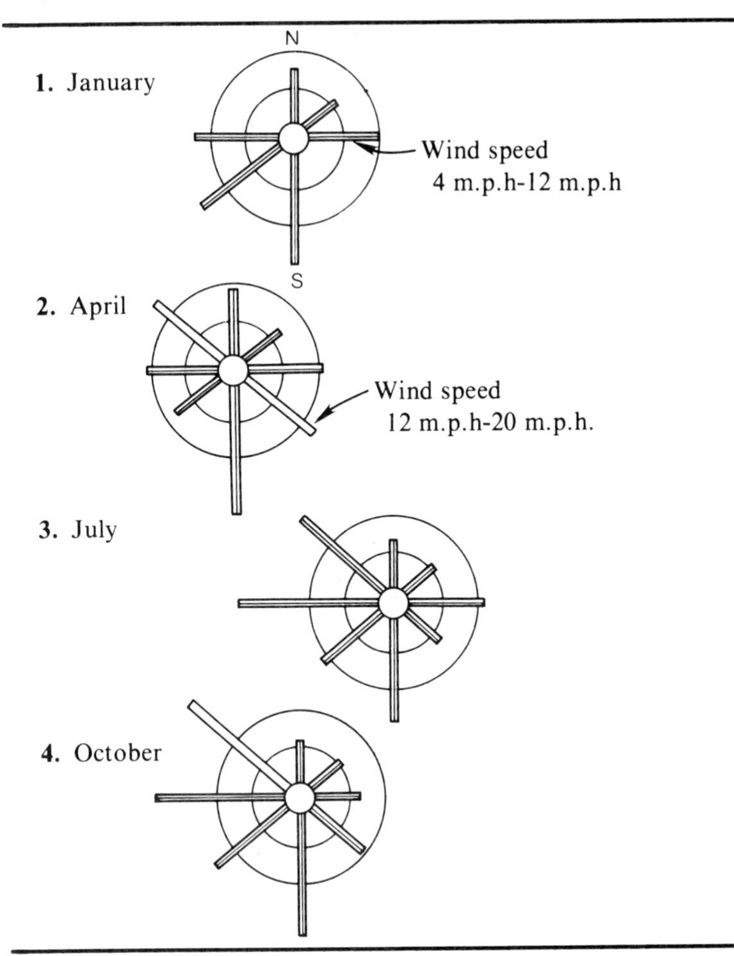

[1]Reference-Atlas of Alberta (1969).

[2]Data for Edmonton, Alberta, Canada: 53° 30′ north latitude/113° 50′ west longitude.

4. What provisions are needed for site drainage?
5. What snow load must be accommodated?
6. Should energy efficient buildings be designed? To what design level?
7. With the range of temperatures and humidity levels indicated, what vapor barrier will be required?
8. Would a solar energy application be feasible?

Ground frost conditions vary from year to year, changing according to winter severity and snow cover. The slope of the site in relation to the sun's rays also makes a considerable difference. Site tests would be needed in April or May and on some shaded slopes in June. With good snow cover all winter, early spring construction may be considered.

The data suggests that May and June could be the wettest months. Drainage provisions would be high on a priority list. Late summer would probably be better for starting, but warm weather construction time would be reduced. Late summer starting would be useful if the general plan included closing-in the building and doing the backfilling in the fall, completing inside work during the winter, and finishing the landscaping the next spring.

TABLE 25-3 RAIN AND SNOW

Edmonton (Mean elevation/altitude: 2,192′ (668.1 m). Area: 680 km^2
1. Mean precipitation: annual 20″ (508 mm).
2. During growing season April-August: 14″ (356 mm).
3. Mean annual snowfall: 50″ to 60″ (1,270 mm to 1,524 mm).
4. One day rainfall 1916-1960 5 year return: 2.5″ (64 mm) 10 year return: 4″ (102 mm).
5. Maximum precipitation date 1921-1950: June 16-30.
6. Hours of thunderstorms 1943-1957: 40 (annual).
7. Hail frequency 1951-1960: 6 per section of land.
Regional (Alberta)
8. Dry month August 1961: 0.5″ (13 mm).
9. Wet month May 1964: 5″ (127 mm).
10. Moisture index: dry subhumid (average surplus 1″ to 4″ [25 mm to 102 mm]).
11. Average yearly flow of North Saskatchewan River at Edmonton: 5,704,750 acre feet with a maximum 48,000 cubic feet per second.

TABLE 25-4 SUNSHINE[1]

1. Hours of sunshine per year: 2,200
2. Percentage of sky covered by cloud during daylight hours:
   a. March—65
   b. August—60

[1]Data for Edmonton and Eastern Alberta Plains.

Local codes should be read with respect to velocity of wind gusts, snow load provisions, and drainage provisions. Mean temperatures, degree-days below 65° F (18.3° C), coldest month, and cold spell data may indicate a need for energy efficient construction. Table 25-4 suggests use of solar energy.

These are some general observations. Actual site study and research with respect to energy resources of the area would be needed. Reading of local technical data (charts, graphs, indexes, and papers) would help decide about viable trends in using less natural gas and more solar, less natural gas and more solar and more fuel wood, or natural gas and electrical.

Table 25-5 indicates that Edmonton may draw from regional sources of fuel and power. All of the fuels/power are available in Edmonton. For good reasons oil, coal, and propane are seldom used as fuels within the city. Natural gas and electrical power are accessible and the techniques are such that they are acceptable. In case of supply problems in cold weather, moves are under way to provide more than one system to a building. Solar energy in energy efficient structures is being tested. Long-term cost analysis is needed for each energy form.

TABLE 25-5 AVAILABLE FUELS AND POWER

Alberta
1. Natural Gas
2. Oil
3. Coal
4. Electrical power a. Coal b. Gas c. Hydro (No nuclear)

Table 25-6 describes soil and tree growth characteristics of the area and region.

TABLE 25-6 SOIL AND TREE GROWTH

Edmonton Area
1. Aspen, poplar forest land now more than 50% cultivated.
2. Soils: black (parkland), 60%-75% arable.
Region
3. Trees: jack pine, lodgepole pine, balsam fir, black spruce, white spruce, alpine larch, balsam poplar, trembling aspen, water and white birch.
4. Forest Products: Coniferous plywood logs Poplar plywood logs Fuelwood Railway ties Round timber Lath Pulp chips

As the soil in the vicinity is mainly black arable land, laying of gas lines, water/sewer lines, and underground electrical cables is accomplished without problems of rock formations. Topographical maps show generally flat surface features, except where the river and creeks cut through.

Soft plywoods and some dimension lumber are regionally available. Additional wood products are brought by rail and trucks from the West Coast, or inland from it.

Data in Table 25-7 indicates that, for the region, materials are available to produce the ingredients for concrete. Cement is manufactured in Edmonton. Sand, gravel, and water are available regionally.

Current data is always needed with respect to building materials and to why and how they may be used in an area.

TABLE 25-7 MINERALS

Alberta
1. Marl
2. Sodium sulphite
3. Limestone
4. Dolomite
5. Bentonite
6. Pumicide
7. Clays and shales
8. Silica sand
9. Gypsum
10. Iron-rich deposits
11. Salt

25.3 SOCIAL ENVIRONMENT

Continuing with a general description of the particular urban construction environment, one might study the needs of the people and their acceptance of current building designs. Particular attention would be given to engineering/technical data at hand for planning. Such information would indicate the level of community awareness and readiness for change and the money spent on planning.

Land use studies show where people work and live. Types of residential buildings show how they live—single dwellings, multiple dwellings, high-rise buildings—at city center or in the suburbs. Shopping center locations indicate the degree of decentralization from midcity.

Transportation studies will reveal how people cope with "to and from work" travel and what the expected changes are. Changes would indicate where people will live in the future, that is, where building construction may take place.

Designs of new public buildings may reflect the interest a community has in buildings generally.

Tables 25-8 through 25-15 contain census data which describe population trends.

TABLE 25-8 POPULATION 1976-1981 AS OF JUNE 30

Year	Alberta	% Inc.	Edmonton Metro*	% Inc.	Edmonton Proper	% Inc.	% Edm. Metro of Alberta
1976	1,838,037		597,964		461,559		32.0
1977	1,868,407	1.7	616,768	3.1	471,474	2.2	32.2
1978	1,950,000	4.4	634,098	2.8	478,066	1.4	32.3
1979	2,050,000	5.1	653,780	3.1	491,359	2.8	31.9
1980	2,093,940	2.1	676,964	3.8	505,773	2.9	32.4
1981	2,180,990	4.2	702,299	3.4	521,205	3.1	32.2
Average		3.5		3.2		2.5	32.2

Source: Alberta Municipal Affairs—Official Population List.
*Compiled by Edmonton Regional Planning Commission.

TABLE 25-9 POPULATION GROWTH OF EDMONTON METRO (in thousands)

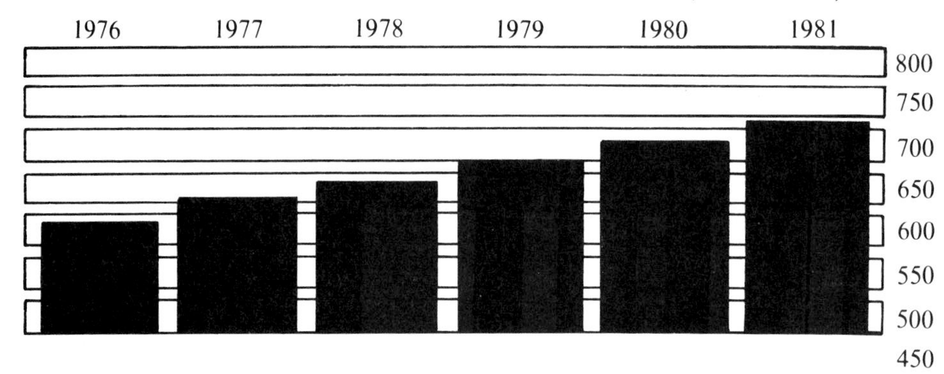

TABLE 25-10 EDMONTON METRO POPULATION BY AGE AND SEX—1981

Age*	Male	% of Total	Female	% of Total	Total	% of Total	Cumulative Total	% Cumulative
0-4	30,418	8.5	28,988	8.4	59,406	8.5	59,406	8.5
5-14	59,047	16.5	56,135	16.3	115,182	16.4	174,588	24.9
15-24	78,012	21.8	74,731	21.7	152,743	21.7	327,331	46.6
25-44	109,146	30.5	101,593	29.5	210,739	30.0	538,070	76.6
45-64	57,257	16.0	55,446	16.1	112,703	16.1	650,773	92.7
65+	23,976	6.7	27,550	8.0	51,526	7.3	702,299	100.0
Total	357,856	100.0	344,443	100.0	702,299	100.0		

*Average age is approximately 31.

Source: Estimated by Business Development Department, City of Edmonton from the Alberta Health Care Insurance Plan—Annual Report 1980.

Further study of the population and labor force may be done with local labor reviews, intercity cost of living comparisons, and a forecast of selected economic indicators for population, households, building permits, retail trade and manufacturing shipments.

TABLE 25-11 PERSONAL INCOME—1976–1981 ($ millions)

Year	Alberta	% Inc.	Edmonton Metro*	% Inc.	% Edm. Metro of Alberta
1976	12,829		4,105		32.0
1977	14,527	13.2	4,678	13.9	32.2
1978	16,348	12.5	5,280	12.8	32.3
1979	19,493	19.2	6,218	17.8	31.9
1980	22,655	16.2	7,343	17.8	32.4
1981	26,053	15.0	8,467	15.0	32.5
Average		15.2		15.5	32.2

Source: Alberta Bureau of Statistics—Economic Accounts, 1980.
*Estimated by Business Development Department, City of Edmonton.

TABLE 25-12 PERSONAL DISPOSABLE INCOME—1976–1981 ($ millions)

Year	Alberta	% Inc.	Edmonton Metro*	% Inc.	% Edm. Metro of Alberta
1976	10,292		3,406		33.1
1977	11,733	14.0	3,813	11.9	32.5
1978	13,505	15.1	4,416	15.8	32.7
1979	16,066	17.2	5,223	18.2	33.0
1980	18,628	15.9	6,037	15.5	32.9
1981	21,363	14.7	6,943	15.0	32.5
Average		15.4		15.3	32.8

Source: Alberta Bureau of Statistics—Economic Accounts, 1980.
*Estimated by Business Development Department, City of Edmonton.

TABLE 25-13 WAGES AND SALARIES—1976–1981 ($ millions)

Year	Alberta	% Inc.	Edmonton Metro*	% Inc.	% Edm. Metro of Alberta
1976	8,284		3,157		38.1
1977	9,521	14.9	3,721	17.8	39.0
1978	10,529	10.6	4,203	12.9	39.9
1979	12,552	19.2	4,809	14.4	38.6
1980	14,828	18.1	5,568	15.8	37.5
1981	17,156	15.7	6,414	15.2	37.4
Average		15.7		15.2	38.4

Source: Alberta Bureau of Statistics—Economic Accounts, 1980.
*Estimated by Business Development Department, City of Edmonton.

25.4 BUSINESS ENVIRONMENT

The current economic climate determines, to a large extent, the volume and kind of building construction. Incentives to build may be local, regional, or national in character and should be sorted with care. When

TABLE 25-14 EDMONTON METRO 1981 BASE LABOR FORCE BY AGE, BY EDUCATION (in thousands)

	15 Years & Older	Age 15-24	Age 25-44	Age 45-64	Age 65+
Population 15 Years & Older	527	153	211	112	52
Less than Grade 9	79	5	19	27	28
Grade 9 to 13	237	93	84	46	13
Post-Secondary (Nonuniversity)	110	31	57	18	5
University	101	24	51	21	5

Source: Estimated by Business Development Department, City of Edmonton from 1976 Census of Canada.

TABLE 25-15 ALBERTA JOURNEYMEN & APPRENTICES 1980

Location	Journey-men	% of Total	Appren-tices	% of Total
Edmonton	23,563	46	12,117	47
Calgary	12,684	25	6,537	25
Lethbridge	4,282	8	1,679	7
Red Deer	3,584	7	1,599	6
Grande Prairie	810	2	559	2
Fort McMurray	3,064	6	1,022	4
Other	3,297	6	2,207	9
Total	51,284		25,720	

Source: Alberta Advanced Education & Manpower.

the market is down, builders turn to renovations and additions. When regional or national needs are to be satisfied, building projects are to be undertaken. Visits to the respective local offices will keep one in communication with offered support programs. Trends may be studied with respect to research results, homeowner grants, and changes in the national building codes which must later be reflected in regional regulations and local by-laws.

Tables 25-16 and 25-17 provide recent data with respect to the value of building permits (Edmonton proper) 1980-1981 and Dwelling Starts and Completions (Alberta and Metro-Edmonton) 1976-1981.

The number of dwelling starts indicates overall construction activity. The data needs further breakdown to light and heavy construction (of high-rise condominiums, for example). More data is also needed with respect to the number of contractors and the sizes of their respective organizations.

25.5 INCREASING KNOWLEDGE OF THE LIGHT BUILDING CONSTRUCTION INDUSTRY

To meet their respective goals, governments and organizations offer educational and training programs. National and regional governments take initiatives to keep the labor force tuned to satisfy needs. They may do this directly through publicly owned schools, colleges, or

TABLE 25-16 VALUE OF BUILDING PERMITS, EDMONTON PROPER (1980-1981)

Category	Year	No. of Permits	No. of Permits Increase	Dollar Value	Total $ Value Increase	Total Value % Increase
Commercial	1981	2,063	478	$513,428,713	$151,581,179	41.9
	1980	1,585		361,847,534		
Industrial	1981	84	3	55,918,368	7,736,705	16.1
	1980	81		48,181,663		
Institutional	1981	167	15	54,301,300	-53,317,394	-49.5
	1980	152		107,618,694		
Residential	1981	7,145	1,170	523,901,682	178,907,505	51.9
	1980	5,975		344,994,177		
Misc.	1981	1,838	-253	9,488,230	1,069,744	-10.1
	1980	2,091		10,557,974		
TOTAL	1981	11,297	1,413	1,157,038,293	283,838,251	32.5
	1980	9,884		873,200,042		

Source: Building Inspection Branch, Bylaw Enforcement, City of Edmonton.

TABLE 25-17 DWELLING STARTS & COMPLETIONS (1976-1981)

	Alberta		Edmonton Metro*		% Edm. Metro Starts of Alberta	% Edm. Metro Compl. of Alberta
Year	Starts	Compl.	Starts	Compl.		
1976	38,771	25,858	14,945	10,260	38.6	39.7
1977	38,075	37,879	13,135	13,830	34.5	36.5
1978	47,925	43,025	18,549	16,921	38.7	39.3
1979	39,947	44,492	12,298	15,567	30.8	34.9
1980	32,031	34,717	10,717	10,748	33.4	30.9
1981	37,648	31,000	12,901	12,101	34.3	39.0

Source: Statistics Canada, Cat. #64-002.
*Estimated by Business Development Department, City of Edmonton.

universities or they may reimburse industry for costs of training. Labor organizations meet specific needs of the members.

To know what is being offered, one needs to know something of each organization and how it functions, where the local offices are, and what communication patterns are in place.

Other organizations require study and first-hand experiences. Banking must be understood and personal accounting practices developed to the point of being able to manage not only one's own business affairs but the affairs of an organization. Business law should be studied and the principles observed.

When working with small groups on a construction basis, one learns to appreciate and employ the skills and knowledge of individuals.

Formal organizations usually require careful reporting by their members. Reporting skills become an asset, not only to the reporter, but to the organization. Good reports strengthen an organization by improving communication levels. Learning how formal organizations function and how to help them do so will be beneficial in a light construction industry of any size. If one actually helps an organization function, one is accepting responsibility and giving leadership.

25.6 RESEARCH

1. Obtain a copy of the latest urban study for your area. Visit the urban office, chamber of commerce office, or public library. Review similar studies from other centers.
2. Select current data to describe the climate for light building construction work in your area. Organize the data in useful sets. (If you cannot find the answers, seek more information from local architects, engineers, and builders. Visit local colleges or universities for additional help. Find the persons who are interested in helping you.)
3. As one who is interested in contracting for a light construction building project, write a set of questions that you wish to have answered.
4. Obtain and study charts which show the formal organization of your local and regional governments. Identify those parts of the organization which directly affect light construction operations. Visit the respective offices to become acquainted not only with their locations, but their directive printed materials such as codes, laws, regulations, and by-laws. Talk to some of the people working there to learn of their points of view in contracting for light construction.
5. Visit a library to find a geography book which explains the *Koppen Classification of Climates*. From the information given, describe the climate for your area.

25.7 REFERENCES

25.7.1 Government of Alberta and the University of Alberta, Edmonton. *Atlas Alberta (1969).*

25.7.2 *1981 Edmonton Annual Economic Report,* Edmonton, Alberta.

APPENDIX

ON-SITE MEASUREMENTS

A.1 BATTER BOARDS

Figure A-1 illustrates a building layout on a rectangular lot. For excavating purposes, the exact location of the building's corners must be determined and marked off. The corner positions are also important later when the forms are set for the foundation footings. Driving stakes into the ground to mark the corner positions will not be sufficient, however, as the excavation work will likely disturb the placement of the stakes.

To solve this problem, batter boards can be set back approximately 4′ (1.2 m) from the excavation lines (Figure A-1). Use 2″ × 4″ (38 mm × 89 mm) lumber for the stakes and 1″ × 6″ (38 mm × 140 mm) lumber for the ledger boards. The ledger boards are nailed to the stakes

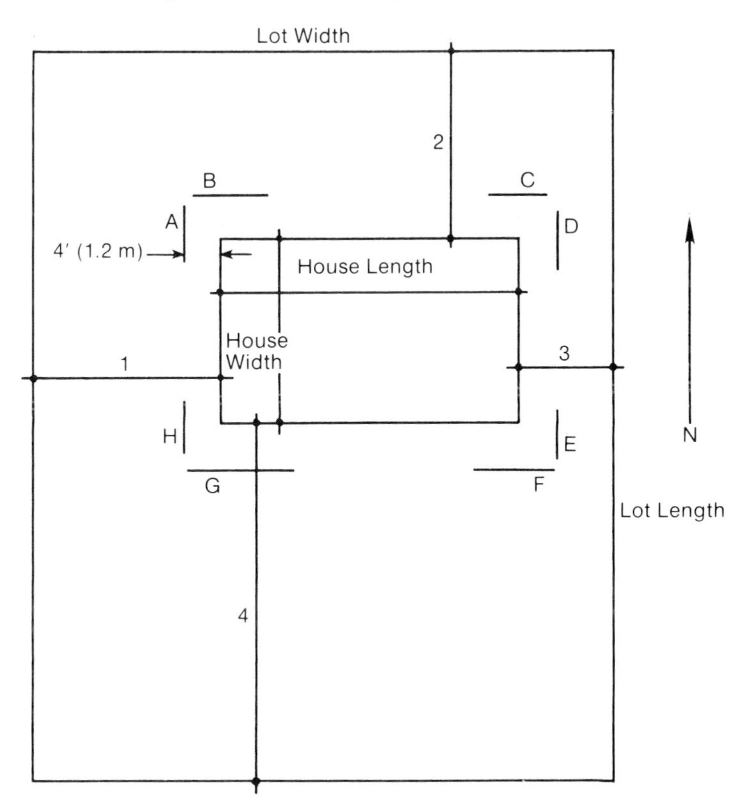

FIGURE A-1. Building layout with batter boards A through H.

at a convenient working height, usually just above the top of the foundation. The batter boards should be driven into the ground level with each other.

Taut lines made from heavy cord or fine wire are then stretched across opposite batter boards and adjusted so they align with the nails of the corner stakes. Use a plumb bob for exact placement of these lines. Now mark the location of the lines on each ledger and make a small kerf cut. Be sure that the lines remain taut while making each of the kerf cuts (Figure A-2).

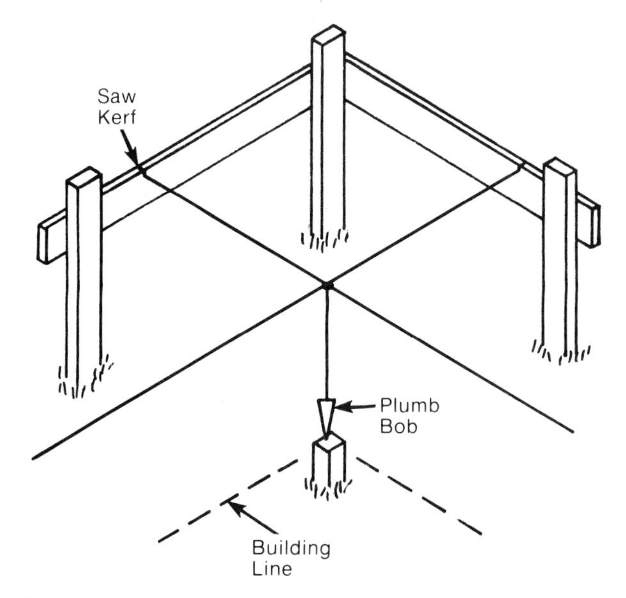

FIGURE A-2. Making a kerf cut.

A.2 BATTER BOARD LEVELING

Before making any kerf cuts, it is important that the batter boards be level and at a reasonable distance above the ground. They should also be securely staked and, if necessary, braced (Figure A-3). Bracing is recommended if the ground is exceptionally loose.

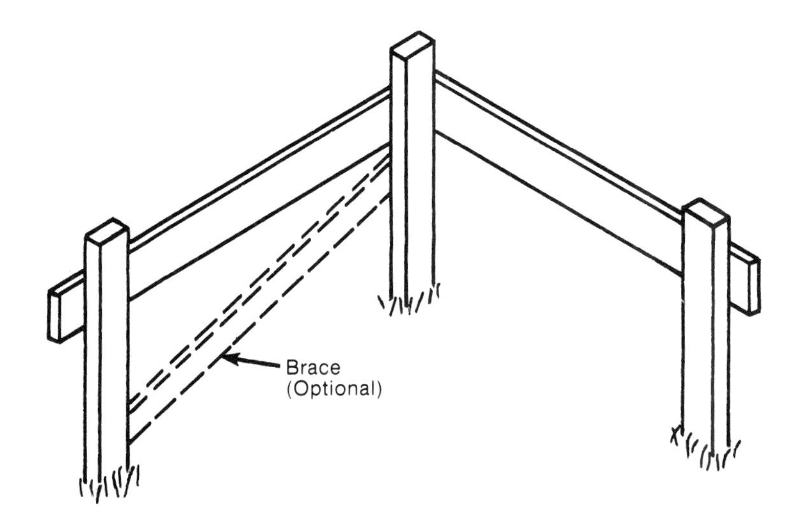

FIGURE A-3. Bracing on batter boards.

To level the first batter board, use a carpenter's level. To set the other ledger boards, attach a line carrying a line level from the first batter board (Figure A-4). When using a line level, keep the line taut. Check its accuracy by taking one level reading and then reversing it on the line; the second reading should be the same as the first. If all readings are taken from a central peg or post, be sure that the pivot end stays flush with the top of the peg or post.

A homemade water level is a very useful instrument, not only for leveling batter boards but also for checking footing forms, foundation wall forms, and general landscaping. It is portable, easy to set up at any height, and easy to use. Always be sure to read the bottom of the meniscus in the reading tube (Figure A-5).

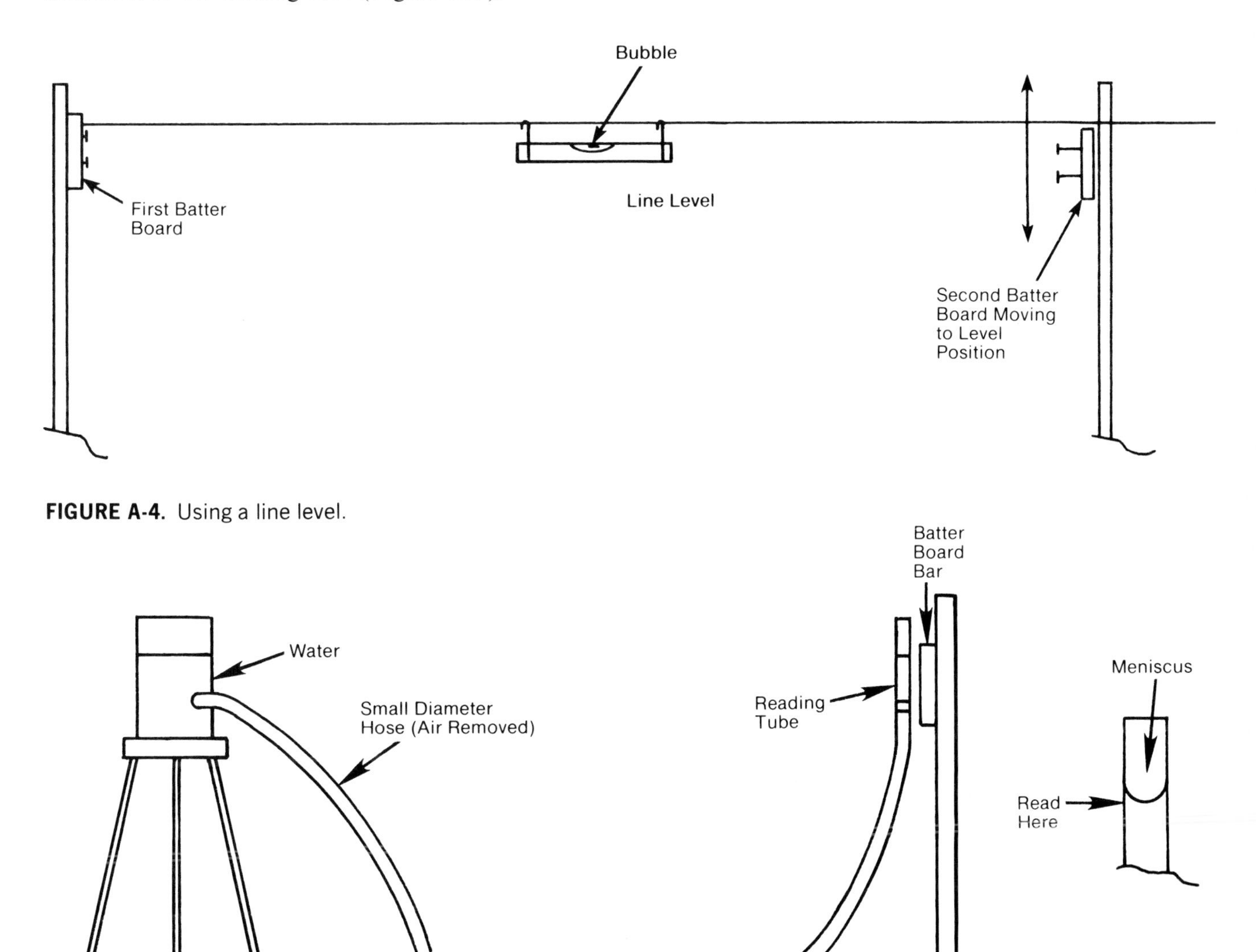

FIGURE A-4. Using a line level.

FIGURE A-5. Homemade water level.

A.3 LOCATING A BUILDING CORNER

When the excavating is complete, it is necessary to locate the building corners. String the building lines tightly between each pair of batter boards. With all of the lines in position, check the length and width measurements and the squareness of each corner. At the points of intersection, drop a plumb bob to the nail heads driven into the stakes at the exact corner positions. Footing forms may now be constructed.

A.4 CHECKING FOR SQUARENESS

It is important that all corners are square and all lines parallel. To make sure that the corners are square, measure the diagnonals; they will be equal in length if the corners are square (Figure A-6). Another way to check for squareness is the 3-4-5 method. Multiples of these figures are used to measure the sides and diagonal of a right triangle. For example, if one side measures 3′ and the adjacent side measures 4′, the diagonal will measure 5′. For greater accuracy, multiples of these numbers are used in Figure A-7. This method can also be used to lay out square corners in the first place.

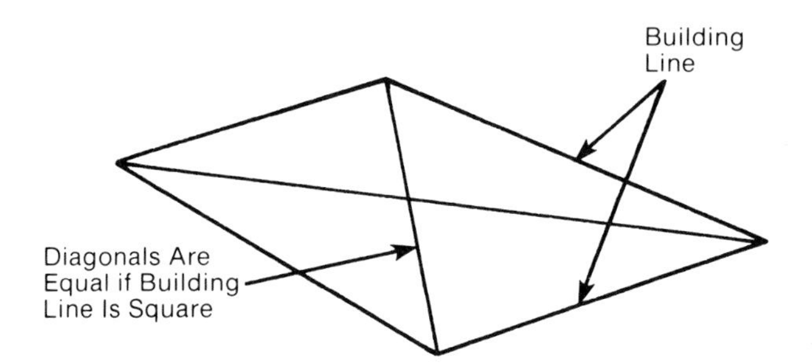

FIGURE A-6. Measuring diagonals to check squareness.

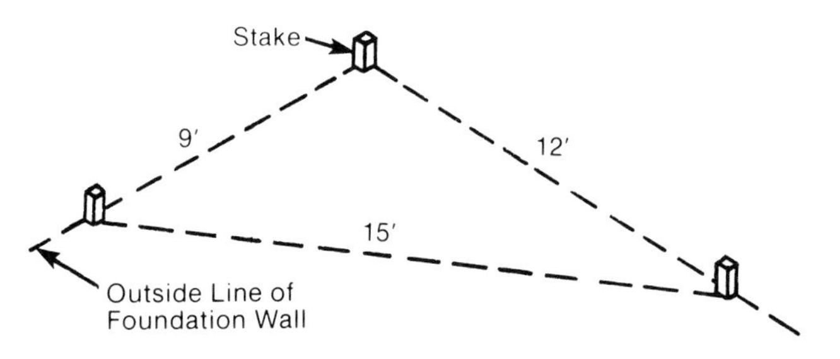

FIGURE A-7. The 3-4-5 method of checking squareness.

INDEX